The Most Successful Way of JEE Preparation

DPP

Daily Practice Problems

JEE Main & Advanced

Rotational Motion & Properties of Matter

with Gravitation

PHYSICS Vol-3

The Most Successful Way of JEE Preparation

DPP

Daily Practice Problems

JEE Main & Advanced

Rotational Motion & Properties of Matter

with Gravitation

PHYSICS Vol-3

Er PRS Murthy (Raghava)
M.Tech, HOD Physics
Jain College, Bengaluru

arihant
ARIHANT PRAKASHAN (SERIES), MEERUT

DPP

Daily Practice Problems

arihant

ARIHANT PRAKASHAN (SERIES), MEERUT

Administrative & Production Offices

Corporate Office 'Ramchhaya' 4577/15, Agarwal Road, Darya Ganj, New Delhi -110002
Tele: 011- 47630600, 43518550; Fax: 011- 23280316

Head Office Kalindi, TP Nagar, Meerut (UP) - 250002
Tel: 0121-2401479, 2512970, 4004199; Fax: 0121-2401648

Sales & Support Offices

Agra, Ahmedabad, Bengaluru, Bareilly, Chennai, Delhi, Guwahati, Hyderabad, Jaipur, Jhansi, Kolkata, Lucknow, Meerut, Nagpur & Pune

ISBN : 978-93-13193-33-3

Published by Arihant Publications (I) Ltd.

Production Team

Publishing Managers	Mahendra Singh Rawat & Keshav Mohan	*Inner Designer*	Mazher Chaudhary
Project Head	Sorabh Chaudhary	*Page Layouting*	Sunil Sharma, Hemant
Cover Designer	Shanu Mahad	*Proof Reading*	Ashish

For further information about the products from Arihant, log on to www.arihantbooks.com or email to info@arihantbooks.com

/arihantpub /@arihantpub Arihant Publications /arihantpub

ABOUT DPP

Daily Practice Problems

Indian Institutes of Technology better known as IITs have always been the dream destination of each science group students who want to pursue their career as Engineers. These are the institutes which have got very high value not even in India but in the whole world. To get the admission in such institutes one need to face entrance Examinations; JEE Main & JEE Advanced, out of which JEE Main is considered as the preliminary exam & JEE Advanced is the final exam to get into the IITs. Approximately per year 12 lakh students enroll themselves for JEE Main and only the top 2.25 lakh in JEE Main merit can attempt for JEE Advanced, the fact clearly express the degree of toughness of JEE Main & JEE Advanced.

To have success in such an exam one needs to be very planned & systematic in its preparation. A very popular wise saying 'Rome was not built in a day' suggests that to achieve a big target like IITs we need to work continuously. For continuous work, one need to have proper planning that how much work he/she has to do on regular basis, i.e., on daily basis. Each student preparing for JEE has many books but unfortunately none of them gives the students a plan for daily studies, which is the most effective success key for JEE.

Keeping the daily practice needs of the students in mind we have come up with DPP Series (Daily Practice Problems) having approximately 26 books. The primary focus of this series is on achieving success through practice & proper planning. For that we have divided the whole syllabus of JEE Main & JEE Advanced; Physics, Chemistry & Mathematics, into different volumes (around 8 volumes of Mathematics & 9 volumes each of Physics & Chemistry), each volume has Daily Practice Problems Sheets having specific questions on various Topics of the individual chapter of the volume ensuring the complete Practice & Assessment of the topic.

DPP
Daily Practice Problems

In this way we have tried to give you an effective daily planning for JEE keeping in mind the daily hours that a student can devote to his practice, you are just required to practice one sheet of each subject on daily basis. This habit of daily practice will keep you away from exam stress and thoughts like; how much I have prepared/am I preparing properly or not/ do I need to study more on daily basis or what I am doing is ok, etc., which regularly tease your mind and can give you panic at any stage of your preparation.

Some special features that we have tried to incorporate in our Daily Practice Problems (DPP) series are discussed below

1. Micro Level coverage of all the topics of a Chapter/ Unit given in JEE Main & Advanced Syllabus.
2. Each DPP has various types of questions; Subjective Questions, Single/ Multiple Correct Options, Statements type, Comprehension based, Integer Value Answer type, Matching, etc., making you practice & ready for all the question formats of JEE.
3. New & innovative questions in each DPP challenging & improving your problem solving skills.
4. Along with Topical Coverage, Revisal DPPs for JEE Main & JEE Advanced with each Chapter, covering multidisciplinary questions proving comprehensive practice.
5. JEE Main & JEE Advanced Archive (collection of last 13 years' exams questions) with each chapter.
6. Complete Solutions for each DPP.

We hope that DPP series will prove to be very beneficial for the students preparing for JEE, by giving them daily practice in a systematic & planned manner. The suggestions/criticism from the students, teachers & parents about DPP series will be welcomed whole heartedly and we promise to update this series accordingly from time to time.

DPP
Daily Practice Problems

DAILY TASKS...

DPP
Daily Practice Problems

Dedicated To My Mother

Late Smt. Pavani. Subbarathnamma,
whose Lullaby I want to listen
atleast one more time before
my Departure

1. Centre of Mass

DPP

DAILY PRACTICE PROBLEMS

DPP-1 Centre of Mass

Only One Option Correct Type

Working Space

Direction (Q. Nos. 1-8) *This section contains 8 multiple choice questions. Each question has four choices (a), (b), (c) and (d), out of which ONLY ONE is correct.*

1 For the figure shown, a rod of mass 10 kg (of length 100 cm) with some point masses tied to it at different positions. Find the distance of the point (from A) at which if the rod is picked over a knife edge, it will be in equilibrium about that knife edge.

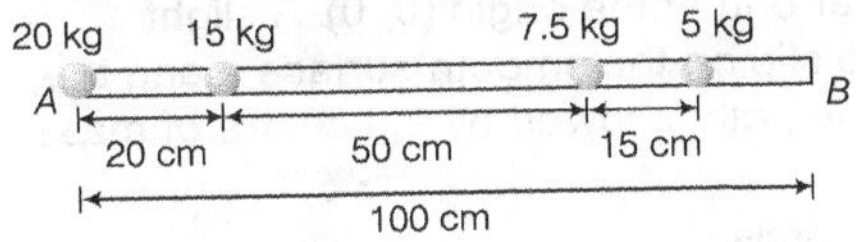

a. 26.32 cm
b. 30.43 cm
c. 28.72 cm
d. 32.50 cm

2 Three point masses are present at the vertices of a right angled triangle as shown in the figure. With respect to the coordinate system shown, the coordinates of centre of mass (X_{COM}, Y_{COM}) of the three point mass system are

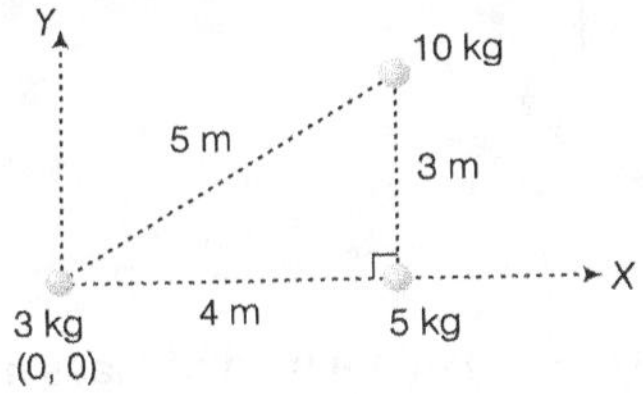

a. (3.33, 1.67) m
b. (1.67, 3.33) m
c. (3.67, 1.33) m
d. (1.33, 3.67) m

3 A particle P of mass m is released from the top of a wedge Q of mass M. Wedge is free to move on a smooth horizontal surface. There is no information regarding the nature of surface of wedge. The speed of particle when it reaches B is v_2 and speed of wedge at this instant is v_1.

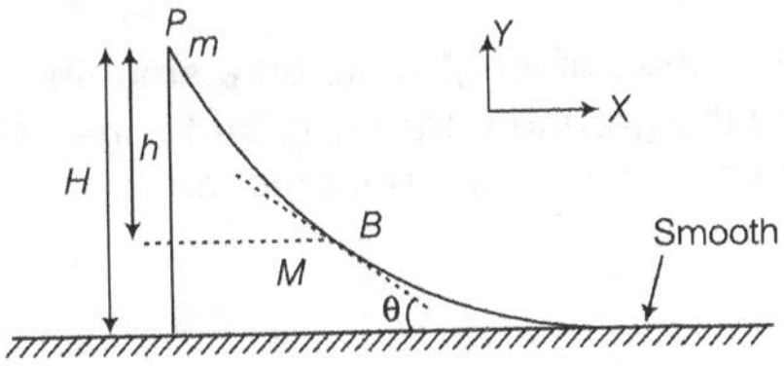

The centre of mass of the system comprising P and Q will be moving in which direction?

a. 
b. 
c. 
d. 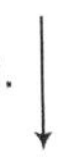

Working Space

4 A block of mass m moves with a velocity v_0 towards a stationary block of mass M (connected to a spring) on a smooth horizontal surface (see figure). If spring constant is k, what is the maximum compression in the spring?

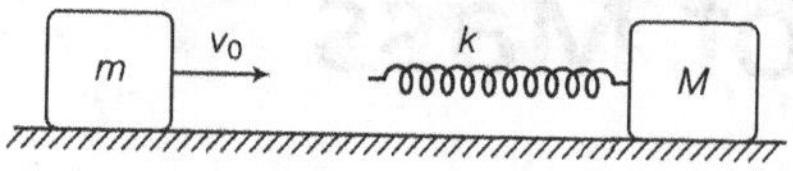

a. $v_0\sqrt{\dfrac{mM}{2k(m+M)}}$

b. $v_0\sqrt{\dfrac{mM}{k(m+M)}}$

c. $v_0\sqrt{\dfrac{2mM}{k(m+M)}}$

d. None of the above

5 A uniform thin rod of mass m and length L is standing vertically along the Y-axis on a smooth horizontal surface with its lower end at the origin (0, 0). A slight disturbance at $t = 0$ causes the lower end to slip on the smooth surface along the positive X-axis and the rod starts falling. The path followed by the centre of mass during its fall, is

a. parabola
b. circle
c. straight line
d. ellipse

6 For the above problem, the equation of trajectory of a point on the rod located at distance r from the lower end is

a. $\dfrac{x^2}{r^2}+\dfrac{y^2}{r^2}=1$

b. $\dfrac{x^2}{\left(\dfrac{L}{2}-r\right)^2}+\dfrac{y^2}{r^2}=1$

c. $\dfrac{x^2}{\left(\dfrac{L}{2}\right)^2}+\dfrac{y^2}{r^2}=1$

d. $\dfrac{x^2}{\left(\dfrac{L}{2}\right)^2}-\dfrac{y^2}{r^2}=1$

7 A rope thrown over a pulley has ladder (of mass $M-m$) with a man of mass m on one of its ends and a counterbalancing mass M on its other end. The man climbs with a velocity v_r relative to a ladder. Pulley and rope are ideal. The velocity of centre of mass of the system is

a. $\dfrac{m}{M}v_r$
b. $\dfrac{m}{2M}v_r$
c. $\dfrac{M}{m}v_r$
d. $\dfrac{2M}{m}v_r$

8 Two masses P and Q connected with an ideal string of length L lie on a smooth horizontal plane. Q is given a velocity v along the ground perpendicular to line PQ (see figure). Find the tension in the string during their subsequent motion.

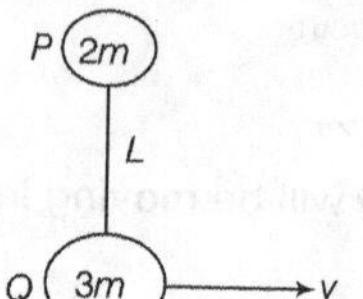

a. $\dfrac{mv^2}{L}$
b. $\dfrac{3mv^2}{L}$
c. $\dfrac{6}{5}\cdot\dfrac{mv^2}{L}$
d. $\dfrac{8}{5}\cdot\dfrac{mv^2}{L}$

Working Space

One or More than One Options Correct Type

Direction (Q. Nos. 9 and 10) This section contains 2 multiple choice questions. Each question has four choices (a), (b), (c) and (d), out of which ONE or MORE THAN ONE are correct.

9 Three particles each of mass m are welded to a rigid frame consisting of three massless rods, each of length L. The rods are welded to each other at point O. The frame is placed on a smooth horizontal floor. Three forces start acting as shown in the figure. Then, just at the initial moment,

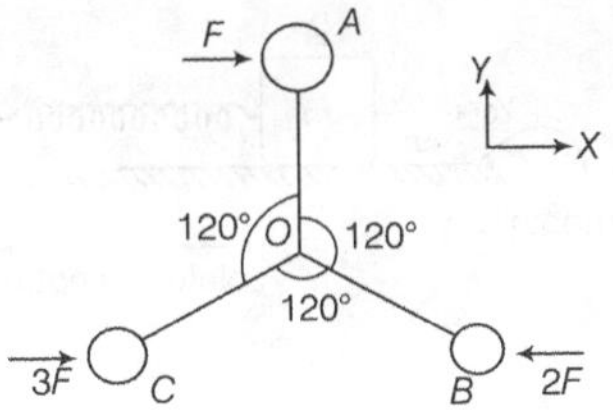

a. acceleration of O is $\bar{a}_O = \dfrac{2F}{3m}\hat{i}$

b. acceleration of A is $\bar{a}_A = \dfrac{5F}{6m}\hat{i}$

c. acceleration of B is $\bar{a}_B = \dfrac{7F}{12m}\hat{i} - \dfrac{F}{4\sqrt{3}}\hat{j}$

d. magnitudes of accelerations of B and C are equal

10 A block of mass m slides down on an inclined plane of a wedge of same mass m (see figure). Friction is absent everywhere. Which of the following statement(s) is/are correct?

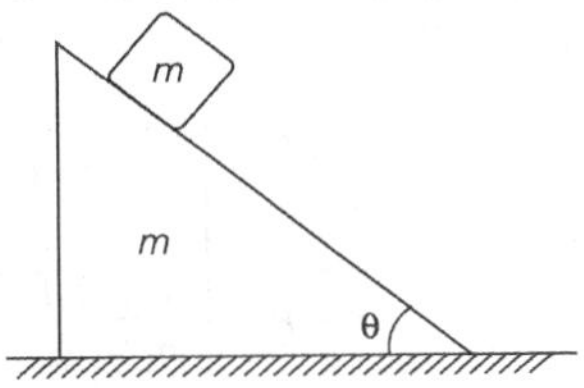

a. Vertical component of acceleration of block is $\dfrac{g\cos^2\theta}{1+\sin^2\theta}$ downwards

b. Acceleration of centre of mass of (block + wedge) is $\dfrac{g\sin^2\theta}{1+\sin^2\theta}$ downwards

c. Acceleration of centre of mass of (block + wedge) is $\dfrac{g\cos\theta}{1+\sin^2\theta}$ downwards

d. Vertical component of acceleration of block is $\dfrac{2g\sin^2\theta}{1+\sin^2\theta}$ downwards

One Integer Value Correct Type

Direction (Q. Nos. 11-14) This section contains 4 questions. When worked out will result in an integer from 0 to 9 (both inclusive).

11 Both the blocks shown in the given arrangement are given together a horizontal velocity towards right. If x ms^{-2} be the subsequent acceleration of centre of mass of the system of blocks, find the minimum value of x (in ms^{-2}). (Given, $m_A = 2m_B$ and $g = 10$ ms^{-2}, all in SI units)

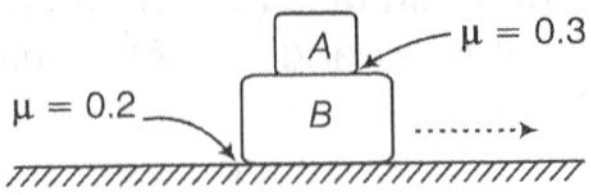

12 A bullet of mass 10 g moving horizontally with speed of 400 ms^{-1} and embeded into a block of mass 390 g initially at rest. If ratio of final KE of centre of mass to initial KE of bullet is r, then find the value of $200r$.

13 A small particle of mass m starts sliding down from rest along the smooth surface of a fixed hollow hemisphere of mass $M(4m)$. The distance of centre of mass of (particle + hemisphere) from centre O of hemisphere, when the particle separates from the surface of hemisphere is $\dfrac{\sqrt{69}}{5\alpha}R$. Find the value of α.

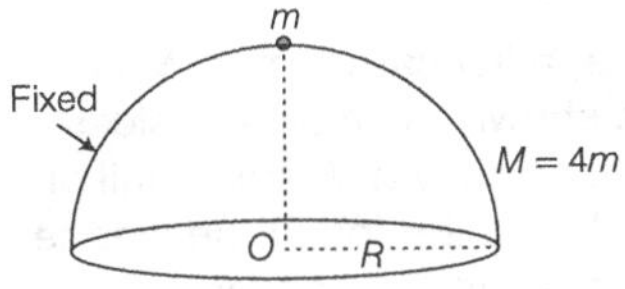

Working Space

14 In the given figure there are two blocks A and B. They are given initial impulses as shown. The maximum stretch in the spring is $\sqrt{\frac{m}{k}}\beta\, v_0$. Find the value of β.

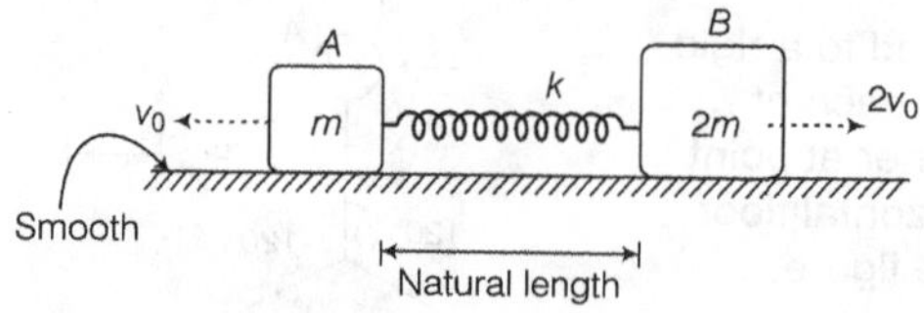

Subjective Questions

Direction (Q. Nos. 15-19) *These questions are subjective in nature, need to be solved completely on notebook.*

15 From a solid sphere of radius R and mass M, a spherical portion of radius r is removed as shown in the figure. With respect to the coordinate system shown, find out the position of centre of mass of residual body.

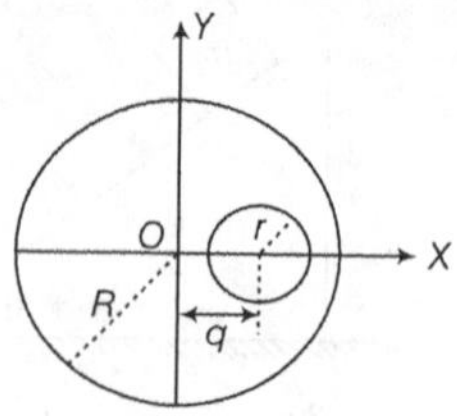

16 For the figure there is an arc of a thin ring, that subtends at an angle θ at the centre. The mass of the arc is m and its radius is R. Locate the centre of mass of the arc.

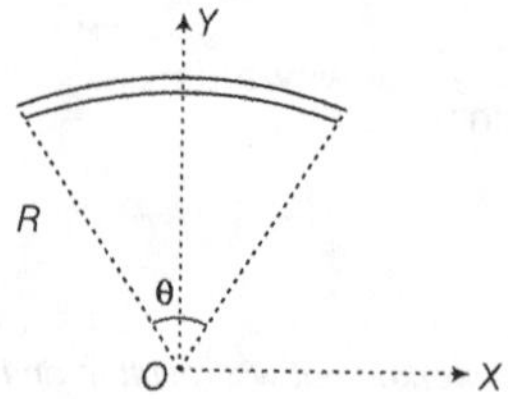

17 For the figure shown, a thin disc (of sector shape) which is part of a circular disc of radius R. The disc has an angle θ. If the mass of the shown disc is m, then locate its centre of mass.

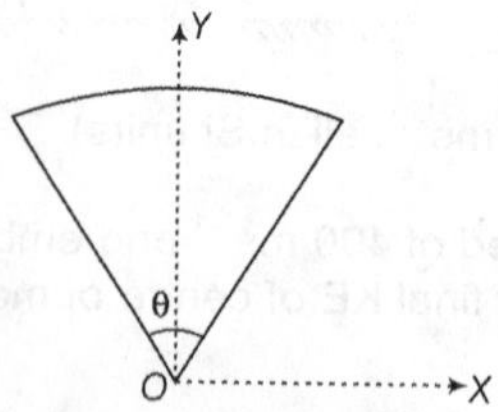

18 For the given figure, a flat car of mass M on a frictionless road is shown. A small massless wedge is fitted on it as shown. A small ball of mass m is released from the top of the wedge. It slides over it and falls into the hole at a horizontal distance L from the initial position of the ball. Find the distance travelled by the flat car till the ball gets into the hole from its initial position.

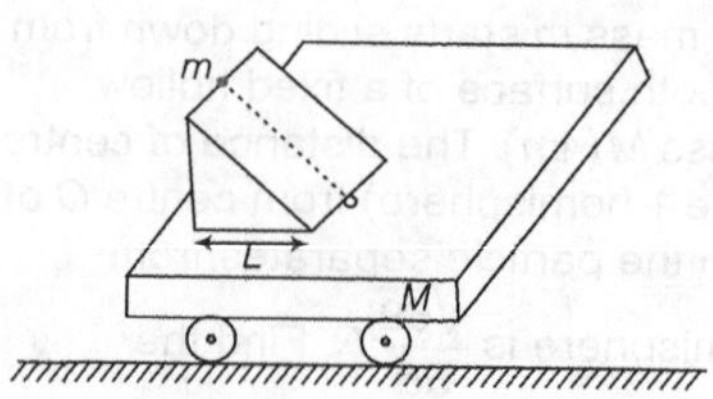

19 Two small discs A and B are connected by an inextensible string of length $5L$. Mass of ball A is m and that of ball B is $2m$. The balls are resting on a frictionless horizontal surface with the distance between them $3L$. In this position, ball B is given a horizontal velocity u perpendicular to the wire joining two balls.

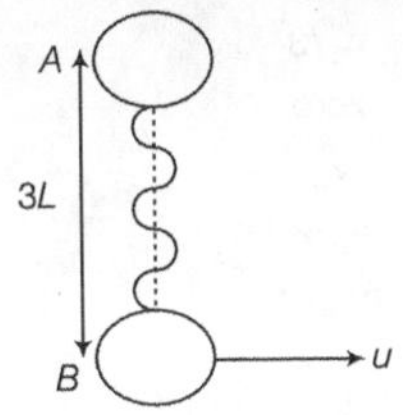

i. Find the speeds of ball A and ball B just after the string gets taut.
ii. Find the impulse of the tension in the string when the string becomes taut.
iii. Find the steady tension in the string much after the string gets taut.

Matching List Type

Direction (Q. Nos. 20 and 21) Choices for the correct combination of elements from Column I and Column II are given as options (a), (b), (c) and (d), out of which one is correct.

20 For the figure shown, two blocks A and B are given. They are given initial impulses as shown. Match the following columns and select the correct option from the codes below.

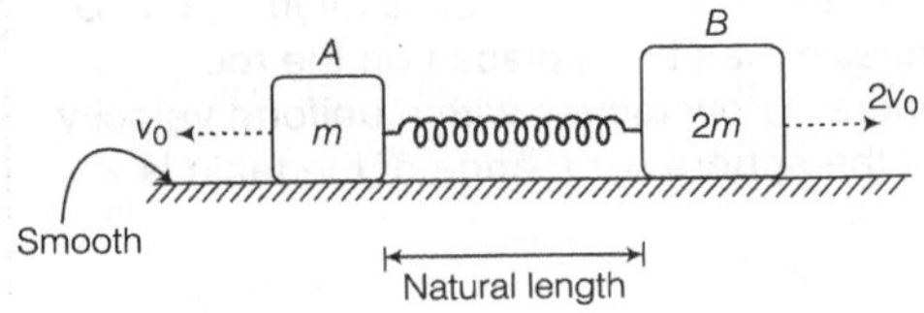

	Column I		Column II
i.	$\lvert v_{A,\max}\rvert$	p.	$\frac{3}{2}v_0$
ii.	$\lvert v_{A,\min}\rvert$	q.	$2v_0$
iii.	$\lvert v_{B,\max}\rvert$	r.	zero
iv.	$\lvert v_{B,\min}\rvert$	s.	$3v_0$
v.	When $v_A = 0$, $v_B = ?$	t.	$\frac{5}{2}v_0$

Codes

	i	ii	iii	iv	v
a.	s	r	q	r	p
b.	r	s	q	p	s
c.	q	p	r	s	t
d.	s	r	q	p	t

21 Two blocks A and B of masses $2m$ and $3m$ placed on a smooth horizontal surface are connected with a light spring. The two blocks are given velocities as shown in the figure when the spring is in its natural length.

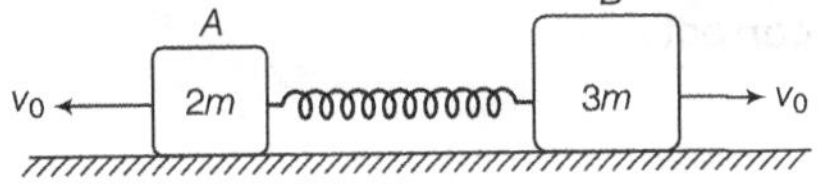

Match the following columns and select the correct option from the codes given below.

Working Space

	Column I		Column II
i.	$\|v_{A,\,min}\|$	p.	$7v_0/5$
ii.	$\|v_{A,\,max}\|$	q.	$v_0/3$
iii.	$\|v_{B,\,min}\|$	r.	zero
iv.	$\|v_{B,\,max}\|$	s.	v_0
v.	$v_A = ?$ when $v_B = 0$	t.	$v_0/2$
vi.	$v_B = ?$ when $v_A = 0$	u.	$5v_0/3$

Codes

	i	ii	iii	iv	v	vi
a.	r	p	r	s	t	q
b.	r	s	q	p	s	t
c.	q	p	r	s	t	u
d.	s	r	q	p	t	u

Comprehension Type

Direction (Q. Nos. 22-24) *This section contains a paragraph, which describing theory, experiments, data, etc. Three questions related to the paragraph have been given. Each question has only one correct answer among the four given options (a), (b), (c) and (d).*

Passage

A rod of mass m and length $2L$ is placed on a smooth table with its middle point B on the edge of the table. A sphere A of mass m has been placed on the rod exactly at the middle point B of it. The sphere A now moves with a uniform velocity on the rod leftwards. The displacement of the sphere w.r.t. edge of the table is x at a certain time.

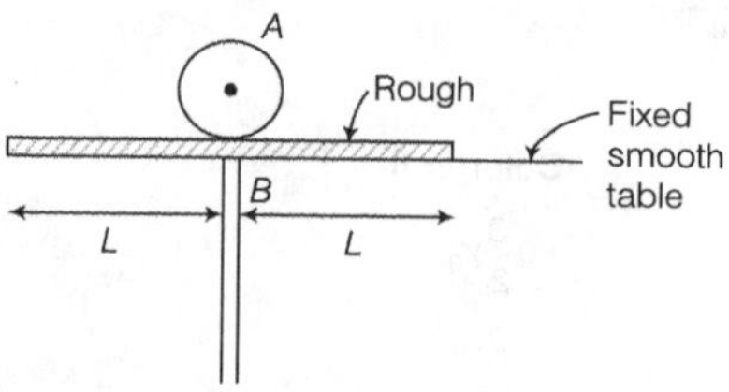

22 For $x < \frac{L}{2}$, which of the following options is correct?

a. Only sphere will topple down
b. Only rod will fall down
c. The system will topple down
d. The system will not topple down

23 For $x = \frac{L}{2}$, which of the following options is correct?

a. Only sphere will topple down
b. Only rod will fall down
c. The system will topple down
d. The system will not topple down

24 For $x > \frac{L}{2}$, which of the following options is correct?

a. Only sphere will topple down
b. Only rod will fall down
c. The system will topple down
d. The system will not topple down

DPP-2 Momentum and Its Conservation

Only One Option Correct Type

Working Space

Direction (Q. Nos. 1-4) This section contains 4 multiple choice questions. Each question has four choices (a), (b), (c) and (d), out of which ONLY ONE is correct.

1 The graph shows the variation of force acting on a body with time t. Assuming that body to start from rest, the variation of its momentum with time is best represented by

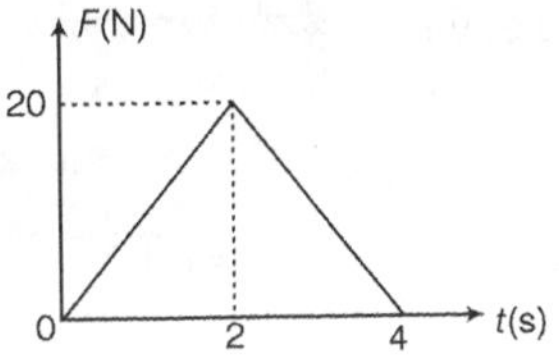

a. 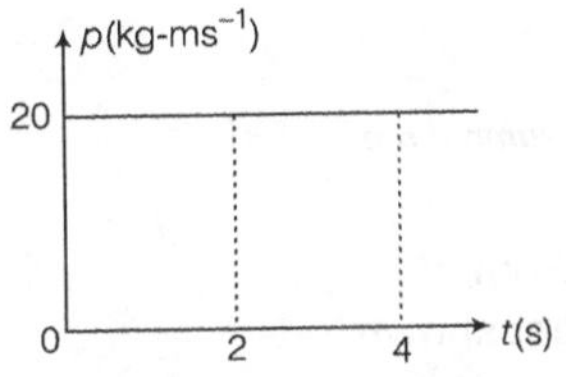

b. 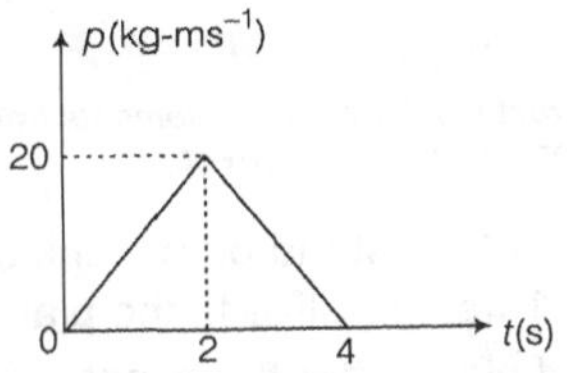

c. 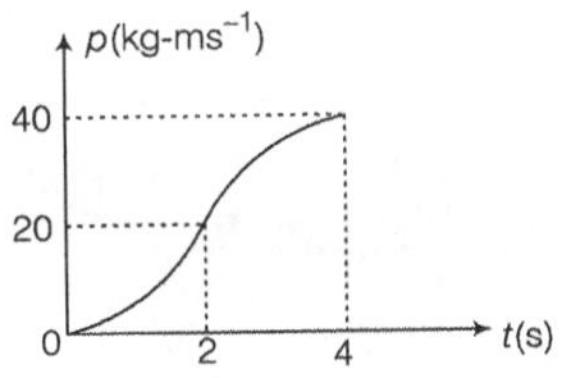

d. 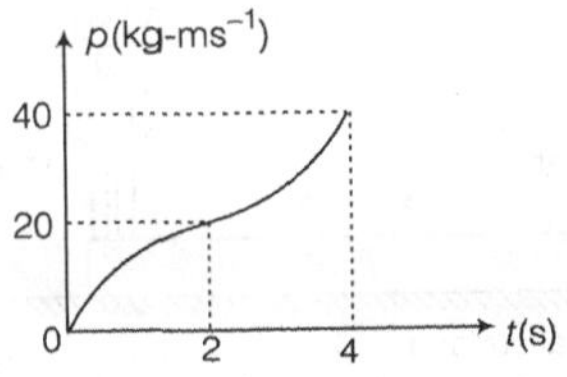

2 A small ring of mass m is attached at one end of a light string of length $l = 0.6$ m. The other end of the string is tied to a small block B of mass $2m$. The ring is free to move on a fixed smooth horizontal rod. The block B is released from rest from the position shown. The velocity (in ms^{-1}) of the ring when the string becomes vertical is (Given, $g = 10\ ms^{-2}$)

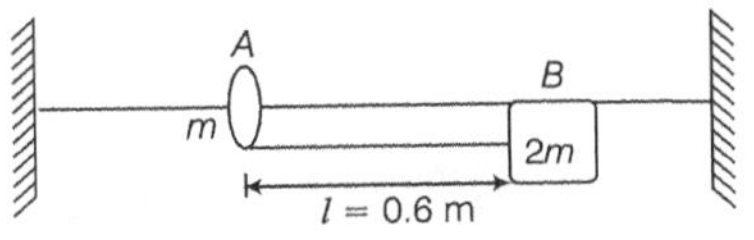

a. 4
b. 3
c. 3.5
d. 2.5

Working Space

3 A platform of mass m and counterweight of $M + m$ (where, $M = 3m$) are connected by a light cord which passes over two smooth pulleys (see figure). A man of mass M is standing on the platform which is at rest. If the man leaps vertically upwards (and out of platform) with velocity $10\ ms^{-1}$, find the distance through which the platform will descend?

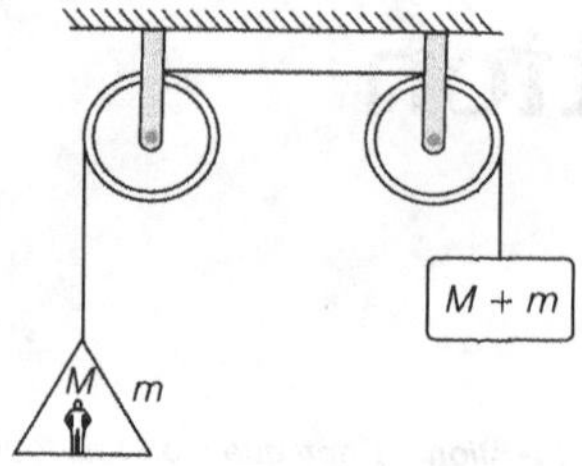

a. 1 m b. 2 m c. 3 m d. 4 m

4 A hemisphere of radius R and of mass $4m$ is free to move with its base on a smooth horizontal table. A particle of mass m is placed on the top of the hemisphere. The angular velocity of the particle relative to hemisphere at an angular displacement θ when velocity of hemisphere has became v is

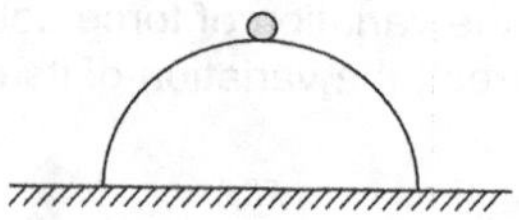

a. $\frac{2v}{R\cos\theta}$ b. $\frac{3v}{R\sin\theta}$ c. $\frac{5v}{R\cos\theta}$ d. $\frac{5v}{R\sin\theta}$

Matching List Type

Direction (Q. No. 5) Choices for the correct combination of elements from Column I and Column II are given as options (a), (b), (c) and (d), out of which one is correct.

5 Two men of masses 60 kg and 80 kg stand on a plank of mass 20 kg. Both of them can jump with a velocity $1\,ms^{-1}$ relative to the plank. For each event shown in Column I, find the velocity of plank after the event.

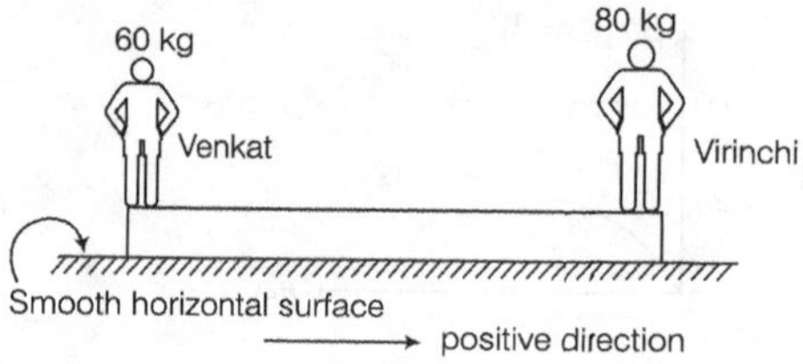

Match the Column I with Column II and select the correct option from the codes given below.

	Column I		Column II
i.	Venkat alone jumps to the left.	p.	$-\frac{1}{2}\,ms^{-1}$
ii.	Virinchi alone jumps to the right.	q.	$-\frac{1}{8}\,ms^{-1}$
iii.	Venkat jumps to the left and Virinchi jumps to the right simultaneously.	r.	$+\frac{3}{8}\,ms^{-1}$
iv.	Venkat jumps to the left after that Virinchi jumps to right.	s.	$-\frac{1}{40}\,ms^{-1}$

Codes

	i	ii	iii	iv
a.	r	p	q	s
b.	p	q	r	s
c.	p	s	q	r
d.	q	p	s	r

One Integer Value Correct Type

Direction *(Q. Nos. 6 and 7) This section contains 2 questions. When worked out will result in an integer from 0 to 9 (both inclusive).*

6 A block of mass m is kept on a rough plank of mass $2m$ which is resting on a smooth horizontal surface. The plank is sufficiently long. Now, the upper block is given an initial velocity u as shown in the figure.The total work done by friction is $-\frac{1}{k}mu^2$. Find the value of k.

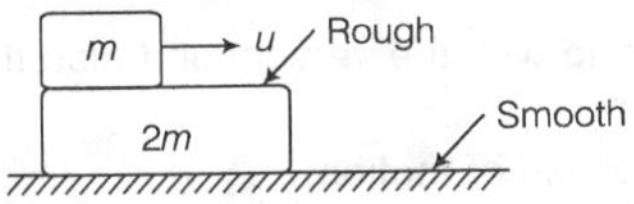

7 for the figure shown, the wedge + particle is resting at a vertical wall. The particle of mass $m\left(\frac{M}{2}\right)$ is released from rest from the position shown. The wedge has mass M. The maximum height reached by the particle on the circular groove is $\frac{2}{\alpha}R$. Find the value of α.

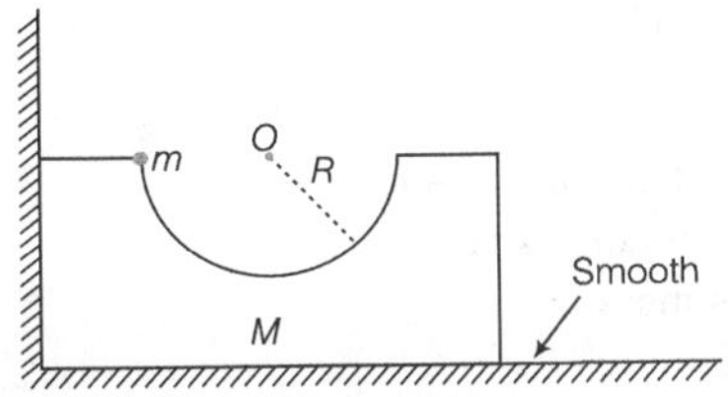

Subjective Questions

Direction *(Q. Nos. 8-12) These questions are subjective in nature, need to be solved completely on notebook.*

8 A man of mass M falls from an aeroplane which is moving horizontally with a speed u at a height h. He removes his jacket of mass m and throws it with a velocity u_r with respect to him (see figure). If he is able to just skip the land under him and is able to fall in the lake. Find out the distance s. Neglect the air resistance.

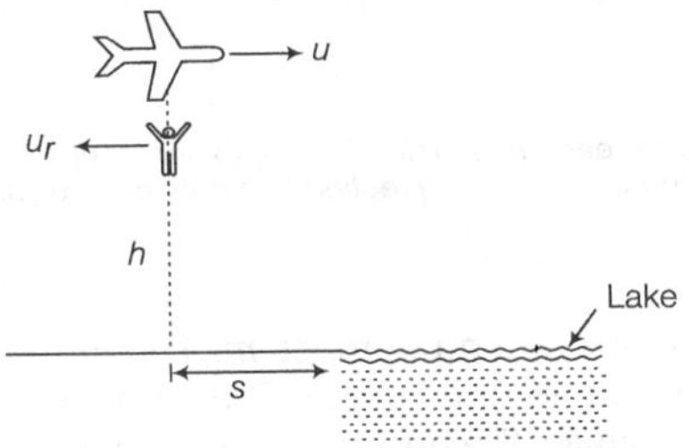

9 Two trucks of mass M each are moving in opposite directions on adjacent parallel tracks with same speed u. One is carrying wheat bags and the other is carrying rice bags. Each bag of wheat has mass m_1 and that of rice has mass m_2 (included in the mass of truck M). When trucks get close to each other while passing, the drivers exchange one bag with each other. Find the speeds of the trucks after the exchange.

Working Space

10 For the figure shown, a wedge of mass M having semi-circular groove with radius R. A particle of mass $m = \frac{M}{2}$ is released from P. It slides on the smooth circular track and starts climbing on the right face.

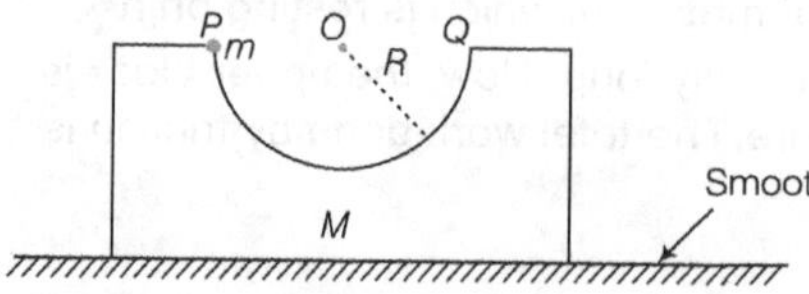

i. Find the maximum value of θ which it can subtend with the vertical and also find the distance displaced by wedge at this position.
ii. Find the maximum velocity of wedge during process of motion.

11 A particle of mass 1 kg is attached to an ideal string of length 5 m. The string is attached to a fixed point O (see figure). It is released from the position shown in the figure. (Given, $g = 10\ \text{ms}^{-2}$) Calculate

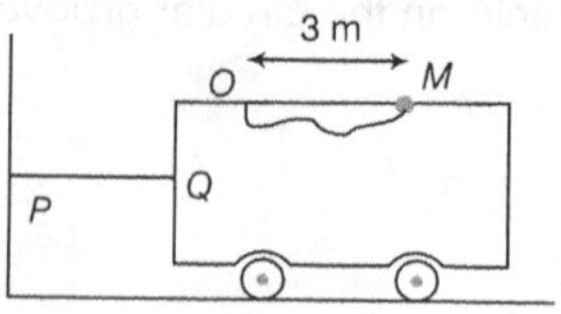

i. the impulse developed in the string when it just becomes taut.
ii. the velocity of the particle just after the string becomes taut.
iii. the impulse developed in the string PQ at this instant.
iv. the impulse given by the earth at this instant.

12 Three particles A, B and C of respective masses m_1, m_2 and m_3 lie on a smooth horizontal surface and are fastened to two light inextensible strings as shown in the figure. The particle A is imparted an impulse J along BA. Find the initial speed of each particle. In the figure, $\theta = 45°$.

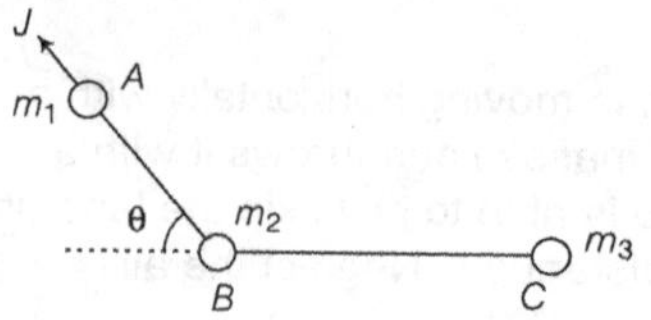

Comprehension Type

Direction (Q. Nos. 13-20) This section contains 3 paragraphs, each describing theory, experiments, data, etc. Eight questions related to the paragraphs have been given. Each question has only one correct answer among the four given options (a), (b), (c) and (d).

Passage I

A horizontal frictionless string is threaded through a bead of mass m. The string is tied to two vertical opposite sides of a cart of mass M (see figure). The length of the cart is L and the radius of the bead is r [$r << L$]. Initially, the bead is at the right edge of the cart. The cart is struck and as a result moves with a velocity v_0. Assuming that no energy is lost in any collision.

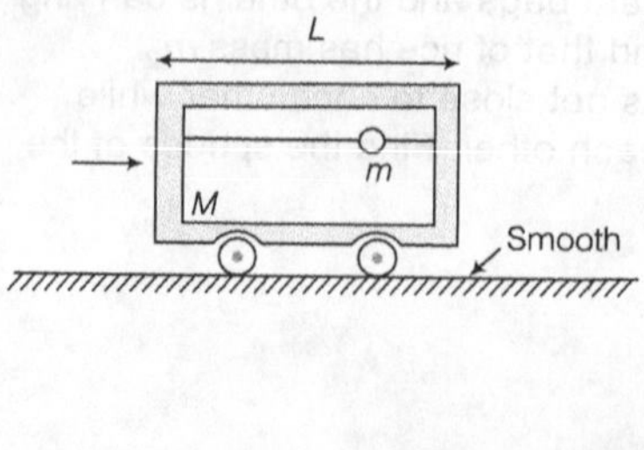

13 The velocity of centre of mass of the cart and the bead is

a. $\dfrac{mv_0}{m+M}$ b. $\dfrac{Mv_0}{m+M}$ c. $\dfrac{Mv_0}{M-m}$ d. $\dfrac{mv_0}{M-m}$

14 The first collision of the bead with the wall of the cart takes place after a time t_1, given by

a. $\dfrac{L}{v_0}$ b. $\dfrac{L}{v_{COM}}$ c. $\dfrac{2L}{v_{COM}}$ d. $\dfrac{L}{2v_0}$

15 The velocity of the bead after the first collision in the centre of mass frame is

a. $\dfrac{2Mv_0}{m+M}$ b. $\dfrac{2m\,v_0}{m+M}$ c. $\dfrac{mv_0}{m+M}$ d. $\dfrac{Mv_0}{m+M}$

Passage II

Block A (mass 1 kg) is placed on smooth horizontal surface and connected to a block B (mass 2 kg) as shown in the figure. A bullet of mass 250 g strikes the block horizontally with a speed $u = 200\ \text{ms}^{-1}$. The bullet penetrates through the block A and comes out with velocity $100\ \text{ms}^{-1}$. (Given, $g = 10\ \text{ms}^{-2}$)

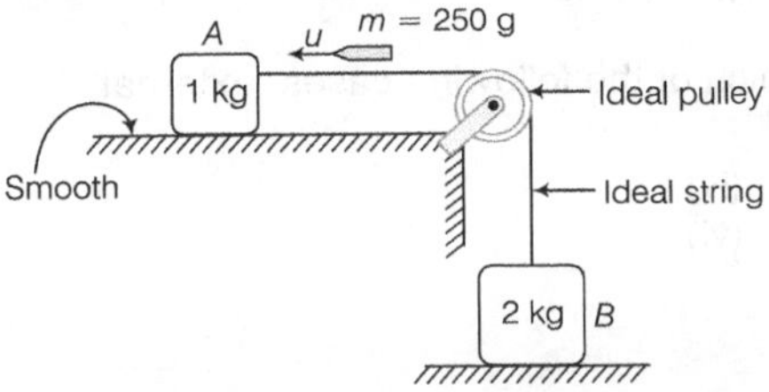

16 Velocity of 2 kg block just after bullet comes out of block A is

a. $8\ \text{ms}^{-1}$ b. $5\ \text{ms}^{-1}$

c. $\dfrac{25}{3}\ \text{ms}^{-1}$ d. $\dfrac{25}{7}\ \text{ms}^{-1}$

17 Maximum displacement of block A in left direction is approximately

a. 2.2 m b. 3.2 m c. 4.2 m d. 5.2 m

Passage III

An initially stationary box on a frictionless horizontal floor explodes into two pieces, piece 1 with mass m_1 and piece 2 with mass m_2. Two pieces then move across the floor along X-axis. Graphs of position *versus* time for the two pieces are given below.

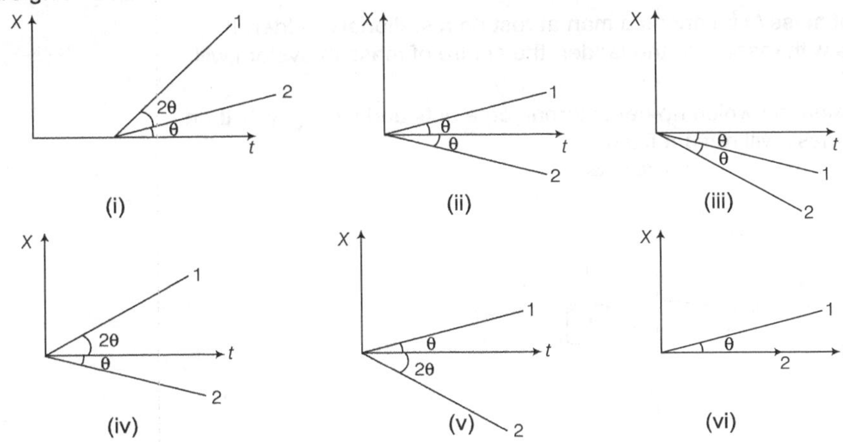

Working Space

18 Which of the graphs shown represent physically possible explosions? (Column II has graph numbers)

a. (ii), (iv) and (v)
b. (i), (iii) and (iv)
c. (i), (iii) and (v)
d. (ii), (iii) and (vi)

19 Based on the above question, match Column I with Column II (Column II has graph numbers).

	Column I		Column II
i.	$m_1 = m_2$	p.	(i)
ii.	$m_1 > m_2$	q.	(ii)
iii.	$m_1 < m_2$	r.	(iii)
		s.	(iv)
		t.	(v)
		u.	(vi)

Codes

	i	ii	iii		i	ii	iii
a.	u	r	p	b.	q	t	s
c.	q	s	s	d.	u	t	s

20 If following graphs are possible, then in which of the following cases, external impulse must be acting on the box?

a. (ii)
b. (iv)
c. (v)
d. (vi)

Statement Type

Direction *(Q. Nos. 21 and 22) This section is based on Statement I and Statement II. Select the correct answer from the codes given below.*

a. Both Statement I and Statement II are correct and Statement II is the correct explanation of Statement I
b. Both Statement I and Statement II are correct but Statement II is not the correct explanation of Statement I
c. Statement I is correct but Statement II is incorrect
d. Statement II is correct but Statement I is incorrect

21 **Statement I** If linear momentum of a system of discrete particles is zero, the kinetic energy of the system of discrete particles will be zero.

Statement II If kinetic energy of a system of discrete particles is zero, the linear momentum of the system of discrete particles will be zero.

22 **Statement I** A block of mass M balances a man at rest on a stationary ladder. If man moves upwards with respect to the ladder, the centre of mass of system will not move.

Statement II For a system on which no net external force acts and if $u_{COM} = 0$, then position of centre of mass will remain fixed.

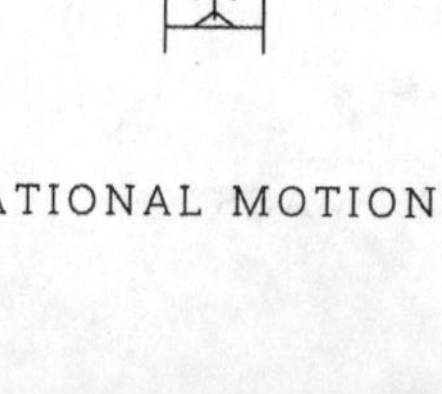

DPP-3 Variable Mass

Subjective Questions

Working Space

Direction (Q. Nos. 1-5) *These questions are subjective in nature, need to be solved completely on notebook.*

1 Derive an expression for the velocity of a rocket after a time t.
Given, u = initial velocity, m_0 = initial mass of the rocket,
$\left[-\frac{dm}{dt}\right]$ = mass of the gases rejected per unit time,
v_r = exhaust velocity of gases and m = mass of rocket at t.

2 A mass m_1 is connected by a weightless cable passing over a frictionless pulley to a container of water, whose mass is $m_0\ (< m_1)$ at $t = 0$. If the container ejects water in downward direction at a constant rate $\mu\ \text{kgs}^{-1}$ with a velocity v_0 relative to the container, then determine the acceleration of m_1 as a function of time.

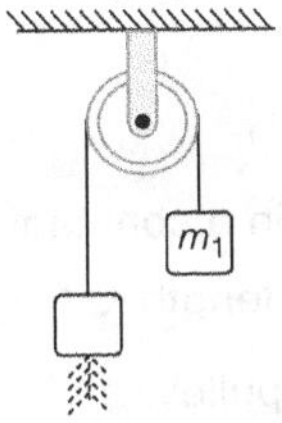

3 A railroad car of length L and mass m_0, when empty, is moving freely on a horizontal track while being loaded with sand from a stationary hopper at a rate $\frac{dm}{dt} = q$. Assuming that the car was approaching the hopper at a speed v_0, then determine

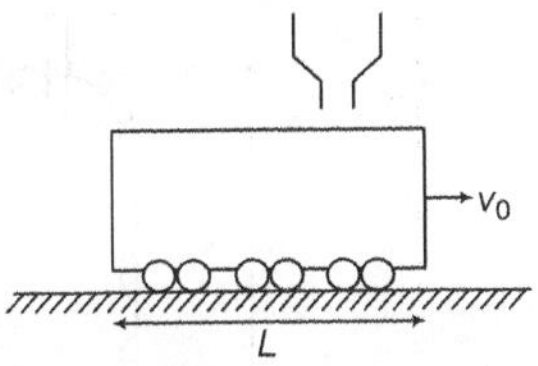

i. the mass of the car and its load after the car has cleared the hopper.
ii. the speed of the car at that time.

4 A chain of length L and mass m lies in a pile on the floor. If its end A is raised vertically at a constant speed v_0, then express in terms of the length y of the chain, which is off the floor at any given instant

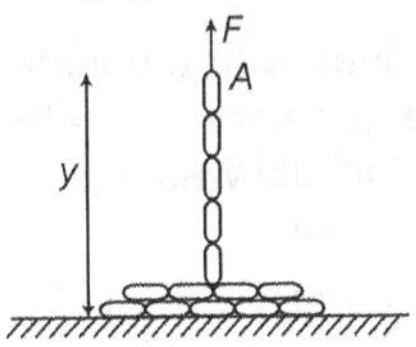

i. the magnitude of the force F applied to end A.
ii. energy lost during the lifting of chain.

5 A rocket is moving vertically upwards against gravity. Its mass at time t is $m = m_0 - \mu t$ and it expels burnt fuel at a speed u vertically downwards relative to the rocket. Here, m_0 and μ are constants. Derive the equation of motion of rocket. You need not solve it.

One Integer Value Correct Type

Working Space

Direction (Q. No. 6) *This section contains 1 question. When worked out will result in an integer from 0 to 9 (both inclusive).*

6 A cylindrical solid of mass 10^{-2} kg and cross-sectional area 10^{-4} m^2 is moving parallel to its axis (the X-axis) with a uniform speed of 10^3 ms^{-1} in the positive direction. At $t = 0$, its front face passes the plane $x = 0$. The region to the right of this plane is filled with stationary dust particles of uniform density 10^{-3} kg m^{-3}. When a dust particle collides with the face of the cylinder, it sticks to its surface. Assuming that the dimensions of the cylinder remain practically unchanged and that the dust sticks only to the front face of the cylinder, the x-coordinate of the front of the cylinder at $t = 150$ s is 10^{β} m. Find the value of β.

Only One Option Correct Type

Direction (Q. Nos. 7-12) *This section contains 6 multiple choice questions. Each question has four choices (a), (b), (c) and (d), out of which ONLY ONE is correct.*

7 A rocket with an initial mass of 1000 kg is launched vertically upwards from rest under gravity. The rocket burns the fuel at the rate of 10 kg s^{-1}. The burnt matter is ejected vertically downwards with a speed of 2000 ms^{-1} relative to the rocket. If burning is complete after 1 min, find the maximum velocity of rocket. (Given, $g = 10$ ms^{-2} and $\ln 2.5 = 0.916$)

a. 1232 ms^{-1}
b. 1000 ms^{-1}
c. 1423 ms^{-1}
d. 1523 ms^{-1}

8 Determine the force F required to give the open link chain a constant velocity $v = \dfrac{dy}{dt}$. The chain has a total length L and mass per unit length λ. Neglect the small size and mass of the pulley and any friction in the pulley.

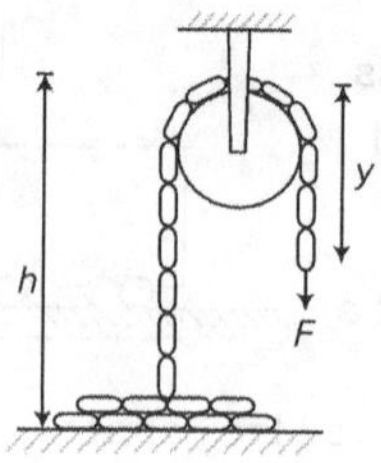

a. $\dfrac{\lambda v^2}{2} + \lambda g(h - y)$
b. $\lambda v^2 + \lambda g(h - y)$
c. $\lambda v^2 + \dfrac{\lambda g(h - y)}{2}$
d. $\dfrac{\lambda v^2 + \lambda g(h - y)}{2}$

9 Find the mass of a rocket as a function of time, if it moves with a constant acceleration α, in the absence of external forces. The gas escapes with a constant velocity u relative to the rocket and its mass initially was m_0.

a. $m = m_0 e^{-(\alpha/u)t}$
b. $m = m_0 e^{-(2\alpha/u)t}$
c. $m = m_0 e^{-(u/\alpha t)}$
d. $m = m_0 e^{-(2u/\alpha t)}$

10 A 4200 kg rocket is travelling in outer space with a velocity of 150 ms^{-1} towards the sun. It wishes to alter its course by 30° and can do this by shooting its rockets chiefly in a direction perpendicular to its original motion. If the small rockets are expelled at a speed 2700 ms^{-1}, what mass has to be expelled?

a. 92 kg
b. 214 kg
c. 134 kg
d. 312 kg

Working Space

11 A chain of mass m and length L is held vertical such that its lower end just touches the floor. It is released from rest. Find the force exerted by the chain on the table when upper end is about to hit the floor.

a. $2mg$ b. $3mg$ c. $\frac{5}{2}mg$ d. $\frac{3}{2}mg$

12 For the above problem, what is the force on the table by the chain when half of the chain has fallen?

a. mg b. $\frac{mg}{2}$ c. $\frac{5}{2}mg$ d. $\frac{3}{2}mg$

One or More than One Options Correct Type

Direction (Q. Nos. 13 and 14) This section contains 2 multiple choice questions. Each question has four choices (a), (b), (c) and (d), out of which ONE or MORE THAN ONE are correct.

13 A car of mass m_0 starts moving to the right due to a constant horizontal force F. At the same time sand starts falling into the car from a stationary hopper. The rate of loading is constant and equal to $\mu\,kgs^{-1}$. Mark the correct option(s).

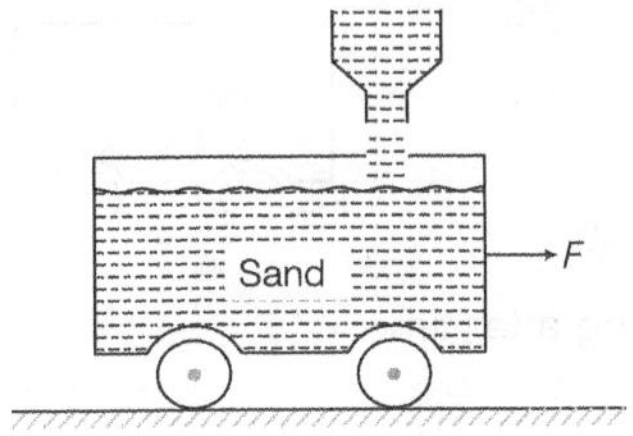

a. velocity of the car at time t is $\frac{Ft}{(m_0 + \mu t)}$

b. velocity of the car at time t is $\frac{Ft}{\sqrt{2}\,(m_0 + \mu t)}$

c. acceleration of the car at time t is $\frac{F m_0}{2(m_0 + \mu t)^2}$

d. acceleration of the car at time t is $\frac{F m_0}{(m_0 + \mu t)^2}$

14 A cart loaded with sand moves along a horizontal floor due to a constant force F, coinciding in direction with the velocity of cart. In the process, the sand spills through a hole in the bottom with a constant rate $\mu\,kgs^{-1}$. If at the initial moment $t = 0$, the cart with loaded sand had the mass m_0 and its velocity was equal to zero (neglect friction). Then,

a. the acceleration of the cart at time t is $\frac{F}{m_0 - \mu t}$

b. the velocity of the cart at time t is $\frac{F}{\mu}\ln\left[\frac{m_0}{m_0 - \mu t}\right]$

c. the acceleration of the cart at time t is $\frac{F}{m_0 + \mu t}$

d. the velocity of the cart at time t is $\frac{F}{2\mu}\ln\left[\frac{m_0}{m_0 - \mu t}\right]$

Comprehension Type

Direction (Q. Nos. 15-20) This section contains 2 paragraphs, each describing theory, experiments, data, etc. Six questions related to the paragraphs have been given. Each question has only one correct answer among the four given options (a), (b), (c) and (d).

Passage I

A rocket set for vertical firing weighs 50 kg and contains 450 kg of fuel. It can have a maximum exhaust velocity of 2 kms^{-1}. (Given, $g = 10\ ms^{-2}$)

15 What should be the minimum rate of fuel consumption of the rocket to just lift it off the launching pad?

a. 1.5 $kg\,s^{-1}$ b. 2.5 $kg\,s^{-1}$ c. 3 $kg\,s^{-1}$ d. 3.5 $kg\,s^{-1}$

Working Space

16 What should be the minimum rate of fuel consumption of the rocket to have an acceleration of 20 ms^{-2}?

a. 3.5 $kg\,s^{-1}$ b. 4.5 $kg\,s^{-1}$
c. 7.5 $kg\,s^{-1}$ d. 10 $kg\,s^{-1}$

17 What will be the speed of the rocket when the rate of consumption of fuel is 10 kgs^{-1} after the entire fuel is consumed? (Given, $\ln 10 = 2.302$)

a. 4156 ms^{-1} b. 5146 ms^{-1}
c. 3146 ms^{-1} d. 2782 ms^{-1}

Passage II

A car has total mass 50 kg. Gases are ejected from it backwards with relative velocity 20 ms^{-1}. The rate of ejection of gases is 2 kg s^{-1}. Total mass of gas is 20 kg. The coefficient of friction between car and road is $\mu = 0.1$ and $g = 10\ ms^{-2}$.

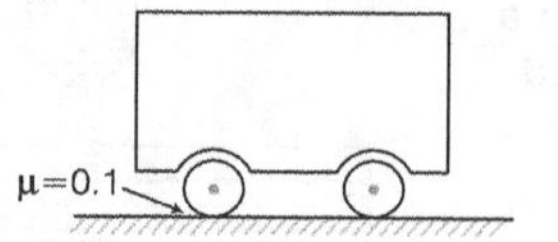

18 Car will start moving after time

a. 4 s b. 10 s c. 5 s d. 8 s

19 Maximum speed of the car will be $\left(\text{Given, } \ln \frac{4}{3} = 0.28\right)$

a. 0.6 ms^{-1} b. 0.8 ms^{-1} c. 1 ms^{-1} d. 1.2 ms^{-1}

20 Car will stop after a time (from the start)

a. 12.2 s b. 15.8 s c. 9.3 s d. 10.6 s

Statement Type

Direction (Q. Nos. 21 and 22) *This section is based on Statement I and Statement II. Select the correct answer from the codes given below.*

a. Both Statement I and Statement II are correct and Statement II is the correct explanation of Statement I
b. Both Statement I and Statement II are correct but Statement II is not the correct explanation of Statement I
c. Statement I is correct but Statement II is incorrect
d. Statement II is correct but Statement I is incorrect

21 **Statement I** A chain of mass per unit length λ begins to fall through a hole in the ceiling as shown in the figure. The thrust force F_t on the hanging part is upwards.

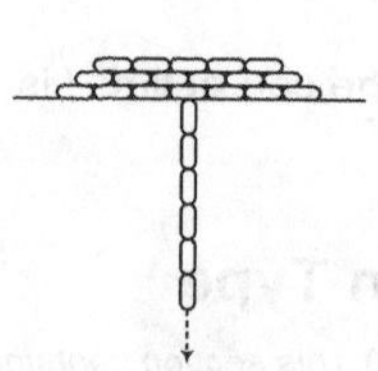

Statement II If relative velocity of the mass joining (leaving) is v_r, then F_t is parallel to v_r if mass is increasing and F_t is opposite to v_r if mass is decreasing.

22 **Statement I** A chain of mass per unit length λ is piled on a platform. Its end is lifted vertically up as shown in the figure. The thrust force F_t on the vertical portion is downwards.

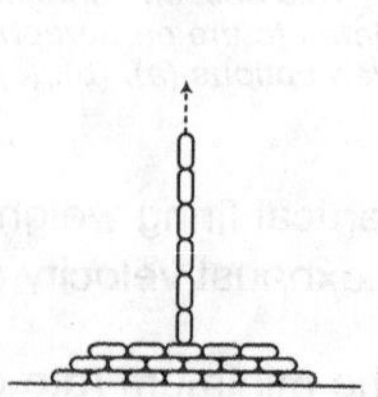

Statement II If the relative velocity of the mass joining (leaving) is v_r, then F_t is parallel to v_r if mass is increasing and F_t is opposite to v_r if mass is decreasing.

DPP-4 Collisions

Only One Option Correct Type

Working Space

Direction (Q. Nos. 1-12) *This section contains 12 multiple choice questions. Each question has four choices (a), (b), (c) and (d), out of which ONLY ONE is correct.*

1 A piece of wood of mass 0.03 kg is dropped from the top of a building 100 m high. At the same time, a bullet of mass 0.02 kg is fired vertically upward with a velocity of 100 ms^{-1} from the ground. The bullet gets embedded in the wooden piece after striking. Find the height to which the combination rises above the building before it starts falling.

a. 46.5 m
b. 41.6 m
c. 49.7 m
d. 52.3 m

2 Two balls B_1 and B_2 having different but unknown masses collide. B_1 is initially at rest and B_2 has a speed u. After collision, B_2 has a speed $\frac{u}{2}$ and moves at right angles to its original motion. Find the direction (with respect to initial direction of B_2) in which ball B_1 moves after collision.

a. $\tan^{-1}\left(\frac{1}{2}\right)$
b. $\tan^{-1}(2)$
c. $\tan^{-1}\left(\frac{1}{4}\right)$
d. $\tan^{-1}(4)$

3 A ball of mass m is projected with speed u into the barrel of a spring gun of mass M initially at rest on a frictionless surface. The mass m sticks in the barrel at the point of maximum compression of the spring. No energy is lost in friction. What fraction of the initial kinetic energy of the ball is stored in the spring?

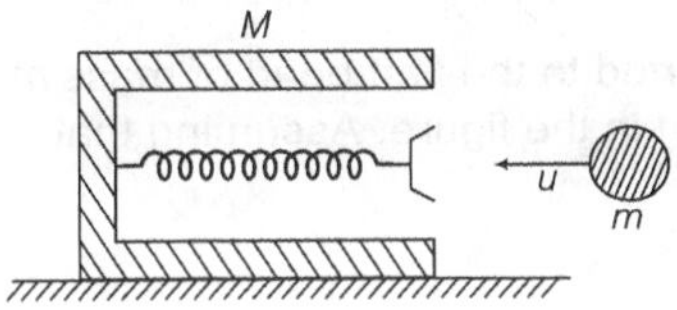

a. $\frac{m}{\sqrt{2}(m+M)}$
b. $\frac{M}{\sqrt{2}(m+M)}$
c. $\frac{m}{m+M}$
d. $\frac{M}{m+M}$

4 A 4 kg block moving with a velocity of 10 ms^{-1} collided with a 6 kg block. After collision, both blocks are moving together. If both of them blocks are made up of a material of specific heat of 2 $Jkg^{-1}K^{-1}$ and there are no radiative loss in the form of heat and sound, then increase in temperature of both blocks (combinedly) will be

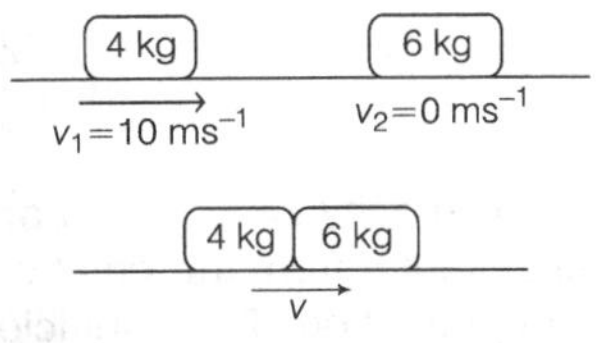

a. 4° C
b. 10° C
c. 6° C
d. 40° C

5 The bob of a pendulum of length 980 cm is released from rest with its string making an angle 60° with the vertical. It collides with a ball of mass 20 g resting on a smooth horizontal surface at the position of rest of pendulum. If the collision is perfectly elastic, with what velocity will the bob move immediately after the collision? Mass of the bob is 10 g.

a. $490\ \text{cms}^{-1}$ b. $\frac{2}{3} \times 980\ \text{cms}^{-1}$

c. $\frac{980}{4}\ \text{cms}^{-1}$ d. $\frac{980}{3}\ \text{cms}^{-1}$

6 A bullet of mass 20 g pierces through a plate of mass $M_1 = 1\ \text{kg}$ and then comes to rest inside a second plate of mass $M_2 = 2.98\ \text{kg}$ as shown in the figure. The plates are on a smooth horizontal surface and bullet moves along the surface. It is found that the two plates, initially at rest, now move with equal velocities. Find the percentage loss in the initial velocity of the bullet when it is in between M_1 and M_2. Neglect any loss of material of the plates.

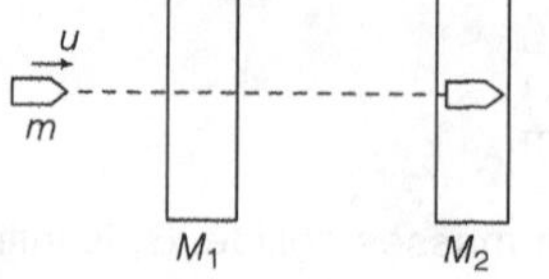

a. 30% b. 25% c. 35% d. 22.5%

7 A ball moving translationally collides elastically with another, stationary, identical ball. At the moment of impact, the angle between the straight line passing through the centres of the balls and the direction of the initial motion of the striking ball is equal to ϕ. Assuming that balls to be smooth, find the fraction of the kinetic energy of the striking ball that is stored as the deformation potential energy at the moment of maximum deformation.

a. $\frac{1}{2}\cos\phi$ b. $\frac{1}{2}\sin\phi$ c. $\frac{1}{2}\cos^2\phi$ d. $\frac{1}{2}\sin^2\phi$

8 n beads are resting on a smooth horizontal wire which is circular at the end with radius R as shown. The masses of the beads are $m, \frac{m}{2}, \frac{m}{4}, \ldots, \frac{m}{2^{n-1}}$, respectively.

Find the minimum velocity that should be imparted to the first bead of mass m such that the nth bead will fall in the tank shown in the figure. Assuming that elastic collisions.

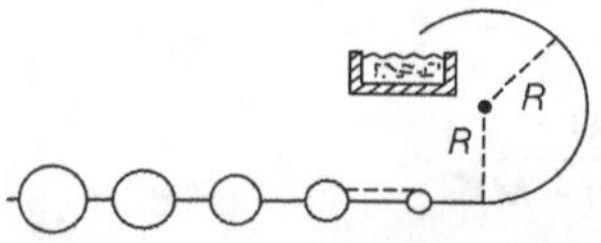

a. $\left(\frac{3}{4}\right)^{n-1}\sqrt{5gR}$ b. $\left(\frac{4}{3}\right)^{n-1}\sqrt{5gR}$

c. $\left(\frac{3}{4}\right)^{n-1} 2\sqrt{gR}$ d. $\left(\frac{4}{3}\right)^{n-1} 2\sqrt{gR}$

9 A ball is projected with velocity $v_0 = \sqrt{gh}$ at an angle θ to the horizontal towards a vertical smooth wall distant d from the point of projection. After collision the ball returns to the point of projection. The coefficient of restitution is

a. $\frac{d}{h\sin\theta - d}$ b. $\frac{d}{2h\sin\theta - d}$

c. $\frac{d}{h\sin 2\theta - d}$ d. $\frac{d}{\sin 2\theta + 2d}$

Working Space

10 A smooth ring is fixed horizontally on a smooth horizontal surface. From a point A of the ring, a particle is projected at an angle θ to the radius vector at A. If the coefficient of restitution between the ring and the particle is 0.5 and the particle returns to the point of projection A after two reflections, then $\cot^2 \theta$ is equal to

a. 10
b. 12
c. 11
d. 14

11 A small ball is projected from point A on the floor towards a vertical wall as shown in the figure. It hits the wall when its velocity is horizontal. Ball reaches point A after one bounce on the floor. If the coefficient of restitution is the same for the collisions, find its value.

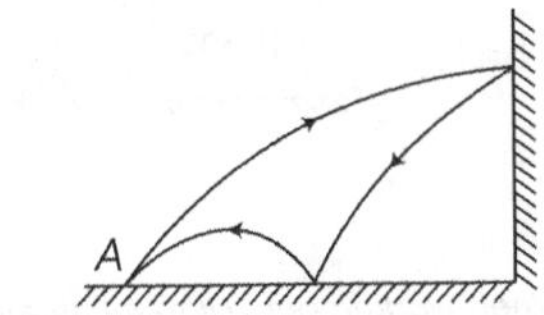

a. 0.5
b. 0.25
c. 0.75
d. 0.33

12 Two equal discs initially at rest are in contact on a smooth horizontal table. A third disc of same mass but of double radius strikes them symmetrically and comes to rest after impact. The coefficient of restitution is

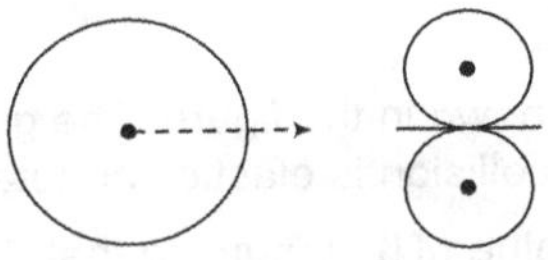

a. $\frac{3}{4}$
b. $\frac{9}{16}$
c. $\frac{3}{16}$
d. $\frac{4}{9}$

Subjective Questions

Direction (Q. Nos. 13-16) *These questions are subjective in nature, need to be solved completely on notebook.*

13 Particle 1 experiences a perfectly elastic collision with a stationary particle 2. Determine the ratio of their masses if

i. after a head on collision the particles fly apart in the opposite directions with equal speeds.
ii. the particles fly apart symmetrically relative to the initial motion, the direction of particle 1 with the angle of divergence $\theta = 60°$.

14 A ball of mass m is shot with a velocity u at an angle θ with the horizontal which strikes the box of M resting at the edge of a smooth table (see figure). After striking the box horizontally, the ball falls down vertically. Find the value of u such that the box is raised to the point P shown in the figure. Also, find the horizontal distance l from which the shot was fired. All surfaces are smooth.

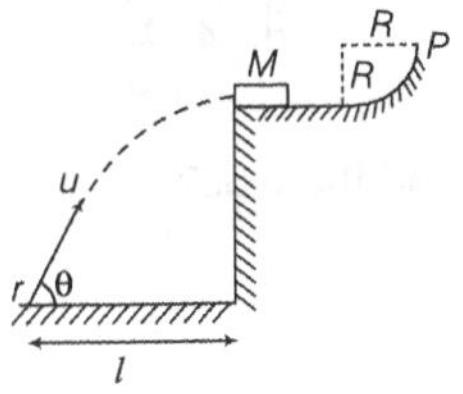

Working Space

15 A smooth ball of mass m strikes a smooth floor with a velocity u at an angle θ with the normal to the floor. If the coefficient of restitution is e, then determine

i. the rebound speed of the particle.

ii. the angle of reflection with the normal.

iii. the impulse on the ball.

iv. the loss in KE of the ball.

16 A smooth ball is dropped from a height h on a smooth incline as shown in the figure. After collision, the velocity of the ball becomes horizontal.

i. Find the coefficient of restitution.

ii. If the collision is elastic, what is the impulse on the ball?

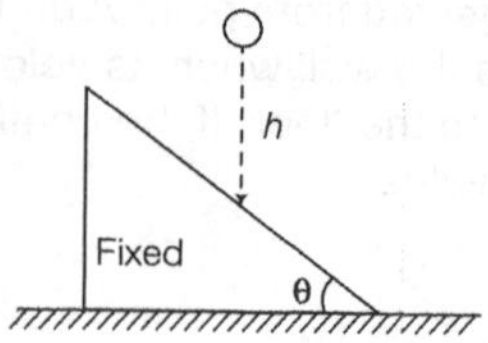

One Integer Value Correct Type

Direction (Q. Nos. 17 and 18) This section contains 2 questions. When worked out will result in an integer from 0 to 9 (both inclusive).

17 A ball is projected with velocity u at an angle θ to the horizontal towards a smooth wall which approaches the ball with velocity u'. After collision the ball retraces its path to the point of projection. The time t taken by the ball from the instant of projection to the point of impact is $\dfrac{(u\cos\theta + ku')\,u\sin\theta}{g(u\cos\theta + u')}$. Find the value of k.

18 A block of mass m is projected with velocity u as shown in the figure. The ground is smooth but there is friction between A and B. If collision is elastic, the total energy dissipated in friction is $\dfrac{4}{3\beta}mu^2$. Find the value of β. Assuming that A does not fall off B.

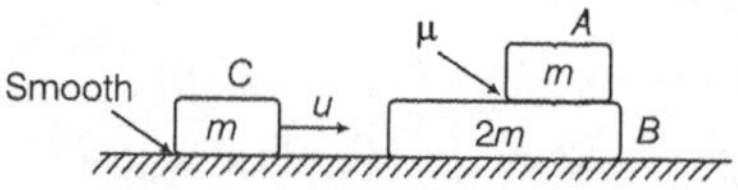

Comprehension Type

Direction (Q. Nos. 19-23) This section contains 2 paragraphs, each describing theory, experiments, data, etc. Five questions related to the paragraphs have been given. Each question has only one correct answer among the four given options (a), (b), (c) and (d).

Passage I

A particle is thrown with a velocity u from a height h horizontally towards a vertical wall which is moving away with a speed $\dfrac{u}{4}$ as shown in the figure. The particle returns to the point of projection after suffering two elastic collisions (one with the wall and the other with the ground).

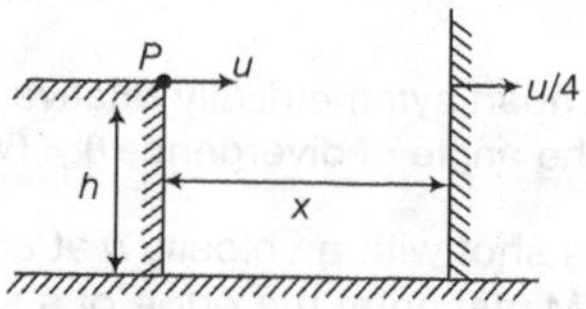

19 The total time of flight is

a. $2\sqrt{\dfrac{2h}{g}}$ b. $2\sqrt{\dfrac{h}{g}}$ c. $\sqrt{\dfrac{2h}{g}}$ d. $4\sqrt{\dfrac{2h}{g}}$

20 What is the initial separation x between the particle and the wall?

a. $2u\sqrt{\dfrac{h}{2g}}$ b. $u\sqrt{\dfrac{h}{2g}}$ c. $u\sqrt{\dfrac{2h}{g}}$ d. $u\sqrt{\dfrac{h}{g}}$

Working Space

Passage II

Two smooth balls A and B, each of mass m and radius R, have their centres at $(0, 0, R)$ and at $(5R, -R, R)$, respectively in a coordinate system shown. Ball A, moving along the X-axis, collides with ball B. Just before collision, speed of the ball A is 4 ms^{-1} and B is stationary. The collision between the balls is elastic.

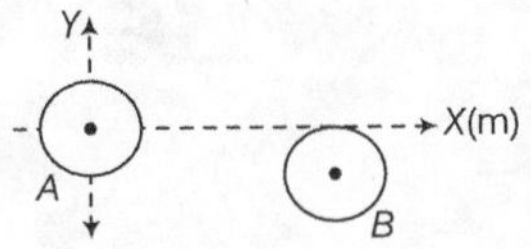

21 Velocity of the ball A just after the collision is (in ms^{-1})

a. $(\hat{i} + \sqrt{3}\,\hat{j})$ b. $(\hat{i} - \sqrt{3}\,\hat{j})$ c. $(2\hat{i} + \sqrt{2}\,\hat{j})$ d. $(2\hat{i} + 2\hat{j})$

22 Impulse of the force exerted by A on B during the collision (in kg-ms^{-1}) is

a. $(\sqrt{3}m\hat{i} + 3m\hat{j})$ b. $\left(\frac{\sqrt{3}}{2}m\hat{i} - 3m\hat{j}\right)$

c. $(3m\,\hat{i} - \sqrt{3}m\hat{j})$ d. $(2\sqrt{2}m\hat{i} + 3m\hat{j})$

23 If coefficient of restitution for the above collision is 0.5 keeping all other parameters unchanged, what will be the velocity of the ball B after collision? (in ms^{-1})

a. $\frac{1}{2}(3\sqrt{3}\,\hat{i} + 9\hat{j})$ b. $\frac{1}{4}(9\hat{i} - 3\sqrt{3}\,\hat{j})$

c. $(6\hat{i} - 3\sqrt{2}\,\hat{j})$ d. $(6\hat{i} + 3\sqrt{3}\,\hat{j})$

Matching List Type

Direction *(Q. No. 24) Choices for the correct combination of elements from Column I and Column II are given as options (a), (b), (c) and (d), out of which one is correct.*

24 A box B of mass M, having a frictionless spherical empty space of radius R (as shown) is placed on a smooth horizontal floor. A ball C of mass m is placed at the bottom position of the spherical space of box B. A ball A of mass M_0 strikes elastically with the box with velocity v, while moving on the same floor. h is the height up to which ball C, inside the box, will rise after the impact.

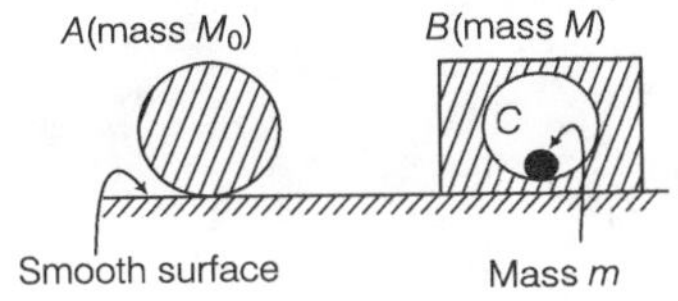

Match the Column I with Column II and select the correct option from the codes given below.

	Column I		Column II
i.	If $M = 2m$ and $M_0 = 3m$	p.	$h = 0$
ii.	If $M = 3m$ and $M_0 = 4m$	q.	$h = \frac{v^2}{4g}$
iii.	For $M_0 = M + m$, the value of M so that ball C of mass m rises to a height kR is	r.	$(k+1)m$
		s.	None of the above

Codes

	i	ii	iii
a.	p	p	s
b.	p	q	r
c.	r	p	s
d.	q	p	r

SET - 1

Revisal Problems for JEE Main

Only One Option Correct Type

Direction *(Q. Nos. 1-20) This section contains 20 multiple choice questions. Each question has four choices (a), (b), (c) and (d), out of which ONLY ONE is correct.*

Working Space

1 A thin rod of length $3L$ is bent at right angles at a distance L from one end. Locate the centre of mass of the arrangement w.r.t. the corner (see figure). (Given, $L = 1.2$ m)

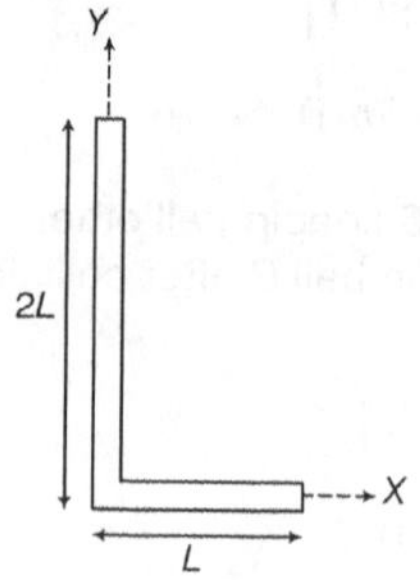

a. $\mathbf{r}_{COM} = (0.2\,\hat{\mathbf{i}} + 0.8\,\hat{\mathbf{j}})$ m
b. $\mathbf{r}_{COM} = (0.4\hat{\mathbf{i}} + 0.6\hat{\mathbf{j}})$ m
c. $\mathbf{r}_{COM} = (0.5\,\hat{\mathbf{i}} + \hat{\mathbf{j}})$ m
d. $\mathbf{r}_{COM} = (0.2\hat{\mathbf{i}} + 0.6\hat{\mathbf{j}})$ m

2 Find the coordinates of centre of mass of the uniform lamina shown in the figure.

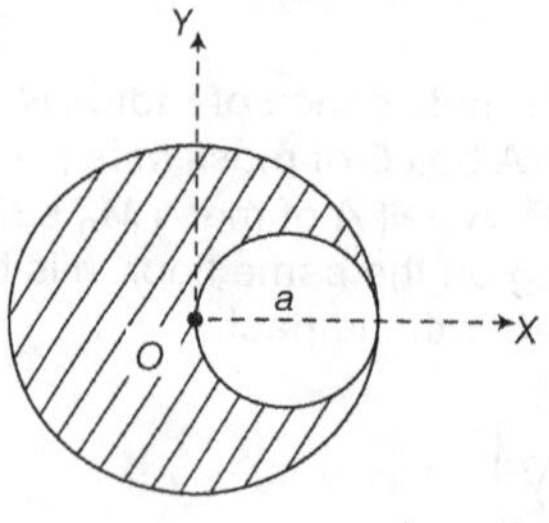

a. $\left(\frac{a}{6}, 0\right)$
b. $\left(-\frac{a}{6}, 0\right)$
c. $\left(-\frac{a}{4}, 0\right)$
d. $\left(\frac{a}{4}, 0\right)$

3 Two particles A and B of masses 1 kg and 2 kg respectively are projected in the directions shown in the figure with speeds $u_A = 200$ ms^{-1} and $u_B = 55$ ms^{-1}. Initially, they were 90 m apart. Find the maximum height attained by the centre of mass of the particles. Assume g is constant and is equal to 10 ms^{-2}.

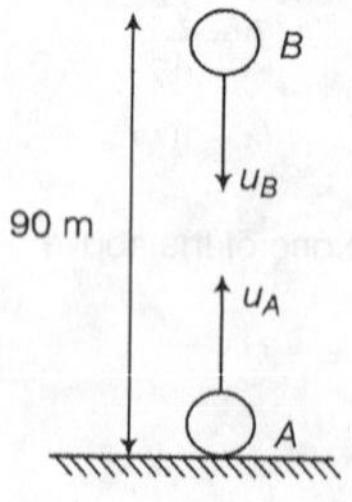

a. 125 m
b. 145 m
c. 115 m
d. 105 m

4 A wooden plank of mass 20 kg is resting on a smooth horizontal floor. A man of mass 60 kg starts moving from one end of the plank to the other end. The length of the plank is 10 m. Find the displacement of the plank over the floor when the man reaches the other end of the plank.

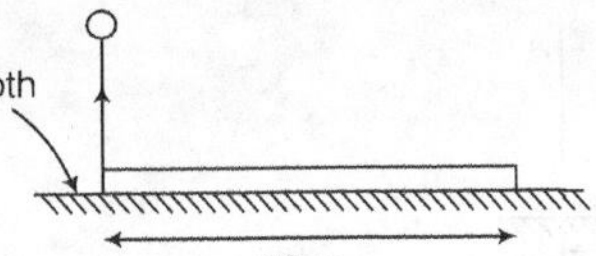

a. 5 m
b. 6.5 m
c. 2.5 m
d. 7.5 m

5 A liquid has been poured into a cylindrical vessel of mass m (the mass of the vessel bottom can be ignored) and height h. The linear density of the liquid which is, the ratio of the mass of the liquid column to its height is denoted by α. Calculate the height x of the column of the liquid at which the common centre of mass of (the liquid + vessel) system is in the lowest position. Assume the centre of mass of the vessel to be at $\frac{h}{2}$ from the ground.

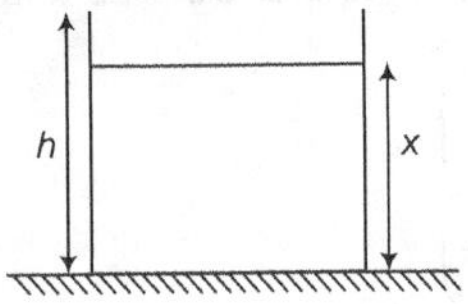

a. $\frac{\sqrt{m(m+h\alpha)}+m}{\alpha}$
b. $\frac{\sqrt{m(m+h\alpha)}}{\alpha}$
c. $\frac{\sqrt{m(m+h\alpha)}-m}{\alpha}$
d. None of these

6 A small particle of mass m slides down a circular path of radius R cut into a large block of mass M as shown in the figure. M rests on a table and both block and particle move without friction. The particle and the block are initially at rest and m starts from the top of the path. The horizontal distance from the bottom of the block to where the particle hits the table is

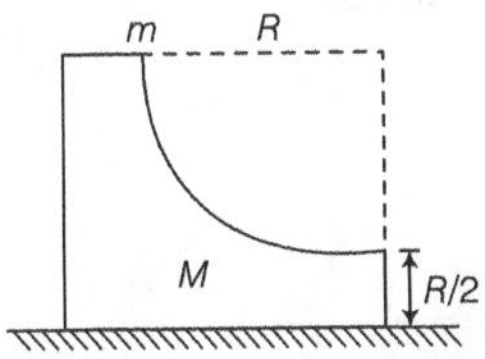

a. $R\sqrt{\frac{2(M+m)}{M}}$
b. $R\sqrt{\frac{2(M+m)}{m}}$
c. $R\sqrt{\frac{M+m}{M}}$
d. $R\sqrt{\frac{M+m}{2m}}$

7 A wagon of mass M can move without friction along horizontal rails. A simple pendulum consisting of a sphere of mass m is suspended from the ceiling of the wagon by an ideal string of length l. At the initial moment, the wagon and the pendulum are at rest and the string is making angle α with the vertical. The velocity of wagon when the pendulum passes through its mean position is

a. $m\sin\left(\frac{\alpha}{2}\right)\sqrt{\frac{gl}{(M+m)M}}$

b. $m\sin\left(\frac{\alpha}{2}\right)\sqrt{\frac{2gl}{(M+m)M}}$

c. $2m\sin\left(\frac{\alpha}{2}\right)\sqrt{\frac{2gl}{(M+m)M}}$

d. $2m\sin\left(\frac{\alpha}{2}\right)\sqrt{\frac{gl}{(M+m)M}}$

8 A ball of negligible size and mass m is given a velocity v_0 when it is at the centre of a cart of mass M. Cart is originally at rest. The coefficient of restitution between the walls A and B and the ball is e. The time taken by the ball to hit first A and B and then to reach the centre of the cart is

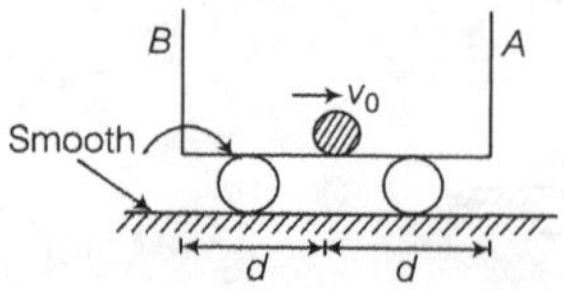

a. $\frac{d}{v_0}\left(1+\frac{1}{e}+\frac{1}{e^2}\right)$

b. $\frac{d}{v_0}\left(1+\frac{2}{e}+\frac{1}{e^2}\right)$

c. $\frac{d}{v_0}\left(1+\frac{2}{e}+\frac{2}{e^2}\right)$

d. $\frac{d}{2v_0}\left(\frac{1}{e}+\frac{1}{e^2}+1\right)$

9 A ball of mass m is dropped from a height h to hit the ground and then rebound to a certain height. The process carries on and the ball continues to hit the ground time and again. The coefficient of restitution between the ground and the ball is e. The total distance travelled by the ball before coming to rest is

a. $\frac{h}{2}\left(\frac{1+e^2}{1-e^2}\right)$

b. $h\left(\frac{1+2e^2}{1-2e^2}\right)$

c. $h\left(\frac{1+e^2}{1-e^2}\right)$

d. $h\left(\frac{1-2e^2}{1+2e^2}\right)$

10 For the above problem, what is the average speed?

a. $\sqrt{\frac{gh}{2}}\frac{(1+e^2)}{(1+e)^2}$

b. $\sqrt{gh}\left(\frac{1+e^2}{1-e^2}\right)$

c. $\sqrt{2gh}\left(\frac{1+e^2}{1-e^2}\right)$

d. $2\sqrt{gh}\left(\frac{1+e^2}{1-e^2}\right)$

11 A small ball of mass m is placed on the top of a super ball of mass M. The two balls are dropped to the floor from height h. How high does the small ball rise after the collision? Assume that collisions are elastic and neglect the size of balls.

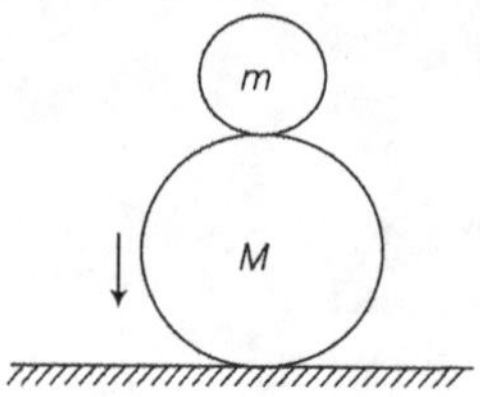

a. $6h$

b. $8h$

c. $9h$

d. $12h$

12 A point P is fixed at a height h above a perfectly inelastic smooth horizontal plane. An ideal string of length $L\,(> h)$ has one end attached to a fixed point P and the other is attached to a heavy particle. The particle is held at the level of P with the string just taut and released from rest. Find the height of the particle above the plane when it is next instantaneously at rest.

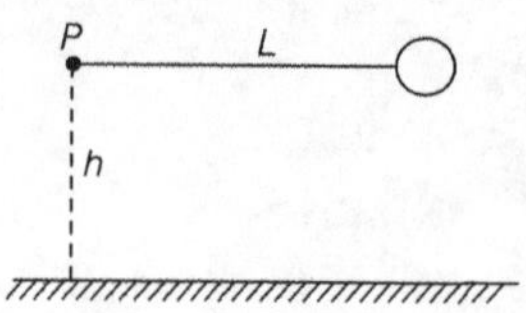

a. $\frac{h^7}{L^6}$

b. $\frac{h^4}{L^3}$

c. $\frac{h^2}{L}$

d. $\frac{h^5}{L^4}$

13 A sphere of mass m is moving with a velocity $4\hat{\mathbf{i}} - \hat{\mathbf{j}}$. It hits a wall and rebounds with velocity $\hat{\mathbf{i}} + 3\hat{\mathbf{j}}$. The coefficient of restitution between the sphere and the wall is

a. $\frac{3}{4}$ b. $\frac{9}{16}$

c. $\frac{4}{15}$ d. $\frac{2}{3}$

14 A sphere of mass m, impinges obliquely on a sphere of mass M which is at rest. The directions of motion of the spheres make an angle x with each other after impact. If $m = eM$ (where, e is coefficient of restitution), then x is

a. $\frac{\pi}{3}$ b. $\frac{\pi}{2}$

c. $\frac{\pi}{6}$ d. $\frac{\pi}{4}$

15 A rocket has a mass of 21000 kg of which 15000 kg is fuel. The rocket engine can exhaust fuel at the rate of 190 kg s^{-1} with an exhaust velocity 2500 ms^{-1} relative to the rocket. If the rocket is fired vertically upwards, its final velocity at the burn out is (in ms^{-1})
[Given, ln (3.5) = 1.253]

a. 2735 b. 2935

c. 2535 d. 1975

16 A pile of loose link chain of mass per unit length λ lies on a rough surface with coefficient of kinetic friction μ_k. One end of the chain is being pulled horizontally along the surface by a constant force F. Determine the acceleration of chain in terms of x and $\frac{dx}{dt} = v$.

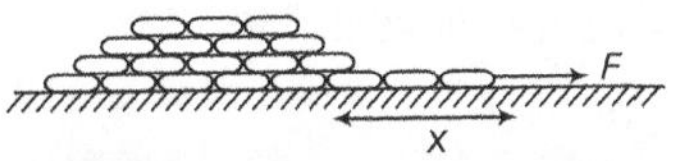

a. $\frac{F}{\lambda x} - \mu_k g + \frac{v^2}{x}$

b. $\frac{F}{\lambda x} - \mu_k g - \frac{v^2}{\lambda x}$

c. $\frac{F}{\lambda x} - \mu_k g - \frac{v^2}{2x}$

d. $\frac{F}{\lambda x} - \mu_k g + \frac{v^2}{2x}$

17 A sphere moving with velocity v strikes elastically with a wall moving towards the sphere with velocity u. If the mass of the wall is infinitely large, the work done by the wall during collision will be

a. $mu(u + v)$ b. $2mu(u + v)$

c. $\frac{2}{3}mu(u + v)$ d. $\frac{3}{2}mu(u + v)$

18 A particle of mass m, initially at rest is acted upon by a variable force F for a brief interval of time T_0. It is moving with a velocity u after the force stops acting F as a function of t is a semi-circle as shown. Then,

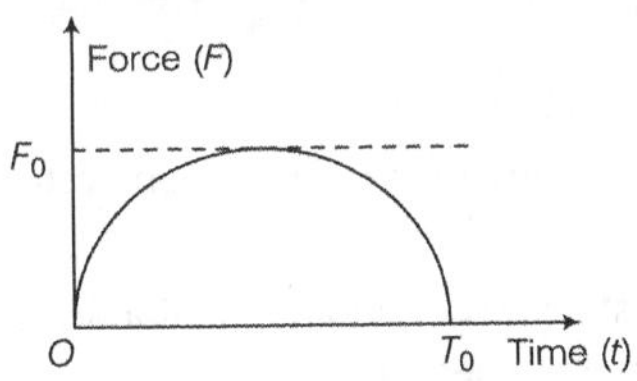

a. $\frac{\pi F_0^2}{2m}$ b. $\frac{\pi F_0 T_0}{2m}$

c. $\frac{\pi F_0 T_0}{4m}$ d. $\frac{F_0 T_0}{2m}$

19 Two identical balls *A* and *B* lie on a smooth horizontal surface, which gradually merge into a curve to a height 3.2 m. Ball *A* is given a velocity 10 ms^{-1} to collide head on with ball *B*, which then takes up the curved path. The minimum coefficient of restitution *e* for the collision between *A* and *B*, in order that of *B* reaches the highest point *P* of curve is (Given, $g = 10\ ms^{-2}$)

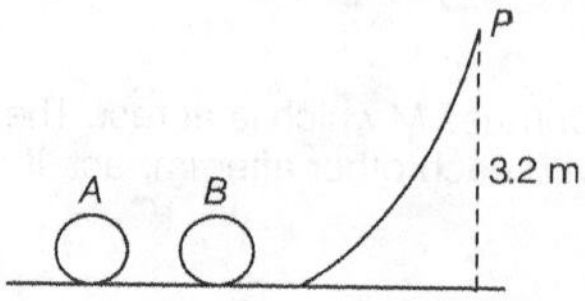

a. 1/2
b. 3/5
c. 1/4
d. 3/4

20 A ball is projected with initial velocity *u* at an angle θ with the horizontal. Then, horizontal displacement covered by the ball as it collides third time with the ground would be (coefficient of restitution is *e*)

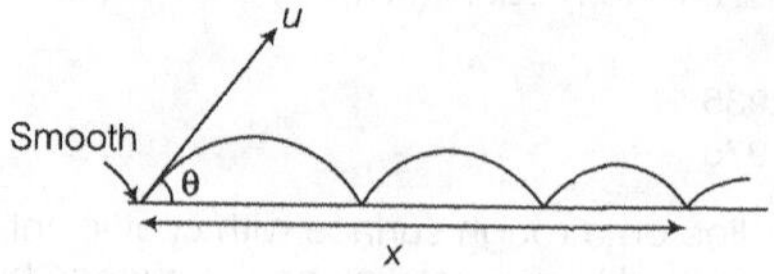

a. $(1+e)\dfrac{u^2 \sin 2\theta}{g}$
b. $(1+e+e^2)\dfrac{u^2 \sin 2\theta}{2g}$
c. $e\dfrac{u^2 \sin 2\theta}{g}$
d. $(1+e+e^2)\dfrac{u^2 \sin 2\theta}{g}$

Statement Type

Direction (Q. Nos. 21-23) ***This section is based on Statement I and Statement II. Select the correct answer from the codes given below.***

a. Both Statement I and Statement II are correct and Statement II is the correct explanation of Statement I
b. Both Statement I and Statement II are correct but Statement II is not the correct explanation of Statement I
c. Statement I is correct but Statement II is incorrect
d. Statement II is correct but Statement I is incorrect

21 **Statement I** A particle of mass *m* strikes a smooth wedge of mass *M* as shown in the figure. Linear momentum of particle along the surface of wedge is conserved during collision.

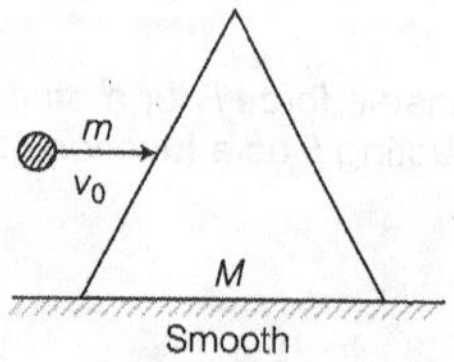

Statement II Wedge exerts a force perpendicular to inclined face of wedge on particle during collision.

22 **Statement I** A body of mass m_1 collides head on elastically with another stationary body of mass m_2. After the collision, velocity of m_2 is maximum, when $m_1 >> m_2$.

Statement II Velocity of second body is always maximum, when its mass m_2 is greater than mass of the hitting body.

23 **Statement I** If no external force acts on a system of particles, then the centre of mass will not move in any direction.

Statement II If net external force is zero, then the linear momentum of the system remains constant.

SET - 2

Revisal Problems
for JEE Advanced

Only One Option Correct Type

Working Space

Direction (Q. Nos. 1-11) *This section contains 11 multiple choice questions. Each question has four choices (a), (b), (c) and (d), out of which ONLY ONE is correct.*

1 A small sphere of radius R is held against the inner surface of larger sphere of radius $6R$ (as shown in the figure). The masses of large and small spheres are $4M$ and M, respectively. This arrangement is placed on a horizontal table. There is no friction between any surfaces of contact. The small sphere is now released. Find the coordinates of the centre of the large sphere, when the smaller sphere reaches the opposite extreme position.

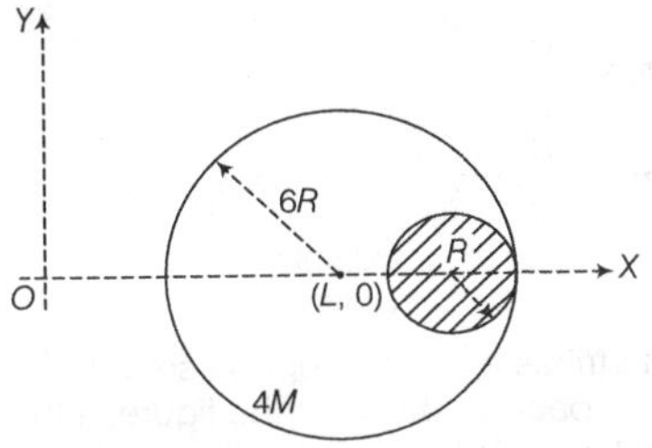

a. $(L + 2R, 0)$
b. $(L + R, 0)$
c. $(L - 2R, 0)$
d. $(L - R, 0)$

2 Two blocks each of mass m are connected by an ideal string and are placed on a smooth wedge as shown. Pulley shown is ideal. Find the magnitude of acceleration of centre of mass of two block system.

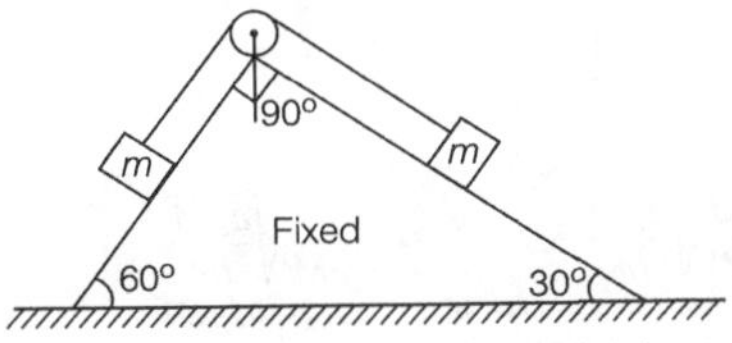

a. $g\left(\frac{\sqrt{3}-1}{2\sqrt{2}}\right)$
b. $g\left(\frac{\sqrt{3}-1}{\sqrt{2}}\right)$
c. $g\left(\frac{\sqrt{3}-1}{4\sqrt{2}}\right)$
d. $\frac{g}{\sqrt{2}}$

3 A ladder PQ of length 5 m inclined to a vertical wall is slipping over a horizontal surface with a velocity of 2 ms^{-1}, when Q is at a distance 3 m from the ground. Calculate the velocity of centre of mass of the rod at this moment.

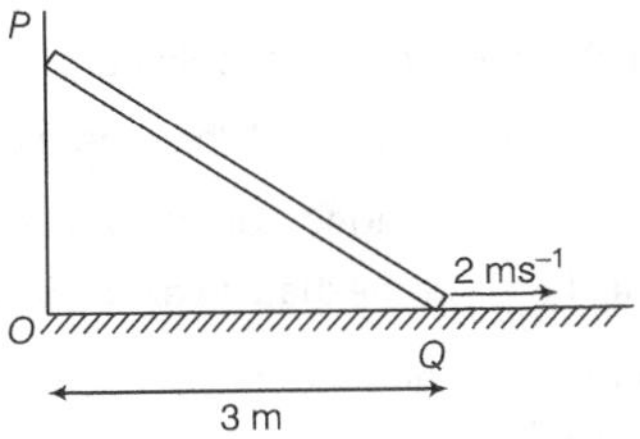

a. 2.5 ms^{-1}
b. 1.25 ms^{-1}
c. 1.75 ms^{-1}
d. 2.75 ms^{-1}

Working Space

4 A wedge of mass m_2 is placed at rest on a scale as shown in the figure. A small block of mass m_1 slides down the frictionless incline of the wedge. Find the scale reading while the block slides. (m_2 does not move)

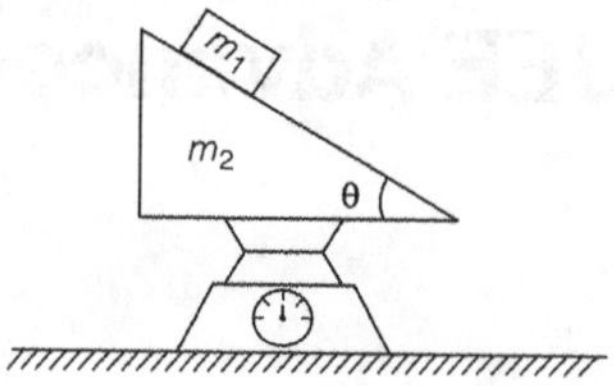

a. $(m_1 + m_2)\cos^2 \theta g$
b. $(m_1 + m_2)\sin^2 \theta g$
c. $(m_1 \sin^2 \theta + m_2)g$
d. $(m_1 \cos^2 \theta + m_2)g$

5 A plank of mass 5 kg is placed on a frictionless horizontal plane. Further a block of mass 1 kg is placed over the plank. A massless spring of natural length 2 m is fixed to the plank at its one end. The other end of the spring (not fixed to block) is compressed by the block by half of spring's natural length. The system is now released from rest. If stiffness constant of the spring is 100 Nm^{-1}, the velocity of the plank when the block leaves the plank is (in ms^{-1})

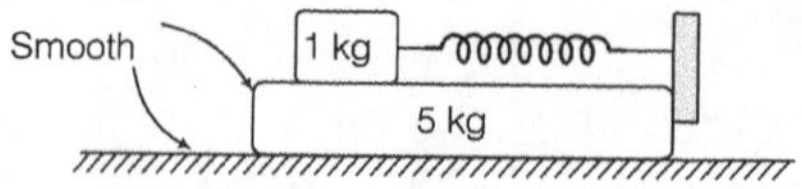

a. $\sqrt{\frac{20}{3}}$
b. $\sqrt{\frac{10}{3}}$
c. 3
d. 3.5

6 A ball of mass m when dropped from certain height strikes a wedge kept on smooth horizontal surface and moves horizontally just after impact as shown in the figure. If the ball strikes the ground at a distance d from its initial line of fall, determine the amplitude of oscillation of the wedge after being hit by the ball.

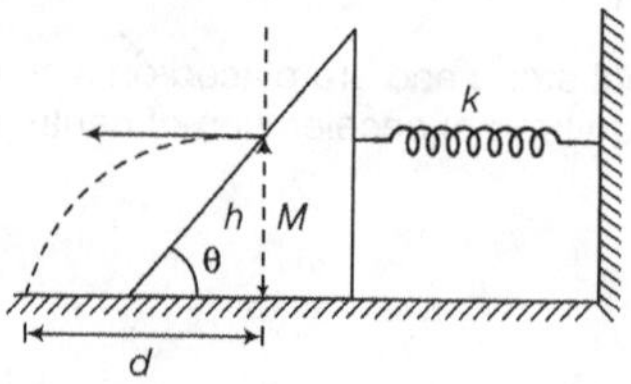

a. $\frac{md}{M}\sqrt{\frac{gM}{2hk}}$
b. $\frac{md}{M}\sqrt{\frac{gM}{hk}}$
c. $\frac{md}{M}\sqrt{\frac{2gM}{hk}}$
d. $\frac{md}{2M}\sqrt{\frac{2gM}{hk}}$

7 Three particles A, B and C of masses m, $2\,m$ and $3\,m$ respectively lie on a smooth horizontal table at the vertices of an equilateral triangle. A and B as well as B and C are connected by ideal strings. C is given a velocity v_0 parallel to AB as shown in the figure. Eventually, A begins to move with speed.

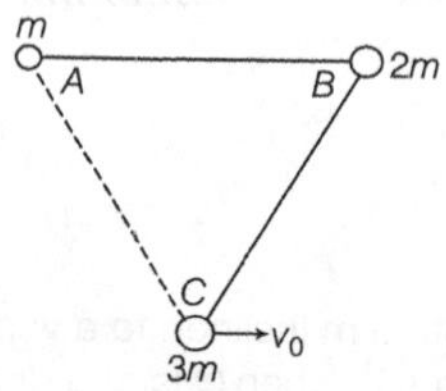

a. $\frac{2v_0}{17}$
b. $\frac{v_0}{14}$
c. $\frac{3v_0}{17}$
d. $\frac{2v_0}{19}$

8 A double stage rocket has an initial mass M_i. Gas is exhausted from the rocket at a constant rate of $\frac{dm}{dt}$ and with an exhaust velocity u relative to the rocket. When the mass of the rocket reaches the value μ, the first state of mass m is disengaged from the rocket then the rocket continues the second stage at the same rate $\left(\frac{dm}{dt}\right)$ and exhaust velocity (u) as in the first stage until it reaches a mass M_f. If gravity is neglected and the rocket has started from rest, the velocity of the rocket at the end of first stage is

a. $u \ln\left(\frac{M_i}{\mu}\right)$
b. $u \ln\left(\frac{M_i - \mu}{M_i}\right)$
c. $u \ln\left(\frac{M_i + \mu}{M_i}\right)$
d. $u \ln\left(\frac{\mu}{M_i}\right)$

Working Space

9 In the above problem, what is the velocity of rocket after second stage?

a. $u \ln \left[\frac{M_i(\mu + m)}{M_f(\mu - m)}\right]$

b. $u \ln \left(\frac{M_i M_f}{\mu(\mu - m)}\right)$

c. $u \ln \left(\frac{M_i (\mu - m)}{\mu M_f}\right)$

d. $u \ln \left(\frac{M_i(\mu - m)}{M_f^2}\right)$

10 A thin hoop of mass M and radius r is placed on a smooth horizontal plane vertically. At the initial instant, the hoop is at rest. A small washer of mass m with zero initial velocity slides from the upper point of the hoop smoothly along the hoop. Determine velocity u of the centre of the hoop at the moment when the washer is at a certain point A of the hoop, whose radius vector forms an angle ϕ with the vertical.

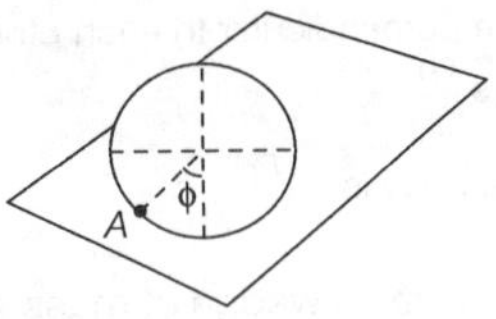

a. $m \cos \phi \sqrt{\frac{gr(1 + \cos \phi)}{(M + m)(M + m \sin^2 \phi)}}$

b. $m \cos \phi \sqrt{\frac{2gr(1 + \cos \phi)}{(M + m)(M + m \sin^2 \phi)}}$

c. $m \cos \phi \sqrt{\frac{2gr(1 + \sin \phi)}{(M + m)(M + m \cos^2 \phi)}}$

d. $m \cos \phi \sqrt{\frac{gr(1 + \sin \phi)}{(M + m)(M + m \cos^2 \phi)}}$

11 A ball B falls vertically downwards onto a wedge with velocity v_1. The wedge is moving horizontally with velocity v_2 as shown in the figure. The ball strikes the wedge elastically. What must be the ratio $\frac{v_1}{v_2}$, so that the ball bounces back in vertically upwards direction relative to wedge?

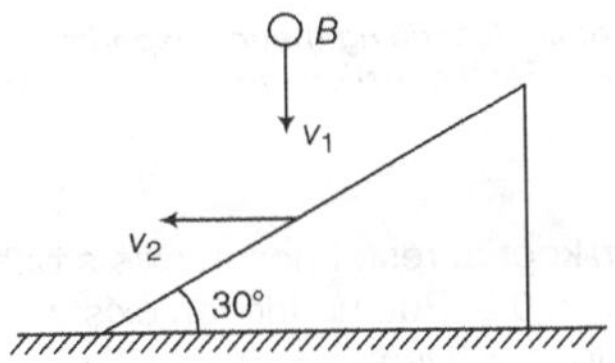

a. $\sqrt{3}$

b. $1/\sqrt{3}$

c. 2

d. 1/2

One or More than One Options Correct Type

Direction (Q. Nos. 12-15) *This section contains 4 multiple choice questions. Each question has four choices (a), (b), (c) and (d), out of which ONE or MORE THAN ONE are correct.*

12 A block B of mass m starts from the rest at the top of the wedge A of mass M. All surfaces are smooth. A can slide on the ground. B slides down onto the ground, moves along with it with a speed v, has an elastic collision with the wall and climbs back on to A. Then,

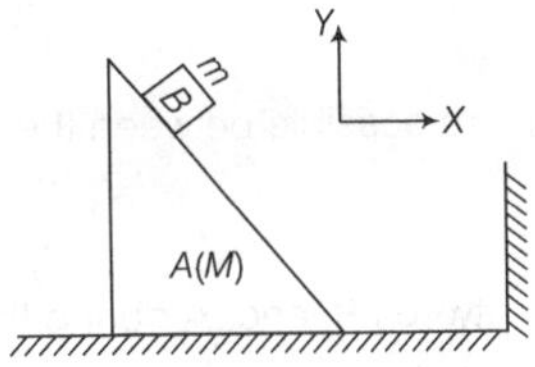

a. B will reach the top of A again

b. from the beginning, till the collision with the wall, the x-coordinate of centre of mass of $(B + A)$ system is unaltered

c. immediately after the collision, the centre of mass of $(B + A)$ system moves with a velocity $\frac{2mv}{M + m}$

d. when B reaches its highest point on A, the speed of A is $\frac{2mv}{M + m}$

13 Two friends P and Q (each weighing 40 kg) are sitting on a frictionless platform at some distance d apart. P rolls a ball of mass 4 kg towards Q which Q catches. Then, Q rolls the ball towards P and P catches it. The ball keeps on moving back and forth between P and Q. The ball has a fixed speed 5 ms^{-1} on the platform. Choose the correct option(s).

a. Speed of P after he catches the ball for the first time is $\frac{10}{11}$ ms^{-1}

b. Speed of P after he catches the ball for the first time is $\frac{10}{9}$ ms^{-1}

c. The centre of mass of the system remains stationary irrespective of the direction of motion of the ball

d. P can roll the ball only for 6 times

14 A small particle travelling with a velocity v collides elastically with a spherical body of equal mass and radius r initially kept at rest. The centre of this spherical body is located at a distance $\rho (< r)$ from the direction of motion of the particle (see figure). After collision,

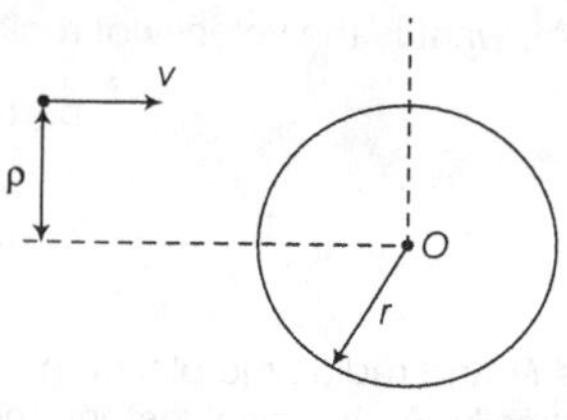

Working Space

a. particle stops and spherical body moves in the direction of particle
b. both of them move perpendicular to each other
c. speed of particle is $\frac{v\rho}{r}$
d. speed of spherical body is $\frac{v}{r}\sqrt{r^2 - \rho^2}$

15 A ball of mass 1 kg strikes a wedge of mass 4 kg horizontally with a velocity of 10 ms^{-1}. Just after collision, the velocity of wedge becomes 4 ms^{-1}. Friction is absent everywhere and collision is elastic. Then,

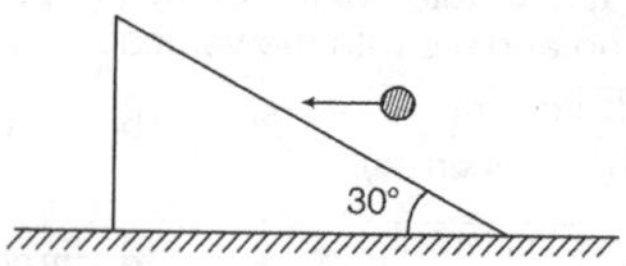

a. speed of ball after collision is 6 ms^{-1}
b. speed of ball after collision is 8 ms^{-1}
c. impulse between ball and wedge during collision is 16 N-s
d. impulse between ball and wedge during collision is 32 N-s

Comprehension Type

Direction (Q. Nos. 16-21) *This section contains 2 paragraphs, each describing theory, experiments, data, etc. Six questions related to the paragraphs have been given. Each question has only one correct answer among the four given options (a), (b), (c) and (d).*

Passage I

A block of mass 2 m collides elastically with a mass m kept at rest. Friction exists between the block B_2 and the surface HH'. Coefficient of friction is 0.3. But, no friction exists between B_1 and HH'. HH' is horizontal and sufficiently long. (Given, $g = 10$ ms^{-2})

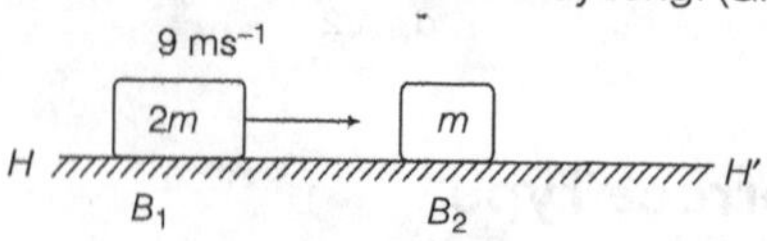

16 The blocks will again collide after a time?

a. 6 s
b. 4 s
c. 8 s
d. ∞

17 How many collisions are possible between the blocks?

a. 2
b. 4
c. 27
d. ∞

18 Maximum distance between B_1 and B_2 during the time interval between first collision and second collision is

a. 12.5 m
b. 13.5 m
c. 20 m
d. 9 m

Passage II

In Bohr's atomic model, energy levels of the hydrogen atom are given by $E_n = \frac{-E_1}{n^2}$ where, $n = 1, 2, 3, \ldots$ and $n = 1$ corresponds to ground state and $E_1 = 13.6$ eV.
Let m_H denotes mass of H-atom.
Consider a collision between a moving H-atom (in its ground state) and a stationary H-atom (in its ground state).

19 Smallest energy associated with moving H-atom for excitation of H-atom at rest is

a. E_1
b. $\frac{2}{3}E_1$
c. $\frac{3}{4}E_1$
d. $\frac{4}{3}E_1$

20 Maximum velocity v_0 of moving H-atom which results in an elastic collision is

a. $\sqrt{\frac{3E_1}{m_H}}$ b. $\sqrt{\frac{2E_1}{m_H}}$

c. $\sqrt{\frac{E_1}{m_H}}$ d. $\sqrt{\frac{5E_1}{m_H}}$

21 At $v = v_0$, the collision is

a. purely elastic b. purely inelastic
c. either elastic or inelastic d. cannot be determined

Matching List Type

Direction ***(Q. No. 22) Choices for the correct combination of elements from Column I and Column II are given as options (a), (b), (c) and (d), out of which one is correct.***

22 For the arrangement shown in the figure, match the following columns and select the correct option from the codes given below.

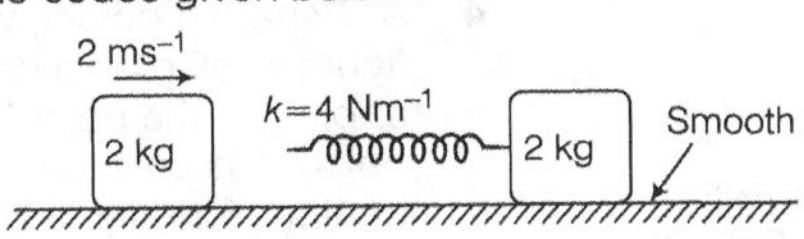

	Column I		Column II
i.	Velocity of COM (in ms^{-1}).	p.	2
ii.	Velocity of combined mass (in ms^{-1}) when compression in the spring is maximum.	q.	1
iii.	Maximum compression (in m) in the spring.	r.	4
iv.	Maximum potential energy (in J) stored in the spring.	s.	0.5
		t.	None of the above

Codes

	i	ii	iii	iv		i	ii	iii	iv
a.	q	q	q	p	b.	p	q	r	s
c.	p	r	t	s	d	t	p	s	q

One Integer Value Correct Type

Direction ***(Q. Nos. 23 and 24) This section contains 2 questions. When worked out will result in an integer from 0 to 9 (both inclusive).***

23 A shell of mass 1 kg is projected with velocity 20 ms^{-1} at an angle 60° with horizontal. It collides inelastically with a ball of mass 1 kg which is suspended through a thread of length 1m. The other end of the thread is affected to the ceiling of a trolley of mass 18 kg as shown in the figure. Initially, the trolley is stationary and it is free to move along horizontal rails without any friction. The maximum deflection of the thread with vertical is θ, such that $\cos\theta = \frac{1}{\alpha}$. Taking, $g = 10 \, ms^{-2}$, find the value of α.

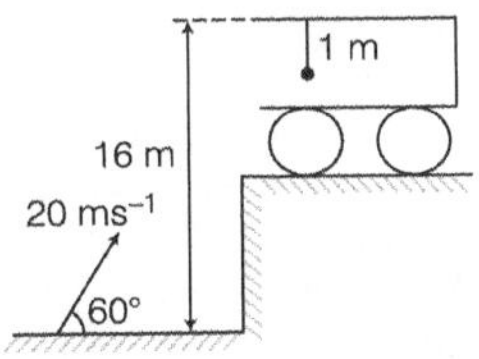

24 After perfectly inelastic collision between two identical particles moving with same speed in different directions, the speed of the particles becomes half of the initial speed. If the angle between the velocities just before impact is ϕ (in degree), find the value of $\frac{\phi}{20}$.

JEE Main & AIEEE Archive

(Compilation of Last 13 Years' Questions)

1 A thin uniform rod of length l and mass m is swinging freely about a horizontal axis passing through its end. Its maximum angular speed is ω. Its centre of mass rises to a maximum height of **(AIEEE 2009)**

a. $\frac{1}{3} \cdot \frac{l^2\omega^2}{g}$

b. $\frac{1}{6} \cdot \frac{l\omega}{g}$

c. $\frac{1}{2} \cdot \frac{l^2\omega^2}{g}$

d. $\frac{1}{6} \cdot \frac{l^2\omega^2}{g}$

2 A thin rod of length L is lying along the X-axis with its ends at $x = 0$ and $x = L$. Its linear density (mass/length) varies with x as $k\left(\frac{x}{L}\right)^n$, where n can be zero or any positive number. If the position x_{CM} of the centre of mass of the rod is plotted against n, which of the following graphs best approximates the dependence of x_{CM} on n? **(AIEEE 2008)**

a. 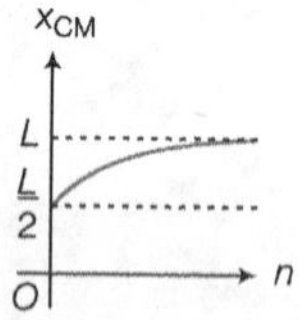

b. 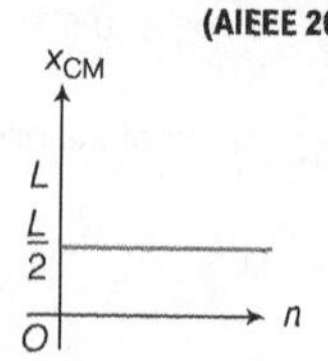

c. 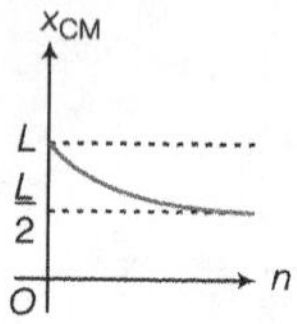

d. 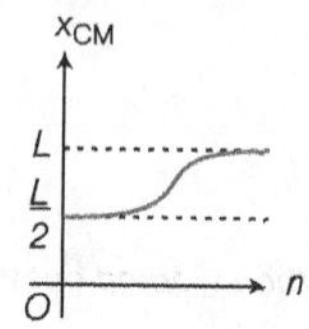

3 Consider a two particle system with particles having masses m_1 and m_2. If the first particle is pushed towards the centre of mass through a distance d, by what distance should the second particle be moved, so as to keep the centre of mass at the same position? **(AIEEE 2006)**

a. $\frac{m_2}{m_1} d$

b. $\frac{m_1}{m_1 + m_2} d$

c. $\frac{m_1}{m_2} d$

d. d

4 A T shaped object with dimensions shown in the figure is lying on a smooth floor. A force **F** is applied at the point P parallel to AB such that the object has only the translational motion without rotation. Find the location of P with respect to C. **(AIEEE 2005)**

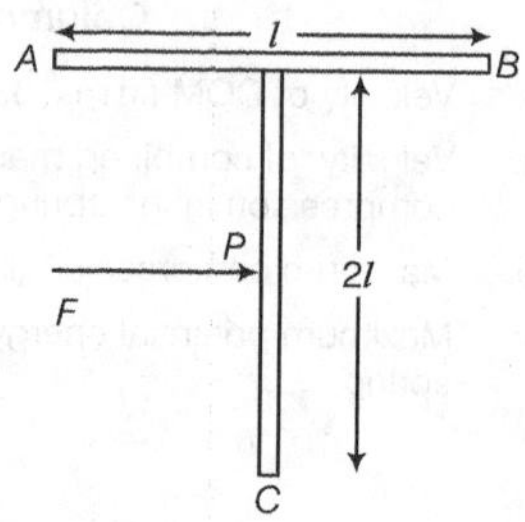

a. $\frac{2}{3} l$

b. $\frac{3}{2} l$

c. $\frac{4}{3} l$

d. l

5 Two identical particles move towards each other with velocity $2v$ and v, respectively. The velocity of centre of mass is **(AIEEE 2002)**

a. v

b. $\frac{v}{3}$

c. $\frac{v}{2}$

d. zero

JEE Advanced & IITJEE Archive

(Compilation of Last 10 Years' Questions)

1 Three objects A, B and C are kept in a straight line on a frictionless horizontal surface. These have masses m, $2m$ and m, respectively. The object A moves towards B with a speed 9 ms^{-1} and makes an elastic collision with it. Thereafter, B makes completely inelastic collision with C. All motions occur on the same straight line. Find the final speed (in ms^{-1}) of the object C.

(2009, Integer Type)

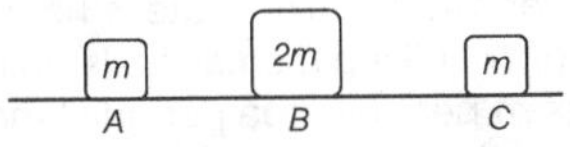

2 Two small particles of equal masses start moving in opposite directions from a point A in a horizontal circular orbit. Their tangential velocities are v and $2v$ respectively, as shown in the figure. Between collisions, the particles move with constant speeds. After making how many elastic collisions, other than that at A, these two particles will again reach the point A?

(2009, Only One Option Correct Type)

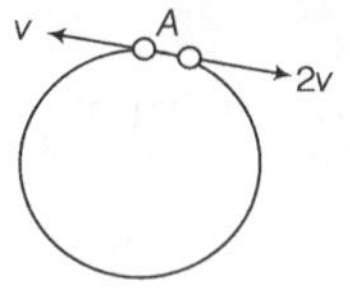

a. 4 b. 3
c. 2 d. 1

3 Look at the drawing given in the figure, which has been drawn with ink of uniform line-thickness. The mass of ink used to draw each of the two inner circles and each of the two line segments is m. The mass of the ink used to draw the outer circle is $6m$. The coordinates of the centres of the different parts are : outer circle (0, 0), left inner circle $(-a, a)$, right inner circle (a, a), vertical line (0, 0) and horizontal line $(0, -a)$. The y-coordinate of the centre of mass of the ink in this drawing is

(2009, Only One Option Correct Type)

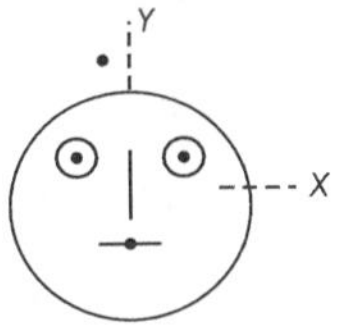

a. $\frac{a}{10}$ b. $\frac{a}{8}$
c. $\frac{a}{12}$ d. $\frac{a}{3}$

4 If the resultant of all the external forces acting on a system of particles is zero, then from an inertial frame, one can surely say that

(2009, Only One Option Correct Type)

a. linear momentum of the system does not change in time
b. kinetic energy of the system does not change in time
c. angular momentum of the system does not change in time
d. potential energy of the system does not change in time

Passage for Q. Nos. (5-7)

A small block of mass M moves on a frictionless surface of an inclined plane as shown in figure. The angle of the incline suddenly changes from 60° to 30° at point B. The block is initially at rest at A. Assume that collisions between the block and the incline are totally inelastic.
(Given, $g = 10 \text{ ms}^{-2}$).

(2008, Comprehension Type)

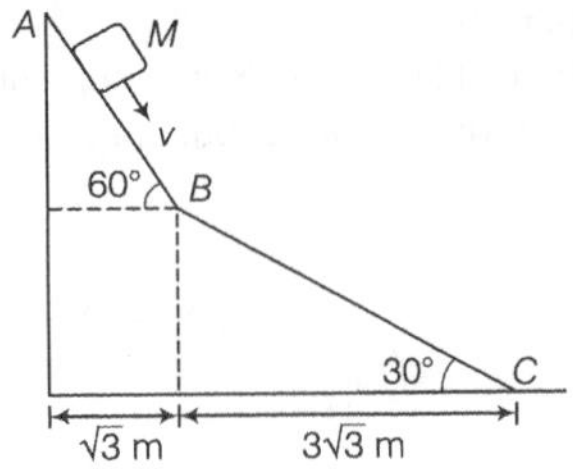

5 The speed of the block at point B immediately after it strikes the second incline is

a. $\sqrt{60} \text{ ms}^{-1}$ b. $\sqrt{45} \text{ ms}^{-1}$
c. $\sqrt{30} \text{ ms}^{-1}$ d. $\sqrt{15} \text{ ms}^{-1}$

6 The speed of the block at point C, immediately before it leaves the second incline is

a. $\sqrt{120} \text{ ms}^{-1}$
b. $\sqrt{105} \text{ ms}^{-1}$
c. $\sqrt{90} \text{ ms}^{-1}$
d. $\sqrt{75} \text{ ms}^{-1}$

7 If collision between the block and the incline is completely elastic, then the vertical (upward) component of the velocity of the block at point B, immediately after it strikes the second incline is

a. $\sqrt{30} \text{ ms}^{-1}$
b. $\sqrt{15} \text{ ms}^{-1}$
c. zero
d. $-\sqrt{15} \text{ ms}^{-1}$

8 Two balls having linear momenta $\mathbf{p}_1 = p\hat{\mathbf{i}}$ and $\mathbf{p}_2 = -p\hat{\mathbf{i}}$, undergo a collision in free space. There is no external force acting on the balls. Let $\mathbf{p}'_1$ and $\mathbf{p}'_2$ be their final momenta. The following option(s) is(are) not allowed for any non-zero values of $p, a_1, a_2, b_1, b_2, c_1$ and c_2.

(2008, One or More than One Options Correct Type)

a. $\mathbf{p}'_1 = a_1\hat{\mathbf{i}} + b_1\hat{\mathbf{j}} + c_1\hat{\mathbf{k}}$, $\mathbf{p}'_2 = a_2\hat{\mathbf{i}} + b_2\hat{\mathbf{j}}$

b. $\mathbf{p}'_1 = c_1\hat{\mathbf{k}}$, $\mathbf{p}'_2 = c_2\hat{\mathbf{k}}$

c. $\mathbf{p}'_1 = a_1\hat{\mathbf{i}} + b_1\hat{\mathbf{j}} + c_1\hat{\mathbf{k}}$, $\mathbf{p}'_2 = a_2\hat{\mathbf{i}} + b_2\hat{\mathbf{j}} - c_1\hat{\mathbf{k}}$

d. $\mathbf{p}'_1 = a_1\hat{\mathbf{i}} + b_1\hat{\mathbf{j}}$, $\mathbf{p}'_2 = a_2\hat{\mathbf{i}} + b_1\hat{\mathbf{j}}$

9 A particle moves in the *XY*-plane under the influence of a force such that its linear momentum is $\mathbf{p}(t) = A[\hat{\mathbf{i}}\cos(kt) - \hat{\mathbf{j}}\sin(kt)]$, where A and k are constants. The angle between the force and the momentum is **(2007, Only One Option Correct Type)**

a. 0° b. 30°

c. 45° d. 90°

Direction (Q. Nos. 10 and 11) *This section is based on Statement I and Statement II. Select the correct answer from the codes given below.*

a. Both Statement I and Statement II are correct and Statement II is the correct explanation of Statement I

b. Both Statement I and Statement II are correct but Statement II is not the correct explanation of Statement I

c. Statement I is correct but Statement II is incorrect

d. Statement II is correct but Statement I is incorrect

10 **Statement I** In an elastic collision between two bodies, the relative speed of the bodies after collision is equal to the relative speed before the collision.

Statement II In an elastic collision, the linear momentum of the system is conserved.

(2007, Statement Type)

11 **Statement I** If there is no external torque on a body about its centre of mass, then the velocity of the centre of mass remains constant.

Statement II The linear momentum of an isolated system remains constant.

(2007, Statement Type)

12 There is a rectangular plate of mass M kg of dimensions $(a \times b)$. The plate is held in horizontal position by striking n small balls uniformly each of mass m per unit area per unit time. These are striking in the shaded half region of the plate. The balls are colliding elastically with velocity v. What is the value of v?

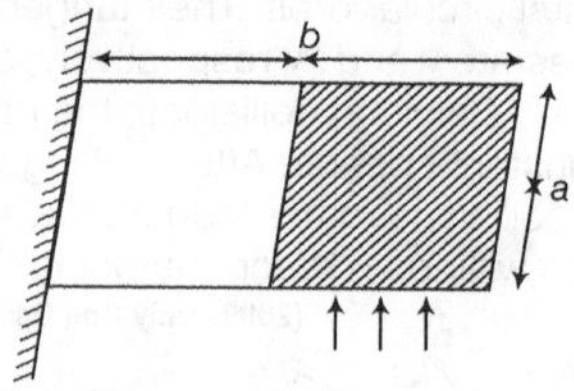

Given, $n = 100$, $M = 3$ kg, $m = 0.01$ kg, $b = 2$ m, $a = 1$ m, $g = 10$ ms^{-2}. **(2006, Integer Type)**

2. Rotational Motion

DPP-1 Rotational Kinematics

Only One Option Correct Type

Working Space

Direction (Q. Nos. 1-11) *This section contains 11 multiple choice questions. Each question has four choices (a), (b), (c) and (d), out of which ONLY ONE is correct.*

1 Find the angular speed of second's hand in a clock.

a. $\frac{\pi}{40}$ rad s^{-1}
b. $\frac{\pi}{20}$ rad s^{-1}
c. $\frac{\pi}{60}$ rad s^{-1}
d. $\frac{\pi}{30}$ rad s^{-1}

2 A body rotates about a fixed axis with an angular acceleration 1 rad s^{-2}. Through what angle does it rotate during the time in which its angular velocity increases from 5 rad s^{-1} to 15 rad s^{-1}?

a. 100 rad
b. 180 rad
c. 200 rad
d. 80 rad

3 A wheel starting from rest is uniformly accelerated at 4 rad s^{-2} for 10 s. It is allowed to rotate at constant angular speed for the next 10 s and is finally brought to rest in the next 10 s. Find the total angle rotated by the wheel?

a. 600 rad
b. 700 rad
c. 800 rad
d. 1000 rad

4 A body rotating at 20 rad s^{-1} is acted upon by a constant torque providing it a deceleration of 2 rad s^{-2}. At what time will the body have kinetic energy as the initial value if the torque continues to act?

a. 20 s
b. 15 s
c. 10 s
d. 25 s

5 A particle is located at (3 m, 4 m) and moving with velocity $\mathbf{v} = (4\hat{\mathbf{i}} - 3\hat{\mathbf{j}})$ ms^{-1}. Find its angular velocity about origin at this instant.

a. $-\frac{1}{2}\hat{\mathbf{k}}$
b. $-\frac{3}{4}\hat{\mathbf{k}}$
c. $-\hat{\mathbf{k}}$
d. $-\frac{2}{3}\hat{\mathbf{k}}$

6 The angular displacement (in radian) of an accelerated flywheel is given by $\theta = 9t^3/32$. The linear velocity and acceleration of a point at a distance of 0.75 m from the axis of rotation at the instant, when magnitudes of its tangential acceleration and normal acceleration are equal respectively, are

a. $\frac{7}{8}$ ms^{-1}, $\frac{1}{2}\sqrt{6}$ ms^{-2}
b. $\frac{5}{8}$ ms^{-1}, $\frac{2}{5}\sqrt{6}$ ms^{-2}
c. $\frac{9}{8}$ ms^{-1}, $\frac{3}{4}\sqrt{6}$ ms^{-2}
d. $\frac{11}{8}$ ms^{-1}, $\frac{2}{3}\sqrt{6}$ ms^{-2}

7 In a system of gears load L rotates a pulley of radius r and a gear wheel of radius r_1. This gear wheel is geared with a second gear wheel of radius r_2. If the load starts from rest and moves down with a constant acceleration (a), find the equation for angular displacement of second gear wheel.

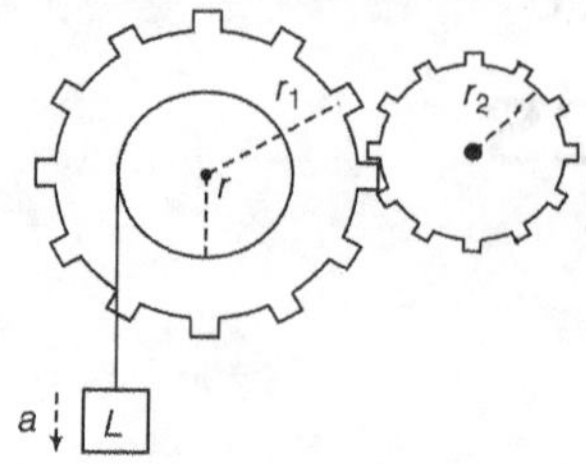

a. $\theta = \left(\frac{ar_1}{rr_2}\right)t^2$

b. $\theta = \left(\frac{ar_1}{2rr_2}\right)t^2$

c. $\theta = \left(\frac{ar_1}{\sqrt{2}\, rr_1}\right)t^2$

d. $\theta = \left(\frac{2ar_1}{rr_2}\right)t^2$

8 A table fan rotating at a speed of 2400 rpm is switched OFF and the resulting variation of rpm with time is shown in the figure. The total number of revolutions of the fan before it comes to rest is

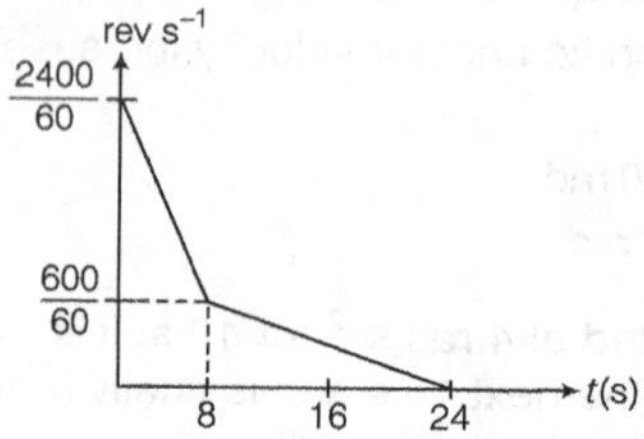

a. 420

b. 280

c. 190

d. 380

9 The angular acceleration of a fly wheel is given by $\alpha = 12 - t$, where α is in rad s^{-2} and t in second. If the angular velocity of the wheel is 60 rad s^{-1} at the end of 4 s, find the angular velocity at the end of 6 s and the number of revolutions that take place in these 6 s, respectively are

a. 54 rad s^{-1}, 37.8

b. 84 rad s^{-1}, 57.8

c. 74 rad s^{-1}, 47.8

d. 80 rad s^{-1}, 50

10 A flywheel rotates about its axis. Due to friction at the axis, it experiences an angular retardation proportional to its angular velocity. If its angular velocity gets halved while it makes n rotations, how many more rotations will it make before coming to rest?

a. $\frac{n}{2}$

b. n

c. $\frac{2n}{3}$

d. $\frac{n}{3}$

11 A wheel rotates around a stationary axis so that the rotation angle θ varies with time as $\theta = kt^2$, where $k = 1.5$ rad s^{-2}. Find the magnitude of net acceleration of a point P on the rim at the moment $t = 2/3$ s, if the linear velocity of the point P at this moment is $v = 2$ ms^{-1}.

a. 3 ms^{-2}

b. 4 ms^{-2}

c. 5 ms^{-2}

d. 6 ms^{-2}

Working Space

One or More than One Options Correct Type

Direction (Q. Nos. 12-16) *This section contains 5 multiple choice questions. Each question has four choices (a), (b), (c) and (d), out of which ONE or MORE THAN ONE are correct.*

12 A disc starts rotating with constant angular acceleration of π rad s^{-2} about a fixed axis perpendicular to its plane and through its centre. Choose the correct option(s).

a. Angular velocity of the disc after 4 s is 4π rad s^{-1}

b. Angular displacement of the disc after 4 s is 4π rad

c. Number of turns accomplished by the disc in 4 s is 4

d. Angular displacement of the disc after 4 s is 8π rad

13 A flywheel of radius 20 cm starts from rest and has a constant angular acceleration of 60 rad s^{-2}. Choose the correct option(s).

a. The magnitude of net linear acceleration of a point on the rim after 0.15 s is 20.0 ms^{-2}

b. The magnitude of net linear acceleration of a point on the rim after 0.15 s is 40.4 ms^{-2}

c. The number of revolutions completed in 0.25 s is 0.6

d. The number of revolutions completed is 0.25 s is 0.3

14 A wheel rotates with an angular acceleration given by $\alpha = 4at^3 - 3bt^2$, where t is the time and a and b are positive constants. If the initial angular speed of the wheel is ω_0, then

a. the equation for angular speed is $\omega = \omega_0 + at^3 - bt^2$

b. the equation for angular speed is $\omega = \omega_0 + at^4 - bt^3$

c. the equation for angular displacement is $\theta = \omega_0 t + \frac{1}{4}at^5 - bt^4$

d. the equation for angular displacement is $\theta = \omega_0 t + \frac{1}{5}at^5 - \frac{1}{4}bt^4$

15 A solid body rotates about a stationary axis according to the law $\theta = k_1 t - k_2 t^2$, where $k_1 = 6$ rad s^{-1} and $k_2 = 2$ rad s^{-2}. Over the time interval between $t = 0$ and the time, when the body comes to rest,

a. the mean value of angular velocity is 3 rads^{-1}

b. the mean value of angular velocity is 5.2 rad s^{-1}

c. the mean value of angular acceleration is -4 rads^{-2}

d. the mean value of angular acceleration is -4.9 rads^{-2}

16 The angular speed of a motor wheel is increased from 1200 rpm to 3120 rpm in 16 s. Its angular acceleration is uniform. Then, choose the correct option(s).

a. Its angular acceleration is 2π rad s^{-2}

b. Its angular acceleration is 4π rads^{-2}

c. Number of revolutions it makes in the given time is 576

d. Number of revolutions it makes in the given time is 288

Comprehension Type

Working Space

Direction (Q. Nos. 17-19) *This section contains a paragraph, which describing theory, experiments, data, etc. Three questions related to the paragraph have been given. Each question has only one correct answer among the four given options (a), (b), (c) and (d).*

Passage

The step pulley shown, starts from rest and accelerates at 2 rad s^{-2}.

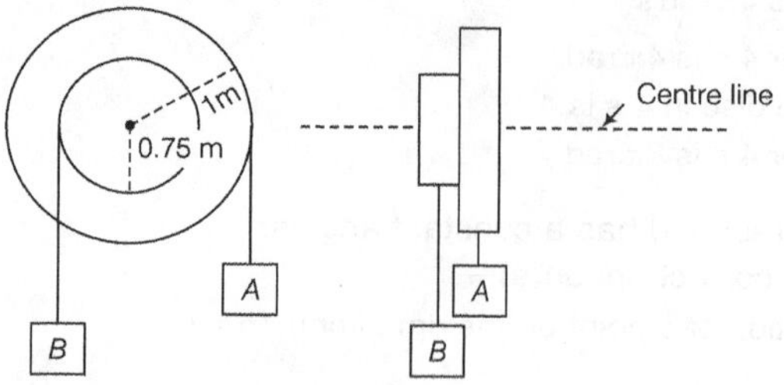

17 What time t is required for block A to move 20 m?

a. 4.47s b. 3.47s

c. 5.47s d. 6.47s

18 The velocity of A at time t (mentioned in the above question) (in ms^{-1}) is

a. 9.94 b. 8.94

c. 7.94 d. 6.94

19 The velocity of B at time t (mentioned in the above question) (in ms^{-1}) is

a. 4.7 b. 3.7

c. 5.7 d. 6.7

Matching List Type

Direction (Q. No. 20) *Choices for the correct combination of elements from Column I and Column II are given as options (a), (b), (c) and (d), out of which one is correct.*

20 A gyroscope is spinning at angular velocity ω_0. The power is shut off and the gyroscope begins to slow down with its angular velocity described by $\omega = \omega_0 e^{-t/T}$, where t is the time and T is a positive constant. Now, match the following columns and select the correct option from the codes given below.

	Column I		Column II
i.	The time in which the gyroscope reaches half of its initial angular velocity.	p.	$\frac{\omega_0 T}{4\pi}$
ii.	Initial angular acceleration of the gyroscope	q.	$-\omega_0/(2T \ln 2)$
iii.	The average angular acceleration between the time, the power is cut-off and the instant it reaches half speed.	r.	$-\omega_0/T$
iv.	Number of revolutions the gyroscope turns between the time the power is cut-off and the instant it reaches half speed.	s.	$T \ln 2$

Codes

	i	ii	iii	iv
a.	s	r	q	p
b.	p	q	r	s
c.	q	r	s	p
d.	p	r	q	s

DPP-2 Moment of Inertia

Only One Option Correct Type

Working Space

Direction (Q. Nos. 1-15) *This section contains 15 multiple choice questions. Each question has four choices (a), (b), (c) and (d), out of which ONLY ONE is correct.*

1 Find the moment of inertia of a cylinder of mass M, radius R and length L about an axis passing through its centre and perpendicular to its length.

a. $\frac{MR^2}{4} + \frac{ML^2}{12}$ b. $\frac{MR^2}{2} + \frac{ML^2}{12}$

c. $\frac{MR^2}{2} + \frac{ML^2}{6}$ d. $\frac{MR^2}{4} + \frac{ML^2}{6}$

2 Find the moment of inertia of the triangular lamina of mass M about the axis of rotation AA' shown in the figure.

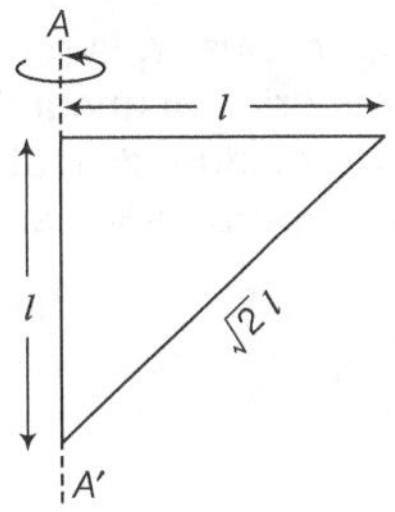

a. $\frac{Ml^2}{3}$

b. $\frac{Ml^2}{6}$

c. $\frac{2}{3}Ml^2$

d. $\frac{1}{3}Ml^2$

3 Find the moment of inertia of a hemisphere of mass M (shown in the figure) about an axis AA' passing through its centre of mass.

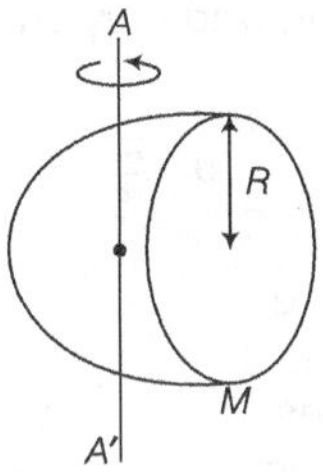

a. $\frac{69}{240}MR^2$ b. $\frac{89}{320}MR^2$

c. $\frac{74}{243}MR^2$ d. $\frac{83}{320}MR^2$

4 On the flat surface of a disc of radius R, a small circular hole of radius r is made with its centre at a distance d from the centre of the disc. If mass of the whole uncut disc is M, calculate the moment of inertia of the residual disc about an axis passing through centre of the hole and perpendicular to the plane of the disc.

a. $M\left[R^2 + 2d^2 - \frac{r^4}{R^2}\right]$ b. $2M\left[R^2 + 2d^2 - \frac{r^4}{R^2}\right]$

c. $\frac{M}{2}\left[R^2 + 2d^2 - \frac{r^4}{R^2}\right]$ d. $\frac{M}{2}\left[R^2 + d^2 - \frac{r^4}{R^2}\right]$

Working Space

5 Find the moment of inertia of a spherical ball of mass m and radius r attached at the end of a thin straight rod of mass M and length L about the axis AA' shown in the figure.

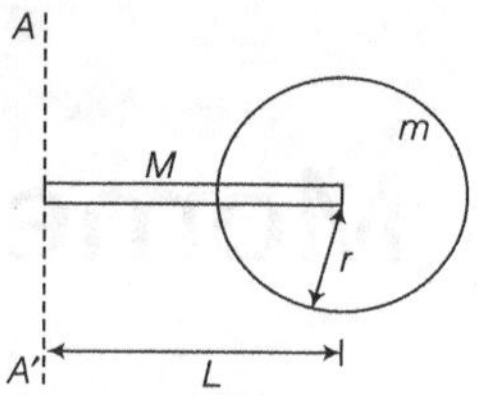

a. $\frac{L^2}{2}\left(m+\frac{M}{3}\right)+\frac{1}{5}mr^2$ b. $L^2\left(m+\frac{M}{3}\right)+\frac{2}{5}mr^2$

c. $\frac{2L^2}{3}\left(m+\frac{M}{3}\right)+\frac{2}{5}mr^2$ d. None of these

6 A rod of length L is pivoted at an end. The linear mass density of the rod λ varies with the distance x from this end as $\lambda = ax^2 + b$ kgm^{-1}, where a and b are positive constants. Find the moment of inertia of the rod about the axis passing through this end and perpendicular to its length.

a. $\frac{2aL^5}{5}+\frac{2bL^3}{3}$ b. $\frac{2aL^5}{3}+\frac{2bL^3}{5}$

c. $\frac{aL^5}{7}+\frac{bL^3}{3}$ d. $\frac{aL^5}{5}+\frac{bL^3}{3}$

7 From a semi-circular disc of mass M and radius R_2, a semi-circular disc of radius R_1 is removed as shown in the figure. If the mass of original uncut disc is M, find the moment of inertia of residual disc about an axis passing through centre O and perpendicular to the plane of the disc.

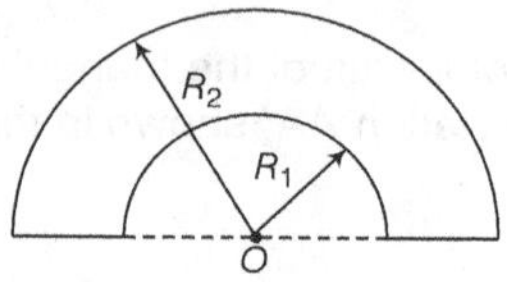

a. $\frac{M}{2R_2^2}(R_2^4 - R_1^4)$ b. $\frac{M}{R_2^2}(R_2^4 - R_1^4)$

c. $\frac{M}{R_2^2}\left(\frac{R_2^4}{4} - R_1^4\right)$ d. $\frac{M}{2R_2^2}\left(R_2^4 - \frac{R_1^4}{4}\right)$

8 If I_1 is the moment of inertia of a thin rod about an axis perpendicular to its length and passing through its centre of mass and I_2 is the moment of inertia of the ring formed by the same rod about an axis tangent to the ring and perpendicular to the plane of the ring, then find the ratio $I_1/_2$.

a. $\frac{\pi^2}{3}$ b. $\frac{\pi^2}{6}$ c. $\frac{\pi^2}{2}$ d. $\frac{\pi^2}{5}$

9 Three identical thin rods each of length L and mass m are welded perpendicular to one another as shown in the figure. The assembly is rotated about an axis that passes through the end of one rod and is parallel to another. Find moment of inertia of this structure.

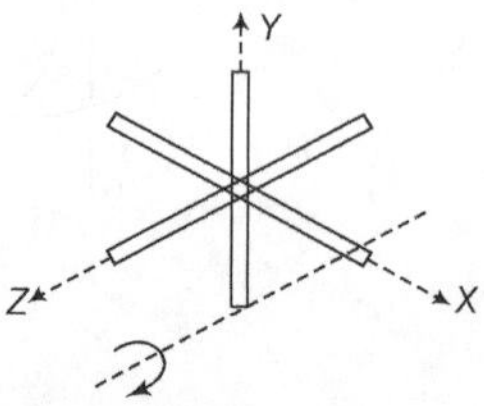

a. $\frac{11}{12}mL^2$ b. $\frac{9}{16}mL^2$

c. $\frac{8}{11}mL^2$ d. $2mL^2$

10 Three identical thin rods, each of mass m and length L are joined to form an equilateral triangle. Find the moment of inertia of the triangle about one of its sides.

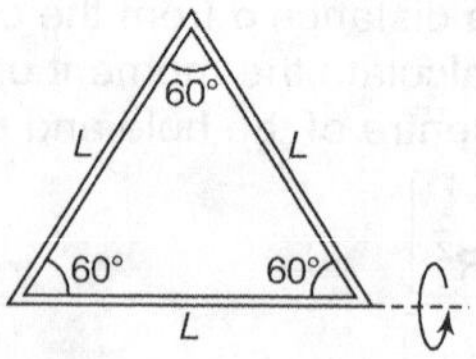

a. $\frac{mL^2}{6}$ b. $\frac{3mL^2}{2}$

c. $\frac{mL^2}{3}$ d. $\frac{mL^2}{2}$

Working Space

11 A symmetrical lamina of mass M consists of a square shape with a semi-circular section over each edge of the square (see figure). The side of the square is $2a$. The moment of inertia of the lamina about an axis through its centre of mass and perpendicular to the plane is $1.6Ma^2$. The moment of inertia of the lamina about the tangent AB in the plane of the lamina is

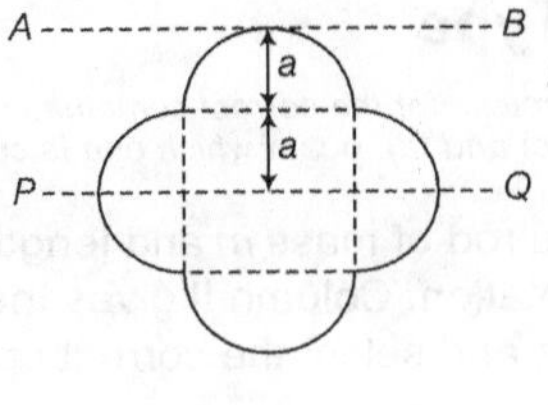

a. $3.6\,Ma^2$ b. $4.2Ma^2$
c. $4.8Ma^2$ d. $5.4Ma^2$

12 A disc of radius R has mass $9m$. A hole of radius $\frac{R}{3}$ is cut from it as shown in the figure. What is the moment of inertia of the remaining part about an axis passing through centre O of the disc and perpendicular to the plane of the disc?

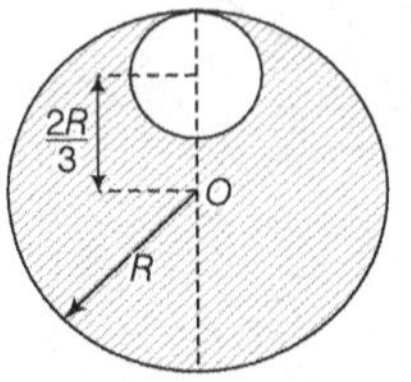

a. $6\ mR^2$ b. $4\ mR^2$
c. $2\ mR^2$ d. $3.5mR^2$

13 A circular disc X of radius R is made from an iron plate of thickness t and another disc Y of radius $4R$ is made from an iron plate of thickness $\frac{t}{4}$. Then, what is the ratio of their moments of inertia $\frac{I_Y}{I_X}$ equal to? (Both moments of inertia are about the axis passing through the centre of mass and perpendicular to plane of disc.)

a. 64 b. 81
c. 49 d. 36

14 Four point masses lie at the corners of a rectangle of sides 3 m and 4 m as shown in the figure. Take, $M = 1$ kg. The diagonal joining $4M$ and $2M$ axis 1. The other diagonals is axis 2. Then,

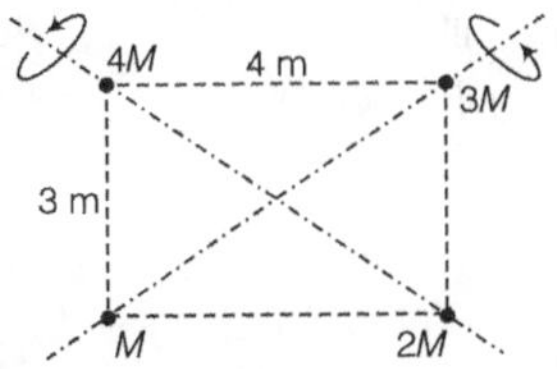

a. moment of inertia of the system about axis 1 is 23 kg-m^2
b. moment of inertia of the system about axis 1 is 26 kg-m^2
c. moment of inertia of the system about axis 2 is 44 kg-m^2
d. moment of inertia of the system about axis 2 is 34 kg-m^2

15 Four particles with equal masses (each m) are connected by rods of negligible mass. Moment of inertia, about the indicated axis, of (i) the cross in Fig. (i) and (ii) the square in Fig. (ii), respectively and I_1 and I_2. Then,

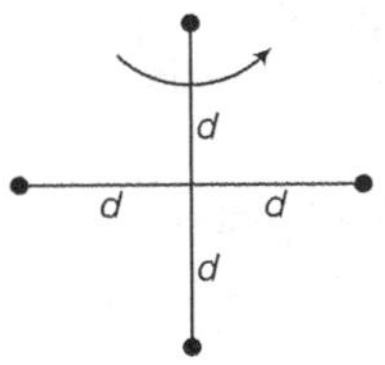

Fig. (i)

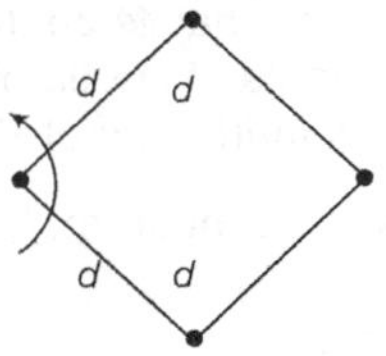

Fig. (ii)

a. $I_1 = 6md^2$ b. $I_1 = 8md^2$
c. $I_2 = 6md^2$ d. $I_2 = 4md^2$

Matching List Type

Working Space

Direction (Q. No. 16) *Choices for the correct combination of elements from Column I and Column II are given as options (a), (b), (c) and (d), out of which one is correct.*

16 Column I shows a rod of mass m and length L, axis of rotation and its orientation with the axis of rotation. Column II gives the moments of inertia. Match the following columns and select the correct option from the codes given below.

	Column I		Column II
i.	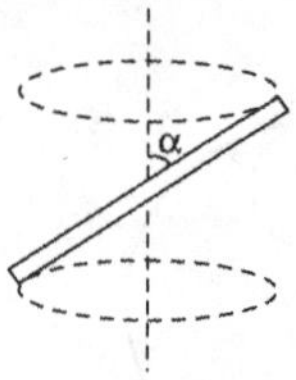	p.	$\frac{1}{3} mL^2 \sin^2 \alpha$
ii.	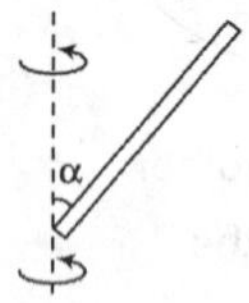	q.	$\frac{1}{3} mL^2$
iii.	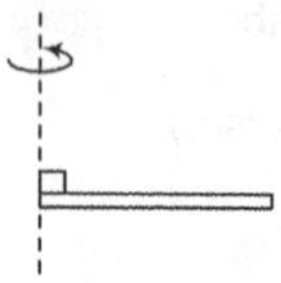	r.	$\frac{1}{6} mL^2 \sin^2 \alpha$
		s.	$\frac{1}{12} mL^2 \sin^2 \alpha$

Codes

	i	ii	iii		i	ii	iii
a.	s	p	q	b.	p	q	r
c.	r	s	p	d.	q	r	s

Statement Type

Direction (Q. Nos. 17 and 18) *This section is based on Statement I and Statement II. Select the correct answer from the codes given below.*

a. Both Statement I and Statement II are correct and Statement II is the correct explanation of Statement I
b. Both Statement I and Statement II are correct but Statement II is not the correct explanation of Statement I
c. Statement I is correct but Statement II is incorrect
d. Statement II is correct but Statement I is incorrect

17 **Statement I** A uniform sphere is placed such that its centre is at the origin of coordinate system. If I_x and I_y are moments of inertia of the sphere about X-axis and Y-axis respectively, then moment of inertia about Z-axis is $I_x + I_y$.

Statement II According to perpendicular axes theorem, $I_z = I_x + I_y$ when object is lying on XY-plane.

18 **Statement I** Minimum moment of inertia of a uniform body is I about an axis. The passing axis must be through centre of mass of the body.

Statement II Moment of inertia depends on distribution of mass about axis of rotation.

Working Space

Subjective Questions

Direction (Q. Nos. 19-22) *These questions are subjective in nature, need to be solved completely on notebook.*

19 Find the moment of inertia of the half cylinder of mass M (shown in the figure) about the axis AA'.

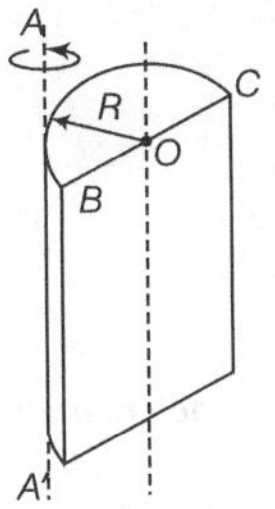

20 Find out the moment of inertia of a solid cone of mass M and base radius R, about its axis of symmetry.

21 Find out the moment of inertia of a thin hollow cone of mass M and radius R, about its axis of symmetry.

22 The density of a sphere of mass M, radius R varies with distance r from its centre as $\rho = \rho_0\left(1 + \frac{r}{R}\right)$, where ρ_0 is a positive constant. Calculate the moment of inertia of this sphere about its diameter.

Comprehension Type

Direction (Q. Nos. 23-25) *This section contains a paragraph, which describing theory, experiments, data, etc. Three questions related to the paragraph have been given. Each question has only one correct answer among the four given options (a), (b), (c) and (d).*

Passage

A thin square lamina has side L. Its moment of inertia about an axis passing through its centre of mass (O) and perpendicular to its plane is I_0.

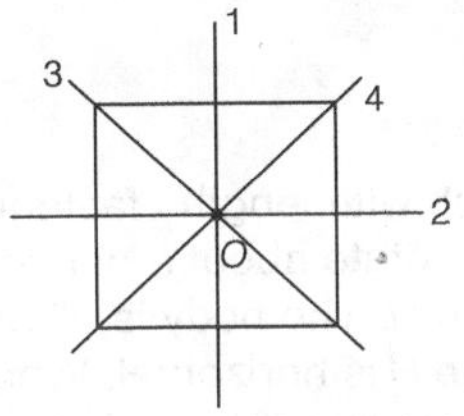

Axes 1,2,3 and 4 are on the plane of the lamina as shown and moments of a inertia about them are I_1, I_2, I_3 and I_4, respectively.

23 $I_1 + I_2$ is equal to

a. $\frac{I_0}{2}$ b. $2I_0$ c. I_0 d. $\sqrt{2}\,I_0$

24 $I_3 + I_4$ is equal to

a. $\frac{I_0}{2}$ b. $2I_0$ c. I_0 d. $\sqrt{2}\,I_0$

25 $I_2 + I_3$ is equal to

a. $\frac{I_0}{2}$ b. $2I_0$ c. I_0 d. $\sqrt{2}\,I_0$

DPP-3 Rotation about a Fixed Axis

Only One Option Correct Type

Working Space

Direction (*Q. Nos. 1-8*) *This section contains 8 multiple choice questions. Each question has four choices (a), (b), (c) and (d), out of which ONLY ONE is correct.*

1 A uniform rod of length L is pivoted at point A. It is struck by a horizontal force which delivers an impulse J at a distance x from point A as shown in the figure. The impulse delivered by pivot is zero when x equals

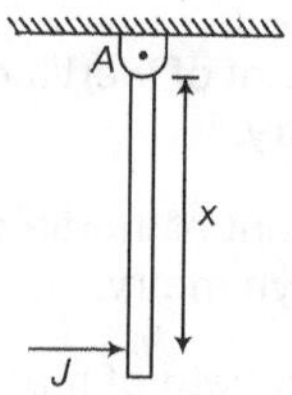

a. $\frac{L}{2}$ b. $\frac{L}{3}$

c. $\frac{2L}{3}$ d. $\frac{3L}{4}$

2 A block of mass $m = 4$ kg is attached to a spring of constant, $k = 32\ \text{Nm}^{-1}$ by a rope that hangs over a pulley of mass $M = 8$ kg. If the system starts from rest with the spring unstretched, find the speed of the block after it falls 1 m. Treat the pulley as a disc.

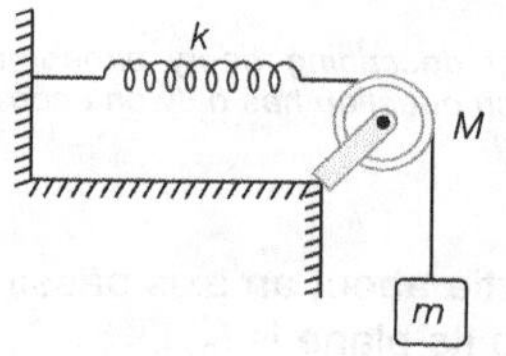

a. $1.2\ \text{ms}^{-1}$ b. $2.8\ \text{ms}^{-1}$

c. $2.4\ \text{ms}^{-1}$ d. $1.8\ \text{ms}^{-1}$

3 A rigid body is made of three identical thin rods, each with length l fastened together in the form of a letter H. The body is free to rotate about a horizontal axis that runs along the length of one of the lengths of the H. The body is allowed to fall from rest from a position in which the plane of the H is horizontal. What is the angular speed of the body when the plane of the H is vertical?

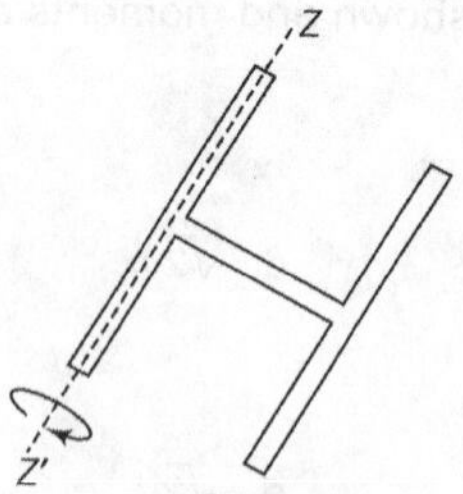

a. $\frac{2}{3}\sqrt{\frac{g}{l}}$ b. $\frac{1}{3}\sqrt{\frac{g}{l}}$

c. $\frac{3}{2}\sqrt{\frac{g}{l}}$ d. $\frac{4}{3}\sqrt{\frac{g}{l}}$

Working Space

4 A uniform spherical shell of mass M and radius R rotates about a vertical axis on frictionless bearing [see figure]. An ideal cord passes around the equator of the shell, over a pulley of rotational inertia I and radius r and is attached to a small object of mass m, that is otherwise free to fall under the influence of gravity. There is no friction in the axle of pulley, the cord does not slip. What is the speed of the object after it has fallen a distance h from rest?

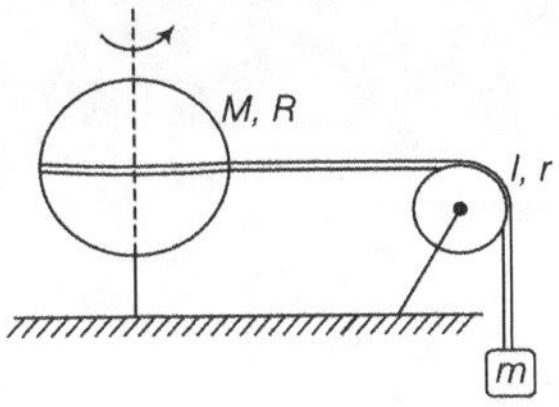

a. $\left(\dfrac{mgh}{m+\dfrac{I}{r^2}+M}\right)^{1/2}$ b. $\left(\dfrac{mgh}{2m+\dfrac{I}{2r^2}+\dfrac{M}{5}}\right)^{1/2}$

c. $\left(\dfrac{mgh}{m+\dfrac{I}{2r^2}+\dfrac{M}{3}}\right)^{1/2}$ d. $\left(\dfrac{mgh}{\dfrac{m}{2}+\dfrac{I}{2r^2}+\dfrac{M}{3}}\right)^{1/2}$

5 A uniform cylinder of radius R is spinned about its axis to the angular velocity ω_0 and then placed into a corner (see figure). The coefficient of friction between the corner walls and the cylinder is equal to μ. How many turns does the cylinder accomplish before it stops?

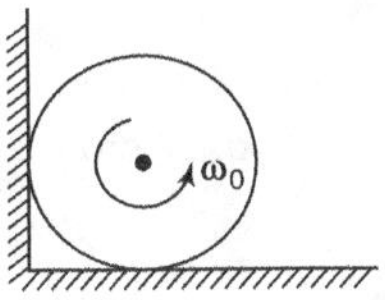

a. $\dfrac{\omega_0^2 R(1+\mu^2)}{8\pi \mu g(1+\mu)}$ b. $\dfrac{\omega_0^2 R(1+\mu^2)}{4\pi \mu g(1+\mu)}$

c. $\dfrac{\omega_0^2 R(1+\mu^2)}{2\pi \mu g(1+\mu)}$ d. $\dfrac{\omega_0^2 R(1+\mu^2)}{6\pi \mu g(1+\mu)}$

6 Three forces are applied to a square plane as shown in the figure. Find the modulus, direction and the point of application of the resultant force, if this point is taken on BC.

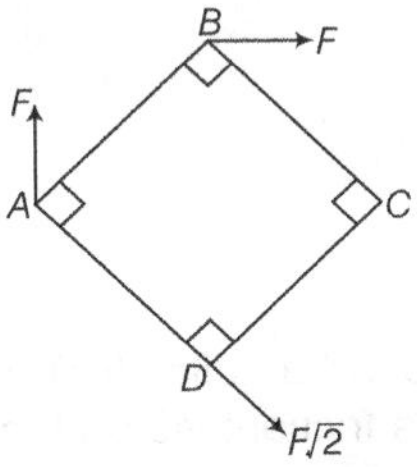

a. $2F$, parallel to diagonal AC, at a distance $\frac{1}{3}BC$ from B

b. $2F$, parallel to diagonal AC, at the mid-point of side BC

c. $2\sqrt{2}F$, parallel to diagonal AC, at a distance $\frac{2}{3}BC$ from B

d. $\sqrt{2}F$, parallel to diagonal AC, at a distance $\frac{1}{3}BC$ from B

Working Space

7 A uniform rod AB of mass $m = 2$ kg and length $L = 1$m is placed on a sharp support O such that $a = 0.4$ m and $b = 0.6$ m. A spring of force constant $k = 600$ Nm^{-1} is attached to the end B. To keep the rod horizontal, its end A is tied with a thread such that the spring is elongated by 1 cm. When the thread is burnt, what will be the reaction of the support O? (Given, $g = 10$ ms^{-2})

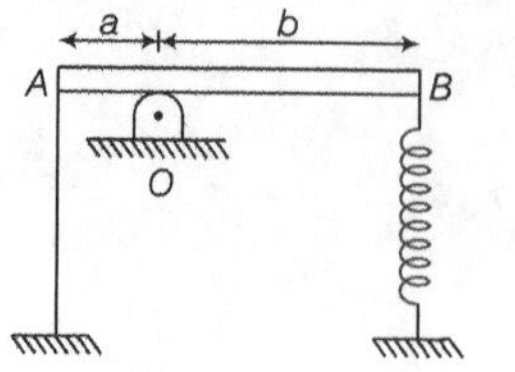

a. 15 N
b. 20 N
c. 25 N
d. 30 N

8 A cylinder of mass $m = 2$ kg and radius $r = 0.1$m is placed such that, it is in contact simultaneouly with a vertical surface and a horizontal surface as shown in the figure. For both the surfaces, the coefficient of static friction is $\mu = \frac{1}{3}$. Find the distance d from the centre of the cylinder at which a force $F = 40$ N should be applied so that the cylinder just starts rotating in the anti-clockwise direction. (Given, $g = 10$ ms^{-2})

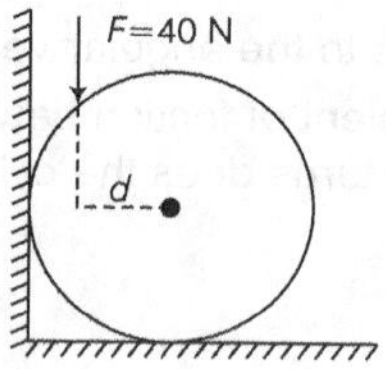

a. 4 cm
b. 6 cm
c. 8 cm
d. 7 cm

One or More than One Options Correct Type

Direction (Q. Nos. 9-13) *This section contains 5 multiple choice questions. Each question has four choices (a), (b), (c) and (d), out of which ONE or MORE THAN ONE are correct.*

9 A uniform rod of mass m and length L is held horizontally by two vertical strings of neligible mass as shown in the figure. Immediately after the string is cut, then

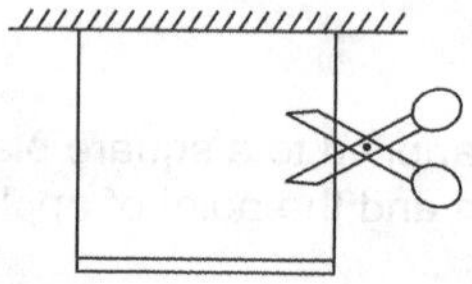

a. linear acceleration of free end of the rod is $\frac{3}{2}g$

b. linear acceleration of centre of mass of the rod is $\frac{3}{4}g$

c. tension in the left string is $\frac{mg}{4}$

d. tension in the left string is $\frac{mg}{2}$

10 An object of mass m is tied to a light string wound around a pulley that has a moment of inertia I and radius R. The wheel bearing is frictionless and the string does not slip on the rim. Choose the correct option(s).

a. Tension in the string is $\dfrac{mg}{1+\dfrac{mR^2}{I}}$

b. Acceleration of the object is $\dfrac{g}{1+\dfrac{mR^2}{I}}$

c. Tension in the string is $\dfrac{mg}{1+\dfrac{I}{mR^2}}$

d. Acceleration of the object is $\dfrac{g}{1+\dfrac{I}{mR^2}}$

Working Space

11 Two blocks are connected by an ideal string that passes over a pulley of radius R and moment of inertia I. The block of mass m_1 slides on a frictionless, horizontal surface and the block of mass m_2 is suspended from the string (see figure). Assume that the string does not slip on the pulley. Choose the correct option(s).

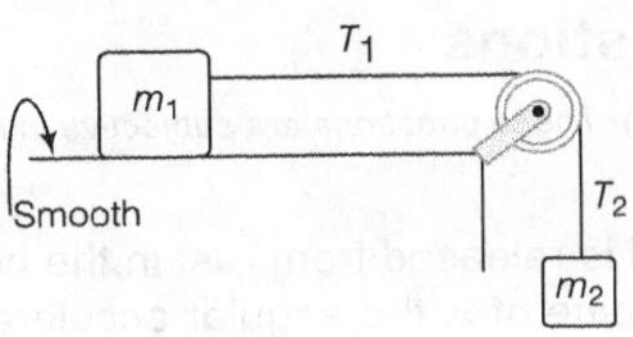

a. Acceleration of blocks is $\dfrac{2m_2}{m_1 + m_2 + \dfrac{I}{R^2}} g$

b. Acceleration of blocks is $\dfrac{m_2}{m_1 + m_2 + \dfrac{I}{R^2}} g$

c. Tension T_1 is $\dfrac{m_1 m_2 g}{m_1 + m_2 + \dfrac{I}{R^2}}$

d. Tension T_2 is $\dfrac{\left(m_1 + \dfrac{I}{R^2}\right) m_2 g}{m_1 + m_2 + \dfrac{I}{R^2}}$

12 Figure shows two blocks m_1 and m_2 connected by ideal string, supported on two identical pulleys of moment of inertia I and radius R. There is no slipping between string and pulleys. Then,

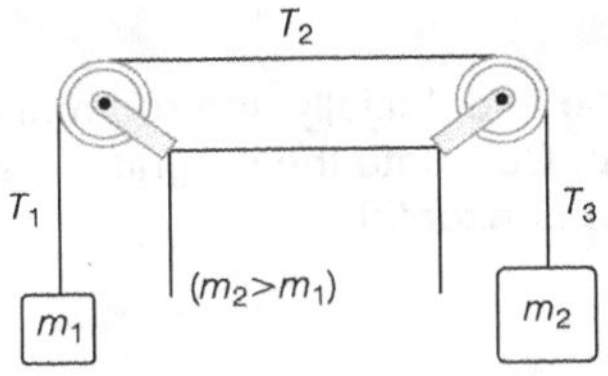

a. $T_1 = \dfrac{2m_1 g\left[m_2 + \dfrac{I}{R^2}\right]}{\left(m_1 + m_2 + \dfrac{2I}{R^2}\right)}$

b. $T_2 = \dfrac{\left[(m_1 + m_2)\dfrac{I}{R^2} + 2m_1 m_2\right] g}{\left(m_1 + m_2 + \dfrac{2I}{R^2}\right)}$

c. $T_3 = \dfrac{\left(m_1 + \dfrac{I}{R^2}\right) 2m_2 g}{\left(m_1 + m_2 + \dfrac{2I}{R^2}\right)}$

d. acceleration of blocks is $\dfrac{(m_2 - m_1) g}{\left(m_1 + m_2 + \dfrac{2I}{R^2}\right)}$

13 A cord is wrapped around the rim of a solid cylinder of radius 0.25 m and a constant force of 40 N is exerted on the cord. The cylinder is mounted on frictionless bearings and its moment of inertia is 4 kg-m^2.

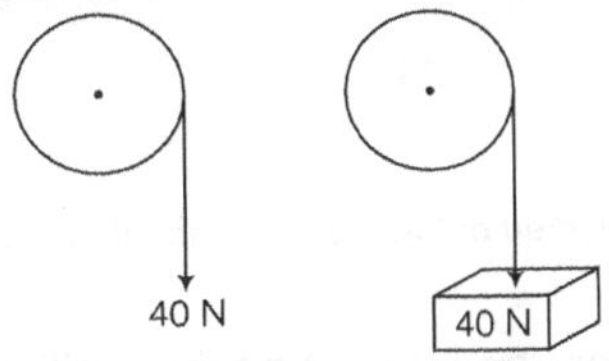

The angular velocity of the cylinder after 5 m of cord has been unwound, is ω_1. If the 40 N force is replaced by a 40 N weight, the angular velocity of the cylinder after 5 m of cord has been unwound is ω_2 .(Given, $g = 10\ \text{ms}^{-2}$). Then,

a. $\omega_1 = 9.16\ \text{rad s}^{-1}$

b. $\omega_2 = 9\ \text{rad s}^{-1}$

c. $\omega_1 = 8.16\ \text{rad s}^{-1}$

d. $\omega_2 = 8\ \text{rad s}^{-1}$

Subjective Questions

Working Space

Direction (Q. Nos. 14-17) *These questions are subjective in nature, need to be solved completely on notebook.*

14 A uniform thin rod is released from rest in the horizontal position as shown in the figure. For what value of x, the angular acceleration is maximum? Determine the corresponding angular acceleration α.

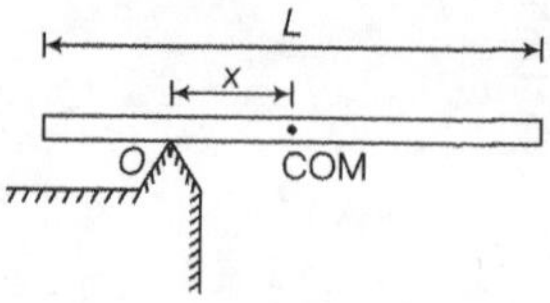

15 A uniform stick of length L and mass m hinged at one end is released from rest at an angle θ_0 with the vertical. Show that when the angle with the vertical is θ, the hinge exerts a force F_r (along the stick) and F_t (perpendicular to the stick) given by $F_r = \frac{1}{2}mg\,(5\cos\theta - 3\cos\theta_0)$ and $F_t = \frac{1}{4}mg\sin\theta$.

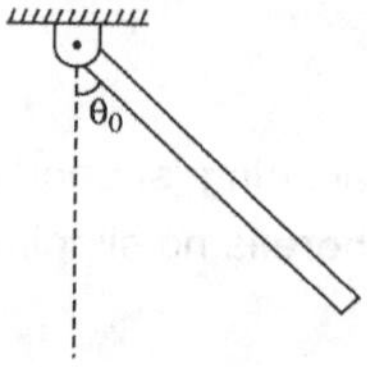

16 A thin rod of mass m and length L is pivoted at one end. Initially, the rod makes an angle of 60° with the vertical and is then released. Calculate the magnitude and direction of the force on the pivot, when the rod is horizontal.

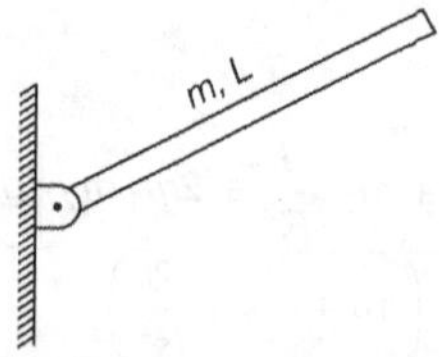

17 A uniform rod AB of length of $2l$ and mass $2m$ has a particle of mass m attached at B. The rod is free to rotate in a vertical plane about a horizontal axis through A. When the rod is hanging at rest with B below A it is given an angular velocity, $\omega_0 = \frac{7}{2}\sqrt{\frac{g}{l}}$.

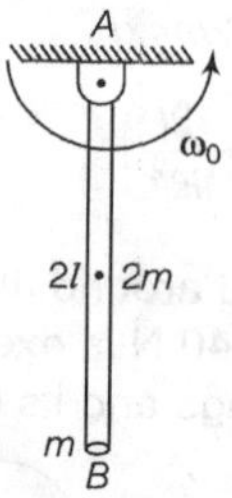

Find the reaction at the axis, when the rod becomes horizontal first. (Given, $g = 10\ ms^{-2}$)

One Integer Value Correct Type

Direction (Q. No. 18) *This section contains 1 question. When worked out will result in an integer from 0 to 9 (both inclusive).*

18 A uniform rod of length $2a$ is placed horizontally on the edge of a table. Initially, the centre of mass of the rod is at a distance $\frac{a}{3}$ from the edge. The rod is released from rest. If the rod slips after it has turned through an angle θ, the coefficient of friction between the rod and the table is $\beta\tan\theta$. Find the value of β.

Comprehension Type

Direction (Q. Nos. 19-23) This section contains 2 paragraphs, each describing theory, experiments, data, etc. Five questions related to the paragraphs have been given. Each question has only one correct answer among the four given options (a), (b), (c) and (d).

Passage I

A uniform rod of length L and mass m is pivoted freely at one end as shown in the figure. Answer the following questions.

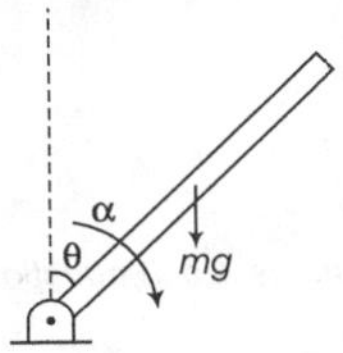

19 The angular acceleration of the rod when it is at an angle θ to the vertical is

a. $\frac{2g\sin\theta}{3L}$ b. $\frac{g\sin\theta}{3L}$

c. $\frac{4g\sin\theta}{3L}$ d. $\frac{3g\sin\theta}{2L}$

20 Assuming the rod to start from the vertical position, the angular velocity as a function of θ is

a. $\omega=\sqrt{\frac{3g}{L}(1-\sin\theta)}$ b. $\omega=\sqrt{\frac{2g}{L}(1-\sin\theta)}$

c. $\omega=\sqrt{\frac{3g}{L}(1-\cos\theta)}$ d. $\omega=\sqrt{\frac{2g}{L}(1-\cos\theta)}$

21 The tangential linear acceleration of the free end when the rod is horizontal, is

a. $\frac{2g}{3}$ b. $\frac{3g}{2}$

c. $\frac{4}{3}g$ d. $\frac{g}{2}$

Passage II

Three particles A, B and C, each of mass m are connected to one another by three massless rigid rods to form a rigid, equilateral triangular body of side l. This body is placed on a horizontal frictionless table (XY-plane) and is hinged to it at the point A so that it can move without friction about the vertical axis through A with a constant angular velocity ω.

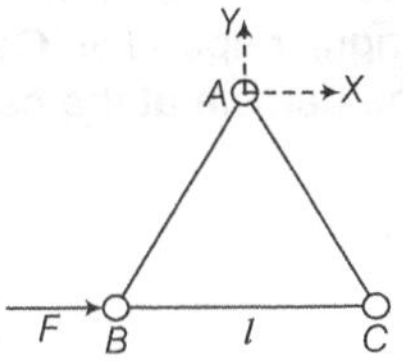

22 The magnitude of horizontal force exerted by the hinge on the body is

a. $\frac{\sqrt{3}}{2}ml\,\omega^2$ b. $\frac{\sqrt{3}}{4}ml\,\omega^2$

c. $\sqrt{3}\,ml\,\omega^2$ d. $2\sqrt{3}\,ml\,\omega^2$

23 At time T, when the side BC is parallel to X-axis, a force F is applied on B along BC (see figure). The X-component and Y-component of the force exerted by the hinge on the body, at this moment T, respectively are

a. $-\frac{3F}{4}, \sqrt{3}\,ml\,\omega^2$ b. $-\frac{2F}{3}, \frac{\sqrt{3}}{2}ml\,\omega^2$

c. $-\frac{F}{3}, \sqrt{3}\,ml\,\omega^2$ d. $-\frac{F}{4}, \frac{\sqrt{3}}{2}ml\,\omega^2$

DPP-4 Angular Momentum and Its Conservation

Only One Option Correct Type

Direction (Q. Nos. 1-13) *This section contains 13 multiple choice questions. Each question has four choices (a), (b), (c) and (d), out of which ONLY ONE is correct.*

Working Space

1 A circular disc of mass m and radius R is set into motion on a horizontal floor with a linear speed v in the forward direction and an angular speed $\omega = \frac{v}{R}$ in clockwise direction as shown in the figure. Find the magnitude of the total angular momentum of the disc about the bottom most point O of the disc.

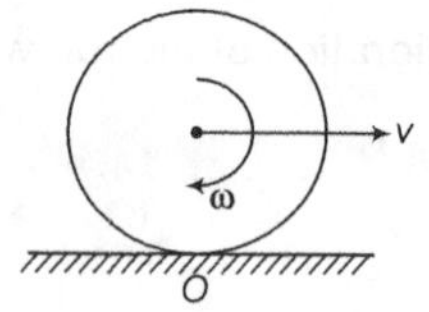

a. $\frac{1}{2}mvR$ b. $\frac{2}{3}mvR$

c. $\frac{3}{2}mvR$ d. $\frac{4}{3}mvR$

2 In the above problem, suppose the disc starts rotating anti-clockwise, then in this new situation, what will be the angular momentum of the disc about point O?

a. $\frac{1}{2}mvR$ b. $\frac{2}{3}mvR$

c. $\frac{3}{2}mvR$ d. $\frac{4}{3}mvR$

3 Two particles, each of mass m are attached to a light rod of length d, one at its centre and the other at a free end. The rod is fixed at the other end and is rotated in a plane at an angular speed ω. Calculate the angular momentum of the particle at the end w.r.t. the particle at the centre.

a. $\frac{1}{2}m\omega d^2$ b. $\frac{1}{4}m\omega d^2$

c. $\frac{1}{3}m\omega d^2$ d. $\frac{2}{3}m\omega d^2$

4 A wheel is rotating at an angular speed ω about its axis which is kept vertical. An identical wheel initially at rest is gently dropped into the same axle and the two wheels start rotating with a common angular speed. Find this common angular speed.

a. ω b. $\frac{\omega}{2}$ c. $\frac{\omega}{3}$ d. $\frac{\omega}{4}$

5 A uniform circular disc of mass m and radius a is rotating with constant angular velocity ω in a horizontal plane about the vertical axis passing through its centre and perpendicular to its plane. A particle P of mass $2m$ is placed gently on the disc at a point, distant $\frac{a}{2}$ from the centre. If the particle does not slip on the disc, find the new angular velocity of the rotating system.

a. ω b. $\frac{\omega}{2}$ c. $\frac{\omega}{3}$ d. $\frac{\omega}{4}$

Working Space

6 The polar ice caps contain about 2.3×10^{19} kg of ice. This contributes essentially nothing to the moment of inertia of the earth because it is located at the poles, close to the axis of rotation. Estimate the change in length of the day to be expected if the polar ice caps melt, distributing the water uniformly over the surface of the earth. Given, mass of the earth = 5.94×10^{24} kg and the radius of the earth = 6.37×10^{6} m.

a. time period increases by 0.9 s
b. time period decreases by 0.9 s
c. time period increases by 0.6 s
d. time period decreases by 0.6 s

7 A child is standing with folded hands at the centre of a platform rotating about its central axis. The kinetic energy of the system is K. The child now stretches his arms so that the moment of inertia of the system doubles. What is the kinetic energy of the system, now?

a. $\frac{K}{4}$
b. $\frac{K}{3}$
c. $\frac{K}{5}$
d. $\frac{K}{2}$

8 A boy is standing at the centre on a platform which is free to rotate about its axis. The boy holds an open umbrella in his hand. The axis of the umbrella coincides with that of the platform. The moment of inertia of "The platform of the boy system" is 3×10^{-3} kg m^{-2} and that of the umbrella is 2×10^{-3} kg m^{-2}. The boy starts spinning the umbrella about the axis at an angular speed of 2 rad s^{-1} w.r.t. himself. Find the angular velocity imparted to the platform. (Note that the values in this problem are not practical).

a. 0.7 rad s^{-1}
b. 0.8 rad s^{-1}
c. 0.9 rad s^{-1}
d. 1 rad s^{-1}

9 A kid of mass M stands at the edge of a platform of radius R which can be freely rotated about its axis. The moment of inertia of the platform is I. The system is at rest when a friend throws a ball of mass m and the kid catches it. If the velocity of the ball is horizontally along the tangent to the edge of the platform when it was caught by the kid, find the angular speed of the platform after the event.

a. $\frac{mvR}{I+(m+M)R^2}$
b. $\frac{mvR}{2[I+(m+M)R^2]}$
c. $\frac{2mvR}{I+(m+M)R^2}$
d. $\frac{mvR}{I+\frac{(m+M)R^2}{2}}$

10 Suppose the platform of the previous problem is brought to rest with the ball in the hand of the kid standing on the rim. The kid now throws the ball horizontally to his friend in a direction tangential to the rim with a speed v as seen by his friend. Find the angular velocity with which the platform will start rotating.

a. $\frac{mvR}{(M+m)R^2+I}$
b. $\frac{mvR}{(M-m)R^2+I}$
c. $\frac{mvR}{MR^2+I}$
d. $\frac{mvR}{mR^2+I}$

11 Suppose the platform with the kid in the previous problem is rotating in anti-clockwise direction at an angular speed ω. The kid starts walking along the rim with a speed v relative to the platform also in anti-clockwise direction. Find the new angular speed of the platform.

a. $\omega+\frac{MvR}{I+MR^2}$
b. $\omega-\frac{mvR}{I+MR^2}$
c. $\omega+\frac{mvR}{I+MR^2}$
d. $\omega-\frac{MvR}{I+MR^2}$

12 A circular platform is free to rotate in a horizontal plane about a vertical axis passing through its centre. A tortoise is sitting at the edge of the platform. Now, the platform is given an angular velocity ω_0. When the tortoise moves along a chord of the platform with a constant velocity (w.r.t. the platform), the angular velocity of the platform $\omega(t)$ will vary with time t as

a. 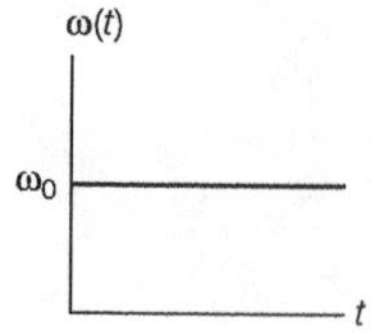

b. 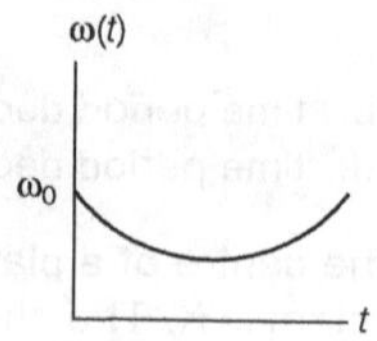

c. 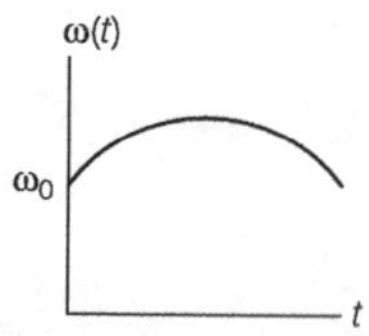

d. 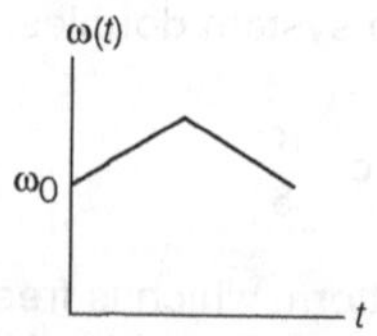

13 A rod of length L and mass M is hinged at point O. A small bullet of mass m hits the rod as shown in the figure. The bullet gets embedded in the rod. Find the angular velocity of the system just after impact.

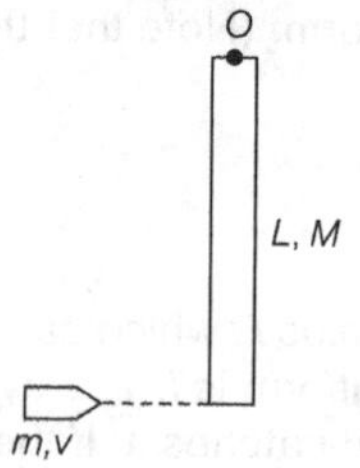

a. $\dfrac{2mv}{L(3m+M)}$ b. $\dfrac{3mv}{L(3m+M)}$

c. $\dfrac{2mv}{L(2m+M)}$ d. None of these

One or More than One Options Correct Type

Direction *(Q. Nos. 14-17) This section contains 4 multiple choice questions. Each question has four choices (a), (b), (c) and (d), out of which ONE or MORE THAN ONE are correct.*

14 Two small balls A and B, each of mass m, are attached rigidly to the ends of a light rod of length d. The structure rotates about the perpendicular bisector of the rod at an angular speed ω. Choose the correct option(s).

a. The angular momentum of individual ball about the axis is $\frac{1}{4}m\,\omega d^2$

b. The angular momentum of individual ball about the axis is $\frac{1}{2}m\,\omega d^2$

c. The angular momentum of the system about the axis is $\frac{1}{2}m\,\omega d^2$

d. The angular momentum of the system about the axis is $m\omega d^2$

Working Space

15 The angular momentum of a disc about Z-axis in Fig. (i) is l_1 and in Fig (ii) is l_2. Then,

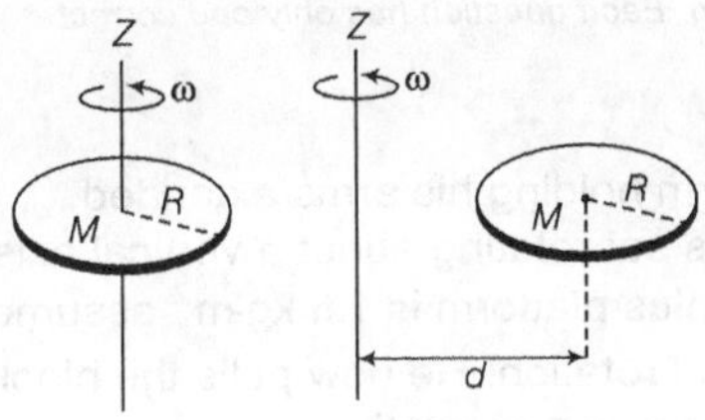

Fig. (i) Fig. (ii)

a. $l_1 = \frac{MR^2\omega}{2}\hat{\mathbf{k}}$

b. $l_2 = M\left(\frac{R^2}{2}+d^2\right)\omega\hat{\mathbf{k}}$

c. $l_1 = \frac{MR^2\omega}{4}\hat{\mathbf{k}}$

d. $l_2 = M\left(\frac{R^2}{2}+\frac{d^2}{2}\right)\omega\hat{\mathbf{k}}$

16 The angular momentum of the disc shown about the origin in Fig. (i) is l_1 and in Fig (ii) is l_2. Then,

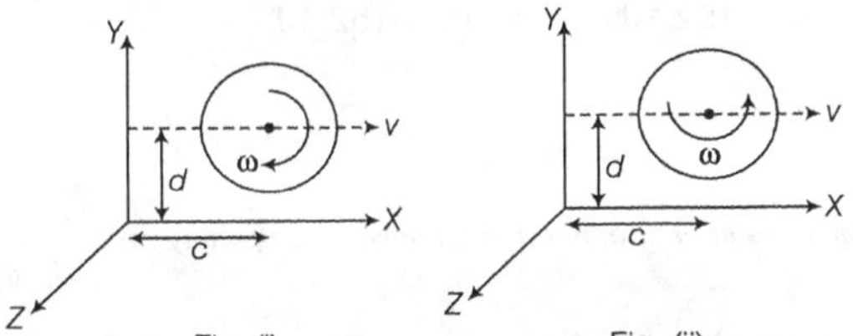

Fig. (i) Fig. (ii)

a. $\left(\frac{MR^2\omega}{4}+Mvd\right)(-\hat{\mathbf{k}})$

b. $\left(-\frac{MR^2\omega}{4}+Mvd\right)(-\hat{\mathbf{k}})$

c. $\left(\frac{MR^2\omega}{2}+Mvd\right)(-\hat{\mathbf{k}})$

d. $\left(-\frac{MR^2\omega}{2}+Mvd\right)(-\hat{\mathbf{k}})$

17 A horizontally oriented uniform disc of mass M and radius R rotates freely about a stationary vertical axis passing through its centre. The disc has a radial guide along which can slide, without friction, a small body of mass m. A light thread running down through the hollow axle of the disc, is tied up to the body. Initially, the body was located at the edge of the disc and the whole system rotated with an angular velocity $\omega_0 (> 0)$. Then, by means of force F applied to the lower end of the thread, the body was slowly pulled to the rotation axis. Choose the correct option(s).

a. The angular velocity of the system in its final state is $\left(1+\frac{m}{2M}\right)\omega_0$

b. The angular velocity of the system in its final state is $\left(1+\frac{2m}{M}\right)\omega_0$

c. Work performed by the force F is $\frac{1}{2}mR^2\omega_0^2\left(1+\frac{2m}{M}\right)$

d. Work performed by the force F is $\frac{1}{2}mR^2\omega_0^2\left(1+\frac{m}{2M}\right)$

Comprehension Type

Working Space

Direction (Q. Nos. 18-20) *This section contains a paragraph, which describing theory, experiments, data, etc. Three questions related to the paragraph have been given. Each question has only one correct answer among the four given options (a), (b), (c) and (d).*

Passage

A man stands at the centre of a circular platform holding his arms extended horizontally with 4 kg block in each hand. He is set rotating about a vertical axis at 0.5 rev s^{-1}. The moment of inertia of the man plus platform is 1.6 kg-m^2 assumed constant. The blocks are 90 cm from the axis of rotation. He now pulls the blocks in towards his body until they are 15 cm from the axis of rotation.

18 His new angular velocity (in rad s^{-1}) is

a. 12.3 b. 13.3 c. 14.3 c. 15.3

19 The initial KE and final KE of (man + platform) respectively are

a. 39.9 J, 182 J
b. 49.9 J, 172 J
c. 39.9 J, 192 J
d. 49.9 J, 192 J

20 Work done by the man is

a. −142.1 J b. +142.1 J c. +152.1 J d. −152.1 J

Subjective Questions

Direction (Q. Nos. 21-23) *These questions are subjective in nature, need to be solved completely on notebook.*

21 Two uniform rods *A* and *B* of length 0.6 m each and of masses 0.01 kg and 0.02 kg, respectively are rigidly joined end to end. The combination is pivoted at the higher end *P* shown in figure, such that it can freely rotate about point *P* in a vertical plane. A small object of mass 0.05 kg moving horizontally hits the lower end of the combination and sticks to it. What should be the velocity of the object so that the system could just be raised to the horizontal position?

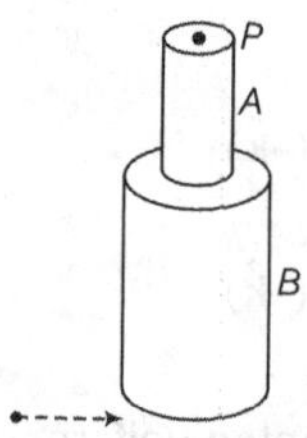

22 A man of mass $m = 80$ kg runs at a speed $u = 4\ ms^{-1}$ along the tangent to a disc shaped platform of mass $M = 160$ kg and radius $R = 2$ m. The platform is initially at rest and can rotate freely about an axis through its centre. Find the angular velocity of the man after the man jumps on to the disc.

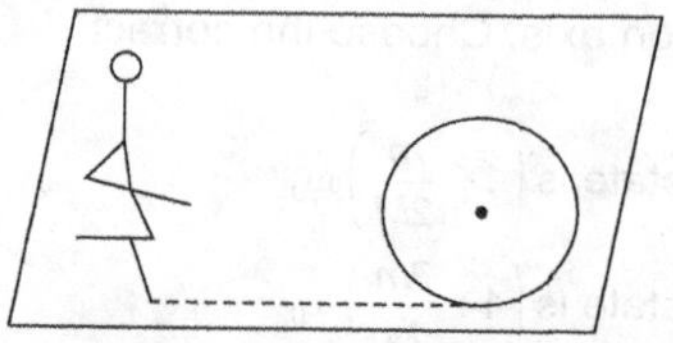

23 A homogeneous rod *AB* of length $L = 1.8$ m and mass *M* is pivoted at the centre *O* in such a way that it can rotate freely in a vertical plane (see figure). The rod is initially in the horizontal position. An insect *S* of the same mass *M* falls vertically with speed *v* on the point *C* mid-way between *O* and *B*. Immediately after falling, the insect moves towards the end *B* such that the rod rotates with a constant angular velocity ω.

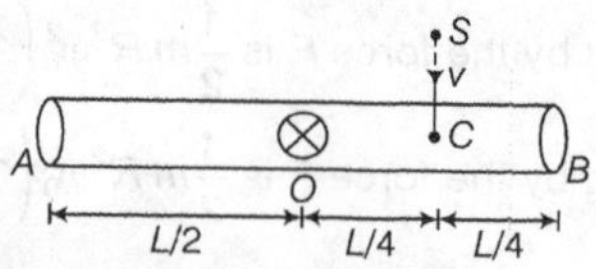

i. Determine the angular velocity in terms of *v* and *L*.
ii. If the insect reaches the end *B* when the rod has turned through an angle of 90°, determine the value of *v*. (Given, $g = 10\ ms^{-2}$)

DPP-5 Combined Translation and Rotation

Only One Option Correct Type

Working Space

Direction (Q. Nos. 1-3) *This section contains 3 multiple choice questions. Each question has four choices (a), (b), (c) and (d), out of which ONLY ONE is correct.*

1 A rigid massless rod of length L joins two particles each of mass m. The rod lies on a frictionless table and is struck by a particle of mass m and velocity v_0 moving as shown. After the collision, the projectile moves straight back. Find the angular velocity about its centre of mass after collision assuming that mechanical energy is conserved.

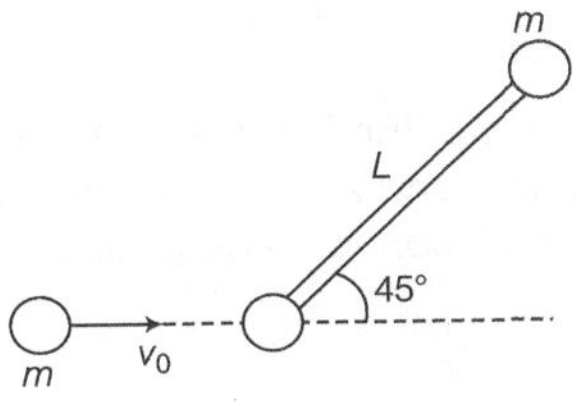

a. $\frac{2\sqrt{2}}{7} \cdot \frac{v_0}{L}$

b. $\frac{3\sqrt{2}}{7} \cdot \frac{v_0}{L}$

c. $\frac{\sqrt{2}}{7} \cdot \frac{v_0}{L}$

d. $\frac{4\sqrt{2}}{7} \cdot \frac{v_0}{L}$

2 A uniform 50 kg pole ABC of length 6 m is balanced in the vertical position (see figure.) A 500 N horizontal force is suddenly applied at B. If the coefficient of kinetic friction between the pole and the ground is 0.3, determine the initial acceleration of point A. (Given, $g = 10 \text{ ms}^{-2}$)

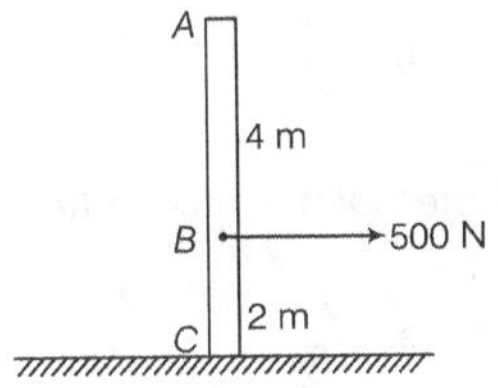

a. 4 ms^{-2} towards right

b. 4 ms^{-2} towards left

c. 6 ms^{-2} towards right

d. 6 ms^{-2} towards left

3 A wedge of mass $4m$ is placed at rest on a smooth horizontal surface. A uniform solid sphere of mass m and radius r is placed at rest on the flat portion of the wedge at a point Q as shown in the figure. A sharp horizontal impulse J is given to the sphere at a point below $h = 0.4\ r$ from the centre of the sphere. The radius of curvature of curved portion on the wedge is R. The coefficient of friction to the left side of point Q is μ and to the right side of point Q is zero. The maximum height to which the centre of mass of the sphere will climb on the curved portion of the wedge is

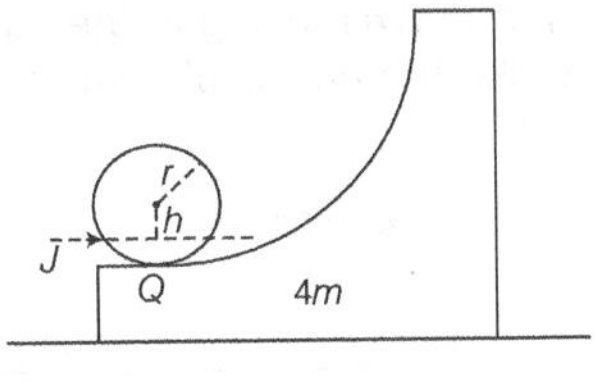

a. $\frac{2J^2}{5m^2g}$

b. $\frac{J^2}{5m^2g}$

c. $\frac{J^2}{2m^2g}$

d. $\frac{3J^2}{5m^2g}$

One or More than One Options Correct Type

Working Space

Direction (Q. No. 4) *This section contains 1 multiple choice question. This question has four choices (a), (b), (c) and (d), out of which ONE or MORE THAN ONE are correct.*

4 A rod leans against a stationary cylindrical body as shown in the figure. Now, its right end slides to the right on the floor with a constant speed v. Choose the correct option(s).

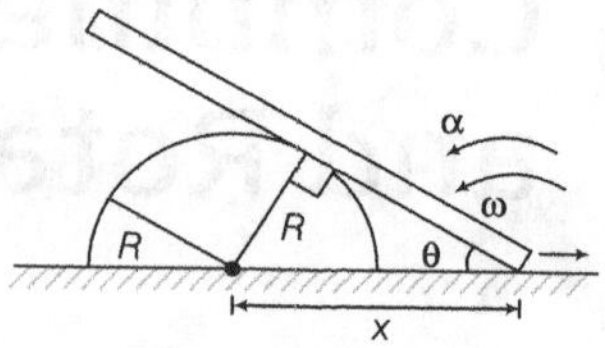

a. The angular acceleration α is $-2Rv^2(2x^2-R^2)/[x^2(x^2-R^2)^{3/2}]$

b. The angular speed ω is $2Rv/[x(x^2-R^2)^{1/2}]$

c. The angular speed ω is $Rv/[x(x^2-R^2)^{1/2}]$

d. The angular acceleration α is $-Rv^2(2x^2-R^2)/[x^2(x^2-R^2)^{3/2}]$

Comprehension Type

Direction (Q. Nos. 5-22) *This section contains 5 paragraphs, each describing theory, experiments, data, etc. Eighteen questions related to the paragraphs have been given. Each question has only one correct answer among the four given options (a), (b), (c) and (d).*

Passage I

A rod AB of mass M and length L is lying on a horizontal frictionless surface. A particle of mass m travelling along the surface hits the end A of the rod with a velocity v_0 in a direction perpendicular to AB. The collision is elastic. After collision, the particle comes to rest.

5 Find the ratio $\frac{m}{M}$.

a. $\frac{2}{3}$ b. $\frac{1}{5}$ c. $\frac{1}{4}$ d. $\frac{2}{5}$

6 A point P on the rod is at rest immediately after collision. Find the distance AP.

a. $\frac{L}{6}$ b. $\frac{5L}{6}$ c. $\frac{2L}{3}$ d. $\frac{L}{3}$

7 Find the speed of the point P after a time $\frac{\pi L}{3v_0}$ after the collision.

a. $\frac{v_0}{3\sqrt{2}}$ b. $\frac{v_0}{\sqrt{2}}$ c. $\frac{v_0}{2}$ d. $\frac{v_0}{2\sqrt{2}}$

Passage II

A thin uniform bar of length $6a$ and mass $8m$ lies on a smooth horizontal table. Two point masses m and $2m$ moving in the same horizontal plane with speeds $2v$ and v respectively strike the bar as shown in the figure and stick to it after collision.

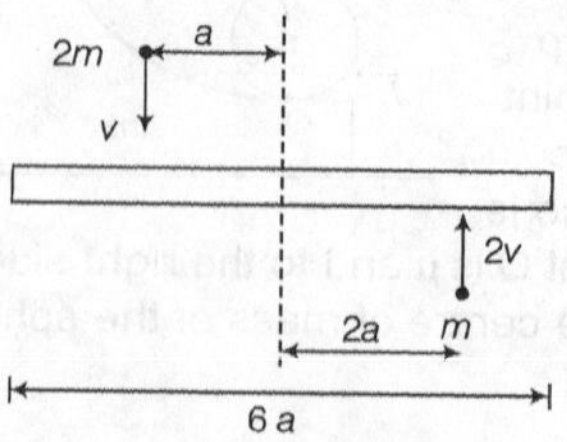

8 After collision, the velocity of centre of mass is

a. zero b. $\frac{v}{5}$ c. $\frac{v}{2}$ d. $\frac{v}{2\sqrt{2}}$

Working Space

9 After collision, the angular velocity of the rod about centre of mass is

a. $\frac{v}{3a}$ b. $\frac{v}{5a}$

c. $\frac{v}{a}$ d. $\frac{v}{2\sqrt{2}a}$

10 After collision, the total energy is

a. $\frac{1}{3}mv^2$ b. $\frac{2}{3}mv^2$

c. $\frac{2}{5}mv^2$ d. $\frac{3}{5}mv^2$

Passage III

A thin rod of length L and mass m moves on a smooth horizontal surface at a velocity $\mathbf{v}_0$ in the direction of its length. The end of the rod hits the end of a second rod of the same length and mass, which is initially at rest, perpendicular to the first rod. At the moment they touch, the two poles stick together while the angle between them remains $\frac{\pi}{2}$. Take the point of contact as origin and directions of X and Y-axes as shown.

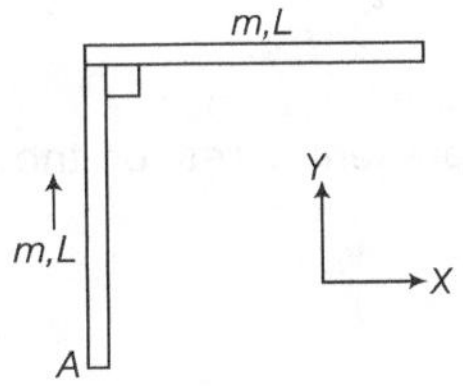

11 The velocity of centre of mass of the system is

a. $\frac{v_0}{2}\hat{\mathbf{j}}$ b. $v_0\,\hat{\mathbf{j}}$

c. $\frac{v_0}{3}\hat{\mathbf{j}}$ d. $\frac{2v_0}{3}\hat{\mathbf{j}}$

12 Just before collision, the centre of mass is located at

a. $\frac{1}{3}L(\hat{\mathbf{i}}-\hat{\mathbf{j}})$ b. $\frac{1}{2}L(\hat{\mathbf{i}}+\hat{\mathbf{j}})$

c. $\frac{1}{4}L(\hat{\mathbf{i}}-\hat{\mathbf{j}})$ d. $\frac{1}{4}L(\hat{\mathbf{i}}+\hat{\mathbf{j}})$

13 Angular velocity of the system after collision about a perpendicular axis through centre of mass is

a. $-\frac{3}{2}\frac{v_0}{L}\hat{\mathbf{k}}$ b. $-\frac{3}{5}\frac{v_0}{L}\hat{\mathbf{k}}$

c. $-\frac{3}{4}\frac{v_0}{L}\hat{\mathbf{k}}$ d. $-\frac{2}{3}\frac{v_0}{L}\hat{\mathbf{k}}$

14 The maximum velocity attained by the end of the rod denoted by A, after the collision is

a. $\frac{1}{2}v_0\left[1+\frac{3}{\sqrt{5}}\right]$ b. $\frac{1}{2}v_0\left[3+\frac{2}{\sqrt{5}}\right]$

c. $\frac{1}{2}v_0\left[2+\frac{3}{\sqrt{5}}\right]$ d. $\frac{1}{2}v_0\left[1+\frac{3}{\sqrt{10}}\right]$

15 Maximum velocity is attained for the first time after a time.

a. $\frac{5L}{3v_0}\tan^{-1}(3)$ b. $\frac{5L}{4v_0}\tan^{-1}(2)$

c. $\frac{3L}{4v_0}\tan^{-1}(3)$ d. $\frac{2L}{v_0}\tan^{-1}(2)$

Passage IV

A rigid zero mass rod of length $2L$ is placed on a smooth horizontal plane and is free to move on it. At $t < 0$, the rod is at rest on the Y-axis ($-L$ to L). Two identical small bullets, each of mass m hit the rod at $t = 0$. The first one hits the lower end A with a velocity $v\hat{\mathbf{i}}$ and the second bullet hits the rod at its upper end B with a velocity $-v\hat{\mathbf{i}}$ as depicted in the figure. The two bullets stick to the rod simultaneously at $t = 0$ and move with it.

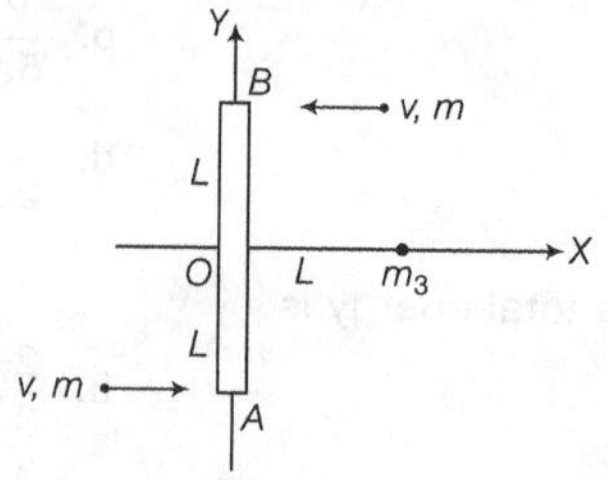

16 Angular speed of the system composed of the rod and the two bullets is

a. $\frac{v}{2L}$ b. $\frac{v}{L}$ c. $\frac{2v}{L}$ d. $\frac{v}{4L}$

17 A mass $m_3(m)$ is present at rest on the point $(L, 0)$. After what time, does the rod hit it?

a. $\frac{\pi L}{2v}$ b. $\frac{\pi L}{3v}$

c. $\frac{\pi L}{v}$ d. $\frac{\pi L}{\sqrt{2}v}$

18 If immediately after collision m_3 sticks to the point of the rod which hits m_3, what is the new angular velocity of the whole system about the centre of mass?

a. $\frac{3v}{2L}$ b. $\frac{2v}{3L}$

c. $\frac{3v}{4L}$ d. $\frac{v}{4L}$

19 Linear velocity of the whole system is

a. zero b. v c. $\frac{v}{2}$ d. $\frac{v}{3}$

Passage V

A thin uniform bar lies on a frictionless horizontal surface and is free to move in any way on the surface. Its mass is 0.16 kg and length is $\sqrt{3}$ m. Two particles, each of mass 0.08 kg are moving on the same surface and towards the bar in a direction perpendicular to the bar, one with a velocity 10 ms^{-1} and the other with 6 ms^{-1} as shown in the figure. The first particle strikes the bar at point A and the other at point B. Each of A and B is at a distance of 0.5 m from the centre of the bar. The particles strike the bar at the same instant of time and stick to the bar after collision.

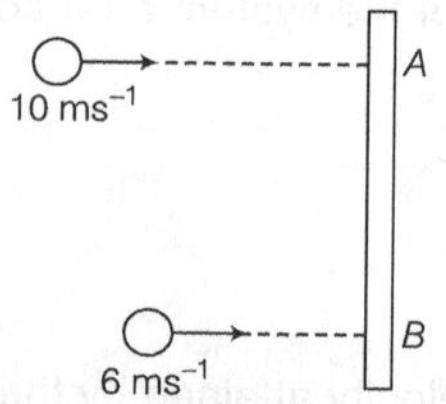

20 The velocity of centre of mass of the system just after impact (in ms^{-1}) is

a. 1 b. 2

c. 3 d. 4

21 The angular velocity of the system just after impact (in rad/s) is

a. 8 b. 4

c. 2 d. 1

22 The loss in kinetic energy of the system in the above collision process is

a. 2.72 J b. 1.72 J

c. 3.36 J d. 4.36 J

DPP-6 Instantaneous Axis of Rotation and Instantaneous Centre

Only One Option Correct Type

Working Space

Direction (Q. Nos. 1-8) *This section contains 8 multiple choice questions. Each question has four choices (a), (b), (c) and (d), out of which ONLY ONE is correct.*

1 A rotating disc (with radius r) moves in the positive direction of the X-axis. Find the equation $y(x)$ describing the position of the instantaneous axis of rotation, if at the initial moment, the centre C of the disc was located at the point O after which it moved with constant velocity v, while the disc started rotating counter clockwise with a constant angular acceleration α. Assume the initial angular velocity to be zero.

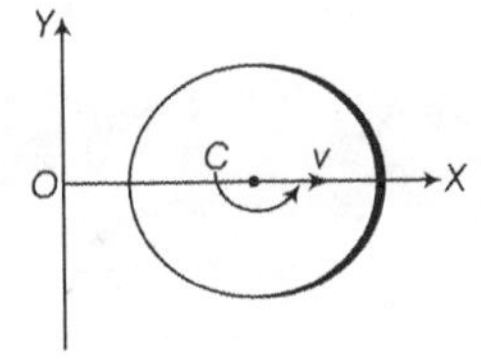

a. $xy = \dfrac{v^2}{\alpha}$

b. $xy = v\alpha$

c. $y = v^2\alpha + x$

d. $y = v^2\alpha - x$

2 A uniform rod of length L and mass m stands vertically touching a vertical wall (Y-axis). When slightly displaced, its lower end begins to slide along the floor (X-axis). Obtain an expression for the angular velocity (ω) as a function of (θ). Friction is absent everywhere.

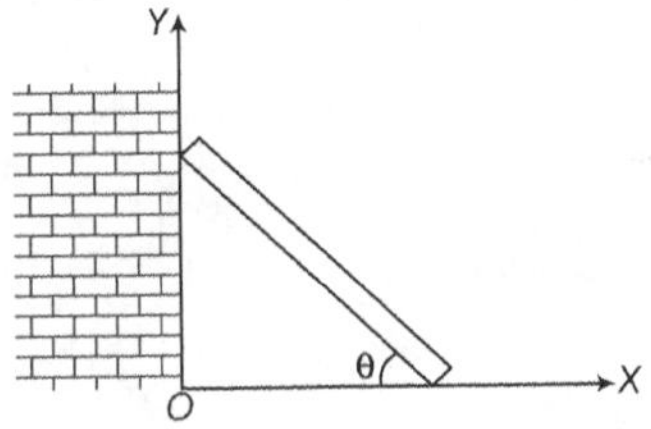

a. $\sqrt{\dfrac{3g\,(1-\sin\theta)}{L}}$

b. $\sqrt{\dfrac{3g\,(1-\cos\theta)}{L}}$

c. $\sqrt{\dfrac{3g\,(1+\sin\theta)}{L}}$

d. None of these

3 For the above problem, determine the distance moved by the lower end by the instant at which the rod no longer touches the vertical wall.

a. $\dfrac{\sqrt{5}}{2}L$

b. $\dfrac{\sqrt{5}}{7}L$

c. $\dfrac{\sqrt{5}}{3}L$

d. $\dfrac{\sqrt{5}}{4}L$

Working Space

4 A hoop rolls without slipping on a horizontal ground with its centre of mass having a speed v_0. What is the speed of a particle P on the circumference of the hoop at an angle θ (see figure)?

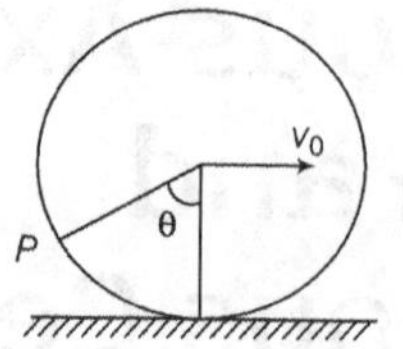

a. $2v_0 \sin\theta$
b. $2v_0 \cos\theta$
c. $2v_0 \cos\dfrac{\theta}{2}$
d. $2v_0 \sin\dfrac{\theta}{2}$

5 A rod of length L is given two velocities v_1 and v_2 (at its ends A and B respectively) in opposite directions at right angles to the length. What is the distance of IAOR from end A?

a. $\left(\dfrac{v_1}{v_1+v_2}\right)L$
b. $\left(\dfrac{v_1}{v_1-v_2}\right)L$
c. $\left(\dfrac{v_1+v_2}{v_1}\right)L$
d. $\left(\dfrac{v_2}{v_1+v_2}\right)L$

6 A disc is rolling (without slipping) on a horizontal surface. C is its centre and A and B are two points equidistant from C. Let v_A, v_B and v_C be the magnitudes of velocities of points A, B, and C, respectively. Arrange them in increasing order.

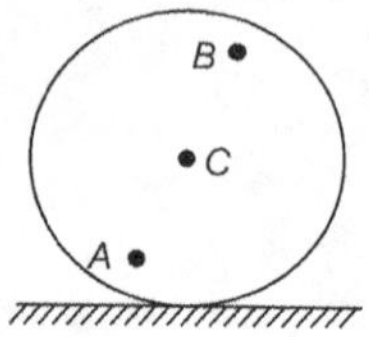

a. $v_A < v_B < v_C$
b. $v_A < v_C < v_B$
c. $v_B < v_C < v_A$
d. $v_C < v_B < v_A$

7 A uniform rod OA of length L resting on smooth horizontal surface is slightly disturbed from its vertical position. P is a point on the rod whose locus is a circle during the subsequent motion of the rod. What is the distance OP?

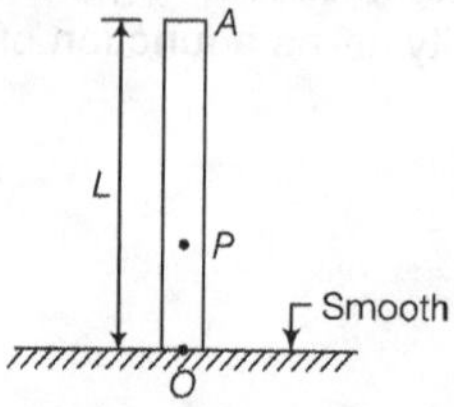

a. $\dfrac{L}{4}$
b. $\dfrac{L}{3}$
c. $\dfrac{2L}{3}$
d. $\dfrac{L}{2}$

8 A disc of radius 20 cm is rolling with slipping on a flat horizontal surface. At a certain instant the velocity of its centre is 4 ms^{-1} and its angular velocity is 10 rad/s. The lowest contact point is O. What is the distance between point O and instantaneous centre at this moment?

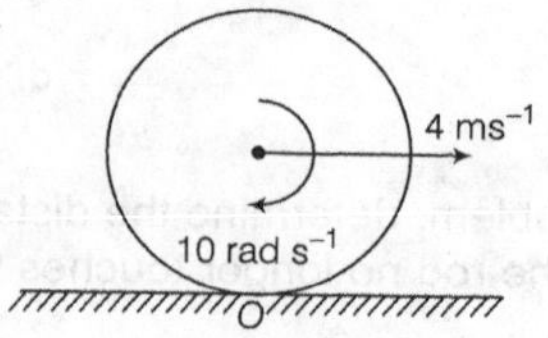

a. 0.2 m below O
b. 0.2 m above O
c. 0.1 m below O
d. 0.1 m above O

Statement Type

Direction (Q. No. 9) *This section is based on Statement I and Statement II. Select the correct answer from the codes given below.*

a. Both Statement I and Statement II are correct and Statement II is the correct explanation of Statement I
b. Both Statement I and Statement II are correct but Statement II is not the correct explanation of Statement I
c. Statement I is correct but Statement II is incorrect
d. Statement II is correct but Statement I is incorrect

9 **Statement I** A disc is rolling on an inclined plane without slipping. The velocity of centre of mass is v_0. The other points of the disc having same speed (i.e. v_0) lie on a circular arc.

Statement II When a disc is rolling on an inclined plane, the magnitudes of velocities of all the points at distance equal to radius from contact point are same.

One Integer Value Correct Type

Direction (Q. No. 10) *This section contains 1 question. When worked out will result in an integer from 0 to 9 (both inclusive).*

10 A constant force F is exerted on a rod of mass M. The rod is supported by a frictionless wall. If the rod starts from a position of rest when $\theta = 45°$ (see figure). The angular speed when the end A has moved a distance 1.5 m is $\left(\frac{F}{KM} - 1.6\right)^{1/2}$.

Find the value of K.

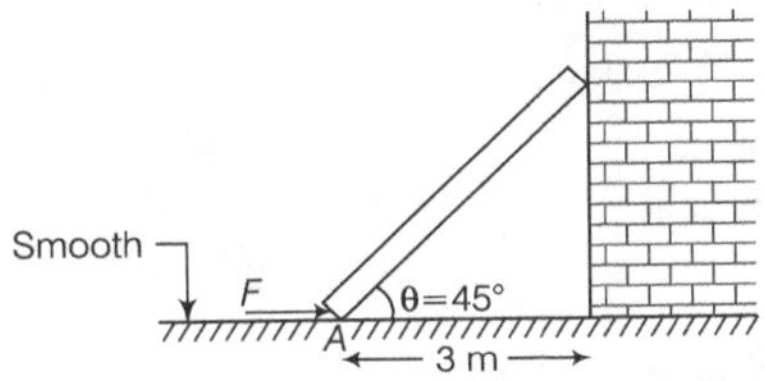

Subjective Questions

Direction (Q. Nos. 11-14) *These questions are subjective in nature, need to be solved completely on notebook.*

11 A uniform thin rod of mass m and length L is standing on a smooth horizontal surface. A slight disturbance causes the lower end to slip on the smooth surface and the rod starts falling by slipping. Describe the trajectory of centre of mass. Find the speed of centre of mass of the rod at the instant, when the rod makes an angle θ with horizontal.

12 In the above problem, what is the velocity of point B when the rod makes an angle θ with the horizontal?

13 For the above problem, what is the velocity of the point C when the rod touches the ground?

14 A uniform rod of length $2a$ is held with one end resting on a smooth horizontal table making an angle α with the vertical. The rod is released from rest. What is its angular velocity when it makes an angle θ with the vertical?

Comprehension Type

Working Space

Direction (Q. Nos. 15-20) This section contains 2 paragraphs, each describing theory, experiments, data, etc. Six questions related to the paragraphs have been given. Each question has only one correct answer among the four given options (a), (b), (c) and (d).

Passage I

The disc of mass m, radius R is confined to roll without slipping at A and B as shown in the figure.

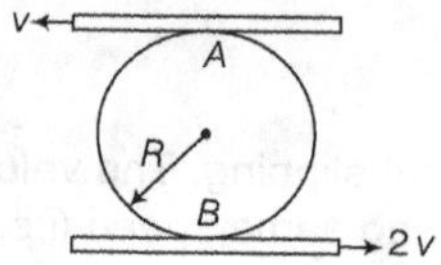

15 If the plates have velocities shown, then find out the distance of instantaneous centre from the point A.

a. $\frac{2R}{3}$ b. $\frac{R}{3}$

c. $\frac{4R}{3}$ d. $\frac{R}{2}$

16 Find out the angular velocity of the disc.

a. $\frac{2v}{3R}$ b. $\frac{3v}{2R}$

c. $\frac{v}{2R}$ d. $\frac{4v}{3R}$

17 What is the velocity of centre of mass?

a. v b. $\frac{v}{3}$

c. $\frac{v}{2}$ d. $\frac{2v}{3}$

18 What is the total energy of the disc?

a. $\frac{9}{16}mv^2$ b. $\frac{13}{16}mv^2$

c. $\frac{7}{8}mv^2$ d. $\frac{11}{16}mv^2$

Passage II

A rod AB of length 3 m remains in contact with a horizontal floor and a vertical wall. At the instant shown, the point A has a velocity 2 ms^{-1} and an acceleration of 1.6 ms^{-2} rightwards.

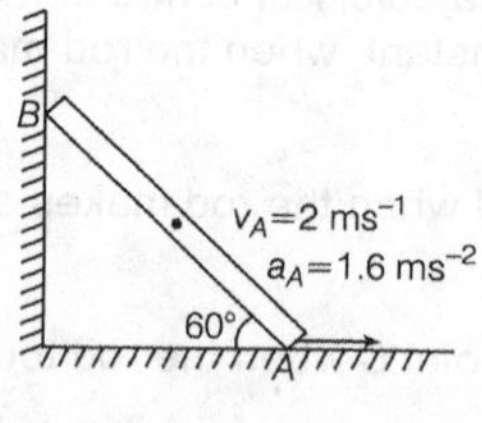

19 The velocity of B at the instant shown is

a. 1.72 ms^{-1} b. 1.16 ms^{-1}

c. 1.98 ms^{-1} d. 1.54 ms^{-1}

20 The acceleration of B at the instant shown is

a. 3.72 ms^{-2} b. 3.43 ms^{-2}

c. 2.21 ms^{-2} d. 2.98 ms^{-2}

DPP-7 Rolling without Slipping and Rolling with Slipping

Only One Option Correct Type

Working Space

Direction *(Q. Nos. 1-7) This section contains 7 multiple choice questions. Each question has four choices (a), (b), (c) and (d), out of which ONLY ONE is correct.*

1 A solid cylindrical wheel of mass M and radius R is pulled by a force F applied to the centre of the wheel at 37° to the horizontal. If the wheel is to roll without slipping, what is the maximum value of $|F|$? The coefficients of static and kinetic friction are $\mu_s = 0.40$ and $\mu_k = 0.30$ and $\sin 37° = \frac{3}{5}$.

a. $0.79\,Mg$ b. $0.98\,Mg$ c. $0.6\,Mg$ d. $0.49\,Mg$

2 A $Yo-Yo$ of mass M has an axle of radius r and a spool of radius R. Its moment of inertia can be taken to be $\frac{MR^2}{2}$.

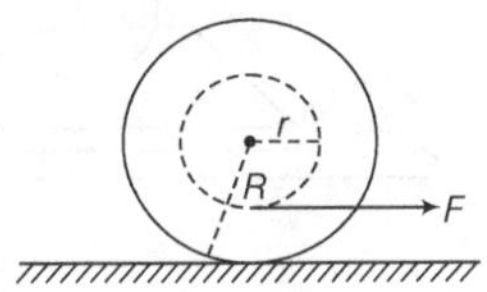

The $Yo-Yo$ is placed upright on a table and the string is pulled with a horizontal force as shown. The coefficient of friction between $Yo-Yo$ and the table is μ. What is the maximum value of F for which the $Yo-Yo$ will roll without slipping?

a. $\frac{2\mu MgR}{R+2r}$ b. $\frac{3\mu MgR}{R+2r}$

c. $\frac{\mu MgR}{R+r}$ d. $\frac{\mu MgR}{R+2r}$

3 The arrangement shown in the figure consists of two identical uniform solid cylinders, each of mass M, radius R on which two light threads are wound symmetrically. Find the tension in each thread in the process of motion. The friction in the axle of the upper cylinder is ignored.

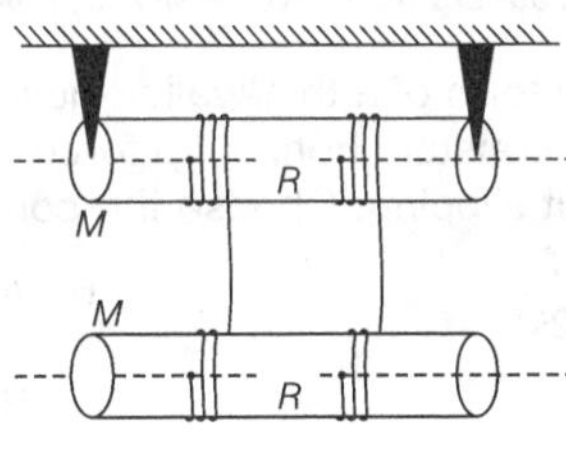

a. $\frac{Mg}{10}$ b. $\frac{Mg}{8}$

c. $\frac{Mg}{12}$ d. $\frac{Mg}{6}$

4 A thin spherical shell of mass m and radius R lying on a rough horizontal surface is hit sharply and horizontally by a cue. Where should it hit, so that the shell does not slip on the surface from the next moment?

a. At a height $\frac{1}{3}R$ from the centre b. At a height $\frac{2}{3}R$ from the centre

c. At a height $\frac{2}{5}R$ from the centre d. At a height $\frac{1}{4}R$ from the centre

Working Space

5 A billiard ball is struck by a cue.The line of action of the applied impulse is horizontal and passes through the centre of the ball. The initial velocity of the ball is v_0. If R is the radius, m is the mass of the ball and μ is the coefficient of friction between the ball and the floor, find how far the ball moves before it ceases to slip on the floor.

a. $\frac{12v_0^2}{49\mu g}$ b. $\frac{6v_0^2}{49\mu g}$ c. $\frac{12v_0^2}{7\mu g}$ d. $\frac{6v_0^2}{7\mu g}$

6 A billiard ball, initially at rest, is given a sharp impulse by a cue. The cue is held horizontally a distance h above the centre line as shown in the figure. The ball leaves the cue with a speed v_0 and because of its backward slipping, eventually acquires a final speed $\frac{9}{7}v_0$. If R is the radius of the ball, then h is

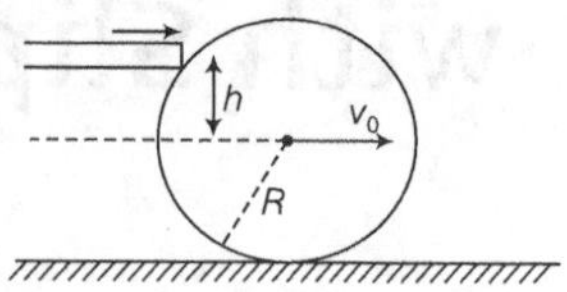

a. $\frac{2R}{5}$ b. $\frac{3R}{5}$ c. $\frac{4R}{5}$ d. $\frac{R}{5}$

7 A sphere of mass m and radius R is placed at rest on a plank of mass M which is placed on a smooth horizontal surface as shown in the figure. The coefficient of friction between the sphere and the plank is μ. At $t = 0$, a horizontal velocity v_0 is given to the plank. Find the time after which, the sphere starts pure rolling. The plank is long enough.

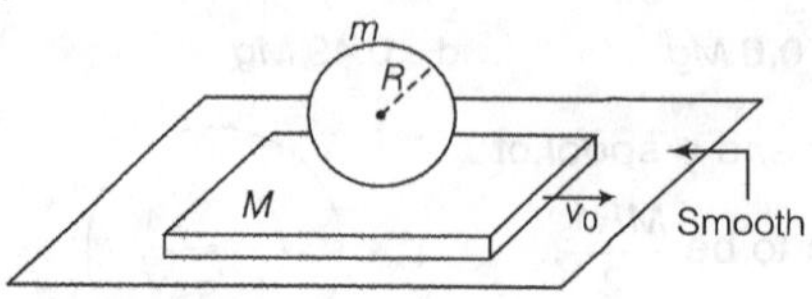

a. $\frac{2Mv_0}{(7M + 2m)\mu g}$ b. $\frac{Mv_0}{(7M + 2m)\mu g}$

c. $\frac{2Mv_0}{(5M + 2m)\mu g}$ d. $\frac{Mv_0}{(5M + 2m)\mu g}$

One or More than One Options Correct Type

Direction (Q. Nos. 8-10) This section contains 3 multiple choice questions. Each question has four choices (a), (b), (c) and (d), out of which ONE or MORE THAN ONE are correct.

8 A lawn roller in the form of a thin-walled hollow cylinder of mass M is pulled horizontally with a constant horizontal force F applied by a handle attached to the axle. It rolls without slipping. Choose the correct option(s).

a. Acceleration is $\frac{F}{2M}$ b. Acceleration is $\frac{F}{M}$

c. Friction is $\frac{F}{2}$ d. Friction is $\frac{F}{3}$

9 For the figure, the cylinder of mass 10 kg and radius 10 cm has a tape wrapped around it. The pulley weighs 100 N and has a radius 5 cm. When the system is released, the 5 kg mass comes down and the cylinder rolls without slipping. Choose the correct option(s).

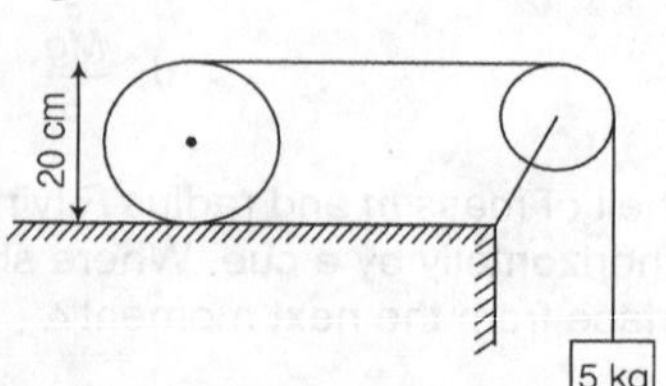

a. Acceleration of mass is $\frac{3g}{11}$ b. Acceleration of mass is $\frac{4g}{11}$

c. Velocity of mass is $\frac{3gt}{11}$ d. Velocity of mass is $\frac{4gt}{11}$

Working Space

10 The acceleration of the cylinder and tension in each string in the system shown are a and T, respectively. Assume that the cylinder remains horizontal and the strings are same. Then,

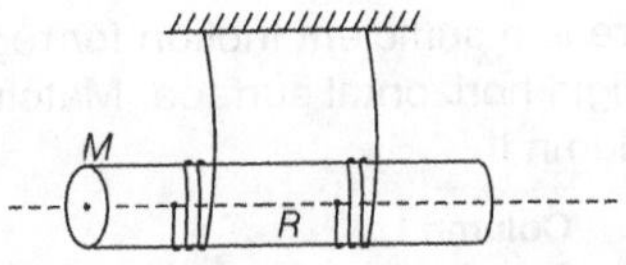

a. $a = \frac{2g}{3}$
b. $a = \frac{g}{3}$
c. $T = \frac{Mg}{3}$
d. $T = \frac{Mg}{6}$

Statement Type

Direction (Q. Nos. 11-13) *This section is based on Statement I and Statement II. Select the correct answer from the codes given below.*

a. Both Statement I and Statement II are correct and Statement II is the correct explanation of Statement I
b. Both Statement I and Statement II are correct but Statement II is not the correct explanation of Statement I
c. Statement I is correct but Statement II is incorrect
d. Statement II is correct but Statement I is incorrect

11 **Statement I** If a spherical body rolls on a surface without slipping, no frictional force acts on it.

Statement II In case of rolling without slipping, points of contact are relatively at rest.

12 **Statement I** A ring moving down on a smooth inclined plane will be in slipping motion.

Statement II Translational work done by friction in pure rolling motion is zero.

13 **Statement I** The force of friction in the case of a disc rolling without slipping down an inclined plane is $\frac{1}{3} g \sin \alpha$.

Statement II When the disc rolls without slipping, friction is required because for rolling condition, velocity of point of contact is zero.

Matching List Type

Direction (Q. Nos. 14 and 15) *Choices for the correct combination of elements from Column I and Column II are given as options (a), (b), (c) and (d), out of which one is correct.*

14 A disc rolls without slipping on a horizontal ground such that velocity of centre of mass of the disc is v_0. A point P on the circumference of the disc at an angle θ with the vertical (see figure) has a speed v_P.

Match the following columns and select the correct option from the codes given below.

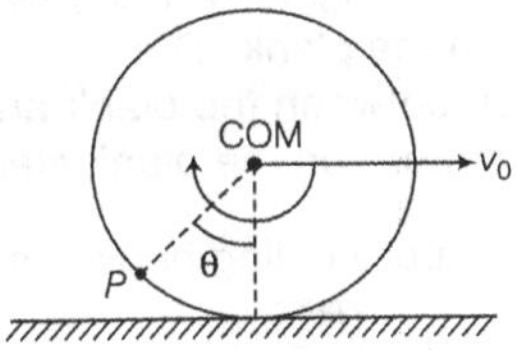

	Column I		Column II
i.	If $\theta = 60°$	p.	$v_P = \sqrt{2} v_0$
ii.	If $\theta = 90°$	q.	$v_P = v_0$
iii.	If $\theta = 120°$	r.	$v_P = 2v_0$
iv.	If $\theta = 180°$	s.	$v_P = \sqrt{3} v_0$

Codes

	i	ii	iii	iv		i	ii	iii	iv
a.	q	p	s	r	b.	r	s	p	q
c.	p	q	v	r	d.	s	p	v	q

Working Space

15 In each case, there is a sufficient friction for regular rigid uniform disc to undergo pure rolling on a rigid horizontal surface. Match the situations in Column I to the Statements in Column II.

	Column I		Column II
i.	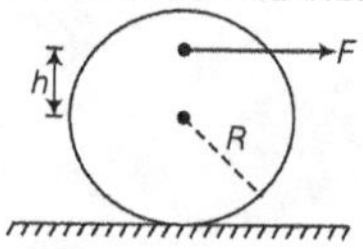	p.	The direction of static friction may be forward or backward or static friction may be zero.
ii.	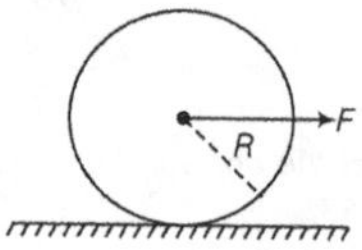	q.	The direction of static friction is backwards.
iii.	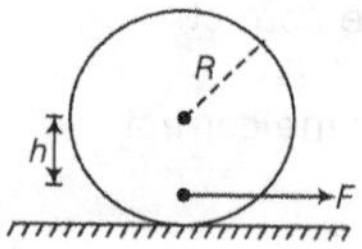	r.	The angular acceleration will be clockwise.
iv.	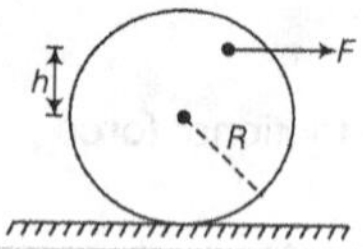	s.	Acceleration of the centre of mass will be along the direction of F.

Codes

	i	ii	iii	iv
a.	p,r,s	q,r,s	q,r,s	p,r,s
b.	q,r,s	p,r,s	q,r,s	p,r,s
c.	p,r,s	q,r,s	p,r,s	p,r,s
d.	q,r,s	q,r,s	p,r,s	p,r,s

Comprehension Type

Direction *(Q. Nos. 16-18) This section contains a paragraph, which describing theory, experiments, data, etc. Three questions related to the paragraph have been given. Each question has only one correct answer among the four given options (a), (b), (c) and (d).*

Passage

A long horizontal plank of mass m is lying on a smooth horizontal surface. A sphere of same mass m and radius R is spinned about its own axis with angular velocity ω_0 and gently placed on the plank. The coefficient of friction between the plank and the sphere is μ. After some time, the pure rolling of the sphere on the plank starts.

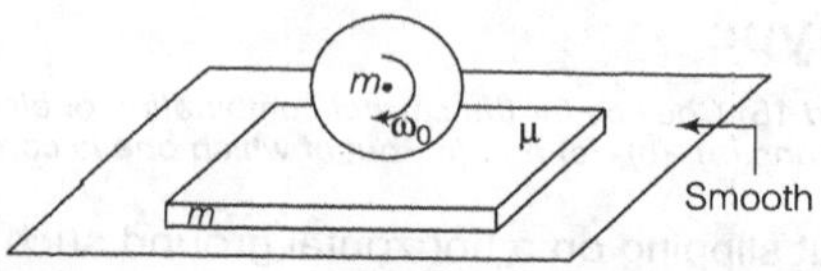

16 The time after which pure rolling begins is

a. $\dfrac{R\omega_0}{9\mu g}$ b. $\dfrac{2R\omega_0}{9\mu g}$ c. $\dfrac{R\omega_0}{3\mu g}$ d. $\dfrac{2R\omega_0}{\mu g}$

17 The velocity of the sphere after pure rolling begins is

a. $\dfrac{2}{9}R\omega_0$ b. $\dfrac{R\omega_0}{9}$ c. $\dfrac{R\omega_0}{3}$ d. $\dfrac{2}{3}R\omega_0$

18 The displacement of the plank till the sphere starts pure rolling is

a. $\dfrac{R^2\omega_0^2}{81\mu g}$ b. $\dfrac{2}{27}\left[\dfrac{R^2\omega_0^2}{\mu g}\right]$ c. $\dfrac{4}{81}\left[\dfrac{R^2\omega_0^2}{\mu g}\right]$ d. $\dfrac{2}{81}\left[\dfrac{R^2\omega_0^2}{\mu g}\right]$

Working Space

Subjective Questions

Direction (Q. Nos. 19-23) *These questions are subjective in nature, need to be solved completely on notebook.*

19 A cylinder of mass M and radius R is lying on a rough horizontal floor. It has a plank lying at its top as shown in the figure. A force F is applied on the plank such that the plank moves and causes the cylinder to roll. The plank always remains horizontal. There is no slipping at any point of contact. Find the acceleration of the cylinder and the frictional forces at the contacts.

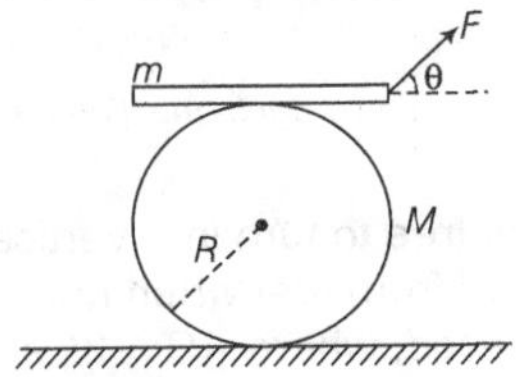

20 Consider a hoop of radius R having centre of mass velocity v_0. Let the hoop roll on a surface without sliding. Assume a point P (see figure) on the hoop to be at the origin initially. Find the velocity of this point as a function of time. Also, find the acceleration of this point P.

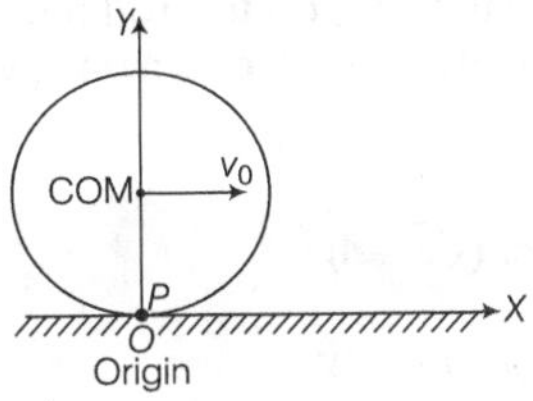

21 For the above problem, prove that the total distance travelled by point P in one complete cycle is $8R$.

22 Two thin circular discs of mass 2 kg and radius 10 cm each are joined by a rigid massless rod of length 20 cm. The axis of the rod is along the perpendicular to the planes of the discs through their centres. This object is kept on a truck in such a way that the axis of the object is horizontal and perpendicular to the direction of motion of the truck. Its friction with the floor of the truck is large enough so that the object can roll on the truck without slipping. Take X-axis as the direction of motion of the truck and Z-axis as the vertically upward direction. If the truck has an acceleration 9 ms^{-2}. Calculate

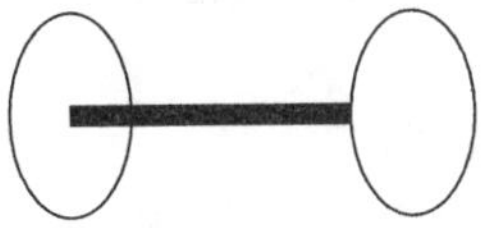

i. the force of friction on each disc.

ii. the magnitude and direction of the friction torque acting on each disc about the centre of mass O of the object. Express the torque in the vector form in terms of unit vectors $\hat{\mathbf{i}}$, $\hat{\mathbf{j}}$ and $\hat{\mathbf{k}}$ in X, Y and Z-directions.

23 A solid cylinder of mass m and radius R is set in rotation about its axis with an angular velocity ω_0, then lowered with its lateral surface onto a horizontal plane and released. The coefficient of friction between the cylinder and the plane is μ. Find

i. how much time the cylinder move with sliding?

ii. total work performed by friction.

DPP-8 Angular Impulse

Only One Option Correct Type

Working Space

Direction *(Q. Nos. 1-10) This section contains 10 multiple choice questions. Each question has four choices (a), (b), (c) and (d), out of which ONLY ONE is correct.*

1 A uniform rod AB of mass $3m$ and length $4l$, which is free to turn in a vertical plane about a smooth horizontal axis through A, is released from rest when horizontal. When the rod first becomes vertical, a point C of the rod, where $AC = 3l$, strikes a fixed peg. Find the impulse exerted by the peg on the rod if the rod is brought to rest by the peg.

a. $\frac{4}{3}m\sqrt{3gl}$ b. $\frac{5}{3}m\sqrt{3gl}$ c. $\frac{1}{3}m\sqrt{3gl}$ d. $\frac{8}{3}m\sqrt{3gl}$

2 For the above problem, find the impulse exerted by the peg on the rod if the rod rebounds and next comes to instantaneous rest inclined to the downward vertical at an angle $\frac{\pi}{3}$ rad.

a. $\frac{4}{3}m\sqrt{6gl}\,(\sqrt{2}+1)$ b. $\frac{2}{3}m\sqrt{6gl}\,(\sqrt{2}+1)$

c. $\frac{1}{3}m\sqrt{6gl}\,(\sqrt{2}+1)$ d. $\frac{5}{3}m\sqrt{6gl}\,(\sqrt{2}+1)$

3 A dumb bell consists of two identical small balls of mass $\frac{1}{2}$ kg each connected to the two ends of a 50 cm long light rod. The dumb bell is rotating about a fixed axis through the centre of the rod and perpendicular to it at an angular speed 10 rad s^{-1}. An impulsive force of average magnitude 5 N acts on one of the masses in the direction of its velocity for 0.1 s. Find the angular velocity of the system.

a. 10 rad s^{-1} b. 12 rad s^{-1} c. 8 rad s^{-1} d. 14 rad s^{-1}

4 Consider a body shown in the figure consisting of two identical balls, each of mass M connected by a light rigid rod. If an impulse $J = Mv$ is imparted to the body at one of its ends what would be its angular velocity?

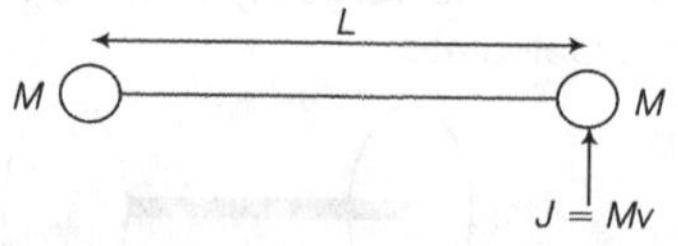

a. $\frac{2v}{L}$ b. $\frac{v}{2L}$ c. $\frac{v}{L}$ d. $\sqrt{2}\frac{v}{L}$

5 A uniform sphere of mass m and radius R starts rolling without slipping down an inclined plane. Find the time dependence of the angular momentum of the sphere relative to the point of contact at the initial moment. The angle of inclination of the plane is θ. How will the result be affected in the case of a perfectly smooth inclined plane?

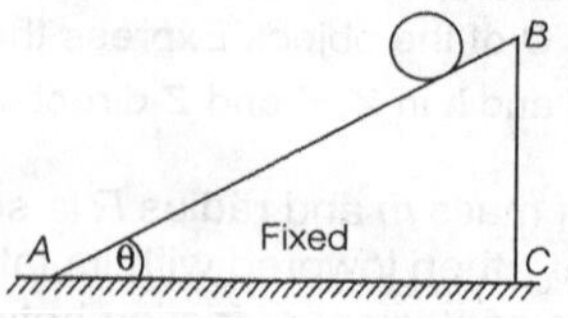

a. $\frac{2}{7}(mg\sin\theta)Rt$ b. $L = mg\sin\theta\frac{Rt^2}{2}$

c. $L = \sqrt{2}mg\sin\theta Rt$ d. $\frac{2}{5}(mg\sin\theta)Rt$

6 A uniform rod AB of mass $3m$ and length $2l$ is lying at rest on a smooth horizontal table with a smooth vertical axis through the end A. A particle of mass $2m$ moves with speed $2u$ across the table and strikes the rod at its mid-point C. If the impact is perfectly elastic, then find the speed of the particle after impact if it strikes the rod normally.

a. $\frac{7u}{3}$ b. $\frac{2u}{3}$ c. $\frac{u}{3}$ d. $\frac{4u}{3}$

7 For the above problem, find the speed of the particle after impact if its path before impact was inclined at angle 60° to AC.

a. $\frac{2}{\sqrt{3}}u$ b. $\frac{2}{3}u$ c. $\frac{\sqrt{2}}{3}u$ d. $\sqrt{\frac{2}{3}}u$

8 A cubical block of side a is moving with velocity v on a horizontal smooth plane as shown in the figure. It hits a ridge at point O. What is the angular speed of the block after it hits O?

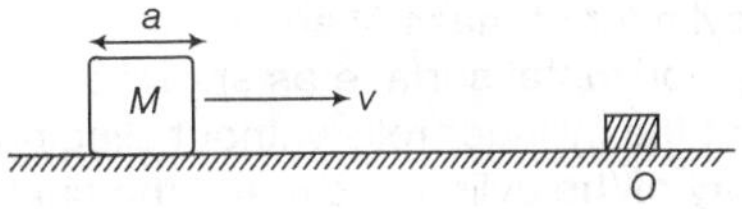

a. $\frac{3v}{4a}$ b. $\frac{2v}{3a}$ c. $\frac{3v}{5a}$ d. $\frac{2v}{5a}$

9 An ice cube of mass M and with sides of length a each is sliding without friction across a counter top with a speed v_0 when it hits a ridge O at the edge of the counter (see figure). This collision causes the cube to tilt as shown. The minimum value of v_0 needed for the cube to fall off the table is given by

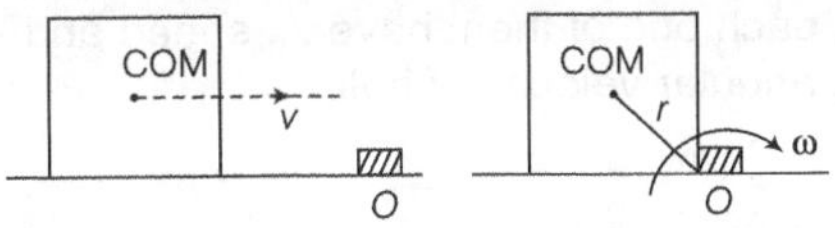

a. $v_0 = \sqrt{2.1ag}$ b. $\sqrt{3.1ag}$ c. $v_0 = \sqrt{1.1ag}$ d. $\sqrt{1.8ag}$

10 A uniform rod AB of length $2l$ and mass m is rotating in a horizontal plane about a vertical axis through A with angular velocity ω. When the mid-point of the rod strikes a fixed nail, it is brought to rest immediately. Find the impulse exerted by the nail.

a. $\frac{2}{3}m\omega l$ b. $\frac{1}{3}m\omega l$ c. $m\omega l$ d. $\frac{4}{3}m\omega l$

One or More than One Options Correct Type

Direction (Q. Nos. 11-15) This section contains 5 multiple choice questions. Each question has four choices (a), (b), (c) and (d), out of which ONE or MORE THAN ONE are correct.

11 A pulley in the form of a uniform disc of mass $2m$ and radius r is free to rotate in a vertical plane about a fixed horizontal axis through its centre. A light inextensible string has one end fastened to a point on the rim of the pulley and is wrapped several times round the rim. The portion of string not wrapped round the pulley is of length $8r$ and carries a particle of mass m at its free end. The particle is held close to the rim of the pulley and level with its centre. If the particle is released from this position, then initial angular velocity of the pulley is ω and the impulse of the sudden tension in the string when it becomes taut is J, then

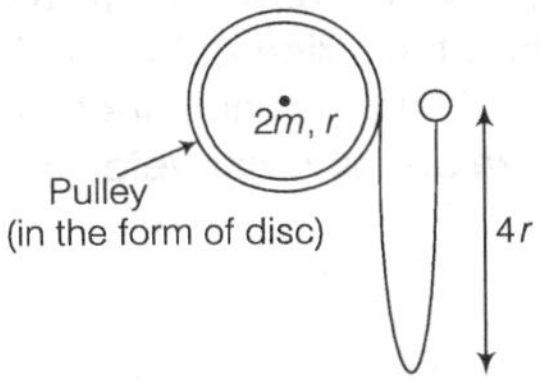

a. $\omega = 2\sqrt{\frac{g}{r}}$ b. $\omega = \sqrt{\frac{2g}{r}}$ c. $J = 2m\sqrt{gr}$ d. $J = m\sqrt{2gr}$

12 A uniform rod of length L rests on a frictionless horizontal surface. The rod is pivoted about a fixed frictionless axis at one end. The rod is initially at rest. A bullet travelling parallel to the horizontal surface and perpendicular to the rod with speed v_0 strikes the rod at its centre and becomes embedded in it. The mass of the bullet is one-sixth of the mass of the rod. The final angular velocity of the rod is ω and the ratio of the kinetic energy of the system after collision to the kinetic energy of the bullet before collision is x. Then,

a. $x = \frac{2}{9}$ b. $x = \frac{1}{9}$

c. $\omega = \frac{v_0}{9L}$ d. $\omega = \frac{2v_0}{9L}$

13 A bullet of mass m moving with velocity u just grazes the top of a solid cylinder of mass M and radius R resting on a rough horizontal surface as shown in the figure. Assume that the cylinder rolls without slipping. The angular velocity of the cylinder is ω and the final velocity of the bullet is v, then

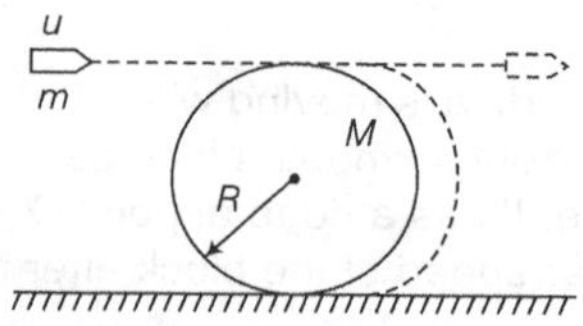

a. $\omega = \frac{4mu}{(8m + 3M)R}$ b. $\omega = \frac{2mu}{(8m + 3M)R}$

c. $v = \frac{2mu}{(8m + 3M)}$ d. $v = \frac{8mu}{(8m + 3M)}$

14 Two solid spherical ball of radius R collide with a rough horizontal surface as shown in the figure. At the time of collision, each one of them have v_0 speed and ω_0 angular speed as shown. After collision, angular velocity of ball,

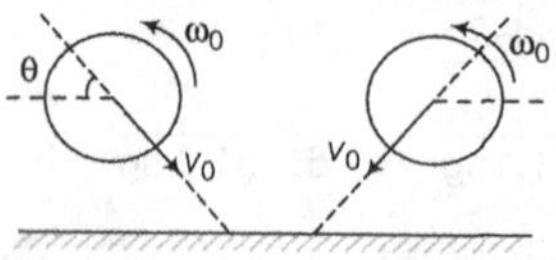

Case (i) Case (ii)

a. decreases in case (i)
b. increases in case (i)
c. decreases in case (ii)
d. cannot be predicted in case (ii)

15 A string is wrapped several times on a cylinder of mass M and radius R. The cylinder is pivoted about its axis of symmetry. A block of mass m tied to the string rests on a support positioned so that the string has no slack. The block is carefully lifted vertically a distance h and the support is removed as shown. Now, the block is released. Just before the string gets taut, kinetic energy of the system is E_0. Just after taut kinetic energy of the system, velocity of m and angular velocity of cylinder are E_1, v_1 and ω_1 respectively. Then,

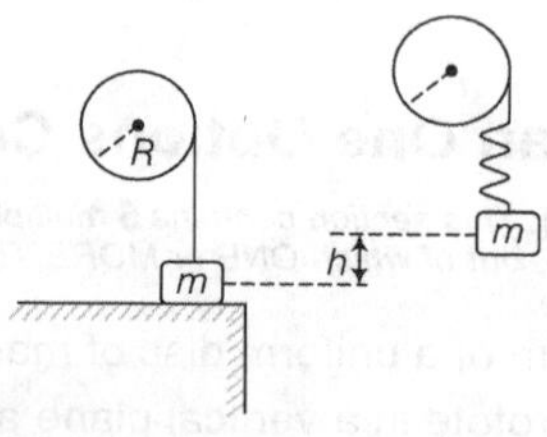

a. $E_0 = mgh$ b. $E_1 = \frac{E_0}{1 + \frac{M}{2m}}$

c. $v_1 = \frac{\sqrt{2gh}}{1 + \frac{M}{2m}}$ d. $\omega_1 = \frac{\sqrt{2gh}}{R\left(1 + \frac{M}{2m}\right)}$

Working Space

Comprehension Type

Direction (*Q. Nos. 16-18*) *This section contains a paragraph, which describing theory, experiments, data, etc. Three questions related to the paragraph have been given. Each question has only one correct answer among the four given options (a), (b), (c) and (d).*

Passage

A uniform flat disc of mass M and radius R rotates about a horizontal axis through its centre with angular speed ω_0.

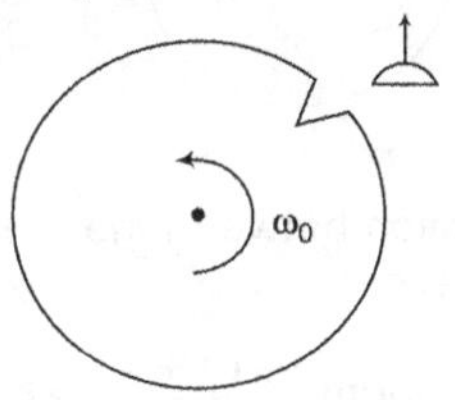

16 What is its angular momentum?

a. $\frac{1}{2}MR^2\omega_0$ b. $MR^2\omega_0$

c. $2MR^2\omega_0$ d. $\frac{2}{3}MR^2\omega_0$

17 A chip of mass m breaks off the edge of the disc at an instant such that the chip rises vertically above the point at which it breaks off. What is the maximum height the chip rises to from the location of its breaking? (no impulsive internal force acts on the chip)

a. $\frac{R^2\omega_0^2}{g}$ b. $\frac{2R^2\omega_0^2}{g}$

c. $\frac{R^2\omega_0^2}{2g}$ d. $\frac{R^2\omega_0^2}{3g}$

18 What are the final angular momentum and energy of the disc?

a. $MR^2\omega_0, \frac{1}{2}MR^2\omega_0^2$ b. $mR^2\omega_0, \frac{1}{2}mR^2\omega_0^2$

c. $\left(\frac{M}{2}-m\right)R^2\omega_0, \left(\frac{M}{2}-m\right)R^2\omega_0^2$ d. $\left(\frac{M}{2}-m\right)R^2\omega_0, \frac{1}{2}\left[\left(\frac{M}{2}-m\right)R^2\omega_0^2\right]$

One Integer Value Correct Type

Direction (*Q. No. 19*) *This section contains 1 question. When worked out will result in an integer from 0 to 9* (*both inclusive*).

19 A spherical ball of radius R and mass m collides with a plank of mass M kept on a smooth horizontal surface. Just before impact, the centre of the ball has a velocity v_0 and angular velocity ω_0 as shown in the figure. After the impact, the normal velocity is reversed with same magnitude and the ball stops rotating after the impact. The coefficient of friction between the ball and the plank is μ. Assume that the plank is large enough. The distance on the plank between first two impacts of the ball is $\frac{4v_0R\omega_0}{\beta g}\left(1+\frac{m}{M}\right)$. Find the value of β.

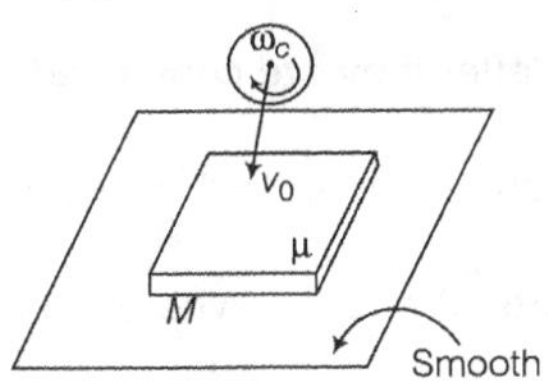

Subjective Questions

Working Space

Direction (Q. Nos. 20-23) *These questions are subjective in nature, need to be solved completely on notebook.*

20 Figure shows two cylinders of radii r_1 and r_2 having moments of inertia I_1 and I_2 about their respective axes. Initially, the cylinders rotate about their axes with angular speeds ω_1 and ω_2 as shown in the figure. The cylinders are moved closer to touch each other keeping the axes parallel. The cylinders first slip over each other at the contact, but the slipping finally ceases due to friction between them. Find the angular speeds of the cylinders after the slipping ceases.

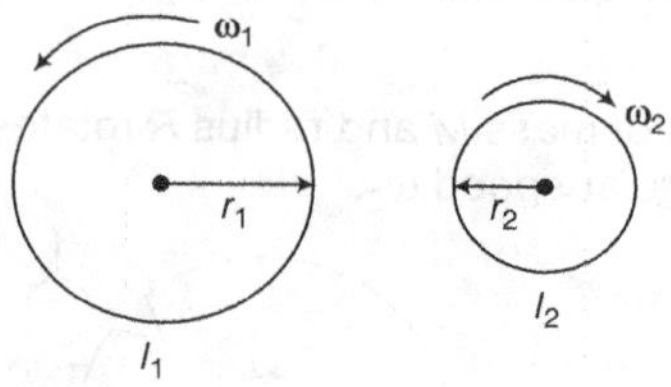

21 A thin rod AB of length a has variable mass per unit length $\frac{\rho_0 (a + x)}{a}$, where x is the distance measured from A and ρ_0 is a constant. The rod is pivoted at A and is hanging vertically in equilibrium, when it is struck by a horizontal impulse of magnitude J at the point B. Show that if B passes through a point vertically above A, then $J > \frac{\sqrt{70\, a^3 \rho_0^2 g}}{6}$.

22 Four thin rods, each of mass M and length d = 1m are rigidly connected in the form of a positive sign. The entire assembly rotates in a horizontal plane around a vertical axle at the centre with initial (clockwise) angular velocity $\omega_i = -2\text{rads}^{-1}$. A mud ball with mass m and initial speed $v_i = 12\text{ms}^{-1}$ is thrown at and sticks to the end of one rod. Let $M = 3m$. What is the final angular velocity ω_f of the "plus sign + mud ball" system, if the initial path of the mud ball is each of the four paths shown in figure.

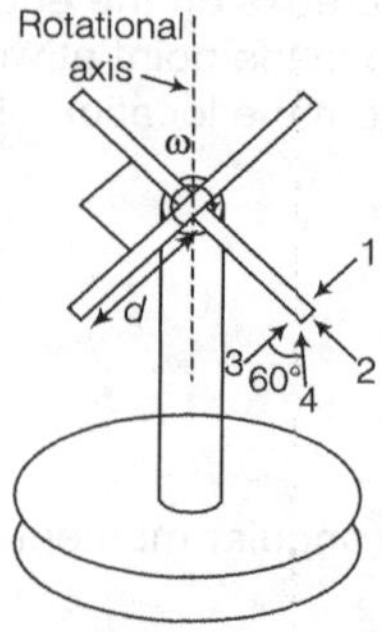

Path 1 Contact is made when the velocity of the ball is perpendicular to the rod.

Path 2 Radial contact.

Path 3 Perpendicular contact.

Path 4 Contact is made 60° to the perpendicular. Note that all 4 paths are horizontal.

23 Two skaters, each of mass 50 kg, approach each other along parallel paths separated by 3 m. They have equal and opposite velocities of 10 ms^{-1}. The first skater carries a long light pole 3 m long and the second skater grabs it as he passes (assume frictionless ice).

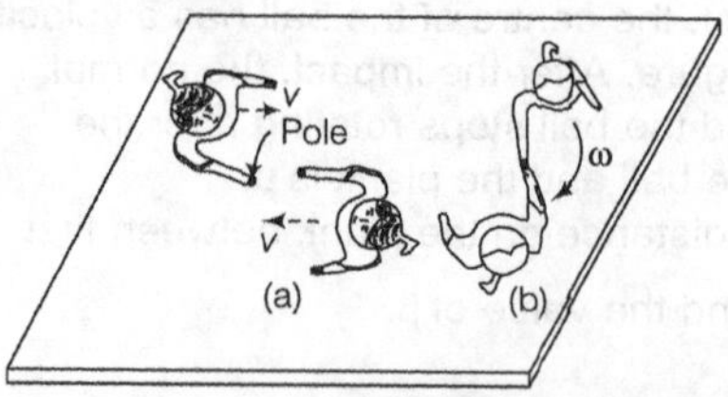

i. Describe quantitatively, the motion of the skaters after they are connected by the pole.

ii. By pulling on the pole, the skaters reduce their distance apart to 1 m. What is their motion then?

iii. Compare the kinetic energies of the system in parts (i) and (ii). Where does the change come from?

DPP-9 Toppling and Rolling on an Inclined Plane

Only One Option Correct Type

Working Space

Direction *(Q. Nos. 1-7) This section contains 7 multiple choice questions. Each question has four choices (a), (b), (c) and (d), out of which ONLY ONE is correct.*

1 A cubical block of side 30 cm is placed on a rough horizontal surface of coefficient of friction 0.6. At what minimum height above the ground, a horizontal force must be applied on the block so that it starts toppling before slipping?

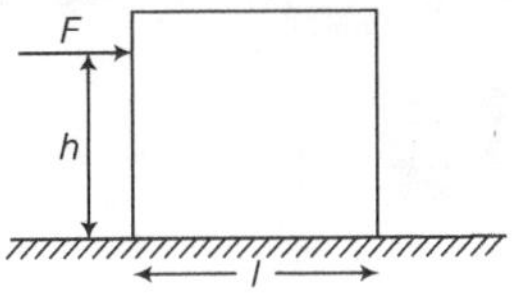

a. $h \geq 25$ cm b. $h \geq 23$ cm c. $h \geq 20$ cm d. $h \geq 15$ cm

2 A wedge in the form of equilateral triangle is placed on a rough horizontal surface as shown in the figure. For what value of coefficient of friction, can the block topple without slipping?

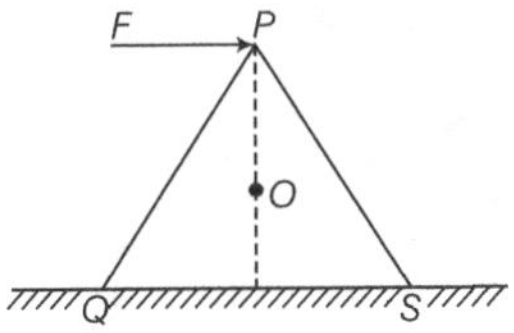

a. $\mu \geq \frac{1}{\sqrt{7}}$ b. $\mu \geq \frac{1}{\sqrt{5}}$

c. $\mu \geq \frac{1}{\sqrt{3}}$ d. $\mu \geq \frac{1}{2}$

3 A spool of thread of mass M is placed on an inclined smooth plane set at an angle θ with the horizontal. The free end of the thread is attached to the wall as shown in the figure. The moment of inertia of spool about its axis is I and radius of wound thread layer is r. Find the acceleration of centre of mass of the spool.

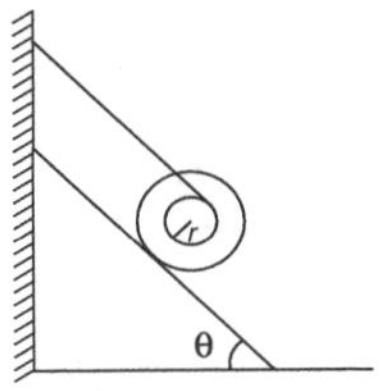

a. $\dfrac{g\sin\theta}{1+\dfrac{I}{Mr^2}}$ b. $\dfrac{g\sin\theta}{1+\dfrac{2I}{MR^2}}$

c. $\dfrac{2g\sin\theta}{1+\dfrac{2I}{Mr^2}}$ d. $\dfrac{2g\sin\theta}{1+\dfrac{I}{Mr^2}}$

Working Space

4 A solid cylinder of mass m and radius r starts rolling down an inclined plane of inclination θ. Friction is enough to prevent slipping. Find the speed of its centre of mass, when its centre of mass has fallen a height h.

a. $\sqrt{\frac{2gh}{3}}$ b. $\sqrt{gh}$

c. $\sqrt{\frac{gh}{3}}$ d. $\sqrt{\frac{4gh}{3}}$

5 A block of mass m and a cylinder of mass $2m$ are released on a rough inclined plane, inclined at an angle $\theta = 45°$ with the horizontal. The coefficient of friction for all the contact surfaces is 0.5. Find the accelerations of the block and the cylinder. Assume pure rolling.

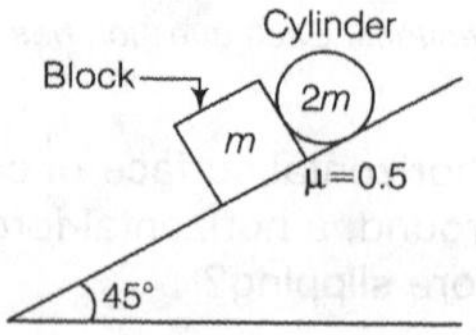

a. $\frac{3g}{2\sqrt{2}}$ b. $\frac{3g}{4\sqrt{2}}$

c. $\frac{3g}{5\sqrt{2}}$ d. $\frac{3g}{\sqrt{2}}$

6 In the figure shown, a light ring with three rods, each of mass m is welded on this ring. The rods form an equilateral triangle. The rigid assembly is released on a rough fixed inclined plane. Determine the minimum value of the coefficient of static friction, that will allow pure rolling of the assembly.

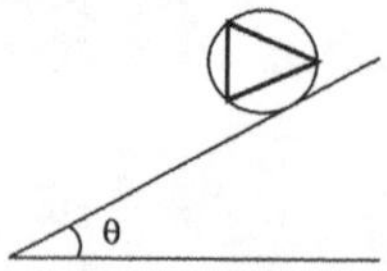

a. $\frac{2}{3}\tan\theta$ b. $\frac{1}{3}\tan\theta$

c. $\tan\theta$ d. $\frac{4}{3}\tan\theta$

7 A cylinder of mass m is rotated about its axis by an angular velocity ω and lowered gently on an inclined plane as shown in the figure. Then,

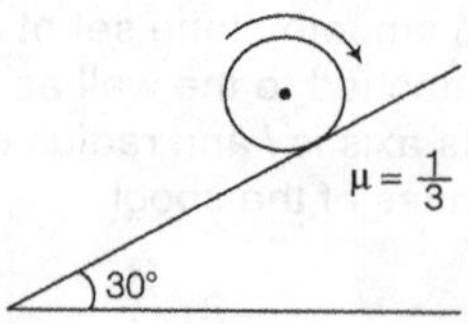

a. it will stay there permanently
b. it will first go upwards and then moves downwards
c. it will go downwards just after it is released
d. it will never go upwards

Working Space

One or More than One Options Correct Type

Direction (Q. Nos. 8-10) This section contains 3 multiple choice questions. Each question has four choices (a), (b), (c) and (d), out of which ONE or MORE THAN ONE are correct.

8 A uniform cylinder of height h and radius r is placed with its circular face on a rough inclined plane and the inclination of the plane to the horizontal is gradually increased. If μ is the coefficient of friction, then the cylinder will

a. slide before toppling if $\mu < \frac{r}{2h}$ b. topple before sliding if $\mu > \frac{r}{2h}$

c. slide before toppling if $\mu < \frac{2r}{h}$ d. topple before sliding if $\mu > \frac{2r}{h}$

9 For the arrangement shown in the figure, a spool of mass m, moment of inertia I is placed on a plank of mass M. There is no friction between the plank and the surface inclined at an angle θ. The friction between spool and plank is sufficient enough to prevent slipping.

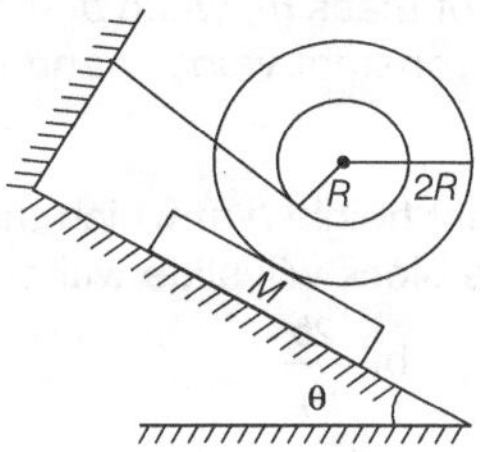

a. If $\frac{m}{M} > 1$, then plank will ascend

b. If $\frac{m}{M} < 1$, then plank will descend

c. If $\frac{m}{M} = 1$, then plank will remain stationary

d. Angular acceleration of spool is $\left[\frac{(m - M)\, g \sin\theta}{mR^2 + MR^2 + I}\right] R$

10 A block A of mass 0.5 kg is held by a long massless string on a frictionless inclined plane of inclination 30° to the horizontal. The string is wound on a uniform solid cylindrical drum P of mass 2 kg and radius 0.2 m as shown in the figure. The drum is given an initial angular velocity such that the block A starts moving up the plane. Given, $g = 9.8\ \text{ms}^{-2}$. The tension in the string is T. At a certain instant of time, the magnitude of the angular velocity of P is 10 rad s^{-1}. The distance travelled by A from this instant of time until it comes to rest is d. Then,

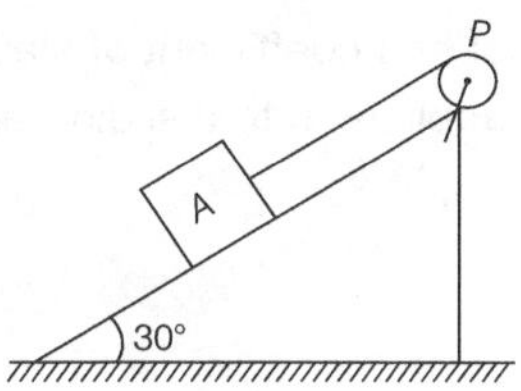

a. $T = 1.63$ N b. $d = 2.22$ m

c. $s = 1.22$ m d. $T = 2.32$ N

Comprehension Type

Direction (Q. Nos. 11-16) This section contains 2 paragraphs, each describing theory, experiments, data, etc. Six questions related to the paragraphs have been given. Each question has only one correct answer among the four given options (a), (b), (c) and (d).

Passage I

A block of mass m, height $2h$ and width $2b$ rests on a flat car which moves horizontally with constant acceleration a (see figure).

(Given, $b = 0.6$ m, $h = 0.9$ m, $\mu = 0.5$ and $g = 9.8\ \text{ms}^{-2}$)

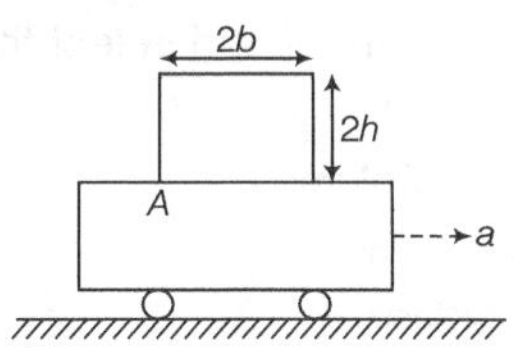

11 Determine the value of acceleration at which slipping of the block on the car starts if the coefficient of friction is μ.

a. 4.9 ms^{-2} b. 2.45 ms^{-2}

c. 7.35 ms^{-2} d. 3.02 ms^{-2}

12 The value of the acceleration at which block topples about point A assuming sufficient friction to prevent slipping is

a. 5.53 ms^{-2} b. 6.53 ms^{-2}
c. 7.23 ms^{-2} d. 4.52 ms^{-2}

13 The shortest distance in which it can be stopped from a speed of 72 kmh^{-1} with constant deceleration so that the block is not disturbed is

a. 25.23 m b. 35.28 m c. 40.82 m d. 52.32 m

Passage II

A horizontal force F is applied to a homogeneous rectangular block of mass m, width b and height H. The block moves with constant velocity and the coefficient of friction is μ_k.

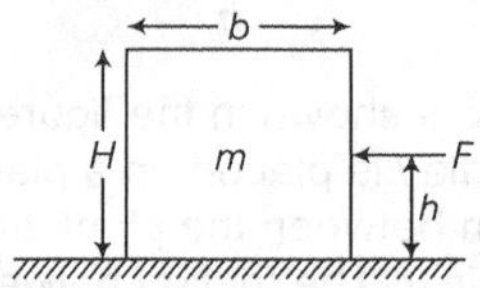

14 What is the greatest height h at which the force F can be applied so that the block will slide without toppling?

a. $\frac{b}{\mu_k}$ b. $\frac{2b}{\mu_k}$ c. $\frac{b}{\sqrt{2}\,\mu_k}$ d. $\frac{b}{2\mu_k}$

15 At what distance (from the left end) on the bottom face of the block will the normal reaction act when F acts at the height $\frac{H}{2}$? (Here, assume no toppling happens)

a. $\frac{\mu_k H}{2}$ b. $\frac{b-\mu_k H}{2}$ c. $b-\frac{\mu_k H}{2}$ d. $\frac{H}{2\mu_k}$

16 If the block is at rest and coefficient of static friction is μ_s, what is the height at which F has to be applied so that sliding and toppling are equally likely to occur?

a. $\frac{b}{2\mu_s}$ b. $\frac{b}{\mu_s}$ c. $\frac{b}{\sqrt{2}\,\mu_s}$ d. $2b\mu_s$

Matching List Type

Direction (Q. No. 17) Choices for the correct combination of elements from Column I and Column II are given as options (a), (b), (c) and (d), out of which one is correct.

17 A sphere, a ring and a disc of same mass and same radius are allowed to roll down three similar, sufficiently rough inclined planes as shown from the same height.

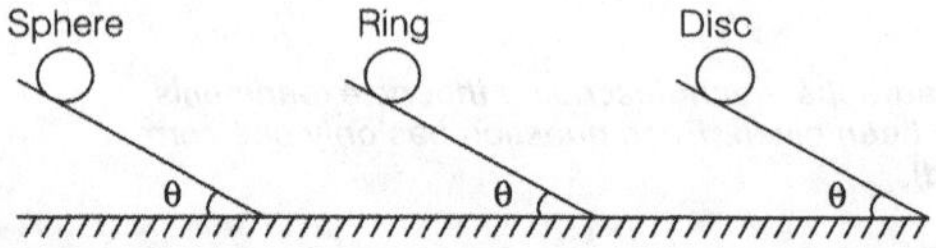

Match the following columns and select the correct option from the codes given below.

	Column I		Column II
i.	Final kinetic energy	p.	Sphere > disc > ring
ii.	Final linear velocity	q.	Sphere < disc < ring
iii.	Frictional force acting during rolling	r.	Disc = ring = sphere
iv.	Time taken to reach the bottom	s.	Ring > sphere > disc

Codes

	i	ii	iii	iv		i	ii	iii	iv
a.	r	p	q	q	b.	p	q	s	r
c.	r	p	s	q	d.	q	r	p	s

Working Space

Subjective Questions

Direction (Q. Nos. 18-20) *These questions are subjective in nature, need to be solved completely on notebook.*

18 A car of mass m travelling at speed v moves on a horizontal track. The centre of mass of the car describes a circle of radius r. If $2a$ is the separation between the inner wheels and outer wheels and h is the height of the centre of mass above the ground, show that the limiting speed beyond which the car will overturn is given by $v = \sqrt{\frac{gra}{h}}$. (Assume that the car overturns before skidding as v increases)

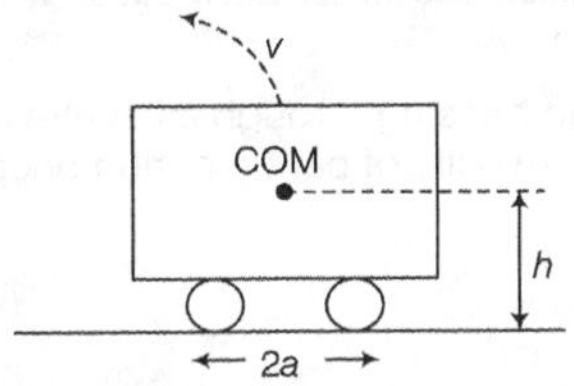

19 A solid sphere is released from a height of 49 m from horizontal on an inclined plane as shown in the figure. All the surfaces are smooth except AB which has $\mu = \frac{2}{3}$ and length of AB is 49 m. After passing over the rough surface, the sphere climbs to the right portion BD of incline. (Given, $g = 10 \text{ ms}^{-2}$)

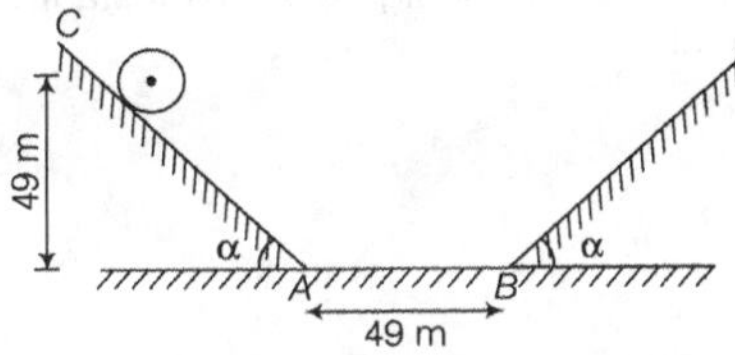

i. Find the height from horizontal to which it climbs on BD.

ii. Now, the sphere returns and goes up the incline AC. Find the linear velocity of sphere at A during the leftward journey.

20 A uniform circular cylinder of mass m and radius R is given an initial angular velocity ω_0 and no initial translational velocity. It is placed in contact with a plane inclined at an angle θ to the horizontal. If there is a coefficient of friction μ for sliding between the cylinder and the plane, find the distance the cylinder moves up the plane before sliding stops. Also, calculate maximum distance it travels up the plane. Assume $\mu > \tan\theta$.

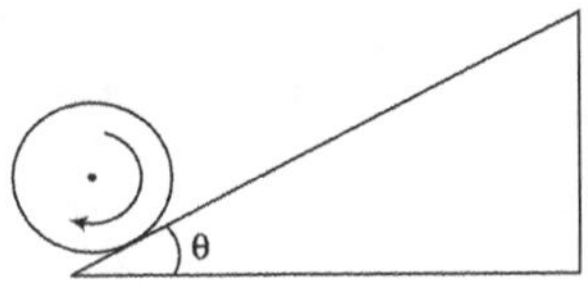

SET - 1

Revisal Problems for JEE Main

Only One Option Correct Type

Direction *(Q. Nos. 1-20) This section contains 20 multiple choice questions. Each question has four choices (a), (b), (c) and (d), out of which ONLY ONE is correct.*

Working Space

1 An impulse J is applied on a ring of mass m along a line passing through its centre O. The ring is placed on a rough horizontal surface. The linear velocity of centre of ring once it starts rolling without slipping is

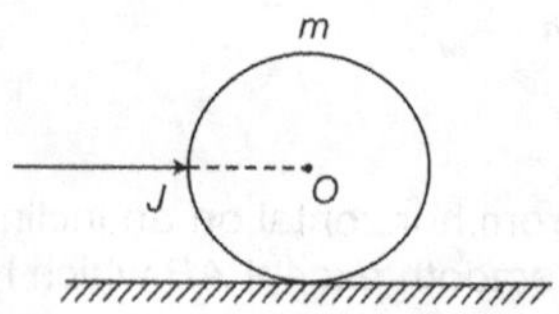

a. $\frac{J}{m}$ b. $\frac{J}{2m}$ c. $\frac{J}{4m}$ d. $\frac{J}{3m}$

2 A rod of length l slides down along an inclined wall as shown in the figure. At the instant shown, the speed of end A is v, then speed of end B will be

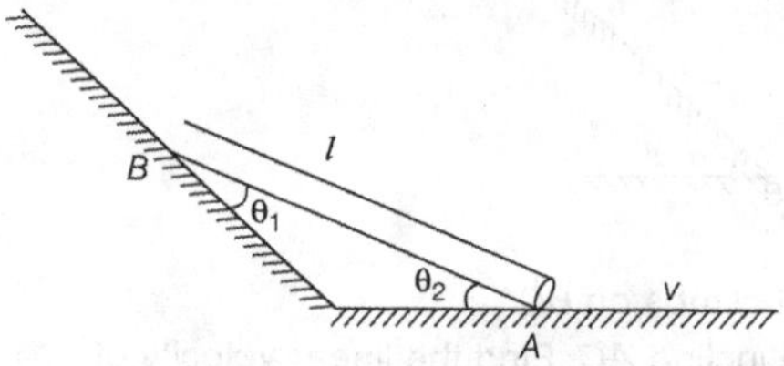

a. $\frac{v \sin \theta_1}{\sin \theta_2}$ b. $\frac{v \sin \theta_2}{\sin \theta_1}$ c. $\frac{v \cos \theta_1}{\cos \theta_2}$ d. $\frac{v \cos \theta_2}{\cos \theta_1}$

3 Two rods OA and OB of equal length and equal mass are lying on XY-plane as shown in the figure. Let I_x, I_y and I_z be the moment of inertia of both the about X, Y and Z-axes respectively. Then,

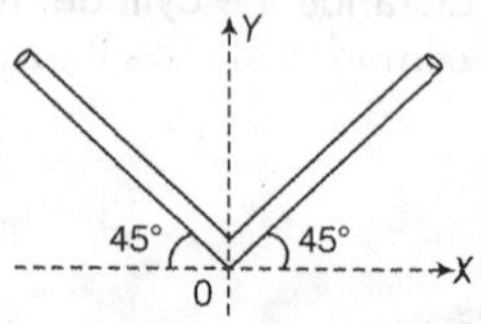

a. $I_x = I_y > I_z$ b. $I_x = I_y < I_z$
c. $I_x > I_y > I_z$ d. $I_z > I_y > I_x$

4 AB and CD are two identical rods, each of length L and mass m joined to form a cross. The moment of inertia of these two rods about a bisector of the angle between the rods is

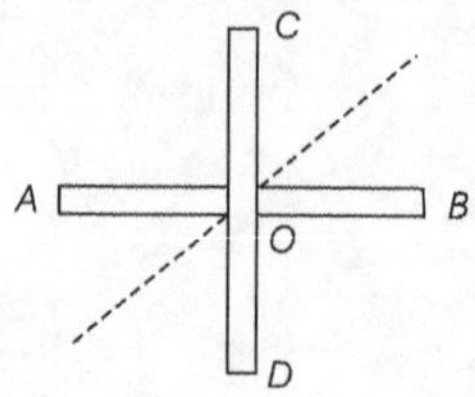

a. $\frac{mL^2}{6}$ b. $\frac{mL^2}{3}$ c. $\frac{mL^2}{12}$ d. $\frac{2mL^2}{3}$

Working Space

5 Four holes, each of radius R, are cut from a thin square plate of side $4R$ and mass M. The moment of inertia of the remaining portion about Z-axis is

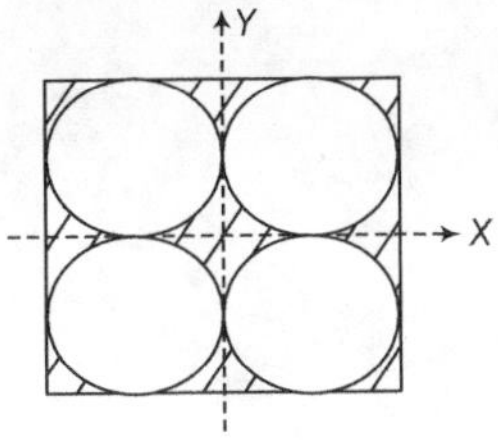

a. $\frac{\pi}{12} MR^2$

b. $\left(\frac{4}{3} - \frac{\pi}{4}\right) MR^2$

c. $\left(\frac{8}{3} - \frac{10\pi}{6}\right) MR^2$

d. $\left(\frac{4}{3} - \frac{\pi}{6}\right) MR^2$

6 A wheel is rolling without slipping on a horizontal surface. The centre of the wheel moves with a constant speed v_0. Consider a point P on the rim which is at the top at time $t = 0$. The square of the speed of point P is plotted against time t. If R is the radius of the wheel, then correct plot is

a.

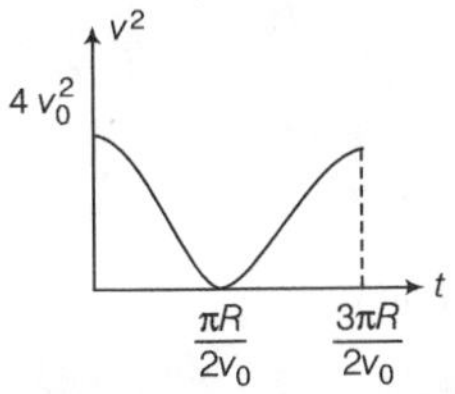

b.

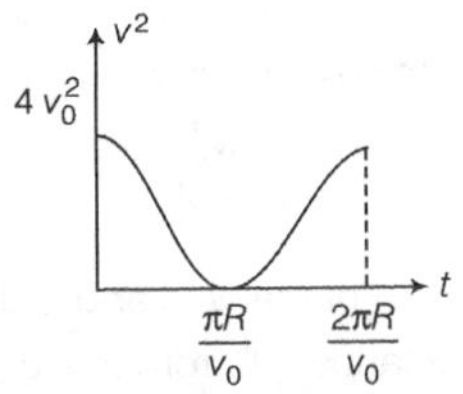

c.

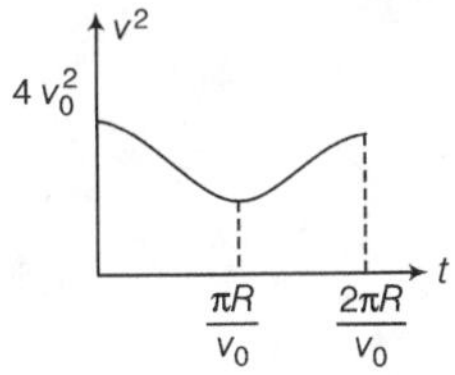

d.

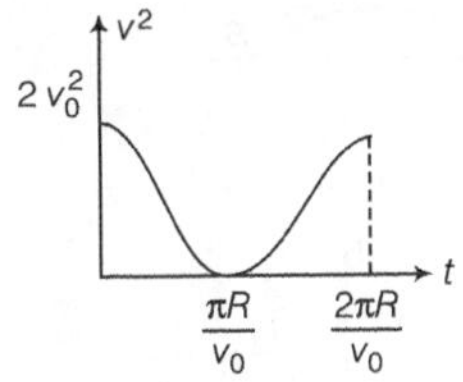

7 A time varying force $F = 2t$ is applied on a spool as shown in the figure. The angular momentum of the spool at time t, about bottom most point is

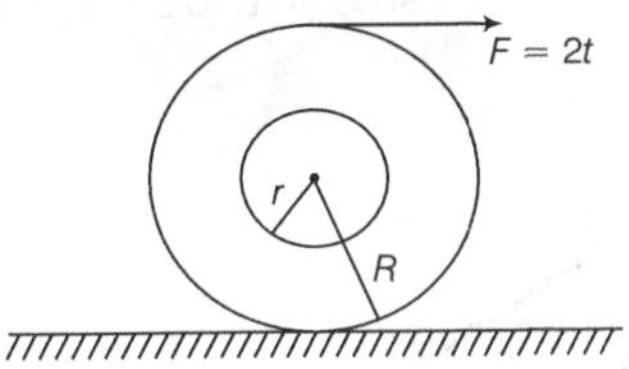

a. $\frac{r^2t^2}{R}$

b. $\frac{(R+r)^2t^2}{r}$

c. Rt^2

d. $\frac{R^2t^2}{r}$

8 A disc of radius R, having angular velocity ω rolls without slipping along positive X-axis. The velocity of point P at the instant shown in the figure is

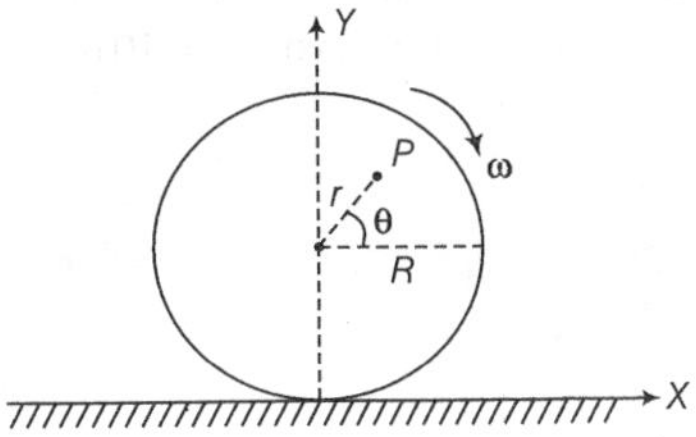

a. $\mathbf{v}_p = \omega\,[r \sin\theta\,\hat{i} + r\cos\theta\,\hat{j}]$

b. $\mathbf{v}_p = \omega\,[r \sin\theta\,\hat{i} - r\cos\theta\,\hat{j}]$

c. $\mathbf{v}_p = \omega\,[(R + r\sin\theta)\,\hat{i} + r\cos\theta\,\hat{j}]$

d. $\mathbf{v}_p = \omega\,[(R + r\sin\theta)\,\hat{i} - r\cos\theta\,\hat{j}]$

Working Space

9 A solid sphere of radius R is resting on a smooth horizontal surface. A constant force F is applied at a height h from the bottom. Choose the correct option.

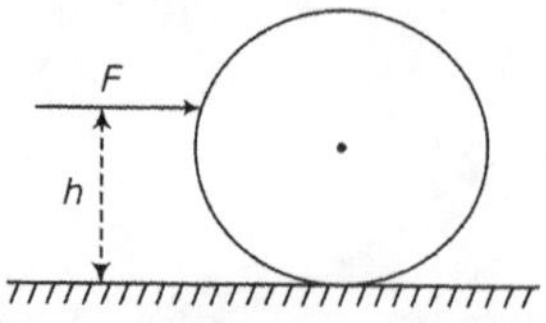

a. Whatever may be the value of h, sphere always slides
b. Sphere will roll without slipping for $2R > h \geq 1.4\,R$
c. Sphere will roll without slipping when $h = 1.4\,R$
d. None of the above

10 A ring of mass m and radius R has three particles attached to the ring as shown in the figure. The centre of the ring has speed v_0. Assume slipping to be absent. The kinetic energy of the system is

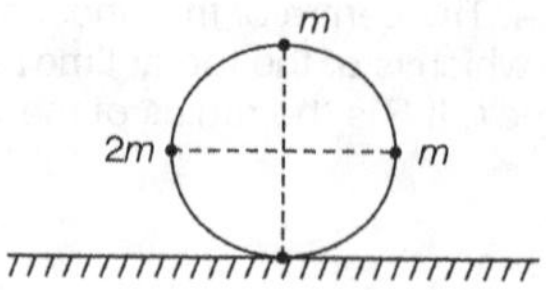

a. $4\,mv_0^2$
b. $6\,mv_0^2$
c. $8\,mv_0^2$
d. $12\,mv_0^2$

11 The radius of gyration of a solid hemisphere of mass m and radius R about an axis (AB in the figure) parallel to the diameter at a distance $\frac{3}{4}R$ from this diameter is

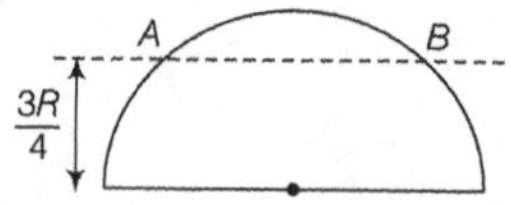

a. $\frac{3R}{\sqrt{10}}$
b. $\frac{5R}{4}$
c. $\frac{5R}{8}$
d. $\sqrt{\frac{2}{5}}R$

12 In both the cases shown, B is the mid-point of A and C and all other factors are same. But in case (i), AB is rough and BC is smooth and in case (ii), AB is smooth and BC is rough. The kinetic energy of the ball on reaching the bottom

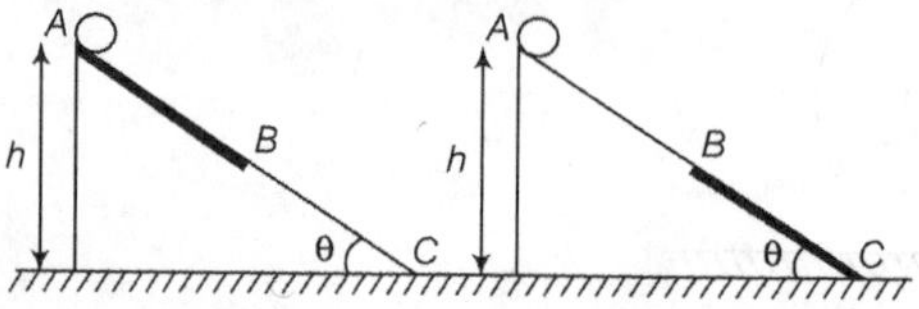

a. is same in both the cases
b. is greater in case (i)
c. is greater in case (ii)
d. None of these

13 A disc of radius 0.1 m rolls without sliding on a horizontal surface with a velocity $6\,ms^{-1}$. Then, it ascends a smooth continuous track as shown in the figure. Given, $g = 10\,ms^{-2}$, the height upto which it will ascend is

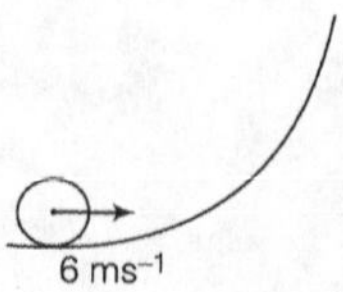

a. 0.9 m
b. 1.8 m
c. 2.4 m
d. 2.7 m

Working Space

14 A small ball of mass *m*, radius *r* is released from rest from point 1 to roll inside a hemispherical shell of radius *R* as shown in the figure. The angular velocity of centre of the ball in point 2 about the centre of the shell is

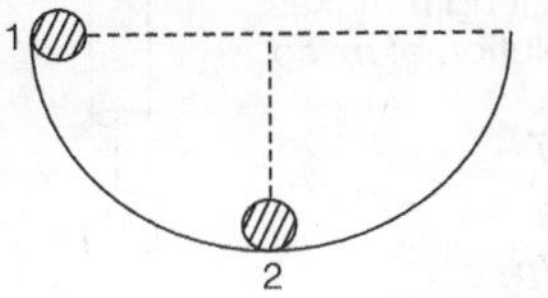

a. $2\sqrt{\dfrac{g}{5(R-r)}}$ b. $\sqrt{\dfrac{10g}{7(R-r)}}$

c. $\sqrt{\dfrac{2g}{5(R-r)}}$ d. $\sqrt{\dfrac{5g}{2(R-r)}}$

15 Two particles *A* and *B* situated at a distance $d = 2$m apart are launched in a gravity free space. The particle *A* is launched with a velocity 10 ms^{-1} at an angle 60° (see figure) and particle *B* is launched with a velocity *v* at an angle 30°, such that distance *d* between *A* and *B* remains constant. The angular velocity of *B* with respect to *A* (in rad s^{-1}) is

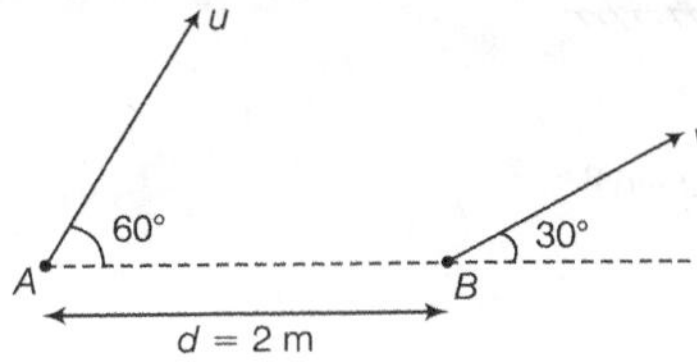

a. $10\sqrt{3}$ b. $\dfrac{10}{\sqrt{3}}$

c. $5\sqrt{3}$ d. $\dfrac{5}{\sqrt{3}}$

16 A plank *P* is placed on a thin hollow cylinder *C*, which rolls on a horizontal surface as shown in the figure. No slippage is there at any of the surfaces in contact. Both Plank and cylinder have same mass *M* and velocity of COM of cylinder is v_0. The ratio of kinetic energy of the plank to that of the cylinder is

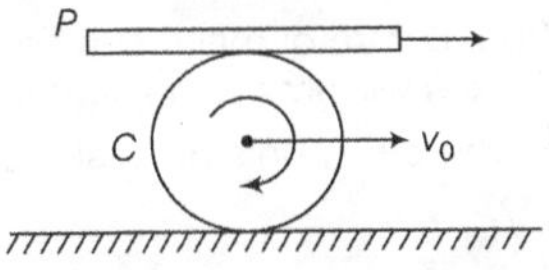

a. 1 : 1 b. 2 : 1

c. 3 : 8 d. 8 : 11

17 A uniform cylinder of mass *M* and radius *R*, rolls without slipping down an incline making an angle θ with the horizontal. The cylinder is connected to a spring of force constant *k* at the centre, the other side of which is connected to a fixed support *S*. The cylinder is released when the spring is unstretched, the force of friction (*f*) is (during downward motion)

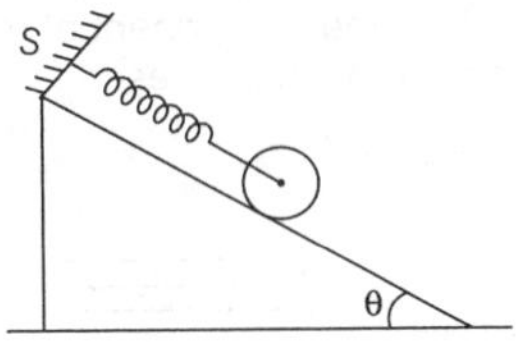

a. always down the incline
b. always up the incline
c. initially down the incline and then up the incline
d. initially up the incline and then down the incline

18 A horizontal turn table in the form of a disc of radius *R* carries a gun *G* and rotates with angular velocity ω_0 about a vertical axis passing through centre *O*. The increase in angular velocity of the system, if the gun fires a bullet of mass *m*, with a tangential velocity *v* relative to the gun is (where, moment of inertia of gun + table about *O* is I_0)

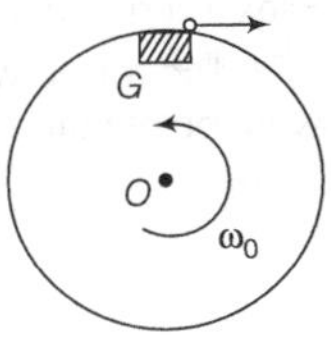

a. $\dfrac{v}{2R}$ b. $\dfrac{mvR}{2I_0}$

c. $\dfrac{2mvR}{I_0 + mR^2}$ d. $\dfrac{mvR}{I_0 + mR^2}$

Working Space

19 Two masses m and M hang at the two ends of a string that passes through a smooth tube as shown in the figure. The mass m moves in a circular path which lies in a horizontal plane. The length of the string from m to the top of the tube is l and θ is the angle this length makes with the vertical. What should be the frequency of revolution of m, so that M remains stationary?

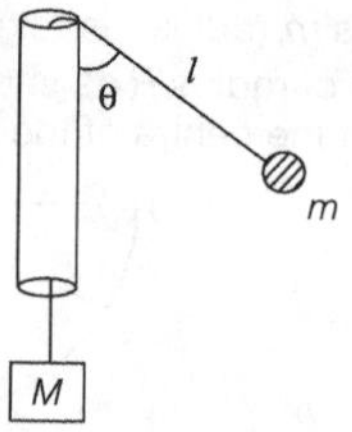

a. $\frac{1}{2\pi}\sqrt{\frac{ml}{Mg}}$ b. $\frac{1}{\pi}\sqrt{\frac{ml}{Mg}}$

c. $\frac{1}{\pi}\sqrt{\frac{Mg}{ml}}$ d. $\frac{1}{2\pi}\sqrt{\frac{Mg}{ml}}$

20 The accleeration a of the plank P required to keep the centre C of a cylinder, which rolls without slipping on the plank, in a fixed position during the motion is

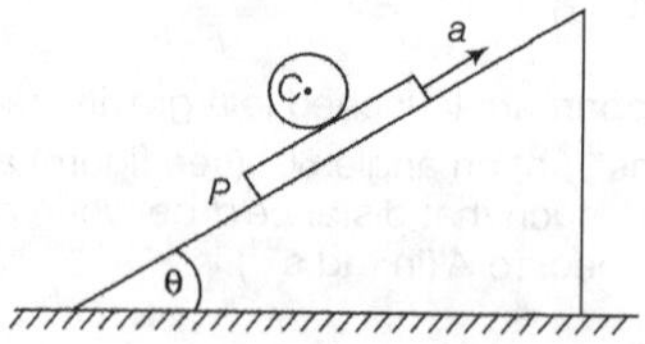

a. $\frac{g}{2}\sin\theta$ b. $g\sin\theta$

c. $2g\sin\theta$ d. $\sqrt{2}\,g\sin\theta$

Statement Type

Direction *(Q. Nos. 21-23) **This section is based on Statement I and Statement II. Select the correct answer from the codes given below.***

a. Both Statement I and Statement II are correct and Statement II is the correct explanation of Statement I
b. Both Statement I and Statement II are correct but Statement II is not the correct explanation of Statement I
c. Statement I is correct but Statement II is incorrect
d. Statement II is correct but Statement I is incorrect

21 **Statement I** If two different axes are at same distance from centre of mass of a rigid body, then moment of inertia of the given body about both the axes will always be same.

Statement II From parallel axes theorem $I = I_{COM} + mr^2$, where symbols have same meaning.

22 **Statement I** A uniform thin rod of length L is hinged about one of its ends and is free to rotate about the hinge without friction. A force F is applied at a distance x from the hinge on the rod such that force is always perpendicular to the rod. Neglect gravity. As the value of x increases from O to L, the component of reaction force by hinge on the rod perpendicular to length of rod increases.

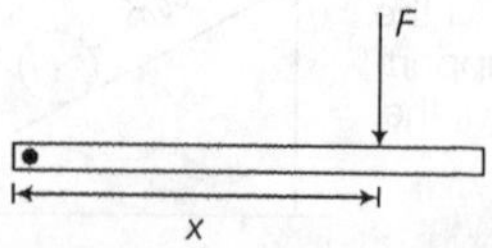

Statement II As x increase from O to L, angular acceleration of rod increases.

23 **Statement I** A rigid disc rolls without slipping on a fixed rough horizontal surface with uniform angular velocity. Then, the acceleration of the lowest point is zero.

Statement II For a rigid disc rolling without slipping on a fixed rough horizontal surface, the velocity of the lowest point in the disc is always zero.

SET - 2

Revisal Problems for JEE Advanced

Only One Option Correct Type

Working Space

Direction (Q. Nos. 1-10) *This section contains 10 multiple choice questions. Each question has four choices (a), (b), (c) and (d), out of which ONLY ONE is correct.*

1 One end of a uniform rod of length L and mass m is hinged at A. It is released from rest from horizontal position AB as shown in the figure. The force exerted by the rod on the hinge, when it becomes vertical is

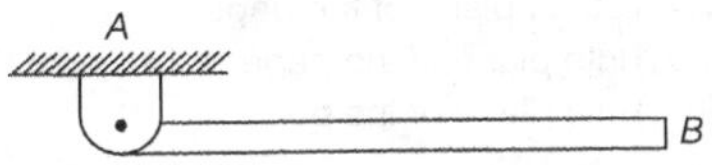

a. $\frac{3}{2}mg$ b. $3mg$ c. $5mg$ d. $\frac{5}{2}mg$

2 In the figure shown, the spherical body and block, each have a mass m. The moment of inertia of the spherical body about centre of mass is $2mR^2$. The spherical body rolls without slipping on the horizontal surface. The ratio of kinetic energy of the spherical body to that of block is

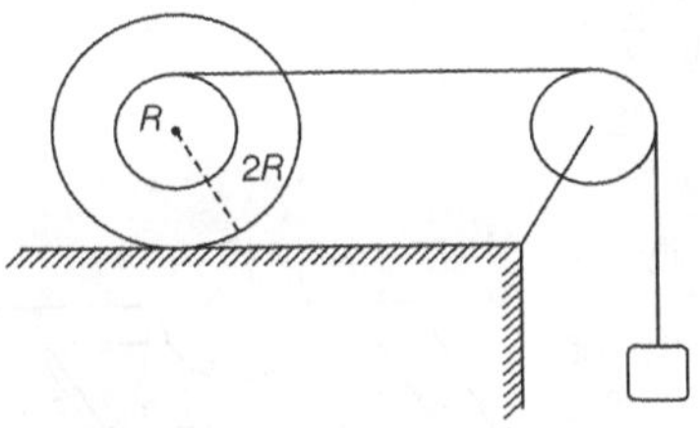

a. $\frac{1}{3}$ b. $\frac{1}{2}$ c. $\frac{2}{3}$ d. $\frac{3}{4}$

3 A uniform cylindrical disc of radius R and mass M is pulled over a horizontal frictionless surface by a constant force. The force is applied by means of a string wound around the disc as shown in the figure. If it starts from rest at $t = 0$, then the linear displacement and angular displacement respectively at time t are

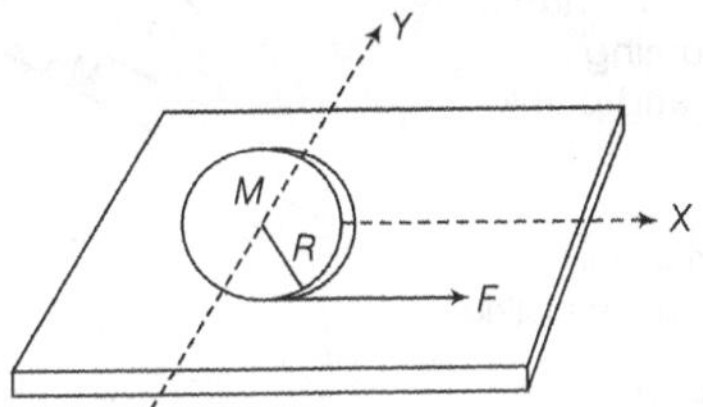

a. $\left(\frac{F}{M}\right)t^2, \left(\frac{F}{M}\right)t^2$ b. $\left(\frac{F}{2M}\right)t, \left(\frac{F}{2MR}\right)t^2$ c. $\left(\frac{F}{2M}\right)t^2, \left(\frac{F}{MR}\right)t^2$ d. $\left(\frac{2F}{M}\right)t^2, \left(\frac{2F}{MR}\right)t^2$

4 A small block is shot into each of the four smooth tracks as shown in the figure. Each of the tracks rises to the same height. The speed with which the block enters the track is the same in all cases. At the highest point of the track, the normal reaction is maximum in

a. 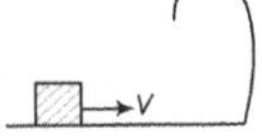

b. 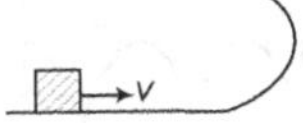

c. 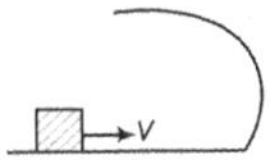

d. 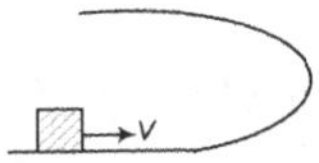

Working Space

5 A uniform sphere of radius R is placed on a rough horizontal surface and given a linear velocity v_0 and angular velocity ω_0 as shown in the figure. For the sphere to come to rest after moving some distance to the right, we have (both rotation and translation cease at the same time)

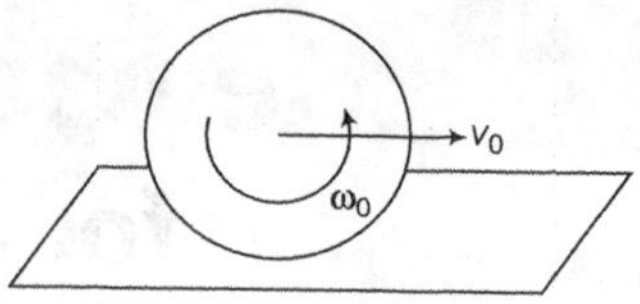

a. $v_0 = R\omega_0$ b. $5v_0 = 2R\omega_0$
c. $2v_0 = 5R\omega_0$ d. $2v_0 = R\omega_0$

6 In the figure shown, the lines of action and moment arms of two forces about the origin O. Imagining these forces to be acting on a rigid body pivoted at O, all vectors shown being in the plane of the figure, the resultant torque will be

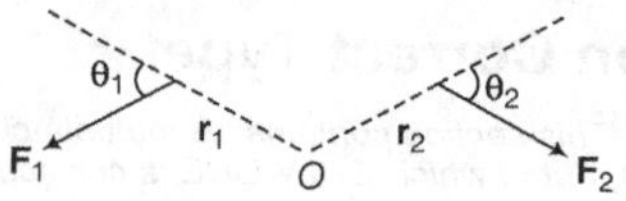

a. $(F_2 r_2 \sin\theta_2 - F_1 r_1 \sin\theta_1)$ out of plane of the page
b. $(F_1 r_1 \sin\theta_1 - F_2 r_2 \sin\theta_2)$ into plane of the page
c. $(F_2 r_2 \sin\theta_2 - F_1 r_1 \sin\theta_1)$ into plane of the page
d. zero

7 Two masses m and M are connected by an ideal string that passes through a smooth hole at O at the centre of a table. Mass m lies on the table and M hangs vertically. m is moved around in a horizontal circle with O as the centre. If l is the length of the string from O to m, then the frequency with which m should revolve so that M remains stationary, is

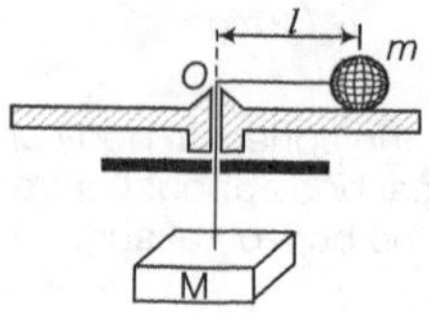

a. $\frac{1}{2\pi}\sqrt{\frac{Mg}{ml}}$ b. $\frac{1}{\pi}\sqrt{\frac{Mg}{ml}}$
c. $\frac{1}{2\pi}\sqrt{\frac{Ml}{mg}}$ d. $\frac{1}{\pi}\sqrt{\frac{Ml}{mg}}$

8 A sphere S of mass m is given a finite angular velocity about a horizontal axis through its centre. Now, it is placed gently on a plank P of same mass. The coefficient of friction between S and P is μ and the plank rests on a smooth horizontal surface as shown in the figure. The initial acceleration of the sphere relative to the planck will be

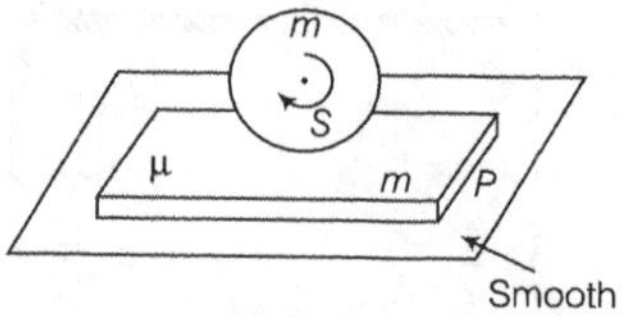

a. μg b. $2\mu g$
c. $\frac{2}{5}\mu g$ d. zero

9 A plank of mass M is placed on a smooth inclined plane and a sphere of mass m is then placed on the plank. Assuming sufficient friction to be present between the sphere and plank, the frictional force on sphere when the plank and sphere are released from rest is

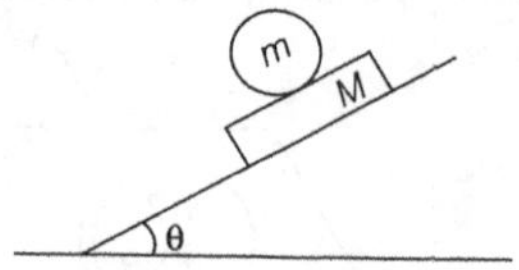

a. zero b. horizontal
c. up the incline d. down the incline

10 Two identical discs of same radius R are rotating about their axes in opposite directions with the same constant angular speed ω. The discs are in the same horizontal plane. At time $t = 0$, the points P and Q are facing each other as shown in the figure. The magnitude of the relative speed between the points P and Q is v_r, which as a function of time t is best represented by

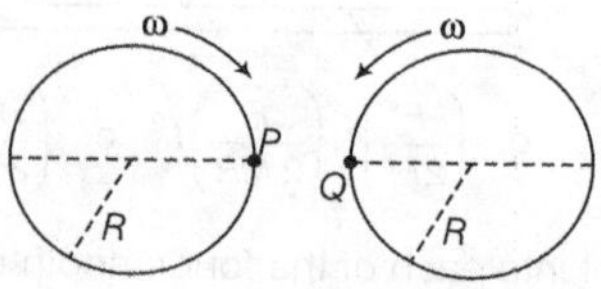

a.
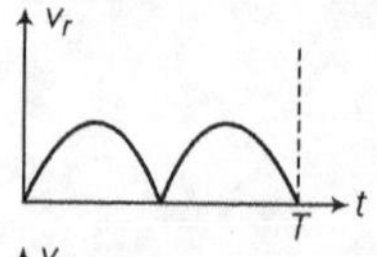

b.
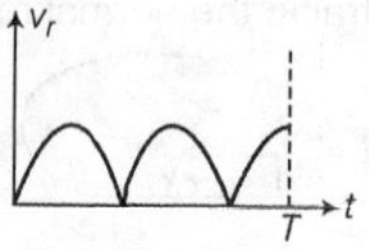

c.
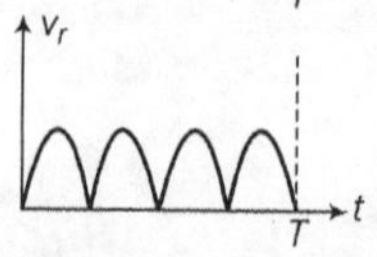

d.
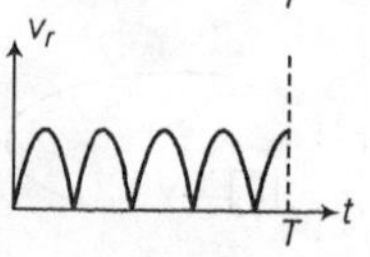

One or More than One Options Correct Type

Working Space

Direction *(Q. Nos. 11-15) This section contains 5 multiple choice questions. Each question has four choices (a), (b), (c) and (d), out of which ONE or MORE THAN ONE are correct.*

11 A uniform rod of mass m and length L is held vertically on a smooth horizontal surface. When the rod is released, choose the correct statement(s).

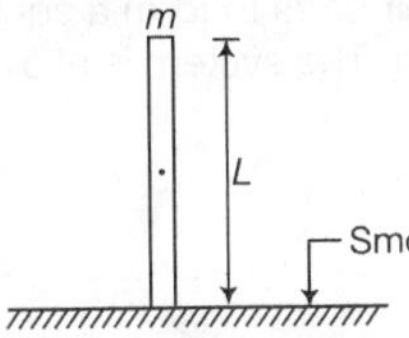

a. The COM of the rod accelerates in vertical direction
b. Initially the magnitude of normal reaction is mg
c. When the rod becomes just horizontal, the magnitude of normal reaction is $mg/2$
d. When the rod becomes just horizontal, the magnitude of normal reaction is $mg/4$

12 A spool of wire rests on a horizontal surface as shown in the figure. As the wire is pulled, the spool does not slip at contact point A. On separate trials, each one of the forces F_1, F_2, F_3 and F_4 is applied to the spool. For each one of these forces, the spool

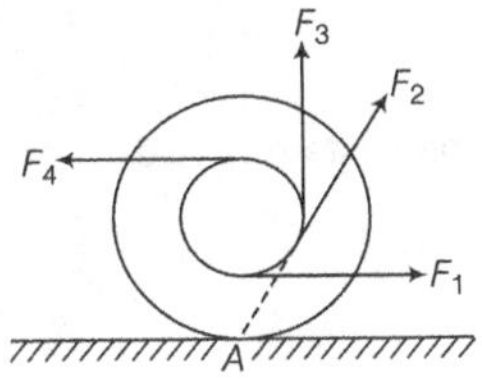

a. will rotate anti-clockwise if F_1 is applied
b. will rotate anti-clockwise if F_3 is applied
c. will rotate clockwise if F_4 is applied
d. will not rotate if F_2 is applied

13 The end B of the rod AB, which makes an angle θ with the floor is being pulled with a constant velocity v_0 as shown in the figure. The length of the rod is L. Which of the following statements is/are correct?

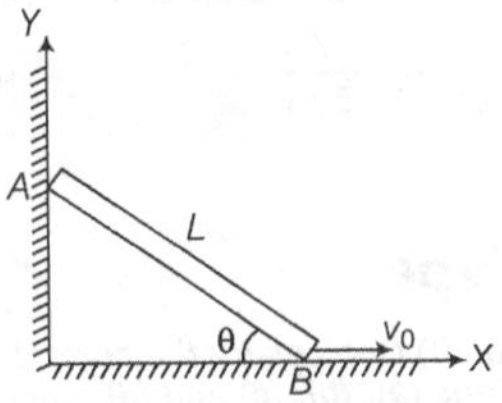

a. Velocity of end A is always same
b. Angular velocity of rod is always same
c. Velocity of end A is $\frac{4}{3}v_0$ downwards when $\theta = 37°$
d. Angular velocity of the rod is $\frac{5v_0}{3L}$ when $\theta = 37°$

14 A solid cylinder of mass m and radius R is rolled horizontally on a rough surface as shown in the figure. Choose the correct statement(s).

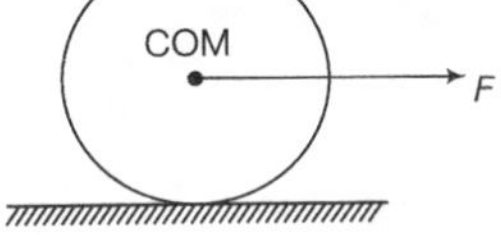

Rough

a. The acceleration of COM is $\frac{F}{m}$
b. The acceleration of COM is $\frac{2}{3}\cdot\frac{F}{m}$
c. The friction on the cylinder acts backwards
d. The friction is $\frac{F}{3}$

15 Two forces F_1 and F_2 are acting on a rod abc as shown in the figure. Select correct statement(s). [τ_a means (τ_{net}) about a]

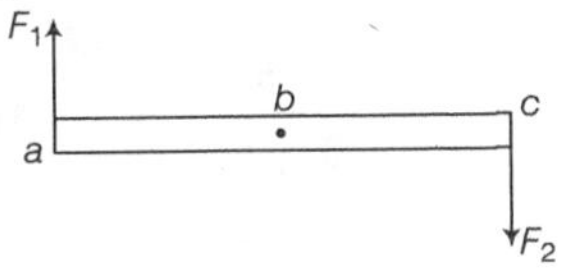

a. If $F_1 \neq F_2$, then $\tau_a \neq \tau_b \neq \tau_c$
b. If $F_1 \neq F_2$, then $\tau_a = \tau_b \neq \tau_c$
c. if $F_1 = F_2$, then $\tau_a = \tau_b = \tau_c$
d. If $F_1 = F_2$, then $\tau_a = \tau_b \neq \tau_c$

Comprehension Type

Working Space

Direction (Q. Nos. 16-18) *This section contains a paragraph, which describing theory, experiments, data, etc. Three questions related to the paragraph have been given. Each question has only one correct answer among the four given options (a), (b), (c) and (d).*

Passage

Four identical rods each of mass $M = 6$ kg, are welded at their ends to form a square and then welded to a ring of mass $m = 4$ kg, having radius $R = 1$ m. The system is allowed to roll down an inclined plane with $\theta = 30°$.

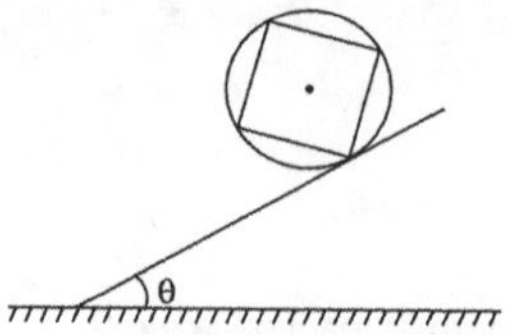

16 The moment of inertia of the system about the centre of the ring will be

a. 10 kg-m^2 b. 20 kg-m^2
c. 25 kg-m^2 d. 40 kg-m^2

17 The acceleration of the system will be

a. $\frac{g}{8}$ b. $\frac{g}{4}$
c. $\frac{7g}{24}$ d. $\frac{5g}{23}$

18 The minimum value of friction coefficient to prevent slipping is

a. $\frac{5}{7}$ b. $\frac{7}{5\sqrt{3}}$ c. $\frac{5\sqrt{3}}{7}$ d. $\frac{5}{12\sqrt{3}}$

Matching List Type

Direction (Q. Nos. 19 and 20) *Choices for the correct combination of elements from Column I and Column II are given as options (a), (b), (c) and (d), out of which one is correct.*

19 A smooth ball of mass m moving with a uniform velocity v_0 strikes a smooth uniform rod AB of equal mass m, lying on a smooth horizontal table. The ball strikes the rod at one end A, perpendicular to the rod as shown in the figure. The collision is perfectly elastic.

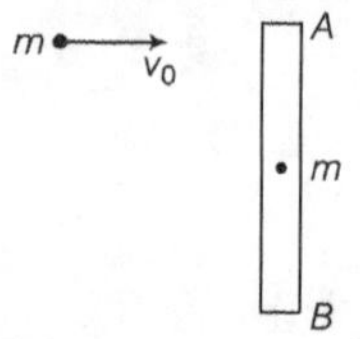

Match the following columns and select the correct option from the codes given below.

	Column I		Column II
i.	$\frac{\text{Final KE of ball}}{\text{Initial KE of ball}}$	p.	$\frac{2}{5}$
ii.	$\frac{\text{Impulse delivered to the rod}}{\text{Initial momentum of ball}}$	q.	$\frac{3}{5}$
iii.	$\frac{\text{Angular momentum of rod about its centre}}{\text{Initial angular momentum of the ball about centre of the rod}}$	r.	$\frac{9}{25}$
iv.	$\frac{\text{Final KE of rotation of rod}}{\text{Final KE of translation of rod}}$	s.	3

Codes

	i	ii	iii	iv		i	ii	iii	iv
a.	r	p	p	s	b.	p	q	r	s
c.	q	p	s	r	d.	r	s	p	q

Working Space

20 The figure shows a uniform smooth solid cylinder A of radius $R = 4$m rolling without slipping on the 8 kg plank, which in turn is supported by a fixed smooth horizontal surface. The blocks B and C accelerate downwards with $6\,ms^{-2}$ and $2\,ms^{-2}$, respectively.

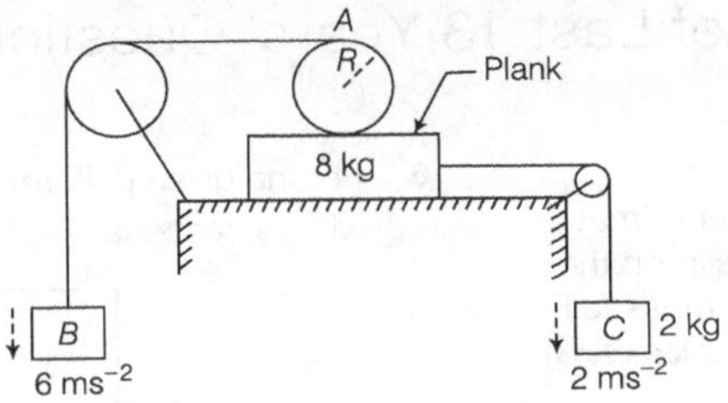

The angular acceleration of the cylinder A is α, the acceleration of COM of the cylinder is a, the ratio of mass of cylinder A to the mass of block B is $\frac{m_A}{M_B}$. The length of the unwrapped thread between the cylinder and the block B at $t = 0$ is $l_0 = 20$m and at $t = 2$s is l. Assume the system to be released from rest. ($g = 10\,ms^{-2}$). Match the following columns and select the correct option from the codes given below.

	Column I		Column II
i.	α (in rad s^{-2})	p.	8
ii.	$2a$ (in ms^{-2})	q.	2
iii.	$\frac{m_A}{m_B}$	r.	4
iv.	$(l - l_0)$ (in m)	s.	1

Codes

	i	ii	iii	iv		i	ii	iii	iv
a.	s	r	q	p	b.	p	q	r	s
c.	r	s	q	p	d.	q	r	s	p

One Integer Value Correct Type

Direction (*Q. Nos. 21-23*) ***This section contains 3 questions. When worked out will result in an integer from 0 to 9 (both inclusive).***

21 A small solid ball (mass = 0.1kg) rolls without slipping along the track shown in the figure. The radius of the circular track is R. If the ball starts from rest at a height $8R$ above the bottom, the the horizontal force acting on it at point P is $5x$ newton. Find the value of x. (Given, $g = 10\,ms^{-2}$)

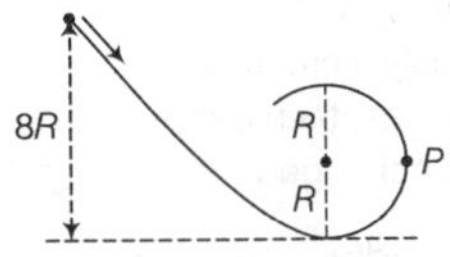

22 A rigid sculpture, consisting of thin hoop (of mass m and radius R) and two thin rods (each of mass m and length $L = 2R$), is arranged as shown in the figure. The sculpture is pivoted and can rotate in a horizontal axis in the plane of the hoop and passing through centre of the hoop. The sculptures rotational inertia about this axis is approximately $K\,mR^2$. Find the value of K.

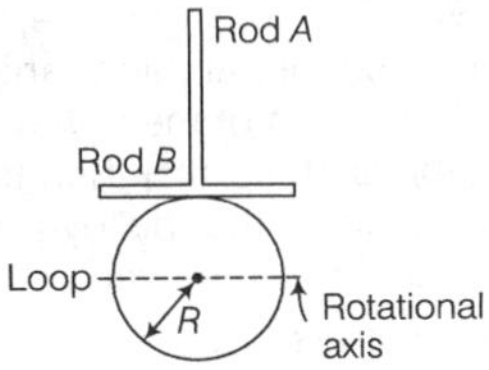

23 A block of mass M is moving with a velocity $v_1 = v_0$ and a frictionless surface as shown in the figure. It passes over a cylinder of radius R and mass M, which has a fixed axis and is initially at rest, when the block first makes contact with the cylinder, it slips on the cylinder, but the friction is large enough so that slipping ceases before it losses position contact with the cylinder. Finally, it goes to the dotted with velocity $v_2 = \frac{2}{\beta} v_0$. Find the value of β.

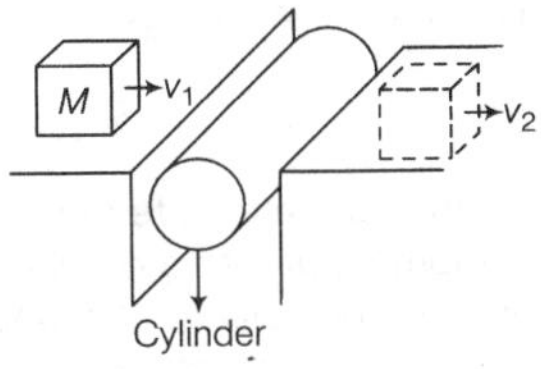

JEE Main & AIEEE Archive

(Compilation of Last 13 Years' Questions)

1 A mass m supported by a massless string wound around a uniform hollow cylinder of mass m and radius R. If the string does not slip on the cylinder, with what acceleraton will the mass fall on release? **(JEE Main 2014)**

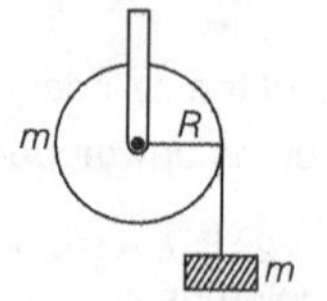

a. $\frac{2g}{3}$ b. $\frac{g}{2}$ c. $\frac{5g}{6}$ d. g

2 A hoop of radius r and mass m rotating with an angular velocity ω_0 is placed on a rough horizontal surface. The initial velocity of the centre of the hoop is zero. What will be the velocity of the centre of the hoop when it ceases to slip? **(JEE Main 2013)**

a. $\frac{r\omega_0}{4}$ b. $\frac{r\omega_0}{3}$ c. $\frac{r\omega_0}{2}$ d. $r\omega_0$

3 A thin horizontal circular disc is rotating about a vertical axis passing through its centre. An insect is at rest at a point near the rim of the disc. The insect now moves along a diameter of the disc to reach its other end. During the journey of the insect, the angular speed of the disc **(AIEEE 2011)**

a. continuously decreases
b. continuously increases
c. first increases, then decreases
d. remains unchanged

4 A pulley of radius 2 m is rotated about its axis by a force $F = (20t - 5t^2)$ N (where, t is measured in seconds) applied tangentially. If the moment of inertia of the pulley about its axis of rotation is 10 kg-m^2, then the number of rotations made by the pulley before its direction of motion is reserved, is **(AIEEE 2011)**

a. more than 3 but less than 6
b. more than 6 but less than 9
c. more than 9
d. less than 3

5 Consider a uniform square plate of side a and mass m. The moment of inertia of this plate about an axis perpendicular to its plane and passing through one of its corners is **(AIEEE 2008)**

a. $\frac{5}{6}ma^2$ b. $\frac{1}{12}ma^2$ c. $\frac{7}{12}ma^2$ d. $\frac{2}{3}ma^2$

6 For the given uniform square lamina $ABCD$, whose centre is O **(AIEEE 2007)**

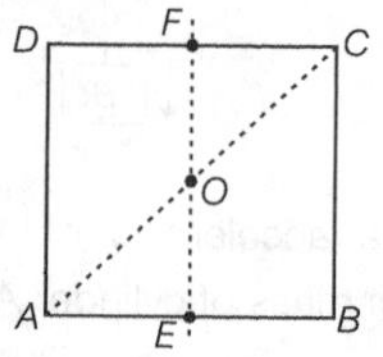

a. $\sqrt{2}\, I_{AC} = I_{EF}$ b. $I_{AD} = 3I_{EF}$ c. $I_{AD} = 4I_{EF}$ d. $I_{AC} = \sqrt{2}\, I_{EF}$

7 A circular disc of radius R is removed from a bigger circular disc of radius $2R$, such that the circumference of the discs coincide. The centre of mass of the new disc is $\frac{\alpha}{R}$ from the centre of the bigger disc. The value of α is **(AIEEE 2007)**

a. $\frac{1}{3}$ b. $\frac{1}{2}$ c. $\frac{1}{6}$ d. $\frac{1}{4}$

8 A round uniform body of radius R, mass M and moment of inertia I, rolls down (without slipping) an inclined plane making an angle θ with the horizontal. Then, its acceleration is **(AIEEE 2007)**

a. $\frac{g \sin\theta}{1 + I/MR^2}$ b. $\frac{g \sin\theta}{1 + MR^2/I}$ c. $\frac{g \sin\theta}{1 - I/MR^2}$ d. $\frac{g \sin\theta}{1 - MR^2/I}$

9 Angular momentum of the particle rotating with a central force is constant due to **(AIEEE 2007)**

a. constant force
b. constant linear momentum
c. zero torque
d. constant torque

10 Four point masses, each of value m, are placed at the corners of a square $ABCD$ of side l. The moment of inertia of this system about an axis passing through A and parallel to BD is **(AIEEE 2006)**

a. $2ml^2$ b. $\sqrt{3}ml^2$ c. $3ml^2$ d. ml^2

11 A force of $-F\hat{k}$ acts on O, the origin of the coordinate system. The torque about the point $(1, -1)$ is **(AIEEE 2006)**

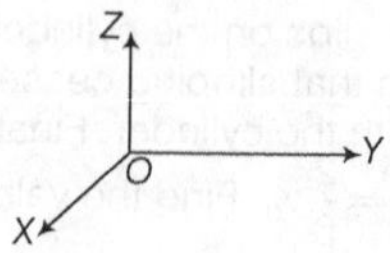

a. $F(\hat{i} - \hat{j})$ b. $-F(\hat{i} + \hat{j})$ c. $F(\hat{i} + \hat{j})$ d. $-F(\hat{i} - \hat{j})$

12 A thin circular ring of mass m and radius R is rotating about its axis with a constant angular velocity ω. Two objects each of mass M are attached gently to the opposite ends of a diameter of the ring. The ring now rotates with an angular velocity ω' is equal to **(AIEEE 2006)**

a. $\frac{\omega(m+2M)}{m}$ b. $\frac{\omega(m-2M)}{(m+2M)}$

c. $\frac{\omega m}{(m+M)}$ d. $\frac{\omega m}{(m+2M)}$

13 The moment of inertia of uniform semi-circular disc of mass M and radius r about a line perpendicular to the plane of the disc through the centre is **(AIEEE 2005)**

a. $\frac{1}{4}Mr^2$ b. $\frac{2}{5}Mr^2$

c. Mr^2 d. $\frac{1}{2}Mr^2$

14 One solid sphere A and another hollow sphere B are of same mass and same outer radii. Their moment of inertia about their diameters are respectively I_A and I_B, such that **(AIEEE 2004)**

a. $I_A = I_B$ b. $I_A > I_B$

c. $I_A < I_B$ d. $\frac{I_A}{I_B} = \frac{d_A}{d_B}$

where, d_A and d_B are their densities.

15 A solid sphere is rotating in free space. If the radius of sphere is increased keeping mass same, which one of the following will not be affected? **(AIEEE 2004)**

a. Moment of inertia
b. Angular momentum
c. Angular velocity
d. Rotational kinetic energy

16 A circular disc X of radius R is made from an iron plate of thickness t and another disc Y of radius $4R$ is made from an iron plate of thickness $t/4$. Then, the relation between the moment of inertia I_X and I_Y is **(AIEEE 2003)**

a. $I_Y = 32 I_X$ b. $I_Y = 16 I_X$

c. $I_Y = I_X$ d. $I_Y = 64 I_X$

17 A particle performing uniform circular motion has angular momentum L. If its angular frequency is doubled and its kinetic energy halved, then the new angular momentum is **(AIEEE 2003)**

a. $\frac{L}{4}$ b. $2L$

c. $4L$ d. $L/2$

18 Let **F** be the force acting on a particle having position vector **r** and τ be the torque of this force about the origin. Then, **(AIEEE 2003)**

a. $\mathbf{r}\cdot\tau = 0$ and $\mathbf{F}\cdot\tau \neq 0$
b. $\mathbf{r}\cdot\tau \neq 0$ and $\mathbf{F}\cdot\tau = 0$
c. $\mathbf{r}\cdot\tau \neq 0$ and $\mathbf{F}\cdot\tau \neq 0$
d. $\mathbf{r}\cdot\tau = 0$ and $\mathbf{F}\cdot\tau = 0$

19 Initial angular velocity of a circular disc of mass M is ω_1. Then, two small spheres of mass m are attached gently to two diametrically opposite points on the edge of the disc. What is the final angular velocity of the disc? **(AIEEE 2002)**

a. $\left(\frac{M+m}{M}\right)\omega_1$

b. $\left(\frac{M+m}{m}\right)\omega_1$

c. $\left(\frac{M}{M+4m}\right)\omega_1$

d. $\left(\frac{M}{M+2m}\right)\omega_1$

20 A particle of mass m moves along line PC with velocity v as shown in the figure. What is the angular momentum of the particle about O? **(AIEEE 2002)**

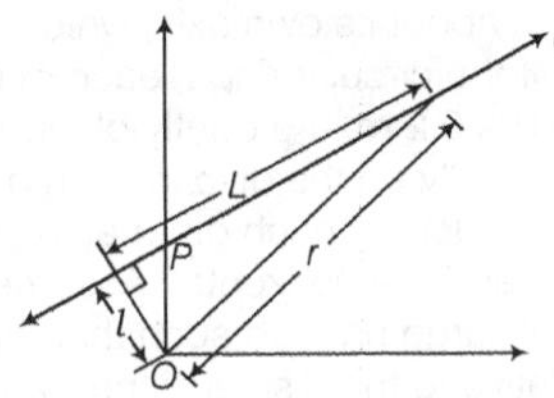

a. mvL b. mvl

c. mvr d. Zero

21 Moment of inertia of a circular wire of mass M and radius R about its diameter, is **(AIEEE 2002)**

a. $\frac{MR^2}{2}$ b. MR^2

c. $2MR^2$ d. $\frac{MR^2}{4}$

22 A solid sphere, a hollow sphere and a ring are released from top of an inclined plane (frictionless), so that they slide down the plane. Then, maximum acceleration down the plane is for (no rolling) **(AIEEE 2002)**

a. solid sphere
b. hollow sphere
c. ring
d. All same

JEE Advanced & IIT JEE Archive

(Compilation of Last 10 Years' Questions)

1 A horizontal circular platform of radius 0.5 m and mass 0.45 kg is free to rotate about its axis. Two massless spring toy-guns, each carrying a steel ball of mass 0.05 kg are attached to the platform at a distance 0.25 m from the centre on its either sides along its diameter (see figure). Each gun simultaneously fires the balls horizontally and perpendicular to the diameter in opposite directions. After leaving the platform, the balls have horizontal speed of 9 ms^{-1} with respect to the ground. The rotational speed of the platform (in rad s^{-1}) after the balls leave the platform is **(2014 Adv., Integer Type)**

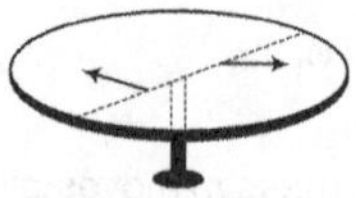

2 A uniform circular disc of mass 50 kg and radius 0.4 m is rotating with an angular velocity of 10 rad/s about its own axis, which is vertical. Two uniform circular rings, each of mass 6.25 kg and radius 0.2 m, are gently placed symmetrically on the disc in such a manner that they are touching each other along the axis of the disc and are horizontal. Assume that the friction is large enough such that the rings are at rest relative to the disc and the system rotates about the original axis. The new angular velocity (in rad s^{-1}) of the system is **(2013 Adv., Integer Type)**

3 Two solid cylinders P and Q of same mass and same radius start rolling down a fixed inclined plane from the same height at the same time. Cylinder P has most of its mass concentrated near its surface, while Q has most of its mass concentrated near the axis. Which statement(s) is(are) correct?

(2012, One or More than One Options Correct Type)

a. Both cylinders P and Q reach the ground at the same time

b. Cylinder P has larger linear acceleration than cylinder Q

c. Both cylinders reach the ground with same translational kinetic energy

d. Cylinder Q reaches the ground with larger angular speed

4 The figure shows a system consisting of (i) a ring of outer radius $3R$ rolling clockwise without slipping on a horizontal surface with angular speed ω and (ii) an inner disc of radius $2R$ rotating anti-clockwise with angular speed $\omega/2$. The ring and disc are separated by frictionless ball bearings. The system is in the XZ-plane. The point P on the inner disc is at a distance R from the origin, where OP makes an angle of 30° with the horizontal. Then, with respect to the horizontal surface,

(2012, One or More than One Options Correct Type)

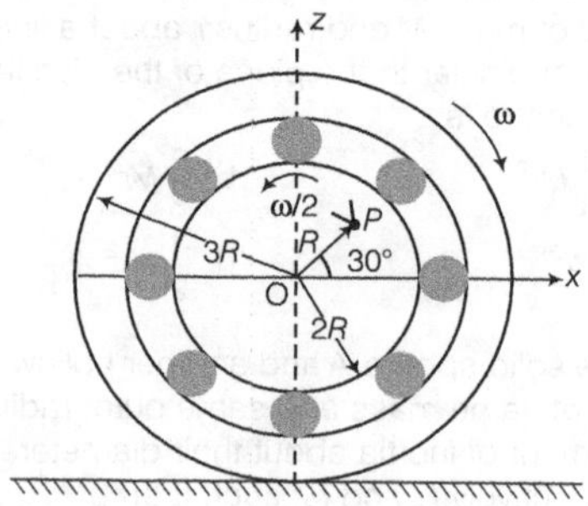

a. the point O has a linear velocity $3R\omega\hat{i}$

b. the point P has a linear velocity $\frac{11}{4}R\omega\hat{i} + \frac{\sqrt{3}}{4}R\omega\hat{k}$

c. the point P has a linear velocity $\frac{13}{4}R\omega\hat{i} - \frac{\sqrt{3}}{4}R\omega\hat{k}$

d. the point P has a linear velocity $\left(3 - \frac{\sqrt{3}}{4}\right)R\omega\hat{i} + \frac{1}{4}R\omega\hat{k}$

5 A thin uniform rod pivoted at O is rotating in the horizontal plane with constant angular speed ω as shown in the figure. At time $t = 0$, a small insect starts from O and moves with constant speed v with respect to the rod towards the other end. It reaches the end of the rod at $t = T$ and stops. The angular speed of the system remains ω throughout.

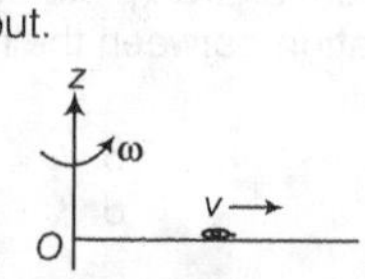

The magnitude of the torque $|\tau|$ on the system about O as a function of time is best represented by which plot? **(2012, Only One Option Correct Type)**

a. $|\tau|$

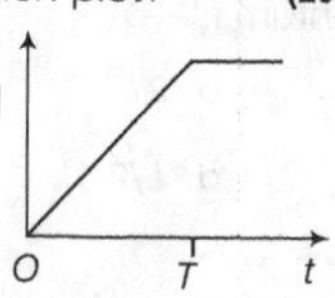

b. $|\tau|$

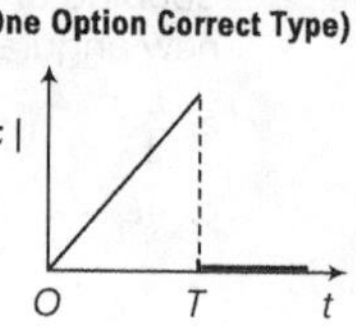

c. $|\tau|$

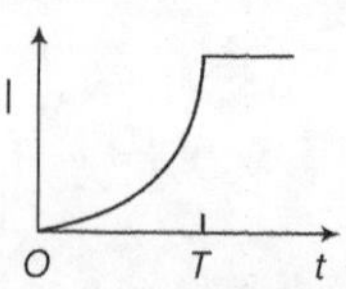

d. $|\tau|$

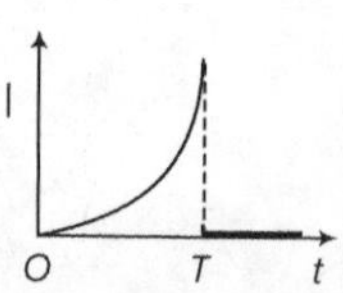

6 Two identical discs of same radius R are rotating about their axes in opposite directions with the same constant angular speed ω. The discs are in the same horizontal plane.

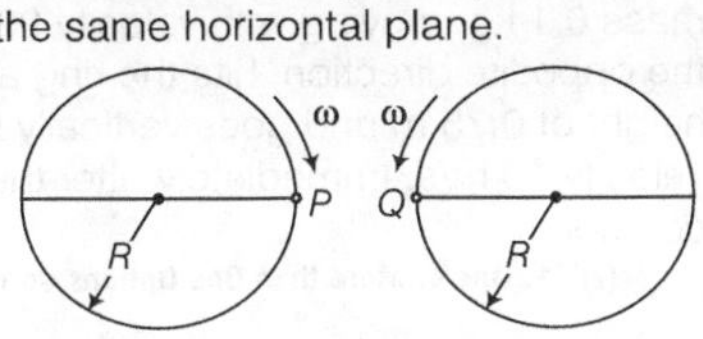

At time $t = 0$, the points P and Q are facing each other as shown in the figure. The relative speed between the two points P and Q is v_r. In one time period (T) of rotation of the discs, v_r as a function of time is best represented by

(2012, Only One Option Correct Type)

a. 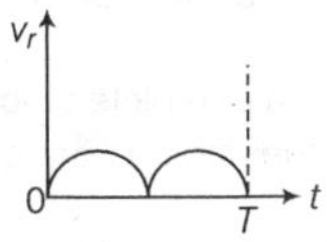

b. 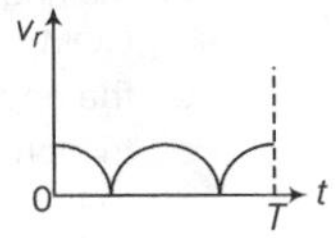

c. 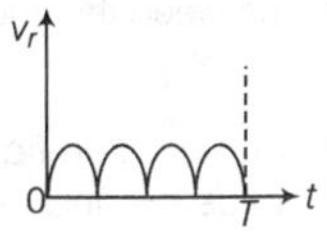

d. 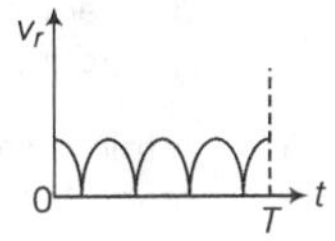

7 A small mass m is attached to a massless string whose other end is fixed at P as shown in the figure. The mass is undergoing circular motion in the XY-plane with centre at O and constant angular speed ω. If the angular momentum of the system, calculated about O and P are denoted by $\mathbf{L}_O$ and $\mathbf{L}_P$ respectively, then **(2012, Only One Option Correct Type)**

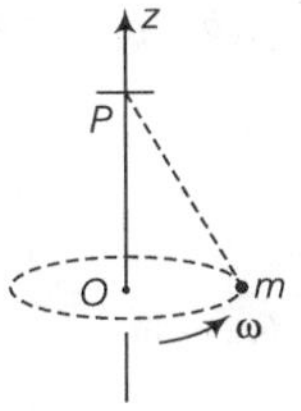

a. $\mathbf{L}_O$ and $\mathbf{L}_P$ do not vary with time
b. $\mathbf{L}_O$ varies with time while $\mathbf{L}_P$ remains constant
c. $\mathbf{L}_O$ remains constant while $\mathbf{L}_P$ varies with time
d. $\mathbf{L}_O$ and $\mathbf{L}_P$ both vary with time

8 Consider a disc rotating in the horizontal plane with a constant angular speed ω about its centre O. The disc has a shaded region on one side of the diameter and an unshaded region on the other side as shown in the figure.

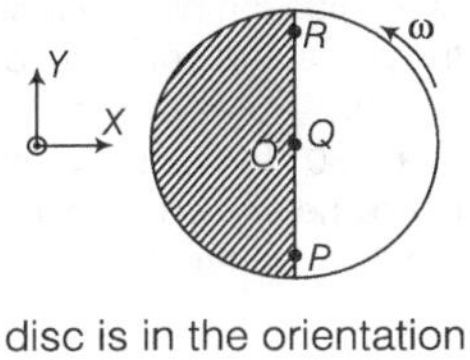

When the disc is in the orientation as shown, two pebbles P and Q are simultaneously projected at an angle towards R. The velocity of projection is in the YZ-plane and is same for both pebbles with respect to the disc. Assume that

(i) they land back on the disc before the disc has completed $\frac{1}{8}$ rotation
(ii) their range is less than half the disc radius
(iii) ω remains constant throughout. Then,

(2012, Only One Option Correct Type)

a. P lands in the shaded region and Q in the unshaded region
b. P lands in the unshaded region and Q in the shaded region
c. both P and Q land in the unshaded region
d. both P and Q land in the shaded region

Passage for Q. Nos. (9-10)

The general motion of a rigid body can be considered to be a combination of (i) a motion of its centre of mass about an axis, and (ii) its relation about an instantaneous axis passing through the centre of mass. These axes need not be stationary. Consider, for example, a thin uniform disc welded (rigidly fixed) horizontally at its rim to a massless stick as shown in the figure. When the disc-stick system is rotated about the origin on a horizontal frictionless plane with angular speed ω, then motion at any instant can be taken as a combination of (i) a rotation of the centre of mass of the disc about the Z-axis, and (ii) a rotation of the disc through an instantaneous vertical axis passing through its centre of mass (as seen from the changed orientation of points P and Q). Both these motions have the same angular speed ω in this case.

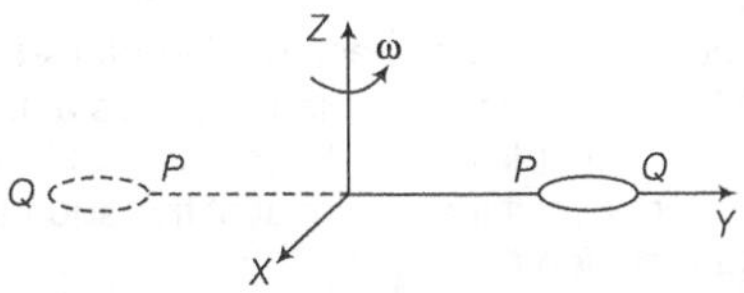

Now, consider two similar systems as shown in the figure : Case (i). the disc with its face vertical and parallel to XZ-plane; Case (ii). the disc with its face making an angle of 45° with XY-plane and its horizontal diameter parallel to X-axis. In both the cases, the disc is welded at point P and the systems are rotated with constant angular speed ω about the Z-axis.

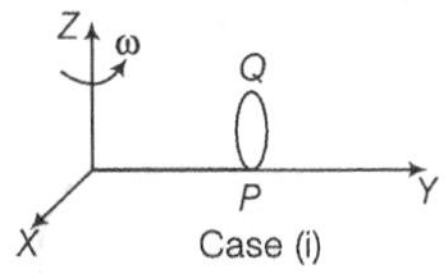

Case (i)

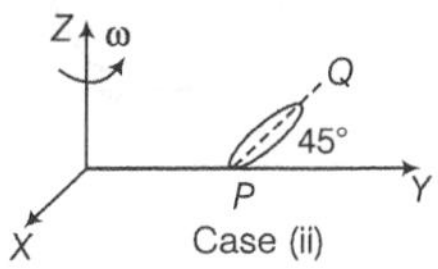

Case (ii)

(2012, Comprehension Type)

9 Which of the following statements regarding the angular speed about the instantaneous axis (passing through the centre of mass) is correct?

a. It is $\sqrt{2}\,\omega$ for both the cases

b. It is ω for case (i) and $\frac{\omega}{\sqrt{2}}$ for case (ii)

c. It is ω for case (i) and $\sqrt{2}\,\omega$ for case (ii)

d. It is ω for both the cases

10 Which of the following statements about the instantaneous axis (passing through the centre of mass) is correct?

a. It is vertical for both the cases (i) and (ii)

b. It is vertical for case (i) and is at 45° to the *XY*-plane and lies in the plane of the disc for case (ii)

c. It is horizontal for case (i) and is at 45° to the *XZ*-plane and is normal to the plane of the disc for case (ii)

d. It is vertical for case (i) and is at 45° to the *XZ*-plane and is normal to the plane of the disc for case (ii)

11 A lamina is made by removing a small disc of diameter $2R$ from a bigger disc of uniform mass density and radius $2R$ as shown in the figure. The moment of inertia of this lamina about axes passing through O and P is I_O and I_P, respectively. Both these axes are perpendicular to the plane of the lamina. The ratio $\frac{I_P}{I_O}$ to the nearest integer is **(2012, Integer Type)**

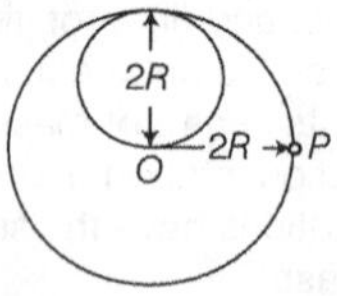

12 Four solid spheres each of diameter $\sqrt{5}$ cm and mass 0.5 kg are placed with their centres at the corners of a square of side 4 cm. The moment of inertia of the system about the diagonal of the square is $N \times 10^{-4}$ kg-m^2, then N is **(2011, Integer Type)**

13 A boy is pushing a ring of mass 2 kg and radius 0.5 m with a stick as shown in the figure. The stick applies a force of 2 N on the ring and rolls it without slipping with an acceleration of 0.3 m/s^2. The coefficient of friction between the ground and the ring is large enough that rolling always occurs and the coefficient of friction between the stick and the ring is $\left(\frac{P}{10}\right)$. The value of P is

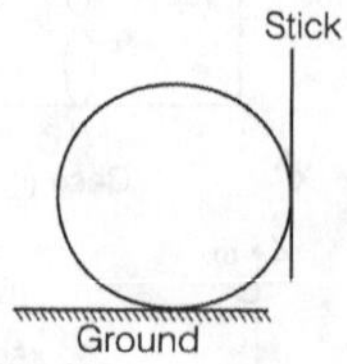

(2011, Integer Type)

14 A thin ring of mass 2 kg and radius 0.5 m is rolling without slipping on a horizontal plane with velocity 1 m/s. A small ball of mass 0.1 kg, moving with velocity 20 m/s in the opposite direction, hits the ring at a height of 0.75 m and goes vertically up with velocity 10 m/s. Immediately after the collision

(2011, One or More than One Options Correct Type)

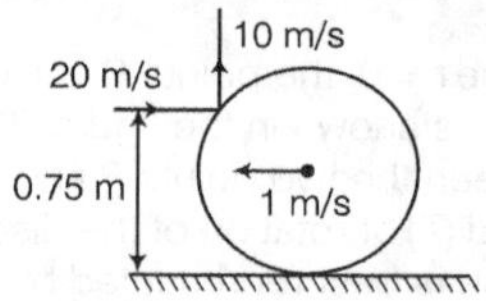

a. the ring has pure rotation about its stationary COM

b. the ring comes to a complete stop

c. friction between the ring and the ground is to the left

d. there is no friction between the ring and the ground

15 A thin uniform annular disc (see figure) of mass M has outer radius $4R$ and inner radius $3R$. The work required to take a unit mass from point P on its axis to infinity is

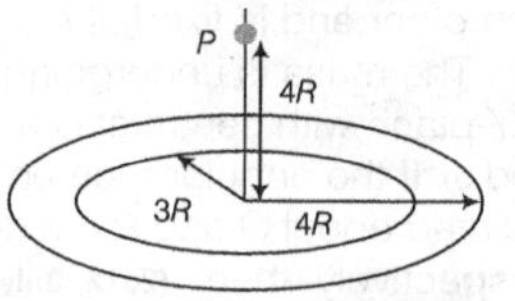

(2010, Only One Option Correct Type)

a. $\frac{2GM}{7R}(4\sqrt{2}-5)$

b. $-\frac{2GM}{7R}(4\sqrt{2}-5)$

c. $\frac{GM}{4R}$

d. $\frac{2GM}{5R}(\sqrt{2}-1)$

16 A block of base 10 cm × 10 cm and height 15 cm is kept on an inclined plane. The coefficient of friction between them is $\sqrt{3}$. The inclination θ of this inclined plane from the horizontal plane is gradually increased from 0°. Then, **(2009, Only One Option Correct Type)**

a. at $\theta = 30°$, the block will start sliding down the plane

b. the block will remain at rest on the plane up to certain θ and then it will topple

c. at $\theta = 60°$, the block will start sliding down the plane and continue to do so at higher angles

d. at $\theta = 60°$, the block will start sliding down the plane and on further increasing θ, it will topple at certain θ

17 A sphere is rolling without slipping on a fixed horizontal plane surface. In the figure, A is the point of contact. B is the centre of the sphere and C is its topmost point. Then,

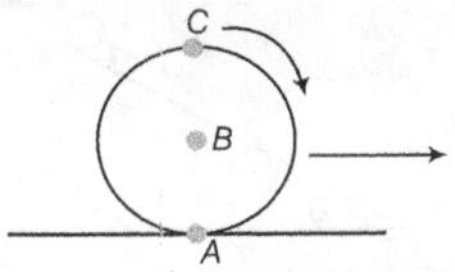

(2009, One or More than One Options Correct Type)

a. $\mathbf{v}_C - \mathbf{v}_A = 2(\mathbf{v}_B - \mathbf{v}_C)$ b. $\mathbf{v}_C - \mathbf{v}_B = \mathbf{v}_B - \mathbf{v}_A$
c. $|\mathbf{v}_C - \mathbf{v}_A| = 2|\mathbf{v}_B - \mathbf{v}_C|$ d. $|\mathbf{v}_C - \mathbf{v}_A| = 4|\mathbf{v}_B|$

18 **Statement I** Two cylinders, one hollow (metal) and the other solid (wood) with the same mass and identical dimensions are simultaneously allowed to roll without slipping down an inclined plane from the same height. The hollow cylinder will reach the bottom of the inclined plane first.

Statement II By the principle of conservation of energy, the total kinetic energies of both the cylinders are identical when they reach the bottom of the incline. **(2008, Statement Type)**

a. Both Statement I and Statement II are correct and Statement II is the correct explanation of Statement I
b. Both Statement I and Statement II are correct but Statement II is not the correct explanation of Statement I
c. Statement I is correct but Statement II is incorrect
d. Statement II is correct but Statement I is incorrect

Passage for Q. Nos. (19-21)

A uniform thin cylindrical disc of mass M and radius R is attached to two identical massless springs of spring constant k which are fixed to the wall as shown in the figure. The springs are attached to the axle of the disc diammetrically on either side at a distance d from its centre.

The axle is massless and both the springs and the axle are in a horizontal plane. The unstretched length of each spring is L.

The disc is initially at its equilibrium position with its centre of mass at a distance L from the wall. The disc rolls without slipping with velocity $\mathbf{v}_0 = v_0\hat{\mathbf{i}}$. The coefficient of friction is μ.

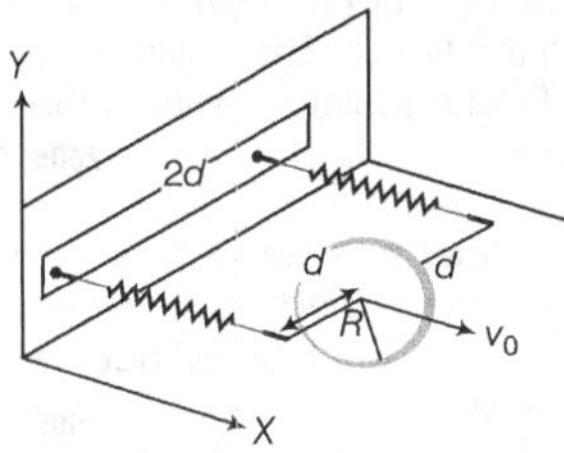

(2008, Comprehension Type)

19 The net external force acting on the disc when its centre of mass is at displacement x with respect to its equilibrium position is

a. $-kx$ b. $-2kx$
c. $-\frac{2kx}{3}$ d. $-\frac{4kx}{3}$

20 The centre of mass of the disc undergoes simple harmonic motion with angular frequency ω equal to

a. $\sqrt{\frac{k}{M}}$ b. $\sqrt{\frac{2k}{M}}$
c. $\sqrt{\frac{2k}{3M}}$ d. $\sqrt{\frac{4k}{3M}}$

21 The maximum value of v_0 for which the disc will roll without slipping is

a. $\mu g\sqrt{\frac{M}{k}}$ b. $\mu g\sqrt{\frac{M}{2k}}$
c. $\mu g\sqrt{\frac{3M}{k}}$ d. $\mu g\sqrt{\frac{5M}{2k}}$

Passage for Q. Nos. (22-24)

Two discs A and B are mounted coaxially on a vertical axle. The discs have moments of inertia I and $2I$ respectively about the common axis. Disc A is imparted an initial angular velocity 2ω using the entire potential energy of a spring compressed by a distance x_1. Disc B is imparted an angular velocity ω by a spring having the same spring constant and compressed by a distance x_2. Both the discs rotate in the clockwise direction. **(2007, Comprehension Type)**

22 The ratio $\frac{x_1}{x_2}$ is

a. 2 b. 1/2
c. $\sqrt{2}$ d. $\frac{1}{\sqrt{2}}$

23 When disc B is brought in contact with disc A, they acquire a common angular velocity in time t. The average frictional torque on one disc by the other during this period is

a. $\frac{2I\omega}{3t}$ b. $\frac{9I\omega}{2t}$
c. $\frac{9I\omega}{4t}$ d. $\frac{3I\omega}{2t}$

24 The loss of kinetic energy during the above process is

a. $\frac{I\omega^2}{2}$ b. $\frac{I\omega^2}{3}$
c. $\frac{I\omega^2}{4}$ d. $\frac{I\omega^2}{6}$

25 A small object of uniform density rolls up a curved surface with an initial velocity v. It reaches up to a maximum height of $\frac{3v^2}{4g}$ with respect to the initial position. The object is

(2007, Only One Option Correct Type)

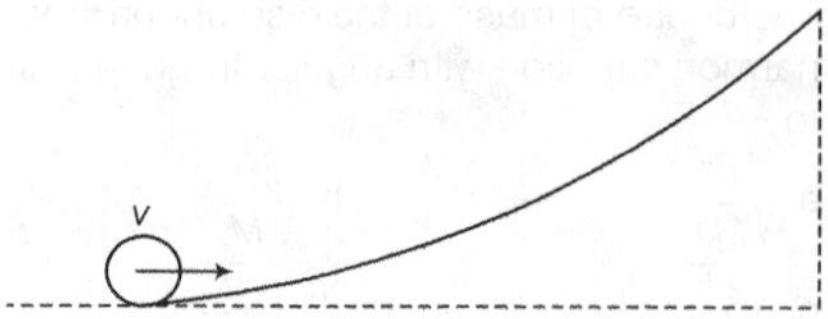

a. ring
b. solid sphere
c. hollow sphere
d. disc

26 A ball moves over a fixed track as shown in the figure. From A to B, the ball rolls without slipping. If surface BC is frictionless and K_A, K_B and K_C are kinetic energies of the ball at A, B and C respectively, then

(2006, Only One Option Correct Type)

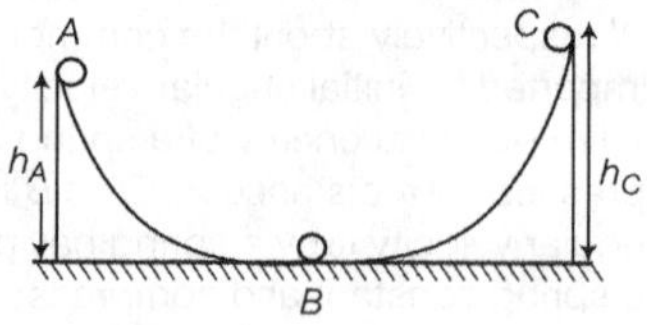

a. $h_A > h_C$; $K_B > K_C$
b. $h_A > h_C$; $K_C > K_A$
c. $h_A = h_C$; $K_B = K_C$
d. $h_A < h_C$; $K_B > K_C$

27 A solid sphere of radius R has moment of inertia I about its geometrical axis. It is melted into a disc of radius r and thickness t. If its moment of inertia about the tangential axis (which is perpendicular to plane of the disc), is also equal to I, then the value of r is equal to **(2006, Only One Option Correct Type)**

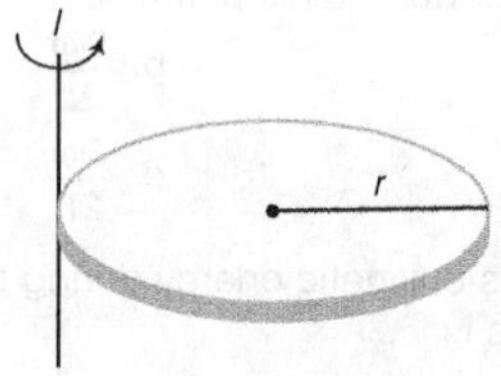

a. $\frac{2}{\sqrt{15}}R$ b. $\frac{2}{\sqrt{5}}R$
c. $\frac{3}{\sqrt{15}}R$ d. $\frac{\sqrt{3}}{\sqrt{15}}R$

28 A solid sphere is in pure rolling motion on an inclined surface having inclination θ. Which of the following statement(s) is/are correct?

(2006, One or More than One Options Correct Type)

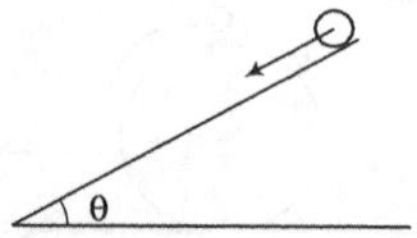

a. Frictional force acting on sphere is $f = \mu mg \cos\theta$
b. f is dissipative force
c. Friction will increase its angular velocity and decrease its linear velocity
d. If θ decreases, friction will decrease

29 From a circular disc of radius R and mass $9M$, a small disc of radius $R/3$ is removed from the disc. The moment of inertia of the remaining disc about an axis perpendicular to the plane of the disc and passing through O is

(2005, Only One Option Correct Type)

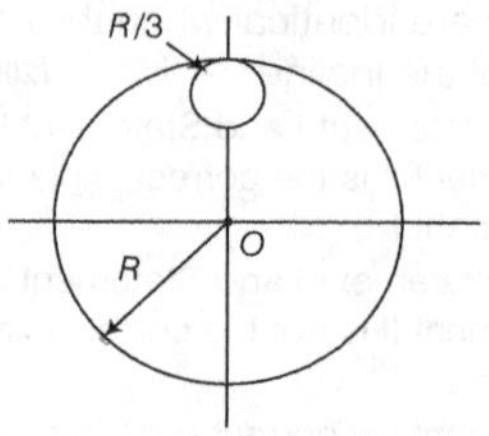

a. $4MR^2$ b. $\frac{40}{9}MR^2$
c. $10MR^2$ d. $\frac{37}{9}MR^2$

30 A particle moves in a circular path with decreasing speed. Choose the correct statement. **(2005, Only One Option Correct Type)**

a. Angular momentum remains constant
b. Acceleration (**a**) is towards the centre
c. Particle moves in a spiral path with decreasing radius
d. The direction of angular momentum remains constant

31 A rod of length L and mass M is hinged at point O. A small bullet of mass m hits the rod as shown in the figure. The bullet gets embedded in the rod. Find angular velocity of the system just after impact. **(2005, Subjective Type)**

32 A solid cylinder rolls without slipping on an inclined plane inclined at an angle θ. Find the linear acceleration of the cylinder. Mass of the cylinder is M. **(2005, Subjective Type)**

3. Gravitation

DPP

DAILY PRACTICE PROBLEMS

DPP-1 Newton's Universal Law of Gravitation and Gravitational Field

Only One Option Correct Type

Working Space

Direction (Q. Nos. 1-9) *This section contains 9 multiple choice questions. Each question has four choices (a), (b), (c) and (d), out of which ONLY ONE is correct.*

1 Spheres of the same material and same radius R are touching each other. The gravitational force between them is proportional to

a. R b. R^3

c. R^4 d. None of these

2 The field strength at a point along the axis of a thin rod of length L and mass m at a distance r from one end is

a. $\dfrac{Gm}{r(L+r)}$ b. $\dfrac{2Gm}{r(L+r)}$

c. $\dfrac{Gm}{2r(L+r)}$ d. $\dfrac{Gm}{\sqrt{2}r(L+r)}$

3 Mass m is distributed uniformly along a wire of length $2L$. A particle of mass m_0 is at a point that is at a distance r above the centre on its perpendicular bisector (point P in the figure). The gravitational force on the particle is

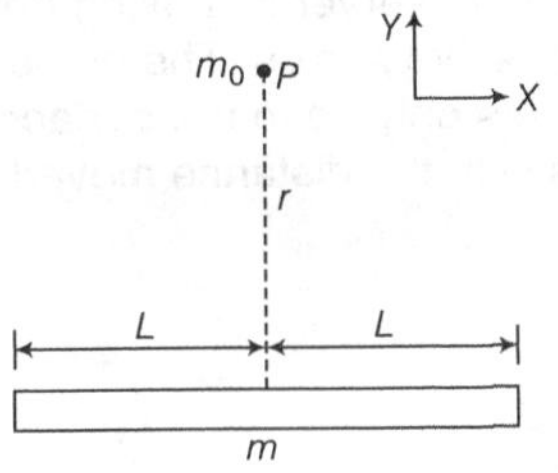

a. $-\dfrac{Gmm_0}{2r\sqrt{r^2+L^2}}\hat{\mathbf{j}}$ b. $-\dfrac{Gmm_0}{r\sqrt{r^2+L^2}}\hat{\mathbf{j}}$

c. $\dfrac{Gmm_0}{\sqrt{r^2+L^2}}(-r\hat{\mathbf{j}}+L\hat{\mathbf{i}})$ d. zero

4 The gravitational field at the centre of a uniform thin wire in the form of an arc of radius R is (the arc subtends an angle α at the centre and λ is the mass per unit length)

a. $\dfrac{\sqrt{2}G\lambda}{R}\sin\left(\dfrac{\alpha}{2}\right)$ b. $\dfrac{G\lambda}{\sqrt{2}R}\sin\left(\dfrac{\alpha}{2}\right)$

c. $\dfrac{(1+\sqrt{2})G\lambda}{R}\sin\left(\dfrac{\alpha}{2}\right)$ d. $\dfrac{2G\lambda}{R}\sin\left(\dfrac{\alpha}{2}\right)$

Working Space

5 For the arrangement shown below, find the magnitude and direction of the net gravitational force acting on the central particle at O, with respect to the axes shown is

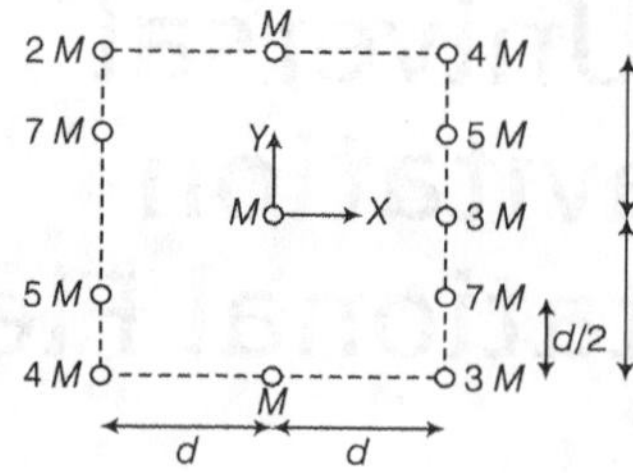

a. $\frac{3GM^2}{d^2}\hat{i}$ b. $\frac{3GM^2}{d^2}\hat{j}$ c. $\frac{3GM^2}{d^2}(\hat{i}+\hat{j})$ d. $\frac{3GM^2}{d^2}(-\hat{i}-\hat{j})$

6 A sphere of radius R has its centre at the origin. It has a uniform mass density ρ_0 except that there is a spherical hole of radius $r = \frac{R}{2}$ whose centre is at $x = \frac{1}{2}R$ as shown in the figure. The magnitude of gravitational field at a point distant x on the X-axis, where $x > R$ is

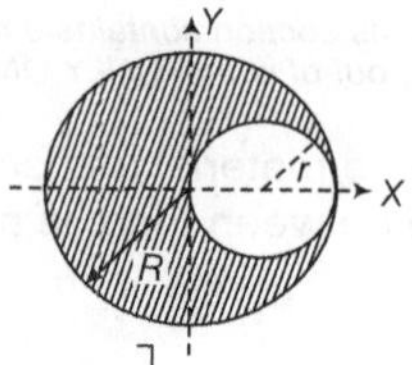

a. $\frac{4G\pi R^3\rho_0}{3}\left[\frac{1}{x^2} - \frac{1}{8\left(x-\frac{R}{2}\right)^2}\right]$ b. $\frac{4G\pi R^3\rho_0}{3}\left[\frac{2}{x^2} - \frac{1}{8\left(x-\frac{R}{2}\right)^2}\right]$

c. $\frac{2G\pi R^3\rho_0}{3}\left[\frac{1}{x^2} - \frac{1}{8\left(x-\frac{R}{2}\right)^2}\right]$ d. $\frac{G\pi R^3\rho_0}{3}\left[\frac{1}{x^2} - \frac{1}{8\left(x-\frac{R}{2}\right)^2}\right]$

7 A ring of radius R, mass m and a solid sphere of same mass m and same radius R are placed with their centres on positive X-axis. The observer is moving from some finite distance on negative X-axis towards positive X-axis. The plane of the ring is perpendicular to X-axis. If the observer moves only up to the surface of the solid sphere, then the net gravitational field varies with the distance moved along X-axis as

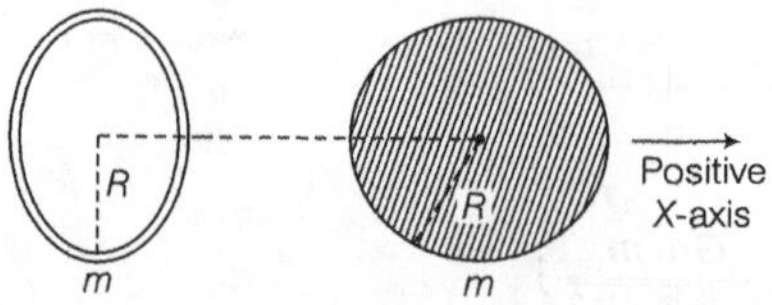

a.
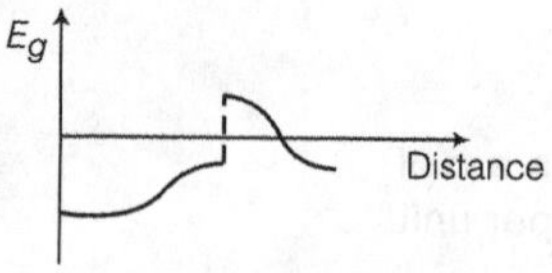

b.
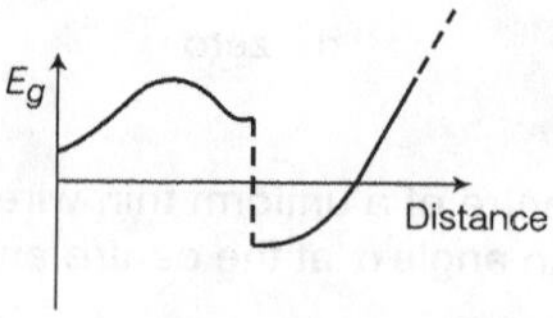

c.
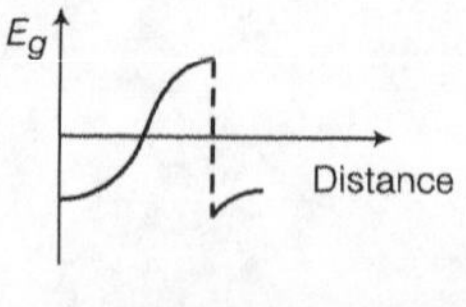

d.
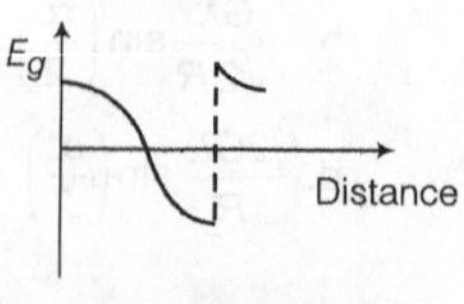

Working Space

8 The gravitational field in a region is given by $\mathbf{E}_g = (4\hat{\mathbf{i}} + \hat{\mathbf{j}})\,\text{Nkg}^{-1}$. Work done by this field is zero when a particle is moved along the line. Choose the correct equation.

a. $y + 4x = 2$ b. $4y + x = 6$
c. $x + y = 5$ d. None of these

9 Three objects, each of mass m are at the vertices of an equilateral triangle and these vertices are on the circumference of a circle of radius R. These three objects are rotating on the circle under their mutual gravitational force of attraction. What is the speed of each mass?

a. $\left(\frac{Gm}{\sqrt{5}R}\right)^{1/2}$ b. $\left(\frac{Gm}{\sqrt{6}R}\right)^{1/2}$
c. $\left(\frac{Gm}{\sqrt{2}R}\right)^{1/2}$ d. $\left(\frac{Gm}{\sqrt{3}R}\right)^{1/2}$

One or More than One Options Correct Type

Direction (Q. Nos. 10-14) *This section contains 5 multiple choice questions. Each question has four choices (a), (b), (c) and (d), out of which ONE or MORE THAN ONE are correct.*

10 Three particles, each of mass m are placed at the three corners of an equilateral triangle of side L. The force exerted by the system on a particle of mass m_0 is F. Then,

a. $F =$ zero, if m_0 is at the centre of the triangle
b. $F = \frac{2Gmm_0}{3L^2}$, if m_0 is at the centre of the triangle
c. $F = \frac{2Gmm_0}{3L^2}$, if m_0 is at the mid-point of a side
d. $F = \frac{4Gmm_0}{3L^2}$, if m_0 is at the mid-point of a side

11 There are two concentric shells of masses m_1 and m_2 and radii R_1 and R_2 $(R_2 > R_1)$. The force on a particle of mass m_0 kept at a distance r from the common centre is F. Then,

a. $F = \frac{G(m_1 + m_2)m_0}{r^2}$, when $r > R_2$
b. $F = \frac{Gm_1m_0}{r^2}$, when $R_1 < r < R_2$
c. $F = 0$, when $r < R_1$
d. $F = \frac{Gm_2m_1}{r^2}$

12 The magnitudes of gravitational field at distances r_1 and r_2 from the centre of a uniform sphere of radius R and mass m are E_{g_1} and E_{g_2}, respectively. Then,

a. $\frac{E_{g_1}}{E_{g_2}} = \frac{r_1}{r_2}$ if $r_1 < R$ and $r_2 < R$
b. $\frac{E_{g_1}}{E_{g_2}} = \frac{r_2^2}{r_1^2}$ if $r_1 > R$ and $r_2 > R$
c. $\frac{E_{g_1}}{E_{g_2}} = \frac{r_1}{r_2}$ if $r_1 > R$ and $r_2 > R$
d. $\frac{E_{g_1}}{E_{g_2}} = \frac{r_2^2}{r_1^2}$ if $r_1 < R$ and $r_2 < R$

Working Space

13 A tunnel is dug along a chord of earth at a perpendicular distance $\frac{R}{2}$ from the earth's centre. The wall of the tunnel may be assumed to be frictionless. A particle is released from one end of the tunnel. The pressing force by the particle on the wall and the acceleration of the particle vary with x (distance of the particle from the centre) as

a. 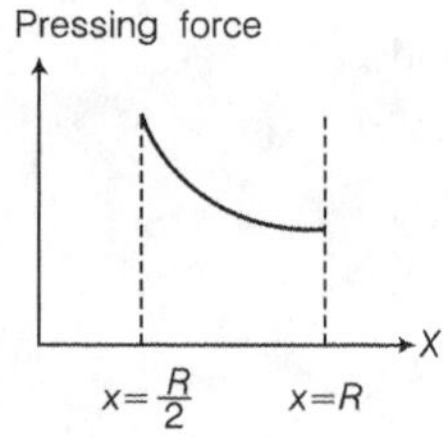

b. 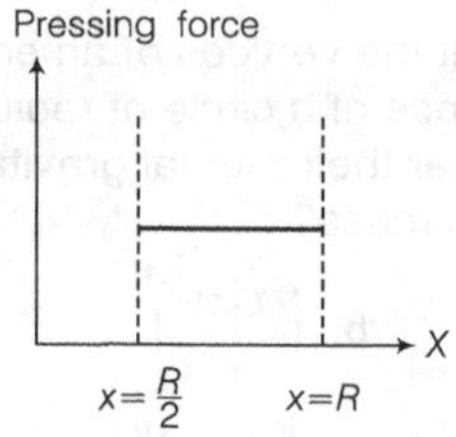

c. 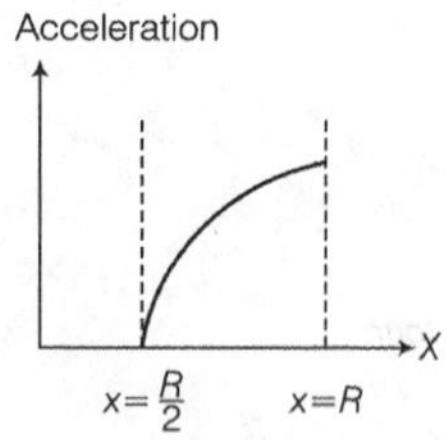

d. 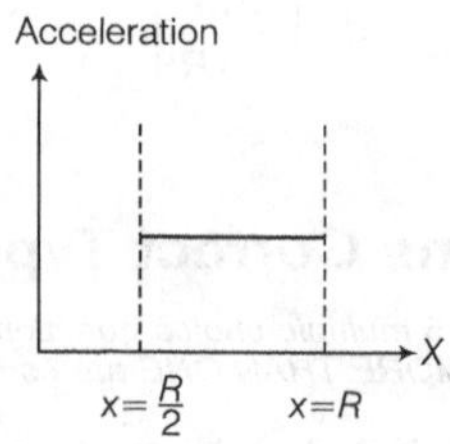

14 Two masses m_1 and m_2 ($m_1 < m_2$) are released from rest from a finite distance. They start moving towards each other under their mutual gravitational attraction. Which of the following statement(s) is/are correct?

a. Acceleration of m_1 is more than that of m_2
b. Acceleration of m_2 is more than that of m_1
c. Centre of mass of system will remain at rest in all the reference frames
d. Total energy of the system remains constant

Statement Type

Direction (Q. No. 15) *This section is based on Statement I and Statement II. Select the correct answer from the codes given below.*

a. Both Statement I and Statement II are correct and Statement II is the correct explanation of Statement I
b. Both Statement I and Statement II are correct but Statement II is not the correct explanation of Statement I
c. Statement I is correct but Statement II is incorrect
d. Statement II is correct but Statement I is incorrect

15 **Statement I** The force of gravitation between sphere of mass M_1 and a rod of mass M_2 is $F = \frac{GM_1M_2}{r^2}$.

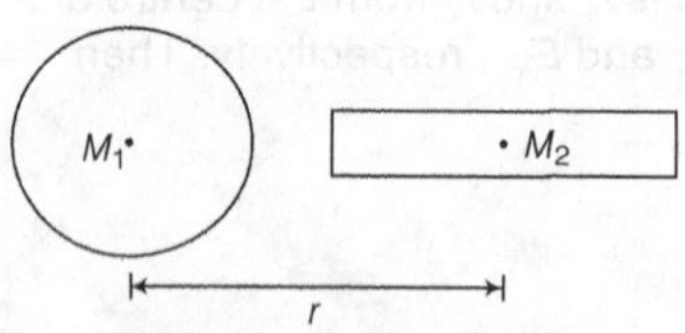

Statement II Newton's law of gravitation holds good for point masses.

One Integer Value Correct Type

Direction (Q. Nos. 16 and 17) This section contains 2 questions. When worked out will result in an integer from 0 to 9 (both inclusive).

16 There are two bodies of masses 10^3 kg and 10^5 kg separated by a distance of 1 km. At a distance $\frac{p}{11}$ km from the smaller body, the intensity of gravitational field is zero. Find the value of p.

17 A mass m is placed at a distance h along the normal through the centre O of a thin circular ring of mass M and radius r. The mass m is removed further away such that OP becomes $2h$.

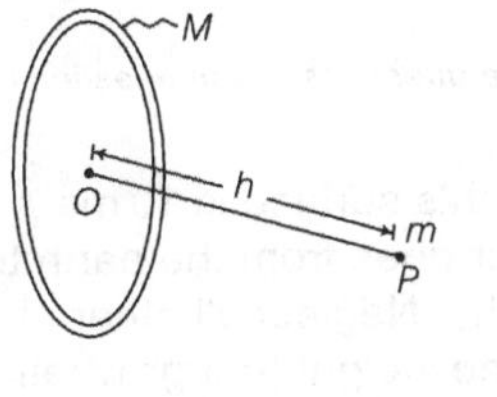

If $h = r$ (radius of ring), then force of gravitation decreases by a factor of $\frac{k\sqrt{10}}{25}$.

Find the value of k.

Comprehension Type

Direction (Q. Nos. 18-20) This section contains a paragraph, which describing theory, experiments, data, etc. Four questions related to the paragraphs have been given. Each question has only one correct answer among the four given options (a), (b), (c) and (d).

Passage

A non-homogeneous sphere of radius R has the following density variation

$$\rho = \rho_0 \text{ for } r \le R/3$$

$$\rho = \frac{\rho_0}{2} \text{ for } \frac{R}{3} < r \le \frac{3R}{4} \text{ and}$$

$$\rho = \frac{\rho_0}{8} \text{ for } \frac{3R}{4} < r \le R$$

18 The gravitational field due to sphere at $r = \frac{R}{4}$ is

a. $\frac{3\pi G R\rho_0}{9}$
b. $\frac{6\pi G R\rho_0}{9}$
c. $\frac{2\pi G R\rho_0}{9}$
d. $\frac{3\pi G R\rho_0}{48}$

19 The gravitational field due to the sphere at $r = \frac{R}{2}$ is

a. $\frac{43\pi G R\rho_0}{50}$
b. $\frac{23\pi G R\rho_0}{100}$
c. $\frac{18\pi G R\rho_0}{25}$
d. $\frac{35\pi G R\rho_0}{81}$

20 The gravitational field due to the sphere at $r = \frac{5R}{6}$ is

a. $0.98\,\pi\, GR\rho_0$
b. $0.48\,\pi\, GR\rho_0$
c. $0.23\,\pi\, GR\rho_0$
d. $0.52\,\pi\, GR\rho_0$

DPP-2 Acceleration due to Gravity

Only One Option Correct Type

Working Space

Direction (Q. Nos. 1-12) This section contains 12 multiple choice questions. Each question has four choices (a), (b), (c) and (d), out of which ONLY ONE is correct.

1 Suppose the acceleration due to gravity at the earth's surface is 10 ms^{-2} and at the surface of mars it is 4 ms^{-2}. A 60 kg passenger goes from the earth to the mars in a spaceship moving with a constant velocity. Neglect all other objects in the sky. Which part of the figure best represents the weight (net gravitational force) of the passenger as a function of time?

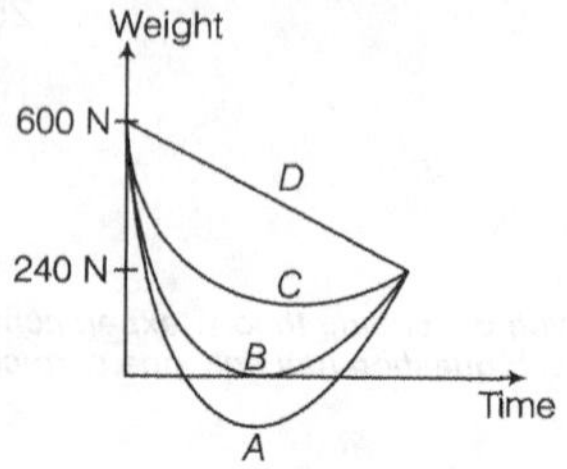

a. A b. B
c. C d. D

2 Let ω be the angular velocity of the earth's rotation about its axis. The acceleration due to the gravity at the surface of earth near the poles is g. An object weighed at the equator gives the same reading as a reading taken at a depth d below earth's surface at a pole ($d << R$). The value of d is

a. $\frac{\omega^2 R^2}{g}$

b. $\frac{\omega^2 R^2}{2g}$

c. $\frac{2\omega^2 R^2}{g}$

d. None of the above

3 Suppose the earth increases its speed of rotation. At what new time period will the weight of a body on the equator becomes zero? (Given, $g = 10\ ms^{-2}$ and radius of the earth, $R = 6400$ km)

a. $\frac{9}{4}\pi$ b. $\frac{4}{9}\pi$
c. $\frac{9}{5}\pi$ d. $\frac{5}{9}\pi$

4 Infinite masses, each of mass 3 kg are placed 1 m, 2 m, 4 m, 8 m, ... marks respectively on X-axis. Acceleration of a mass of 2 kg placed at origin will be

a. $3G$ b. $2G$
c. $4G$ d. ∞

Working Space

5 Rate of change of weight near the earth's surface varies with height h as

a. h b. h^0

c. h^{-1} d. $h^{1/2}$

6 If the radius of the earth were increased by a factor of 2 keeping the mass constant, then the factor by which its density has to be changed to keep g, the same is

a. $\frac{1}{8}$ b. $\frac{1}{4}$

c. 4 d. $\frac{1}{2}$

7 The acceleration due to gravity at the surface of moon is only one-sixth of that of the earth. If the earth and moon are assumed to have same density, then ratio of the radius of the moon to radius of the earth is

a. $\frac{1}{6}$ b. $\frac{1}{(6)^{1/3}}$

c. $\frac{1}{36}$ d. $\frac{1}{(6)^{2/3}}$

8 The radius of the earth is 6400 km and $g = 10\ \text{ms}^{-2}$. In order that the body of 6 kg weighs zero at the equator, the angular speed of the earth should be (in rad s^{-1}) around

a. $\frac{1}{80}$ b. $\frac{1}{400}$

c. $\frac{1}{800}$ d. $\frac{1}{1600}$

9 A straight smooth tunnel is dug through a spherical planet whose mass density ρ_0 is constant. The tunnel passes through the centre of the planet and is perpendicular to the planet's axis of rotation, which is fixed in space. The planet rotates with the angular velocity ω so that objects in the tunnel have no acceleration relative to the tunnel. Find the value of ω.

a. $\sqrt{\frac{4}{3}\pi G\rho_0}$ b. $\sqrt{\frac{2}{3}\pi G\rho_0}$

c. $\sqrt{\frac{1}{3}\pi G\rho_0}$ d. $2\sqrt{\frac{2}{3}\pi G\rho_0}$

10 The density of the core of a planet is ρ_1 and that of the outer shell is ρ_2. The radii of the core and that of the planet are R and $2R$, respectively. If the acceleration due to gravity at the surface is same as at a depth R, then ratio $\frac{\rho_1}{\rho_2}$ is

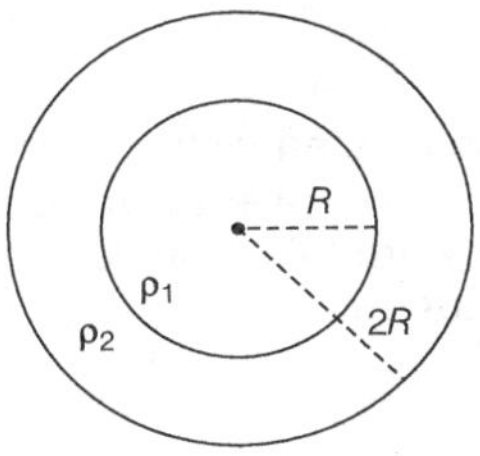

a. $\frac{5}{3}$ b. $\frac{7}{3}$

c. $\frac{7}{5}$ d. 3

Working Space

11 The percentage change in the acceleration of the earth from the situation of total eclipse of the sun to the situation where the moon is on a side of earth directly opposite to the sun is (Given, r_1 = average distance between the sun and earth, r_2 = average distance between the moon and the earth, M_m, M_s, M_e are masses of moon, sun and earth respectively)

a. $\frac{M_s}{M_m}\left(\frac{r_2}{r_1}\right) \times 100$

b. $\frac{M_s}{M_m}\left(\frac{r_2}{r_1}\right)^2 \times 100$

c. $\left(\frac{r_1}{r_2}\right)^2 \frac{M_m}{M_s} \times 100$

d. $2\left(\frac{r_1}{r_2}\right)^2 \frac{M_m}{M_s} \times 100$

12 For an experiment, a boy draws graph between v^2 (on Y-axis) and a^2 (on X-axis) where v = velocity and a = acceleration for a simple pendulum. The graph is found to be a straight line of negative slope making an angle of 30° (with X-axis) when the experiment is performed on the ground and 60° when experiment is done at a height h above the ground. Then, h must be (Given, R = radius of the earth)

a. $0.5R$

b. $0.24R$

c. $0.73R$

d. R

Statement Type

Direction (Q. Nos. 13 and 14) *This section is based on Statement I and Statement II. Select the correct answer from the codes given below.*

a. Both Statement I and Statement II are correct and Statement II is the correct explanation of Statement I

b. Both Statement I and Statement II are correct but Statement II is not the correct explanation of Statement I

c. Statement I is correct but Statement II is incorrect

d. Statement II is correct but Statement I is incorrect

13 **Statement I** If the earth stops rotating about its axis, then the value of acceleration due to gravity will become same at all places on earth.

Statement II The value of acceleration due to gravity is depend on rotational speed of earth.

14 **Statement I** The length of the day is slowly increasing.

Statement II The dominant effect causing a slow down in the rotation of the earth is the viscous force between the earth and the atmosphere around it.

Comprehension Type

Direction (Q. Nos. 15 and 16) *This section contains a paragraph, which describing theory, experiments, data, etc. Two questions related to the paragraph have been given. Each question has only one correct answer among the four given options (a), (b), (c) and (d).*

Passage

A solid sphere of uniform density and radius 4 m is located with its centre at the origin O of the coordinate system (see figure). Two spheres of equal radius 1 m with their cavities at $A(-2, 0, 0)$ and $B(2, 0, 0)$ respectively are taken out leaving behind spherical cavities. The mass of each sphere taken out is M.

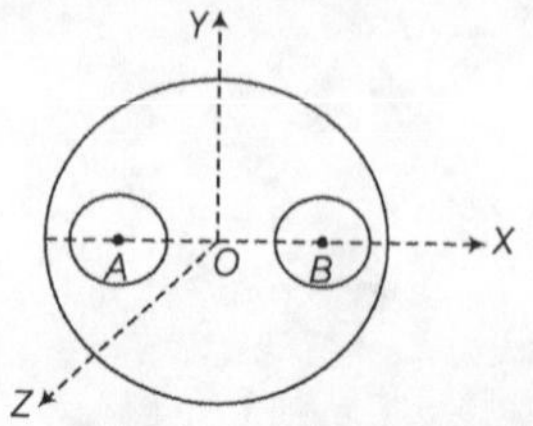

Working Space

15 If a particle of mass m is released at point B, then its acceleration would be

a. GM

b. $\frac{21GM}{5}$

c. $\frac{31}{8}GM$

d. $\frac{31}{16}GM$

16 The gravitational field due to the sphere (with holes) shown at point O is

a. zero

b. $\frac{21GM}{5}$

c. $\frac{7GM}{5}$

d. $\frac{33}{7}GM$

One Integer Value Correct Type

Direction *(Q. Nos. 17-20) This section contains 4 questions. When worked out will result in an integer from 0 to 9 (both inclusive).*

17 If the radius of the earth were to shrink by 2% its mass remaining the same, by how much percentage would the acceleration due to gravity on the earth's surface would increase?

18 In order to simulate different values of g, aspiring astronauts are put on a plane which dives in a parabola given by the equation $x^2 = 500y$, where x is horizontal and y is vertically upwards, both being measured in metre. The x-component of velocity of the plane is constant and has the value of 360 km h^{-1}. The effective value of g experienced by an astronaut on the plane is βg. Find the value of β. (Given, $g = 10\ ms^{-2}$)

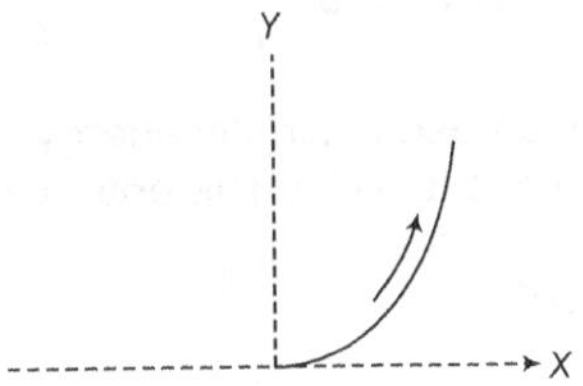

19 The difference between acceleration due to gravity at the poles and equator due to rotational motion of the earth about its own axis is $(p + 30) \times 10^{-3}\ ms^{-2}$. Find the value of p.
(Given, time period of earth = 24 h and radius of earth = 6.4×10^6 m).

20 A smooth tunnel is dug along the radius of earth that end at the centre. A ball is released from the surface of earth along the tunnel. The coefficient of restitution for collision between the centre of the earth and ball is 0.5. The distance travelled by ball till before second collision with the centre is βR. (Given, mass of the earth is M and radius of the earth is R) Find the value of β.

DPP-3 Potential, Potential Energy and Self Energy

Only One Option Correct Type

Working Space

Direction (Q. Nos. 1-9) *This section contains 9 multiple choice questions. Each question has four choices (a), (b), (c) and (d), out of which ONLY ONE is correct.*

1 Two spheres of masses m_1 and m_2 are placed at a distance d apart. The potential at the position, where the gravitational field due to them is zero, is

a. $-\frac{G}{2d}(m_1 + m_2 + 2\sqrt{m_1 m_2})$
b. $-\frac{G}{2d}(m_1 + m_2 - 2\sqrt{m_1 m_2})$
c. $-\frac{G}{d}(2m_1 + m_2 + 2\sqrt{m_1 + m_2})$
d. $-\frac{G}{d}(m_1 + m_2 + 2\sqrt{m_1 m_2})$

2 The gravitational potential of two homogeneous spherical shells A and B of same surface density at their respective centres are in the ratio 3 : 4. If the two shells coalesce into single one such that surface density remains same, then the ratio of potential at an internal point of the new shell to that of shell A is equal to

a. $\frac{3}{2}$
b. $\frac{5}{3}$
c. $\frac{4}{3}$
d. $\frac{5}{7}$

3 A person brings a mass of 2 kg from A to B. The increase in kinetic energy of the mass is 4 J and work done by the person on the mass is -10 J. The potential difference between B and A (i.e. $V_B - V_A$) is

a. $-7\ \text{Jkg}^{-1}$
b. $-3\ \text{Jkg}^{-1}$
c. $3\ \text{Jkg}^{-1}$
d. $7\ \text{Jkg}^{-1}$

4 What is the gravitational self energy of a thin hollow sphere of mass M and radius R?

a. $-\frac{GM^2}{R}$
b. $-\frac{GM^2}{2R}$
c. $-\frac{GM^2}{\sqrt{2}R}$
d. $-\frac{2GM^2}{R}$

5 What is the gravitational self energy of a solid sphere of mass M and radius R?

a. $-\frac{3}{5}\cdot\frac{GM^2}{R}$
b. $-\frac{2}{5}\cdot\frac{GM^2}{R}$
c. $-\frac{1}{5}\cdot\frac{GM^2}{R}$
d. $-\frac{4}{5}\cdot\frac{GM^2}{R}$

6 Find the gravitational potential energy of a point mass m_0 and a thin uniform rod of mass m_0 and length l, if they are located along a straight line at a distance d measured from the centre of the rod.

a. $-\frac{Gm_0^2}{l}\log_e\left(\frac{d+r}{d-r}\right)$
b. $-\frac{Gm_0^2}{l}\log_e\left(\frac{d+2l}{d-2l}\right)$
c. $-\frac{Gm_0^2}{l}\log_e\left(\frac{2d+l}{2d-l}\right)$
d. $-\frac{Gm_0^2}{2l}\log_e\left(\frac{d+l}{d-l}\right)$

7 A particle of mass m_0 is placed at the centre of a fixed, uniform semicircular ring of radius R and mass m as shown in the figure. The work required to displace the particle from centre of the ring to infinity is

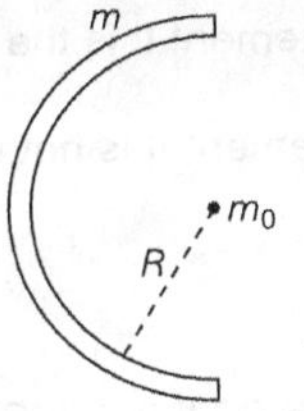

a. $\frac{Gm m_0}{2R}$ b. $\frac{Gm m_0}{R}$ c. $\frac{Gm m_0}{\sqrt{2}R}$ d. $\frac{2Gm m_0}{R}$

8 A particle of mass m_0 is transferred from the centre of the base of a uniform solid hemisphere of mass M and radius R to infinity. The work done in this process by the gravitational force exerted on the particle by the hemisphere, is

a. $-\frac{3}{2}\cdot\frac{GMm_0}{R}$ b. $-\frac{5}{2}\cdot\frac{Gm m_0}{R}$ c. $-\frac{Gm m_0}{2R}$ d. None of these

9 Three solid spheres, each of mass m and radius R are released from the position shown in the figure. The speed of any sphere at the time of collision will be

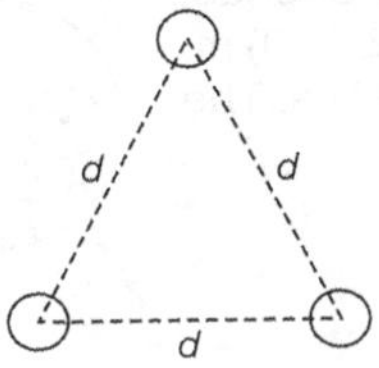

a. $\sqrt{Gm\left(\frac{1}{d}-\frac{3}{R}\right)}$ b. $\sqrt{Gm\left(\frac{3}{d}-\frac{1}{R}\right)}$

c. $\sqrt{Gm\left(\frac{2}{R}-\frac{1}{d}\right)}$ d. $\sqrt{Gm\left(\frac{1}{R}-\frac{2}{d}\right)}$

One or More than One Options Correct Type

Direction (Q. No. 10) This section contains 1 multiple choice question. This question has four choices (a), (b), (c) and (d), out of which ONE or MORE THAN ONE are correct.

10 There are two thin wire rings, each of radius R, whose axes coincide. The masses of the rings are m_1 and m_2. Their centres are separated by a distance d. Which of the following statement(s) is/are correct?

a. The gravitational potential difference between centres is

$$\frac{G(m_1-m_2)}{2}\left(\frac{1}{R}-\frac{1}{\sqrt{R^2+d^2}}\right)$$

b. The gravitational potential difference between the centres is

$$G(m_1-m_2)\left(\frac{1}{R}-\frac{1}{\sqrt{R^2+d^2}}\right)$$

c. The work done to move a mass m_0 from centre of 1st ring to centre of 2nd ring is

$$\frac{Gm_0(m_1-m_2)}{2}\left(\frac{1}{R}-\frac{1}{\sqrt{R^2+d^2}}\right)$$

d. The work done to move a mass m_0 from centre of 1st ring to centre of 2nd ring is

$$Gm_0(m_1-m_2)\left(\frac{1}{R}-\frac{1}{\sqrt{R^2+d^2}}\right)$$

Statement Type

Working Space

Direction (Q. No. 11) *This section is based on Statement I and Statement II. Select the correct answer from the codes given below.*

a. Both Statement I and Statement II are correct and Statement II is the correct explanation of Statement I
b. Both Statement I and Statement II are correct but Statement II is not the correct explanation of Statement I
c. Statement I is correct but Statement II is incorrect
d. Statement II is correct but Statement I is incorrect

11 **Statement I** Gravitational potential of earth at every place on it is negative.

Statement II Everybody on earth is bound by the attraction of the earth.

Comprehension Type

Direction (Q. Nos. 12-16) *This section contains 2 paragraphs, each describing theory, experiments, data, etc. Five questions related to the paragraphs have been given. Each question has only one correct answer among the four given options (a), (b), (c) and (d).*

Passage I

A solid sphere of uniform density and radius 4 m is located with its centre at the origin O of coordinates. Two spheres of equal radius 1m with their cavities at $A(-2, 0, 0)$ m and $B(2, 0, 0)$ m, respectively are taken out leaving behind spherical cavities. The mass of each sphere taken out is M.

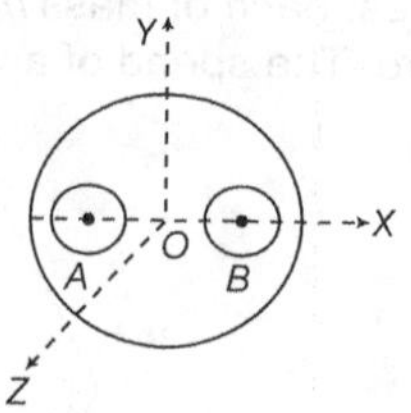

12 The gravitational potential at any point on the circle $y^2 + z^2 = 36$ is

a. $GM\left(\frac{\sqrt{10}}{100} - \frac{32}{3}\right)$ b. $GM\left(\frac{\sqrt{5}}{100} - \frac{32}{3}\right)$

c. $GM\left(\frac{\sqrt{10}}{50} - \frac{32}{3}\right)$ d. $-GM\left(\frac{32}{3}\right)$

13 The gravitational potential at any point on the circle $y^2 + z^2 = 4$ is

a. $\left(\frac{\sqrt{2}}{2} - 22\right)GM$ b. $\left(\frac{\sqrt{2}}{2} - 11\right)GM$

c. $\left(\frac{\sqrt{2}}{2} + 22\right)GM$ d. $-22\,GM$

Passage II

A solid sphere of mass M and radius R is surrounded by a spherical shell of same mass M and radius $2R$ as shown in the figure. A small particle of mass m_0 is released from rest from a height $h(<< R)$ above the shell. There is a hole in the shell.

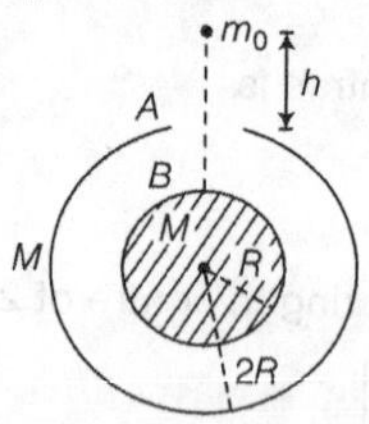

14 The time taken by the particle just to enter the hole at A is

a. $2\sqrt{\frac{hR^2}{GM}}$ b. $\sqrt{\frac{2hR^2}{GM}}$ c. $\sqrt{\frac{hR^2}{GM}}$ d. $\sqrt{\frac{hR^2}{GM^3}}$

Working Space

15 The time taken by the particle to reach the surface of solid sphere from the hole t_{AB} is

a. $\dfrac{R^2}{\sqrt{GMh}} = t_{AB}$ b. $\dfrac{R^2}{\sqrt{GMh}} > t_{AB}$

c. $\dfrac{R^2}{\sqrt{GMh}} < t_{AB}$ d. particle never reaches solid sphere

16 The approximate speed with which the particle hits the solid sphere is

a. $\sqrt{\dfrac{2GM}{R}}$ b. $\sqrt{\dfrac{GM}{2R}}$

c. $\sqrt{\dfrac{3GM}{2R}}$ d. $\sqrt{\dfrac{GM}{R}}$

Matching List Type

Direction (Q. Nos. 17-19) Choices for the correct combination of elements from Column I and Column II are given as options (a), (b), (c) and (d), out of which one is correct.

17 $P(r)$ is magnitude of physical quantity as a function of r (distance from centre of spherical distribution of radius R). Match the following columns and select the correct option from the codes given below.

	Column I		Column II
i.	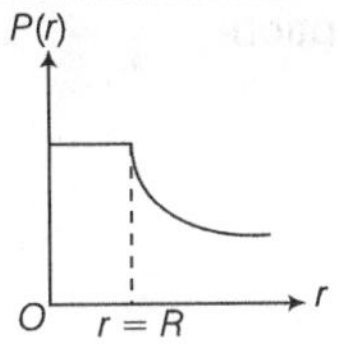	p.	$P(r)$ is variation of gravitational potential for a uniform volumetric spherical distribution of mass.
ii.	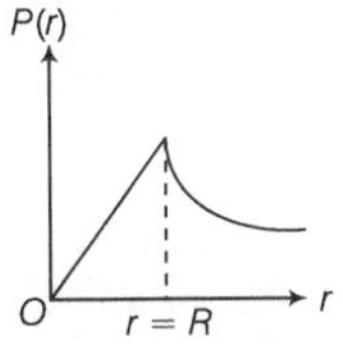	q.	$P(r)$ is variation of gravitational potential for a uniform spherical shell.
iii.	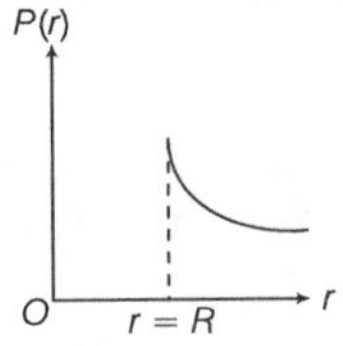	r.	$P(r)$ is variation of gravitational field intensity for a uniform volumetric spherical distribution of mass.
iv.	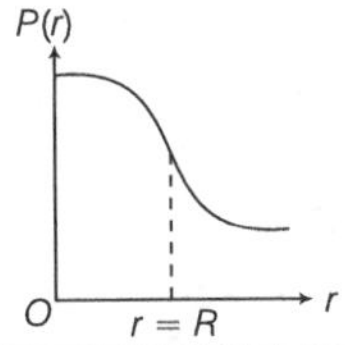	s.	$P(r)$ is variation of gravitational field intensity for a uniform spherical shell.

Codes

	i	ii	iii	iv
a.	s	q	p	r
b.	p	q	s	r
c.	q	r	s	p
d.	r	s	q	p

Working Space

18 Two concentric spherical shells are as shown in the figure.

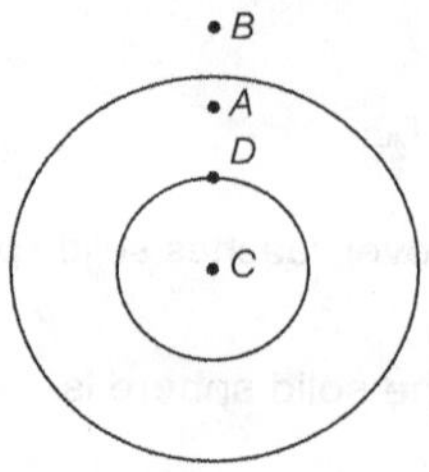

Match the following columns and select the correct option from the codes given below.

	Column I		Column II
i.	Potential at A	p.	Greater than B
ii.	Gravitational field at A	q.	Less than B
iii.	As one moves from C to D	r.	Potential remains constant
iv.	As one moves from D to A	s.	Gravitational field decreases
		t.	None of the above

Codes

	i	ii	iii	iv		i	ii	iii	iv
a.	q	t	r	s	b.	q	p	s	t
c.	r	q	p	s	d.	t	s	r	q

19 On the surface of earth, the gravitational field is E_g and gravitational potential is V (R = radius of earth). Match the following columns and select the correct option from the codes given below.

	Column I		Column II
i.	At a height $h = R$, value of E_g	p.	Decreases by a factor $\frac{1}{4}$
ii.	At a depth $d = \frac{R}{2}$, value of E_g	q.	Decreases by a factor $\frac{1}{2}$
iii.	At a height $h = R$, value of V	r.	Increases by a factor $\frac{11}{8}$
iv.	At a depth $d = \frac{R}{2}$, value of V	s.	Increases by a factor 2
		t.	Decreases by a factor $\frac{11}{8}$

Codes

	i	ii	iii	iv		i	ii	iii	iv
a.	q	s	r	p	b.	r	q	p	s
c.	p	q	s	t	d.	q	p	t	s

One Integer Value Correct Type

Direction (Q. No. 20) *This section contains 1 question. When worked out will result in an integer from 0 to 9 (both inclusive).*

20 Inside a fixed sphere of radius R and uniform density ρ, there is a spherical cavity of radius $\frac{R}{2}$ such that, the surface of cavity passes through the centre of the sphere as shown in the figure. A particle of mass m_0 is released from rest at centre B of the cavity. The velocity with which particle strikes the centre A of sphere is $\frac{\sqrt{2\pi G\rho R^2}}{\sqrt{\beta}}$. Find the value of β.

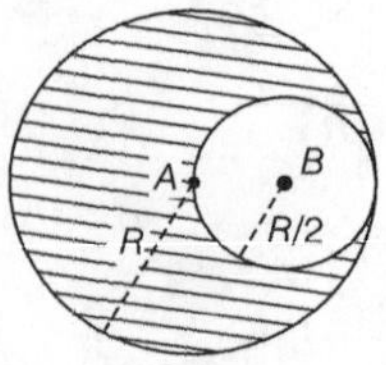

DPP-4 Escape Velocity and Binding Energy

Only One Option Correct Type

Working Space

Direction *(Q. Nos. 1-5) This section contains 5 multiple choice questions. Each question has four choices (a), (b), (c) and (d), out of which ONLY ONE is correct.*

1 A projectile is fired straight up from the surface of the earth with an initial velocity v_i. Neglecting air resistance, the ratio of radius of the earth and the maximum height up to which the projectile rises is

a. $1-\dfrac{v_i^2}{2gR}$

b. $\dfrac{v_i^2}{2gR}$

c. $\dfrac{v_i^2}{2mgR}$

d. $\dfrac{v_i^2}{mgR}$

2 An object is released from rest at an altitude h above the surface of earth's mass M_E radius R_E. Its speed at a distance r from centre of earth is

a. $\sqrt{2gh}$

b. $2\sqrt{gh}$

c. $\sqrt{2GM_E\left(\dfrac{1}{r}-\dfrac{1}{R_E+h}\right)}$

d. $2\sqrt{GM_E\left(\dfrac{1}{R_E+h}-\dfrac{1}{r}\right)}$

3 The masses and radii of the earth and the moon are M_1, R_1 and M_2, R_2, respectively. Their centres are at a distance d apart. What is the minimum speed with which a particle of mass m, should be projected from a point mid-way between the two centres so as to escape to infinity?

a. $\sqrt{\dfrac{G(M_1+M_2)}{d}}$

b. $2\sqrt{\dfrac{G(M_1+M_2)}{d}}$

c. $\sqrt{\dfrac{G(M_1+M_2)}{2d}}$

d. $\sqrt{\dfrac{2G(M_1+M_2)}{d}}$

4 Velocity of a point on the equator of a rotating spherical planet is v. The angular velocity of the planet is such that, the value of g at the equator is half of g at the pole. Determine the escape velocity for a polar particle on the planet as a function of v.

a. $6v$

b. $2v$

c. $4v$

d. $3v$

5 A body is projected vertically upwards from the surface of the earth with a velocity sufficient to carry it to infinity. The time taken by it to reach height h is

(Given, R = radius of the earth)

a. $\dfrac{1}{3}\sqrt{\dfrac{2R}{g}}\left[\left(1+\dfrac{h}{R}\right)^{3/2}-1\right]$

b. $\sqrt{\dfrac{R}{g}}\left[\left(1+\dfrac{h}{R}\right)^{3/2}-1\right]$

c. $\dfrac{1}{2}\sqrt{\dfrac{R}{g}}\left[\left(1+\dfrac{h}{R}\right)^{3/2}-1\right]$

d. $\dfrac{1}{3}\sqrt{\dfrac{R}{g}}\left[\left(1+\dfrac{h}{R}\right)^{3/2}-1\right]$

One or More than One Options Correct Type

Working Space

Direction (Q. Nos. 6-8) *This section contains 3 multiple choice questions. Each question has four choices (a), (b), (c) and (d), out of which ONE or MORE THAN ONE are correct.*

6 An orbiting satellite will escape if

a. its speed is made $\sqrt{1.5}$ times of its initial value
b. its speed is increased by 41%
c. its kinetic energy is doubled
d. it stops moving in the orbit

7 Two spherical planets P_1 and P_2 have the same uniform density ρ, masses M_1 and M_2 and surface areas A and $4A$, respectively. A spherical planet P_3 also has uniform density ρ and its mass is $(M_1 + M_2)$. The escape velocities from the planets P_1, P_2 and P_3 are v_1, v_2 and v_3, respectively. Then,

a. $v_2 > v_3 > v_1$
b. $v_3 > v_2 > v_1$
c. $\frac{v_3}{v_1} = 3$
d. $\frac{v_1}{v_2} = \frac{1}{2}$

8 Two spherical planets have the same mass but their densities are in the ratio 1:8. For the planets,

a. the accelerations due to gravity will be in the ratio 4 : 1
b. the accelerations due to gravity will be in the ratio 1 : 4
c. the escape velocities from their surfaces will be in the ratio $\sqrt{2}:1$
d. the escape velocities from their surfaces will be in the ratio $1:\sqrt{2}$

Statement Type

Direction (Q. Nos. 9-11) *This section is based on Statement I and Statement II. Select the correct answer from the codes given below.*

a. Both Statement I and Statement II are correct and Statement II is the correct explanation of Statement I
b. Both Statement I and Statement II are correct but Statement II is not the correct explanation of Statement I
c. Statement I is correct but Statement II is incorrect
d. Statement II is correct but Statement I is incorrect

9 **Statement I** The value of escape velocity from the surface of earth at 30° is $v_1 = 2v_e$ and at 60° is $v_2 = \frac{2}{\sqrt{3}}v_e$.

Statement II The value of total initial energy of a projected particle is independent of angle of projection.

10 **Statement I** Escape velocity from surface of a particular planet is v. If a tunnel is made from the surface, then escape velocity from a point inside the tunnel must be greater than v.

Statement II Gravitational force is a conservative force.

11 **Statement I** The escape speed of a particle of mass m is v_e. The escape speed of another particle of mass $2m$ for the same planet is v_e.

Statement II The escape speed of a body for a given planet is independent of mass of the body.

Matching List Type

Direction (Q. No. 12) *Choices for the correct combination of elements from Column I and Column II are given as options (a), (b), (c) and (d), out of which one is correct.*

12 Consider a planet of mass M and radius R. A satellite is launched from a height $h(<< R)$ from its surface with a speed v. For different values of v. Match the corresponding paths taken by the satellite.

	Column I		Column II
i.	$v > \sqrt{\frac{2GM}{R}}$	p.	Ellipse
ii.	$v = \sqrt{\frac{2GM}{R}}$	q.	Hyperbola
iii.	$v < \sqrt{\frac{2GM}{R}}$	r.	Circle
iv.	$v = \sqrt{\frac{GM}{R}}$	s.	Parabola

Codes

	i	ii	iii	iv
a.	r	s	q	p
b.	q	s	p	r
c.	p	r	q	s
d.	s	q	p	r

Comprehension Type

Direction (Q. Nos. 13-15) *This section contains a paragraph, which describing theory, experiments, data, etc. Three questions related to the paragraph have been given. Each question has only one correct answer among the four given options (a), (b), (c) and (d).*

Passage

If a tunnel is dug across the earth passing through centre and a particle is dropped from the surface, it performs SHM inside tunnel with time period given by $T = 2\pi\sqrt{\frac{R}{g}}$ where, R = radius of earth.

13 Assuming that a tunnel is dug across the earth passing through its centre. Find the time, a particle takes to reach the centre of the earth, if it is projected into the tunnel from the surface of earth with speed needed for it to escape the gravitational field of earth.

a. $\left[\sin^{-1}\left(\frac{1}{\sqrt{3}}\right)\right]\sqrt{\frac{R}{2g}}$
b. $\left[\sin^{-1}\left(\frac{1}{\sqrt{3}}\right)\right]\sqrt{\frac{2R}{g}}$
c. $\left[\sin^{-1}\left(\frac{1}{\sqrt{3}}\right)\right]\sqrt{\frac{R}{g}}$
d. None of these

14 If a particle is dropped from a height $h (h << R)$ into the tunnel, then choose the correct option.

a. particle will perform SHM
b. particle will not perform SHM
c. particle will perform periodic motion
d. Both (b) and (c)

15 For the above problem, find the time period of motion.

a. $2\pi\sqrt{\frac{R}{g}} + 4\sqrt{\frac{2h}{g}}$
b. $2\pi\sqrt{\frac{R}{g}} + 2\sqrt{\frac{2h}{g}}$
c. $\pi\sqrt{\frac{R}{g}} + 2\sqrt{\frac{2h}{g}}$
d. $\pi\sqrt{\frac{R}{g}} + 2\sqrt{\frac{h}{g}}$

One Integer Value Correct Type

Working Space

Direction (Q. No. 16-20) *This section contains 5 questions. When worked out will result in an integer from 0 to 9 (both inclusive).*

16 A particle is fixed vertically upwards with a speed of 15 kms^{-1}. The speed of the particle (in kms^{-1}), when it goes out of the earth's gravitational pull is 5α. Find the value of α. [Given, $v_e^2 = 125\ (kms^{-1})^2$]

17 Gravitational acceleration on the surface of a planet is $\frac{\sqrt{6}}{11}g$, where g is the acceleration due to gravity on the surface of earth. The average mass density of the planet is $\frac{2}{3}$ times that of the earth. If the escape speed on the surface of the earth is taken to be 11kms^{-1}, then find the escape speed on the surface of the planet (in kms^{-1}).

18 A projectile of mass m is fired from the surface of earth at an angle $\alpha = 60°$ from the vertical. The initial speed of the projectile is $u = \sqrt{\frac{GM}{R}}$, where M and R are mass and radius of the earth. The projectile rises to a height $\frac{R}{\psi}$. Find the value of ψ. Neglect air resistance and the rotation of earth.

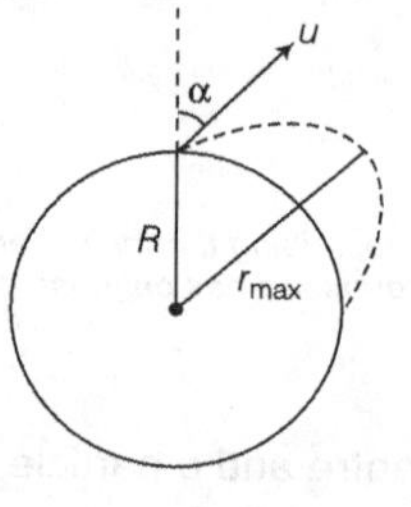

19 An artificial satellite is moving in a circular orbit around the earth with a speed equal to half the magnitude of escape velocity from the earth. If the satellite is stopped suddenly in its orbit and allowed to fall freely onto the earth, then speed (in ms^{-1}) with which it hits the surface of the earth is $\sqrt{kgR}$. Find the value of k. (Given, $g = 9.8\ ms^{-2}$ and $R = 6.4 \times 10^6$ m)

20 A stone is thrown from moon towards earth at an speed which is twice the lunar escape speed (Lunar escape speed is speed required to move infinitely far away from a stationary moon alone in universe). That stone strikes the surface of earth with speed

$$v = \left[pG\left(\frac{qM_m}{R_m} + \frac{M_m - M_E}{R_{ME}} + \frac{M_E}{R_E}\right)\right]^{1/2}$$

where p, q are constants,

G = universal gravitational constant

M_E = mass of earth

M_m = mass of moon

R_E = Radius of earth

R_m = Radius of moon

R_{ME} = Radius of moon's orbit relative to earth

Then, find the value of $2 \times \frac{q}{p}$.

DPP-5 Motion of Satellites and Planets

Only One Option Correct Type

Working Space

Direction (Q. Nos. 1-6) *This section contains 6 multiple choice questions. Each question has four choices (a), (b), (c) and (d), out of which ONLY ONE is correct.*

1 A magnetic storm from sun can disrupt a satellite as well as move it either away from or towards the earth radially. Ground based engineers start it back in a new circular orbit at the new position. Due to the storm,

a. the period of a satellite displaced further away is more than previous period
b. the mechanical energy of a satellite displaced towards earth is more than the previous energy
c. the speed of a satellite displaced further away is more than the previous speed
d. the angular momentum of a satellite displaced towards earth is more than the previous angular momentum

2 Figure shows the variation of energy with the orbit radius r of a satellite in a circular motion. Mark the correct statement.

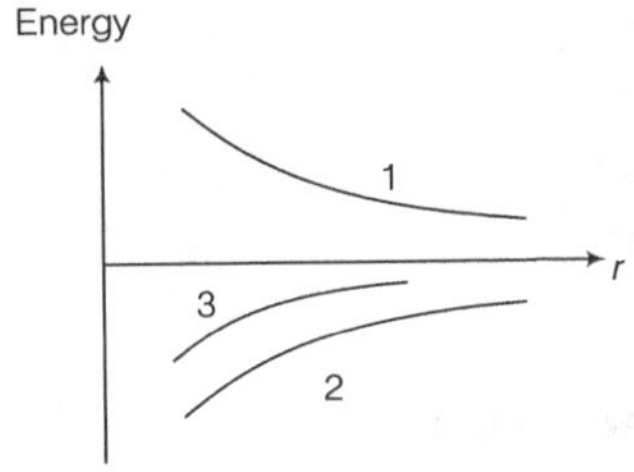

a. 3 shows the total energy, 2 the kinetic energy and 1 the potential energy of the satellite
b. 1 shows the kinetic energy, 2 the total energy and 3 the potential energy of the satellite
c. 1 and 2 show kinetic and potential energies, respectively and 3 the total energy of the satellite
d. 3 and 1 show kinetic and potential energies, respectively and 2 the total energy of the satellite

3 A satellite can be in a geostationary orbit around earth in an orbit of radius r. If the angular velocity of earth about its axis doubles, a satellite can now be in a geostationary orbit around earth of radius

a. $\frac{r}{2}$
b. $\frac{r}{2\sqrt{2}}$
c. $\frac{r}{(4)^{1/3}}$
d. $\frac{r}{(2)^{1/3}}$

4 The radius of a planet is n times the radius of earth R. A satellite revolves around it in a circle of radius $4nR$ with an angular velocity ω. The acceleration due to gravity on planet's surface is

a. $R\omega^2$ b. $16R\omega^2$

c. $32nR\omega^2$ d. $64nR\omega^2$

5 An artificial satellite of a planet revolves in a circular orbit whose radius exceeds the radius of the planet 7 times. In the process of motion, the satellite experiences slight resistance due to cosmic dust. Assuming the resistive force depends on the speed of the satellite as $F = kv^2$ (where k is a constant), find how long the satellite will stay in orbit until it falls on to the planet's surface.
(Given, m = mass of the satellite, g = acceleration due to gravity at the surface of planet and R = radius of the planet)

a. $\frac{m}{k\sqrt{2gR}}(\sqrt{7}-1)$

b. $\frac{m}{k\sqrt{gR}}(\sqrt{7}-1)$

c. $\frac{2m}{k\sqrt{2gR}}(\sqrt{7}-1)$

d. $\frac{m}{2k\sqrt{gR}}(\sqrt{7}-1)$

6 A double star is a system of two stars moving around the centre of mass of the system due to gravitation. The distance between the components of the double star is (Given, the total mass is M and time period is T)

a. $\left(\frac{GMT^2}{\pi^2}\right)^{1/3}$ b. $\left(\frac{GMT^2}{4\pi^2}\right)^{1/3}$

c. $\left(\frac{GMT^2}{2\pi^2}\right)^{1/3}$ d. $\left(\frac{GMT}{3\pi^2}\right)^{1/3}$

One or More than One Options Correct Type

Direction (Q. Nos. 7-11) This section contains 5 multiple choice questions. Each question has four choices (a), (b), (c) and (d), out of which ONE or MORE THAN ONE are correct.

7 A geostationary satellite is a height h above the surface of earth. If radius of earth is R (θ is the colatitude), then

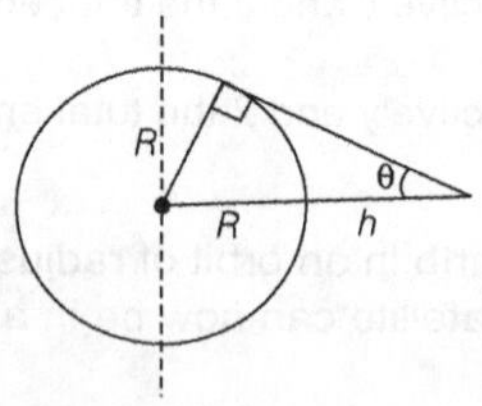

a. the minimum colatitude on earth up to which the satellite can be used for communication is $\sin^{-1}\left(\frac{R}{R+h}\right)$

b. the maximum colatitude on earth up to which the satellite can be used for communication is $\sin^{-1}\left(\frac{R}{R+h}\right)$

c. the area on earth escaped from this satellite is given as $2\pi R^2(1+\sin\theta)$

d. the area on earth escaped from this satellite is given as $2\pi R^2(1+\cos\theta)$

Working Space

8 For a satellite to orbit around the earth, which of the following must be true?

a. It cannot pass over the poles at any time

b. It must be above the equator at sometime

c. Its period of rotation must be greater than $2\pi\sqrt{\frac{R}{g}}$

d. Its height above the surface of earth cannot exceed 36000 km

9 Two satellites A and B of the same mass are orbiting the earth at altitudes R and $3R$, respectively, where R is the radius of the earth. Let their orbits are circular K and U represent kinetic and potential energies, respectively. Choose the correct option(s).

a. $\frac{K_A}{K_B} = 2$

b. $\frac{U_A}{U_B} = 2$

c. $\frac{K_A}{K_B} = \frac{1}{2}$

d. $\frac{U_A}{U_B} = \frac{1}{2}$

10 A satellite of mass m is orbiting a planet of mass M at a radial distance r from the centre of the planet. The satellite explodes by expelling very rapidly a small amount of its mass Δm in opposite direction to its orbital velocity. The immediate recoil velocity of the satellite is v_r (additional to the velocity already possessed by the satellite). The maximum value of v_r for which the satellite remains in the gravitational field (assuming $\Delta m << m$) is x. Then,

a. $x = (\sqrt{2} + 1)\sqrt{\frac{GM}{R}}$

b. $x = (\sqrt{2} - 1)\sqrt{\frac{GM}{R}}$

c. Δm is not negligible, the value of x gets modified

d. Δm is not negligible, the value of x does not get modified

11 Two satellites S_1 and S_2 are to be set into the orbits of heights $\frac{R}{4}$ and $\frac{R}{6}$ (above the surface of earth), respectively. They revolve around the earth in coplanar circular orbits in opposite sense (where, R = radius of earth). Which of the following statement(s) is/are correct?

a. The ratio of their time periods is $\left(\frac{15}{14}\right)^{3/2}$

b. The ratio of their time periods is $\left(\frac{14}{15}\right)^{3/2}$

c. The ratio of their speeds of projection from the surface of earth is $\sqrt{\frac{23}{15}}$

d. The ratio of their speeds of projection from the surface of earth is $\sqrt{\frac{21}{20}}$

Statement Type

Working Space

Direction (Q. No. 12) *This section is based on Statement I and Statement II. Select the correct answer from the codes given below.*

a. Both Statement I and Statement II are correct and Statement II is the correct explanation of Statement I
b. Both Statement I and Statement II are correct but Statement II is not the correct explanation of Statement I
c. Statement I is correct but Statement II is incorrect
d. Statement II is correct but Statement I is incorrect

12 **Statement I** Orbital velocity of a satellite is greater than its escape velocity.

Statement II Orbit of a satellite is within the gravitational field of earth, whereas escaping is beyond the gravitational field of earth.

Matching List Type

Direction (Q. No. 13) *Choices for the correct combination of elements from Column I and Column II are given as options (a), (b), (c) and (d), out of which one is correct.*

13 A satellite is revolving around the earth in a circular orbit of radius x with a velocity v_0. A particle of mass m_0 is projected from the satellite in forward direction with relative velocity $v = \left(\sqrt{\frac{5}{4}} - 1\right)v_0$. During subsequent motion of satellite, match the following columns and select the correct option from the codes given below. (M = mass of earth).

	Column I		Column II
i.	Total energy of particle	p.	$-\frac{3GMm_0}{8x}$
ii.	Minimum distance of particle from the earth	q.	$\frac{5}{8}\cdot\frac{GMm_0}{x}$
iii.	Maximum distance of particle from the earth	r.	$\frac{5x}{3}$
iv.	The kinetic energy of the particle	s.	x
		t.	$\frac{3}{8}\cdot\frac{GMm_0}{x}$
		u.	None of the above

Codes

	i	ii	iii	iv		i	ii	iii	iv
a.	p	r	s	t	b.	p	s	r	q
c.	t	q	u	p	d.	t	u	q	p

Comprehension Type

Direction (Q. Nos. 14-19) *This section contains 3 paragraphs, each describing theory, experiments, data, etc. Six questions related to the paragraphs have been given. Each question has only one correct answer among the four given options (a), (b), (c) and (d).*

Passage I

Two satellites S_1 and S_2 revolve around a planet in coplanar circular orbits in the same sense around a planet. Their periods of revolution are 1h and 8h, respectively. The radius of the orbit of S_1 is 10^4 km.

14 The speed of S_2 relative to S_1, when they are closest (in kmh^{-1}) is

a. $10^4\pi$
b. $2\times10^4\pi$
c. $\frac{10^4}{2}\pi$
d. $4\times10^4\pi$

Working Space

15 The angular speed of S_2, as observed by an astronaut in S_1 is (in rad h^{-1})

a. $\frac{\pi}{2}$ b. $\frac{\pi}{3}$

c. $\frac{\pi}{4}$ d. $\frac{\pi}{6}$

Passage II

An artificial satellite is moving in a circular orbit around the earth with a speed equal to half the escape speed from the earth. Now, the satellite is stopped suddenly in its orbit and allowed to fall freely onto the earth. The radius of the earth is R and acceleration due to gravity on earth's surface is g.

16 The height of the satellite above the surface of the earth is

a. $\frac{R}{2}$ b. $\frac{2R}{3}$

c. R d. $2R$

17 The speed with which it hits the surface of the earth is

a. $\sqrt{gR}$ b. $\sqrt{2gR}$

c. $\sqrt{\frac{gR}{2}}$ d. $\frac{\sqrt{gR}}{2}$

Passage III

A sky lab of mass 2×10^3 kg is first launched from the surface of earth in a circular orbit of radius $2R$ (from the centre of earth) and is then shifted from this circular orbit to another circular orbit of radius $3R$. (Given, $R = 6400$ km, $g = 10\,ms^{-2}$)

18 The minimum energy required to place the lab in first orbit is

a. $\frac{GMm}{R}$ b. $\frac{3GMm}{4R}$

c. $\frac{2GMm}{R}$ d. $\frac{4GMm}{3R}$

19 The minimum energy required to shift the lab from first orbit to the second orbit is

a. $\frac{mgR}{2}$ b. $\frac{2mgR}{3}$

c. mgR d. $\frac{mgR}{12}$

One Integer Value Correct Type

Direction (Q. No. 20) This section contains 1 question. When worked out will result in an integer from 0 to 9 (both inclusive).

20 If G is universal gravitational constant and ρ is the uniform density of a spherical planet, the shortest possible period of rotation of the planet can be $\sqrt{\frac{\beta\pi}{G\rho}}$. Find the value of β.

DPP-6 Kepler's Laws

Only One Option Correct Type

Direction *(Q. Nos. 1-5) This section contains 5 multiple choice questions. Each question has four choices (a), (b), (c) and (d), out of which ONLY ONE is correct.*

Working Space

1 A planet revolves around the sun in elliptical orbit. The areal velocity $\left(\frac{dA}{dt}\right)$ of the planet is $4 \times 10^{16}\ m^2s^{-1}$. The least distance between the planet and the sun is 2×10^{12} m. Then, the maximum speed of the planet (in km s^{-1}) is

a. 10
b. 20
c. 30
d. 40

2 The moon revolves around the earth 13 times in a year. If the ratio of the distance of the earth from the sun to the distance of the moon from the earth is 392, then ratio of mass of the sun to the mass of the earth is

a. $\frac{(392)^3}{(13)^2}$
b. $\left(\frac{392}{13}\right)^{3/2}$
c. $\frac{(392)^2}{(13)^3}$
d. $\frac{(13)^3}{(392)^2}$

3 A satellite revolves around a planet in an elliptical orbit. Its maximum and minimum distances from the planet are 1.5×10^7 m and 0.5×10^7 m respectively. If the speed of the satellite at the farthest point is $5 \times 10^3\ ms^{-1}$, then its speed at the nearest point is

a. $4.5 \times 10^4\ ms^{-1}$
b. $3.5 \times 10^4\ ms^{-1}$
c. $2.5 \times 10^4\ ms^{-1}$
d. $1.5 \times 10^4\ ms^{-1}$

4 Imagine a light planet revolving around a very massive star in a circular orbit of radius r with a period of revolution T. On what power of r will the square of time period depend if the gravitational force of attraction between the planet and the star is proportional to $r^{-5/2}$?

a. $r^{5/2}$
b. $r^{9/2}$
c. $r^{7/2}$
d. $r^{3/2}$

5 Halley's comet has a period of T years and in the year 1986, had a distance of closest approach to the sun equal to r. If the mass of the sun is M, then comet's farthest distance from the sun is

a. $\left(\frac{T^2GM}{4\pi^2}\right)^{1/3} - r$
b. $2\left(\frac{T^2GM}{4\pi^2}\right)^{1/3} - r$
c. $r^3 - \frac{T^2GM}{4\pi^2}$
d. $r - \frac{T^2GM}{4\pi^2}$

Working Space

One or More than One Options Correct Type

Direction (Q. Nos. 6-8) *This section contains 3 multiple choice questions. Each question has four choices (a), (b), (c) and (d), out of which ONE or MORE THAN ONE are correct.*

6 A planet mass m is moving in an elliptical orbit with the sun of mass M at is focus. The perihelion and aphelion distances are r_2 and r_1 respectively, Now, choose the correct option(s).

a. Angular momentum of planet relative to sun is

$$J = \sqrt{\frac{2GMm\, r_1 r_2}{r_1 + r_2}}$$

b. Total mechanical energy depends only on the semi-major axis of the orbit

c. Total mechanical energy depends on both semi-major and semi-minor axes of the orbit

d. If J denotes angular momentum of planet, then

$$\frac{J}{2m}\left(\frac{1}{r_1^2} - \frac{1}{r_2^2}\right) = GM_s m\left(\frac{1}{r_1} - \frac{1}{r_2}\right)$$

7 An earth's satellite is revolving in a circular orbit of radius a with a velocity v_0. A gun is in the satellite and is aimed directly towards the earth. A bullet is fired from the gun with a muzzle velocity $\frac{v_0}{2}$. Neglecting resistance offered by cosmic dust and recoil of gun during subsequent motion of bullet from the centre of the earth. Choose the correct option(s).

a. maximum distance of bullet is $3a$

b. minimum distance of bullet is $\frac{3a}{4}$

c. maximum distance of bullet is $2a$

d. minimum distance of bullet is $\frac{2a}{3}$

8 A planet A at a certain moment of time was at a distance r_0 from the sun and was travelling with a velocity v_0. The angle between $\mathbf{r}_0$ and $\mathbf{v}_0$ is ϕ. Then,

($K = \frac{r_0^2 v_0^2}{GM}$, where M is mass of the sun)

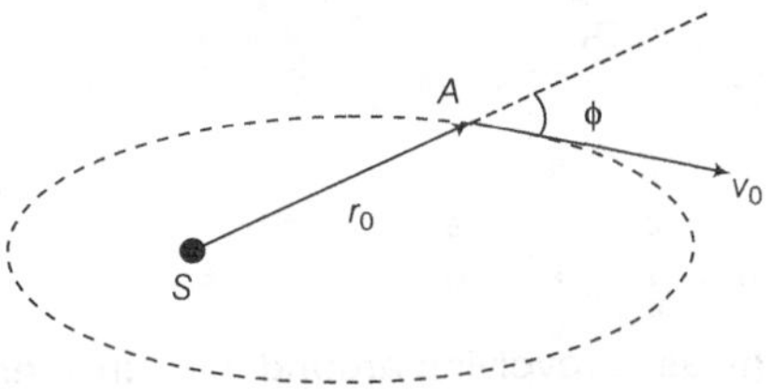

a. minimum distance of the planet from the sun is

$$\frac{r_0}{2-K}\left(1 - \sqrt{1 + K(2-K)\sin^2\phi}\right)$$

b. minimum distance of the planet from the sun is

$$\frac{r_0}{2-K}\left(1 - \sqrt{1 - K(2-K)\sin^2\phi}\right)$$

c. maximum distance of the planet from the sun is

$$\frac{r_0}{2-K}\left(1 + \sqrt{1 - K(2-K)\sin^2\phi}\right)$$

d. maximum distance of the planet from the sun is

$$\frac{r_0}{2-K}\left(1 + \sqrt{1 + K(2-K)\sin^2\phi}\right)$$

Statement Type

Working Space

Direction (Q. Nos. 9 and 10) *This section is based on Statement I and Statement II. Select the correct answer from the codes given below.*

a. Both Statement I and Statement II are correct and Statement II is the correct explanation of Statement I
b. Both Statement I and Statement II are correct but Statement II is not the correct explanation of Statement I
c. Statement I is correct but Statement II is incorrect
d. Statement II is correct but Statement I is incorrect

9 **Statement I** For the planets orbiting around the sun, angular speed, linear speed and KE change with time but angular momentum remains constant.

Statement II This is because no torque is acting on the rotating planet.

10 **Statement I** When a planet approaches the point which is farthest from the sun, its orbital speed decreases.

Statement II Work done on the planet by the gravitational force exerted by the sun is negative.

Matching List Type

Direction (Q. Nos. 11 and 12) *Choices for the correct combination of elements from Column I and Column II are given as options (a), (b), (c) and (d), out of which one is correct.*

11 Let a satellite of mass m is revolving in a circular orbit of radius r around a planet of mass M such that the system is bound. If E, U, K and B denote total energy, potential energy, kinetic energy and binding energy respectively, then match the following columns and select the correct option from the codes given below.

	Column I		Column II
i.	E	p.	$\frac{GMm}{2r}$
ii.	U	q.	$\frac{GMm}{r}$
iii.	K	r.	$-\frac{GMm}{r}$
iv.	B	s.	$\frac{GMm}{2r}$

Codes

	i	ii	iii	iv			i	ii	iii	iv
a.	s	r	p	p		b.	p	q	r	s
c.	r	p	q	s		d.	q	r	s	p

12 Match the following columns for a planet of mass m revolving around the sun in an elliptical orbit of semi-major axis a and semi-minor axis b. T is time period of planet.

	Column I		Column II
i.	$T^2 \propto r^3$	p.	Kepler's first law
ii.	Areal velocity is constant	q.	Kepler's second law
iii.	Orbit of planet is elliptical	r.	Kepler's third law
iv.	$\frac{2\pi mab}{T}$ = constant	s.	None of the above

Codes

	i	ii	iii	iv			i	ii	iii	iv
a.	s	r	p	q		b.	q	r	p	s
c.	p	s	q	r		d.	s	q	p	q

Comprehension Type

Working Space

Direction (Q. Nos. 13-18) This section contains 2 paragraphs, each describing theory, experiments, data, etc. Six questions related to the paragraphs have been given. Each question has only one correct answer among the four given options (a), (b), (c) and (d).

Passage I

It can be assumed that orbits of earth and mars are nearly circular around the sun. It is proposed to launch an artificial planet around the sun such that its apogee is at the orbit of mars, while its perigee is at the orbit of earth. Let T_e and T_m be periods of revolution of earth and mars. Further the variables are assigned the meanings as follows:

M_e, M_m = Masses of earth and mars

L_e, L_m = Angular momenta of earth and mars (each about the sun)

a_e, a_m = Semi-major axes of earth and mars orbits

m = Mass of the artificial planet

E_e, E_m = Total energies of earth and mars

13 Time period of revolution of the artificial planet about the sun will be (neglect gravitational effects of earth and mars)

a. $\frac{T_e + T_m}{2}$ b. $\sqrt{T_e T_m}$

c. $\frac{2T_e T_m}{T_e + T_m}$ d. None of these

14 The total energy of artificial planet's orbit will be

a. $\frac{2m}{M_e}\left(\frac{a_e E_e}{a_e + a_m}\right)$ b. $\frac{2m}{M_m}\left(\frac{a_e E_e}{a_e + a_m}\right)$

c. $\frac{2E_e m}{M_m}\left(\frac{a_e + a_m}{a_m}\right)$ d. $\frac{2E_e m}{M_e}\left(\frac{a_e + a_m}{\sqrt{a_e^2 + a_m^2}}\right)$

15 Areal velocity of the artificial planet around the sun will be

a. less than that of earth

b. more than that of mars

c. more than that of earth

d. same as that of earth

Passage II

An earth station receives data transmitted back in time from a future intergalactic expedition. The table summarises the data for the moons of a planet of a distance galaxy.

	Radius	Gravitational acceleration at the surface	Orbital radius	Orbital period
Moon I			5×10^7m	
Moon II			9×10^7m	3×10^5 s
Moon III	2×10^5 m	0.2 ms^{-2}		2×10^6s

Based on above information, answer the following questions :

16 Mass of the planet is closest to (in kg)

a. 1.2×10^{20} b. 1.7×10^{21}

c. 2.4×10^{22} d. 4.8×10^{24}

Working Space

17 Mass of moon III is closest to (in kg)

a. 2.4×10^{19} b. 4.8×10^{19}
c. 1.2×10^{20} d. 2.4×10^{21}

18 The centripetal acceleration of moon II due to orbital motion is closest to (in ms^{-2})

a. 0.2 b. 0.04
c. 0.6 d. 8.0

One Integer Value Correct Type

Direction (*Q. Nos.* 19 *and* 20) *This section contains 2 questions. When worked out will result in an integer from* 0 *to* 9 (*both inclusive*).

19 Suppose earth's orbital motion around the sun is suddenly stopped. The time (in days) taken by the earth to fall into the sun is $\frac{130}{k}$. Find the value of k.

20 A missile, which missed its target, went into orbit around the earth at a mean radius 3 times as great as the parking orbit of the satellite. The period of missile is $\sqrt{3}x$ days. Find the value of x.

SET - 1

Revisal Problems for JEE Main

Only One Option Correct Type

Working Space

Direction (Q. Nos. 1-19) *This section contains 19 multiple choice questions. Each question has four choices (a), (b), (c) and (d), out of which ONLY ONE is correct.*

1 The height at which the gravitational field of the earth becomes one-fourth the field at the surface is (where, R = radius of the earth)

a. $2R$ b. $\frac{R}{2}$

c. R d. $\frac{2R}{3}$

2 Assuming the earth to be a sphere of uniform mass density, how much would a body weight half way down the centre of the earth, if it weighed 100 N on the surface? (Given, $g = 10\ \text{ms}^{-2}$)

a. 75 N b. 50 N

c. 60 N d. 62 N

3 A system consists of a thin ring of radius R and a very long uniform wire oriented along axis of the ring with one of its ends coinciding with the centre of the ring. If mass of ring is m and the linear mass density of the wire is λ, then the interaction force between the ring and the wire is

a. $\frac{Gm\lambda}{R}$ b. $\frac{Gm\lambda}{2R}$

c. $\frac{Gm\lambda}{\sqrt{2}R}$ d. $\frac{\sqrt{2}Gm\lambda}{R}$

4 A thick spherical shell has an inner radius R_1 and an outer radius R_2. It has mass m and a uniform density. If r is the distance from the centre, then gravitational field E_g for $R_1 < r < R_2$, is

a. $\frac{Gm}{\sqrt{2}r^2}\left[\frac{r^3 - R_1^3}{R_2^3 - R_1^3}\right]$ b. $\frac{Gm}{2r^2}\left[\frac{r^3 - R_1^3}{R_2^3 - R_1^3}\right]$

c. $\frac{\sqrt{2}Gm}{r^2}\left[\frac{r^3 - R_1^3}{R_2^3 - R_1^3}\right]$ d. $\frac{Gm}{r^2}\left[\frac{r^3 - R_1^3}{R_2^3 - R_1^3}\right]$

5 Find the gravitational potential energy of a system of eight identical masses, each of mass m, placed at the corners of a cube of side L.

a. $-\frac{Gm^2}{L}\left(12 + 6\sqrt{2} + \frac{4}{\sqrt{3}}\right)$ b. $-\frac{Gm^2}{L}\left(12\sqrt{2} + \frac{4}{\sqrt{3}}\right)$

c. $-\frac{Gm^2}{L}\left(6 + 10\sqrt{2} + \frac{2}{\sqrt{3}}\right)$ d. $-\frac{Gm^2}{L}\left(12 + 10\sqrt{2} + \frac{4}{\sqrt{3}}\right)$

6 Calculate the gravitational potential energy of a point mass m_0 and a uniform thin rod of mass m and length L. m_0 is at a distance a from one end of the rod on the axis line of the rod as shown in figure.

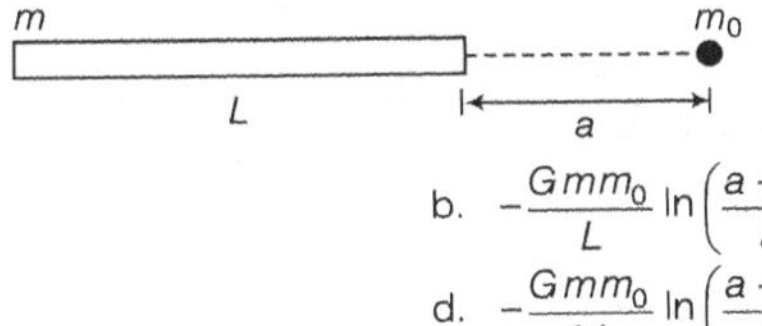

a. $-\frac{Gmm_0}{L}\ln\left(\frac{a+L}{2a}\right)$ b. $-\frac{Gmm_0}{L}\ln\left(\frac{a+L}{a}\right)$

c. $-\frac{Gmm_0}{\sqrt{2}L}\ln\left(\frac{a+L}{a}\right)$ d. $-\frac{Gmm_0}{2L}\ln\left(\frac{a+L}{a}\right)$

Working Space

7 A uniform spherical shell of mass m and radius R has a particle of mass m placed at its centre. The gravitational potential at a distance $\frac{R}{2}$ from the centre is

a. $-\frac{2GM}{R}$ b. $-\frac{GM}{R}$ c. $-\frac{4GM}{R}$ d. $-\frac{3GM}{R}$

8 In a region of space, the gravitational potential is represented by $V = 2x + 3y - z$. Then, gravitational field strength is

a. $2\hat{i} + 3\hat{j} - \hat{k}$ b. $2\hat{i} - 3\hat{j} + \hat{k}$
c. $-2\hat{i} - 3\hat{j} + \hat{k}$ d. $-2\hat{i} - 3\hat{j} - \hat{k}$

9 A particle is projected from the surface of the earth with an initial speed of 4 km s^{-1}. The maximum height attained by the particle is (Given, radius of the earth = 6400 km, $g = 9.8\ ms^{-2}$)

a. 835 km b. 935 km
c. 735 km d. 1035 km

10 Given a thin homogeneous disc of radius R and mass m_1. A particle of mass m_2 is placed at a distance L from the disc on its axis of symmetry. Initially, both are motionless in free space but they ultimately collide because of gravitational attraction between them. The relative velocity at the time of collision is ($L >> R$)

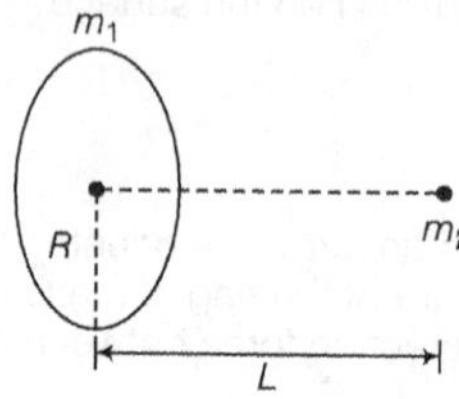

a. $\sqrt{2G(m_1 + m_2)\left(\frac{2}{R} - \frac{1}{L}\right)}$ b. $\sqrt{2G(m_1 + m_2)\left(\frac{2}{R} + \frac{1}{L}\right)}$
c. $\sqrt{2G(m_1 + m_2)\left(\frac{1}{R} + \frac{2}{L}\right)}$ d. $\sqrt{2G(m_1 + m_2)\left(\frac{1}{R} - \frac{2}{L}\right)}$

11 Distance between the centres of two stars is $10x$. The masses of these stars are m and $16m$ and their radii x and $2x$, respectively. A body of mass m_0 is fired straight from the surface of larger star towards the smaller star. Its minimum speed to reach the surface of smaller star is

a. $\frac{5\sqrt{5}}{2}\sqrt{\frac{Gm}{x}}$ b. $\frac{3\sqrt{5}}{2}\sqrt{\frac{Gm}{x}}$
c. $\frac{\sqrt{5}}{2}\sqrt{\frac{Gm}{x}}$ d. $2\sqrt{5}\sqrt{\frac{Gm}{x}}$

12 A planet of mass M moves along an ellipse around the sun. Its maximum and minimum distances from the sun are r_1 and r_2, respectively. Mass of the sun is M_s. The angular momentum of this planet relative to the centre of the sun is

a. $\frac{M}{2}\sqrt{\frac{2GM_s r_1 r_2}{(r_1 + r_2)}}$ b. $\frac{M}{2}\sqrt{\frac{GM_s r_1 r_2}{(r_1 + r_2)}}$
c. $M\sqrt{\frac{2GM_s r_1 r_2}{(r_1 + r_2)}}$ d. $M\sqrt{\frac{GM_s r_1 r_2}{(r_1 + r_2)}}$

13 The velocity of a satellite travelling in an elliptical orbit, when it reaches the end of the semi-major axis is (R = radius of the earth, g = acceleration due to gravity)

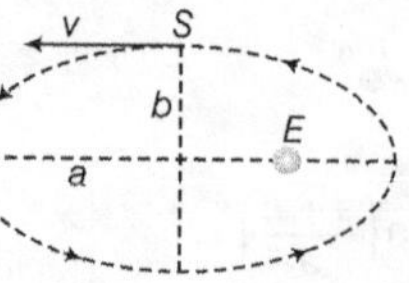

a. $v = 2R\sqrt{\frac{g}{a}}$ b. $v = 2R\sqrt{\frac{g}{b}}$
c. $v = R\sqrt{\frac{g}{b}}$ d. $v = R\sqrt{\frac{g}{a}}$

Working Space

14 Two concentric spherical shells have masses m_1 and m_2 and radii R_1 and R_2 ($R_2 > R_1$). The force exerted by this system on a particle of mass m_0, if it is placed at a distance r ($R_1 < r < R_2$) from the centre is

a. $\frac{Gm_1m_0}{r^2}$
b. $\frac{Gm_2m_0}{r^2}$
c. $\frac{Gm_1m_2}{m_0r^3}R_1$
d. $\frac{Gm_1m_2}{m_0r^3}R_2$

15 A spaceship is launched into a circular orbit close to the earth's surface. The additional velocity now to be imparted to the spaceship in the orbit to overcome the gravitational pull is (in ms^{-1})

a. $\sqrt{2}gR$
b. $(\sqrt{2}+1)\sqrt{gR}$
c. $\sqrt{2gR}$
d. $(\sqrt{2}-1)\sqrt{gR}$

16 The minimum and maximum distances of a satellite from the centre of the earth are $2R$ and $4R$ respectively, where R is the radius of the earth and M is mass of the earth. The radius of curvature of the orbit at the point of minimum distance is

a. $\frac{5R}{3}$
b. $\frac{7R}{3}$
c. $\frac{8R}{3}$
d. $3R$

17 Three concentric spherical shells of masses m_1, m_2 and m_3 having radii a, b and c respectively are shown in the figure. Find the force on a particle of mass m_0 present at point P as shown below.

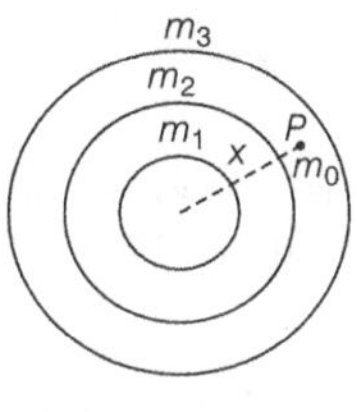

a. $\frac{Gm_0}{x^2}(m_1+m_2+m_3)$
b. $\frac{Gm_0}{x^2}\left(\frac{am_1}{b}+\frac{bm_2}{c}+\frac{cm_3}{a}\right)$
c. $\frac{Gm_0}{x^2}(m_1+m_2)$
d. $\frac{Gm_0}{x^2}\left(\frac{am_1+bm_2}{c}\right)$

18 If the radius of the earth contracts to half of its present value without change in its mass, the new duration of day will be

a. 18 h
b. 12 h
c. 8 h
d. 6 h

19 An infinite number of masses, each of 1 kg, are placed on the positive X-axis at distance of 1 m, 2 m, 4 m, 8 m, 16 m, ... ∞ from the origin. The magnitude of the gravitational field at the origin due to this distribution is

a. 2G
b. $\frac{4G}{3}$
c. $\frac{3G}{4}$
d. ∞

Statement Type

Direction (Q. No. 20-22) *This section is based on Statement I and Statement II. Select the correct answer from the codes given below.*

a. Both Statement I and Statement II are correct and Statement II is the correct explanation of Statement I
b. Both Statement I and Statement II are correct but Statement II is not the correct explanation of Statement I
c. Statement I is correct but Statement II is incorrect
d. Statement II is correct but Statement I is incorrect

20 **Statement I** A particle is at a height R from the surface of the earth. (R is radius of the earth). If mass of the particle is m, then its Gravitational Potential Energy is mgR.

Statement II If a particle is slowly lifted above the surface of the earth, then work is to be done by external energy. Work done by external agent is stored in the form of potential energy.

21 **Statement I** If the product of density and radius of a planet is less than the same product value for the earth, then the acceleration due to gravity at the planet will be more than the earth.

Statement II $g = \frac{GM}{R^2}$.

22 **Statement I** A balloon filled with hydrogen will fall with an acceleration $\frac{g}{6}$ on the moon.

Statement II Moon has no atmosphere.

SET - 2

Revisal Problems for JEE Advanced

Only One Option Correct Type

Working Space

Direction (Q. Nos. 1-8) *This section contains 8 multiple choice questions. Each question has four choices (a), (b), (c) and (d), out of which ONLY ONE is correct.*

1 A spaceship approaches the moon (mass = M_m, radius = R_m) along a parabolic path which is almost tangential to its surface (very close to it). At the moment of closest approach, a brake rocket is fired to convert the spaceship into a satellite of the moon. The change in speed of spaceship is

a. $\frac{1}{(\sqrt{2}-1)}\sqrt{\frac{GM_m}{R_m}}$ b. $(\sqrt{3}-1)\sqrt{\frac{GM_m}{R_m}}$

c. $(\sqrt{2}-1)\sqrt{\frac{GM_m}{R_m}}$ d. $(\sqrt{5}-1)\sqrt{\frac{GM_m}{R_m}}$

2 A satellite of mass m is moving in a circular orbit of radius r around earth (mass M). Its angular momentum with respect to the centre of the orbit is

a. $(GM^2mr)^{1/2}$ b. $(GMm^2r)^{1/2}$

c. $\left(\frac{GM^2mr}{2}\right)^{1/2}$ d. $\left(\frac{GMm^2r}{2}\right)^{1/2}$

3 A solid sphere of mass m and radius r is initially placed at a distance $5r$ from the point mass m_0 as shown in the figure. Now, it is shifted to a position at a distance $3r$ from the point mass. During displacement, the point mass m_0 is also uniformly expanded to a radius $2r$ so that its density decreases uniformly throughout its volume. The work required in this process is

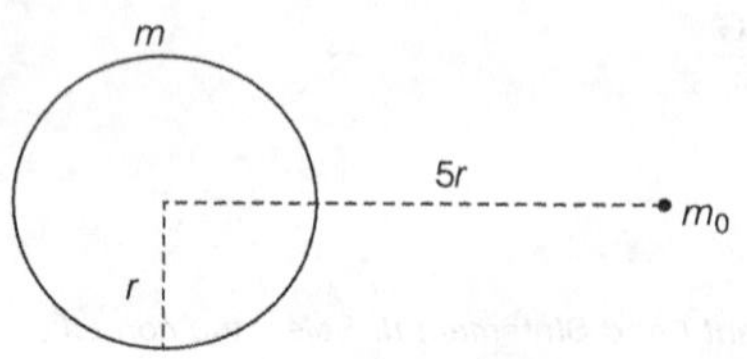

a. $\frac{Gm}{30r}(9m-4m_0)$ b. $\frac{Gm}{25r}(9m-4m_0)$

c. $\frac{Gm}{30r}(19m-9m_0)$ d. $\frac{Gm}{20r}(5m-2m_0)$

4 A satellite is orbiting around the earth in an orbit in equatorial plane of radius $2R_e$, where R_e is the radius of the earth. The area on the earth, which this satellite covers for communication purpose in its complete revolution is

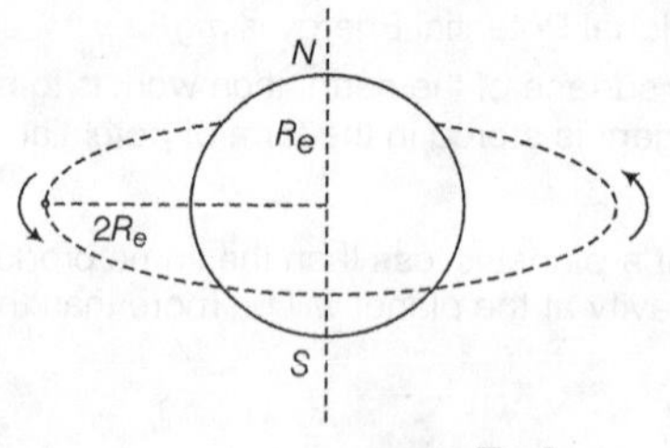

a. $3\sqrt{3}R_e^2$ b. $4\sqrt{3}R_e^2$

c. $5\sqrt{3}R_e^2$ d. $2\sqrt{3}R_e^2$

Working Space

5 A satellite is revolving around the earth in a circular orbit in a plane containing earth's axis of rotation. If the angular speed of satellite is equal to that of earth, the time it takes to move from a point above North pole to a point above equator is

a. 12 h
b. 8 h
c. 6 h
d. 10 h

6 A cosmic body B coming from infinity with a velocity v_0 is approaching the sun (of mass M_s), with its line of motion at a distance l_0 from the sun as shown in the figure. When it gets closest to the sun, i.e. at P, its distance from the sun is

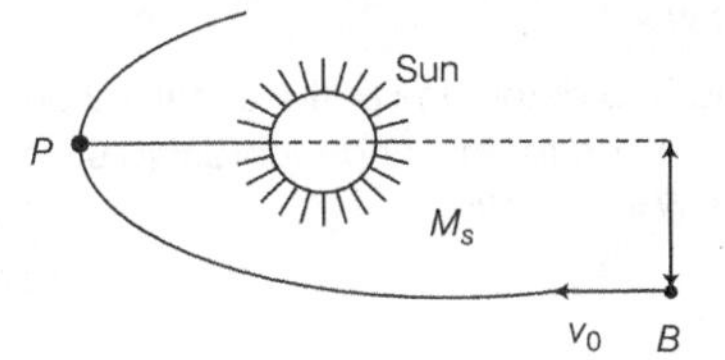

a. $\frac{GM_s}{v_0^2}\left[\sqrt{1+\left(\frac{v_0^2 l_0}{GM_s}\right)^2}-1\right]$
b. $\frac{GM_s}{2v_0^2}\left[\sqrt{1+\left(\frac{v_0^2 l_0}{GM_s}\right)}-1\right]$
c. $\frac{GM_s}{\sqrt{2}v_0^2}\left[\sqrt{1+\left(\frac{v_0^2 l_0}{GM_s}\right)^2}+1\right]$
d. $\frac{GM_s}{v_0^2}\left[\sqrt{1+\left(\frac{v_0^2 l_0}{2GM_s}\right)^2}-1\right]$

7 A particle of mass m_0 ascends from the earth's surface with zero initial velocity due to the action of two forces as shown in the figure. The force varies with height y as $\mathbf{F} = -2\,mg\,(1-ky)$, where k is a positive constant. The work performed by this force $\mathbf{F}$ over the first half (distance) of the ascent is

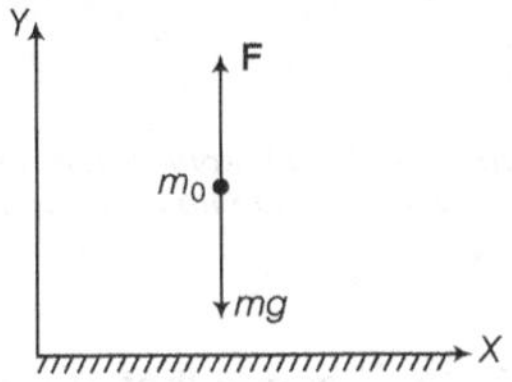

a. $\frac{mg}{2k}$
b. $\frac{2mg}{k}$
c. $\frac{3mg}{4k}$
d. $\frac{2mg}{3k}$

8 Two identical spheres are kept at some distance as shown in the figure. The potential energy when a mass particle m_0 is taken from the surface of one sphere to the other

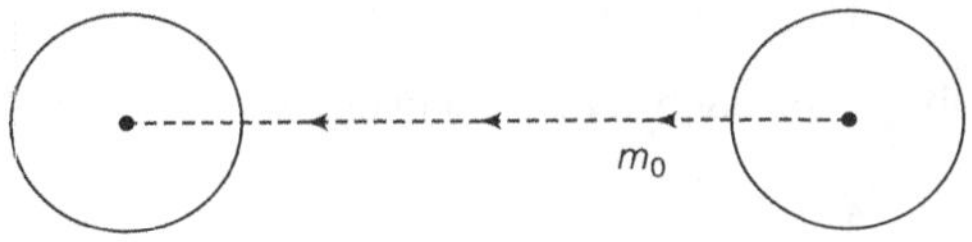

a. increases continuously
b. first increases then decreases
c. decreases continuously
d. first decreases then increases

One or More than One Options Correct Type

Direction **(Q. Nos. 9-13)** ***This section contains 5 multiple choice questions. Each question has four choices (a), (b), (c) and (d), out of which ONE or MORE THAN ONE are correct.***

9 Two tunnels are dug from one side of the, earth's surface to the other side, one along a diameter and the other along a chord. Now, two particles are dropped from one end of each of the tunnels. Both the particles oscillate simple harmonically along the tunnels. Let T_1 and T_2 be the time periods and v_1 and v_2 be the maximum speeds of two particles respectively, then neglecting friction of tunnel walls, correct options are

a. $T_1 = T_2$
b. $T_1 > T_2$
c. $v_1 = v_2$
d. $v_1 > v_2$

Working Space

10 In some region, the gravitational field is zero. The gravitational potential in this region

a. may be zero
b. cannot be zero
c. must be constant
d. may be variable

11 Let V and E be the gravitational potential and gravitational field strength. Then, select the correct option(s). (r is distance from the centre)

a. The plot of V *versus* r is continuous for a spherical shell
b. The plot of E *versus* r is discontinuous for a spherical shell
c. The plot of V *versus* r is continuous for a solid sphere
d. The plot of E *versus* r is discontinuous for a solid sphere

12 A student uses simple pendulum of exactly 1m length to determine g, the acceleration due to gravity. He uses a stopwatch with the least count of 1 s for this and records 40 s for 20 oscillations. Which of the following statement(s) is/are correct?

a. Error ΔT in measuring T, the time period is 0.05 s
b. Error ΔT is measuring T, the time period is 1s
c. Percentage error in the determination of g is 5%
d. Percentage error in the determination of g is 2.5%

13 The potential at the surface of a planet of mass M and radius R is assumed to be zero. Choose the appropriate option(s).

a. The potential at infinity is zero
b. The potential at the centre of the planet is $-\frac{GM}{2R}$
c. The potential at infinity is $\frac{GM}{R}$
d. The potential at the centre of the planet is $\frac{-3GM}{2R}$

Comprehension Type

Direction (Q. Nos. 14-18) *This section contains 2 paragraphs, each describing theory, experiments, data, etc. Five questions related to the paragraphs have been given. Each question has only one correct answer among the four given options (a), (b), (c) and (d).*

Passage I

The mean radius of the earth's orbit around the sun is 1.5×10^{11} m and that of the orbit of mercury is 6×10^{10} m.

14 The mercury will revolve around the sun in nearly

a. $\sqrt{\frac{2}{5}}$ yr
b. $\frac{2}{5}$ yr
c. $\left(\frac{2}{5}\right)^2$ yr
d. $\left(\frac{2}{5}\right)^{3/2}$ yr

15 The ratio of the orbital velocity of mercury to that of the earth is (assuming orbits to be circular)

a. $\sqrt{\frac{10}{4}}$
b. $\frac{4}{\sqrt{10}}$
c. $2\sqrt{10}$
d. $\frac{1}{2}\sqrt{10}$

Passage II

A binary star system consists of two stars of masses m and $2m$. The separation between their centres is L. They revolve about the centre of mass of the system under their mutual gravitational interaction force.

16 Heavier star revolves in an orbit of radius

a. L
b. $\frac{L}{2}$
c. $\frac{L}{3}$
d. $\frac{L}{4}$

Working Space

17 The system revolves with an angular velocity ω about the centre of mass given by

a. $\sqrt{\frac{Gm}{L^3}}$ b. $\sqrt{\frac{2Gm}{L^3}}$

c. $\sqrt{\frac{3Gm}{L^3}}$ d. $2\sqrt{\frac{Gm}{L^3}}$

18 If K_L be the kinetic energy of light body and K_H be the kinetic energy of heavy body, then $K_L + K_H$ is equal to

a. $\frac{mL^2\omega^2}{3}$ b. $\frac{2mL^2\omega^2}{9}$

c. $\frac{2mL^2\omega^2}{3}$ d. $\frac{mL^2\omega^2}{9}$

One Integer Value Correct Type

Direction **(Q. Nos. 19 and 20)** ***This section contains 2 questions. When worked out will result in an integer from 0 to 9 (both inclusive).***

19 Two identical thin rods each of length $2l$ have equal mass m, that is distributed uniformly along their lengths. The rods lie along X-axis with their centres separated by a distance $d > 2l$.

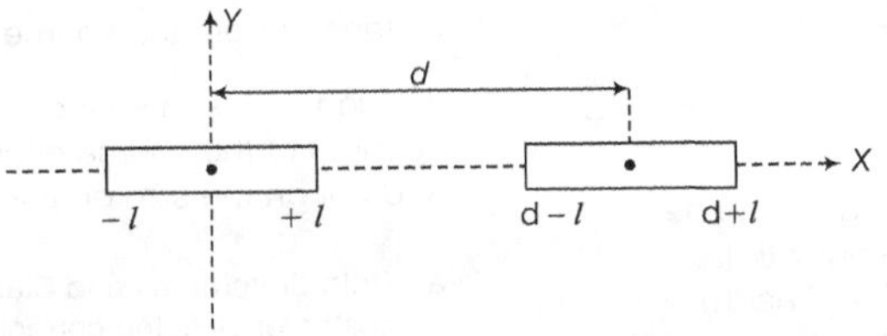

The magnitude of the force exerted by the left rod on the right one is

$$\frac{Gm^2}{4l^2} \ln\left[\frac{d^2}{d^2 - \beta l^2}\right]$$

Find the value of β.

20 A ring of radius $R = 4$ m made of a highly dense material has a mass $m_1 = 5.4 \times 10^9$ kg distributed uniformly over its circumference. A highly dense particle of mass $m_2 = 6 \times 10^8$ kg is placed on the axis of the ring at a distance 3 m from the centre of the ring. The speed of the particle, when the particle is at the centre of the ring is $2\,\alpha\,\text{cms}^{-1}$. Find the value of α. Except mutual gravitational interaction of the two, neglect all other forces.

JEE Main & AIEEE Archive

(Compilation of Last 13 Years' Questions)

1 Four particles, each of mass M and equidistant from each other move along a circle of radius R under the action of their mutual gravitational attraction, the speed of each particle is **(JEE Main 2014)**

a. $\sqrt{\frac{GM}{R}}$ b. $\sqrt{2\sqrt{2}\frac{GM}{R}}$

c. $\sqrt{\frac{GM}{R}(1+2\sqrt{2})}$ d. $\frac{1}{2}\sqrt{\frac{GM}{R}(1+2\sqrt{2})}$

2 What is the minimum energy required to launch a satellite of mass m from the surface of a planet of mass M and radius R in a circular orbit at an altitude of $2R$? **(JEE Main 2013)**

a. $\frac{5GmM}{6R}$ b. $\frac{2GmM}{3R}$

c. $\frac{GmM}{2R}$ d. $\frac{GmM}{3R}$

3 The mass of a spaceship is 1000 kg. It is to be launched from the earth's surface out into free space. The value of g and R (radius of earth) are 10 m/s^2 and 6400 km, respectively. The required energy for this work will be **(AIEEE 2012)**

a. 6.4×10^{11} J b. 6.4×10^{8} J

c. 6.4×10^{9} J d. 6.4×10^{10} J

4 Two particles of equal mass m go around a circle of radius R under the action of their mutual gravitational attraction. The speed of each particle with respect to their centre of mass is **(AIEEE 2011)**

a. $\sqrt{\frac{Gm}{R}}$ b. $\sqrt{\frac{Gm}{4R}}$

c. $\sqrt{\frac{Gm}{3R}}$ d. $\sqrt{\frac{Gm}{2R}}$

5 Two bodies of masses m and $4m$ are placed at a distance r. The gravitational potential at a point on the line joining them, where the gravitational field is zero, is **(AIEEE 2011)**

a. $-\frac{4Gm}{r}$ b. $-\frac{6Gm}{r}$

c. $-\frac{9Gm}{r}$ d. zero

6 The height at which the acceleration due to gravity becomes $\frac{g}{9}$ (where, g is the acceleration due to gravity on the surface of the earth in terms of R, the radius of the earth) is **(AIEEE 2009)**

a. $2R$ b. $\frac{R}{\sqrt{3}}$

c. $\frac{R}{2}$ d. $\sqrt{2}R$

7 A planet in a distant solar system is 10 times more massive than the earth and its radius is 10 times smaller. Given that the escape velocity from the earth is 11 kms^{-1}, the escape velocity from the surface of the planet would be **(AIEEE 2008)**

a. 1.1 kms^{-1} b. 11 kms^{-1}

c. 110 kms^{-1} d. 0.11 kms^{-1}

8 **Statement I** For a mass M kept at the centre of a cube of side a, the flux of gravitational field passing through its sides is $4\pi GM$.

Statement II If the direction of a field due to a point source is radial and its dependence on the distance r from the source is given as $\frac{1}{r^2}$, its flux through a closed surface depends only on the strength of the source enclosed by the surface and not on the size or shape of the surface. **(AIEEE 2008)**

a. Both Statement I and Statement II are correct and Statement II is the correct explanation of Statement I

b. Both Statement I and Statement II are correct but Statement II is not the correct explanation of Statement I

c. Statement I is correct but Statement II is incorrect

d. Statement II is correct but Statement I is incorrect

9 If g_E and g_M are the accelerations due to gravity on the surfaces of the earth and the moon respectively and if Millikan's oil drop experiment could be performed on the two surfaces, one will find the ratio $\frac{\text{electronic charge on the moon}}{\text{electronic charge on the earth}}$ to be **(AIEEE 2007)**

a. 1 b. zero

c. $\frac{g_E}{g_M}$ d. $\frac{g_M}{g_E}$

10 Average density of the earth **(AIEEE 2005)**

a. does not depend on g

b. is a complex function of g

c. is directly proportional to g

d. is inversely proportional to g

11 The change in the value of g at a height h above the surface of the earth is the same as at a depth d below the surface of earth. When both d and h are much smaller than the radius of earth, then which one of the following is correct? **(AIEEE 2005)**

a. $d=\frac{h}{2}$ b. $d=\frac{3h}{2}$

c. $d=2h$ d. $d=h$

12 A particle of mass 10 g is kept on the surface of a uniform sphere of mass 100 kg and radius 10 cm. Find the work to be done against the gravitational force between them to take the particle far away from the sphere, (you may take $G = 6.67 \times 10^{-11}\ Nm^2/kg^2$) **(AIEEE 2005)**

a. 13.34×10^{-10} J b. 3.33×10^{-10} J
c. 6.67×10^{-9} J d. 6.67×10^{-10} J

13 A satellite of mass m revolves around the earth of radius R at a height x from its surface. If g is the acceleration due to gravity on the surface of the earth, then orbital speed of the satellite is **(AIEEE 2004)**

a. gx b. $\frac{gR}{R-x}$
c. $\frac{gR^2}{R+x}$ d. $\left(\frac{gR^2}{R+x}\right)^{1/2}$

14 The time period of an earth satellite in circular orbit is independent of **(AIEEE 2004)**

a. the mass of the satellite
b. radius of its orbit
c. both the mass and radius of the orbit
d. neither the mass of the satellite nor the radius of its orbit

15 If g is the acceleration due to gravity on the earth's surface, then the gain in the potential energy of an object of mass m raised from the surface of the earth to a height equal to the radius R of the earth is **(AIEEE 2004)**

a. $2mgR$ b. $\frac{1}{2}mgR$ c. $\frac{1}{4}mgR$ d. mgR

16 Suppose the gravitational force varies inversely as the nth power of distance. Then, the time period of a planet in circular orbit of radius R around the sun will be proportional to **(AIEEE 2004)**

a. $R^{\left(\frac{n+1}{2}\right)}$ b. $R^{\left(\frac{n-1}{2}\right)}$ c. R^n d. $R^{\left(\frac{n-2}{2}\right)}$

17 The time period of a satellite of earth is 5 h. If the separation between the earth and the satellite is increased to 4 times the previous value, the new time period will become **(AIEEE 2003)**

a. 10 h b. 80 h c. 40 h d. 20 h

18 Two spherical bodies of mass M and $5M$ and radii R and $2R$, respectively are released in free space with initial separation between their centres equal to $12R$. If they attract each other due to gravitational force only, then the distance covered by the smaller body just before collision is **(AIEEE 2003)**

a. $2.5R$ b. $4.5R$
c. $7.5R$ d. $1.5R$

19 The escape velocity for a body projected vertically upwards from the surface of earth is 11 km/s. If the body is projected at an angle of 45° with the vertical, then the escape velocity will be **(AIEEE 2003)**

a. $11\sqrt{2}$ kms^{-1} b. 22 kms^{-1}
c. 11 kms^{-1} d. $\frac{11}{\sqrt{2}}$ ms^{-1}

20 Energy required to move a body of mass m from an orbit of radius $2R$ to $3R$ is **(AIEEE 2002)**

a. $\frac{GMm}{12R^2}$ b. $\frac{GMm}{3R^2}$
c. $\frac{GMm}{8R}$ d. $\frac{GMm}{6R}$

21 The kinetic energy needed to project a body of mass m from the earth's surface (radius R) to infinity is **(AIEEE 2002)**

a. $\frac{mgR}{2}$ b. $2mgR$
c. mgR d. $\frac{mgR}{4}$

22 The escape velocity of a body depends upon mass as **(AIEEE 2002)**

a. m^0 b. m^1
c. m^2 d. m^3

23 If suddenly the gravitational force of attraction between earth and a satellite revolving around it becomes zero, then the satellite will **(AIEEE 2002)**

a. continue to move in its orbit with same velocity
b. move tangentially to the original orbit with the same velocity
c. become stationary in its orbit
d. move towards the earth

JEE Advanced & IIT JEE Archive

(Compilation of Last 10 Years' Questions)

1 A planet of radius $R = \frac{1}{10} \times$ (radius of earth) has the same mass density as earth. Scientists dig a well of depth $\frac{R}{5}$ on it and lower a wire of the same length and of linear mass density 10^{-3} kgm^{-1} into it. If the wire is not touching anywhere, then force applied at the top of the wire by a person holding it in place is (Given, radius of the earth $= 6 \times 10^6$ m and the acceleration due to gravity of earth is 10 ms^{-2}) **(2014 Adv., Only One Option Correct Type)**

a. 96 N b. 108 N c. 120 N d. 150 N

2 Two bodies, each of mass M, are kept fixed with a separation $2L$. A particle of mass m is projected from the mid-point of the line joining their centres perpendicular to the line. The gravitational constant is G. The correct statement(s) is(are)

(2013 Adv., One or More than One Options Correct Type)

a. The minimum initial velocity of the mass m to escape the gravitational field of the two bodies is $4\sqrt{\frac{GM}{L}}$

b. The minimum initial velocity of the mass m to escape the gravitational field of the two bodies is $2\sqrt{\frac{GM}{L}}$

c. The minimum initial velocity of the mass m to escape the gravitational field of the two bodies is $\sqrt{\frac{2GM}{L}}$

d. The energy of the mass m remains constant

3 Two spherical planets P and Q have the same uniform density ρ, masses M_P and M_Q, and surface areas A and $4A$, respectively. A spherical planet R also has uniform density ρ and its mass is $(M_P + M_Q)$. The escape velocities from the planets P, Q and R, are v_P, v_Q and v_R, respectively. Then, **(2012, One or More than One Options Correct Type)**

a. $v_Q > v_R > v_P$ b. $v_R > v_Q > v_P$

c. $v_R/v_P = 3$ d. $v_P/v_Q = \frac{1}{2}$

4 A satellite is moving with a constant speed v in a circular orbit about the earth. An object of mass m is ejected from the satellite such that it just escapes from the gravitational pull of the earth. At the time of its ejection, the kinetic energy of the object is

(2011, Only One Option Correct Type)

a. $\frac{1}{2}mv^2$ b. mv^2 c. $\frac{3}{2}mv^2$ d. $2mv^2$

5 Gravitational acceleration on the surface of a planet is $\frac{\sqrt{6}}{11}g$, where g is the gravitational acceleration on the surface of the earth. The average mass density of the planet is $\frac{2}{3}$ times that of the earth. If the escape speed on the surface of the earth is taken to be 11 kms^{-1}, then escape speed on the surface of the planet (in kms^{-1}) will be **(2010, Integer Type)**

6 A binary star consists of two stars A (mass $2.2\,M_s$) and B (mass $11M_s$), where M_s is the mass of the sun. They are separated by distance d and are rotating about their centre of mass, which is stationary. The ratio of the total angular momentum of the binary star to the angular momentum of star B about the centre of mass is **(2010, Integer Type)**

7 A spherically symmetric gravitational system of particles has a mass density $\rho = \begin{cases} \rho_0 & \text{for } r \le R \\ 0 & \text{for } r > R \end{cases}$ where ρ_0 is a constant. A test mass can undergo circular motion under the influence of the gravitational field of particles. Its speed v as a function of distance r from the centre of the system is represented by

(2008, Only One Option Correct Type)

a. 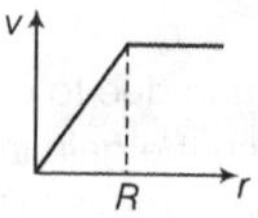

b. 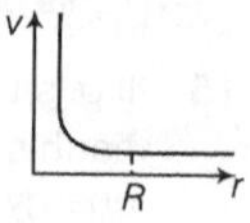

c. 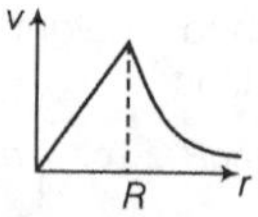

d. 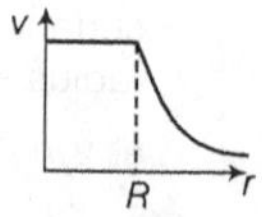

8 **Statement I** An astronaut in an orbiting space station above the earth experiences weightlessness.

Statement II An object moving around the earth under the influence of earth's gravitational force is in a state of 'free-fall'. **(2008, Statement Type)**

a. Both Statement I and Statement II are correct and Statement II is the correct explanation of Statement I

b. Both Statement I and Statement II are correct but Statement II is not the correct explanation of Statement I

c. Statement I is correct but Statement II is incorrect

d. Statement II is correct but Statement I is incorrect

9 A double star system consists of two stars A and B which have time periods T_A and T_B, radius R_A, R_B and mass M_A, M_B. Choose the correct option.

(2006, Only One Option Correct Type)

a. If $T_A > T_B$, then $R_A > R_B$

b. If $T_A > T_B$, then $M_A > M_B$

c. $\left(\frac{T_A}{T_B}\right)^2 = \left(\frac{R_A}{R_B}\right)^3$

d. $T_A = T_B$

4. Elasticity

DPP
DAILY PRACTICE PROBLEMS

DPP-1 Stresses and Strains

Only One Option Correct Type

Working Space

Direction (Q. Nos. 1-6) *This section contains 6 multiple choice questions. Each question has four choices (a), (b), (c) and (d), out of which ONLY ONE is correct.*

1 A steel wire 2 m long is suspended from a ceiling. When a mass is hung at its lower end, then increase in length is 1 cm. Determine the strain in the wire.

a. 0.001 b. 0.005
c. 0.002 d. 0.008

2 Figure shows an elemental rectangular block *ABCD*, whose thickness normal to the plane of the figure is unity. The four faces are acted upon by pure shear stress of intensity σ. The maximum normal stress carried by any plane is

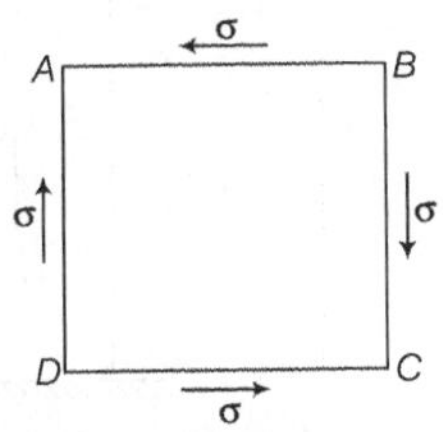

a. σ b. $\sqrt{2}\sigma$
c. 2σ d. $\frac{\sigma}{\sqrt{2}}$

3 An air filled balloon is at a depth of 1 km below the water level in an ocean. The normal stress of the balloon (in Pa) is
(Given, $\rho_{water} = 10^3 \text{ kg m}^{-3}$, $g = 9.8 \text{ ms}^{-2}$ and $p_{atm} = 10^5$ Pa)

a. 10^6 b. 9.9×10^5
c. 9.9×10^7 d. 9.9×10^6

4 An elastic wire is found to be elongated by 1 cm when a mass is gently hung from it. It extends by 3 cm more, when the attached mass is made to move uniformly in a horizontal circle performing 2 rev s^{-1}. The length of the unstretched wire (in cm) is (Given, $g = 9.81 \text{ms}^{-2}$)

a. $\frac{(24-\sqrt{15})}{\sqrt{15}} \times 4$ b. $\frac{24}{\sqrt{15}} \times 4$
c. $\frac{4}{\sqrt{5}}$ d. $\frac{(4-\sqrt{15})}{2}$

5 A load of 10 kN is supported from a pulley, which in turn is supported by a rope of cross-sectional area 10^3 mm^2 and modulus of elasticity 10^3 Nmm^{-2} as shown in the figure. Neglecting friction at the pulley, then downward deflection of the load (in mm) is

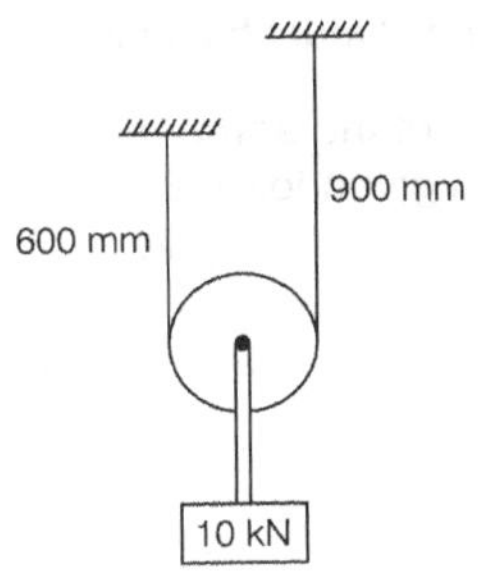

a. 3.75 b. 4.25 c. 2.75 d. 4.00

Working Space

6 A body of mass 3.14 kg is suspended from one end of a wire of length 10 m. The radius of cross-section of the wire is changing uniformly from 5×10^{-4} m at the top (i.e. point of suspension) to 9.8×10^{-4} m at the other end. Young's modulus of the material of the wire is 2×10^{11} Nm^{-2}. The change in length of the wire is

a. 4×10^{-3} m
b. 3×10^{-3} m
c. 10^{-3} m
d. 2×10^{-3} m

One or More than One Options Correct Type

Direction (Q. Nos. 7-12) *This section contains 6 multiple choice questions. Each question has four choices (a), (b), (c) and (d), out of which ONE or MORE THAN ONE are correct.*

7 A vertical bar of uniform cross-section is fixed at both of its ends and a load $w = 5000$N is applied axially at an intermediate section as shown in the figure. Choose the correct option(s).

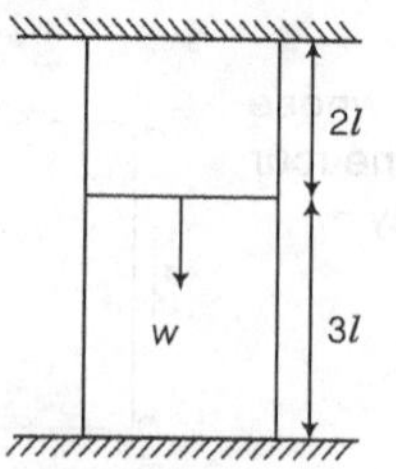

a. Reaction at the top support is 3000 N
b. Reaction at the bottom support is 2000 N
c. Reaction at the top support is 1000 N
d. Reaction at the bottom support is 3000 N

8 Two light wires P and Q shown in the figure are made of same material and have radii r_P and r_Q, respectively. The block between them has a mass m. When the force $F = \frac{mg}{3}$, then one of the wires breaks. Choose the correct option(s).

a. P breaks, if $r_P = r_Q$
b. P breaks, if $r_P < 2r_Q$
c. Either P or Q may break, if $r_P = 2r_Q$
d. To predict, which wire will break, the lengths of P and Q must be known

9 Two wires A and B have equal lengths and are made of the same material, but diameter of wire A is twice that of wire B. Then, for a given load,

a. the extension of B will be four times that of A
b. the extensions of A and B will be equal
c. the strain in B is four times that in A
d. the strains in A and B will be equal

Working Space

10 A light rod of length 2m is suspended from a ceiling horizontally by means of two vertical wires of equal length tied to its ends. One of the wires is made of steel and is of cross-section $10^{-3} m^2$ and the other is of brass of cross-section $2 \times 10^{-3} m^2$. x is the distance from steel wire end, at which a weight may be hung.

$Y_{steel} = 2 \times 10^{11}$ Pa and $Y_{brass} = 10^{11}$ Pa

Which of the following statement(s) is/are correct?

a. $x = 1.2$ m, if the strains of both the wires are to be equal
b. $x = 1.42$m, if the stresses of both the wires are to be equal
c. $x = 1$m, if the strains of both the wires are to be equal
d. $x = 1.33$ m, if the stresses of both the wires are to be equal

11 A steel rod of cross-sectional area 16 cm^2 and two brass rods each of cross-sectional area 10 cm^2 together support a load of 5000 kg as shown in the figure. (Given, $Y_{steel} = 2 \times 10^6$ kg cm^{-2} and $Y_{brass} = 10^6$ kg cm^{-2}). Choose the correct option(s).

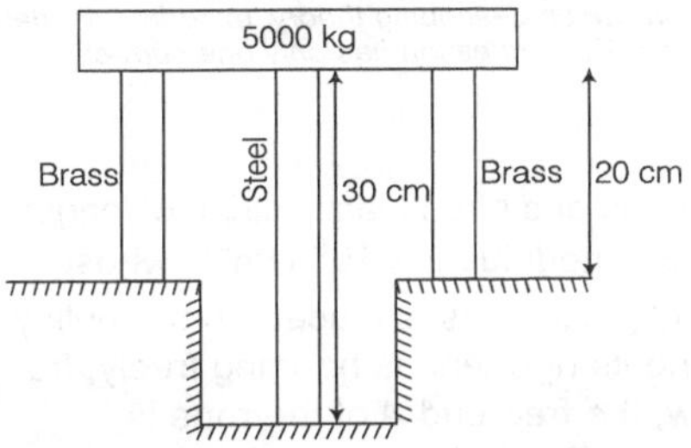

a. Stress in brass rod = 121 kg cm^{-2}
b. Stress in steel rod = 161 kg cm^{-2}
c. Stress in brass rod = 141 kg cm^{-2}
d. Stress in steel rod = 141 kg cm^{-2}

12 A member $PQRS$ is subjected to point loads F_1, F_2, F_3 and F_4 as shown in the figure. Given, $F_1 = 4500$ kg, $F_3 = 45000$ kg, $F_4 = 13000$ kg and modulus of elasticity = 2.1×10^6 kg cm^{-2}. Choose the correct option(s).

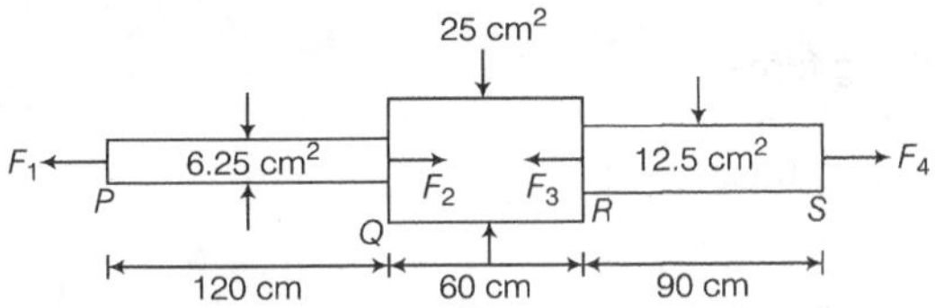

a. For equilibrium, $F_2 = 36500$ kg
b. Total elongation is 0.98 mm
c. For equilibrium, $F_2 = 26500$ kg
d. Total elongation is 0.49 mm

Statement Type

Direction (Q. No. 13) This section is based on Statement I and Statement II. Select the correct answer from the codes given below.

a. Both Statement I and Statement II are correct and Statement II is the correct explanation of Statement I
b. Both Statement I and Statement II are correct but Statement II is not the correct explanation of Statement I
c. Statement I is correct but Statement II is incorrect
d. Statement II is correct but Statement I is incorrect

Working Space

13 **Statement I** A uniform elastic rod lying on smooth horizontal surface is pulled by a constant horizontal force of magnitude F as shown in the figure. Another identical elastic rod is pulled vertically upwards by a constant vertical force of magnitude F as shown in the figure. The extensions in both the rods will be same.

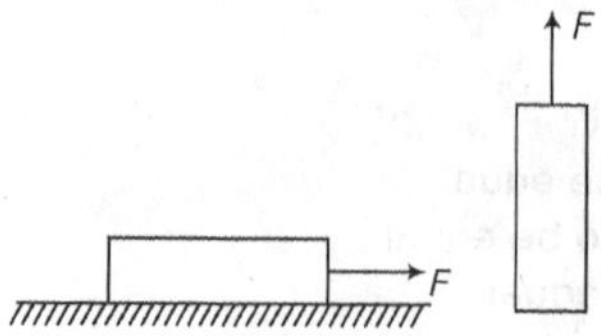

Statement II In a uniform elastic rod, the extension depends only on forces acting at the ends of rods.

Comprehension Type

Direction (Q. Nos. 14-16) *This section contains a paragraph, which describing theory, experiments, data, etc. Three questions related to the paragraph have been given. Each question has only one correct answer among the four given options (a), (b), (c) and (d).*

Passage

The axle of a pulley of mass 1 kg is attached to the end of an elastic string of length 1 m, cross-sectional area 10^{-3} m^3 and Young's modulus 2×10^5 Nm^{-2}, whose other end is fixed to the ceiling. A rope of negligible mass is placed on the pulley such that its left end is fixed to the ground and its right end is hanging freely, from the pulley which is at rest in equilibrium. Now, the free end A of the rope is subjected to pulling with constant force $F = 10$ N. Friction between the rope and the pulley can be neglected.(Given, $g = 10$ ms^{-2})

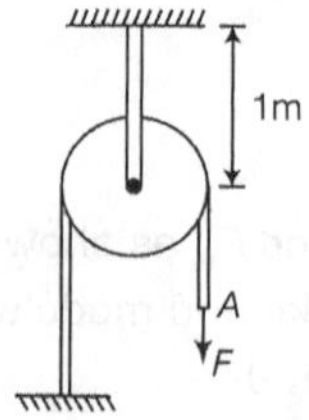

14 The elongation of the string before applying force is

a. 50 cm b. 5 cm
c. 0.5 cm d. 0.05 cm

15 The greatest elongation of the string is

a. 35 cm b. 30 cm
c. 25 cm d. 20 cm

16 The maximum displacement of point A after applying F is

a. 70 cm b. 60 cm
c. 40 cm d. 50 cm

Matching List Type

Direction (Q. No. 17) *Choices for the correct combination of elements from Column I and Column II are given as options (a), (b), (c) and (d), out of which one is correct.*

17 A bar of cross-sectional area A is subjected to equal and opposite tensile forces at its ends. Consider a plane section (PS) of the bar, whose normal makes an angle ϕ with the axis (axis is along the length) of the bar.

Working Space

Match the following columns and select the correct option from the codes given below.

	Column I		Column II
i.	Shearing stress on *PS*	p.	$\frac{F}{A}\cos^2\phi$
ii.	Tensile stress on *PS*	q.	0°
iii.	The tensile stress is maximum for ϕ =	r.	$\frac{F}{A}\sin\phi\cos\phi$
iv.	The shearing stress is maximum for ϕ =	s.	60°
		t.	45°

Codes

	i	ii	iii	iv
a.	r	p	q	t
b.	p	q	r	s
c.	q	r	p	s
d.	s	t	q	p

One Integer Value Correct Type

Direction (*Q. Nos.* 18-20) *This section contains* 3 *questions. When worked out will result in an integer from* 0 *to* 9 (*both inclusive*).

18 A thin ring of radius R is made of a material of density ρ and Young's modulus Y. When the ring is rotated about its centre in its own plane with angular velocity ω, the small increase in its radius is $\Delta R = x\left(\frac{\rho\omega^2R^3}{Y}\right)$. Find the value of x.

19 A sphere of radius 0.1 m and mass 8π kg is attached to the lower end of a steel wire of length 5 m and diameter 10^{-3} m. The wire is suspended from 5.22 m high ceiling of a room. When the sphere is made to swing as a simple pendulum, it just grazes the floor at its lowest point. Young's modulus of steel is 2×10^{11} Nm^{-2}. If the velocity of the sphere at the lowest position is v, then $v^2 = N\times 20.48$. Find the value of N.

20 The breaking stress for a metal is 7.8×10^9 Nm^{-2}. The density of the metal is 7800 kg m^{-3}. If g = 10 N kg^{-1}, the maximum length of the wire made of this metal which may be suspended without breaking is 10^k m. Find the value of k.

DPP-2 Modulii of Elasticity

Only One Option Correct Type

Working Space

Direction (*Q. Nos. 1-9*) *This section contains 9 multiple choice questions. Each question has four choices (a), (b), (c) and (d), out of which ONLY ONE is correct.*

1 A load of 4 kg is suspended from a ceiling through a steel wire of length 20 m and radius 2 mm. It is found that the length of the wire increases by 0.031 mm, as equilibrium is achieved. If $g = 3.1 \times \pi \text{ ms}^{-2}$, the value of Young's modulus of the material of the wire (in Nm^{-2}) is

a. 2×10^{12} b. 4×10^{11}

c. 2×10^{11} d. 0.02×10^{9}

2 A wire of length 1m and radius 1mm is subjected to a load. The extension is x. The wire is melted and then drawn into a wire of square cross-section of side 2 mm. Its extension under the same load will be

a. $\frac{\pi^2 x}{8}$ b. $\frac{\pi^2 x}{16}$

c. $\frac{\pi^2 x}{2}$ d. $\frac{x}{2\pi}$

3 A wire elongates by l units, when a load w is suspended from it. If the wire gets over a pulley (equally on both the sides) and two weights w each are hung at the two ends, the elongation of wire (in units) will be

a. zero b. $\frac{l}{2}$

c. l d. $2l$

4 When the tension on wire is 4 units, its length is l_1 and when the tension is 5 units, its length is l_2. Its natural length is

a. $4l_1 - 3l_2$ b. $5l_1 - 4l_2$

c. $5l_2 - 4l_1$ d. $\frac{l_1 + l_2}{2}$

5 If stress is numerically equal to Young's modulus, then the elongation will be

a. $\frac{1}{4}$th of the original length

b. $\frac{1}{2}$th of the original length

c. equal to the original length

d. twice the original length

6 Maximum stress that can be applied to a wire which supports on elevator is σ. Mass of elevator is m and it is moved upwards with an acceleration of $g/2$. Minimum diameter of wire (Neglecting weight of wire) must be

a. $\sqrt{\frac{2mg}{\pi\sigma}}$ b. $\sqrt{\frac{3mg}{2\pi\sigma}}$

c. $\sqrt{\frac{5mg}{2\pi\sigma}}$ d. $\sqrt{\frac{6mg}{\pi\sigma}}$

Working Space

7 A body is in the form of solid revolution, radius of upper face of the body is r_0 and volumetric density is ρ. The solid body supports a load P. If the compressive stress at all cross-sections should be same, then the radius r as a function of y (see figure) is

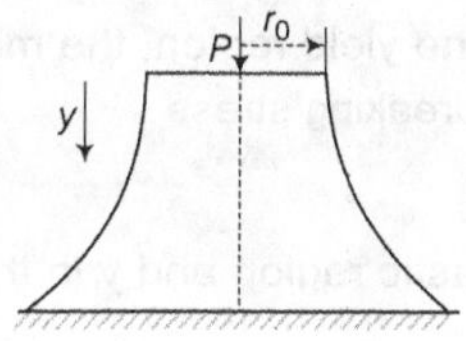

a. $r = \frac{r_0}{2} e^{(\pi r_0^2 \rho y / 2P)}$
b. $r = r_0 e^{(\pi r_0^2 \rho y / 2P)}$
c. $r = r_0 e^{(\pi r_0^2 \rho y / P)}$
d. $r = r_0 e^{(2\pi r_0^2 \rho y / P)}$

8 Figure shows the stress-strain curve of two metals P and Q. From the graph, it can be concluded that

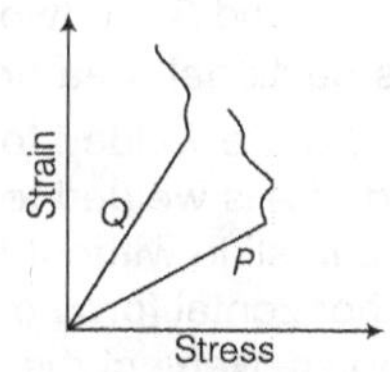

a. P has greater Young's modulus and lesser ductility
b. Q has greater Young's modulus and lesser ductility
c. P has greater Young's modulus and greater ductility
d. Q has greater Young's modulus and greater ductility

9 Depth of sea is maximum at Mariana Trench in West Pacific Ocean. Trench has a maximum depth of about 11 km. At bottom of trench water column above exerts over 1000 atm pressures. Percentage change in density of sea water at such depth will be around.

(Given, $B = 2 \times 10^9\ \text{Nm}^{-2}$ and $p_{atm} = 1 \times 10^5\ \text{Nm}^{-2}$)

a. about 5%
b. about 10%
c. about 3%
d. about 7%

One or More than One Options Correct Type

Direction (Q. No. 10) *This section contains 1 multiple choice question. Each question has four choices (a), (b), (c) and (d), out of which ONE or MORE THAN ONE are correct.*

10 A solid shaft radius R, length L, coefficient of modulus of rigidity G, can transmits power of P watts at constant angular speed ω

Now, choose the correct option(s).

a. Power and torque (T) are related as $P = \omega T$
b. Torque $\tau \propto R^4$
c. Torque $\tau \propto R^3$
d. $\frac{T}{j} = \frac{G\theta}{L} = \frac{\tau}{r}$, where $\tau = \frac{G\theta r}{L}$ and $j = 2\pi \int_0^R r^3\, dr$

Comprehension Type

Direction (Q. Nos. 11-16) *This section contains 2 paragraphs, each describing theory, experiments, data, etc. Six questions related to the paragraphs have been given. Each question has only one correct answer among the four given options (a), (b), (c) and (d).*

Passage I

On gradual loading, stress-strain relationship for a metal wire is as follows.

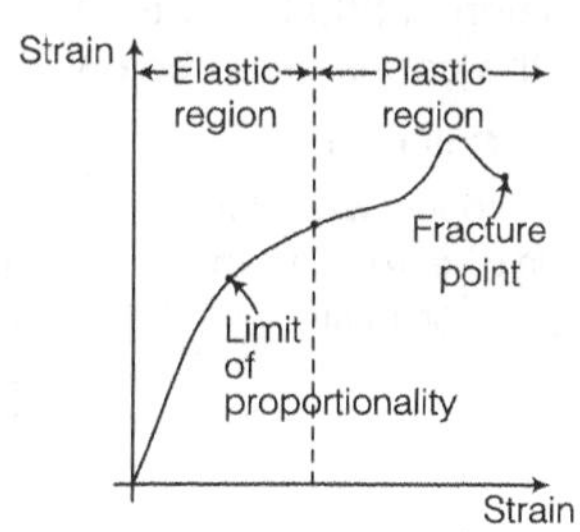

Within proportionality limit, stress $\propto$ strain and so,

$\frac{\text{Stress}}{\text{Strain}}$ = a constant for the material of wire.

11 Two wires of same material have length and radius (L, r) and $\left(2L, \frac{r}{2}\right)$. The ratio of their Young's modulii is

a. 1 : 2
b. 2 : 3
c. 2 : 1
d. 1: 1

12 Just on crossing the yield region, the material will have

a. increased and breaking stress b. reduced and breaking stress
c. constant stress d. None of the above

13 If $\dfrac{\text{Stress}}{\text{Strain}}$ is x in elastic region and y in the region of yield, then

a. $x > y$ b. $x = y$ c. $x < y$ d. $x = 2y$

Passage II

In the figure shown, A and B are two short steel rods each of cross-sectional area 5cm^2. The lower ends of A and B are welded to a fixed plate CD. The upper end of A is welded to an L-shaped piece EFG, which can slide without friction on the upper end of B. A horizontal force of 1200 N is exerted at G as shown. Neglect the weight of EFG.

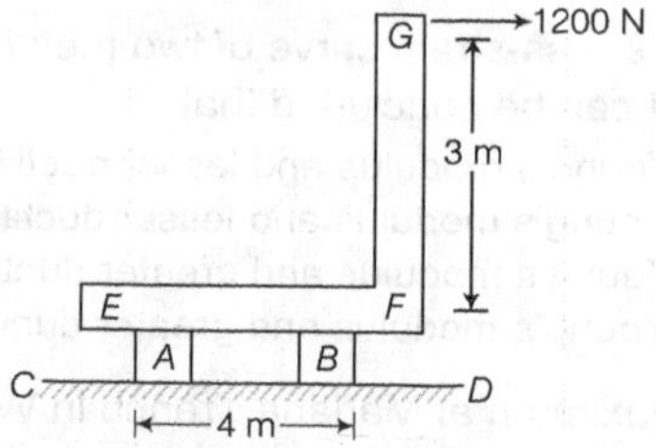

14 Mark out the correct statement(s).

a. Shearing stress in A is zero
b. Shearing stress in B is zero
c. Shearing stress in both A and B is zero
d. Shearing stress in both A and B is non-zero

15 Longitudinal stress in A is

a. tensile in nature and has the magnitude 180Ncm^{-2}
b. tensile in nature and has the magnitude 240Ncm^{-2}
c. compressive in nature and has the magnitude 180Ncm^{-2}
d. compressive in nature and has the magnitude 240Ncm^{-2}

16 Longitudinal stress in B is

a. tensile in nature and has the magnitude 180Ncm^{-2}
b. tensile in nature and has the magnitude 240Ncm^{-2}
c. compressive in nature and has the magnitude 180Ncm^{-2}
d. compressive in nature and has the magnitude 240Ncm^{-2}

Matching List Type

Direction (*Q. Nos.* 17 and 18) *Choices for the correct combination of elements from Column I and Column II are given as options (a), (b), (c) and (d), out of which one is correct.*

17 A copper wire ($Y = 10^{11}\ \text{Nm}^{-2}$) of length 8 m and steel wire ($Y = 2 \times 10^{11}\text{Nm}^{-2}$) of length 4 m and each of cross-sectional area $0.5\ \text{cm}^2$ are fastened end to end and stretched with a tension of 500 N. Match the following columns and select the correct option from the codes given below.

	Column I		Column II
i.	Elongation in copper wire (in mm)	p.	0.2
ii.	Elongation in steel wire (in mm)	q.	1.0
iii.	Total elongation (in mm)	r.	0.8
		s.	0.1
		t.	0.9

Codes

	i	ii	iii		i	ii	iii
a.	r	p	q	b.	s	p	q
c.	r	p	q	d.	p	s	r

18 In Column I, a uniform bar of uniform cross-sectional area under the application of forces is shown in the figure and in Column II, some effects/phenomena are given. Match the following columns and select the correct option from the codes given below.

	Column I		Column II
i.	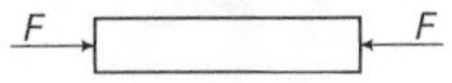	p.	Uniform stresses are developed in the rod.
ii.	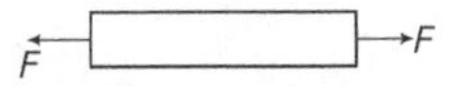	q.	Non-uniform stresses are developed in the rod.
iii.	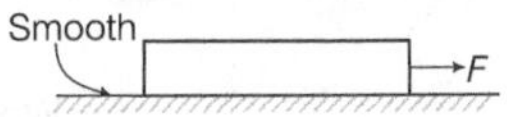	r.	Compressive stresses are developed in the rod.
iv.	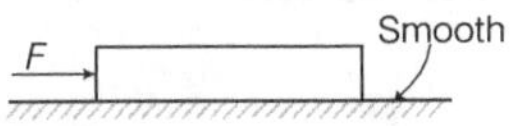	s.	Tensile stresses are developed in the rod.

Codes

	i	ii	iii	iv		i	ii	iii	iv
a.	p,r	p,s	q,s	q,r	b.	p,s	q,s	q,s	p,r
c.	p,r	q,s	p,r	q,r	d.	q,r	p,s	p,r	q,s

One Integer Value Correct Type

Direction (Q. Nos. 19-21) This section contains 3 questions. When worked out will result in an integer from 0 to 9 (both inclusive).

19 The average depth of Indian ocean is about 3000 m. The fractional compression $\frac{\Delta V}{V}$, of water of the bottom of the ocean is $\beta(0.68)\%$. Find out β. Given, bulk modulus of water is 2.2×10^9 Nm^{-2} and density of water = 1000 kgm^{-3}.

20 A solid sphere of radius R made of a material of bulk modulus B is surrounded by a liquid in a cylindrical container. A massless piston of area A (the area of container is also A) floats on the surface of the liquid. When a mass M is placed on the piston to compress the liquid, fractional change in radius of the sphere is $\frac{Mg}{\alpha AB}$. Find the value of α.

21 A 0.1 kg mass is suspended from a wire of negligible mass. The length of the wire is 1 m and its cross-sectional area is 4.9×10^{-7} m^2. If the mass is pulled a little in the vertically downward direction and released, it performs SHM with angular frequency 140 rad s^{-1}. If the Young's modulus of the material of the wire is $p \times 10^9$ Nm^{-2}, find the value of p.

DPP-3 Elastic Potential Energy

Only One Option Correct Type

Working Space

Direction (Q. Nos. 1-9) This section contains 9 multiple choice questions. Each question has four choices (a), (b), (c) and (d), out of which ONLY ONE is correct.

1 Which of these option(s) is correct about elastic forces?

a. Elastic forces are always conservative in nature
b. Elastic forces are considered conservative only within elastic limit
c. Beyond elastic limit, work done by elastic forces appears completely in form of heat
d. Elastic forces can be conservative or non-conservative in nature

2 The graph shows the behaviour of a steel wire in the region for which the wire obeys Hooke's law. The graph is a parabola. The variables X and Y-axes, respectively can be [where, stress (σ), strain (ε) and elastic potential energy (U)]

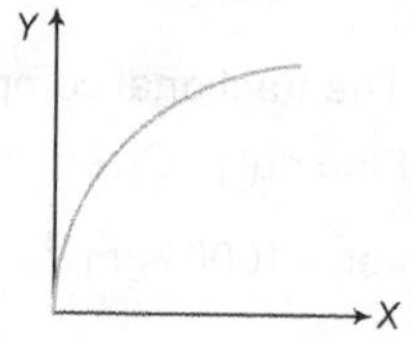

a. U, σ
b. U, ε
c. σ, ε
d. None of these

3 Free end of a light elastic wire is loaded with a mass m and the wire shows an extension h. Elastic potential energy stored by the wire is

a. mgh
b. $\frac{2}{3}mgh$
c. $\frac{1}{2}mgh$
d. zero

4 A stone of mass m is projected from a rubber catapult of length l and51133 cross-sectional area A, stretched by an amount Δl. If Y is the Young's modulus of rubber, then velocity of projection of stone is

a. $\sqrt{\frac{YA(\Delta l)^2}{lm}}$
b. $\sqrt{\frac{YA(\Delta l)^2}{2lm}}$
c. $\sqrt{\frac{2YA(\Delta l)^2}{lm}}$
d. $\frac{1}{2}\sqrt{\frac{YA(\Delta l)^2}{lm}}$

5 A metal wire of length l and cross-sectional area A has mass m.It is stretched by an amount Δl by a load of mass M. If the wire breaks at the point of suspension due to the load, then rise in temperature of the wire ΔT is (where,C is specific heat of the wire)

a. $\Delta T = \frac{Mg(\Delta l)}{4mC}$
b. $\Delta T = \frac{2Mg(\Delta l)}{2mC}$
c. $\Delta T > \frac{Mg(\Delta l)}{2mC}$
d. $\Delta T < \frac{Mg(\Delta l)}{4mC}$

6 If the work done in stretching a wire by 1 mm is 2J, then work necessary for stretching another wire of same material but with double radius of cross-section and half of the length by 1 mm is

a. 8 J b. 16 J
c. 4 J d. 32 J

7 A rubber of volume 2000 cc is alternately subjected to tension and released. The figure shows the stress-strain curve of rubber. Each curve is a quadrant of an ellipse. The amount of energy lost as heat per cycle per unit volume will be

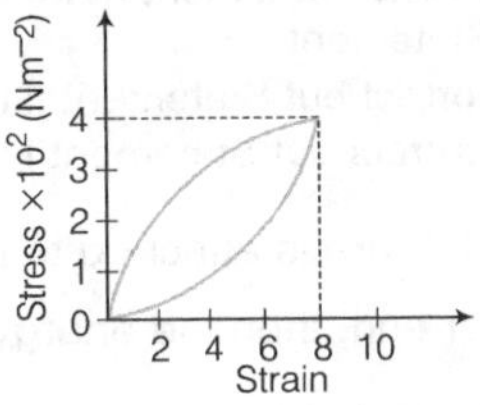

a. $\left(\frac{\pi}{2}-1\right)\times 16\times 10^2$ J

b. $\left(\frac{\pi}{4}-1\right)\times 8\times 10^2$ J

c. $\left(\frac{\pi}{4}-1\right)\times 32\times 10^2$ J

d. $\left(\frac{\pi}{2}-1\right)\times 32\times 10^2$ J

8 Two different types of rubber are found to have the stress-strain curves as shown in the figure.

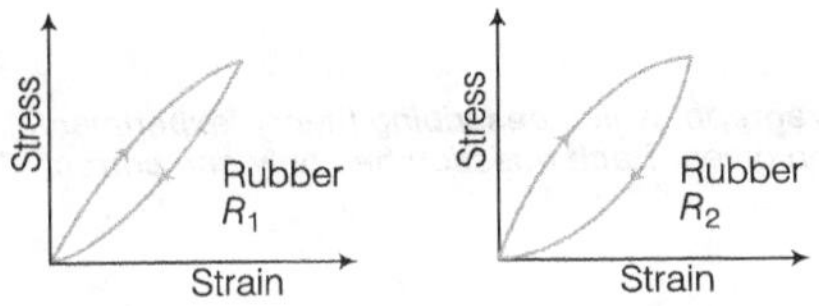

The following are two situations, where rubber is required.

i. A heavy machine is to be installed in a factory. To absorb the vibrations of the machine , a block of rubber, is to be placed between the machinery and the floor.

ii. A rubber tyre for a car has to be manufactured.

Which of the following statements is/are correct?

a. R_2 is suggested for (i) and (ii) b. R_1 is suggested for (i) and (ii)
c. R_1 is suggested for (i) and R_2 for (ii) d. R_1 is suggested for (ii) and R_2 for (i)

9 A smooth uniform string of natural length L_0, cross-sectional area A and Young's modulus Y is pulled along its length by a force F on a horizontal smooth surface. The elastic potential energy stored in the string is

a. $\frac{2F^2L_0}{AY}$ b. $\frac{F^2L_0}{3AY}$
c. $\frac{F^2L_0}{2AY}$ d. $\frac{F^2L_0}{6AY}$

One or More than One Options Correct Type

Direction (Q. No. 10) This section contains 1 multiple choice question. This question has four choices (a), (b), (c) and (d), out of which ONE or MORE THAN ONE are correct.

10 Figure shows the graph of elastic potential energy (U) stored *versus* extension, for a steel wire ($Y = 2\times 10^{11}$ Pa) of volume 200 cc. If area of cross-section A and original length L, then

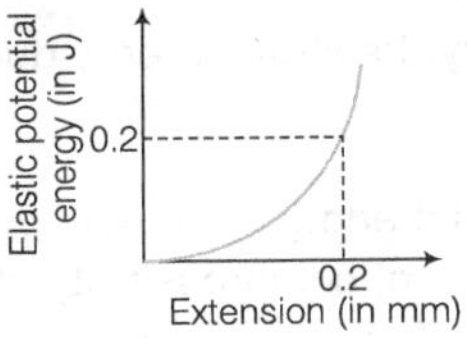

a. $A = 10^{-4}\,\text{m}^2$ b. $A = 10^{-3}\,\text{m}^2$
c. $L = 1.5\,\text{m}$ d. $L = 2\,\text{m}$

Statement Type

Working Space

Direction (Q. Nos. 11 and 12) *This section is based on Statement I and Statement II. Select the correct answer from the codes given below.*

a. Both Statement I and Statement II are correct and Statement II is the correct explanation of Statement I
b. Both Statement I and Statement II are correct but Statement II is not the correct explanation of Statement I
c. Statement I is correct but Statement II is incorrect
d. Statement II is correct but Statement I is incorrect

11 **Statement I** If a metal wire is attached to the ceiling of a room and a mass M is attached to another end, then the energy stored in the stretched wire is $\frac{Mgl}{2}$, where l is the increment in length of the wire.

Statement II In the above statement, loss in gravitational energy is Mgl, while the loss in energy to the surroundings is $\frac{Mgl}{2}$.

12 **Statement I** Identical springs of steel and copper are equally stretched. More work will be done on the steel spring.

Statement II Steel is more elastic than copper.

Comprehension Type

Direction (Q. Nos. 13 and 14) *This section contains a paragraph, which describing theory, experiments, data, etc. Two questions related to the paragraph have been given. Each question has only one correct answer among the four given options (a), (b), (c) and (d).*

Passage

A solid has a volume V_0 when external pressure is p_0.The bulk modulus of the material is B. Now, the pressure is increased to p.

13 What is the decrease in volume of the solid?

a. $\frac{(p-p_0)V_0}{2B}$
b. $\frac{(p-p_0)V_0}{B}$
c. $\frac{2(p-p_0)V_0}{B}$
d. $\frac{(2p-p_0)V_0}{B}$

14 What additional energy per unit volume is now stored in the material?

a. $\frac{1}{2B}(p^2-p_0^2)$
b. $\frac{1}{B}(p^2-p_0^2)$
c. $\frac{2}{B}(p^2-p_0^2)$
d. $\frac{1}{\sqrt{2}B}(p^2-p_0^2)$

One Integer Value Correct Type

Direction (Q. Nos. 15 and 16) *This section contains 2 questions. When worked out will result in an integer from 0 to 9 (both inclusive).*

15 A wire having a length L and cross-sectional area A is suspended at one of its ends from a ceiling. Density and Young's modulus of material of the wire are ρ and Y, respectively. Its strain energy due to its own weight is $\frac{\rho^2 g^2 AL^3}{\alpha Y}$. Find the value of α.

16 The elastic potential energy per unit volume of water at a depth of 1 km is $3x \times 10^3$ Jm^{-3}. Given, compressibility of water $= 5\times10^{-10}$ SI units and density of water $= 1000$ kg m^{-3}. Find the value of x.

DPP-4 Thermal Stresses and Strains

Only One Option Correct Type

Working Space

Direction (Q. Nos. 1-12) *This section contains 12 multiple choice questions. Each question has four choices (a), (b), (c) and (d), out of which ONLY ONE is correct.*

1 A uniform metal rod is fixed at one end to a rigid support. Its temperature is gradually increased. If a plot be made of the stress (σ) *versus* increment in temperature $\Delta\theta$, which curve best represents it?

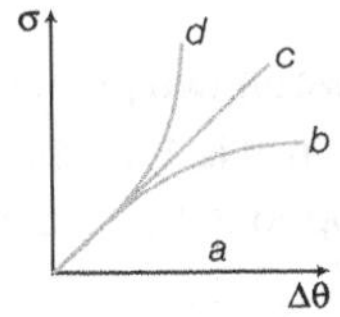

a. *a* b. *b*
c. *c* d. *d*

2 Steel wire of length L at 40° C is suspended from the ceiling and then a mass m is hung from its free end. The wire is cooled down from 40° C and 30° C to regain its original length L. The coefficient of linear thermal expansion of the steel is 10^{-5}mol^{-1}, Young's modulus of steel is 10^{11} Nm^{-2} and the radius of the wire is 1 mm. Assume that $L >>$ diameter of the wire. Then, the value of m is (Given, $g = 10\text{ms}^{-2}$)

a. 3 kg b. 4 kg
c. 5 kg d. 2 kg

3 Two rods of different materials having coefficients of linear expansion α_1 and α_2 and Young's modulii of elasticity Y_1 and Y_2 respectively, are fixed between two rigid massive walls. The rods are heated such that they undergo the same increase in temperature without bending. If $\alpha_1 : \alpha_2 = 2:3$, then the thermal stress developed in the two rods will be equal if $Y_1 : Y_2$ is equal to

a. 2 : 3 b. 3 : 2
c. 1 : 2 d. 2 : 1

4 A steel rod of length 1 m is heated from 25°C to 75°C keeping its length constant. The longitudinal strain developed in the rod is (Given, $\alpha_{\text{steel}} = 12 \times 10^{-6} \text{°C}^{-1}$)

a. -4×10^{-4} b. 6×10^{-4}
c. -6×10^{-4} d. 4×10^{-4}

5 A rod of length 1000 mm and coefficient of linear expansion $\alpha = 10^{-4}$ °C^{-1} is placed symmetrically between fixed walls separated by 1001 mm. The Young's modulus of the rod is 10^{11} Nm^{-2}. If the temperature is increased by 20° C, then the stress developed in the rod is

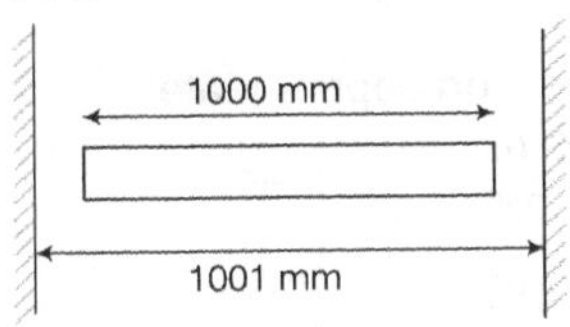

a. 10^{11} Pa b. 10^{10} Pa c. 10^{9} Pa d. 10^{8} Pa

Working Space

6 A steel wire of uniform cross-section $1\,mm^2$ is stretched and its ends are rigidly fixed. Calculate the change in tension of the wire when the temperature falls from 70°C to 35°C. Coefficient of linear expansion of steel is $1.1 \times 10^{-5}\,°C^{-1}$ and Young's modulus is $2 \times 10^{11}\,Nm^{-2}$.

a. 88 N
b. 101 N
c. 92 N
d. 77 N

7 Two bobs of different metals but of same mass are suspended from different metal wires of same cross-sectional area but of different length L_1 and L_2. Thus, the bobs get heated by different temperature $\Delta\theta_1$ and $\Delta\theta_2$. The specific heats of two bobs are related as $\frac{C_1}{C_2} = \frac{\Delta\theta_2}{\Delta\theta_1}$. Find the ratio of the strains in two wires.

a. $\frac{\Delta\theta_1 L_1}{\Delta\theta_2 L_2}$
b. $\frac{C_1 L_1}{C_2 L_2}$
c. $\frac{L_1}{L_2}$
d. $\frac{L_2}{L_1}$

8 A rod is fixed between two points at 20°C. The coefficient of linear expansion of the material of the rod is $1.1 \times 10^{-5}\,°C^{-1}$ and Young's modulus is $1.2 \times 10^{11}\,Nm^{-2}$. The stress developed in the rod, if temperature of the rod becomes 10°C is

a. $1.32 \times 10^7\,Nm^{-2}$
b. $1.1 \times 10^{15}\,Nm^{-2}$
c. $1.32 \times 10^8\,Nm^{-2}$
d. $1.1 \times 10^6\,Nm^{-2}$

9 Two rods of different metals but of equal cross-section and length (1 m each) are joined to make a rod of length 2 m. The metal of one rod has a coefficient of linear thermal expansion $10^{-5}\,°C^{-1}$ and Young's modulus $3 \times 10^{10}\,Nm^{-2}$. The other metal has the values $2 \times 10^{-5}\,°C^{-1}$ and $10^{10}\,Nm^{-2}$, respectively. How much pressure must be applied to the ends of the composite rod to prevent its expansion when the temperature is raised to 100°C?

a. $4.25 \times 10^7\,Nm^{-2}$
b. $3.25 \times 10^7\,Nm^{-2}$
c. $2.25 \times 10^7\,Nm^{-2}$
d. $1.25 \times 10^7\,Nm^{-2}$

10 A steel wire of cross-sectional area $0.5\,mm^2$ is held between two fixed supports. If tension in the wire is negligible and is just taut at 20°C, determine the tension when the temperature falls to 0°C. Young's modulus is 21×10^{11} dyne cm^{-2} and coefficient of linear expansion is $12 \times 10^{-6}\,°C^{-1}$.

a. 25.2 N
b. 15.2 N
c. 35.2 N
d. 27.5 N

11 A copper plate has two steel plates soldered on its either side with rigid plates securing ends. What is the tension in each plate, if temperature is increased by $\Delta\theta$? Cross-sectional area of each interface is A. The coefficients of expansion of copper and steel are α_c and α_s $(\alpha_c > \alpha_s)$ and Young's modulii are Y_c and Y_s, respectively.

a. $\frac{AY_cY_s(\alpha_c - \alpha_s)(\Delta\theta)}{2Y_s + Y_c}$
b. $\frac{2AY_cY_s(\alpha_c - \alpha_s)(\Delta\theta)}{Y_s + Y_c}$
c. $\frac{AY_cY_s(\alpha_c - \alpha_s)(\Delta\theta)}{Y_s + Y_c}$
d. $\frac{AY_cY_s(\alpha_c - \alpha_s)(\Delta\theta)}{2Y_s + 2Y_c}$

12 In the above problem, what is the compressive force on copper plate?

a. $\frac{AY_cY_s(\alpha_c - \alpha_s)(\Delta\theta)}{2Y_s + Y_c}$
b. $\frac{2AY_cY_s(\alpha_c - \alpha_s)(\Delta\theta)}{2Y_s + Y_c}$
c. $\frac{AY_cY_s(\alpha_c - \alpha_s)(\Delta\theta)}{Y_s + Y_c}$
d. $\frac{2AY_cY_s(\alpha_c - \alpha_s)(\Delta\theta)}{2Y_s + 2Y_c}$

Working Space

One or More than One Options Correct Type

Direction (Q. Nos. 13 and 14) *This section contains 2 multiple choice questions. Each question has four choices (a), (b), (c) and (d), out of which ONE or MORE THAN ONE are correct.*

13 A rod is made of uniform material and has non-uniform cross-section. It is fixed at both the ends as shown and heated at mid-section. Which of the following statements are not correct?

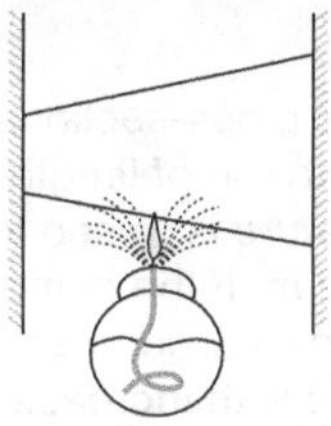

a. Force of compression in the rod will be maximum at mid-section
b. Compressive stress in the rod will be maximum at left end
c. Since the rod is fixed at both the ends, its length will remain unchanged. Hence, no strain will be induced in it
d. Compressive stress in the rod will be maximum at right end

14 The figure shows bimetallic strip used as a thermostat in a circuit. Which will be switched ON when the bimetallic strip becomes hot?

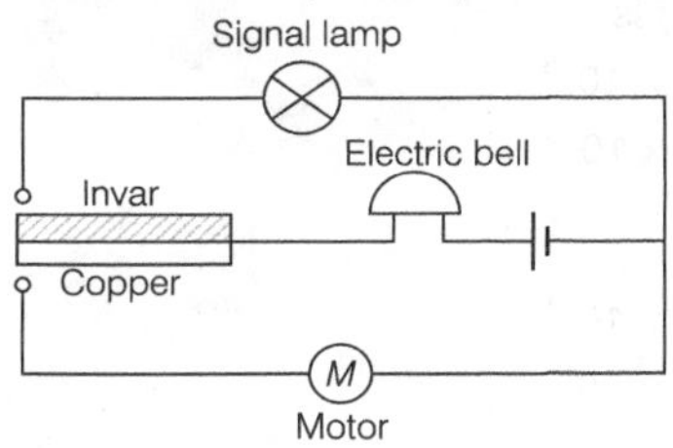

a. Bell   b. Signal lamp
c. Motor   d. All of these

Matching List Type

Direction (Q. No. 15) *Choices for the correct combination of elements from Column I and Column II are given as options (a), (b), (c) and (d), out of which one is correct.*

15 A steel rod of length 6 m and radius 1 mm is fixed between two rigid supports. Given, $Y = 2 \times 10^6$ kg cm^{-2} and $\alpha = 12 \times 10^{-6}$ °C^{-1} (stress is denoted by σ).

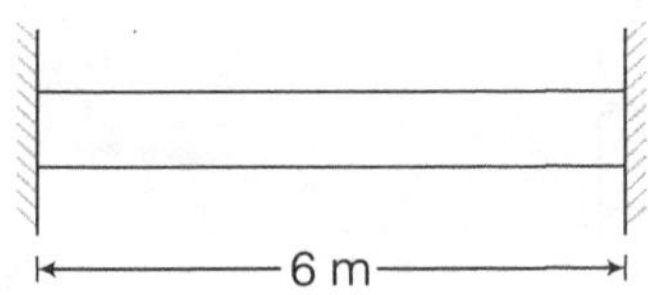

Match the following columns and select the correct option from the codes given below.

	Column I		Column II
i.	When the temperature increases by 80°C and the ends do not yield.	p.	$\sigma = 1920$ kg cm^{-2}
ii.	When the temperature increases by 80°C and the ends yield by 1 mm.	q.	$\sigma = 1587$ kg cm^{-2}
iii.	When the temperature increases by 100°C and the ends do not yield.	r.	$\sigma = 2400$ kg cm^{-2}
iv.	When the temperature increases by 100°C and the ends yield by 2 mm.	s.	$\sigma = 1733$ kg cm^{-2}

Codes

	i	ii	iii	iv		i	ii	iii	iv
a.	p	q	r	s	b.	q	r	p	s
c.	p	r	s	q	d.	q	r	p	s

Comprehension Type

Working Space

Direction (Q. Nos. 16-20) *This section contains 2 paragraphs, each describing theory, experiments, data, etc. Five questions related to the paragraphs have been given. Each question has only one correct answer among the four given options (a), (b), (c) and (d).*

Passage I

Two rods of equal cross-sections, one of copper and the other of steel are joined to form a composite rod of length 2 m at 20°C, the length of copper rod is 0.5 m. When the temperature is raised to 120°C, the length of the composite rod increases to 2.002 m. If the composite rod is fixed between two rigid walls and thus not allowed to expand, then it is found that the length of the component rods also do not change with increase in temperature. Given, Young's modulus of copper is 1.3×10^{11} Pa and coefficient of linear expansion of copper is $1.6 \times 10^{-5}\,{}^\circ C^{-1}$.

16 Coefficient of linear expansion of steel (in $^\circ C^{-1}$) is

a. 0.6×10^{-5}
b. 0.7×10^{-5}
c. 0.8×10^{-5}
d. 1.0×10^{-5}

17 Young's modulus of steel (in Nm^{-2}) is

a. 2.6×10^{15}
b. 2.6×10^{14}
c. 2.6×10^{12}
d. 2.6×10^{11}

Passage II

A steel bolt of cross-sectional area $A_b = 5 \times 10^{-5}$ m^2 is passed through a cylindrical tube made of aluminium. Cross-sectional area of the tube material is $A_t = 10^{-4}$ m^2 and its length is l = 50cm. The bolt is just taut so that there is no stress in the bolt and temperature of the assembly increases through $\Delta\theta = 10^\circ$C. The increase in length of the system is 0.075 mm. Given, coefficient of linear thermal expansion of steel is $\alpha_b = 10^{-5}\,{}^\circ C^{-1}$, Young's modulus of steel is $Y_b = 2 \times 10^{11}$ Nm^{-2}, Young's modulus of aluminium is $Y_t = 10^{11}$ Nm^{-2} and coefficient of linear expansion of Al is $\alpha_t = 2 \times 10^{-5}\,{}^\circ C^{-1}$.

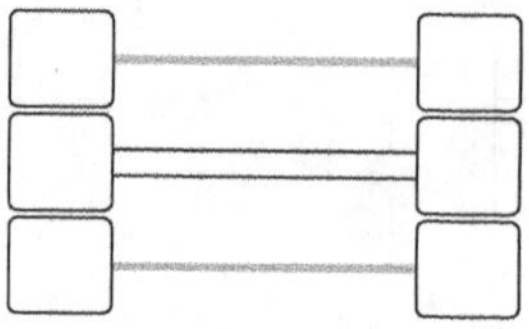

18 The compressive strain in the tube is

a. 10^{-4}
b. 5×10^{-5}
c. 2×10^{-3}
d. 10^{-6}

19 The compressive stress in the tube is

a. 5×10^{6} Nm^{-2}
b. 10^{5} Nm^{-2}
c. 10^{8} Nm^{-2}
d. 10^{3} Nm^{-2}

20 The tensile stress in the bolt is

a. 10^{4} Nm^{-2}
b. 10^{7} Nm^{-2}
c. 2×10^{8} Nm^{-2}
d. 10^{10} Nm^{-2}

Revisal Problems
for JEE Main

Only One Option Correct Type

Working Space

Direction (Q. Nos. 1-20) This section contains 20 multiple choice questions. Each question has four choices (a), (b), (c) and (d), out of which ONLY ONE is correct.

1 A steel wire (original length = 2 m, diameter = 1 mm) and a copper wire (original length = 1m, diameter = 2 mm) are loaded as shown in the figure. Find the ratio of extension of steel wire to that of copper wire. Given, Young's modulus of steel = 2×10^{11} Nm^{-2} and that of copper is 10^{11} Nm^{-2}.

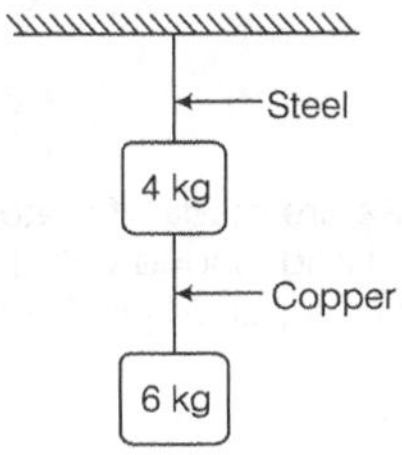

a. $\frac{10}{3}$ b. 5
c. $\frac{20}{3}$ d. $\frac{14}{3}$

2 When the pressure on all sides of a metal cube is increased by 10^7 Pa, its volume decreases by 0.015%. The bulk modulus of the metal is

a. 6.67×10^{10} Pa b. 7.67×10^{10} Pa
c. 5.67×10^{10} Pa d. 4.67×10^{10} Pa

3 A rubber cube is subjected to a pressure of 10^6 Pa on all sides. As a result, each side of the cube decreases by 1%. Then, bulk modulus of rubber is

a. 5.33×10^7 Pa
b. 3.33×10^7 Pa
c. 4.33×10^7 Pa
d. 2.33×10^7 Pa

4 A rubber eraser 3 cm × 1 cm × 8 cm is clamped at one end with 8 cm edge as vertical. A horizontal force $F = 2.1$ N is applied on the free face. Shear modulus of rubber = 1.4×10^{-2} Nm^{-2}. Then, horizontal displacement of the top face is

a. 1 mm b. 2 mm
c. 3 mm d. 4 mm

5 A thick uniform rubber rope of density 1.5 g cm^{-3} and Young's modulus 5×10^6 Nm^{-2} has a length of 8 m. When hung from the ceiling of a room, the increase in length of the rope due to its own weight will be (Given, $g = 10$ ms^{-2})

a. 9.6×10^{-8} m b. 19.2×10^{-3} m
c. 9.6×10^{-3} m d. 9.6 m

6 A wire of length L is stretched by a length l when a force F is applied at one end. If the elastic limit is not exceeded, then the amount of energy stored in the wire is given by

a. Fl b. $\frac{1}{2}Fl$
c. $\frac{Fl^2}{L}$ d. $\frac{1}{2} \cdot \frac{Fl^2}{L}$

Working Space

7 A wire, suspended vertically from one end, is stretched by attaching a weight of 20 N to the lower end. The weight stretches the wire by 1 mm. The energy gained by the wire is

a. 0.01 J
b. 0.02 J
c. 0.04 J
d. 1.0 J

8 A uniform pressure p is exerted on all sides of a solid cube at temperature $t°$ C. By what amount should the temperature of the cube be raised in order to bring its volume back to the value it had before the pressure was applied? The coefficient of volume expansion of the cube is γ and the bulk modulus is B.

a. $\frac{p}{\sqrt{2}\gamma B}$
b. $\frac{p}{2\gamma B}$
c. $\frac{2p}{\gamma B}$
d. $\frac{p}{\gamma B}$

9 The following four wires are made of the same material. Which of these will have the largest extension when the same tension is applied to each?
(length $= L$, diameter $= d$)

a. $L = 0.5\,\text{m}, d = 0.5\,\text{mm}$
b. $L = 1\,\text{m}, d = 1\,\text{mm}$
c. $L = 2\,\text{m}, d = 2\,\text{mm}$
d. $L = 3\,\text{m}, d = 3\,\text{mm}$

10 Two strings of equal lengths and equal cross-sectional areas are made of materials whose Young's modulii are in the ratio of 3 : 2. They are suspended and loaded with the same mass. When stretched and released, they will oscillate with time periods in the ratio of

a. $\sqrt{3} : \sqrt{2}$
b. $\sqrt{2} : \sqrt{3}$
c. $3\sqrt{3} : 2\sqrt{2}$
d. 9 : 4

11 The density of water at the surface of an ocean is ρ. If the bulk modulus of water is B, then what is the density of ocean water at a depth where the pressure is np_0, where p_0 is the atmospheric pressure?

a. $\frac{\rho B}{B + np_0}$
b. $\frac{\rho B}{B + (n-1)p_0}$
c. $\frac{\rho B}{B - np_0}$
d. $\frac{\rho B}{B - (n-1)p_0}$

12 One end of a uniform rod of mass m and cross-sectional area A is suspended from a rigid support and an equal mass m is suspended from the other end. The stress at the mid-point of the rod will be

a. $\frac{2\,mg}{A}$
b. $\frac{3\,mg}{2A}$
c. $\frac{mg}{A}$
d. None of these

13 The potential energy U between two atoms in a diatomic molecule as function of distance x between atoms has been shown in the figure.

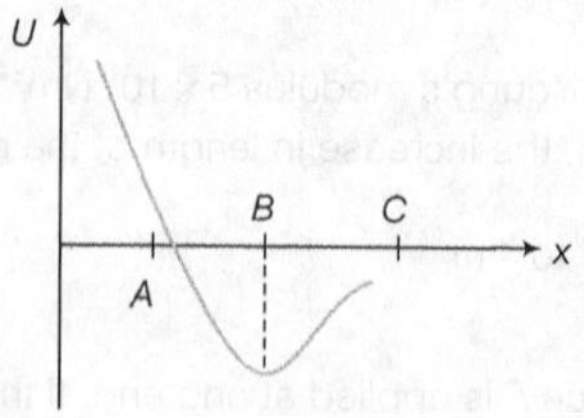

The atoms are

a. attracted when x lies between A and B and repelled when x lies between B and C
b. attracted when x lies between B and C and repelled when x lies between A and B
c. attracted when they reach B
d. None of the above

Working Space

14 The graph shown was obtained from experimental measurements of the period of oscillation T for different masses m placed in the scale pan on the lower end of the spring balance. The most likely reason for the line not passing through the origin is that

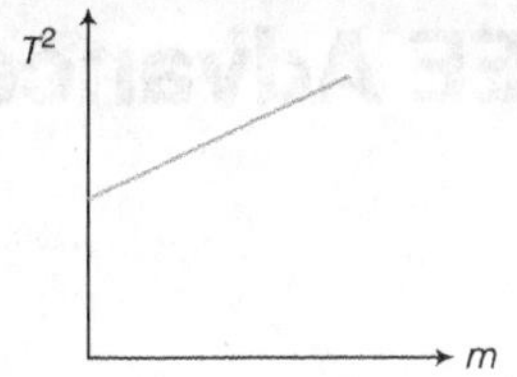

a. spring does not obey Hooke's law
b. amplitude of oscillation was too large
c. clock used needed regulating
d. the mass of the pan was neglected

15 Rigidity modulus of steel is η and its Young's modulus is Y. A piece of steel of cross-sectional area A is stretched into a wire of length L and area $A/10$. Then,

a. Y increases and η decreases
b. Y and η remain the same
c. Y decreases and η increases
d. Y and η decrease

16 A nylon rope 3 cm in diameter, has a breaking strength of 1.5×10^5 N. The breaking strength of a similar rope 1.5 cm in diameter is

a. 75000 N
b. 300000 N
c. 37500 N
d. 60000 N

17 An aluminium rod having Young's modulus 7×10^9 Nm^{-2} has a breaking strain of 0.2%. The minimum cross-sectional area of the rod (in m^2) in order to support a load of 10^4N is

a. 1×10^{-2}
b. 1×10^{-8}
c. 1.4×10^{-2}
d. 7.1×10^{-4}

18 There is no change in volume of a wire due to change in its length on stretching. The Poisson's ratio of the material of the wire is

a. + 0.25
b. −0.25
c. −0.5
d. +0.5

19 Two bars A and B of circular cross-section and of same volume made of the same material are subjected to the same tension. If the diameter of A is half that of B, then the ratio of extension of A to that of B will be

a. 16
b. 15
c. 8
d. 24

20 You are given three wires A, B and C of same length and cross-section. They are each stretched by applying the same force to the ends. The wire A is stretched least and comes back to its original length when the stretching force is removed. The wire B is stretched more than A and also comes back to its original length, when the stretching force is removed. The wire C is stretched most and remains stretched even when the stretching force is removed. The greatest Young's modulus of elasticity (Y) is possessed by the wire

a. A
b. B
c. C
d. All have same Y

SET - 2

Revisal Problems
for JEE Advanced

Only One Option Correct Type

Working Space

Direction *(Q. Nos. 1-9) This section contains 9 multiple choice questions. Each question has four choices (a), (b), (c) and (d), out of which ONLY ONE is correct.*

1 A mass m is hanging from a wire of cross-sectional area A and Length L. Y is the Young's modulus of the wire. An external force F is applied on the wire, which is then slowly further pulled down by Δx from its equilibrium position. The work done by the force that the wire exerts on the mass is

a. $mg\,\Delta x + \frac{YA}{2L}(\Delta x)^2$ b. $mg\Delta x + \frac{YA(\Delta x)^2}{L}$

c. $mg\,\frac{\Delta x}{2} + \frac{YA\,(\Delta x)^2}{L}$ d. $mg\,\frac{\Delta x}{2} + \frac{YA\,(\Delta x)^2}{2L}$

2 A wire of length L and density ρ and Young's modulus Y is hanging from a support. The elongation in the length of the wire at which wire will break is

a. $\frac{L^2\rho g}{2Y}$ b. $\frac{L^2\rho g}{4Y}$

c. $\frac{2L^2\rho g}{Y}$ d. $\frac{L^2\rho g}{Y}$

3 A plank of mass M is suspended horizontally by using two wires as shown in the figure. They have same length (L) and same cross-sectional area (A). Their Young's modulii are Y_1 and Y_2, respectively. The elastic potential energy of the system will be

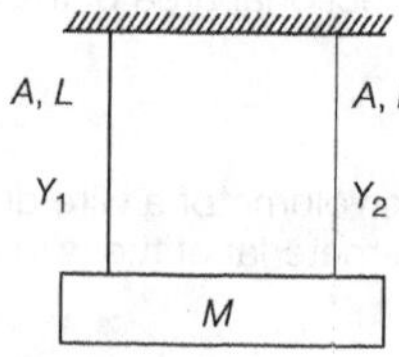

a. $\frac{2M^2g^2L}{A(Y_1+Y_2)}$ b. $\frac{M^2g^2L}{A(Y_1+Y_2)}$

c. $\frac{M^2g^2L}{2A(Y_1+Y_2)}$ d. $\frac{M^2g^2L(Y_1+Y_2)}{2AY_1Y_2}$

4 The product of Young's modulus of the material of the wire with its cross-sectional area is equal to its length. Find the parameters representing X and Y-axes of the curve shown in the figure.

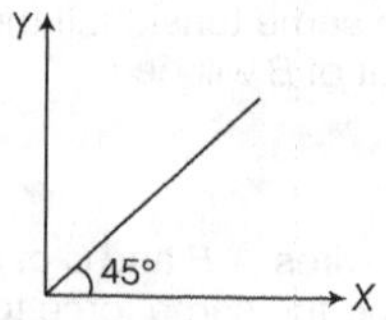

a. Stress and strain
b. Load and increase in length
c. Young's modulus and stress
d. Young's modulus and strain

5 In the above problem, if the ratio of strain to the load is $10^{-1}\,mN^{-1}m^{-1}$, then the length of the wire is

a. 100 cm b. 10 cm c. 1000 cm d. 50 cm

6 A wire of length L and cross-sectional area A is stretched by a distance x, so that work done is W. The Young's modulus is

a. $\frac{LW}{Ax^2}$ b. $\frac{LW}{2Ax^2}$ c. $\frac{2LW}{Ax^2}$ d. $\frac{LW}{\sqrt{2}\,Ax^2}$

7 The length of the wire OE is divided into five equal parts OA, AB, BC, CD and DE. The wire is made to hang from E and its length is $\frac{5\sigma_b}{4\rho g}$, where σ_b is the breaking stress and ρ is the density of the material of the wire. The point at which wire will break, is

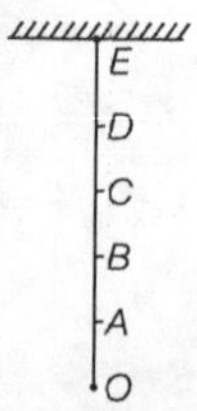

a. A b. B

c. C d. D

Working Space

8 Young's modulus of a rod is $\frac{AgL^2}{2\,(\Delta L)}$ for which elongation is ΔL due to its own weight, when suspended from the ceiling. L is the length of the rod. A is

a. area
b. mass per unit length
c. mass per unit length per unit area
d.. area per unit mass

9 The graph shows the extension (ΔL) of a wire of length 1m suspended from the top of a roof at one end and with a load F connected to the other end. If the cross-sectional area of the wire is 10^{-6}m^2, then Young's modulus of the wire is

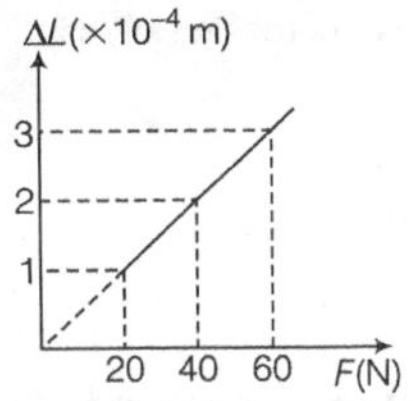

a. $2 \times 10^{11}\ \text{Nm}^{-2}$
b. $2 \times 10^{-11}\ \text{Nm}^{-2}$
c. $3 \times 10^{12}\ \text{Nm}^{-2}$
d. $2 \times 10^{13}\text{Nm}^{-2}$

One or More than One Options Correct Type

Direction (Q. Nos. 10-13) *This section contains 4 multiple choice questions. Each question has four choices (a), (b), (c) and (d), out of which ONE or MORE THAN ONE are correct.*

10 Three vertical wires 1, 2 ane 3 are supporting a rod of mass M in horizontal position as shown in the figure. The wires are of equal length and equal cross-sectional area. It is given that $Y_2 = Y_3$. The wires 1 and 3 are attached at the extreme ends of the rod. Choose the correct option(s).

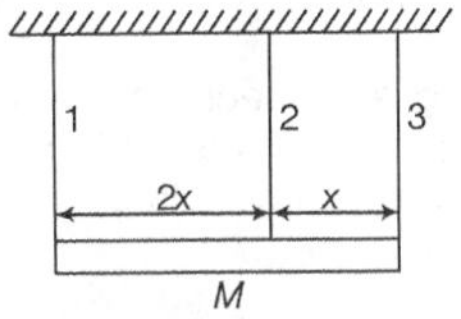

a. $T_1 = 2T_3$
b. $2T_1 = T_3$
c. $T_2 = T_3$
d. $T_1 = \frac{Mg}{5}$

11 A metallic wire of length L is held between two rigid supports. If the wire is cooled through a temperature θ, then fundamental frequency of oscillation is proportional to (where, Y = Young's modulus of elasticity of wire, ρ = density, α = thermal coefficient of linear expansion)

a. $\frac{1}{L}$
b. $\sqrt{Y}$
c. $\sqrt{\alpha/\rho}$
d. $\sqrt{\theta}$

12 For two different materials it is given that $Y_1 > Y_2$ and $B_1 < B_2$. Here, Y is Young's modulus of elasticity and B is the bulk modulus of elasticity. Then, we can conclude that

a. 1 is more ductile
b. 2 is more ductile
c. 1 is more malleable
d. 2 is more malleable

13 A load w is suspended from a wire of length L and area of cross-section A. Change in length of the wire is ΔL. Change in length can be increased to two times by increasing

a. w to two times
b. L to two times
c. A to two times
d. A to four times

Matching List Type

Direction (Q. No. 14 and 15) *Choices for the correct combination of elements from Column I and Column II are given as options (a), (b), (c) and (d), out of which one is correct.*

14 What kind of elasticity is used in various item listed in Column I ? Match with Column II.

	Column I		Column II
i.	Automobile tyre	p.	Bulk
ii.	Automobile drive shaft	q.	Tensile
iii.	Coiled spring	r.	Rigidity
iv.	Rope of lifting crane	s.	None of the above

Codes

	i	ii	iii	iv		i	ii	iii	iv
a.	p	r	r	q	b.	r	s	p	q
c	q	r	s	p	d.	p	q	r	s

15 Match the following columns and select the correct option from the codes given below.

Working Space

	Column I		Column II
i.	Shear stress	p.	Shape changes
ii.	Tensile stress	q.	Shape does not change
iii.	Longitudinal stress	r.	Volume changes
iv.	Compressive stress	s.	Volume does not change

Codes

	i	ii	iii	iv		i	ii	iii	iv
a.	p,q	r,s	p,s	p,r	b.	p,s	q,r	q,r	q,r
c.	q	p,q	r,s	s	d.	s	r,s	p,q	q

One Integer Value Correct Type

Direction (*Q. Nos. 16 and 17*) ***This section contains 2 questions. When worked out will result in an integer from 0 to 9 (both inclusive).***

16 A block of mass m produces an extension of 9 cm in the elastic spring of length 60 cm when it is hung by it and is in equilibrium. The spring is cut into two parts of 30 cm and 30 cm lengths. The same block hangs in equilibrium with the help of these two parts (in parallel). The extension in this case is $\frac{\beta}{4}$ cm. Find the value of β.

17 A body of mass 10 kg is connected to wire of length 0.3 m and of cross-sectional area $10^{-6} m^2$. Its breaking stress is $4.8 \times 10^7 Nm^{-2}$. Find the maximum angular velocity (in rad s^{-1}) with which it can be rotated in a horizontal circle.

Comprehension Type

Direction (*Q. Nos. 18-20*) ***This section contains a paragraph, which describing theory, experiments, data, etc. Three questions related to the paragraph have been given. Each question has only one correct answer among the four given options (a), (b), (c) and (d).***

Passage

In the table, utimalte strength represents the maximum stress applied to a specimen without fracture.

Material	Density ($kg m^{-3}$)	Young's modulus ($10^9 Nm^{-2}$)	Ultimate strength ($10^6 Nm^{-2}$)
Steel	7860	200	400
Aluminium	2710	70	110
Glass	2190	65	50
Concrete	2320	30	40
Wood	525	13	50
Bone	1900	9	170

18 According to the table, which of the following can withstand the greatest amount of stress without breaking?

a. A cylinder of glass 1 m long with a 10 cm radius
b. A human femur 0.5 m long with a 10 cm radius
c. A cylinder of concrete 10 m long with a 20 cm radius
d. A cylinder of aluminum 0.5 m long with a 5 cm radius

19 According to the table, what is the maximum per cent change in length withstood by a steel bar before it will break?

a. 0.125% b. 0.2%
c. 12.5% d. 125%

20 According to the table, which of the following is the most flexible? (expands the most without breaking)

a. Steel b. Glass
c. Wood d. Bone

JEE Main & AIEEE Archive

(Compilation of Last 13 Years' Questions)

1 The pressure that has to be applied to the end of a steel wire of length 10 cm to keep its length constant when its temperature is raised by 100°C is (for steel, Young's modulus is 2×10^{11} Nm^{-2} and coefficient of thermal expansion is 1.1×10^{-5} K^{-1}) **(JEE Main 2014)**

a. 2.2×10^8 Pa
b. 2.2×10^9 Pa
c. 2.2×10^7 Pa
d. 2.2×10^6 Pa

2 A sonometer wire of length 1.5 m is made of steel. The tension in it produces an elastic strain of 1%. What is the fundamental frequency of steel, if density and elasticity of steel are 7.7×10^3 kgm^{-3} and 2.2×10^{11} Nm^{-2}, respectively? **(JEE Main 2013)**

a. 188.5 Hz b. 178.2 Hz
c. 200.5 Hz d. 770 Hz

3 This question has Statement I and Statement II. Of the four choices given after the statements, choose the one that best describes the two statements.
If two springs S_1 and S_2 of force constants k_1 and k_2 respectively are stretched by the same force, it is found that more work is done on spring S_1 than on spring S_2.
Statement I If stretched by the same amount, then work done on S_1 will be more than that on S_2.
Statement II $k_1 < k_2$ **(AIEEE 2012)**

a. Both Statement I and Statement II are correct and Statement II is the correct explanation of Statement I
b. Both Statement I and Statement II are correct but Statement II is not the correct explanation of Statement I
c. Statement I is correct but Statement II is incorrect
d. Statement II is correct but Statement I is incorrect

4 A wooden wheel of radius R is made of two semi-circular parts (see figure). The two parts are held together by a ring made of a metal strip of cross-sectional area S and length L. L is slightly less than $2\pi R$. To fit the ring on the wheel, it is heated so that its temperature rises by ΔT and it just steps over the wheel. As it cools down to surrounding temperature, it presses the semi-circular parts together. If the coefficient of linear expansion of the metal is α and its Young's modulus is Y, then the force that one part of the wheel applies on the other part is **(AIEEE 2012)**

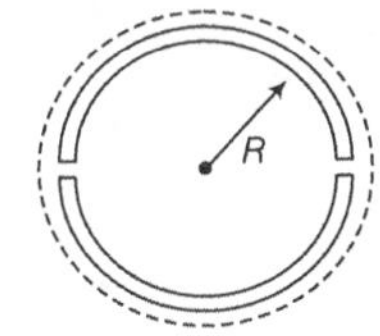

a. $2\pi SY\alpha\Delta T$ b. $SY\alpha\Delta T$
c. $\pi SY\alpha\Delta T$ d. $2SY\alpha\Delta T$

5 A metal rod of Young's modulus Y and coefficient of thermal expansion α is held at its two ends such that its length remains invariant. If its temperature is raised by $\Delta T°$ C, then linear stress developed in it is **(AIEEE 2011)**

a. $\frac{\alpha\Delta T}{Y}$ b. $\frac{Y}{\alpha\Delta T}$
c. $Y\alpha\Delta T$ d. $\frac{1}{Y\alpha\Delta T}$

6 Two identical charged spheres are suspended by strings of equal lengths. The strings make an angle of 30° with each other. When suspended in a liquid of density 0.8 gcm^{-3}, the angle remains same. If density of the material of sphere is 1.6 g cm^{-3}, then dielectric constant of the liquid is **(AIEEE 2010)**

a. 4 b. 3
c. 2 d. 1

7 Two wires are made of the same material and have the same volume. However, wire 1 has cross-sectional area A and wire 2 has cross-sectional area $3A$. If the length of wire 1 increases by Δx on applying force F, then how much force is needed to stretch wire 2 by the same amount? **(AIEEE 2009)**

a. F b. $4F$
c. $6F$ d. $9F$

8 If S is stress and Y is Young's modulus of material of a wire, then the energy stored in the wire per unit volume is **(AIEEE 2005)**

a. $2S^2Y$ b. $\frac{S^2}{2Y}$
c. $\frac{2Y}{S^2}$ d. $\frac{S}{2Y}$

9 A wire fixed at the upper end stretches by length l by applying a force F. The work done in stretching is **(AIEEE 2004)**

a. $\frac{F}{2l}$ b. Fl
c. $2Fl$ d. $\frac{Fl}{2}$

10 A wire suspended vertically from one of its ends is stretched by attaching a weight of 200 N to the lower end. The weight stretches the wire by 1 mm. Then, the elastic energy stored in the wire is **(AIEEE 2003)**

a. 0.2 J b. 10 J
c. 20 J d. 0.1 J

JEE Archive

JEE Advanced & IIT JEE Archive

(Compilation of Last 10 Years' Questions)

1 One end of a horizontal thick copper wire of length $2L$ and radius $2R$ is welded to an end of another horizontal thin copper wire of length L and radius R. When the arrangement is stretched by applying forces at two ends, the ratio of the elongation in the thin wire to that in the thick wire is **(2013 Adv., Only One Option Correct Type)**

a. 0.25
b. 0.50
c. 2.00
d. 4.00

2 A solid sphere of radius R and density ρ is attached to one end of a massless spring of force constant k. The other end of the spring is connected to another solid sphere of radius R and density 3ρ. The complete arrangement is placed in a liquid of density 2ρ and is allowed to reach equilibrium. Which of the following statement(s) is/are correct? **(2013 Adv., One or More than One Options Correct Type)**

a. The net elongation of the spring is $\frac{4\pi R^3 \rho g}{3k}$

b. The net elongation of the spring is $\frac{8\pi R^3 \rho g}{3k}$

c. The light sphere is partially submerged

d. The light sphere is completely submerged

3 The pressure of a medium is changed from 1.01×10^5 Pa to 1.165×10^5 Pa and change in volume is 10% keeping temperature constant. The bulk modulus of the medium is **(2005, Only One Option Correct Type)**

a. 204.8×10^5 Pa
b. 102.4×10^5 Pa
c. 51.2×10^5 Pa
d. 1.55×10^5 Pa

5. Fluid Mechanics

DPP-1 Pressure in Static/ Accelerated Fluids

Only One Option Correct Type

Working Space

Direction (Q. Nos. 1-13) *This section contains 13 multiple choice questions. Each question has four choices (a), (b), (c) and (d), out of which ONLY ONE is correct.*

1 A vertical narrow glass tube of uniform cross-section, closed at one end has air filled in it upto 40 cm of height, with a column of mercury of height 10 cm above it. If the tube is gradually inclined in the vertical plane and brought to a position making an angle of 60° with the vertical, then new length of air column entrapped within the tube is (atmospheric pressure = 76 cm of Hg)

a. 45.47 cm
b. 43.47 cm
c. 42.47 cm
d. 44.47 cm

2 The press plunger of a Bramah (hydraulic) press is $5\ m^2$ in cross-section and is employed to lift a load of mass 500 kg. What minimum force will be required to be applied on the pump plunger if its cross-sectional area is $200\ cm^2$?

a. 39.2 N
b. 9.8 N
c. 29.4 N
d. 19.6 N

3 To what height (h) should a cylindrical vessel of radius (R) be filled with a homogeneous liquid to make the force with which the liquid presses on the curved side wall of the vessel equal to the force exerted by the liquid on the bottom of the vessel?

a. $h = R$
b. $h = \sqrt{2}R$
c. $h = \frac{R}{2}$
d. $h = 2R$

4 Water is contained in a rectangular tank, at rest, to a height H. Which of the following graphs best represents the variation of hydrostatic thrust F per unit width of the side of the vessel (width is measured perpendicular to paper) on the area measured from the top of a level, at a height y from the bottom level?

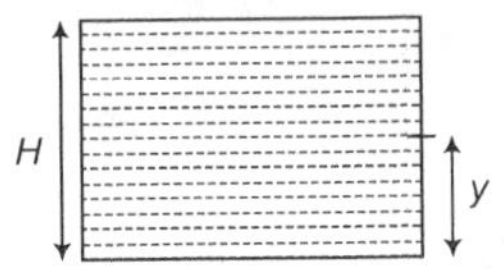

a.
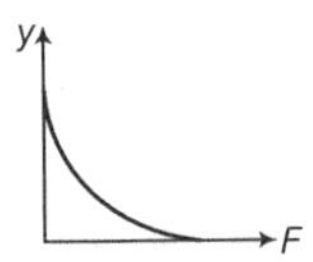

b.
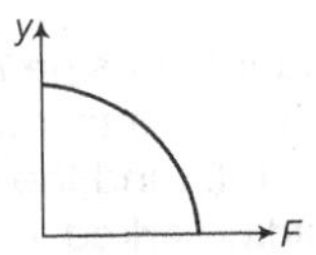

c.
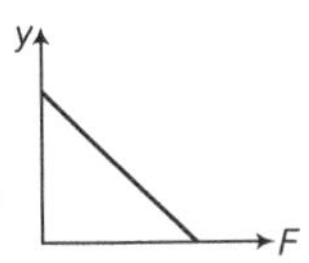

d.
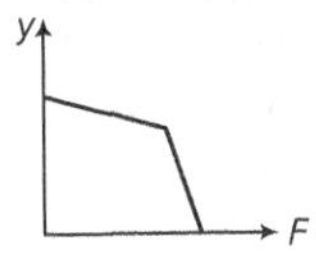

5 A vessel contains non-homogeneous liquid whose density ρ increases linearly with depth h. Which of the following graphs best depicts the variation of liquid pressure p with depth h?

a. 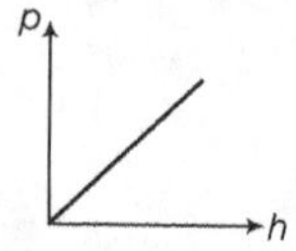

b. 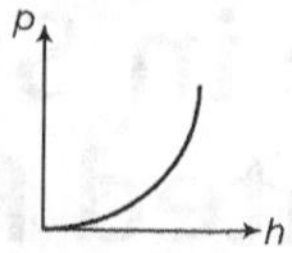

c. 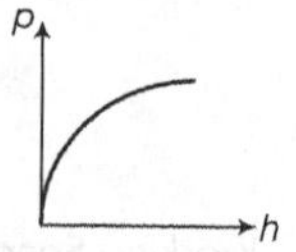

d. 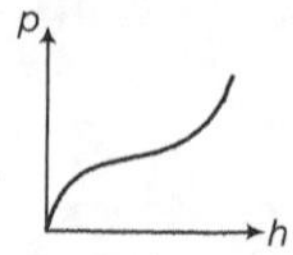

6 A U-tube of length L (see figure) contains liquid and is vertical. It is mounted on a horizontal turn table rotating with an angular speed ω about one of its arms as shown in the figure. The difference in heights (h) between the liquid columns in two vertical arms is

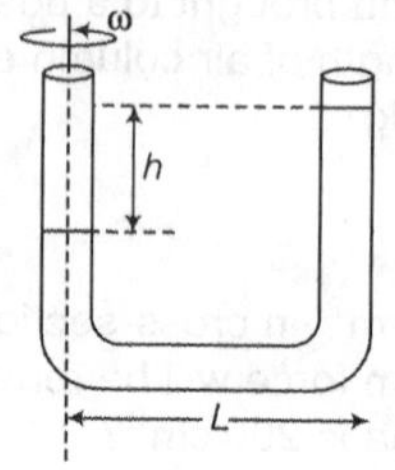

a. $\dfrac{\omega^2 L^2}{2g}$ b. $\dfrac{\omega^2 L^2}{g}$

c. $\dfrac{2\omega^2 L^2}{g}$ d. $\dfrac{\omega^2 L^2}{g}$

7 On a smooth inclined plane (with angle of inclination $\alpha = 37^\circ$), a vessel containing liquid is being pushed upwards with an acceleration $5\ \text{ms}^{-2}$. If $g = 10\ \text{ms}^{-2}$, then angle (θ) that the liquid surface makes with the horizontal is (Given, $\sin 37^\circ = \frac{3}{5}$)

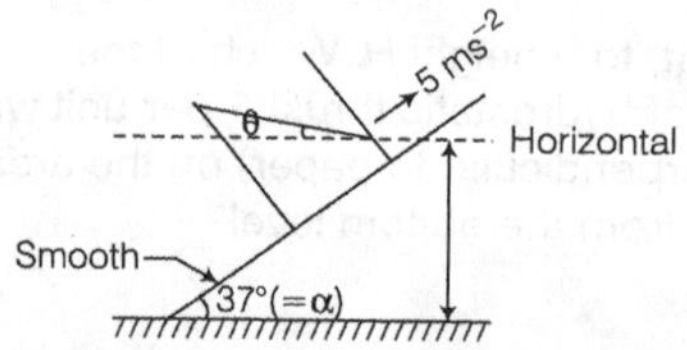

a. 37° b. 53°

c. $\tan^{-1}\left(\dfrac{4}{13}\right)$ d. $\tan^{-1}\left(\dfrac{5}{13}\right)$

8 A closed tube in the form of an equilateral triangle of side l contains equal volumes of three liquids which do not mix and is placed vertically with its lowest side horizontal. Find the value of x (see figure), if the densities of liquids marked 1, 2 and 3 are in arithmetic progression in increasing order.

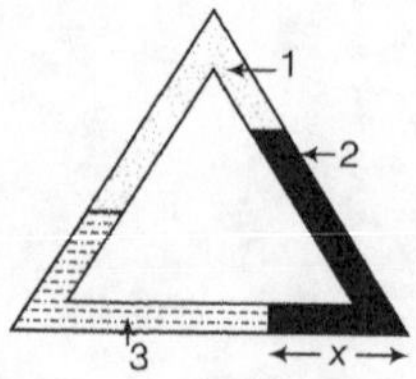

a. $\dfrac{l}{3}$ b. $\dfrac{2l}{3}$

c. $\dfrac{l}{2}$ d. $\dfrac{3l}{4}$

Working Space

Working Space

9 A narrow tube completely filled with a liquid is lying on a series of cylinders as shown in the figure. Assuming pure rolling (no sliding between any surfaces),the value of acceleration (a) of the cylinders for which liquid will not come out of the tube from anywhere is given by

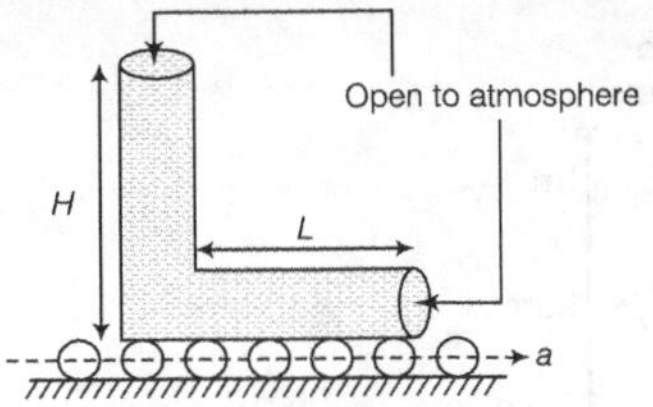

a. $\frac{gH}{\sqrt{2}L}$ b. $\frac{gH}{L}$

c. $\frac{2gH}{L}$ d. $\frac{gH}{2L}$

10 The force acting due to the liquid per unit width (width is measured perpendicular to the plane of figure) on a plane wall inclined at an angle θ with the horizontal (see figure), if the density of liquid is ρ, is

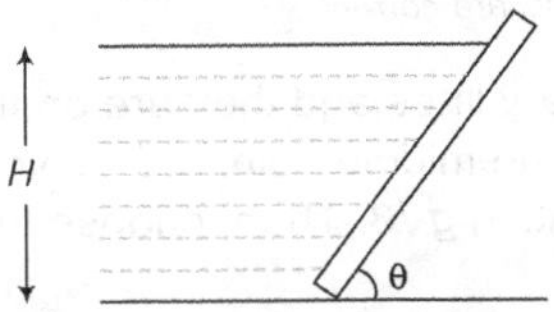

a. $\frac{\rho g H^2}{2\sin\theta}$ b. $\frac{\rho g H^2}{\sin\theta}$

c. $\frac{2}{3}\cdot\frac{\rho g H^2}{\sin\theta}$ d. $\frac{\rho g H^2}{\sqrt{2}\sin\theta}$

11 A closed container completely filled with water is mounted on a cart. The cart moves with an acceleration a_0 on a plane horizontal road. What is the difference in pressure between points B and A shown in the figure?

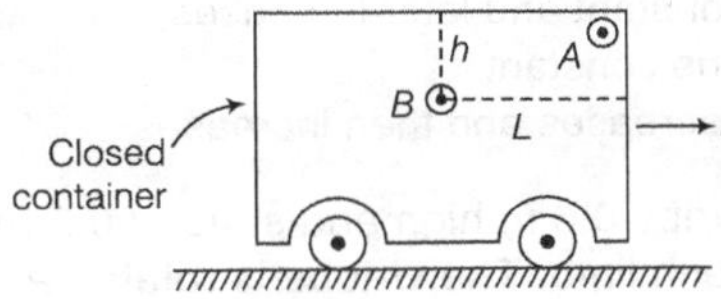

a. ρgh b. $\rho g(h+L)$

c. $\rho g(h-L)$ d. $\rho(gh+a_0L)$

12 A liquid of density ρ is in a bucket that spins with angular velocity ω as shown in the figure. What is the pressure at point A, which is at a radial distance r from the centre as shown in the figure? Atmospheric pressure is p_0.

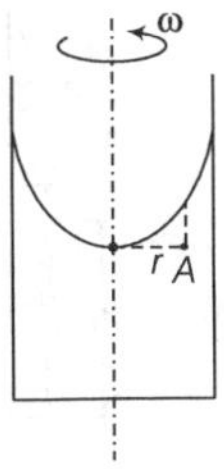

a. $p_0+\frac{\rho\omega^2r^2}{2}$

b. $p_0+\frac{2}{3}\rho\omega^2r^2$

c. $p_0+\frac{1}{3}\rho\omega^2r^2$

d. None of the above

13 A cubical container with side 3 m has a small hole with a cap at point C as shown in the figure. The water level is upto D ($BC = 0.5$ cm and $BD = 2$ m).If container is given a constant acceleration $a_0 = 9\,\text{ms}^{-2}$ and the hole is simultaneously opened, then amount of water that will spill out is (Given, $g = 10\,\text{ms}^{-2}$)

Working Space

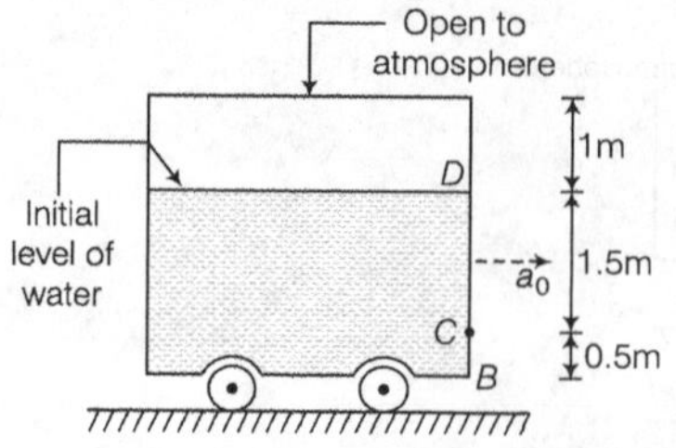

a. 2150 L
b. 4150 L
c. 3150 L
d. 2750 L

One or More than One Options Correct Type

Direction (Q. Nos. 14 and 15) This section contains 2 multiple choice questions. Each question has four choices (a), (b), (c) and (d), out of which ONE or MORE THAN ONE are correct.

14 A vessel contains water. PQR and SQ are imaginary lines and they are on the same plane, which is the plane of the figure. Q is the middle point of PR. Now, the vessel moves rightwards with a constant acceleration $g\sqrt{3}$. Then, choose the correct statement(s).

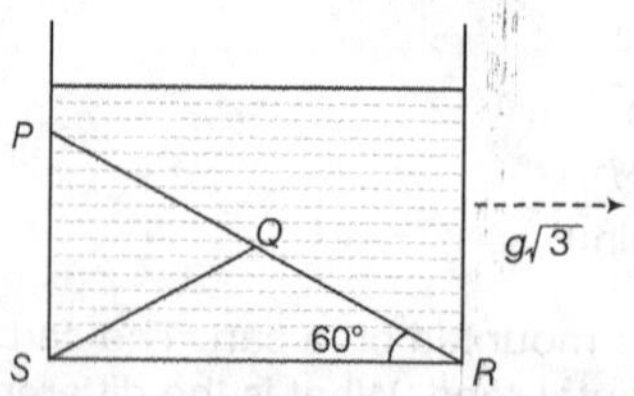

a. If one moves from S to Q to R, pressure first decreases and later remains constant
b. If one moves from P to Q to S, pressure first constant and later increases
c. If one moves from P to Q to R, pressure remains constant
d. If one moves from R to Q to S, pressure first decreases and then increases

15 An upright narrow U-tube manometer with its limbs 0.6 m high and spaced 0.3 m apart contains a liquid to a height of 0.4 m in each limb. The U-tube is rotated at 10 rad s^{-1} about a vertical axis at 0.1 m from one limb (see figure). Then, (Given, $g = 10\text{ms}^{-2}$)

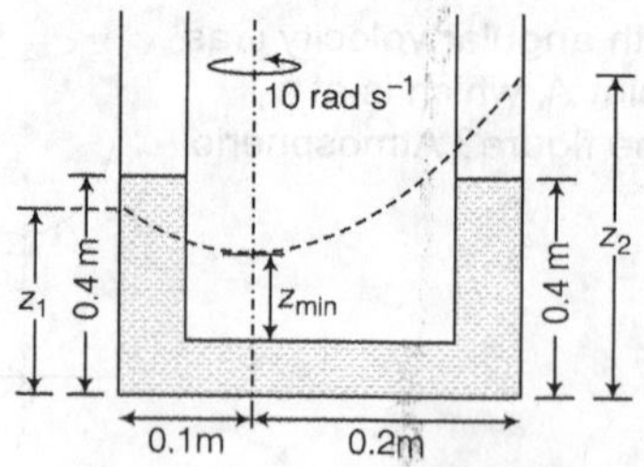

a. $\frac{z_1}{z_2} = 0.684$
b. $\frac{z_1}{z_2} = 0.513$
c. $z_{min} = 0.275$m
d. $z_{min} = 0.1375$ m

Working Space

Comprehension Type

Direction (Q. Nos. 16-18) This section contains a paragraph, which describing theory, experiments, data, etc. Three questions related to the paragraph have been given. Each question has only one correct answer among the four given options (a), (b), (c) and (d).

Passage

A glass full of water, upto a height of 10 cm has a bottom of area 10 cm^2, top of area 30cm^2 and volume 1 L. Atmospheric pressure is 1.01×10^5 Pa. Density of water is 1000 kg m^{-3} and g = 10 ms^{-2}. Note that water completely fills the glass.

16 The force (F_1) exerted by the water on the bottom is

a. 92 N
b. 102 N
c. 98 N
d. 108 N

17 The resultant force (F_R) exerted by the sides of the glass on the water is

a. 198 N
b. 152 N
c. 182 N
d. 211 N

18 If the glass is covered by a jar and air inside the jar is completely pumped out, then F_1 and F_R, respectively are

a. 2 N, 18 N
b. 1N, 9 N
c. 3 N, 27 N
d. 4 N, 36 N

One Integer Value Correct Type

Direction (Q. Nos. 19-21) This section contains 3 questions. When worked out will result in an integer from 0 to 9 (both inclusive).

19 A liquid with varying density is contained in a rectangular container of height 10 cm. The density of the liquid at a depth h cm below the free surface is given by $\rho = \rho_0 (1+h)$, where ρ_0 = 0.8 gcm^{-3}. Its average density is $\frac{8k}{10}$ gcm^{-3}. Find the value of k. (Given, base area = 100 cm^2)

20 A container (see figure) contains a liquid upto a height h. The density of the liquid is ρ. The gauge pressure at point P is $\frac{\rho gh}{x}$. Find the value of x.

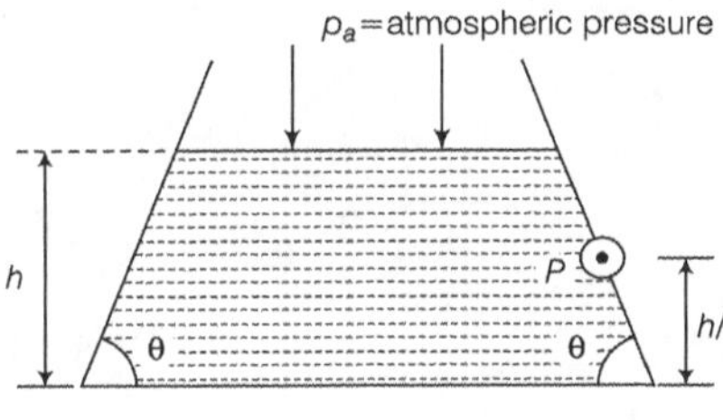

21 In rising from the bottom of a lake to its surface, the temperature of air bubble remains constant but its diameter gets doubled. If h is the barometric height (expressed in metre of mercury of relative density ρ) at the surface of the lake, the depth of the lake is $\beta h\rho$. Find the value of β. (Neglect effects of surface tension)

Subjective Questions

Working Space

Direction (Q. Nos. 22-24) ***These questions are subjective in nature, need to be solved completely on notebook.***

22 A prismatic bar of given dimensions is kept in water. Find the hydrostatic forces on the slant face, vertical face and base of the bar.
(Given, $\rho_{water} = 1000 \text{ kg m}^{-3}$ and $g = 10 \text{ ms}^{-2}$)

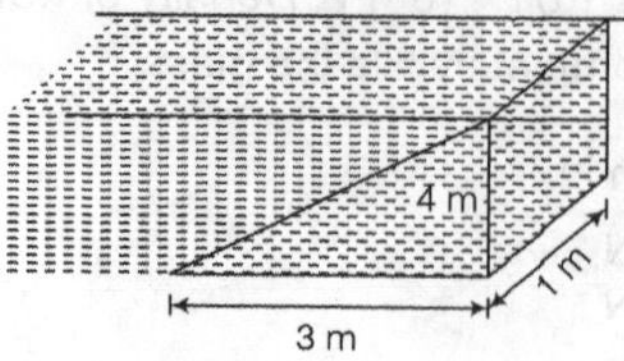

23 A vertical flat gate of area A is kept in equilibrium about the horizontal pivot P by the horizontal force F. ρ is density of water.

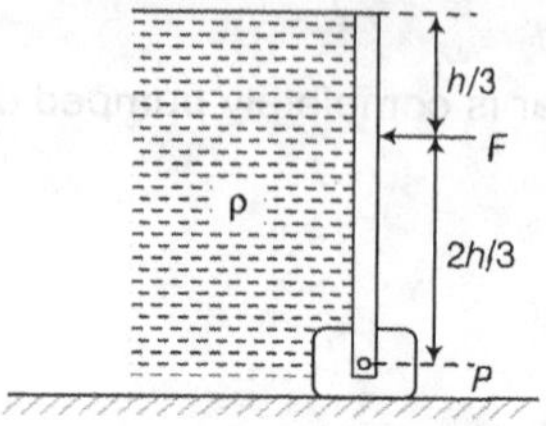

i. Find the value of F.

ii. What is the horizontal reaction force acting at the pivot?

24 A U-tube of base length l filled with same volume of two liquids of densities ρ and 2ρ is moving with an acceleration a_0 on the horizontal plane. If the height difference between the two surfaces (open to atmosphere) becomes zero. What will be the height h?

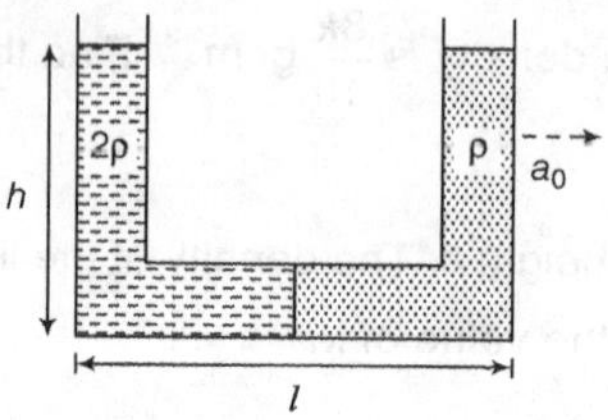

DPP-2 Archimedes' Principle and Buoyant Force

Only One Option Correct Type

Working Space

Direction (*Q. Nos. 1-10*) *This section contains 10 multiple choice questions. Each question has four choices (a), (b), (c) and (d), out of which ONLY ONE is correct.*

1 Density of ice is 900 kg m^{-3}. A piece of ice is floating in water of density 1000 kg m^{-3}. Find the fraction of volume of the piece of the ice outside water.

a. 0.9
b. 0.8
c. 0.1
d. 0.2

2 A piece of ice is floating in a glass vessel filled with water. When ice completely melts,

a. level of water does not change
b. level of water falls down
c. level of water rises
d. None of the above

3 An ornament weighing 50 g in air weighs only 46 g in water. Assuming that some copper is mixed with gold in the preparation of ornament, find the amount of copper in it. Specific gravity of gold is 20 and that of copper is 10.

a. 20 g
b. 30 g
c. 25 g
d. 15 g

4 The tension in a string holding a solid block below the surface of a liquid (of density greater than that of solid) as shown in figure is T_0, when the system is at rest. What will be the tension in the string if the system has an upward acceleration a?

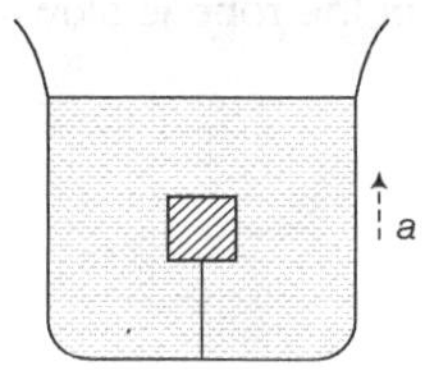

a. $T_0\left(\frac{a}{g}\right)$
b. $T_0\left(1-\frac{a}{g}\right)$
c. $T_0(a+g)$
d. $T_0\left(1+\frac{a}{g}\right)$

5 A vessel contains oil of density 0.8 g cc^{-1} floating over mercury of density 13.6 g cc^{-1}. A homogeneous sphere floats with half of its volume immersed in mercury and the other half in oil. The density of the material of the sphere (in g cc^{-1}) is

a. 7.2
b. 6.4
c. 3
d. 12.8

6 A hollow sphere of inner diameter and outer diameter of 12 cm and 16 cm respectively floats half submerged in water. The specific gravity of material of the sphere is

a. 8.65 b. 6.85
c. 0.865 d. 0.685

7 A small steel ball of mass m is attached to the bottom surface of a wooden cube of mass 200 g. Find the value of m that will just allow the block to float in water. Specific gravity of wood is 0.8 and that of steel is 7.8.

a. 59.3 g b. 57.3 g
c. 56.3 g d. 58.3 g

8 In the above problem, what is the value of m, if the ball is attached to the top surface of the cube, floating with upper surface of cube just immersed in water?

a. 60 g b. 50 g
c. 55 g d. 65 g

9 A cubical block of side a is floating in a fixed and closed cylindrical container of radius $2a$ kept on the ground. Density of the block is ρ, whereas density of the liquid is 2ρ. Container is made up of conducting wall, so that the temperature remains constant. A piston is mounted in the cylinder which can move inside the cylinder without friction. If piston oscillates with a large amplitude A, then

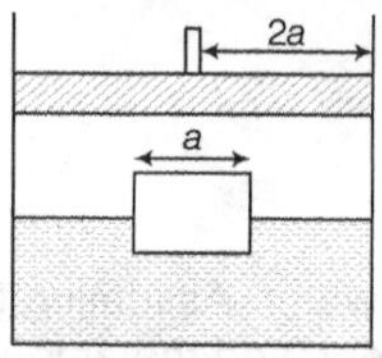

a. the cube will remain stationary
b. the cube will oscillate with very small amplitude in opposite phase with piston
c. the cube will oscillate with very small amplitude in same phase with piston
d. the cube will oscillate with amplitude A

10 A cubical wooden block of side L and relative density 0.5 is floating on the surface of water as shown in the figure. A string is attached by a pulley to it as shown in figure. The graph of tension T in the string *versus* distance d of the bottom of the block from the free surface of water, when the end A of the rope is slowly being pulled up will be (block is moving very slowly)

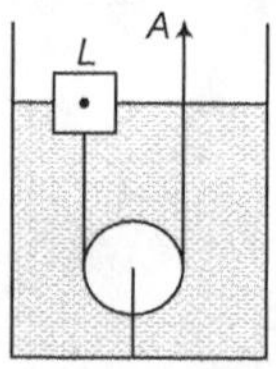

a.
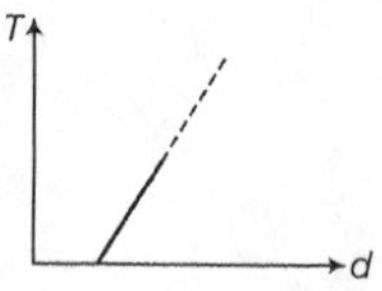

b.
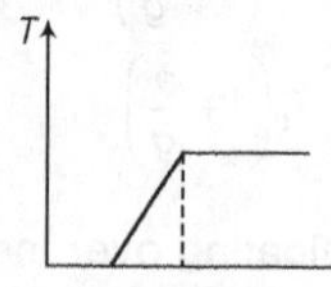

c.
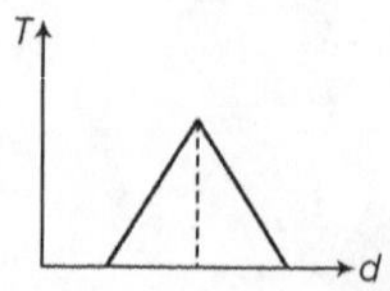

d.
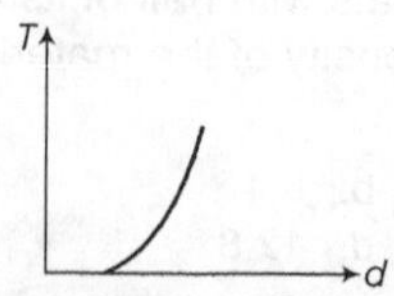

Working Space

Working Space

One or More than One Options Correct Type

Direction (Q. Nos. 11 and 12) This section contains 2 multiple choice questions. Each question has four choices (a), (b), (c) and (d), out of which ONE or MORE THAN ONE are correct.

11 A solid ball of density half that of water falls freely under gravity from a height of 19.6 m and then enters water. (Neglect air resistance and viscosity of water) Choose the correct statement(s).

a. It goes upto a depth of 19.6 m in water
b. After entering water, it takes 4 s to reach surface again
c. It goes upto a depth of 9.8 m in water
d. After entering water, it takes 2 s to reach water surface again

12 A block of mass 1 kg and density $0.8\ \text{g cm}^{-3}$ is held stationary with the help of a string as shown in the figure. The tank is accelerating vertically upwards with a constant acceleration $a_0 = 1\ \text{ms}^{-2}$. Choose the correct statement(s).
(Given, $g = 10\ \text{ms}^{-2}$ and $\rho_{\text{water}} = 1000\ \text{kg m}^{-3}$)

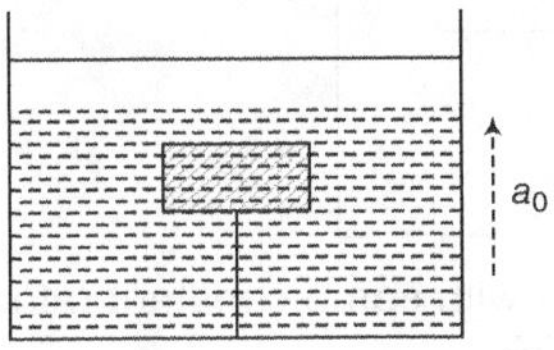

a. The tension in the string is 2.75 N
b. If the string is cut, then acceleration of the block is $1.75\ \text{ms}^{-2}$
c. The tension in the string is 1.75 N
d. If the string is cut, then acceleration of the block is $3.75\ \text{ms}^{-2}$

Comprehension Type

Direction (Q. Nos. 13-18) This section contains 2 paragraphs, each describing theory, experiments, data, etc. Six questions related to the paragraphs have been given. Each question has only one correct answer among the four given options (a), (b), (c) and (d).

Passage I

A cylinder of radius R is kept embedded along the wall of a dam as shown in the figure. Density of water is ρ. Length of cylinder (perpendicular to the plane of paper) is L.

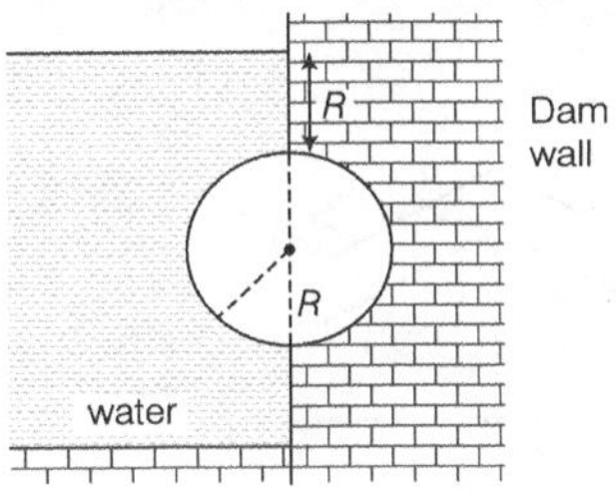

Neglect atmospheric pressure.

13 The vertical force exerted by water on the cylinder is

a. $\rho\pi R^2 Lg$
b. $\dfrac{\rho\pi R^2 Lg}{2}$
c. zero
d. None of these

14 The horizontal force exerted by water on the cylinder is

a. $2R^2\rho gL$
b. $R^2\rho gL$
c. $4R^2\rho gL$
d. $1.61R^2\rho gL$

15 The net torque exerted by water on the cylinder is

a. $\frac{2\rho R^2 Lg}{3}$ b. $\frac{\rho R^3 Lg}{3}$

c. $\frac{\rho R^3 Lg}{2}$ d. zero

Passage II

Two ice cubes, each of side 10 cm have cavity of volume 20 cm^3 at the centre of each cube. These cavities are filled with materials A and B respectively in cube 1 and cube 2. The specific gravity of material of cube A is 4.9 and specific gravity of material of cube B is 1.9. Now, these cubes are placed in two different vessels of same base area as shown in the figure. Before putting the blocks, the water levels in vessels are same.

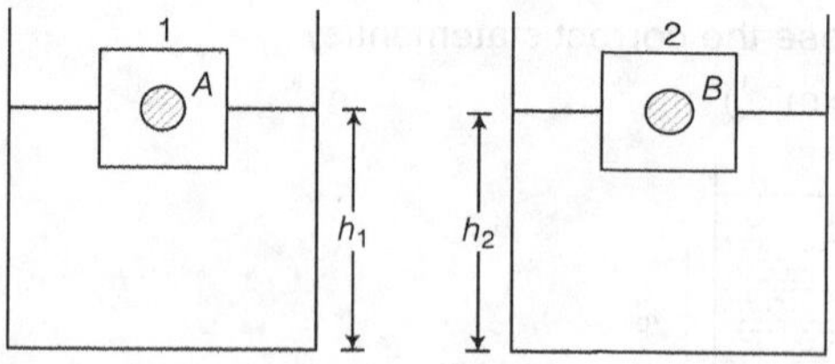

Assume that ice melts uniformly from all sides and with same constant rate in both the vessels. (Given, specific gravity of ice = 0.9)

16 Find the ratio of initial submerged volumes of the blocks containing A and B respectively.

a. $\frac{49}{46}$ b. $\frac{1}{1}$

c. $\frac{49}{19}$ d. $\frac{49}{28}$

17 Choose the correct statement(s).

a. A and B sink simultaneously

b. A and B sink only after complete melting of ice

c. A sinks prior to B

d. B sinks prior to A

18 Choose the correct graph showing the variation of heights of water level in two vessels with time.

a. 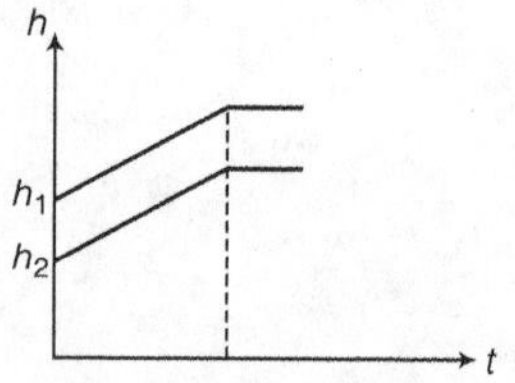

b. 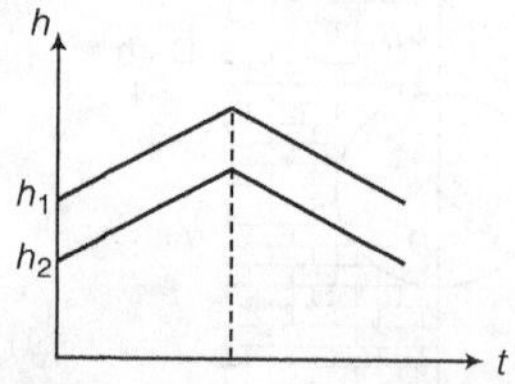

c. 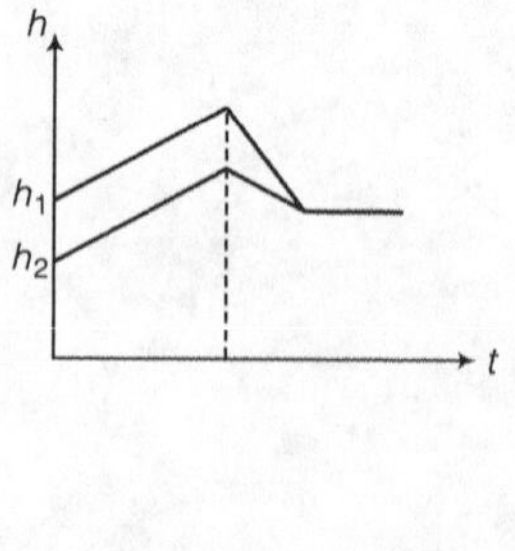

d. 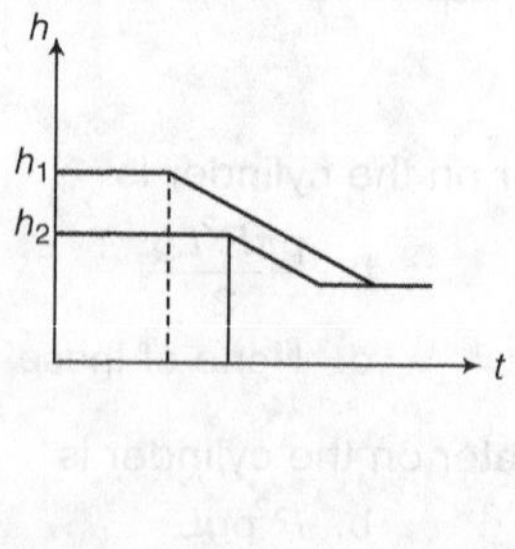

Matching List Type

Direction (Q. No. 19) *Choices for the correct combination of elements from Column I and Column II are given as options (a), (b), (c) and (d), out of which one is correct.*

19 Column I gives different systems as described, with some parameter while Column II gives the change in the parameter. Match the following columns and select the correct option from the codes given below.

	Column I			Column II
i.	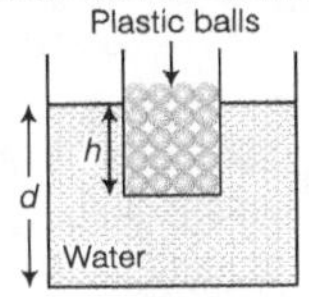	Plastic ball is thrown from the container into the water.	p.	h decreases
ii.	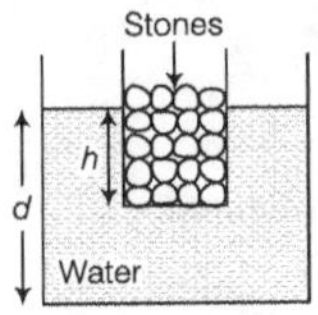	Stone is thrown from the container into the water.	q.	h remains same
iii.	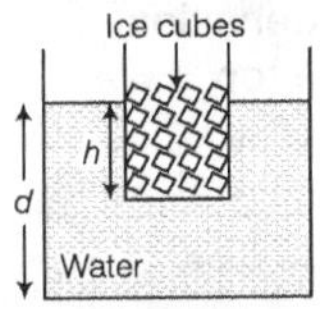	Ice cubes melt and remain in the container.	r.	h increases
iv.	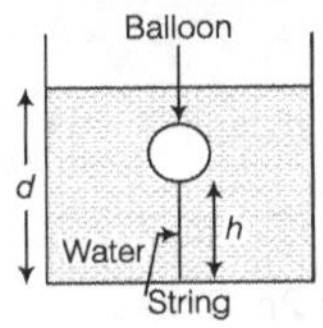	The water is heated slowly $\mu_{air} > \mu_{water}$. Here, h is the tension in the string (μ = coefficient of volume expension).	s.	d decreases
			t.	d remains same

Codes

	i	ii	iii	iv		i	ii	iii	iv
a.	s,q	p	q,p	t	b.	p,t	p,s	q,t	r
c.	q,r	t,s	p	q	d.	r,s	t	p	q,t

One Integer Value Correct Type

Direction (Q. Nos. 20-22) *This section contains 3 questions. When worked out will result in an integer from 0 to 9 (both inclusive).*

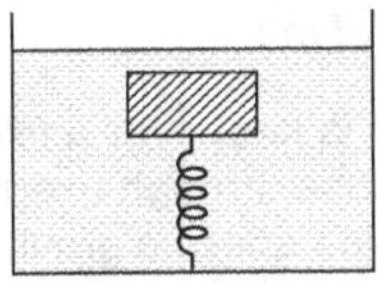

20 A cube of wood of mass 0.5 kg and density 800 kg m^{-3} is fastened to the free end of a vertical spring of spring constant $k = 50$ Nm^{-1}, fixed at the bottom. Now, the entire system is completely submerged in water. The elongation or compression of the spring in equilibrium is $\frac{\beta}{2}$ cm. Find the value of β. (Given, $g = 10$ ms^{-2})

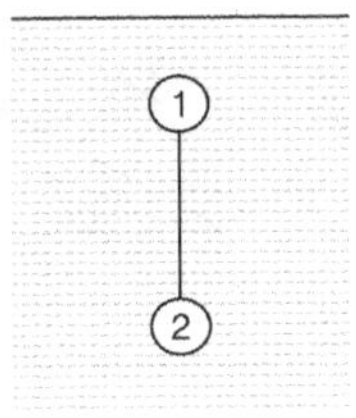

21 Two spheres each of volume 250 cc but of relative densities 0.8 and 1.2 are connected by a string and the combination is immersed in a liquid in vertical position as shown in the figure. The tension in the string is $\frac{k}{18}$ N. Find the value of k. (Given, $g = 10$ ms^{-2})

22 Figure shows a cubical block of side 10 cm and relative density 1.5 suspended by a wire of cross-sectional area $10^{-6} m^2$. The breaking stress of the wire is $7 \times 10^6 \ Nm^{-2}$. The block is placed in a beaker of base area 200 cm² and intially at $t = 0$, the top surfaces of water and block coincide. There is a pump at the bottom corner, which ejects 2 cm³ of water per second. If the time at which wire breaks is $p \times 100$ second, then find the value of p. (Given, $g = 10 \ ms^{-2}$)

Working Space

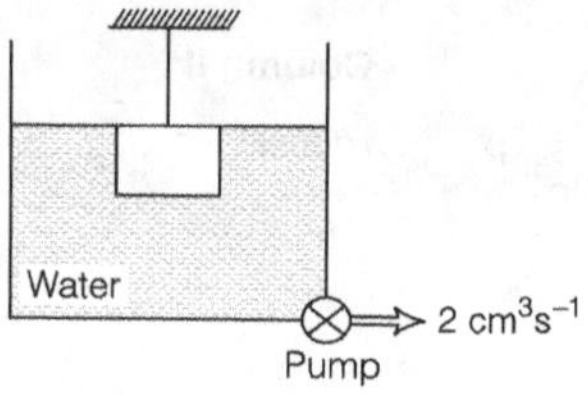

Subjective Questions

Direction *(Q. Nos. 23-26) These questions are subjective in nature, need to be solved completely on notebook.*

23 A conical body of height h is floating in between two liquid layers having densities ρ and 2ρ. If ρ_S be the density of cone, then what will be the height d in the top liquid?

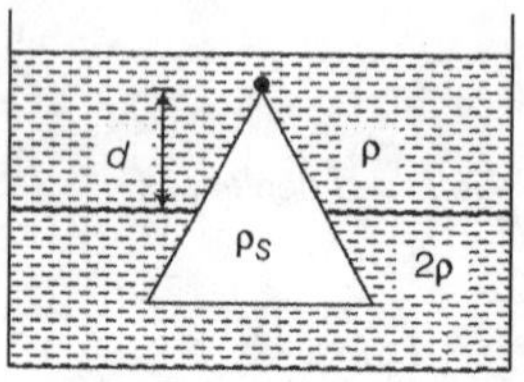

24 A cubical block of wood of edge a and density ρ floats in water of density 2ρ. The lower surface of the cube just touches the free end of a massless spring of force constant k, fixed at the bottom of the vessel. A weight w is put over the block, so that it is completely immersed in water without wetting the weight. Then, find the value of w.

25 A uniform rod of length 6 m and mass 1.5 kg is made of a material of specific gravity 0.6. It is hinged at one end at a depth of 4 m from the free surface of water in a tank. At the other end, a body of mass m is placed on the rod, so that the rod remains in equilibrium with full immersion, with one end just near the free surface of water. Mass m is not immersed. What is the value of m?

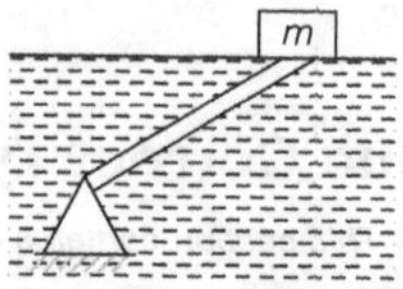

26 A block of mass m is kept over a fixed smooth wedge. Block is attached to a sphere of same mass through fixed ideal pulleys P_1 and P_2. Sphere is dipped in water as shown. The specific gravity of material of sphere is 2. The system is released from rest. Find the acceleration of the sphere.

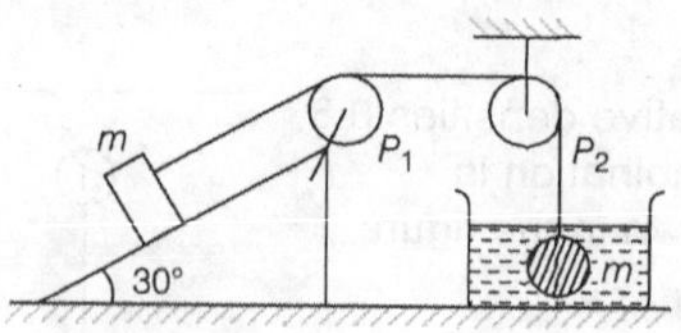

DPP-3 Bernoulli's Principle and Its Applications

Only One Option Correct Type

Working Space

Direction (Q. Nos. 1-9) This section contains 9 multiple choice questions. Each question has four choices (a), (b), (c) and (d), out of which ONLY ONE is correct.

1 Find the force acting on the blade of an undershot wheel (see figure), if the stream after impinging on the blade continues to move with the velocity of the blade. The height of the water head is h, the radius of the wheel is R and angular velocity of the wheel is ω and cross-sectional area of the stream is A.

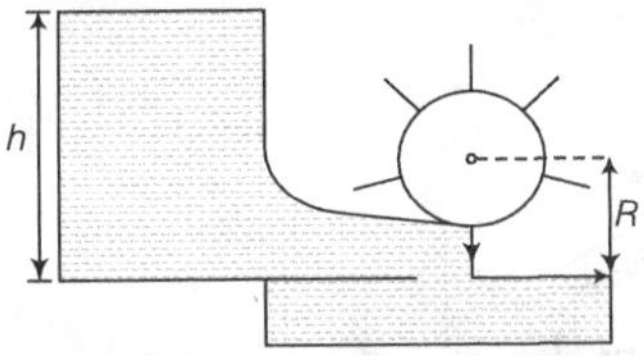

a. $\rho A(\sqrt{2gh} - \omega R)^2$ b. $\rho A(\sqrt{2gh} + \omega R)^2$

c. $\frac{\rho A}{2}(\sqrt{2gh} - \omega R)^2$ d. $\frac{\rho A}{2}(\sqrt{2gh} + \omega R)^2$

2 A water clock (clepsydra) used in ancient Greece is designed as a vessel with a small orifice O (see figure). The time is determined according to the level of water in the vessel. What should be the equation corresponding to the shape of the vessel for the time scale to be uniform? (Vessel is symmetric about vertical axis and a = the area of orifice and v = speed with which water surface lower down)

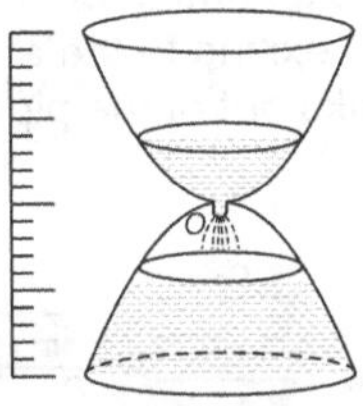

a. $y = kx^4$, where $k = \frac{\pi^2 v^2}{ga^2}$ b. $y = kx^4$, where $k = \frac{2\pi^2 v^2}{ga^2}$

c. $y = kx^4$, where $k = \frac{\pi v^2}{ga^2}$ d. $y = kx^4$, where $k = \frac{\pi^2 v^2}{2ga^2}$

3 A tank with a pipe is fixed to a trolley as shown in the figure. The diameter of the pipe is 5 cm. Water issues from the pipe with velocity $v = 5\ \text{ms}^{-1}$, which makes an angle 45° with the horizontal. The compression in the spring (with stiffness $k = 20\ \text{N cm}^{-1}$) shown is

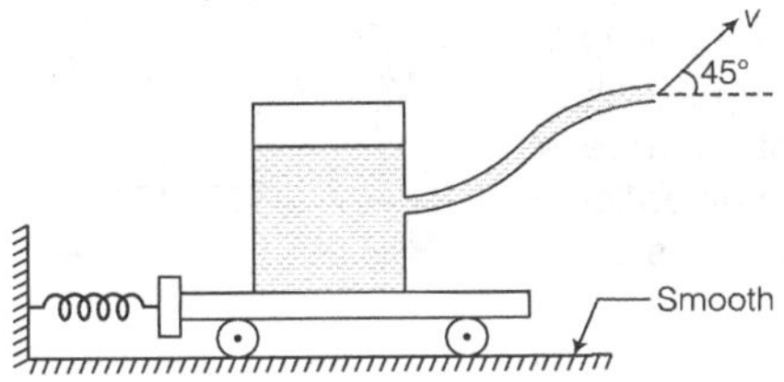

a. 1.12 cm b. 1.74 cm

c. 0.78 cm d. 2.12 cm

Working Space

4 In the figure shown, a light container is kept on a horizontal rough surface of coefficient of friction $\mu = \frac{Ah}{V}$. A very small hole of area A is made at a depth h. Water of volume V is filled in the container. The friction is not sufficient to keep the container at rest. The magnitude of acceleration of the container initially is

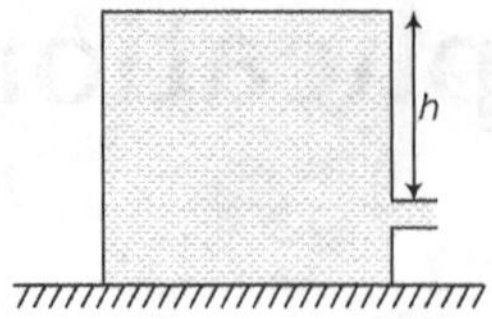

a. $\frac{V}{Ah}g$ b. g c. $\frac{Ah}{V}g$ d. zero

5 An incompressible liquid flows through a horizontal tube as shown in the figure. The velocity v of the fluid is (Given, $v_1 = 3 \text{ ms}^{-1}$ and $v_2 = 1.5 \text{ ms}^{-1}$)

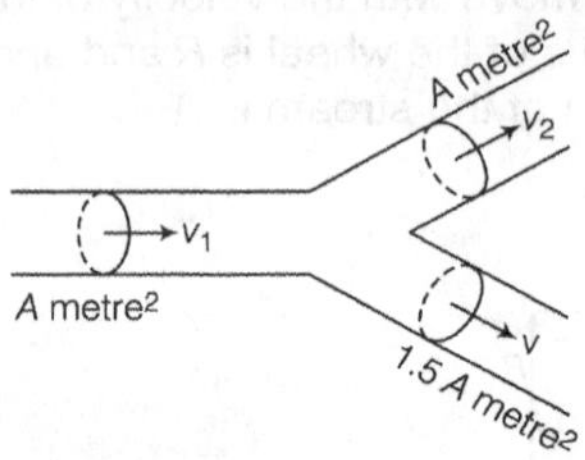

a. 3 ms^{-1} b. 1.5 ms^{-1} c. 1.0 ms^{-1} d. 2.25 ms^{-1}

6 A non-viscous, incompressible costly fluid flows through a pipe of circular cross-section. The initial diameter of the pipe is d_1, the pipe then shrinks to diameter $d_2 < d_1$, expands to diameter $d_3 > d_1$ and then returns to its initial diameter d_1 before exiting to the atmosphere. A thief wants to drill a small hole so that some fluid leaks out of the pipe before the exit. He will be successful if he drills the hole at

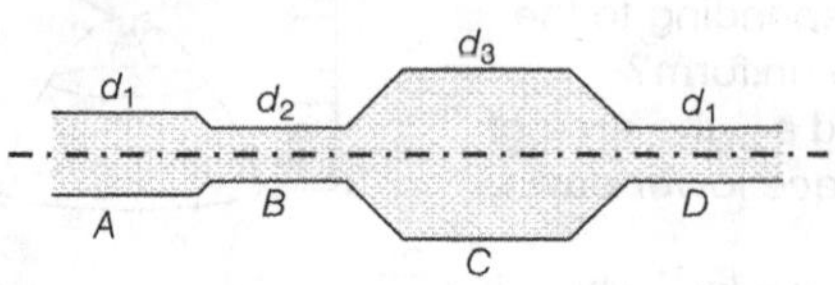

a. point A
b. point B
c. point C
d. any point along the pipe

7 In a horizontal pipeline of uniform area of cross-section, the pressure falls by 8 Nm^{-2} between two points separated by 1 km. What is change in KE per kg of the oil flowing at these points? (Given, $\rho_{oil} = 800 \text{ kgm}^{-3}$)

a. 0.1Jkg^{-1} b. 0.001Jkg^{-1} c. 1Jkg^{-1} d. 0.01Jkg^{-1}

8 A horizontally oriented tube AB of length L rotates with a constant angular velocity ω about a stationary axis OO' passing through the end A as shown in the figure. The tube is filled with an ideal fluid. The end A of the tube is opened and the closed end B has a very small orifice. Find the velocity of the fluid relative to the tube as a function of column height h.

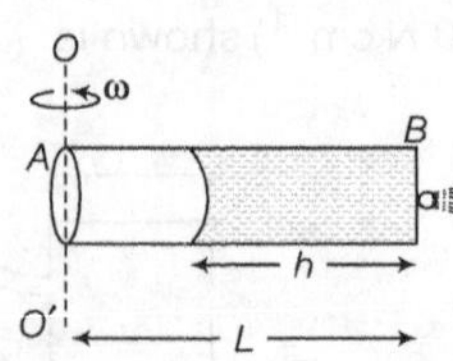

a. $\frac{\omega h}{2}\sqrt{\frac{2L}{h} - 1}$
b. $\omega h\sqrt{\frac{2L}{h} - 1}$
c. $\omega h\sqrt{\frac{L}{h} - 1}$
d. $\frac{\omega h}{2}\sqrt{\frac{2L}{h} - \frac{1}{2}}$

Working Space

9 A container of large uniform cross-sectional area holds two immiscible, non-viscous and incompressible liquids of densities ρ and 4ρ, each of height $\frac{H}{2}$. Determine the position where, a small hole has to be punched, so that the heavier liquid comes out with a maximum range R initially.

a. At a height $\frac{H}{4}$ from the bottom

b. At a height $\frac{5}{8}H$ from the bottom

c. At a height $\frac{5}{16}H$ from the bottom

d. At a height $\frac{7}{18}H$ from the bottom

One or More than One Options Correct Type

Direction (Q. Nos. 10 and 11) *This section contains 2 multiple choice questions. Each question has four choices (a), (b), (c) and (d), out of which ONE or MORE THAN ONE are correct.*

10 Two holes with an area $a = 0.2\ \text{cm}^2$, each are drilled one above the other in the wall of a vertical vessel filled with water. The distance between the holes is $H = 50$ cm. Every second $Q = 140\ \text{cm}^3$ of water is poured into the vessel. The streams flowing out of the holes intersect at a point. x is the horizontal distance of this point from the vessel and y is the vertical distance of this point from the water surface in the tank. x and y remain same as the time passes. Choose the correct option(s).

a. $x = 120$ cm b. $x = 60$ cm

c. $y = 75$ cm d. $y = 130$ cm

11 A U-tube acts as a water siphon. The bend in the tube is 1 m above the water surface. The tube outlet is 7 m below the water surface. The water issues from the bottom of the siphon as a free jet at atmospheric pressure. Speed of jet is v and minimum absolute pressure in the bend is p. (Given, $p_{atm} = 1.01 \times 10^5\ \text{Nm}^{-2}$, $g = 9.8\ \text{ms}^{-2}$ and $\rho_{water} = 1000\ \text{kgm}^{-3}$). Choose the correct option(s).

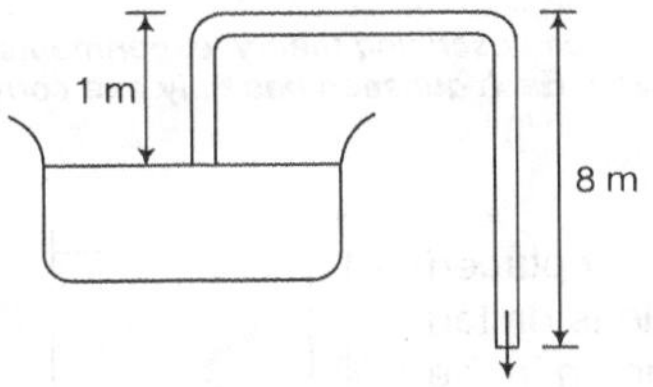

a. $v = 11.7\ \text{ms}^{-1}$ b. $v = 13.2\ \text{ms}^{-1}$

c. $p = 2.27 \times 10^4\ \text{Nm}^{-2}$ d. $p = 3.27 \times 10^4\ \text{Nm}^{-2}$

Statement Type

Direction (Q. Nos. 12 and 13) *This section is based on Statement I and Statement II. Select the correct answer from the codes given below.*

a. Both Statement I and Statement II are correct and Statement II is the correct explanation of Statement I
b. Both Statement I and Statement II are correct but Statement II is not the correct explanation of Statement I
c. Statement I is correct but Statement II is incorrect
d. Statement II is correct but Statement I is incorrect

12 **Statement I** During storms, when the velocity of the wind is very high, it tears off the roofs of the buildings. Two cases can be distinguished

1. if the roof is fastened more firmly at points A and B than at ridge C, the wind will break the roof along ridge C and rise both halves up as shown in Fig. (i).
2. If the roof is secured more firmly at the ridge and less firmly at points A and B, the wind will first lift the roof up and then carry it aside as shown in Fig. (ii).

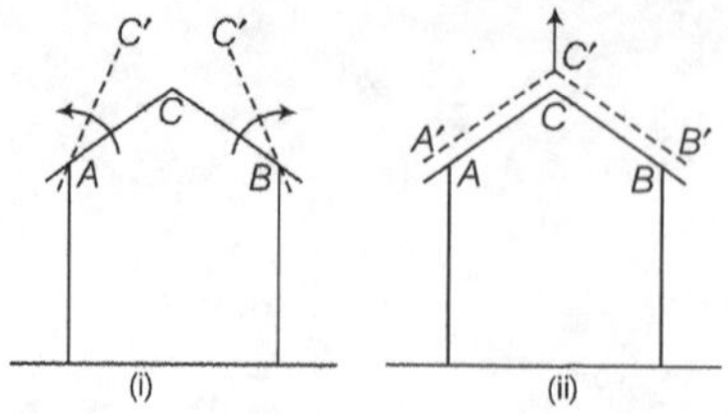

Statement II The pressure of the air streaming over the roof is less than that of air pressure inside. It is this surplus pressure of the stationary air under the roof that causes the described phenomena.

13 **Statement I** The bottom of a broad vessel is provided with a narrow tube through which the water can flow out of the vessel as shown in the figure. A screen is placed between the vessel and the tube. If a light ball is submerged to the bottom of the vessel, then water flows out of it, and the ball will not rise to the surface. If the outflow of the water is stopped, then ball will immediately rise to the surface.

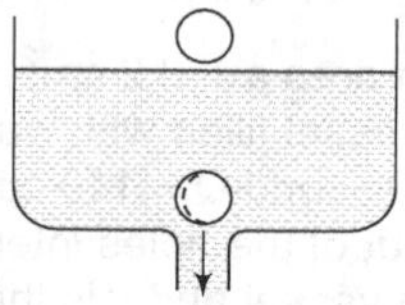

Statement II The pressure diminishes in a stream of a flowing liquid with an increase in its velocity. The velocity with which the water flows in the vessel is much smaller than in the tube and, therefore, the pressure of the water in the vessel is greater than in the tube. The velocity increases at the boundary between the vessel and the tube and the pressure drops. For this reason, the ball is pressed against the screen and does not rise.

Comprehension Type

Direction (Q. Nos. 14-19) This section contains 2 paragraphs, each describing theory, experiments, data, etc. Six questions related to the paragraphs have been given. Each question has only one correct answer among the four given options (a), (b), (c) and (d).

Passage I

Water is filled to a height h in a fixed vertical cylinder placed on a horizontal surface. At time $t = 0$ a small hole is drilled at a height $h/4$ from the bottom of cylinder as shown in the figure. The cross-sectional area of hole is a and the cross-sectional area of cylinder is A and $A >> a$.

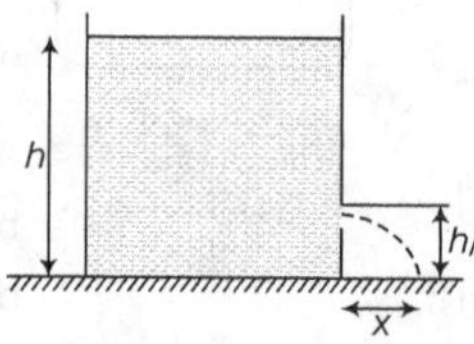

14 Let the value of horizontal distance of point where the water falls on horizontal surface from the bottom of cylinder be x as shown in the figure. Then, from time $t = 0$ till water comes out

a. x decreases with time
b. x increases with time
c. x first increases and then decreases with time
d. x first decreases and increases with time

15 As long as water comes out of hole, the time taken by a water particle starting from the hole to reach the horizontal surface

a. increases
b. decreases
c. remains constant
d. increases and then decreases

Working Space

Working Space

16 The duration of time for which water flows out of hole is

a. $\frac{A}{a}\sqrt{\frac{3h}{2g}}$ b. $\frac{a}{A}\sqrt{\frac{3h}{2g}}$

c. $\frac{A}{a}\sqrt{\frac{2h}{3g}}$ d. $\frac{a}{A}\sqrt{\frac{h}{g}}$

Passage II

Water is filled in a big cylinder vessel upto a height $4h$. Three small orifices, O_1, O_2 and O_3 are made on the wall of the vessel.

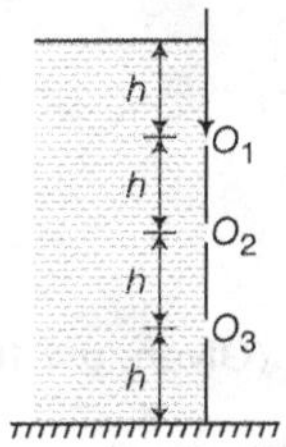

17 If v_1, v_2 and v_3 be the speeds of efflux of water from the three orifices, then

a. $v_1 = v_3 < v_2$ b. $v_1 < v_2 < v_3$

c. $v_1 = v_3 > v_2$ d. $v_1 > v_2 > v_3$

18 Let x_1 be the horizontal distance of the point at which the water stream from orifice O_1 strikes the horizontal surface. The distance x_1 is

a. h b. $2h$

c. $3\sqrt{2}h$ d. $2\sqrt{3}h$

19 The correct graph showing the initial trajectories is

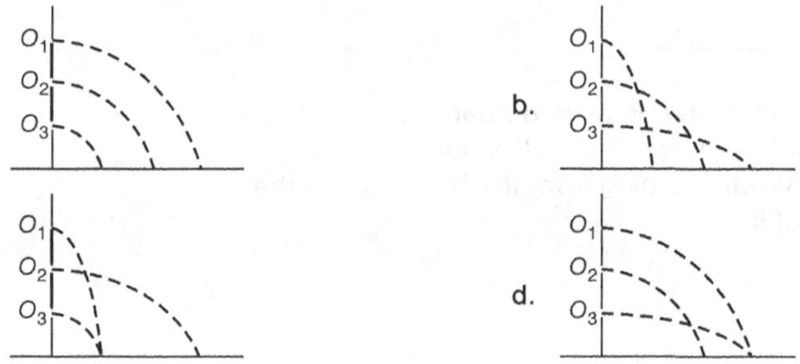

One Integer Value Correct Type

Direction (Q. No. 20) *This section contains 1 question. When worked out will result in an integer from 0 to 9 (both inclusive).*

20 A thin glass plate of area A and mass m is hinged along one of its sides. The speed with which the air should be blown parallel to its upper surface to hold the plate horizontal is $\gamma\sqrt{\frac{mg}{18\rho A}}$. Find the value of γ. The density of air is ρ.

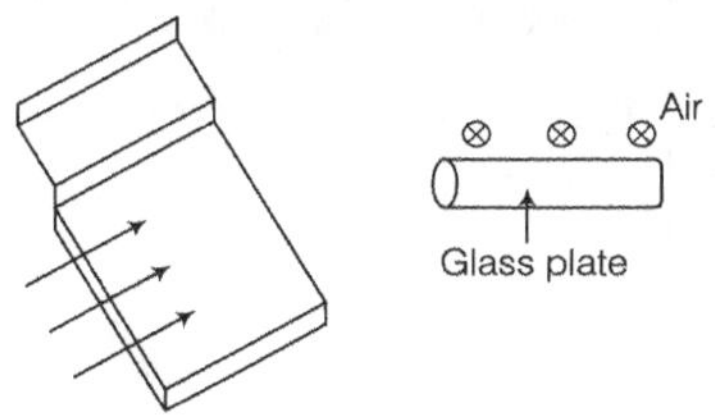

Subjective Questions

Working Space

Direction (Q. Nos. 21-24) *These questions are subjective in nature, need to be solved completely on notebook.*

21 A cylindrical tank 1 m in radius rests on a platform 5 m high. Initially the tank is filled with water to a height of 5 m. A plug whose area is $10^{-4}\ m^2$ is removed from an orifice on the side of the tank at the bottom. Calculate

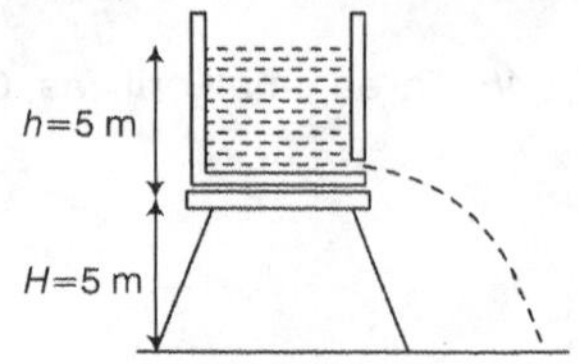

i. initial speed with which the water flows from the orifice.
ii. initial speed with which water strikes the ground.
iii. time taken to empty the tank to half of its original value. (Given, $g = 10\ ms^{-2}$)

22 A long cylindrical tank of cross-sectional area $0.5\ m^2$ is filled with water. It has a small hole at a height 50 cm from the bottom. A movable piston of cross-sectional area almost equal to $0.5\ m^2$ is fitted on the top of the tank such that it can slide in the tank freely. A load of 20 kg is applied on the top water by piston, as shown in the figure. Calculate the speed of the water jet (when it hits the horizontal surface) coming out when the piston is 1 m above the bottom. Piston is massless.

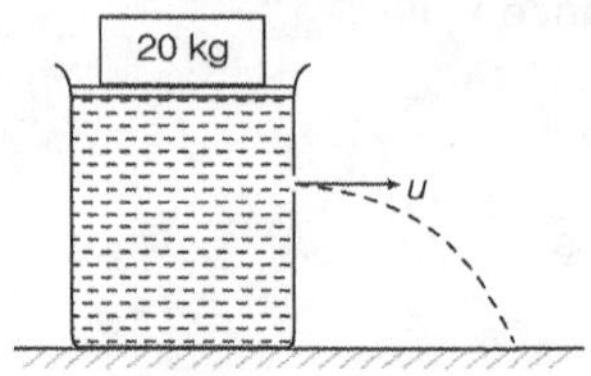

23 A rectangular tank of height 10 m filled with water, is placed near the bottom of a plane inclined at angle 30° with horizontal. At height h from the bottom a small hole is made (see figure) such that the stream coming from the hole, strikes the inclined plane normally. Find the value of h.

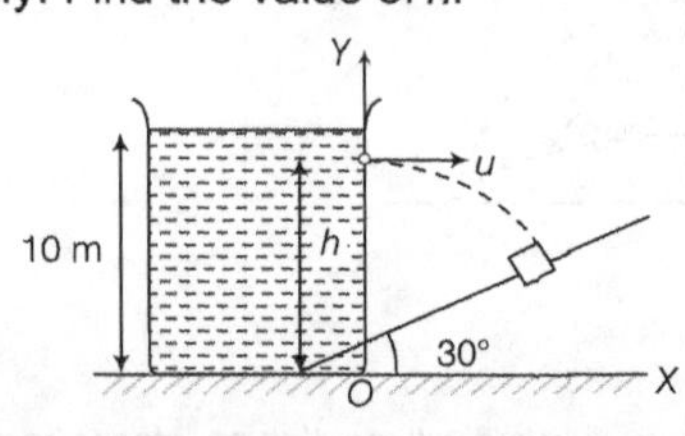

24 What work should be done in order to squeeze all water from a horizontally located cylinder (see figure) during the time t by means of a constant force acting on the piston? The volume of water in the cylinder is equal to V, the cross-sectional area of orifice is a and a is considerably less than the piston area. Density of water is ρ. The friction and viscosity are negligible.

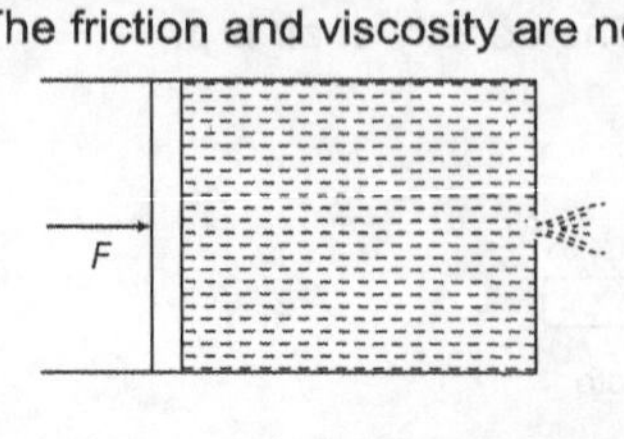

DPP-4 Viscosity

Only One Option Correct Type

Working Space

Direction (Q. Nos. 1-10) *This section contains 10 multiple choice questions. Each question has four choices (a), (b), (c) and (d), out of which ONLY ONE is correct.*

1 Water is flowing through a horizontal tube 8 cm in diameter and 4 km in length at the rate of 20 L s^{-1}. Assuming only viscous resistance, the pressure required to maintain the flow in terms of mercury column is (Given, $\eta_{water} = 0.001$ Pa-s)

a. 69.72 cm
b. 59.72 cm
c. 49.72 cm
d. 39.72 cm

2 Eight spherical raindrops of equal size are falling vertically through air with terminal velocity of 0.10 ms^{-1}. What should be the terminal velocity if these drops were to combine to form one large spherical drop?

a. 0.1 ms^{-1}
b. 0.3 ms^{-1}
c. 0.6 ms^{-1}
d. 0.4 ms^{-1}

3 A solid sphere falls into a viscous liquid from a large height. Which of the graphs shown best represents the variation of velocity of sphere in liquid with time?

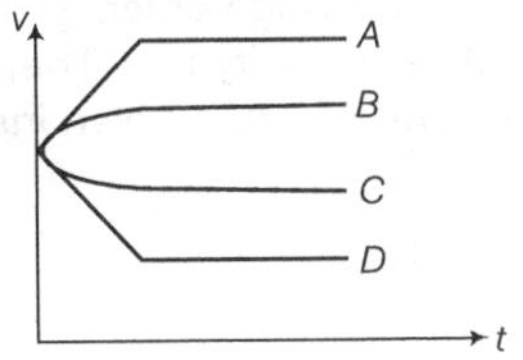

a. *A*
b. *B*
c. *C*
d. *D*

4 A marble of mass x and diameter $2r$ is gently released in a tall cylinder containing honey. If the marble displaces mass $y(< x)$ of the liquid, then the terminal velocity is proportional to

a. $x + y$
b. $x - y$
c. $\frac{x + y}{r}$
d. $\frac{x - y}{r}$

5 A spherical ball is dropped in a long column of viscous liquid. The sequence of given graphs that represent, respectively, the variation of magnitude of

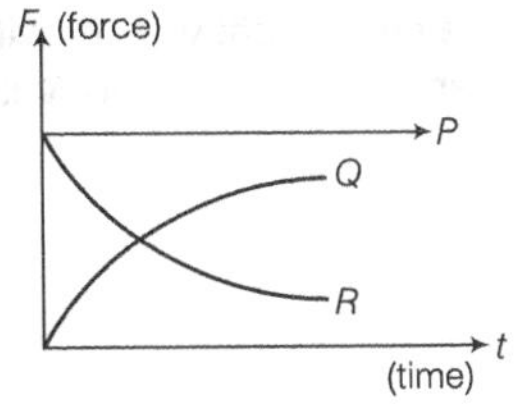

I. gravitational force with time
II. viscous force with time and
III. net force acting on the ball with time.

Choose the correct option(s).

a. *Q, R, P*
b. *R, Q, P*
c. *P, Q, R*
d. *P, R, Q*

Working Space

6 Spherical particles of pollen are shaken up in water and allowed to settle. The depth of water is 2×10^{-2} m. What is the diameter of the largest particles remaining in suspension one hour later?

Given, density of pollen $= 1.8 \times 10^3$ kg m^{-3},

viscosity of water $= 10^{-2}$ Poise

density of water $= 10^3$ kg m^{-3}

and $g = 10$ ms^{-2}

a. 3.54 μm
b. 4.02 μm
c. 5.54 μm
d. 5.02 μm

7 A spherical ball of radius 3×10^{-4} m and density 10^4 kg m^{-3} falls freely under gravity through a distance h before entering a tank of water. If after entering the water, the velocity of the ball does not change, then the value of h is (Given, $\eta_{water} = 9.8 \times 10^{-6}$ Nsm^{-2} and $\rho_{water} = 10^3$ kgm^{-3})

a. 1650 m
b. 165 m
c. 1050 m
d. 105 m

8 A piston of 20 cm diameter and 20 cm long moves down in a cylinder of diameter 20.0628 cm. The oil filling the annular space has a viscosity of 10 Poise and the mass of the piston is 1 kg. Find the speed with which the piston slides down. (Given, $g = 10$ ms^{-2})

a. 15cms^{-1}
b. 2.5 cms^{-1}
c. 2cms^{-1}
d. 3.5cms^{-1}

9 A container filled with viscous liquid is moving vertically downwards with constant speed $3v_0$. At the instant shown, a sphere of radius r is moving vertically downwards, in liquid, has a speed v_0. The coefficient of viscosity is η. There is no relative motion between the liquid and the container. Then at the shown instant, the magnitude of viscous force acting on sphere is

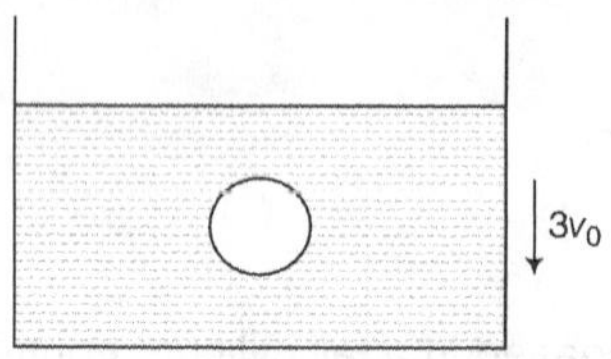

a. $6\pi\eta r v_0$
b. $12\pi\eta r v_0$
c. $18\pi\eta r v_0$
d. $24\pi\eta r v_0$

10 A horizontal capillary tube with an internal radius $r = 1$ mm and a length of $l = 2$cm is inserted into the side of a cylindrical vessel having a radius of $R = 2$cm. The vessel is filled with castor oil whose viscosity is $\eta = 12$gcm^{-1}s^{-1}. The level of the castor oil sinks in the cylindrical vessel with a velocity v. And the height of this level from the capillary tube is h. Then, v is

a. $\dfrac{r^4 \rho g h}{R^2 \eta l}$
b. $\dfrac{r^4 \rho g h}{4R^2 \eta l}$
c. $\dfrac{r^4 \rho g h}{8R^2 \eta l}$
d. $\dfrac{r^4 \rho g h}{16R^2 \eta l}$

Working Space

One or More than One Options Correct Type

Direction *(Q. Nos. 11and 12) This section contains 2 multiple choice questions. Each question has four choices (a), (b), (c) and (d), out of which ONE or MORE THAN ONE are correct.*

11 A liquid is flowing through horizontal pipes shown in the figure. The distances *EB, BG, FC* and *CH* are negligible. Lengths of different pipes has the following ratio. $L_{AB} = L_{CD} = \frac{L_{EF}}{2} = \frac{L_{GH}}{2}$

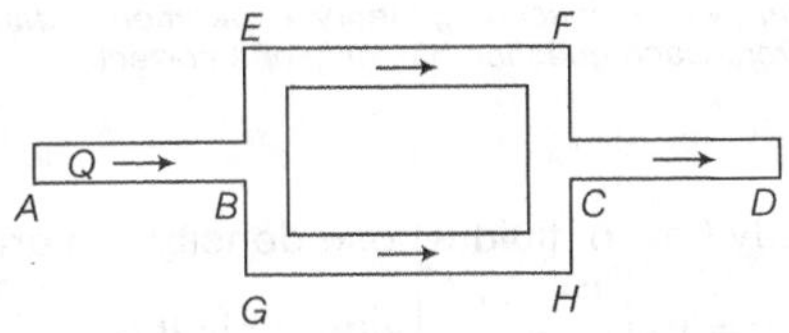

Similarly, radii of the pipes have the ratio $R_{AB} = R_{EF} = R_{CD} = \frac{R_{GH}}{2}$. Pressure at *A* is $2p_0$ and pressure at *D* is p_0. The volume flow rate through the pipe *AB* is *Q*. Then, choose the correct statement(s).

a. Volume flow rate through *EF* is $\frac{Q}{17}$

b. Volume flow rate through *GH* is $\frac{16Q}{17}$

c. Pressure at *E* is nearly 1.53 p_0

d. Pressure at *F* is nearly 1.47 p_0

12 A cubical block of side *a* and density ρ slides over a fixed inclined plane with constant velocity *v*. There is a thin film of viscous fluid of thickness *t* between the plane and the block. Then, the coefficient of viscosity of the thin film cannot be

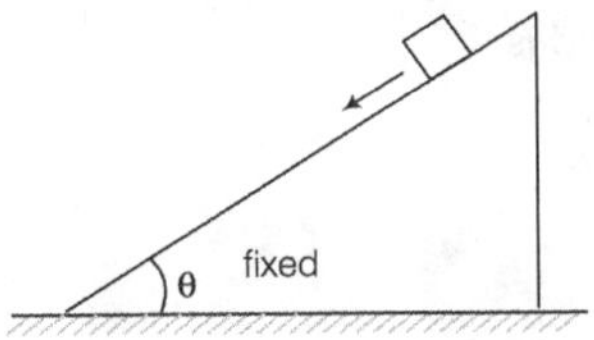

a. $\frac{\rho a g t \sin\theta}{v}$

b. $\frac{\rho a g t}{v \sin\theta}$

c. $\frac{\rho a g t \sin\theta}{\sqrt{2} v}$

d. $\frac{\rho a g t}{\sqrt{2} v \sin\theta}$

Statement Type

Direction *(Q. Nos. 13 and 14) This section is based on Statement I and Statement II. Select the correct answer from the codes given below.*

a. Both Statement I and Statement II are correct and Statement II is the correct explanation of Statement I

b. Both Statement I and Statement II are correct but Statement II is not the correct explanation of Statement I

c. Statement I is correct but Statement II is incorrect

d. Statement II is correct but Statement I is incorrect

13 **Statement I** Machine parts are jammed in winter.

Statement II The viscosity of lubricants used in machine parts increases at low temperatures.

14 **Statement I** All the raindrops hit the surface of the earth with the same constant velocity.

Statement II An object falling through a viscous medium eventually attains a terminal velocity.

Working Space

Comprehension Type

Direction (Q. Nos. 15-17) *This section contains a paragraph, which describing theory, experiments, data, etc. Three questions related to the paragraph have been given. Each question has only one correct answer among the four given options (a), (b), (c) and (d).*

Passage

A tube of length l and radius R carries a steady flow of fluid whose density is ρ and viscosity is η. The velocity of flow is given by $v = v_0\left[\frac{R^2 - r^2}{R^2}\right]$, where r is the distance of the flowing fluid from the axis.

15 Volume of fluid, flowing across the section of the tube, in unit time is

a. $v_0 16\pi R^2$
b. $\frac{v_0 \pi R^2}{2}$
c. $\frac{v_0 \pi R^2}{3}$
d. $\frac{v_0 \pi R^2}{4}$

16 Kinetic energy of the fluid within the volume of the tube is

a. $\frac{\pi l \rho v_0^2 R^2}{2}$
b. $\frac{\pi l \rho v_0^2 R^2}{4}$
c. $\frac{\pi l \rho v_0^2 R^2}{6}$
d. $\frac{\pi l \rho v_0^2 R^2}{3}$

17 The frictional force exerted on the tube by the fluid is

a. $4\pi\eta l v_0$
b. $\pi\eta l v_0$
c. $2\pi\eta l v_0$
d. $3\pi\eta l v_0$

Matching List Type

Direction (Q. No. 18) *Choices for the correct combination of elements from Column I and Column II are given as options (a), (b), (c) and (d), out of which one is correct.*

18 Column II depends on physical quantity/physical law given in Column I. Match the following columns and select the correct option from the codes given below.

	Column I		Column II
i.	Stoke's Law	p.	Radius
ii.	Terminal velocity	q.	Density of material of the body
iii.	Excess pressure inside mercury drop	r.	Coefficient of viscosity
iv.	Viscous force on a plate moving horizontally on a liquid surface	s.	Surface tension
		t.	Velocity gradient

Codes

	i	ii	iii	iv
a.	p,r	p,q,r	p,s	r,t
b.	p,s	q,r	s,t	p,t
c.	p	q,s	t	p,q,r
d.	q	s	p,r	q,s

Working Space

One Integer Value Correct Type

Direction (Q. Nos. 19-21) *This section contains 3 questions. When worked out will result in an integer from 0 to 9 (both inclusive).*

19 The coefficient of viscosity of blood is 2.12×10^{-3} Pa-s and its density is 1.06×10^3 kg m^{-3}. The blood flow rate, corresponding to the largest average velocity in an artery of radius 2×10^{-3} m for the flow to be laminar, is $x\pi \times 10^{-6}$ m^3 s^{-1}. Find the value of x.

20 A plate of area 2m^2 is made to move horizontally with a speed of 2ms^{-1} by applying a horizontal tangential force over the free surface of a liquid. The depth of the liquid is 1 m and the liquid in contact with the bed is stationary. Coefficient of viscosity of liquid = 0.01Poise. The tangential force needed to move the plate is $\alpha \times 10^{-3}$ N. Find the value of α.

21 A space 2.5 cm wide between two large plane surfaces is filled with oil. Force required to drag a very thin plate of area 0.5 m^2 just midway the surfaces at a speed of 0.5 ms^{-1} is 1 N. If the coefficient of viscosity of oil is $5\beta \times 10^{-3}$ kg-sm^{-2}, then find the value of β.

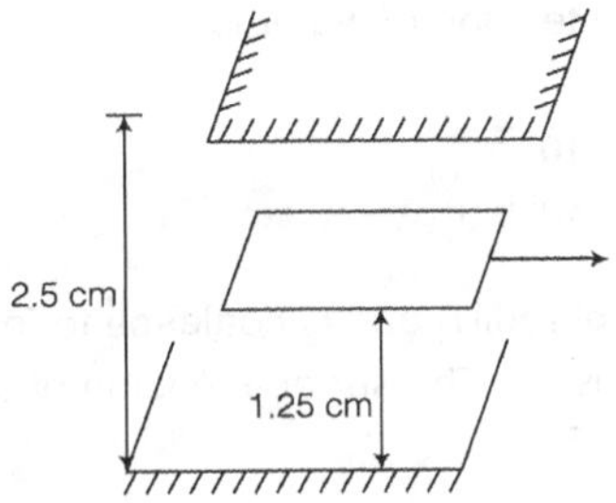

Subjective Questions

Direction (Q. Nos. 22 and 23) *These questions are subjective in nature, need to be solved completely on notebook.*

22 A soap bubble of radius r_1 is blown at the end of a capillary tube of length l and of internal radius r_2. Calculate the time taken by the bubble to reduce radius $r_2 < r_1$. Surface tension of soap is T and coefficient of viscosity of air is η.

23 A metal sphere of radius 1 mm and mass 50 mg falls vertically in glycerine. Density of glycerine = 1260 kg m^{-3} and its coefficient of viscosity at room temperature = 8 Poise. Calculate

i. the viscous force exerted by the glycerine on the sphere when the speed of the sphere is 1 cms^{-1}.

ii. the hydrostatic force exerted by the glycerine on the sphere.

iii. the terminal velocity with which the sphere will move down.

DPP-5 Surface Tension

Only One Option Correct Type

Working Space

Direction (Q. Nos. 1-10) *This section contains 10 multiple choice questions. Each question has four choices (a), (b), (c) and (d), out of which ONLY ONE is correct.*

1 How much work will be done in increasing the diameter of a soap bubble from 2 cm to 5 cm? Surface tension of soap solution is $3 \times 10^{-2}\ Nm^{-1}$.

a. 2.96×10^{-4} J b. 3.96×10^{-4} J

c. 4.96×10^{-4} J d. 1.96×10^{-4} J

2 Calculate the energy released when 1000 small water drops each of same radius 10^{-7} m coalesce to form one large drop. Surface tension of water is $7 \times 10^{-2}\ Nm^{-1}$.

a. 3.9×10^{-12} J b. 6.9×10^{-12} J

c. 4.9×10^{-12} J d. 7.9×10^{-12} J

3 Under isothermal conditions, two soap bubbles of radii r_1 and r_2 coalesce to form a single bubble of radius r. The external pressure is p_0. The surface tension of the soap solution, in terms of the given parameters, is

a. $\dfrac{p_0(r^3 - r_1^3 + r_2^3)}{4(r_1^2 - r_2^2 - r^2)}$

b. $\dfrac{p_0(r^3 + r_1^3 + r_2^3)}{4(r_1^2 + r_2^2 + r^2)}$

c. $\dfrac{p_0(r^3 + r_1^3 - r_2^3)}{2(r_1^2 + r_2^2 + r^2)}$

d. $\dfrac{p_0(r^3 - r_1^3 - r_2^3)}{4(r_1^2 + r_2^2 - r^2)}$

4 There is a rectangular wire frame having a thin film of soap solution. A circular ring of massless thin wire of area of cross-section A is placed on the surface of the film. The radius of the ring is R. Now, inside portion of the ring is pricked. If the surface tension of soap solution is T and Young's modulus of wire is Y, then change in radius of the wire is

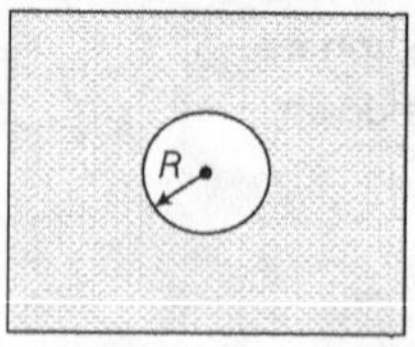

a. $\dfrac{TR^2}{AY}$ b. $\dfrac{2TR^2}{AY}$

c. $\dfrac{TR^2}{3AY}$ d. $\dfrac{TR^2}{2AY}$

Working Space

5 An isolated and charged spherical soap bubble has a radius r and the pressure inside is atmospheric. If T is the surface tension of soap solution, the charge on the drop is

a. $2\sqrt{\frac{2rT}{\varepsilon_0}}$ b. $8\pi r\sqrt{rT\varepsilon_0}$

c. $8\pi r\sqrt{2rT\varepsilon_0}$ d. $8\pi r\sqrt{\frac{2rT}{\varepsilon_0}}$

6 Drops of liquid of density ρ are floating in a liquid of density ρ_L. If the surface tension is T, the radius of the drop will be

a. $\sqrt{\frac{3T}{g(3\rho-\rho_L)}}$ b. $\sqrt{\frac{6T}{g(2\rho-\rho_L)}}$

c. $\sqrt{\frac{3T}{g(2\rho-\rho_L)}}$ d. $\sqrt{\frac{3T}{g(4\rho-3\rho_L)}}$

7 In the figure shown, radii of the limbs (capillary tubes) of a U-tube are r_1 and r_2 ($> r_1$). The surface tension of the liquid is T. If the angle of contact is zero, the difference of the heights of the liquid column, h is equal to (Given, ρ = density of liquid)

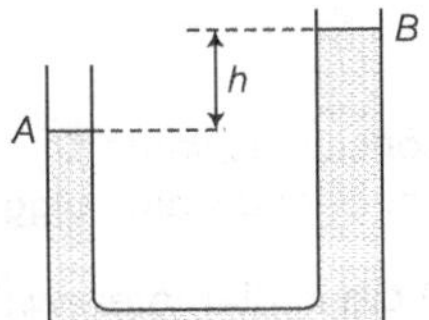

a. $\frac{T}{\rho g}\left[\frac{1}{r_1}+\frac{1}{r_2}\right]$ b. $\frac{T}{\rho g}\left[\frac{1}{r_1}-\frac{1}{r_2}\right]$

c. $\frac{2T}{\rho g}\left[\frac{1}{r_1}+\frac{1}{r_2}\right]$ d. $\frac{2T}{\rho g}\left[\frac{1}{r_2}-\frac{1}{r_1}\right]$

8 Water in a clean aquarium forms a meniscus, as shown in the figure. Calculate the difference in height (h) between the centre and edge of the meniscus. Given, the surface tension of water, $T = 0.073\ \text{Nm}^{-1}$, density of water, $\rho = 1000\ \text{kgm}^{-3}$.

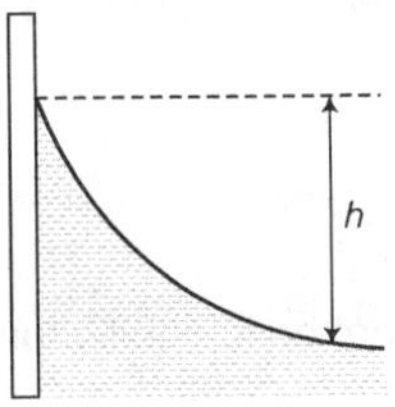

a. $h \simeq 4$mm b. $h \simeq 2$mm

c. $h \simeq 1$mm d. $h \simeq 5$mm

9 The base of an insect's leg is approximately spherical in shape, with a radius of about 2×10^{-5} m. The 0.003 g mass of the insect is supported equally by six legs. Estimate the angle θ (see figure) for an insect on the surface of water. Given, surface tension of water, $T = 0.072\text{Nm}^{-1}$.

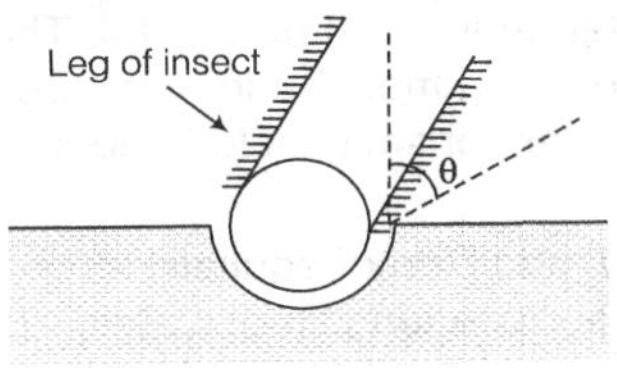

a. $\cos^{-1}(0.54)$ b. $\cos^{-1}(0.47)$

c. $\cos^{-1}(0.63)$ d. $\cos^{-1}(0.23)$

10 A drop of water of volume V is pressed between the two glasses placed, so as to spread to an area A. If T is the surface tension of water, the normal force required to seperate the glass plates is

Working Space

a. $\frac{TA^2}{V}$

b. $\frac{2TA^2}{V}$

c. $\frac{4TA^2}{V}$

d. $\frac{TA^2}{2V}$

One or More than One Options Correct Type

Direction (Q. Nos. 11 and 12) This section contains 2 multiple choice questions. Each question has four choices (a), (b), (c) and (d), out of which ONE or MORE THAN ONE are correct.

11 Two separate soap bubbles (radii 0.002 m and 0.004 m) formed of the same soap solution (surface tension $0.07 Nm^{-1}$) comes together to form a double bubble. The internal film common to both the bubbles has radius R. Then,

a. $R = 0.004$ m
b. $R = 0.003$m
c. internal film is concave towards smaller bubble
d. internal film is concave towards bigger bubble

12 A circular wire, 10 cm in diameter, with a slider wire on it, is in a horizontal plane. A liquid film is formed, bounded by the wires, on the left side of the slider, as shown in figure. The surface tension of the liquid is $0.1 Nm^{-1}$. An applied force of 16 mN, perpendicular to the slider, maintains the film in equilibrium. Ignore the sag in the film. What can be the distance between point P and the slider?

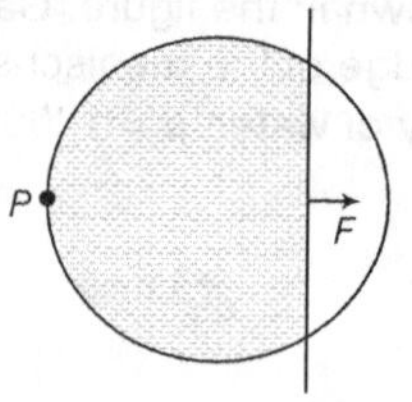

a. 8 cm

b. 2 cm

c. 5 cm

d. Slider cannot be in equilibrium

Comprehension Type

Direction (Q. Nos. 13-18) This section contains 2 paragraphs, each describing theory, experiments, data, etc. Six questions related to the paragraphs have been given. Each question has only one correct answer among the four given options (a), (b), (c) and (d).

Passage I

A prism shaped styrofoam of density $\rho_{styrofoam} < \rho_{water}$ is held completely submerged in water. It lies with its base (i.e. rectangular face) horizontal. The base of foam is at a depth H below water surface and atmospheric pressure is p_0. Surface of water is opened to atmosphere. Styrofoam prism is held in equilibrium by the string attached symmetrically.

(where, $\rho_{styrofoam} = \rho_{sf}$ and $\rho_{water} = \rho_w$, length $= L$ and side of equilateral triangular base $= l$, atmospheric pressure $= p_0$, surface tension of water $= T$ and contact angle $= 0°$)

13 Net force exerted by water on the styrofoam is

a. $\sqrt{2}\rho_w g l^2 L$

b. $2\rho_w g l^2 L$

c. $\rho_w g \frac{l^2 L}{2}$

d. $\rho_w g l^2 L \frac{\sqrt{3}}{4}$

Working Space

14 Magnitude of force on any one of the slant faces of styrofoam is

a. $\left[p_0 + \rho_w g\left(H - \sqrt{\frac{2}{3}}l\right)\right]Ll$ b. $\left[p_0 + \rho_w g\left(H - \frac{l}{\sqrt{2}}\right)\right]Ll$

c. $\left[p_0 - \rho_w g\left(H - \frac{l}{\sqrt{2}}\right)\right]Ll$ d. $\left[p_0 + \rho_w g\left(H - \frac{\sqrt{3}}{4}l\right)\right]Ll$

15 Now string is cut and styrofoam is allowed to come to surface. A point mass is to be placed symmetrically on the upper surface of styrofoam such that it is in equilibrium with its base in horizontal plane. In equilibrium, position styrofoam has half of its slant height submerged. Determine mass m to achieve equilibrium.

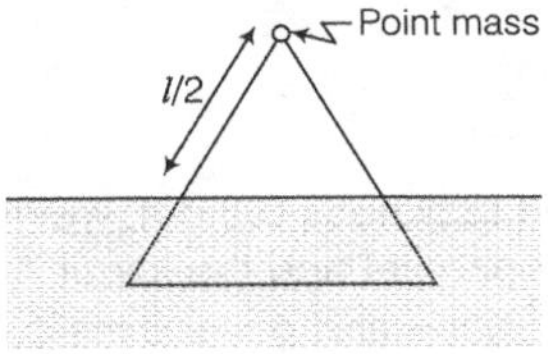

a. $\frac{Ll^2\sqrt{3}}{4}\left[\frac{3}{4}\rho_w - \rho_{sf}\right] - (2L + l)\frac{T}{g}$

b. $\frac{Ll^2\sqrt{3}}{2}\left[\frac{3}{2}\rho_w - \rho_{sf}\right] - (2L + l)\frac{T}{g}$

c. $\frac{Ll^2\sqrt{3}}{4}\left[\frac{3}{2}\rho_w - \rho_{sf}\right] - (L + l)\frac{T}{g}$

d. $\frac{Ll^2\sqrt{3}}{2}\left[3\rho_w - \frac{\rho_{sf}}{2}\right] - (L + l)\frac{T}{g}$

Passage II

An open capillary tube contains a drop of water. When the tube is in its vertical position, the drop forms a column with a length L. The internal diameter of the capillary tube is 1 mm. The radii of upper meniscus and lower meniscus are respectively, R_u and R_l. Consider the wetting to be complete and surface tension of water is 0.075 Nm^{-1}. (Given, $\rho = 1000 kg\,m^{-3}$ and $g = 10 ms^{-2}$)

16 When $L = 2$cm, R_u and R_l respectively are

a. 0.5 mm, 1.5 mm
b. 0.5 mm, 1 mm
c. 1 mm, 1.5 mm
d. 1.5 mm, 1 mm

17 When $L = 4$cm, R_u and R_l respectively are

a. 0.5 mm, 1 mm
b. 0.5 mm, 1.5 mm
c. 1 mm, 1.5 mm
d. 1.5 mm, ∞

18 When $L = 3$ cm, R_u and R_l respectively are

a. 0.5 mm, 1 mm
b. 0.5 mm, 1.5 mm
c. 0.5 mm, ∞
d. 1.5 mm, 1 mm

One Integer Value Correct Type

Working Space

Direction (*Q. Nos. 19 and 20*) *This section contains 2 questions. When worked out will result in an integer from 0 to 9 (both inclusive).*

19 A soap bubble is being blown on a tube of radius 1 cm. The surface tension of the soap solution is 0.05 Nm^{-1} and the bubble makes 60° with the tube as shown in the figure. The excess of pressure over the atmosphere pressure in the tube is 2β Pa. Find the value of β.

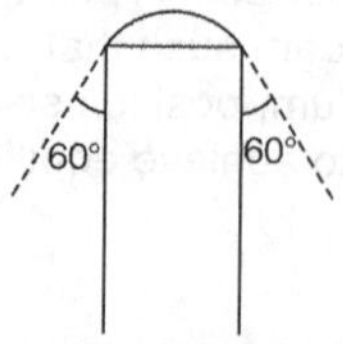

20 A glass plate of length 10 cm, breadth 1.54 cm and thickness 0.2 cm, weight 8.2 g in air. It is held vertically with the long side horizontal and the lower half under water. Surface tension of water is $7.3 \times 10^{-2}\ Nm^{-1}$ and $g = 9.8\ ms^{-2}$. The apparent weight of the plate is $(2.0449)\ \alpha \times 10^{-3}$ kg-wt. Find the value of α.

Subjective Questions

Direction (*Q. Nos. 21-23*) *These questions are subjective in nature, need to be solved completely on notebook.*

21 The end of a capillary tube with radius r is immersed into water. What amount of heat will be evolved when the water rises in the tube? The density of water is ρ and its surface tension is T.

22 There is an air bubble of radius 1 mm in a liquid of surface tension 0.075 Nm^{-1} and density 1000 kgm^{-3}. The bubble is at a depth of 10 cm below the free surface. By what amount is the pressure inside the bubble greater than the atmospheric pressure? (Given, $g = 9.8\ ms^{-2}$)

23 A small water drop falls in air with a uniform velocity. Find the difference (approximately) between the curvature radii of the droplet surface at the upper and lower points of the drop separated by a distance h.

Revisal Problems
for JEE Main

Only One Option Correct Type

Working Space

Direction *(Q. Nos. 1-20) This section contains 20 multiple choice questions. Each question has four choices (a), (b), (c) and (d), out of which ONLY ONE is correct.*

1 Shown in the figure is a conical container of half-apex angle 37° filled with certain quantities of kerosene and water. The force exerted by the water on the kerosene is approximately (Given, $g = 10\ ms^{-2}$)

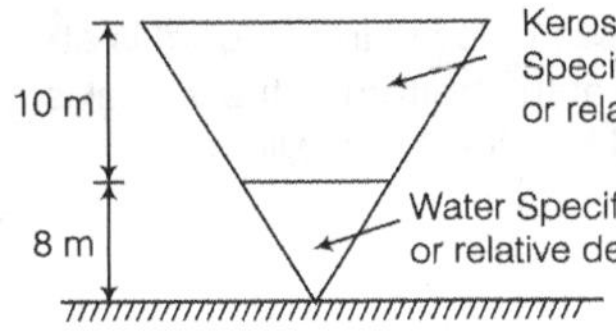

a. 3×10^7 N
b. 4×10^7 N
c. 2×10^7 N
d. 5×10^7 N

2 Two identical discs sit at the bottom of a 3 m pool of water whose surface is exposed to atmospheric pressure. The first acts as a plug to seal the drain as shown in the figure. The second disc covers a container containing nearly a perfect vacuum. If each disc has an area 1 m^2, then what is the approximate difference in the forces necessary to open the discs? (Given, 1 atm = 101300 Pa)

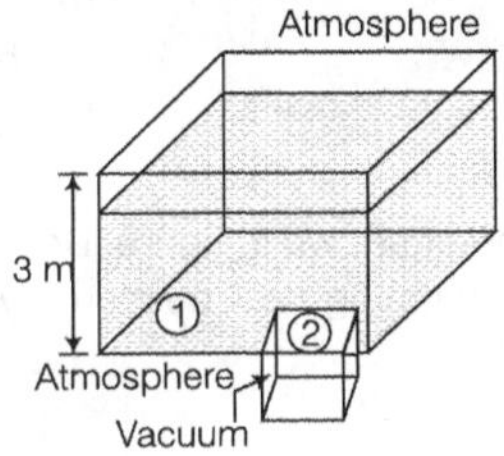

a. There is no difference
b. 3000 N
c. 1013000 N
d. 101300 N

3 A solid cone of uniform density and height $2R$ and base radius R has a conical portion scooped out from its base with the same base radius but height R as shown in the figure. The solid cone is floating in a liquid of density ρ with vertex A touching the fluid surface. If atmospheric pressure is p_0, the force acting on the surface BCD due to liquid is

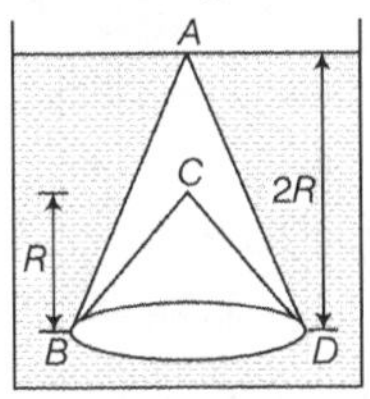

a. $\frac{\pi R^3 \rho g}{3}\left[2 + \frac{2p_0}{R\rho g}\right]$
b. $\frac{\pi R^3 \rho g}{3}\left[5 + \frac{3p_0}{R\rho g}\right]$
c. $\frac{\pi R^3 \rho g}{3}\left[1 + \frac{3p_0}{R\rho g}\right]$
d. $\frac{\pi R^3 \rho g}{3}\left[4 + \frac{3p_0}{R\rho g}\right]$

4 Figure shows a long cylinder of radius R, which is inside a long hollow tube of inner radius R'. A liquid of coefficient of viscosity η is filled in the gap between the cylinder and hollow tube. If the solid cylinder is moved with velocity v parallel to its length, inside the fixed hollow tube, the force required per unit length to be applied on the solid cylinder is (assume uniform velocity gradient in the liquid)

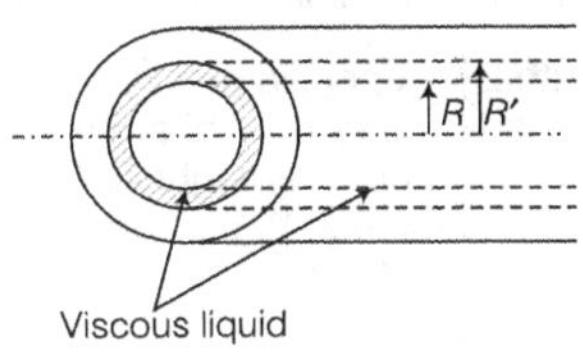

a. $2\pi\eta R\left(\frac{v}{R' - R}\right)$
b. $\frac{2}{3}\pi\eta R\left(\frac{v}{R' - R}\right)$
c. $\frac{1}{3}\pi\eta R\left(\frac{v}{R' - R}\right)$
d. $\frac{1}{2}\pi\eta R\left(\frac{v}{R' - R}\right)$

Working Space

5 A body floats in a liquid contained in a beaker. The whole system as shown in the figure, falls with constant acceleration a_0. The upthrust on the body due to the liquid is

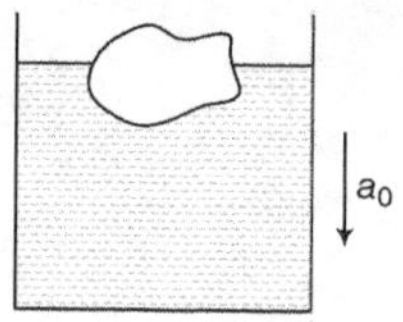

a. zero
b. equal to the weight of the liquid displaced
c. less than the weight of the liquid displaced
d. equal to the weight of the immersed portion of the body

6 The figure shows a pond full of water having the shape of a truncated cone. The depth of the pond is 30 m. The atmospheric pressure (p_0) above the pond is 10^5 Pa. The circular top surface and circular bottom surface of the pond are both parallel to the ground. The magnitude of the force acting on the top surface is the same as the magnitude of the force acting on the bottom surface. If ($R_1 + R_2$) (in metre) is 20α. Find the value of α. (Given, $g = 10 \text{ ms}^{-2}$)

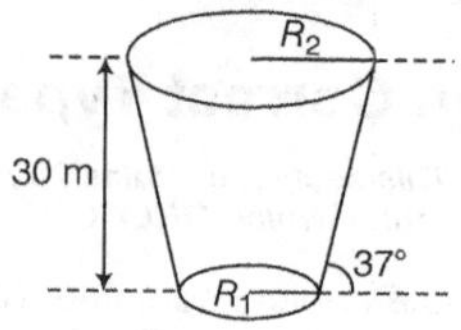

a. 6   b. 5   c. 4   d. 3

7 In a cylindrical vessel containing liquid of density ρ, there are two holes in the side walls at heights of h_1 and h_2 respectively, such that the range of efflux at the bottom of the vessel is same. The height of a hole for which the range of efflux would be maximum, will be

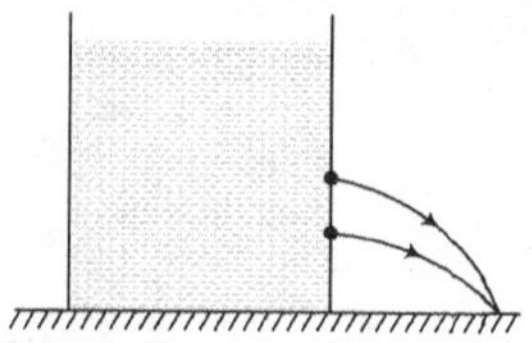

a. $h_2 - h_1$ b. $h_2 + h_1$
c. $\dfrac{h_2 - h_1}{2}$ d. $\dfrac{h_2 + h_1}{2}$

8 Consider the barometer shown in figure. Density of mercury is ρ. A small hole is made at point S as shown in the figure. The mercury comes out from this hole with speed v equal to

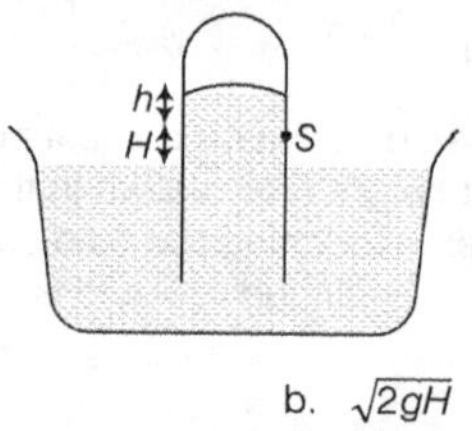

a. $\sqrt{2gh}$ b. $\sqrt{2gH}$
c. $\sqrt{2g(H-h)}$ d. None of these

9 A water proofing agent changes the angle of contact from
a. acute to 90°   b. obtuse to 90°
c. an acute to an obtuse angle   d. an obtuse to an acute angle

10 A and B are two soap bubbles. Bubble A is larger than B. If these are now joined by a tube then,
a. the bubble A gets larger and larger
b. the bubble B gets larger and larger
c. both the bubbles acquire the same size
d. both the bubbles will become smaller and smaller

11 A uniform rod of length 2 m, specific gravity 0.5 and mass 2 kg is hinged at one end to the bottom of a tank of water (specific gravity = 1.0) filled upto a height of 1 m as shown in the figure. Taking the case $\theta \neq 0°$, the force exerted by the hinge on the rod is ($g = 10 \text{ms}^{-2}$)

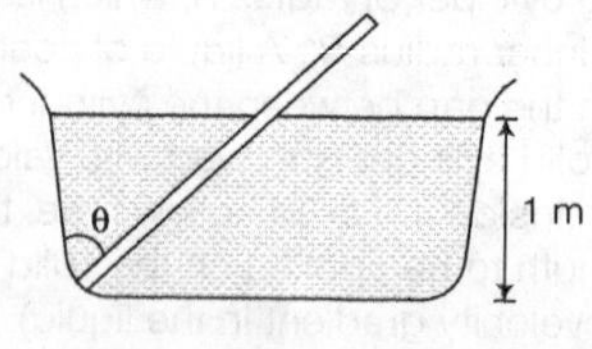

a. 10.2 N upwards
b. 4.2 N upwards
c. 8.3 N downwards
d. 6.2 N downwards

Working Space

12 A cylindrical vessel contains a liquid of density ρ upto a height h. The liquid is closed by a piston of mass m and area of cross-section A. There is a small hole at the bottom of the vessel. The speed v with which the liquid comes out of the hole is

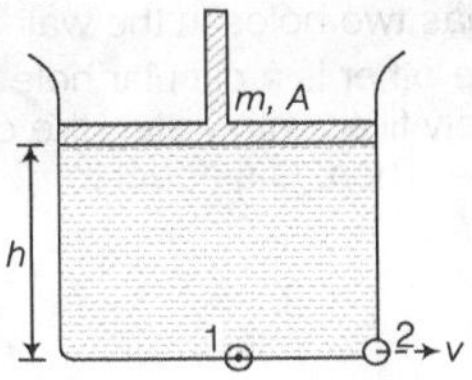

a. $\sqrt{2gh}$
b. $\sqrt{2\left(gh + \frac{mg}{\rho A}\right)}$
c. $\sqrt{2\left(gh + \frac{mg}{A}\right)}$
d. $\sqrt{2gh + \frac{mg}{A}}$

13 The opening near the bottom of the vessel shown in the figure has an area A. A disc is held against the opening to keep liquid, from running out. Let F_1 be the net force on the disc applied by liquid and air in this case. Now, the disc is moved away from the opening a short distance. The liquid comes out and strikes the disc inelastically. Let F_2 be the force exerted by the liquid in this condition. Then, $\frac{F_1}{F_2}$ is equal to

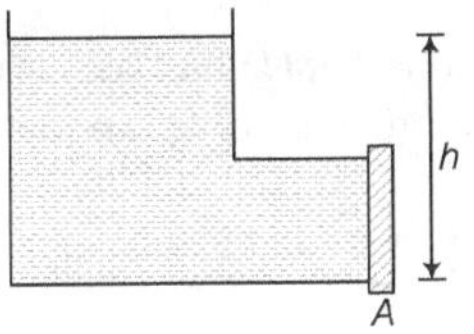

a. $\frac{1}{2}$
b. 1
c. 2
d. $\frac{1}{4}$

14 In the arrangement shown in the figure $\frac{m_A}{m_B} = \frac{2}{3}$ and the ratio of density of block B and of liquid is 2 : 1. The system is released from rest. Then, (neglect viscosity)

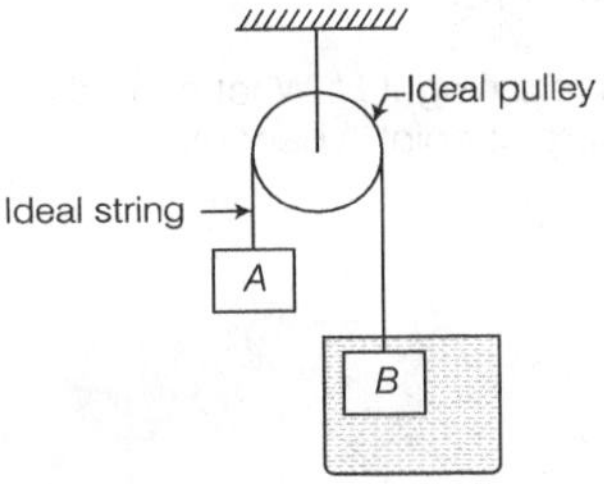

a. block B will oscillate periodically but not simple harmonically
b. block B will oscillate simple harmonically
c. the system will remain in equilibrium
d. None of the above

15 A ball of density ρ is released from deep inside of a viscous liquid of density 2ρ. It will move up

a. with an increasing acceleration
b. with a decreasing acceleration
c. with a constant acceleration
d. with zero acceleration

16 A large tank is filled with water (density = $1000\,kg\,m^{-3}$). A small hole is made at a depth 10 m below water surface. The range of water issuing out of the whole is R on the ground. Approximately, what extra pressure must be applied on the water surface, so that range becomes $2R$? (Given, 1 atm = 10^5 Pa and $g = 10\,ms^{-2}$)

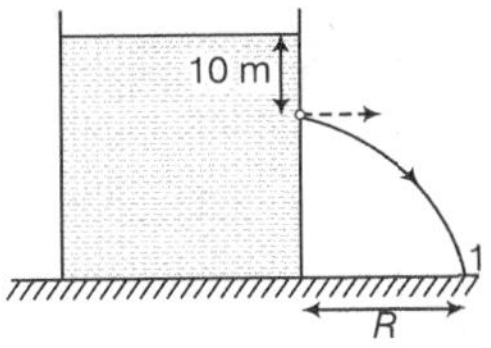

a. 9 atm
b. 4 atm
c. 5 atm
d. 3 atm

Working Space

17 A large open tank has two holes in the wall. One is a square hole of side L at a depth y from the top and the other is a circular hole of radius R at a depth $4y$ from the top. When the tank is completely filled with water, the quantities of water flowing out per second from both holes are same. Then, R is equal to

a. $L/\sqrt{2\pi}$ b. $2\pi L$

c. L d. $\frac{L}{2\pi}$

18 When a block of iron floats in mercury at 0°C, fraction k_1 of its volume is submerged, while at the temperature 60°C, a fraction k_2 is seen to be submerged. If the coefficient of volume expansion of iron is x and that of mercury is y, then the ratio $\frac{k_1}{k_2}$ can be expressed as

a. $\frac{1+60y}{1+60x}$ b. $\frac{1-60y}{1+60x}$

c. $\frac{1+60x}{1+60y}$ d. $\frac{1-60x}{1-60y}$

19 In the figure, find out angle θ. The *zig-zag* tube shown (opened at N) has liquids of densities ρ_1, ρ_2 and ρ_3 and is present in vertical plane. The pressure at M is equal to atmospheric pressure.

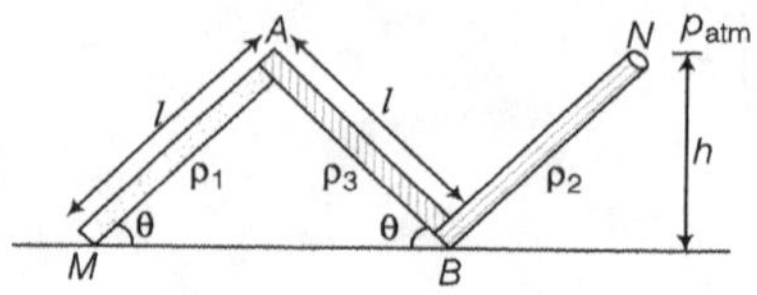

a. $\sin^{-1}\left[\frac{\rho_2 h}{\rho_3 l + \rho_1 l}\right]$ b. $\sin^{-1}\left[\frac{\rho_2 h}{\rho_3 l - \rho_1 l}\right]$

c. $\sin^{-1}\left[\frac{\rho_1 h}{\rho_2 l + \rho_3 l}\right]$ d. $\sin^{-1}\left[\frac{\rho_3 h}{\rho_1 l + \rho_2 l}\right]$

20 Figure shows an L-shaped thin tube filled with a liquid to a height H. What should be the horizontal acceleration a of the tube, so that the pressure at point A becomes atmospheric?

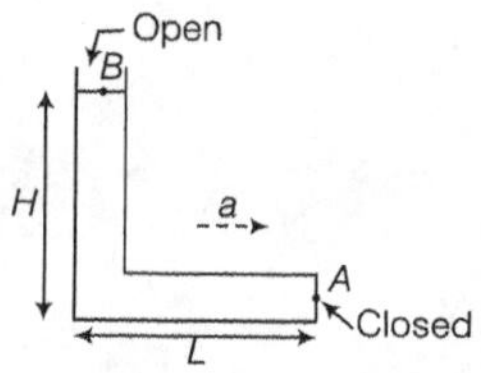

a. $\sqrt{2}\frac{gL}{H}$ b. $\frac{gL}{H}$

c. $\frac{gH}{L}$ d. $\frac{\sqrt{2}gH}{L}$

SET - 2

Revisal Problems for JEE Advanced

Only One Option Correct Type

Working Space

Direction (Q. Nos. 1-10) *This section contains 10 multiple choice questions. Each question has four choices (a), (b), (c) and (d), out of which ONLY ONE is correct.*

1 A simple accelerometer (an instrument for measuring acceleration) can be made in the form of a tube filled with a liquid and bent as shown in the figure. During motion, the level of the liquid in the left arm will be at a height h_1 and in the right arm at a height h_2. Determine the acceleration a_0 of a carriage in which the accelerometer is installed, assuming that the diameter of the tube is much smaller than h_1 and h_2.

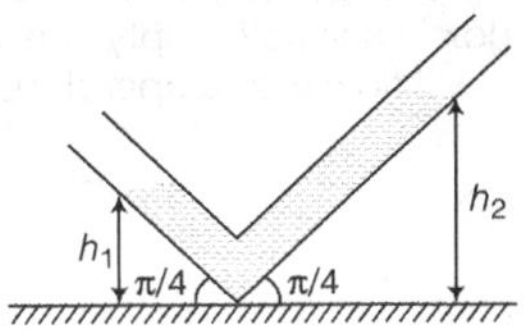

a. $g\left(\frac{h_1 + h_2}{h_2 - h_1}\right)$
b. $g\left(\frac{h_1 + h_2}{h_1}\right)$
c. $g\left(\frac{h_2 - h_1}{h_1 + h_2}\right)$
d. $g\left(\frac{h_1 + h_2}{h_2}\right)$

2 Oil enters the bend of a pipe in the horizontal plane with a velocity $4\,ms^{-1}$ and pressure $280 \times 10^3\,Nm^{-2}$ as shown in the figure. The pressure of oil at point P_2 is (Given, specific gravity of oil is 0.9 and $\sin 37° = 0.6$) (the entire pipe is in the horizontal plane)

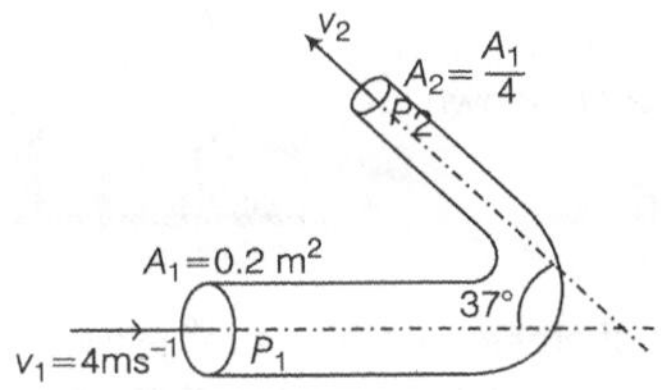

a. $172\,k\,Nm^{-2}$
b. $241\,k\,Nm^{-2}$
c. $221\,k\,Nm^{-2}$
d. $153\,k\,Nm^{-2}$

3 Figure shows a liquid flowing through a tube at the rate of $0.1\,m^3s^{-1}$. The tube is branched into two semicircular tubes of cross-sectional area $\frac{A}{3}$ (at P) and $\frac{2A}{3}$ (at Q). The velocity of liquid at Q is (Given, the cross-section of the main tube is $A = 10^{-2}\,m^2$ and $v_P = 20\,ms^{-1}$)

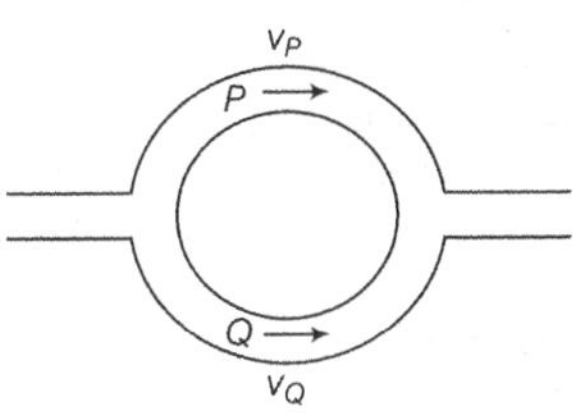

a. $5\,ms^{-1}$
b. $30\,ms^{-1}$
c. $35\,ms^{-1}$
d. $10\,ms^{-1}$

Working Space

4 A cylindrical container of radius R and height h is completely filled with a liquid of density ρ. Two horizontal L-shaped pipes of small cross-sectional area a (each) are connected to the cylinder as shown in the figure. Now, the two pipes are opened and fluid starts coming out of these pipes horizontally in opposite directions. Then, the torque due to ejected liquid on the system is

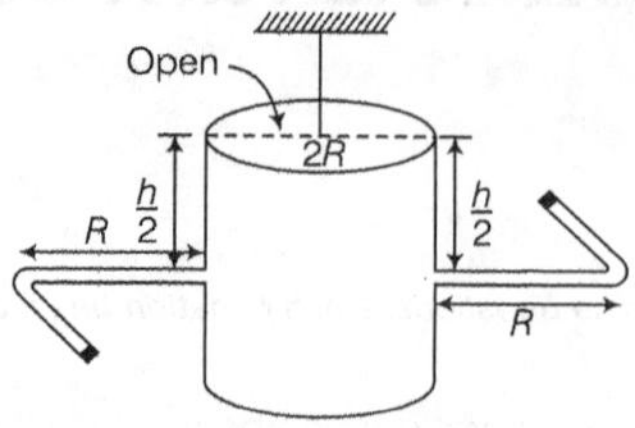

a. $8agh\rho R$ b. $4agh\rho R$
c. $2agh\rho R$ d. $6agh\rho R$

5 A square box of water has a small hole located at one of the bottom corners. When the box is full and sitting on a level surface, complete opening of the hole results in a flow of water with a speed v_0, as shown in the figure. When the box is still half empty, it is tilted by 45°, so that the hole is at the lowest point. Now, the water will flow with a speed v equals to

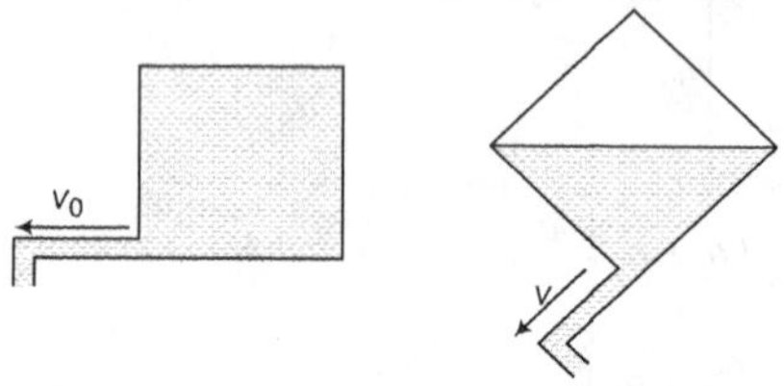

a. v_0 b. $\frac{v_0}{2}$
c. $\frac{v_0}{\sqrt{2}}$ d. $\frac{v_0}{2^{1/4}}$

6 A hollow sphere of mass $m = 50$ kg and radius $r = \left(\frac{3}{40\pi}\right)^{1/3}$ m is immersed in a tank of water (with density $\rho_w = 1000$ kgm^{-3}). The sphere is tied to the bottom of a tank by wires 1 and 2 as shown in the figure. Tension in wire 1 is (Given, $g = 10$ ms^{-2})

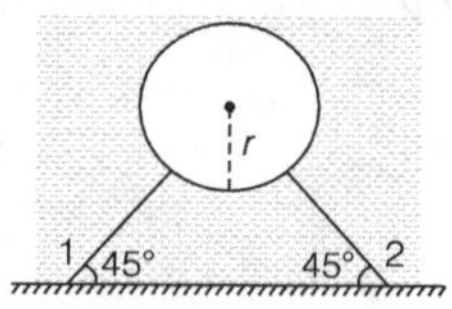

a. $125\sqrt{2}$ N b. 125 N
c. $250\sqrt{2}$ N d. 250 N

7 An empty cylindrical tank of 1.2 m height and of area of cross-section 0.4 m^2 is filled by water starting at $t = 0$ s, flowing with the speed of 2ms^{-1} from a tap of area of cross-section 10^{-4} m^2. There is a small hole of area 0.5×10^{-4} m^2 in the base of the tank. The variation of height h (in metre) of water in tank with time t (in second) is best represented by the following (Given, $g = 10$ ms^{-2})

a. 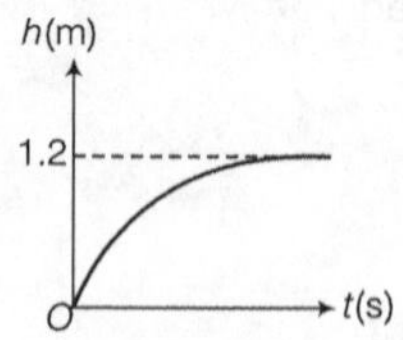

b. 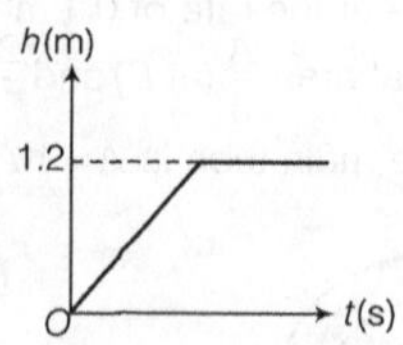

c. 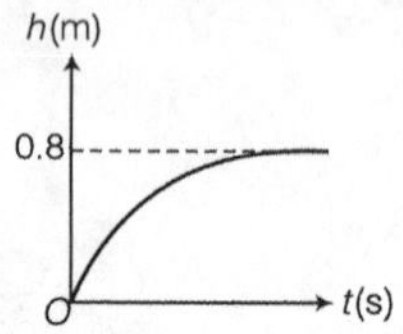

d. 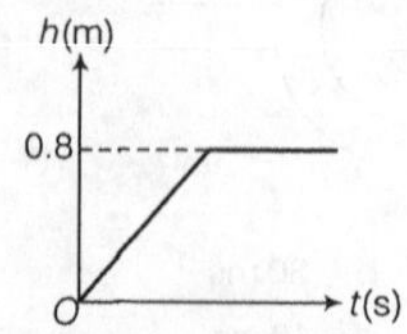

Working Space

8 A non-viscous liquid of density 1000 kg m^{-3} flows in a streamline motion through a tube of non-uniform cross-section. The tube is kept inclined in a vertical plane as shown in the figure. The area of cross-section of tube at points A and B at heights of 2 m and 5 m are $4\times10^{-3}\ m^2$ and $8\times10^{-3}\ m^2$, respectively. Velocity at point A of the liquid is $1\ ms^{-1}$. The work done per unit volume by the pressure, when fluid flows from A to B is

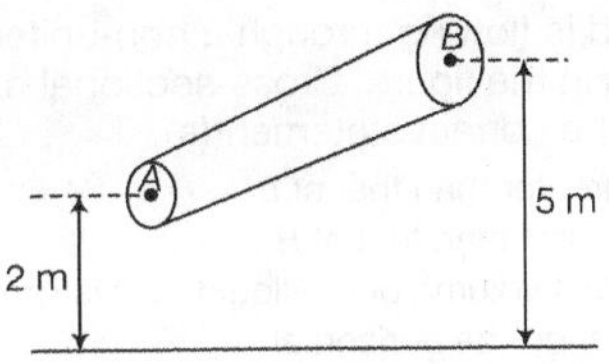

a. $5.8\times10^3\ Jm^{-3}$ b. $2.9\times10^4\ Jm^{-3}$
c. $5.8\times10^4\ Jm^{-3}$ d. $2.9\times10^3\ Jm^{-3}$

9 A hemispherical portion of radius R is removed from the bottom of a cylinder of radius R. The volume of the remaining cylinder is V and its mass is M. It is suspended by a string in a liquid of density ρ, where it stays vertical. The upper surface of the cylinder is at a depth h below the liquid surface. The force on the bottom of the cylinder by the liquid is

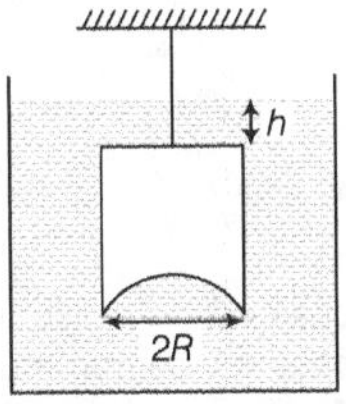

a. Mg
b. $Mg - V\rho g$
c. $Mg + \pi R^2 h\rho g$
d. $\rho g(V + \pi R^2 h)$

10 Water is flowing on a horizontal fixed surface, such that its flow velocity varies with y (vertical direction) as,

$v = k_1\left[\frac{2y^2}{k_2^2} - \frac{y^3}{k_2^3}\right]$, where k_1 and k_2 are positive constants. If coefficient of viscosity for water is η, what will be the shear stress between layers of water at $y = k_2$?

a. $\frac{\eta k_1}{k_2}$ b. $\frac{\eta}{k_1 k_2}$ c. $\frac{\eta k_2}{k_1}$ d. $\frac{\eta k_2}{\sqrt{k_1^2 + k_2^2}}$

One or More than One Options Correct Type

Direction (Q. Nos. 11-14) *This section contains 4 multiple choice questions. Each question has four choices (a), (b), (c) and (d), out of which ONE or MORE THAN ONE are correct.*

11 An ice cube of side 1 cm is floating at the interface of kerosene and water in a beaker of base area $10\ cm^2$. The level of kerosene is just covering the top surface of the ice cube. Specific gravities of kerosene and ice respectively are 0.8 and 0.9. Which of the following statement(s) is/are correct?

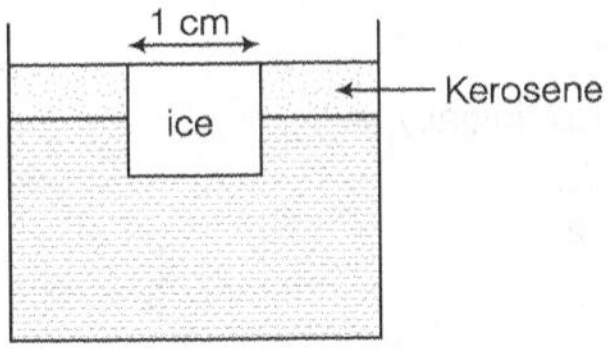

a. Depth of submergence of ice cube in kerosene is 0.5 cm
b. Depth of submergence of ice cube in water is 0.5 cm
c. When the whole ice melts into water, the overall level falls by 0.1 mm
d. When the whole ice melts into water, the overall level raises by 0.1 mm

12 A circular cylinder of radius R and height H is filled with water to a height $\frac{2}{3}H$. It starts rotating about its axis with constantly increasing angular speed. Choose the correct statement(s).

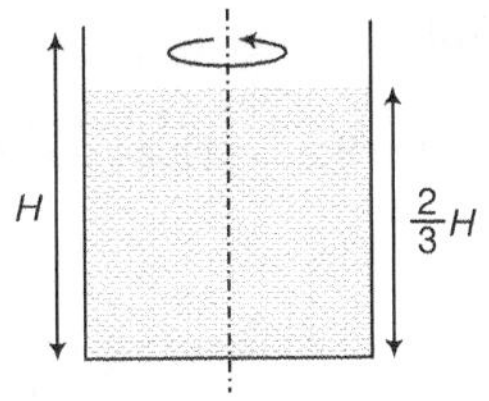

a. At all speeds, shape of the free surface is paraboloid
b. The free surface touches first the brim of the cylinder and then the base of the cylinder
c. The free surface cannot touch the base without spilling water
d. The free surface touches the brim as well as the base at the same instant

Working Space

13 A non-viscous liquid is flowing through a non-uniform pipe from section A to section B as shown in the figure. Cross-sectional area of the pipe at A is less than at B. Choose the correct statement(s).

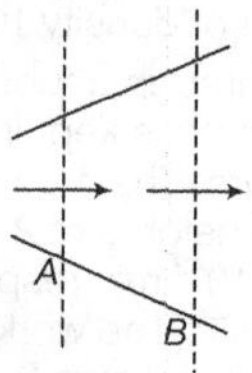

a. Pressure at A is greater than that at B
b. Velocity at A is greater than that at B
c. Total energy per unit volume of the liquid at A is greater than at B
d. Axis of the pipe cannot be horizontal

14 Equal volumes of liquid are poured in the three vessels A, B and C ($h_1 < h_2 < h_3$). All the vessels have same base area. Choose the correct statement(s).

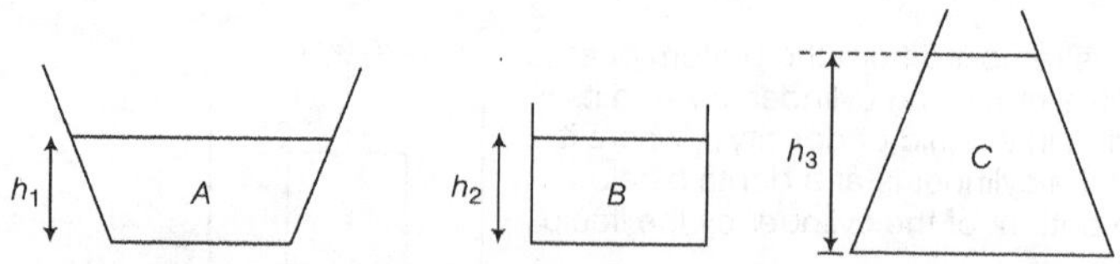

a. The force on the base will be maximum in vessel A
b. The force on the base will be maximum in vessel C
c. Net force exerted by the liquid in all the three vessels is same
d. Net force exerted by the liquid in vessel A is maximum

Comprehension Type

Direction (Q. Nos. 15-17) *This section contains a paragraph, which describing theory, experiments, data, etc. Three questions related to the paragraphs have been given. Each question has only one correct answer among the four given options (a), (b), (c) and (d).*

Passage

In the figure shown, a large closed cylindrical tank containing water. Initially, the air trapped above the water surface has a height h_0 and pressure $2p_0$, where p_0 is the atmospheric pressure. There is a hole in the wall of the tank at a depth h_1 below the top from which water comes out. A long vertical tube is connected as shown.

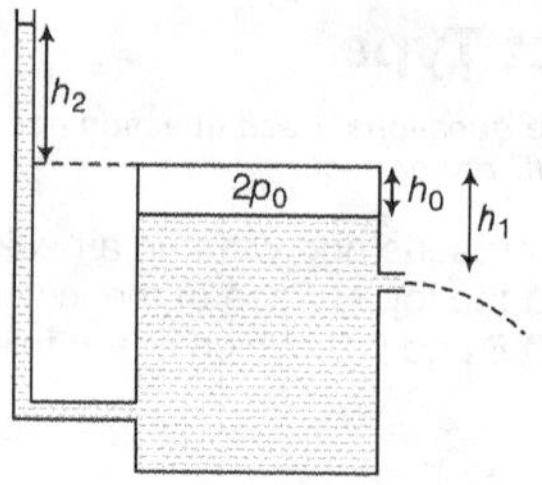

15 The height h_2 of the water in the long tube above the top, initially, is

a. $\frac{3p_0}{\rho g} - \frac{h_0}{3}$
b. $\frac{2p_0}{\rho g} - \frac{h_0}{2}$
c. $\frac{p_0}{\rho g} - h_0$
d. $\frac{p_0}{2\rho g} - 2h_0$

16 The speed with which water comes out of the hole, initially, is

a. $\frac{1}{\rho}[p_0 - \rho g(h_1 - 2h_0)]^{1/2}$
b. $\left\{\frac{2}{\rho}[p_0 + \rho g(h_1 - h_0)]\right\}^{1/2}$
c. $\left\{\frac{3}{\rho}[p_0 + \rho g(h_1 + h_0)]\right\}^{1/2}$
d. $\left\{\frac{4}{\rho}[p_0 - \rho g(h_1 - h_0)]\right\}^{1/2}$

17 The height of the water in the long tube above the top, when the water stops coming out of the hole, is

a. $-2h_0$
b. h_0
c. h_2
d. $-h_1$

Matching List Type

Direction (Q. No. 18) ***Choices for the correct combination of elements from Column I and Column II are given as options (a), (b), (c) and (d), out of which one is correct.***

18 For the value asked in a situation in Column I, match the suitable answer in Column II.

	Column I 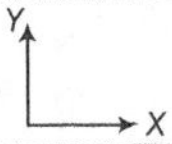		Column II
i.	A cube of mass m and density ρ_S is pulled by a force $\mathbf{F}_x = mg\,\hat{\mathbf{i}}$ in an accelerating liquid of density ρ. The value of $\frac{a_x}{a_y}$ is 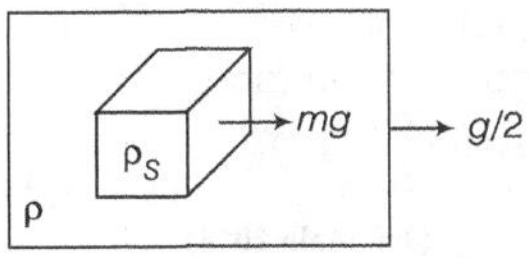	p.	$\frac{\rho}{\rho_s}$
ii.	The sphere of mass m is pulled horizontally by a force $\mathbf{F}_x = mg\,\hat{\mathbf{i}}$ in a non-accelerating liquid. The value of a_x/a_y is 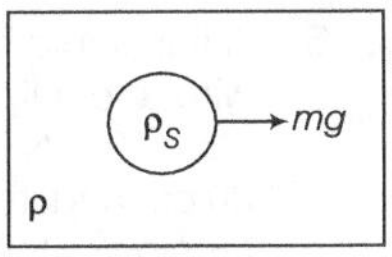	q.	1 : 1
iii.	A closed tube of length L is placed in a rotating liquid. The value of a_x/g, where a_x is the horizontal acceleration of the cube, is 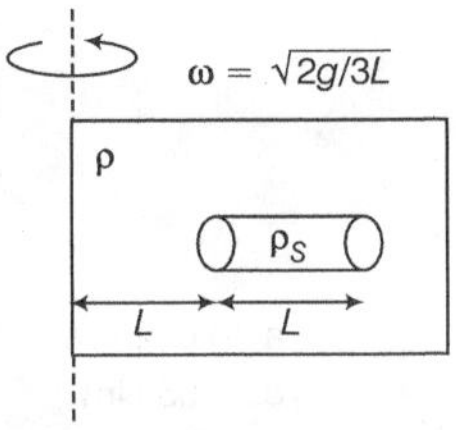	r.	$\frac{\rho_s}{\rho - \rho_s}$
iv.	The string of a pendulum makes an angle θ with vertical when the cart moves with an acceleration a_0. If the cart is filled with liquid of density $\rho < \rho_s$ (ρ_s is density of the bob), then the value of the angle of inclination with the vertical becomes ϕ. Then, $\frac{\tan\theta}{\tan\phi}$ is 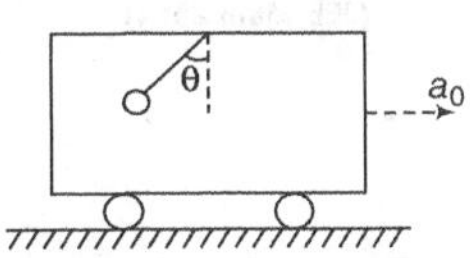	s.	$\frac{2\rho_s + \rho}{2(\rho - \rho_s)}$

Codes

	i	ii	iii	iv		i	ii	iii	iv
a.	s	r	p	q	b.	p	q	r	s
c.	r	p	s	q	d.	q	s	r	p

One Integer Value Correct Type

Direction (Q. Nos. 19 and 20) ***This section contains 2 questions. When worked out will result in an integer from 0 to 9 (both inclusive).***

19 A bowl whose bottom has round holes of diameter 1 mm is filled with water. Assuming that surface tension acts only at holes, find the maximum height (in cm) upto which water can be filled in the vessel without leakage. (Given, surface tension of water $= 75 \times 10^{-3}\ \text{Nm}^{-1}$, $g = 10\ \text{ms}^{-2}$ and density of water $= 1000\ \text{kgm}^{-3}$).

20 A soap film is made by dipping a circular frame of radius b in soap solution. A bubble is formed by blowing air with speed v in the form of a cylinder. Air stops after striking surface of soap bubble. Density of air is ρ. The radius of the bubble is $R >> b$. The radius R of the bubble when the soap bubble separates from the ring is $\frac{\beta T}{\rho v^2}$. Find the value of β. (where, surface tension of liquid is T).

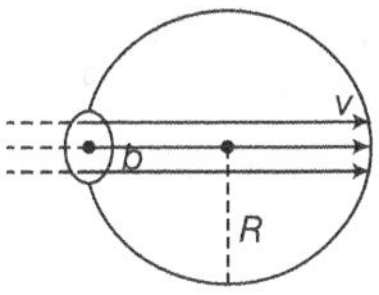

Working Space

JEE Main & AIEEE Archive

(Compilation of Last 13 Years' Questions)

1 There is a circular tube in a vertical plane. Two liquids which do not mix and of densities d_1 and d_2 are filled in the tube. Each liquid subtends 90°angle at centre. Radius joining their interface makes an angle α with vertical. Ratio $\frac{d_1}{d_2}$ is **(JEE Main 2014)**

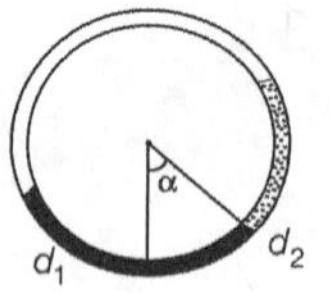

a. $\frac{1+\sin\alpha}{1-\sin\alpha}$ b. $\frac{1+\cos\alpha}{1-\cos\alpha}$

c. $\frac{1+\tan\alpha}{1-\tan\alpha}$ d. $\frac{1+\sin\alpha}{1-\cos\alpha}$

2 On heating water, bubbles being formed at the bottom of the vessel detatch and rise. Take the bubbles to be spheres of radius R and making a circular contact of radius r with the bottom of the vessel. If $r << R$ and the surface tension of water is T, value of r just before bubbles detatch is (density of water is ρ) **(JEE Main 2014)**

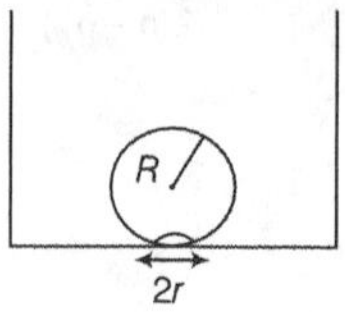

a. $R^2\sqrt{\frac{2\rho_w g}{3T}}$ b. $R^2\sqrt{\frac{\rho_w g}{6T}}$

c. $R^2\sqrt{\frac{\rho_w g}{T}}$ d. $R^2\sqrt{\frac{3\rho_w g}{T}}$

3 A uniform cylinder of length L and mass M having cross-sectional area A is suspended with its length vertical from a fixed point by a massless spring such that it is half submerged in a liquid of density σ at equilibrium position. The extension x_0 of the spring when it is in equilibrium, is **(JEE Main 2013)**

a. $\frac{Mg}{k}$ b. $\frac{Mg}{k}\left(1-\frac{LA\sigma}{M}\right)$

c. $\frac{Mg}{k}\left(1-\frac{LA\sigma}{2M}\right)$ d. $\frac{Mg}{k}\left(1+\frac{LA\sigma}{M}\right)$

4 Assume that a drop of liquid evaporates by decrease in its surface energy, so that its temperature remains unchanged. What should be the minimum radius of the drop for this to be possible? The surface tension is T, density of liquid is ρ and latent heat of vaporisation is L. **(JEE Main 2013)**

a. $\rho L/T$ b. $\sqrt{T/\rho L}$

c. $T/\rho L$ d. $2\,T/\rho L$

5 A thin liquid film formed between a U-shaped wire and a light slider supports a weight of 1.5×10^{-2} N (see figure). The length of the slider is 30 cm and its weight is negligible. The surface tension of the liquid film is **(AIEEE 2012)**

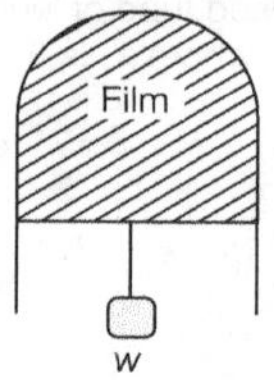

a. 0.0125 Nm^{-1} b. 0.1 Nm^{-1}

c. 0.05 Nm^{-1} d. 0.025 Nm^{-1}

6 Two mercury drops (each of radius r) merge to form a bigger drop. The surface energy of the bigger drop, if T is the surface tension, is **(AIEEE 2011)**

a. $2^{5/3}\pi r^2 T$ b. $4\pi r^2 T$

c. $2\pi r^2 T$ d. $2^{8/3}\pi r^2 T$

7 If a ball of steel (density, $\rho = 7.8$ g cm^{-3}) attains a terminal velocity of 10 cms^{-1} when falling in a tank of water (coefficient of viscosity $\eta_{water} = 8.5\times 10^{-4}$ Pa-s), then its terminal velocity in glycerine ($\rho = 12$ g cm^{-3}, $\eta = 13.2$ Pa-s) would be nearly **(AIEEE 2011)**

a. 1.6×10^{-5} cms^{-1}

b. 6.25×10^{-4} cms^{-1}

c. 6.45×10^{-4} cms^{-1}

d. 1.5×10^{-5} cms^{-1}

8 Work done in increasing the size of a soap bubble from radius of 3 cm to 5 cm is nearly (surface tension of soap solution = 0.03 Nm^{-1}) **(AIEEE 2011)**

a. 0.2π mJ b. 2π mJ

c. 0.4π mJ d. 4π mJ

9 Water is flowing continuously from a tap having an internal diameter 8×10^{-3} m. The water velocity as it leaves the tap is $0.4\,ms^{-1}$. The diameter of the water stream at a distance 2×10^{-1} m below the tap is close to **(AIEEE 2011)**

a. 7.5×10^{-3} m b. 9.6×10^{-3} m
c. 3.6×10^{-3} m d. 5.0×10^{-3} m

10 A ball is made of a material of density ρ, where $\rho_{oil} < \rho < \rho_{water}$ with ρ_{oil} and ρ_{water} representing the densities of oil and water, respectively. The oil and water are immiscible. If the above ball is in equilibrium in a mixture of this oil and water, then which of the following pictures represents its equilibrium position? **(AIEEE 2010)**

a. 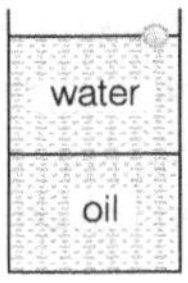

b. 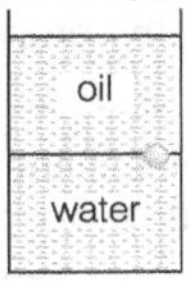

c. 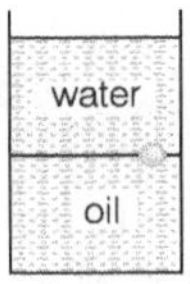

d. 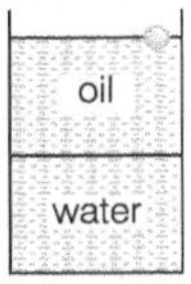

11 A spherical solid ball of volume V is made of a material of density ρ_1. It is falling through a liquid of density $\rho_2 (\rho_2 < \rho_1)$. [Assume that the liquid applies a viscous force on the ball that is proportional to the square of its speed v, i.e. $F_{viscous} = -kv^2 (k > 0)$]. The terminal speed of the ball is **(AIEEE 2008)**

a. $\sqrt{\frac{Vg(\rho_1 - \rho_2)}{k}}$ b. $\frac{Vg\rho_1}{k}$
c. $\sqrt{\frac{Vg\rho_1}{k}}$ d. $\frac{Vg(\rho_1 - \rho_2)}{k}$

12 A jar is filled with two non-mixing liquids 1 and 2 having densities ρ_1 and ρ_2, respectively. A solid ball, made of a material of density ρ_3, is dropped in the jar. It comes to equilibrium in the position shown in the figure.

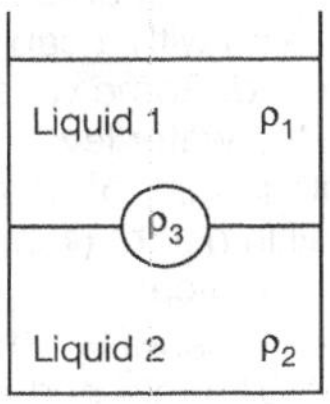

Which of the following is true for ρ_1, ρ_2 and ρ_3? **(AIEEE 2008)**

a. $\rho_3 < \rho_1 < \rho_2$ b. $\rho_1 > \rho_3 > \rho_2$
c. $\rho_1 > \rho_2 > \rho_3$ d. $\rho_1 < \rho_3 < \rho_2$

13 A capillary tube (A) is dipped in water. Another identical tube (B) is dipped in a soap-water solution. Which of the following shows the relative nature of the liquid columns in the two tubes? **(AIEEE 2008)**

a. 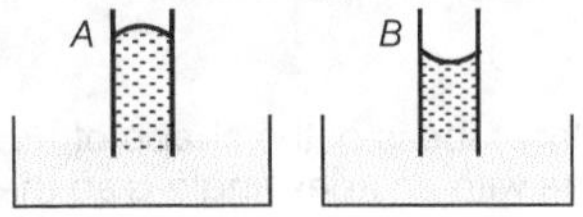

b. 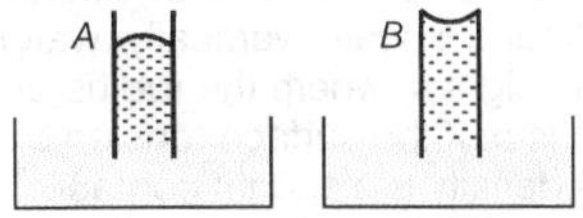

c. 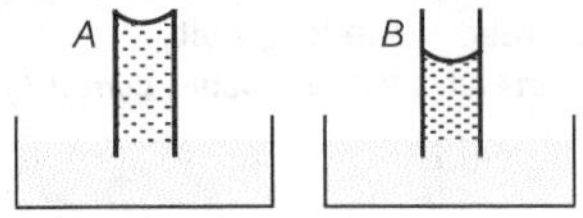

d. 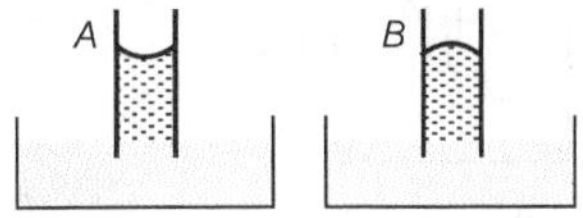

14 If the terminal speed of a sphere of gold (density $= 19.5\,kgm^{-3}$) is $0.2\,ms^{-1}$ in a viscous liquid (density $= 1.5\,kgm^{-3}$), find the terminal speed of a sphere of silver (density $= 10.5\,kgm^{-3}$) of the same size in the same liquid. **(AIEEE 2006)**

a. $0.4\,ms^{-1}$ b. $0.133\,ms^{-1}$
c. $0.1\,ms^{-1}$ d. $0.2\,ms^{-1}$

15 A 20 cm long capillary tube is dipped in water. The water rises upto 8 cm. If the entire arrangement is put in a freely falling elevator, then length of water column in the capillary tube will be **(AIEEE 2005)**

a. 8 cm b. 10 cm
c. 4 cm d. 20 cm

16 Spherical balls of radius R are falling in a viscous fluid of viscosity η with a velocity v. The retarding viscous force acting on the spherical ball is **(AIEEE 2004)**

a. directly proportional to R but inversely proportional to v
b. directly proportional to both radius R and velocity v
c. inversely proportional to both radius R and velocity v
d. inversely proportional to radius R but directly proportional to velocity v

17 If two soap bubbles of different radii are connected by a tube, then **(AIEEE 2004)**

a. air flows from the bigger bubble to the smaller bubble till the sizes become equal
b. air flows from bigger bubble to the smaller bubble till the sizes are interchanged
c. air flows from the smaller bubble to the bigger bubble
d. there is no flow of air

18 A cylinder of height 20 m is completely filled with water. The velocity of efflux of water (in ms^{-1}) through a small hole on the side wall of the cylinder near its bottom is **(AIEEE 2004)**

a. 10 b. 20 c. 25.5 d. 5

JEE Advanced & IIT JEE Archive

(Compilation of Last 10 Years' Questions)

1 A glass capillary tube is of the shape of truncated cone with an apex angle α so that its two ends have cross-sections of different radii. When dipped in water vertically, water rises in it to a height h, where the radius of its cross-section is b. If the surface tension of water is S, its density is ρ and its contact angle with glass is θ, the value of h will be (g is the acceleration due to gravity)

(2014 Adv., Only One Option Correct Type)

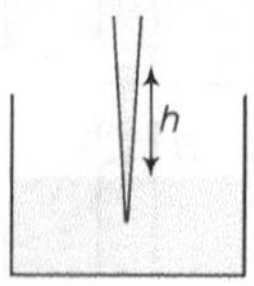

a. $\dfrac{2S}{b\rho g}\cos(\theta-\alpha)$

b. $\dfrac{2S}{b\rho g}\cos(\theta+\alpha)$

c. $\dfrac{2S}{b\rho g}\cos\left(\theta-\dfrac{\alpha}{2}\right)$

d. $\dfrac{2S}{b\rho g}\cos\left(\theta+\dfrac{\alpha}{2}\right)$

2 A thin uniform cylindrical shell, closed at both ends, is partially filled with water. It is floating vertically in water in half-submerged state. If ρ_c is the relative density of the material of the shell with respect to water, then the correct statement is that the shell is

(2012, Only One Option Correct Type)

a. more than half-filled if ρ_c is less than 0.5

b. more than half-filled if ρ_c is more than 1.0

c. half-filled if ρ_c is more than 0.5

d. less than half-filled if ρ_c is less than 0.5

3 Two solid spheres A and B of equal volumes but of different densities d_A and d_B are connected by a string. They are fully immersed in a fluid of density d_F. They get arranged into an equilibrium state as shown in the figure with a tension in the string. The arrangement is possible only if

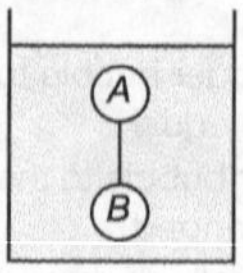

(2011, One or More than One Options Correct Type)

a. $d_A < d_F$

b. $d_B > d_F$

c. $d_A > d_F$

d. $d_A + d_B = 2d_F$

Passage for Q. Nos. (4-6)

When liquid medicine of density ρ is to be put in the eye, it is done with the help of a dropper. As the bulb on the top of the dropper is pressed, a drop forms at the opening of the dropper. We wish to estimate the size of the drop.

We first assume that the drop formed at the opening is spherical because that requires a minimum increase in its surface energy. To determine the size, we calculate the net vertical force due to the surface tension T when the radius of the drop is R. When this force becomes smaller than the weight of the drop, the drop gets detached from the dropper.

(2010, Comprehension Type)

4 If the radius of the opening of the dropper is r, the vertical force due to the surface tension on the drop of radius R (assuming $r << R$) is

a. $2\pi rT$ b. $2\pi RT$

c. $\dfrac{2\pi r^2 T}{R}$ d. $\dfrac{2\pi R^2 T}{r}$

5 If $r = 5\times10^{-4}$ m, $\rho = 10^3$ kg m^{-3}, $g = 10$ ms^{-2}, $T = 0.11$ Nm^{-1}, the radius of the drop when it detaches from the dropper is approximately

a. 1.4×10^{-3} m b. 3.3×10^{-3} m

c. 2.0×10^{-3} m d. 4.1×10^{-3} m

6 After the drop detaches, its surface energy is

a. 1.4×10^{-6} J b. 2.7×10^{-6} J

c. 5.4×10^{-6} J d. 8.1×10^{-9} J

7 Two soap bubbles A and B are kept in a closed chamber, where the air is maintained at pressure 8 Nm^{-2}. The radii of bubbles A and B are 2 cm, respectively. Surface tension of the soap-water used to make bubbles is 0.04 Nm^{-1}. Find the ratio $\dfrac{n_B}{n_A}$, where n_A and n_B are the number of moles of air in bubbles A and B, respectively. (Neglect the effect of gravity). **(2009, Integer Type)**

8 A cylindrical vessel of height 500 mm has an orifice (small hole) at its bottom. The orifice is initially closed and water is filled in it upto height H. Now the top is completely sealed with a cap and the orifice at the bottom is opened. Some water comes out from the orifice and the water level in the vessel becomes steady with height of water column being 200 mm. Find the fall in height (in mm) of water level due to opening of the orifice.

(Given, atmospheric pressure $= 1.0\times10^5$ Nm^{-2}, density of water $= 1000$ kg m^{-3} and $g = 10$ ms^{-2}. Neglect any effect of surface tension.)

(2009, Integer Type)

9 A glass tube of uniform internal radius (r) has a valve separating the two identical ends. Initially, the valve is in a tightly closed position. End 1 has a hemispherial soap bubble of radius r. End 2 has sub-hemispherical soap bubble as shown in the figure. Just after opening the valve,

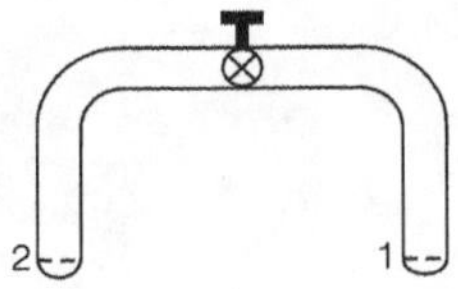

(2008, Only One Option Correct Type)

a. air from end 1 flows towards end 2. No change in the volume of the soap bubbles
b. air from end 1 flows towards end 2. Volume of the soap bubble at end 1 decreases
c. no change occurs
d. air from end 2 flows towards end 1. Volume of the soap bubble at end 1 increases

10 **Statement I** The stream of water flowing at high speed from a garden hose pipe tends to spread like a fountain when held vertically up, but tends to narrow down when held vertically down.

Statement II In any steady flow of an incompressible fluid, the volume flow rate of the fluid remains constant. **(2008, Statement Type)**

a. Both Statement I and Statement II are correct and Statement II is the correct explanation of Statement I
b. Both Statement I and Statement II are correct but Statement II is not the correct explanation of Statement I
c. Statement I is correct but Statement II is incorrect
d. Statement II is correct but Statement I is incorrect

11 Water is filled up to a height h in a beaker of radius R as shown in the figure. The density of water is ρ, the surface tension of water is T and the atmospheric pressure is p_0. Consider a vertical section $ABCD$ of the water column through a diameter of the beaker. The force on water on one side of this section by water on the other side of this section has magnitude

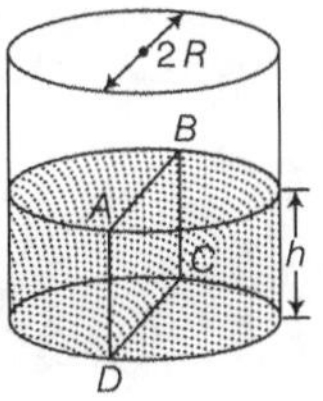

(2007, Only One Option Correct Type)

a. $|2p_0Rh + \pi R^2\rho gh - 2RT|$
b. $|2p_0Rh + R\rho gh^2 - 2RT|$
c. $|p_0\pi R^2 + R\rho gh^2 - 2RT|$
d. $|p_0\pi R^2 + R\rho gh^2 + 2RT|$

Passage for Q. Nos. (12-14)

A wooden cylinder of diameter $4r$, height h and density $\rho/3$ is kept on a hole of diameter $2r$ of a tank, filled with liquid of density ρ as shown in the figure. **(2006, Comprehension Type)**

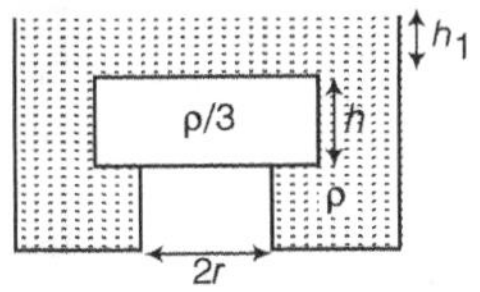

12 Now level of the liquid starts decreasing slowly. When the level of liquid is at a height h_1 above the cylinder, the block starts moving up. At what value of h_1, will the block rise ?

a. $4h/9$
b. $5h/9$
c. $\frac{5h}{3}$
d. Remains same

13 The block in the above question is maintained at the position by external means and the level of liquid is lowered. The height h_2 when this external force reduces to zero, is

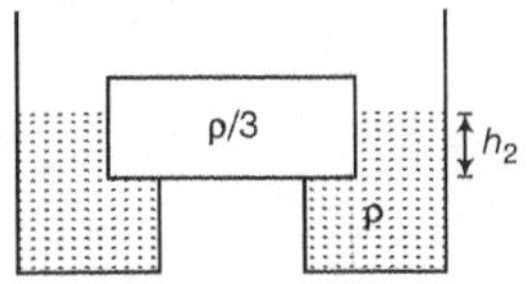

a. $\frac{4h}{9}$
b. $\frac{5h}{9}$
c. remain same
d. $\frac{2h}{3}$

14 If height h_2 of water level is further decreased, then

a. cylinder will not move up and remains at its original position
b. for $h_2 = h/3$, cylinder again starts moving up
c. for $h_2 = h/4$, cylinder again starts moving up
d. for $h_2 = h/5$, cylinder again starts moving up

15 A U-shaped tube contains a liquid of density ρ and it is rotated about the line as shown in the figure. Find the difference in the levels of liquid column. **(2005, Subjective Type)**

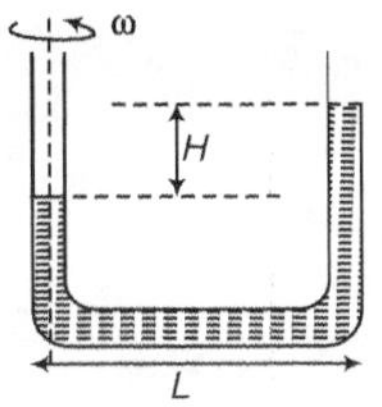

16 Water is filled in a cylindrical container to a height of 3 m. The ratio of the cross-sectional area of the orifice and the beaker is 0.1. The square of the speed of the liquid coming out from the orifice is ($g = 10\ ms^{-2}$)

(2005, Only One Option Correct Type)

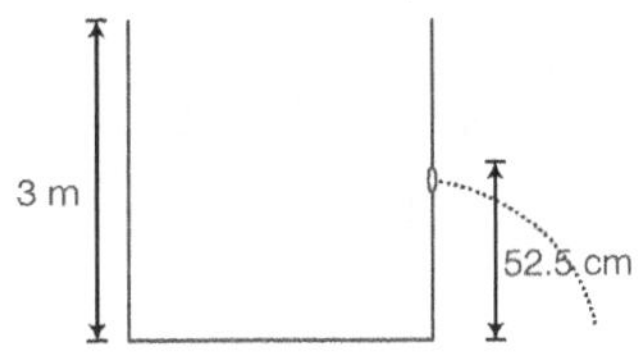

a. $50\ m^2s^{-2}$
b. $50.5\ m^2s^{-2}$
c. $51\ m^2s^{-2}$
d. $52\ m^2s^{-2}$

Answers *with* Explanations

Answers

1. Centre of Mass

DPP-1 Centre of Mass

1. *b* 2. *a* 3. *d* 4. *b* 5. *c* 6. *b* 7. *b* 8. *c* 9. *a,b,c,d*

10. *b,d* 11. 2 12. 5 13. 3 14. 6 15. $\left(\frac{-qr^3}{R^3-r^3}, 0\right)$ 16. $h=\frac{2R\sin(\theta/2)}{\theta}$

17. $h=\frac{4R\sin(\theta/2)}{3\theta}$ 18. $x=\frac{mL}{M+m}$ 19. (i) $\frac{8u}{15}, \frac{12u}{15}$ (ii) $\frac{8mu}{15}$ (iii) $\frac{2u^2}{15L}$ 20. *a* 21. *a*

22. *d* 23. *d* 24. *d*

DPP-2 Momentum and Its Conservation

1. *c* 2. *a* 3. *c* 4. *c* 5. *a* 6. 3 7. 3 8. $\frac{Mu+mu_r}{M}\sqrt{\frac{2h}{g}}$

9. $\frac{Mu-m_1u-m_2u}{M-m_1+m_2}$ and $\frac{Mu-m_1u-m_2u}{M-m_2+m_1}$ 10. (i) $\theta_{max}=90^\circ, \frac{2}{3}R$ (ii) $\sqrt{\frac{gR}{3}}$

11. (i) $\frac{16}{\sqrt5}$ kg-ms^{-1} (ii) $\frac{12}{\sqrt5}$ ms^{-1} (iii) $\frac{48}{5\sqrt5}$ kg-ms^{-1} (iv) $\frac{64}{5\sqrt5}$ kg-ms^{-1}

12. $v_A=\frac{J(m_3+2m_2)}{k}, v_B=\frac{J[(m_3+2m_2)^2+m_3^2]^{1/2}}{k}$ and $v_C=\frac{Jm_2\sqrt2}{k}$, where $k=m_1m_3+2m_2(m_1+m_2+m_3)$

13. *b* 14. *a* 15. *d* 16. *c* 17. *d* 18. *a* 19. *b* 20. *d* 21. *d* 22. *d*

DPP-3 Variable Mass

1. $v=u-gt+v_r\ln\left(\frac{m_0}{m}\right)$ 2. $\frac{(m_1-m_0+\mu t)g+\mu v_0}{(m_1+m_0-\mu t)}$ 3. (i) $m_0e^{\left(\frac{qL}{m_0v_0}\right)}$ (ii) $v_0e^{-\left(\frac{qL}{m_0v_0}\right)}$

4. (i) $F=\frac{m}{L}(gy+v_0^2)$ (ii) $\frac{myv_0^2}{2L}$ 5. $(m_0-\mu t)\frac{d^2y}{dt^2}=\mu u-(m_0-\mu t)g$ 6. 5 7. *a* 8. *b*

9. *a* 10. *c* 11. *b* 12. *d* 13. *a,d* 14. *a,b* 15. *b* 16. *c* 17. *a* 18. *c*
19. *a* 20. *d* 21. *a* 22. *a*

DPP-4 Collisions

1. *b* 2. *a* 3. *d* 4. *c* 5. *d* 6. *b* 7. *c* 8. *c* 9. *c* 10. *d*

11. *a* 12. *b* 13. (i) $\frac{1}{3}$ (ii) 2 14. $u=\frac{M\sqrt{2gR}}{m\cos\theta}$ and $l=\frac{2M^2R\tan\theta}{m^2}$

15. (i) $u\sqrt{\sin^2\theta+e^2\cos^2\theta}$ (ii) $\tan^{-1}\left(\frac{\tan\theta}{e}\right)$ (iii) $mu\cos\theta(1+e)$ (iv) $-\frac{1}{2}mu^2\cos^2\theta(1-e^2)$

16. (i) $\tan^2\theta$ (ii) $2m\sqrt{2gh}\cos\theta$ 17. 2 18. 9 19. *a* 20. *b* 21. *a* 22. *c* 23. *b*

24. *a*

Revisal Problems (JEE Main)

1. *a* 2. *b* 3. *d* 4. *d* 5. *c* 6. *a* 7. *d* 8. *b* 9. *c* 10. *a*
11. *c* 12. *d* 13. *b* 14. *b* 15. *a* 16. *b* 17. *b* 18. *c* 19. *b* 20. *d*
21. *a* 22. *c* 23. *d*

Revisal Problems (JEE Advanced)

1. *a* 2. *c* 3. *b* 4. *d* 5. *b* 6. *a* 7. *d* 8. *a* 9. *c* 10. *b*
11. *b* 12. *b,c,d* 13. *a,c,d* 14. *b,c,d* 15. *a,d* 16. *c* 17. *d* 18. *b* 19. *c* 20. *a*
21. *c* 22. *a* 23. 8 24. 6

JEE Main & AIEEE Archive

1. *d* 2. *a* 3. *c* 4. *c* 5. *c*

JEE Advanced & IIT JEE Archive

1. 4 2. *c* 3. *a* 4. *a* 5. *b* 6. *b* 7. *c* 8. *a,d* 9. *d* 10. *d*
11. *d* 12. 10

2. Rotational Motion

DPP-1 Rotational Kinematics

1. *d* 2. *a* 3. *c* 4. *a* 5. *c* 6. *c* 7. *b* 8. *b* 9. *c* 10. *b*
11. *c* 12. *a,c,d* 13. *a,d* 14. *b,d* 15. *a,c* 16. *b,c* 17. *a* 18. *b* 19. *d* 20. *a*

DPP-2 Moment of Inertia

1. *a* 2. *b* 3. *d* 4. *c* 5. *b* 6. *d* 7. *a* 8. *b* 9. *a* 10. *d*
11. *c* 12. *b* 13. *a* 14. *a,d* 15. *b,d* 16. *a* 17. *d* 18. *b*
19. $I_A = \left(\frac{9\pi - 16}{6\pi}\right) MR^2$ 20. $I_{YY'} = \frac{3}{10} MR^2$ 21. $I_{YY'} = \frac{1}{2} MR^2$ 22. $I_{AA'} = \frac{44}{105} MR^2$ 23. *c*
24. *c* 25. *c*

DPP-3 Rotation about a Fixed Axis

1. *c* 2. *c* 3. *c* 4. *d* 5. *a* 6. *b* 7. *b* 8. *b* 9. *a,b,c* 10. *a,d*
11. *b,c,d* 12. *a,b,c,d* 13. *c,d* 14. $x = \frac{L}{2\sqrt{3}}, \alpha_{\max} = \frac{g\sqrt{3}}{L}$ 16. $F_{\text{net}} = \frac{\sqrt{10}}{4} mg,\ \phi = \tan^{-1}\left(\frac{1}{3}\right)$
17. $F = 442$ mN 18. 2 19. *d* 20. *c* 21. *b* 22. *c* 23. *a*

DPP-4 Angular Momentum and Its Conservation

1. *c* 2. *a* 3. *b* 4. *b* 5. *b* 6. *c* 7. *d* 8. *b* 9. *a* 10. *c*
11. *d* 12. *c* 13. *b* 14. *a,c* 15. *a,b* 16. *c,d* 17. *b,c* 18. *c* 19. *a* 20. *b*
21. $v = 63\ \text{ms}^{-1}$ 22. $\omega = 1\ \text{rad s}^{-1}$ 23. (i) $\omega = \frac{12v}{7L}$ (ii) $v = 3.5\ \text{ms}^{-1}$

DPP-5 Combined Translation and Rotation

1. *d* 2. *c* 3. *a* 4. *c,d* 5. *c* 6. *b* 7. *d* 8. *a* 9. *b* 10. *d*
11. *a* 12. *c* 13. *b* 14. *d* 15. *a* 16. *b* 17. *a* 18. *c* 19. *a* 20. *d*
21. *c* 22. *a*

DPP-6 Instantaneous Axis of Rotation and Instantaneous Centre

1. *a* 2. *a* 3. *c* 4. *d* 5. *a* 6. *b* 7. *a* 8. *a* 9. *a*
10. 2 11. $v_C = \sqrt{\frac{3gL(1-\sin\theta)\cos^2\theta}{(1+3\cos^2\theta)}}$ 12. $v_B = \sqrt{\frac{3gL(1-\sin\theta)\sin^2\theta}{(1+3\cos^2\theta)}}$ 13. $v_C = \sqrt{\frac{3gL}{4}}$
14. $\omega = \left[\frac{6g(\cos\alpha - \cos\theta)}{a(1+3\sin^2\theta)}\right]^{1/2}$ 15. *a* 16. *b* 17. *c* 18. *d* 19. *b* 20. *d*

DPP-7 Rolling without Slipping and Rolling with Slipping

1. *a* 2. *b* 3. *a* 4. *b* 5. *a* 6. *c* 7. *a* 8. *a,c* 9. *b,d*
10. *a,d* 11. *d* 12. *b* 13. *b* 14. *a* 15. *a* 16. *b* 17. *a* 18. *d*
19. $a_2 = \frac{4F\cos\theta}{3M+8m}, f_1 = \frac{3MF\cos\theta}{3M+8m}, f_2 = \frac{MF\cos\theta}{3M+8m}$
20. $v_P = v_0\left[1-\cos\left(\frac{v_0 t}{R}\right)\right]\hat{\mathbf{i}} + v_0 \sin\left(\frac{v_0 t}{R}\right)\hat{\mathbf{j}};\ a_P = \frac{v_0^2}{R}\sin\left(\frac{v_0 t}{R}\right)\hat{\mathbf{i}} + \frac{v_0^2}{R}\cos\left(\frac{v_0 t}{R}\right)\hat{\mathbf{j}}$
22. (i) $\mathbf{f} = (6\hat{\mathbf{i}})$N (ii) $0.6(\hat{\mathbf{k}} - \hat{\mathbf{j}})$ N-m, $0.6(-\hat{\mathbf{j}} - \hat{\mathbf{k}})$ N-m and $|\tau| = 0.85$ N-m
23. (i) $t = \frac{\omega_0 R}{3\mu g}$ (ii) $W = -\frac{1}{6} m\omega_0^2 R^2$

DPP-8 Angular Impulse

1. *d* 2. *a* 3. *b* 4. *c* 5. *a* 6. *b* 7. *a* 8. *a* 9. *c* 10. *d*
11. *a,c* 12. *b,d* 13. *a,d* 14. *a,d* 15. *a,b,c,d* 16. *a* 17. *c* 18. *d* 19. 5
20. $\omega_1' = \frac{I_1\omega_1 r_2 + I_2\omega_2 r_1}{I_2 r_1^2 + I_1 r_2^2}\cdot r_2$ and $\omega_2' = \frac{I_1\omega_1 r_2 + I_2\omega_2 r_1}{I_2 r_1^2 + I_1 r_2^2}\cdot r_1$
22. Path1: $\omega_f = -4\ \text{rad s}^{-1}$, Path2: $\omega_f = -1.6\ \text{rad s}^{-1}$, Path3: $\omega_f = 0.8\ \text{rad s}^{-1}$, Path4: $\omega_f = -0.4\ \text{rad s}^{-1}$
23. (i) $\omega = \frac{20}{3}\ \text{rad s}^{-1}$ (ii) $\omega' = 60\ \text{rad s}^{-1}$ (iii) $\frac{KE_f}{KE_i} = 9$

DPP-9 Toppling and Rolling on an Inclined Plane

1. *a* 2. *c* 3. *a* 4. *d* 5. *c* 6. *b* 7. *d* 8. *c,d* 9. *a,b,c,d*
10. *a,c* 11. *a* 12. *b* 13. *c* 14. *d* 15. *b* 16. *a* 17. *a*
19. (i) 25 m (ii) 9.62 ms^{-1} 20. $\frac{R^2\omega_0^2(\mu\cos\theta-\sin\theta)}{2g(3\mu\cos\theta-\sin\theta)^2}, \frac{R^2\omega_0^2(\mu\cos\theta-\sin\theta)}{4g\sin\theta(3\mu\cos\theta-\sin\theta)}$

Revisal Problems (JEE Main)

1. *b* 2. *d* 3. *b* 4. *c* 5. *c* 6. *b* 7. *c* 8. *d* 9. *c* 10. *b*
11. *d* 12. *b* 13. *b* 14. *b* 15. *d* 16. *b* 17. *d* 18. *d* 19. *d* 20. *c*
21. *d* 22. *d* 23. *d*

Revisal Problems (JEE Advanced)

1. *d* 2. *c* 3. *c* 4. *a* 5. *b* 6. *a* 7. *a* 8. *b* 9. *a* 10. *a*
11. *a,b,d* 12. *b,d* 13. *c,d* 14. *b,c,d* 15. *a,c* 16. *b* 17. *c* 18. *d* 19. *a* 20. *a*
21. 2 22. 6 23. 3

JEE Main & AIEEE Archive

1. *b* 2. *c* 3. *c* 4. *a* 5. *d* 6. *c* 7. *a* 8. *a* 9. *c* 10. *c*
11. *c* 12. *d* 13. *d* 14. *c* 15. *b* 16. *d* 17. *a* 18. *d* 19. *c* 20. *b*
21. *a* 22. *d*

JEE Advanced & IIT JEE Archive

1. 4 2. 8 3. *b,d* 4. *a,b* 5. *b* 6. *a* 7. *c* 8. *a* 9. *d* 10. *a*
11. 3 12. 9 13. 4 14. *a,c* 15. *a* 16. *b* 17. *b,c* 18. *d* 19. *d* 20. *d*
21. *c* 22. *c* 23. *a* 24. *b* 25. *d* 26. *a* 27. *a* 28. *c,d* 29. *a* 30. *d*
31. $\omega=\frac{3mv}{L(3m+M)}$ 32. $a=\frac{2}{3}g\sin\theta$

3. Gravitation

DPP-1 Newton's Universal Law of Gravitation and Gravitational Field

1. *c* 2. *a* 3. *b* 4. *d* 5. *a* 6. *a* 7. *b* 8. *a* 9. *d* 10. *a,d*
11. *a,b,c* 12. *a,b* 13. *b,c* 14. *a,d* 15. *d* 16. 1 17. 4 18. *a* 19. *d* 20. *b*

DPP-2 Acceleration due to Gravity

1. *b* 2. *a* 3. *b* 4. *c* 5. *b* 6. *d* 7. *a* 8. *c* 9. *a* 10. *b*
11. *d* 12. *c* 13. *d* 14. *c* 15. *d* 16. *a* 17. 4 18. 5 19. 4 20. 2

DPP-3 Potential, Potential Energy and Self Energy

1. *d* 2. *b* 3. *a* 4. *b* 5. *a* 6. *c* 7. *b* 8. *a* 9. *d* 10. *b,d*
11. *a* 12. *a* 13. *a* 14. *a* 15. *b* 16. *d* 17. *c* 18. *a* 19. *c* 20. 3

DPP-4 Escape Velocity and Binding Energy

1. *a* 2. *c* 3. *b* 4. *b* 5. *a* 6. *b,c* 7. *b,d* 8. *b,d* 9. *d* 10. *b*
11. *a* 12. *b* 13. *c* 14. *d* 15. *a* 16. 2 17. 3 18. 2 19. 1 20. 3

DPP-5 Motion of Satellites and Planets

1. *a* 2. *c* 3. *c* 4. *d* 5. *b* 6. *b* 7. *a,c* 8. *b,c* 9. *a,b* 10. *b,d*
11. *a,d* 12. *d* 13. *b* 14. *a* 15. *b* 16. *c* 17. *a* 18. *b* 19. *d* 20. 3

DPP-6 Kepler's Laws

1. *d* 2. *a* 3. *d* 4. *c* 5. *b* 6. *a,b* 7. *c,d* 8. *b,c* 9. *a* 10. *a*
11. *a* 12. *d* 13. *d* 14. *a* 15. *c* 16. *d* 17. *c* 18. *b* 19. 2 20. 3

Revisal Problems (JEE Main)

1. *c* 2. *b* 3. *a* 4. *d* 5. *a* 6. *b* 7. *d* 8. *c* 9. *b* 10. *a*
11. *b* 12. *c* 13. *d* 14. *a* 15. *d* 16. *c* 17. *c* 18. *d* 19. *b* 20. *d*
21. *d* 22. *a*

Revisal Problems (JEE Advanced)

1. *c* 2. *b* 3. *a* 4. *d* 5. *c* 6. *a* 7. *c* 8. *b* 9. *a,d* 10. *a,c*
11. *a,b,c* 12. *a,c* 13. *b,c* 14. *d* 15. *a* 16. *c* 17. *c* 18. *a* 19. 4 20. 9

JEE Main & AIEEE Archive

1. *d* 2. *a* 3. *d* 4. *b* 5. *c* 6. *a* 7. *c* 8. *a* 9. *a* 10. *c*
11. *c* 12. *d* 13. *d* 14. *a* 15. *b* 16. *a* 17. *c* 18. *c* 19. *c* 20. *d*
21. *c* 22. *a* 23. *b*

JEE Advanced & IIT JEE Archive

1. *b* 2. *b,d* 3. *b,d* 4. *b* 5. 3 6. 6 7. *c* 8. *a* 9. *d*

4. Elasticity

DPP-1 Stresses and Strains

1. *b* 2. *a* 3. *d* 4. *a* 5. *a* 6. *c* 7. *a,b* 8. *a,b,c* 9. *a,c* 10. *c,d*
11. *a,b* 12. *a,d* 13. *c* 14. *b* 15. *c* 16. *d* 17. *a* 18. 1 19. 5 20. 5

DPP-2 Modulii of Elasticity

1. *a* 2. *b* 3. *c* 4. *b* 5. *c* 6. *d* 7. *b* 8. *a* 9. *a*
10. *a,b,d* 11. *d* 12. *b* 13. *a* 14. *b* 15. *a* 16. *c* 17. *a* 18. *a* 19. 2
20. 3 21. 4

DPP-3 Elastic Potential Energy

1. *d* 2. *b* 3. *c* 4. *a* 5. *c* 6. *b* 7. *d* 8. *d* 9. *d* 10. *a,d*
11. *a* 12. *a* 13. *b* 14. *a* 15. 6 16. 8

DPP-4 Thermal Stresses and Strains

1. *a* 2. *a* 3. *b* 4. *c* 5. *d* 6. *d* 7. *d* 8. *a* 9. *c* 10. *a*
11. *a* 12. *b* 13. *a,c,d* 14. *a,b* 15. *a* 16. *c* 17. *d* 18. *b* 19. *a* 20. *b*

Revisal Problems (JEE Main)

1. *c* 2. *a* 3. *b* 4. *d* 5. *a* 6. *b* 7. *a* 8. *d* 9. *a* 10. *b*
11. *d* 12. *b* 13. *b* 14. *d* 15. *b* 16. *c* 17. *d* 18. *c* 19. *a* 20. *a*

Revisal Problems (JEE Advanced)

1. *a* 2. *d* 3. *c* 4. *b* 5. *c* 6. *c* 7. *d* 8. *c* 9. *a*
10. *b,c,d* 11. *a,b,c,d* 12. *b,c* 13. *a,b* 14. *a* 15. *b* 16. 9 17. 4 18. *b*
19. *b* 20. *d*

JEE Main & AIEEE Archive

1. *a* 2. *b* 3. *d* 4. *d* 5. *c* 6. *c* 7. *d* 8. *b* 9. *d* 10. *d*

JEE Advanced & IIT JEE Archive

1. *c* 2. *a,d* 3. *d*

5. Fluid Mechanics

DPP-1 Pressure in Static/Accelerated Fluids

1. *c* 2. *d* 3. *a* 4. *b* 5. *b* 6. *a* 7. *c* 8. *a* 9. *d* 10. *a*

11. *d* 12. *a* 13. *c* 14. *a,b,c* 15. *a,c* 16. *b* 17. *d* 18. *b* 19. 6 20. 2

21. 7 22. $10^5 \text{N}, 8\times10^4 \text{N}, 12\times10^4 \text{N}$

23. (i) $\frac{\rho gAh}{4}$ (ii) $\frac{\rho gAh}{4}$

24. $\frac{3a_0 l}{2g}$

DPP-2 Archimedes' Principle and Buoyant Force

1. *c* 2. *a* 3. *b* 4. *d* 5. *a* 6. *c* 7. *b* 8. *b* 9. *b* 10. *b*

11. *a,b* 12. *a,d* 13. *b* 14. *c* 15. *d* 16. *a* 17. *c* 18. *d* 19. *b* 20. 5

21. 9 22. 1 23. $\left(2-\frac{\rho_S}{\rho}\right)^{1/3}\cdot h$ 24. $w=a\left[a^2\rho g+\frac{k}{2}\right]$ 25. 0.5 kg 26. 0

DPP-3 Bernoulli's Principle and Applications

1. *a* 2. *d* 3. *b* 4. *c* 5. *c* 6. *c* 7. *b* 8. *b* 9. *c* 10. *a,d*

11. *a,c* 12. *a* 13. *a* 14. *a* 15. *c* 16. *a* 17. *b* 18. *d* 19. *c* 20. 6

21. (i) 10 ms^{-1} (ii) 14.1ms^{-1} (iii) 9200 s

22. 4.51ms^{-1} 23. 8.33 m 24. $\frac{\rho V^3}{2a^2t^2}$

DPP-4 Viscosity

1. *b* 2. *d* 3. *c* 4. *d* 5. *c* 6. *a* 7. *a* 8. *b* 9. *b* 10. *c*

11. *a,b,c,d* 12. *b,c,d* 13. *a* 14. *d* 15. *b* 16. *c* 17. *a* 18. *a* 19. 4

20. 4 21. 5 22. $\frac{2\eta l}{b^4T}(r_1^4-r_2^4)$

23. (i) 1.5×10^{-4} N (ii) 5.2×10^{-5} N (iii) 32.5 ms^{-1}

DPP-5 Surface Tension

1. *b* 2. *d* 3. *d* 4. *b* 5. *c* 6. *c* 7. *d* 8. *a* 9. *a* 10. *b*

11. *a,c* 12. *a,b* 13. *d* 14. *d* 15. *a* 16. *a* 17. *b* 18. *c* 19. 5 20. 4

21. $\frac{2\pi T^2}{\rho g}$ 22. 1130 Nm^{-2} 23. $\frac{\rho gh^3}{8T}$

Revisal Problems (JEE Main)

1. *c* 2. *d* 3. *b* 4. *a* 5. *c* 6. *a* 7. *d* 8. *d* 9. *c* 10. *a*

11. *c* 12. *b* 13. *a* 14. *a* 15. *b* 16. *d* 17. *a* 18. *c* 19. *b* 20. *c*

Revisal Problems (JEE Advanced)

1. *c* 2. *a* 3. *a* 4. *b* 5. *d* 6. *c* 7. *c* 8. *b* 9. *d* 10. *a*

11. *a,b,c* 12. *a,b,c* 13. *a,b,d* 14. *b,c* 15. *c* 16. *b* 17. *d* 18. *a* 19. 3 20. 4

JEE Main & AIEEE Archive

1. *c* 2. *a* 3. *c* 4. *d* 5. *d* 6. *d* 7. *b* 8. *c* 9. *c* 10. *b*

11. *a* 12. *d* 13. *c* 14. *c* 15. *d* 16. *b* 17. *c* 18. *b*

JEE Advanced & IIT JEE Archive

1. *d* 2. *a* 3. *a,b,d* 4. *c* 5. *a* 6. *b* 7. 6 8. 6 9. *b* 10. *a*

11. *b* 12. *c* 13. *a* 14. *a* 15. $H=\frac{\omega^2L^2}{2g}$ 16. *a*

1. Centre of Mass

DPP-1 Centre of Mass

1. (b) We consider A as the origin.

$$\therefore \quad X_{COM} = \frac{(20)(0) + (15)(20) + (10)(50) + (7.5)(70) + (5)(85)}{20 + 15 + 10 + 7.5 + 5}$$

$$= 30.43 \text{ cm}$$

2. (a) Coordinates of point masses are

$$3 \text{ kg} \to (0, 0)$$
$$5 \text{ kg} \to (4, 0)$$
$$10 \text{ kg} \to (4, 3)$$

$$\therefore \quad X_{COM} = \frac{3(0) + 5(4) + 10(4)}{3 + 5 + 10} = 3.33 \text{ m}$$

$$Y_{COM} = \frac{3(0) + 5(0) + 10(3)}{3 + 5 + 10} = 1.67 \text{ m}$$

3. (d) On system of (particle + wedge), net external force is vertically downwards, i.e. force of gravity.

4. (b) $v_{COM} = \dfrac{mv_0 + M(0)}{m + M} = \dfrac{mv_0}{m + M}$ (this is constant)

$\therefore$ In centre of mass reference frame, relative velocity of mass m is

$$u_m = v_0 - \frac{mv_0}{m + M} = \frac{Mv_0}{m + M}$$

and of mass M, $\quad u_M = -\dfrac{mv_0}{m + M}$

$$K_i = \frac{1}{2}mu_m^2 + \frac{1}{2}Mu_M^2$$

$$= \frac{1}{2}m\left(\frac{Mv_0}{m + M}\right)^2 + \frac{1}{2}M\left(\frac{mv_0}{m + M}\right)^2$$

or $\quad K_i = \dfrac{1}{2} \cdot \dfrac{mMv_0^2}{m + M}$

In centre of mass reference frame, at maximum compression, m and M are at rest.

$\therefore$ By conservation of energy, $K_f = 0$

So, $\quad \dfrac{1}{2}kx_{max}^2 = \dfrac{1}{2} \cdot \dfrac{mMv_0^2}{m + M}$

$$x_{max} = \sqrt{\frac{mM}{k(m + M)}}\, v_0$$

5. (c) For COM, $F_x = 0 \Rightarrow a_x = 0$

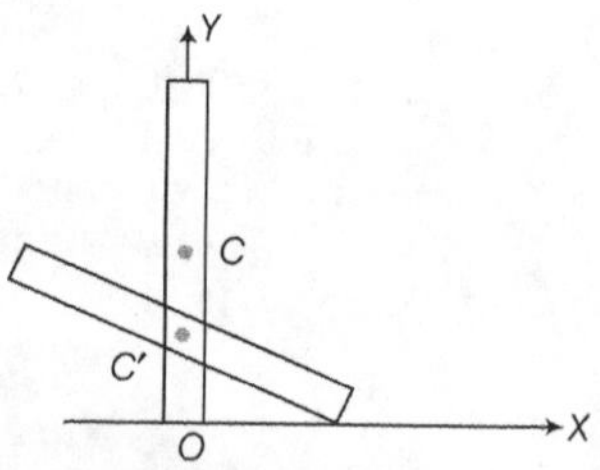

So, COM falls down vertically.

6. (b) For point P,

$$x = \left(\frac{L}{2} - r\right)\cos\theta$$

and $\quad y = r\sin\theta$

$$\therefore \quad \frac{x^2}{\left(\frac{L}{2} - r\right)^2} + \frac{y^2}{r^2} = 1$$

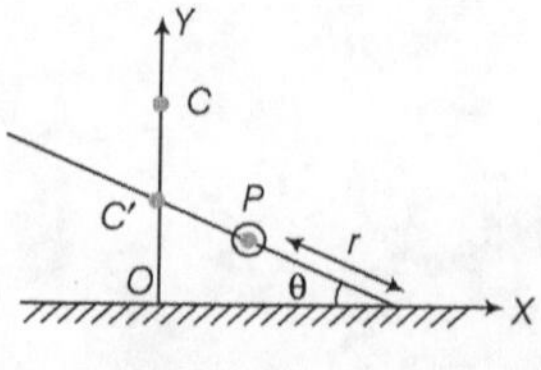

7. (b) F is force between the ladder and the man.

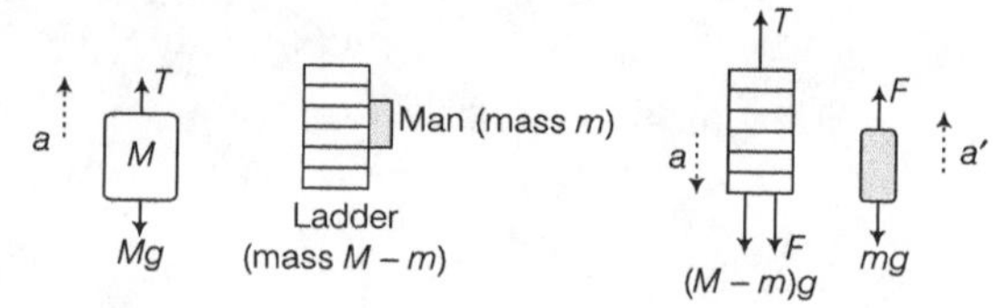

$$T - Mg = Ma \quad \ldots\text{(i)}$$
$$(M - m)g + F - T = (M - m)a \quad \ldots\text{(ii)}$$
$$F - mg = ma' \quad \ldots\text{(iii)}$$

On adding Eqs. (i) and (ii), we get

$$(M - m)g + F - Mg = 2Ma - ma$$
$$\Rightarrow \quad F - mg = (2M - m)a \quad \ldots\text{(iv)}$$

From Eqs. (iii) and (iv), we get

$$a' = \frac{(2M - m)}{m}a$$

$$\therefore \quad v' = \frac{(2M - m)}{m}v$$

$$v_r = v' + v$$

$$\Rightarrow \quad v_r - v = \frac{(2M - m)v}{m}$$

$$\Rightarrow \quad v = v_r \Big/ \left[1 + \frac{2M - m}{m}\right]$$

or $\quad v = v_r \dfrac{m}{2M}$

$$v' = v_r - v = v_r - v_r \cdot \frac{m}{2M} = v_r\left(\frac{2M - m}{2M}\right)$$

$$v_{COM} = \frac{Mv - (M - m)v + mv'}{M + (M - m) + m} = \frac{m(v + v')}{2M}$$

$$= \frac{m}{2M}\left[v_r \cdot \frac{m}{2M} + v_r\left(\frac{2M - m}{2M}\right)\right]$$

$$= \frac{m}{2M}v_r\left[\frac{m}{2M} + 1 - \frac{m}{2M}\right]$$

or $\quad v_{COM} = \dfrac{m}{2M}v_r$

8. (c) In COM reference frame, particles P and Q move along circular path with same angular velocity (ω).

$$Y_{COM} = \frac{(3m \times 0) + (2m)L}{3m + 2m} = \frac{2}{5}L$$

$$v_{COM} = \frac{(3m \times v) + (2m \times 0)}{3m + 2m} = \frac{3}{5}v$$

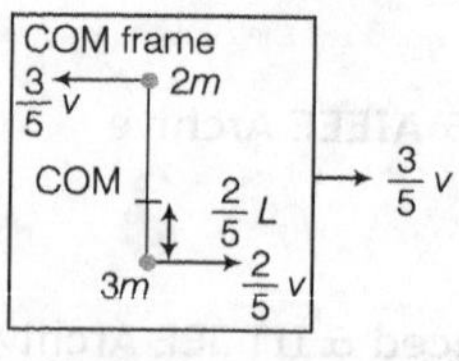

In COM frame Velocity of $3m$ mass $= v - \dfrac{3}{5}v = \dfrac{2}{5}v$

Angular velocity of $3m$ mass is

$$\omega = \frac{(2/5)v}{(2/5)L} = \frac{v}{L}$$

$$T = (3m)\omega^2 r = (3m)\frac{v^2}{L^2} \cdot \frac{2}{5}L = \frac{6mv^2}{5L}$$

9. (a,b,c,d) $\bar{a}_O = \bar{a}_{COM} = \dfrac{F + 3F - 2F}{3m}\hat{i} = \dfrac{2F}{3m}\hat{i}$

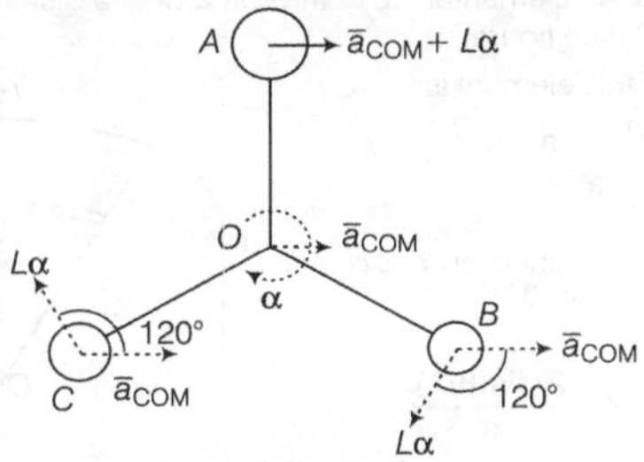

$$\Sigma\tau_O = I\alpha$$

$$FL + 2F\frac{L}{2} - 3F\frac{L}{2} = 3mL^2\alpha$$

$$\alpha = \frac{F}{6mL} \Rightarrow L\alpha = \frac{F}{6m}$$

$$\bar{a}_A = \frac{2F}{3m}\hat{i} + \frac{F}{6m}\hat{i} = \frac{5F}{6m}\hat{i}$$

$$\bar{a}_B = \frac{2F}{3m}\hat{i} + \left[-\frac{F}{6m}\sin 30°\,\hat{i} - \frac{F}{6m}\cos 30°\,\hat{j}\right]$$

$$= \frac{7F}{12m}\hat{i} - \frac{F}{4\sqrt{3}}\hat{j}$$

From the figure, it is clear that $|\bar{a}_B| = |\bar{a}_C|$.

10. (b,d) Let a be the acceleration of wedge leftwards and a_r be the relative acceleration of block down the plane. Then, absolute acceleration of block in horizontal direction will be $(a_r\cos\theta - a)$ rightwards.

$$\bar{F}_{horizontal} = 0 \Rightarrow \bar{a}_{horizontal} = 0$$

$\Rightarrow$ $\quad a_r\cos\theta - a = a$

or $\quad 2a = a_r\cos\theta \quad$ ...(i)

FBD of wedge

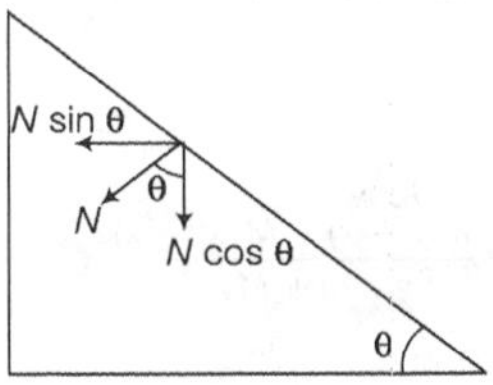

FBD of block w.r.t. wedge

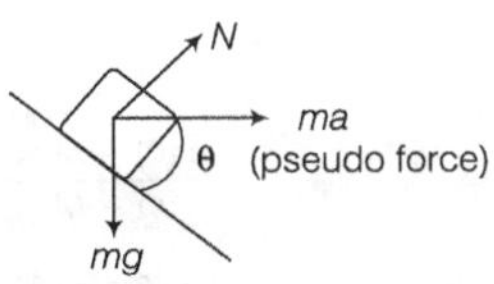

$$N\sin\theta = ma \quad \text{...(ii)}$$

Net force on block perpendicular to plane is zero. So,

$$N + ma\sin\theta = mg\cos\theta \quad \text{...(iii)}$$

On solving Eqs. (i), (ii) and (iii), we get

$$a_r = \frac{2g\sin\theta}{1 + \sin^2\theta}$$

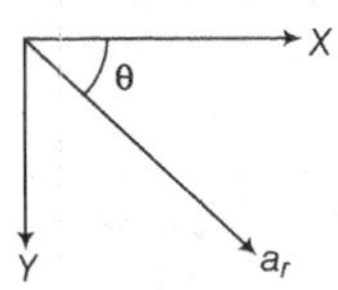

For block, $\quad a_y = a_r\sin\theta = \dfrac{2g\sin^2\theta}{1 + \sin^2\theta}$

$$a_{COM} = \frac{m(0) + m\left[\dfrac{2g\sin^2\theta}{1 + \sin^2\theta}\right]}{m + m} = \frac{g\sin^2\theta}{1 + \sin^2\theta}$$

11. (2) $F_{ext} \geq (0.2)(2m + m)(10) = 6m$

$$\bar{a}_{COM} = \frac{F_{ext}}{2m + m} = \frac{6m}{2m + m} = 2\,\text{ms}^{-2}$$

$\therefore \quad x = 2$

12. (5) Initial KE of bullet, $K_i = \dfrac{1}{2}mu^2 = \dfrac{1}{2} \times 0.01 \times (400)^2$

$$= \frac{1}{2} \times 0.01 \times 16 \times 10^4 = 800\text{ J}$$

Let v_{COM} = velocity of COM before collision, then

$$M_{COM}v_{COM} = \sum_{i=1}^{n} m_iv_i = m_1u_1 + m_2u_2$$

$$= m_1u_1 \quad (\text{as } u_2 = 0)$$

So, $\quad v_{COM} = \dfrac{m_1u_1}{M_{COM}} = \dfrac{m_1u_1}{m_1 + m_2}$

$\Rightarrow \quad v_{COM} = \dfrac{4\text{ kg-ms}^{-1}}{0.4\text{ kg}} = 10\text{ ms}^{-1}$

As there is no external force present in the direction of motion.

$\therefore$ v_{COM} remains constant.

Hence, v_{COM} after the bullet is embedded in the block = 10 ms^{-1}

$\therefore$ Finally KE of COM is

$$K_f = \frac{1}{2}M_{COM} \times v_{COM}^2$$

$$= \frac{1}{2} \times 0.04 \times (10)^2 = 20\text{ J}$$

So, $\quad r = \dfrac{K_f}{K_i} = \dfrac{20}{800} = 0.025$

Hence, $\quad 200r = 200 \times 0.025 = 5$

13. (3) Let the particle is separated when it is at the position θ (see figure).

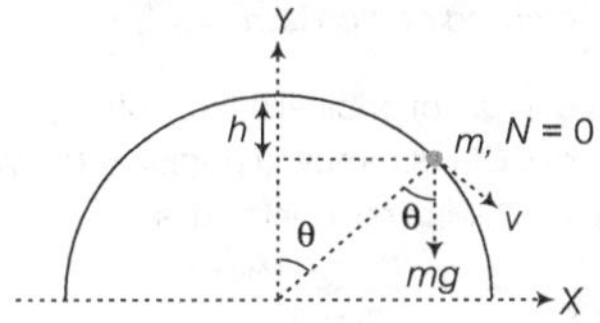

Now, $\quad N = 0$

$$mg\cos\theta = \frac{mv^2}{R}$$

$\Rightarrow \quad v^2 = Rg\cos\theta \quad$ ...(i)

$$h = R(1 - \cos\theta)$$

Decrease in PE = Increase in KE

$\Rightarrow \quad mgR(1 - \cos\theta) = \dfrac{1}{2}mv^2 \quad$ ...(ii)

From Eqs. (i) and (ii), we get

$$\cos\theta = \frac{2}{3}$$

So, $\quad \sin\theta = \dfrac{\sqrt{5}}{3}$

Coordinates of particle, $m = (x_1, y_1) = \left(\dfrac{\sqrt{5}}{3}R, \dfrac{2}{3}R\right)$

Coordinates of COM of hemisphere = $(x_2, y_2) = \left(0, \dfrac{R}{2}\right)$

$\therefore$ x-coordinate of COM of (particle + hemisphere) is

$$X_{COM} = \frac{m\left(\dfrac{\sqrt{5}}{3}R\right) + 4m(0)}{m + 4m} = \frac{R}{3\sqrt{5}}$$

y-coordinate of COM of (particle + hemisphere) is

$$Y_{COM} = \frac{m\left(\dfrac{2}{3}R\right) + 4m\left(\dfrac{R}{2}\right)}{m + 4m} = \frac{8}{15}R$$

Coordinate of point O = (0, 0)

$\therefore$ Required distance $= \sqrt{\left(0 - \dfrac{R}{3\sqrt{5}}\right)^2 + \left(0 - \dfrac{8R}{15}\right)^2} = \dfrac{\sqrt{69}}{15}R$

$\Rightarrow \quad \alpha = 3$

14. (6) $v_{COM} = \dfrac{(2m)(2v_0)\hat{i} + m(-v_0)\hat{i}}{2m + m} = v_0\hat{i}$ (is constant)

Initially $\bar{v}_{A,\,COM} = \bar{v}_{A,\,G} - \bar{v}_{COM,\,G} = -v_0\hat{i} - v_0\hat{i} = -2v_0\hat{i}$

$$\bar{v}_{B,\,COM} = \bar{v}_{B,\,G} - \bar{v}_{COM,\,G} = 2v_0\hat{i} - v_0\hat{i} = v_0\hat{i}$$

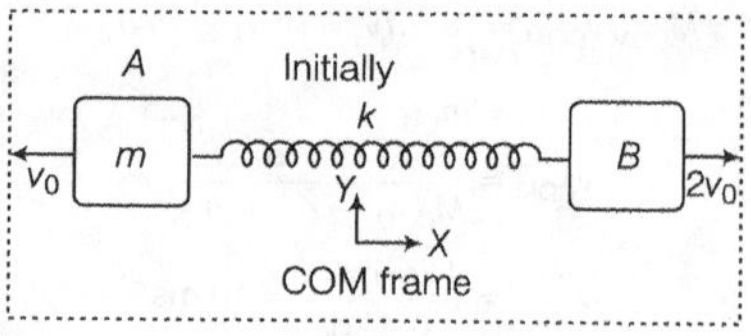

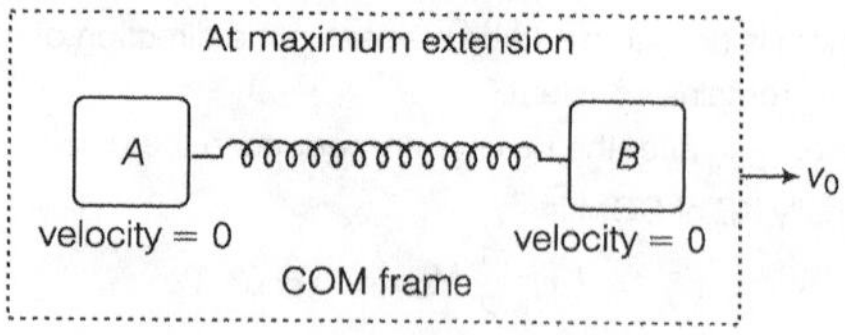

From COM frame at the maximum extension,

$$\frac{1}{2}k\,x_{max}^2 = \frac{1}{2}\mu u_{rel}^2 \Rightarrow \frac{1}{2}k\,x_{max}^2 = \frac{1}{2}\frac{(2m)(m)}{(2m+m)}(3v_0)^2$$

$$\Rightarrow \qquad x_{max} = \sqrt{\frac{6m}{k}}\,v_0 \Rightarrow \beta = 6$$

15. Mass of the original uncut sphere is $m = M$

Mass of the removed portion is $m_1 = \dfrac{M}{R^3}r^3$

Coordinates of COM of original uncut sphere = $(x, y) = (0, 0)$

Coordinates of COM of removed portion = $(x_1, y_1) = (q, 0)$

x-coordinate of COM of residual part is

$$X_{COM} = \frac{mx - m_1x_1}{m - m_1}$$

or $$X_{COM} = \frac{M(0) - \frac{M}{R^3}r^3(q)}{M - \frac{M}{R^3}r^3} = -\frac{r^3q}{R^3}\cdot\frac{R^3}{R^3 - r^3}$$

or $$X_{COM} = -\frac{qr^3}{R^3 - r^3}$$

From symmetry, $\qquad Y_{COM} = 0$

16. From symmetry, it can be said that the COM lies on Y-axis. Mass of the element shown is $dm = \dfrac{m}{\theta}d\alpha$.

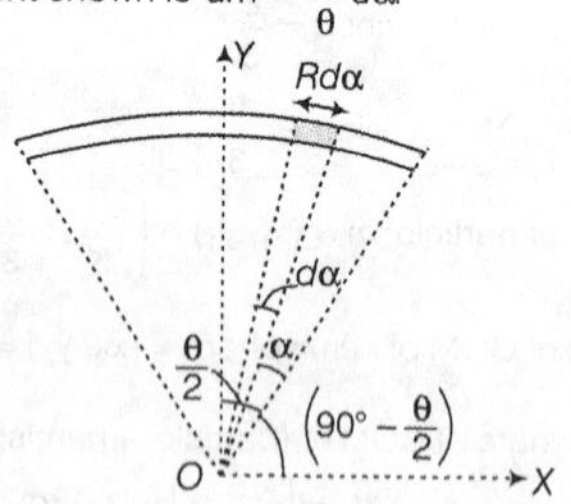

y-coordinate of element is $y = R\sin\left(\alpha + 90° - \dfrac{\theta}{2}\right)$.

$$\therefore Y_{COM} = \frac{\int y\,dm}{\int dm} = \frac{\int_0^\theta R\sin\left(90° + \alpha - \frac{\theta}{2}\right)\frac{m}{\theta}d\alpha}{m}$$

$$= \frac{R}{\theta}\int_0^\theta \cos\left(\alpha - \frac{\theta}{2}\right)d\alpha = \frac{R}{\theta}\left[\sin\left(\alpha - \frac{\theta}{2}\right)\right]_0^\theta = \frac{R}{\theta}\cdot 2\sin\theta/2$$

$\therefore$ COM of the arc is at a height $h = \dfrac{2R\sin(\theta/2)}{\theta}$ from the centre O.

17. From symmetry, the COM should lie on Y-axis .

Consider an elemental arc of thickness dr at a distance r from the centre O (see figure).

Mass of this element is

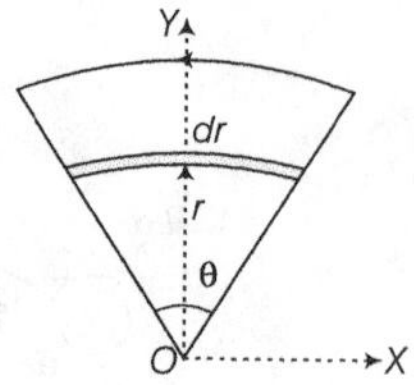

$$dm = \frac{m}{\frac{1}{2}R^2\theta}r\theta dr = \frac{2mr}{R^2}dr$$

COM of element is on Y-axis at a height $= \dfrac{2r\sin(\theta/2)}{\theta}$.

$$\therefore \qquad y = \frac{2r\sin(\theta/2)}{\theta}$$

$$\therefore \qquad Y_{COM} = \frac{\int y\,dm}{\int dm} = \frac{\int_0^R 2r\frac{\sin(\theta/2)}{\theta}\cdot\frac{2mr}{R^2}dr}{m}$$

$$= \frac{4\sin(\theta/2)}{R^2\theta}\cdot\int_0^R r^2dr$$

or $$Y_{COM} = \frac{4R\sin(\theta/2)}{3\theta}$$

18. In the horizontal direction, no external force is acting on the (ball + car) system. So, X_{COM} remains unaltered. Let initial coordinate of COM of M is x_1 and let horizontal distance travelled by wedge is x.

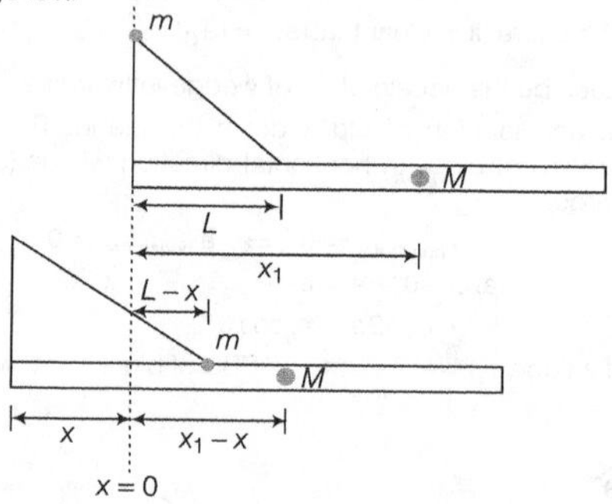

$$(X_i)_{COM} = (X_f)_{COM}$$

$$\frac{m(0) + M\,x_1}{m + M} = \frac{m(L - x) + M(x_1 - x)}{m + M}$$

$$\Rightarrow \qquad mL - mx - Mx = 0 \text{ or } x = \frac{mL}{M + m}$$

19. When the string is jerked, the component of velocity of B along the string and velocity of A along the string will have to be equal (as distance between two points of the string is constant) and linear momentum along the string before and after the jerk is conserved.

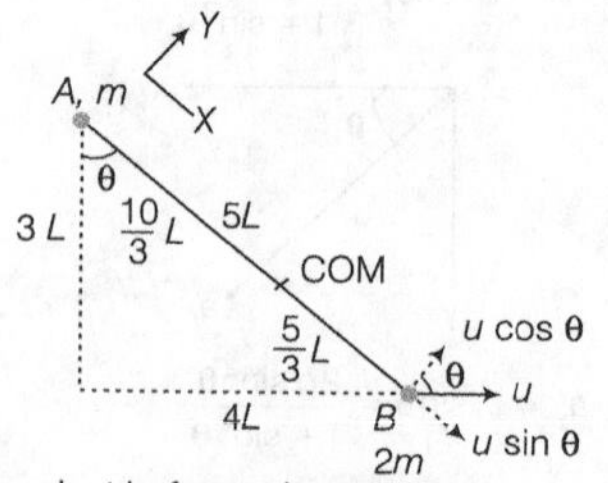

Just before string gets taut

$$(v_{COM})_X = \frac{m(0) + 2m(u\sin\theta)}{m + 2m} = \frac{2}{3}u\sin\theta$$

$$(v_{COM})_Y = \frac{m(0) + 2m(u\cos\theta)}{m + 2m} = \frac{2}{3}u\cos\theta$$

$$X_{COM} = \frac{m(0) + 2m\,(5L)}{3m} = \frac{10}{3}L \qquad \text{(from } A\text{)}$$

Just before string gets taut along the string,

$$p_i = 2m(u\sin\theta) + m(0) = 2mu\sin\theta$$

After the string gets taut along the string,

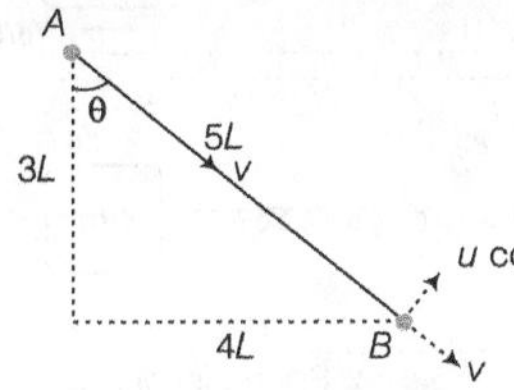

$$p_f = 2m(v) + m(v) = 3mv$$

$$3mv = 2mu\sin\theta$$

$$\Rightarrow \qquad v = \frac{2}{3}u\sin\theta$$

$$\because \sin\theta = \frac{4}{5} \text{ and } \cos\theta = \frac{3}{5}$$

$$\therefore \qquad v = \frac{2}{3}u\left(\frac{4}{5}\right) = \frac{8}{15}u$$

$$u\cos\theta = u\left(\frac{3}{5}\right) = \frac{3}{5}u$$

Just after the string is taut

$$v_A = v = \frac{8}{15}u$$

and $\quad v_B = \sqrt{v^2 + (u\cos\theta)^2}$

or $\quad v_B = \sqrt{\left(\frac{8}{15}u\right)^2 + \left(\frac{3}{5}u\right)^2} = u\sqrt{\frac{64}{225} + \frac{9}{25}} = u\sqrt{\frac{64+81}{225}}$

or $\quad v_B = \frac{\sqrt{145}}{15}u \approx \frac{12}{15}u$

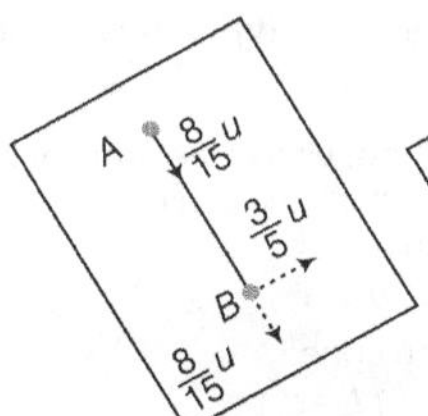

Ground frame after string is taut

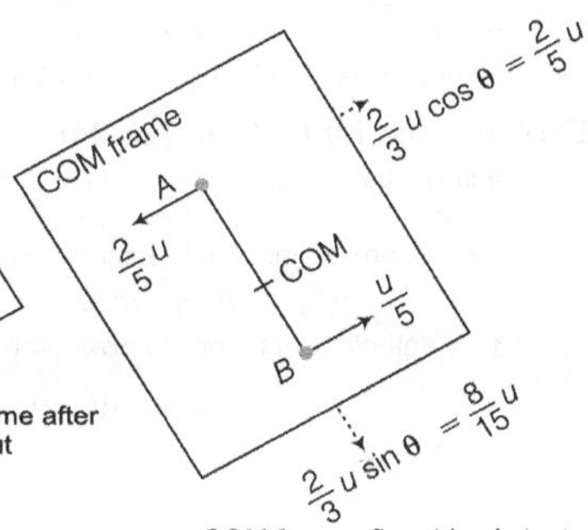

COM frame after string is taut

For B $\quad \omega = \frac{u/5}{5L/3} = \frac{3u}{25L}$

Tension, $T = (2m)\omega^2 r = (2m)\frac{9u^2}{225L^2}\cdot\frac{5}{3}L$

or $\quad T = \frac{30}{225}\cdot\frac{u^2}{L} = \frac{2}{15}\cdot\frac{u^2}{L}$

Check point For A, $\omega = \frac{\frac{2}{5}u}{\frac{10}{3}L} = \frac{3}{25}\cdot\frac{u}{L}$

$$T = (m)\frac{9u^2}{225L^2}\cdot\frac{10}{3}L = \frac{2u^2}{15L}$$

In COM frame

Before string is taut,

$$v'_A = -\frac{2}{3}u\sin\theta\,\hat{i} = -\frac{8}{15}u\,\hat{i}$$

Just after string is taut along the string,

$$v''_A = 0$$

$$\therefore \qquad |J| = m|v'_A - v''_A| = \frac{8}{15}mu$$

Check point

Before string is taut along the string,

$$v'_B = \left(u\sin\theta - \frac{2}{3}u\sin\theta\right)\hat{i}$$

$$= \frac{1}{3}u\sin\theta\,\hat{i} = \frac{4}{15}u\hat{i}$$

Just after the string is taut along the string, $v''_B = 0$

$$\therefore \qquad |J| = 2m|v'_B - v''_B| = \frac{8}{15}mu$$

20. (a) In such problems, the following are the most important points.

In the reference frame of ground, system moves forward with spring getting compressed and relaxed, then extended alternatively.

In the reference frame of COM, both blocks execute SHM with initial position as equilibrium position.

In the ground frame, at maximum extension/maximum compression of the spring both have same velocity.

In the COM frame, at maximum extension/maximum compression of the spring both are momentarily stationary.

Linear momentum is conserved at every moment, in both the frames independently.

$$v_{COM} = \frac{(2m)(2v_0) - (m)(v_0)}{2m + m} = v_0, \text{ take right side as positive.}$$

Let us see how the things look in both the frames.

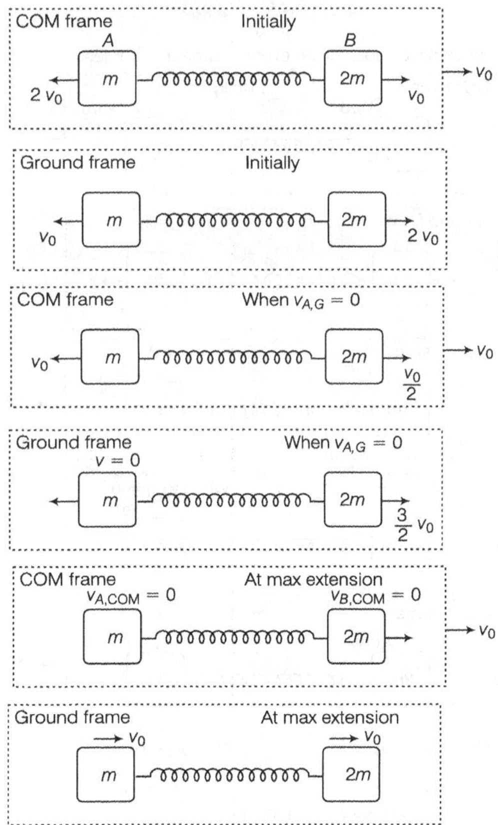

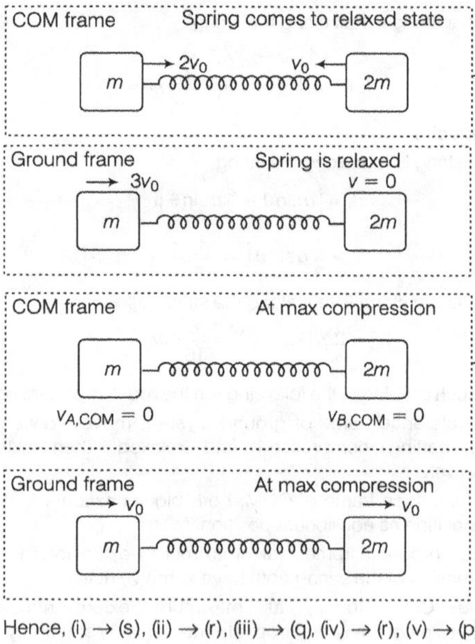

Hence, (i) → (s), (ii) → (r), (iii) → (q), (iv) → (r), (v) → (p)

21. (a) $v_{COM} = \dfrac{(3m)(v_0) - (2m)(v_0)}{3m + 2m} = \dfrac{v_0}{5}$

Total linear momentum in ground frame

$$= (3m)(v_0) - (2m)v_0 = mv_0$$

Total linear momentum in COM frame

$$= -(2m)\left(\frac{6v_0}{5}\right) + (3m)\left(\frac{4v_0}{5}\right) = 0$$

Let us have a look at the events in both the frames.

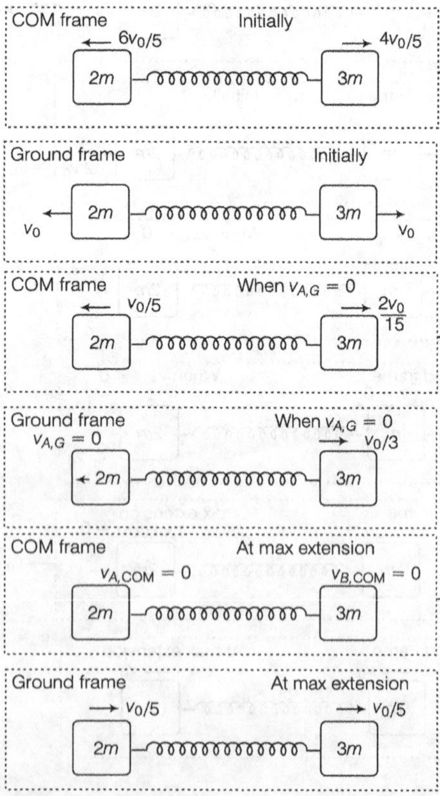

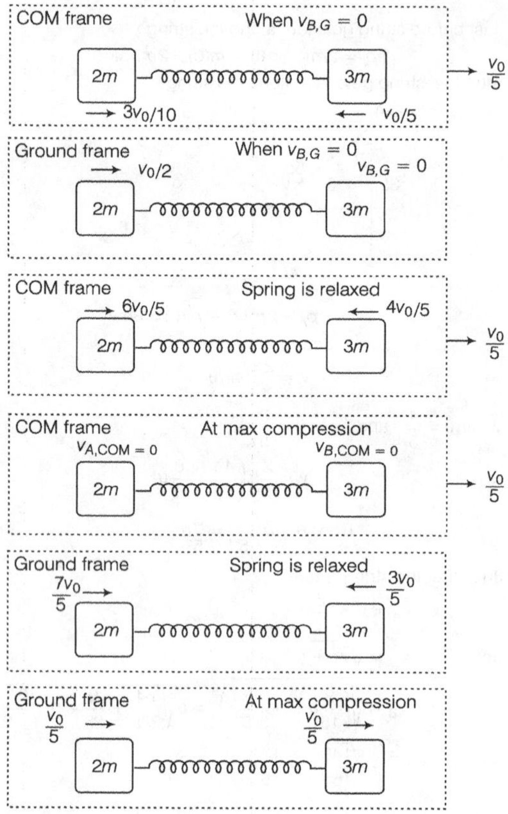

Hence, (i) → (r), (ii) → (p), (iii) → (r), (iv) → s, (v) → (t), (vi) → (q)

Explanations for Q. Nos. (22-24)

Initially, the centre of mass of (A + rod) was at the edge of the table. As there is no external force acting, the centre of mass of the system remains at the same place.

$$m|x| = m|d| \Rightarrow d = x$$

τ_1 = anti-clockwise torque about the edge

$$= mgx + \frac{m}{2L}(L-d)g\left(\frac{L-d}{2}\right)$$

$$= mgx + \frac{mg}{4L}(L-x)^2 = \frac{mgL}{4} + \frac{mgx^2}{4L} + \frac{mgx}{2}$$

τ_2 = clockwise torque about the edge $= \dfrac{m}{2L}(L+d)g\left(\dfrac{L+d}{2}\right)$

$$= \frac{mg}{4L}(L+x)^2 = \frac{mgL}{4} + \frac{mgx^2}{4L} + \frac{mgx}{2}$$

Clearly, $|\tau_1| = |\tau_2|$

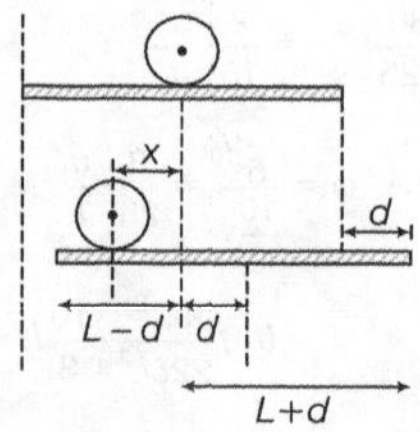

Hence, for any condition i.e. $x < \frac{L}{2}$ or $x = \frac{L}{2}$ or $\frac{L}{2} < x \le L$, the system will not topple.

22. (d) **23.** (d) **24.** (d)

DPP-2 Momentum and Its Conservation

1. (c) For $0 \le t \le 2$,

$$F = 10t$$

Change in momentum,

$$p - 0 = \int_0^t F\,dt = \int_0^t 10t\,dt$$

$$p = 5t^2$$

At $t = 2$ s,

$$p = 20 \text{ kg-ms}^{-1}$$

For $2 \le t \le 4$,

$$F = -10t + 40$$

$$dp = F\,dt$$

$$\therefore \quad p - 20 = \int_2^t (40 - 10t)\,dt = [40t - 5t^2]_2^t$$

$$p - 20 = 40t - 5t^2 - (80 - 20) = -5t^2 + 40t - 60$$

$$p = -5t^2 + 40t - 40$$

At $t = 4$ s,

$$p = -5(16) + 40(4) - 40 = 40 \text{ kg-ms}^{-1}$$

2. (a) Since there is no force in horizontal direction,

$$mv' = 2mv$$

or

$$v' = 2v$$

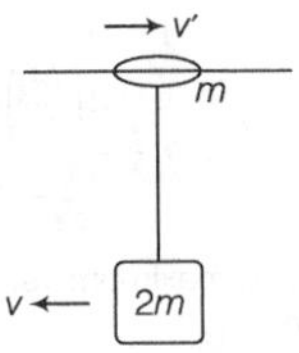

From energy conservation,

$$(2m)gl = \frac{1}{2}mv'^2 + \frac{1}{2}(2m)v^2$$

$$= \frac{1}{2}mv'^2 + \frac{1}{2}(2m)\left(\frac{v'}{2}\right)^2$$

$$2gl = \frac{1}{2}v'^2 + \frac{1}{4}v'^2$$

$$\Rightarrow \quad \frac{3}{4}v'^2 = 2gl$$

$$v' = \sqrt{\frac{8gl}{3}} = \sqrt{\frac{8 \times 10 \times 0.6}{3}} = 4 \text{ ms}^{-1}$$

3. (c) Let v be the velocity of platform in downward direction just after the man jumps up with velocity u. From momentum conservation.

$$Mu - (M + 2m)v = 0$$

$$\Rightarrow \quad v = \frac{Mu}{M + 2m}$$

$$= \frac{3m}{3m + 2m}(10) \qquad (\because M = 3m)$$

$$= 6 \text{ ms}^{-1}$$

After the man leaves, $T - mg = ma$

and $(M + m)g - T = (M + m)a$

$$\therefore \quad a = \frac{Mg}{M + 2m}$$

$$= \frac{3m}{3m + 2m}(10) = 6 \text{ ms}^{-2}$$

$$v^2 = 2ad$$

$$\Rightarrow \quad d = \frac{(6)^2}{2(6)} = 3 \text{ m}$$

4. (c) Let v_r be the relative velocity of the particle at this instant. Then, from conservation of momentum, $4mv = m(v_r \cos\theta - v)$

Absolute components of velocity of particle.

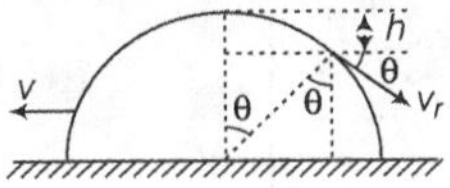

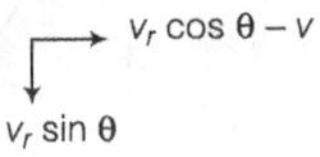

$$\Rightarrow \quad v_r = \frac{5v}{\cos\theta}$$

$$\therefore \quad \omega = \frac{v_r}{R} = \frac{5v}{R\cos\theta}$$

5. (a) (i) $60(1 - v) = 100v \Rightarrow v = \frac{3}{8} \text{ ms}^{-1} (\rightarrow)$

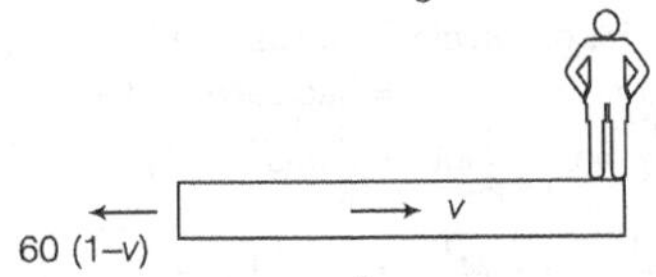

(ii) $80(1 - v) = 80v \Rightarrow v = \frac{1}{2} \text{ ms}^{-1} (\leftarrow)$

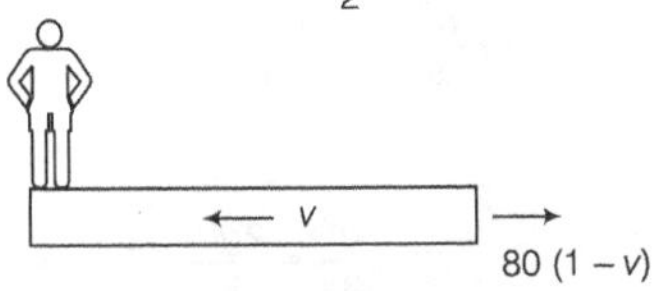

(iii) $-60(1 + v) + 80(1 - v) - 20v = 0$

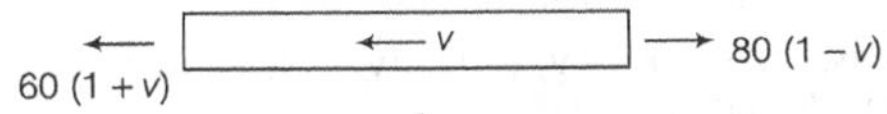

$$\Rightarrow \quad v = \frac{1}{8} \text{ ms}^{-1} (\leftarrow)$$

(iv) $v = \frac{3}{8} \text{ ms}^{-1}$

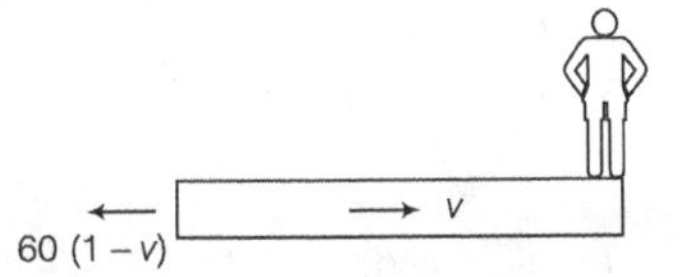

$$p_i = (80 + 20)v = 100 \times \frac{3}{8} = \frac{150}{4} \text{ kg-ms}^{-1}$$

$$p_f = 80(1 - v') - 20v'$$

$$\therefore \quad 80(1 - v') - 20v' = \frac{150}{4}$$

$$\Rightarrow \quad -100v' = \frac{150}{4} - 80$$

$$v' = \frac{10}{400} = \frac{1}{40} \text{ ms}^{-1} (\leftarrow)$$

6. (3) When slipping stops, both will move with common velocity, v (say).

From conservation of momentum,

$$mu + 2m(0) = (m + 2m)v$$

$$v = \frac{u}{3}$$

Work done by friction, $W_f = K_f - K_i = \frac{1}{2}(3m)v^2 - \frac{1}{2}mu^2$

$$= \frac{1}{2}(3m)\frac{u^2}{9} - \frac{1}{2}mu^2 = -\frac{1}{3}mu^2$$

$$\Rightarrow \quad k = 3$$

7. (3) The wedge remains at rest till the particle reaches the bottom of circular track.

Velocity of particle at the lowest point, $u = \sqrt{2gR}$

Now, wedge also starts moving. Let the particle reaches up to a maximum height h.

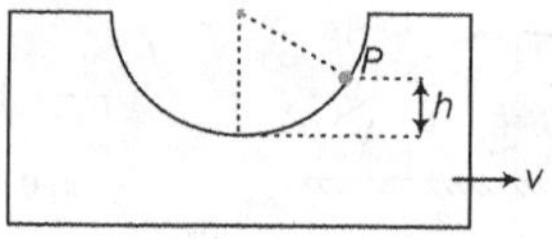

From conservation of momentum,

$$mu = mv + Mv$$

$$\Rightarrow \quad \frac{M}{2}u = \frac{M}{2}v + Mv \Rightarrow v = \frac{u}{3}$$

From energy conservation,

Increase in PE of particle + Increase in KE of wedge

= Decrease in KE of particle

$$mgh + \frac{1}{2}Mv^2 = \frac{1}{2}m(u^2 - v^2)$$

$$\Rightarrow \quad \frac{M}{2}gh + \frac{1}{2}M\frac{u^2}{9} = \frac{1}{2}\cdot\frac{M}{2}\left(u^2 - \frac{u^2}{9}\right)$$

$$gh + \frac{u^2}{9} = \frac{1}{2}\cdot\frac{8u^2}{9}$$

$$\Rightarrow \quad gh = \frac{4u^2}{9} - \frac{u^2}{9} = \frac{u^2}{3}$$

$$h = \frac{u^2}{3g} = \frac{2gR}{3g} = \frac{2}{3}R \Rightarrow \alpha = 3$$

8. Initial momentum of man, $p_i = Mu$

Let velocity of jacket is v_2 (←)and velocity of man is v_1 (→)after he throws the jacket.

$$\therefore \quad -u_r = (-v_2) - (+v_1)$$

$$\Rightarrow \quad v_2 = u_r - v_1$$

$$p_f = (M - m)\,v_1 + m(-v_2) = (M - m)v_1 - m(u_r - v_1)$$

$$= Mv_1 - mu_r$$

$$p_f = p_i \Rightarrow Mv_1 - mu_r = Mu$$

$$\Rightarrow \quad v_1 = \frac{Mu + mu_r}{M}$$

$$t = \sqrt{\frac{2h}{g}} \text{ and } s = v_1 t = \frac{Mu + mu_r}{M}\sqrt{\frac{2h}{g}}$$

9. Here, at the time of exchange of bags, momentum in the direction of individual motion remains conserved.

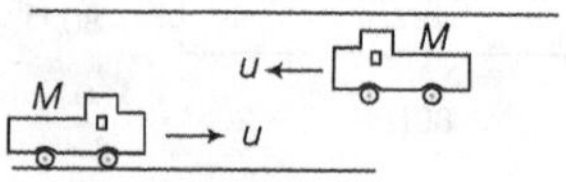

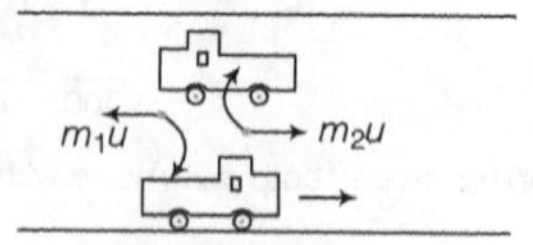

For first truck carrying wheat bags,

$$Mu - m_1u - m_2u = (M - m_1 + m_2)\,v_1$$

or $$v_1 = \frac{Mu - m_1u - m_2u}{M - m_1 + m_2}$$

For second truck carrying rice bags,

$$Mu - m_2u - m_1u = (M - m_2 + m_1)v_2$$

or $$v_2 = \frac{Mu - m_2u - m_1u}{M - m_2 + m_1}$$

10. (i) Initial momentum along X-axis (i.e. horizontal) is $p_i = 0$

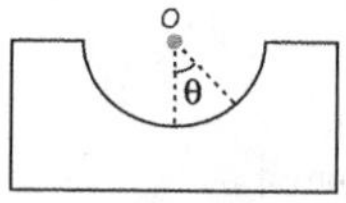

$$\therefore \quad p_f = 0$$

By energy conservation,

Initial potential energy = final potential energy

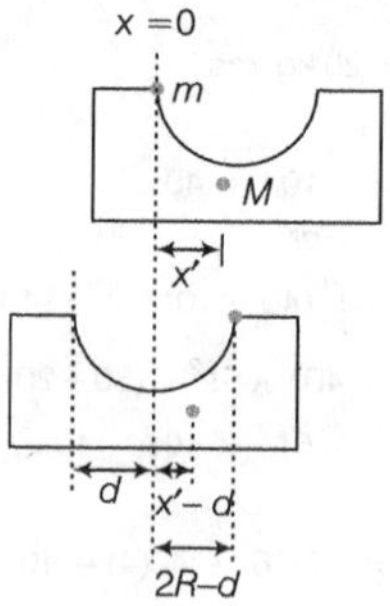

Hence, $$\theta_{max} = 90°$$

$$\Delta x_{COM} = 0$$

$$m(2R - d) = M(d)$$

$$\Rightarrow \quad d\,(m + M) = 2mR$$

or $$d = \frac{2\left(\frac{M}{2}\right)R}{\frac{M}{2} + M} = \frac{2}{3}R$$

(ii) The wedge will have maximum velocity when the particle is at its lowest position.

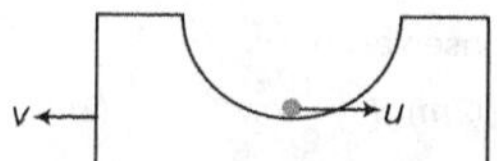

$$p_i = 0 \text{ and } p_f = mu - Mv$$

$$\therefore \quad mu - Mv = 0$$

$$u = \frac{M}{m}v = 2v$$

$$U_i + K_i = U_f + K_f$$

$$\Rightarrow \quad mgR + 0 = 0 + \frac{1}{2}mu^2 + \frac{1}{2}Mv^2$$

$$mgR = \frac{1}{2}m(4v^2) + \frac{1}{2}Mv^2$$

$$\Rightarrow \quad \frac{M}{2}gR = \frac{1}{2}\frac{M}{2}(4v^2) + \frac{1}{2}Mv^2$$

$$3v^2 = gR \Rightarrow v = \sqrt{\frac{gR}{3}}$$

11. From M to M', the ball will fall a distance 4 m, then impulse will be developed in the string OM'.

Velocity of the ball just before the string gets taut is

$$u_0 = \sqrt{2gh}$$

or $$u_0 = \sqrt{2 \times 10 \times 4} = \sqrt{80} \text{ ms}^{-1}$$

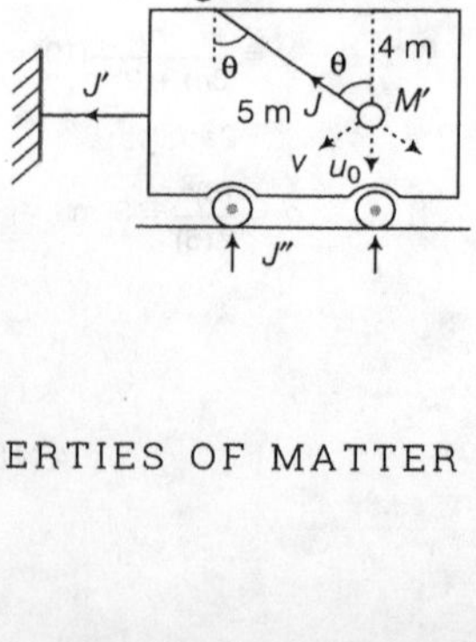

Velocity component of the ball in the direction perpendicular to the impulse will remain the same (just after and just before the impulse developed).

$$\therefore \qquad v = u_0 \sin\theta = (\sqrt{80})\left(\frac{3}{5}\right) = \frac{12}{\sqrt{5}} \text{ ms}^{-1}$$

After the string becomes taut, the velocity (along the string) of the ball becomes zero as point O is fixed.

$\therefore$ Net velocity of the ball just after the string becomes taut $\frac{12}{\sqrt{5}}$ ms^{-1}.

Now, change in momentum along the string

$$= mu_0 \cos\theta - 0 = mu_0 \cos\theta$$

$\therefore$ Impulse (J) developed in the string when it becomes taut is

$$J = mu_0 \cos\theta = (1)(\sqrt{80})\left(\frac{4}{5}\right) = \frac{16}{\sqrt{5}} \text{ kg-ms}^{-1}$$

Impulse in the string PQ,

$$J' = J\sin\theta = \left(\frac{16}{\sqrt{5}}\right)\left(\frac{3}{5}\right) = \frac{48}{5\sqrt{5}} \text{ kg-ms}^{-1}$$

Impulse given by the earth at this instant, $J'' = J\cos\theta$

$$\text{or} \qquad J'' = \left(\frac{16}{\sqrt{5}}\right)\left(\frac{4}{5}\right) = \frac{64}{5\sqrt{5}} \text{ kg-ms}^{-1}$$

12. Just after the impulse is given to particle A, the impulses developed in the strings and the velocities of the particles are as indicated in the following figure. The following equations can be written referring to the figure.

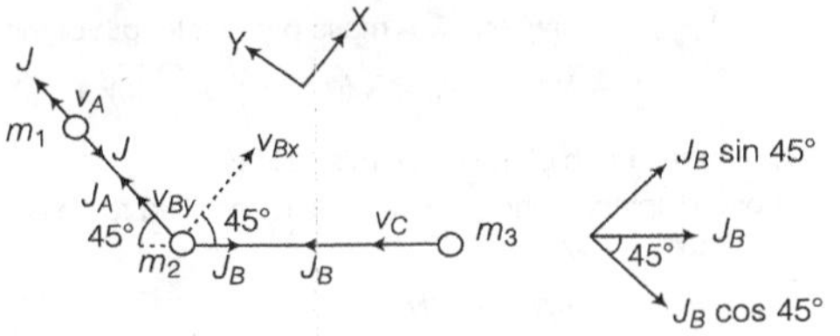

$$J - J_A = m_1 v_A \qquad \ldots\text{(i)}$$

$$J_A - J_B \cos 45° = m_2 v_{By}$$

$$\Rightarrow \qquad J_A - \frac{J_B}{\sqrt{2}} = m_2 v_{By} \qquad \ldots\text{(ii)}$$

$$J_B \sin 45° = m_2 v_{Bx}$$

$$\Rightarrow \qquad \frac{J_B}{\sqrt{2}} = m_2 v_{Bx} \qquad \ldots\text{(iii)}$$

$$J_B = m_3 v_C \qquad \ldots\text{(iv)}$$

Now, $\qquad v_{By} = v_A$

and $v_{By} \cos 45° - v_{Bx} \cos 45° = v_C$

($\because$ Relative velocity along each string is zero)

$$\text{or} \qquad v_{By} = v_A \qquad \ldots\text{(v)}$$

$$\text{and} \qquad v_{By} - v_{Bx} = v_C\sqrt{2} \qquad \ldots\text{(vi)}$$

Solving the above equations, we get

$$v_A = \frac{J(m_3 + 2m_2)}{m_1 m_3 + 2m_2(m_1 + m_2 + m_3)}$$

$$v_{Bx} = \frac{m_3 J}{m_1 m_3 + 2m_2(m_1 + m_2 + m_3)}$$

$$v_{By} = v_A = \frac{J(m_3 + 2m_2)}{m_1 m_3 + 2m_2(m_1 + m_2 + m_3)}$$

$$\text{and} \qquad v_C = \frac{Jm_2\sqrt{2}}{m_1 m_3 + 2m_2(m_1 + m_2 + m_3)}$$

$$v_B = \sqrt{v_{Bx}^2 + v_{By}^2} = \frac{J[(m_3 + 2m_2)^2 + m_3^2]^{1/2}}{m_1 m_3 + 2m_2(m_1 + m_2 + m_3)}$$

13. (b) As, no horizontal force is acting on the bead w.r.t. ground it remains stationary.

$$\therefore \qquad v_{COM} = \frac{m(0) + Mv_0}{M + m} = \frac{Mv_0}{M + m}$$

14. (a) $t_1 = \dfrac{\text{Distance}}{\text{Speed of cart}} = \dfrac{L}{v_0}$

15. (d) From momentum conservation, $Mv_0 = Mv_1 + mv_2$

$$\frac{v_2 - v_1}{v_0} = 1$$

$$\Rightarrow \qquad v_1 = v_2 - v_0$$

$$\therefore \qquad Mv_0 = M(v_2 - v_0) + mv_2$$

$$\text{or} \qquad v_2(M + m) = 2Mv_0$$

$$\text{or} \qquad v_2 = \frac{2Mv_0}{m + M}$$

In COM frame, $\quad v_2' = v_2 - v_{COM} = \dfrac{2Mv_0}{m + M} - \dfrac{Mv_0}{m + M} = \dfrac{Mv_0}{m + M}$

16. (c) Impulse on block A,

$$|J_A| = F\Delta t - T\Delta t$$

where, $\quad F$ = force between bullet and block A

and $\quad T$ = tension in the string

Impulse on bullet,

$$|J_b| = F\Delta t = 0.25(200 - 100)$$

$$F\Delta t = 25 \text{ kg-ms}^{-1}$$

or

Impulse on block B, $\qquad |J_B| = T\Delta t$

Velocities of A and B are equal and say it is v.

For A, $\quad F\Delta t - T\Delta t = 1 \times v$

$$\Rightarrow \qquad T\Delta t = 25 - v$$

For B, $\qquad T\Delta t = 2 \times v$

$$\Rightarrow \qquad 2v = 25 - v$$

$$\Rightarrow \qquad v = \frac{25}{3} \text{ ms}^{-1}$$

17. (d) Total initial energy of $A + B$ is

$$E_i = \frac{1}{2}(1 \times v^2 + 2 \times v^2)$$

$$\text{or} \qquad E_i = \frac{1}{2} \times 3 \times \left(\frac{25}{3}\right)^2 = \frac{625}{6} \text{ J}$$

At maximum displacement of A,

$$E_f = 0 + m_B g\, s_{max}$$

$$\therefore \qquad m_B g\, s_{max} = \frac{625}{6}$$

$$\Rightarrow \qquad s_{max} = \frac{625}{6} \times \frac{1}{2 \times 10}$$

$$\text{or} \qquad s_{max} \simeq 5.2 \text{ m}$$

18. (a) Physically possible explosions are those in which both the particles move in opposite directions, i.e. velocities have opposite sign.

19. (b) If $m_1 = m_2$, magnitudes of velocities are equal $\Rightarrow$ (ii)

If $m_1 > m_2$, then $v_1 < v_2 \ \Rightarrow$ (v)

If $m_1 < m_2$, then $v_1 > v_2 \ \Rightarrow$ (iv)

20. (d) In (vi), the momentum is not conserved. So, force must be acting on it.

21. (d) Linear momentum is a vector and kinetic energy is a scalar. Consider a bomb at rest. Say, it explodes into different fragments. Now, the sum of linear momenta of all the fragments is zero but their total KE is not zero.

22. (d) Let m be the mass of man, then mass of ladder $= M - m$

$$\Delta y_{COM} = \frac{My + (M + m)(-y) + m(h - y)}{2M}$$

$$\Rightarrow \qquad \Delta y_{COM} = \frac{mh}{2M} \text{ (upwards)}$$

DPP-3 Variable Mass

1. At time t, the forces acting on the rocket are

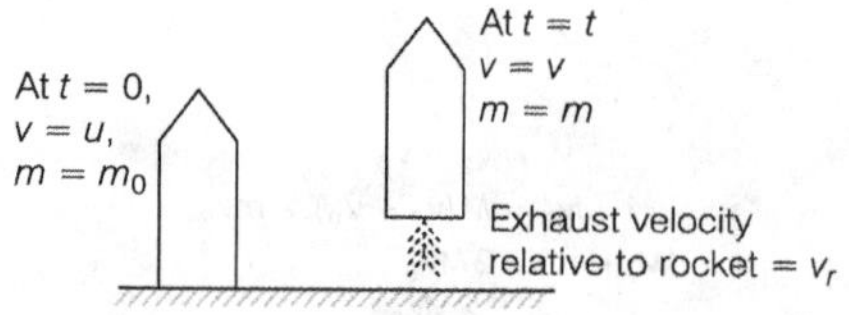

(i) Thrust force on the rocket, $F_t = v_r\left[-\dfrac{dm}{dt}\right]$ (upwards)

(ii) Weight of the rocket = mg (downwards)

Note $F_t = v_r\left[-\dfrac{dm}{dt}\right]$ *is upwards because* v_r *is downwards and mass is decreasing.*

So, at time t $\quad F_{net} = F_t - mg$ (upwards)

or $\quad F_{net} = v_r\left(-\dfrac{dm}{dt}\right) - mg$

or $\quad m\dfrac{dv}{dt} = v_r\left(-\dfrac{dm}{dt}\right) - mg$

or $\quad dv = -v_r\dfrac{1}{m}dm - g\,dt$

$$\int_u^v dv = -v_r\int_{m_0}^{m}\frac{1}{m}dm - g\int_0^t dt$$

$$v - u = v_r \ln\left(\frac{m_0}{m}\right) - gt$$

or $\quad v = u - gt + v_r\ln\left(\dfrac{m_0}{m}\right)$

2. $F_t = v_0\mu$ in the upward direction, since mass is decreasing and v_0 is downwards.

At time t FBD's of both are

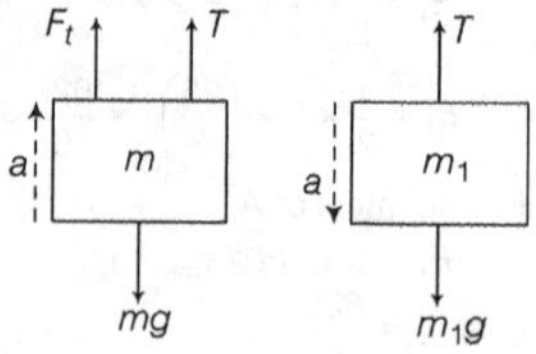

$$T + F_t - mg = ma \qquad \ldots(i)$$

$$m_1g - T = m_1a \qquad \ldots(ii)$$

On adding Eqs. (i) and (ii), we get

$$F_t - mg + m_1g = a(m + m_1)$$

$$F_t = \mu v_0 \quad \text{and} \quad m = m_0 - \mu t$$

$$\mu v_0 - (m_0 - \mu t)g + m_1 g = a(m_0 - \mu t + m_1)$$

$$\therefore \quad a = \frac{(m_1 - m_0 + \mu t)g + \mu v_0}{(m_1 + m_0 - \mu t)}$$

3. At time t, $m = m_0 + qt$

Thrust force is $\quad F_t = v\dfrac{dm}{dt} = qv$ (backwards)

$$a_t = \frac{F_t}{m} = \frac{qv}{m_0 + qt} \Rightarrow -\frac{dv}{dt} = \frac{qv}{m_0 + qt}$$

$$\frac{1}{v}dv = -q\frac{1}{m_0 + qt}dt$$

$$\Rightarrow \quad \int_{v_0}^{v}\frac{1}{v}dv = -q\int_0^t \frac{1}{m_0 + qt}dt$$

$$\ln\left(\frac{v}{v_0}\right) = -\frac{q}{q}\ln\left(\frac{m_0 + qt}{m_0}\right)$$

$$\Rightarrow \quad \frac{v}{v_0} = \frac{m_0}{m_0 + qt} \qquad \ldots(i)$$

$$v = v_0\left(\frac{m_0}{m_0 + qt}\right)$$

$$\Rightarrow \quad \frac{dx}{dt} = v_0\left(\frac{m_0}{m_0 + qt}\right)$$

$$\int_0^L dx = m_0v_0\int_0^t \frac{1}{m_0 + qt}dt$$

$$\Rightarrow \quad L = \frac{m_0v_0}{q}\ln\left(\frac{m_0 + qt}{m_0}\right)$$

Now, $\quad m_0 + qt = m_0\, e^{\left(\frac{qL}{m_0v_0}\right)}$

is the mass of car and its load after the car has cleared the hopper.

From Eq. (i), the speed of the car at this moment is

$$v = \frac{m_0v_0}{m_0e^{(qL/m_0v_0)}} = v_0e^{-\left(\frac{qL}{m_0v_0}\right)}$$

4. (i) When the chain is raised with a constant upward speed, then net force on it is zero. So,

F = Weight of length y of chain + Thrust force

Thrust force,

$$F_t = v_r\frac{dm}{dt} = v_0\frac{\lambda dy}{dt} = \lambda v_0^2$$

(where, λ = mass per unit length of the chain)

$$\therefore \quad F = \lambda yg + \lambda v_0^2 = \lambda(gy + v_0^2) = \frac{m}{L}(gy + v_0^2)$$

(ii) When the length of the chain raised is y.

For a displacement dy, work done by applied force is

$$dW = Fdy$$

$$= \frac{m}{L}(gy + v_0^2)\,dy$$

$$\therefore \quad W = \frac{m}{L}\int_0^y (gy + v_0^2)dy = \frac{m}{L}\left(\frac{gy^2}{2} + v_0^2y\right)$$

and increase in mechanical energy of chain

$$= \left(\frac{m}{L}y\right)g\left(\frac{y}{2}\right) + \frac{1}{2}\left(\frac{m}{L}y\right)v_0^2$$

$$= \frac{mg\,y^2}{2L} + \frac{myv_0^2}{2L}$$

$\therefore$ Energy lost during lifting = Work done by applied force − Increase in mechanical energy of chain

or $\quad -\Delta E = \dfrac{m}{L}\left(\dfrac{gy}{2} + v_0^2\right)y - \dfrac{mgy^2}{2L} - \dfrac{myv_0^2}{2L}$

$$= \frac{mgy^2}{2L} + \frac{myv_0^2}{L} - \frac{mgy^2}{2L} - \frac{myv_0^2}{2L} = \frac{myv_0^2}{2L}$$

5. Thrust force, $\quad F_t = v_r\left(-\dfrac{dm}{dt}\right)$

Given, $\quad m = m_0 - \mu t$

$\Rightarrow \quad \dfrac{dm}{dt} = -\mu$ and given that $v_r = u$

$\therefore \quad F_t = \mu u$

$$F_{net} = F_t - mg = \mu u - (m_0 - \mu t)g$$

If a is the acceleration of the rocket, then

$$F_{net} = ma = m\frac{d^2y}{dt^2}$$

or $\quad F_{net} = (m_0 - \mu t)\dfrac{d^2y}{dt^2}$

$$\therefore \quad (m_0 - \mu t)\left(\frac{d^2y}{dt^2}\right) = \mu u - (m_0 - \mu t)g$$

6. (5) Given, $m_0 = 10^{-2}$ kg,

$A = 10^{-4}\ \text{m}^2,\ v_0 = 10^3\ \text{ms}^{-1}$

and $P_{\text{dust}} = \rho = 10^{-3}\ \text{kg-m}^{-3}$

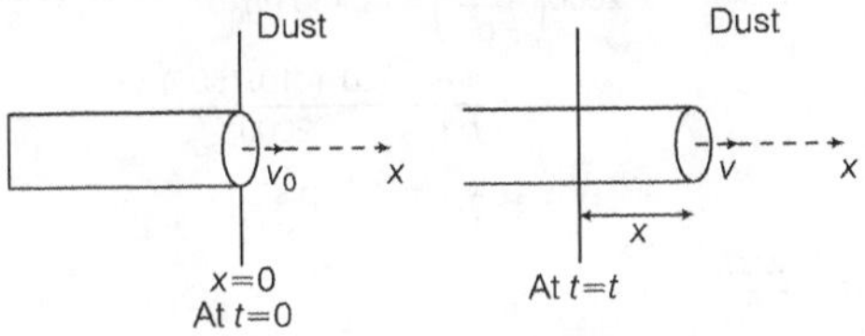

At an instant t, mass of the cylinder is $m = m_0$ + mass of dust collected so far

or $m = m_0 + Ax\rho$

At $t = 0$, the linear momentum is $p_0 = m_0 v_0$

At $t = t$, the linear momentum is

$$p_t = mv = (m_0 + Ax\rho)v$$

From law of conservation of linear momentum, we have

$$p_0 = p_t$$

$$m_0 v_0 = (m_0 + Ax\rho)v$$

$$\Rightarrow \quad m_0 v_0 = (m_0 + Ax\rho)\frac{dx}{dt}$$

$$(m_0 + Ax\rho)dx = m_0 v_0 dt$$

$$\int_0^x (m_0 + Ax\rho)dx = m_0 v_0 \int_0^{150} dt$$

$$\Rightarrow \quad m_0 x + \frac{A\rho x^2}{2} = 150 m_0 v_0$$

Solving this quadratic equation and substituting the values of m_0, v_0, A and ρ, we get

$$x = 10^5\ \text{m} \Rightarrow \beta = 5$$

Note *x can have only positive value.*

7. (a) The maximum velocity is attained after $t = 60$ s. Then,

$m = 1000 - 60(10) = 400$ kg

Initial mass is $m_0 = 1000$ kg and $v_r = 2000\ \text{ms}^{-1}$

$$\therefore \quad v_{\text{max}} = u - gt + v_r \ln\left(\frac{m_0}{m}\right)$$

$$= 0 - 10(60) + 2000 \ln \frac{1000}{400}$$

$$= 1232\ \text{ms}^{-1}$$

8. (b) If v has to be constant, net pulling force on the chain should be zero. This is possible only when

Thrust force + Weight of left side of the chain

= Pull F + Weight of right side of the chain

or $\lambda v^2 + (\lambda h)g = F + (\lambda y)g \quad \left(\because F_t = v\lambda \frac{dy}{dt} = \lambda v^2\right)$

or $F = \lambda v^2 + \lambda g(h - y)$

9. (a) $v_r\left(\frac{-dm}{dt}\right) = m\alpha$ $\quad$ ($\because$ no external forces are acting)

$$u\left(-\frac{dm}{dt}\right) = m\alpha$$

$$\int_{m_0}^{m} \frac{1}{m} dm = -\frac{\alpha}{u}\int_0^t dt$$

$$\Rightarrow \quad \ln \frac{m}{m_0} = -\frac{\alpha}{u}t$$

$$\Rightarrow \quad \ln \frac{m_0}{m} = +\frac{\alpha}{u}t$$

$$\Rightarrow \quad \frac{m_0}{m} = e^{(\alpha/u)t}$$

$$\Rightarrow \quad m = m_0 e^{-(\alpha/u)t}$$

10. (c) Let the mass to be expelled is x.

$$p_i = 4200 \times 150\ \text{kg-ms}^{-1}$$

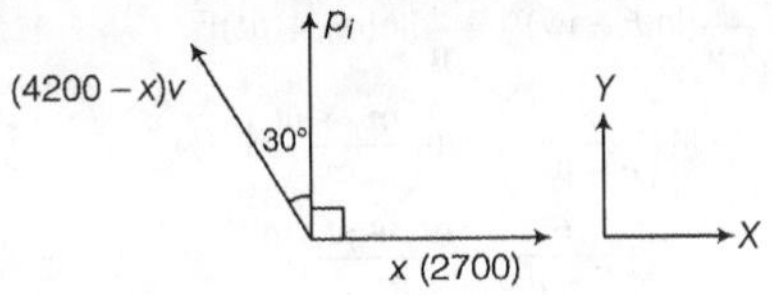

Along Y-axis, $(4200 - x)v \cdot \frac{\sqrt{3}}{2} = 4200 \times 150$ ...(i)

Along X-axis, $(4200 - x)\ v \cdot \frac{1}{2} = x \times 2700$ ...(ii)

On dividing Eq. (i) by Eq. (ii), we get

$$\sqrt{3} = \frac{4200 \times 150}{x \times 2700} \text{ or } x = 134 \text{ kg}$$

11. (b) Force on the table is $F = F_1 + F_2$

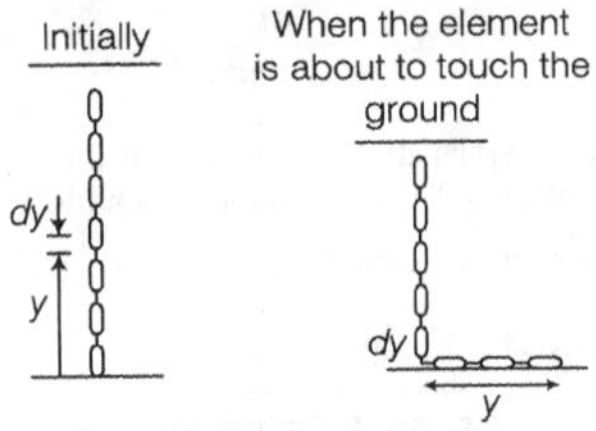

where, F_1 = weight of fallen part = λyg, where λ is mass per unit length

and F_2 = thrust force, $F_t = v_r \frac{dm}{dt} = v\frac{\lambda dy}{dt} = \lambda v^2$

v = speed of element when it is just about to touch the table $= \sqrt{2gy}$

$$\therefore \quad F_2 = \lambda(2gy) = 2\lambda yg$$

$F = \lambda yg + 2\lambda yg = 3\lambda yg$ or $F = 3\lambda yg$ ...(i)

When the upper end is just about to touch the table, $y = L$

$$\therefore \quad F = 3\lambda Lg = 3mg$$

Note *In this, F_t on the table will be vertically downwards because v_r is downwards and mass on table is increasing.*

12. (d) When half of the chain has fallen, $y = \frac{L}{2}$

Substitute this value in the Eq. (i) of previous solution, we get

$$F = 3\lambda \cdot \frac{L}{2} g = \frac{3}{2} mg$$

13. (a,d) Let the velocity of car is v at time t. When the sand particles are just about to touch the sand in the car, their relative velocity w.r.t. car is backwards and is equal to v. Thus, $v_r = v$ (backwards) and mass is increasing. So, F_t is also backwards.

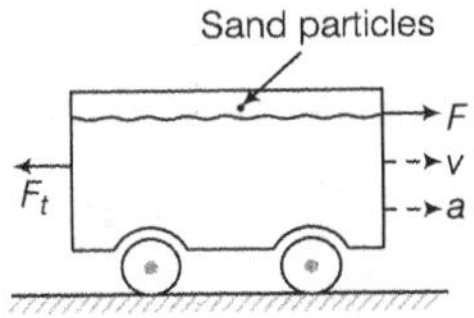

$$\therefore \quad F_{\text{net}} = F - F_t = F - \mu v$$

$\Rightarrow \quad m\frac{dv}{dt} = F - \mu v$ ...(i)

$$\text{or} \quad (m_0 + \mu t)\frac{dv}{dt} = F - \mu v$$

$$\Rightarrow \quad \frac{1}{F - \mu v} dv = \frac{1}{m_0 + \mu t} dt$$

or $\qquad \int_0^v \frac{1}{F-\mu v}dv = \int_0^t \frac{1}{m_0+\mu t}dt$

$\Rightarrow \qquad \frac{1}{-\mu}[\ln(F-\mu v)]_0^v = \frac{1}{\mu}[\ln(m_0+\mu t)]_0^t$

or $\qquad \ln\left(\frac{F}{F-\mu v}\right) = \ln\left(\frac{m_0+\mu t}{m_0}\right)$

$\Rightarrow \qquad \frac{F}{F-\mu v} = \frac{m_0+\mu t}{m_0}$

$$Fm_0 = Fm_0 + F\mu t - \mu m_0 v - \mu^2 vt$$

$\Rightarrow \qquad v(m_0+\mu t) = Ft$

or $\qquad v = \frac{Ft}{m_0+\mu t}$

From Eq. (i), we get

$$a = \frac{F-\mu v}{m} = \frac{F - \mu\left(\frac{Ft}{m_0+\mu t}\right)}{m_0+\mu t}$$

or $\qquad a = \frac{Fm_0}{(m_0+\mu t)^2}$

14. (a,b) The sand spills through a hole in the bottom of the cart. Immediately after spilling, the velocity of sand spilled will be same as that of the cart. So, $v_r = 0$. Thus, $F_t = 0$. $\left[\because F_t = v_r\left(-\frac{dm}{dt}\right)\right]$ and the net force will be F only.

$\therefore \qquad F_{net} = F$ or $m\frac{dv}{dt} = F$

$(m_0 - \mu t)\frac{dv}{dt} = F \qquad (\because$ at time t, mass is $m = m_0 - \mu t)$

or $\qquad dv = F\frac{1}{m_0-\mu t}dt$

$\Rightarrow \qquad \int_0^v dv = F\int_0^t \frac{1}{m_0-\mu t}dt$

or $\qquad v = \frac{F}{-\mu}[\ln(m_0-\mu t)]_0^t$

or $\qquad v = \frac{F}{\mu}\ln\left[\frac{m_0}{m_0-\mu t}\right]$

Acceleration of the cart at time t is

$$a = \frac{F}{m} = \frac{F}{m_0-\mu t}$$

15. (b) $F_t = v_r\left(-\frac{dm}{dt}\right) = 2000\left(-\frac{dm}{dt}\right)$

For just lifting off, $\qquad F_t = m_0 g$

$\therefore \qquad 2000\left(-\frac{dm}{dt}\right) = (50+450)\,10$

$\Rightarrow \qquad \left(-\frac{dm}{dt}\right) = 2.5\ \text{kg s}^{-1}$

16. (c) $F_t = v_r\left(-\frac{dm}{dt}\right) = 2000\left(-\frac{dm}{dt}\right)$

$$F_t - m_0 g = m_0 a$$

$\Rightarrow \qquad 2000\left(-\frac{dm}{dt}\right) = (a+g)m_0$

or $\qquad -\frac{dm}{dt} = \frac{(20+10)\,(500)}{2000}$

or $\qquad -\frac{dm}{dt} = 7.5\ \text{kg s}^{-1}$

17. (a) $t = \frac{450}{10} = 45$ s

$$v = u - gt + v_r\ln\frac{m_0}{m}$$

$$= 0 - 10\times 45 + 2000\ln\frac{500}{50}$$

$$= 4156\ \text{ms}^{-1}$$

18. (c) Mass of the car is decreasing and $v_r = 20\ \text{ms}^{-1}$ is backwards. So, F_t will be forward.

$$F_t = v_r\left(-\frac{dm}{dt}\right) = 20(2) = 40\ \text{N}$$

When this overcomes friction, the car starts moving

So, $\qquad F_t = \mu mg$

$\Rightarrow \qquad 40 = (0.1)\,(50-2t)\,(10)$

or $\qquad t = 5$ s

19. (a) At $t = 5$ s, mass of the car will be 40 kg and it starts moving. Fuel will be consumed in a total time of 10 s. Therefore, maximum velocity will be obtained after a time of $t' = 10 - 5 = 5$ s.

Now, $\qquad$ mass $= 50 - 10(2) = 30$ kg

Comparing with the equation of rocket $\left[v = u - gt + v_r\ln\left(\frac{m_0}{m}\right)\right]$,

g should be replaced by μg and here $t = t' = 5$ s

$\therefore$ The maximum speed of the car is

$$v_{max} = 0 - (0.1)\,(10)\,(5) + 20\ln\left(\frac{40}{30}\right)$$

$$= 0.6\ \text{ms}^{-1}$$

20. (d) At $t = 10$ s, fuel has been exhausted completely. So, the retardation of car will be $\mu g = 1\ \text{ms}^{-2}$ and its speed at this instant is $0.6\ \text{ms}^{-1}$. Let it stops after a time t'' from now. So,

$$v = u + at$$

$\Rightarrow \qquad 0 = 0.6 - 1\times t''$

$\Rightarrow \qquad t'' = 0.6$ s

$\therefore$ Total time is

$$t = 10 + 0.6 = 10.6\ \text{s}$$

21. (a) With respect to hanging part, v_r of mass joining is upwards and mass of hanging part is increasing. So, F_t is upwards.

22. (a) With respect to vertical part, v_r of mass joining is downwards and mass of vertical portion is increasing. So, F_t is downwards.

DPP-4 Collisions

1. (b) $s_1 = \frac{1}{2}gt^2 = 4.9t^2$

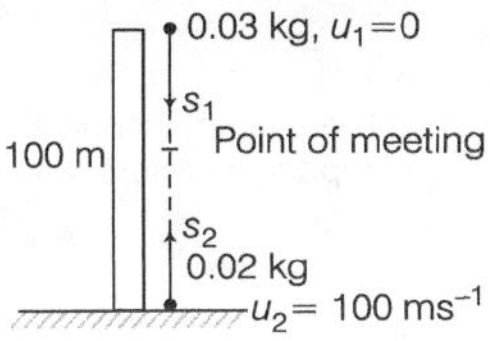

and $s_2 = 100t - \frac{1}{2}gt^2$

$= 100t - 4.9t^2$

$s_1 + s_2 = 100$

$\Rightarrow \quad 100t = 100$

$\Rightarrow \quad t = 1\,\text{s}$

At the point of meeting,

$$v_1 = gt = 9.8(1) = 9.8\ \text{ms}^{-1}$$

$$v_1 = 100 - 9.8(1) = 90.2\ \text{ms}^{-1}$$

Taking downwards direction as negative at the point of meeting

$$p_i = p_f$$

$$-(0.03)(9.8) + (0.02)(90.2) = (0.02 + 0.03)v$$

$\Rightarrow \quad v = 30.2\,\text{ms}^{-1}$

Maximum height attained from the point of meeting is

$$h = \frac{v^2}{2g} = \frac{(30.2)^2}{2(9.8)} = 46.53\ \text{m}$$

Height above the building is

$$h' = h - s_1$$
$$= 46.53 - 4.9$$
$$= 41.63\ \text{m}$$

2. (a) Along X-axis,

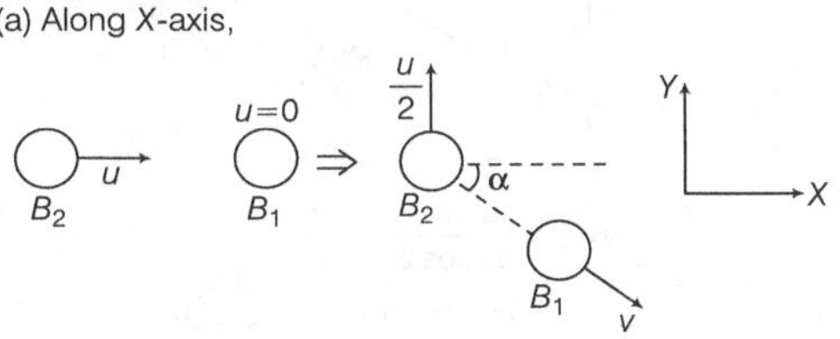

$$m_1v_1\cos\alpha = m_2u \quad \ldots\text{(i)}$$

and along Y-axis,

$$m_1v_1\sin\alpha = m_2\frac{u}{2} \quad \ldots\text{(ii)}$$

On dividing Eq. (ii) by Eq. (i), we get

$$\tan\alpha = \frac{1}{2}$$

$\Rightarrow \quad \alpha = \tan^{-1}\left(\frac{1}{2}\right)$

3. (d) $p_i = p_f$

$\Rightarrow \quad mu = (M + m)v$

$\Rightarrow \quad v = \left(\frac{m}{m + M}\right)u$

$$E_i = E_f$$

$\Rightarrow \quad \frac{1}{2}mu^2 = \frac{1}{2}(m + M)v^2 + \frac{1}{2}kx_{max}^2$

$$\frac{1}{2}kx_{max}^2 = \frac{1}{2}mu^2 - \frac{1}{2}(m + M)\left(\frac{m}{m + M}\right)^2u^2$$

$$= \frac{1}{2}mu^2 - \frac{1}{2}\frac{m^2u^2}{(m + M)}$$

$$= \frac{1}{2}mu^2\left(1 - \frac{m}{m + M}\right)$$

$$\frac{1}{2}kx_{max}^2 = \frac{1}{2}mu^2\left(\frac{M}{m + M}\right)$$

So, required fraction $= \frac{M}{m + M}$

4. (c) Conservation of momentum gives

$$m_1v_1 + m_2v_2 = m_1v_1' + m_2v_2'$$

Here, $v_1' = v_2' = v$ (let) then,

$$v = \frac{m_1v_1 + m_2v_2}{m_1 + m_2}$$

$$= \frac{4 \times 10 + 6 \times 0}{4 + 6} = 4\ \text{ms}^{-1}$$

and loss of energy is ΔK

$$= \text{Initial KE} - \text{Final KE}$$

$$= \frac{1}{2}m_1v_1^2 - \frac{1}{2}(m_1 + m_2)v^2$$

$$= 200 - 80 = 120\ \text{J}$$

Loss of energy appears as heat which rises temperature of blocks.

$\therefore \quad (m_1 + m_2)C \times \Delta\theta = 120$

$\Rightarrow \quad \Delta\theta = \frac{120}{10 \times 2} = 6\ \text{K or } 6\,°\text{C}$

Note *$\Delta\theta$ is same in Kelvin and in degree centigrades.*

5. (d) $h = l(1 - \cos 60°) = \frac{980}{2} = 490$ cm

$$u = \sqrt{2gh} = \sqrt{2 \times 980 \times 490} = 980\ \text{cms}^{-1}$$

Let the speed of the ball is v_1 and that of bob is v_2 after collision

$$p_i = p_f$$

$\Rightarrow \quad (10)(980) = (20)v_1 + (10)v_2$

or $\quad 2v_1 + v_2 = 980 \quad \ldots\text{(i)}$

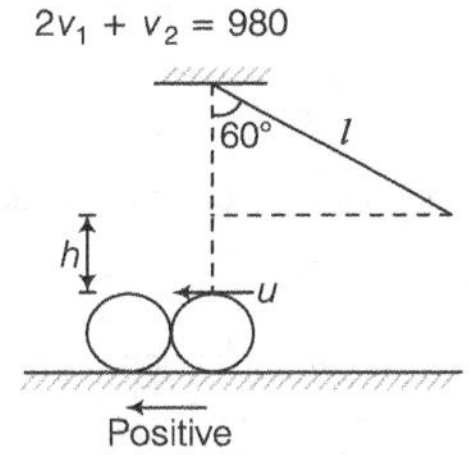

As $e = 1$, speed of separation = speed of approach

$\Rightarrow \quad v_1 - v_2 = u$

or $\quad v_1 - v_2 = 980 \quad \ldots\text{(ii)}$

From Eqs. (i) and (ii), we get

$$v_2 = -\frac{980}{3}\ \text{cms}^{-1}$$

i.e. bob will move backwards with a speed $\frac{980}{3}$ cms^{-1}.

6. (b) Let the initial velocity of the bullet is u and velocities of (M_1 and bullet) after the collision are v_1 and v, respectively. Now, the velocity of (M_2 + bullet) after second collision will be v_1 according to the problem.

For 1st collision,

$\therefore \quad mu + 0 = mv + M_1v_1$

$\Rightarrow \quad 0.02u = 0.02v + 1 \times v_1$

$\Rightarrow \quad u = v + 50v_1 \quad \ldots\text{(i)}$

Now, for 2nd collision,

$$mv + 0 = (m + M_2)v_1$$

$\Rightarrow \quad 0.02v = (0.02 + 2.98)v_1$

$\Rightarrow \quad v = 150v_1 \quad \ldots\text{(ii)}$

From Eqs. (i) and (ii), we get

$$u = v + 50\frac{v}{150} = \frac{4}{3}v$$

or $\quad v = \frac{3}{4}u$

So, percentage loss = 25%

7. (c) At the moment of maximum deformation, both have same velocity components along common normal direction (i.e. line of action of impulses).

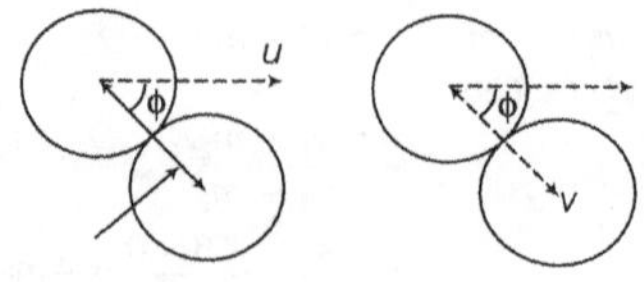

Line of action of impulse $\quad mu\cos\phi = mv + mv \Rightarrow v = \dfrac{u\cos\phi}{2}$

From conservation of energy,

$$\frac{1}{2}m(u\cos\phi)^2 = \frac{1}{2}m\left(\frac{u\cos\phi}{2}\right)^2 \times 2$$

+ Deformation potential energy

or deformation potential energy

$$= \frac{1}{2}mu^2\left[\cos^2\phi - \frac{\cos^2\phi}{2}\right]$$

$$= \frac{1}{2}mu^2\left[\frac{1}{2}\cos^2\phi\right]$$

$\therefore$ Required fraction $= \dfrac{1}{2}\cos^2\phi$

8. (c) Let required minimum velocity of m is x.

Velocity of second bead after collision will be

$$v_2 = \frac{m_2 - m_1}{m_1 + m_2}u_2 + \frac{2m_1}{m_1 + m_2}u_1$$

$$= 0 + \frac{2m}{m + \frac{m}{2}}x$$

or $\quad v_2 = \dfrac{4}{3}x$

Similarly, you get velocity of third bead after collision as $v_1 = \left(\dfrac{4}{3}\right)^2 x$.

So, velocity of nth bead after collision will be

$$v_n = \left(\frac{4}{3}\right)^{n-1} x$$

But $\quad (v_n)_{\min} = \sqrt{4gR}$

$$\sqrt{4gR} = \left(\frac{4}{3}\right)^{n-1} x$$

$\Rightarrow \quad x = \left(\dfrac{3}{4}\right)^{n-1}\sqrt{4gR}$

Note *At the maximum height in the limiting case, the nth bead can be momentarily stationary.*

$\therefore \quad \dfrac{1}{2}m_n(v_n)^2_{\min} = m_n g \times 2R$

$$(v_n)_{\min} = 2\sqrt{gR}$$

9. (c) Time taken to reach the wall $= \dfrac{d}{v_0\cos\theta}$

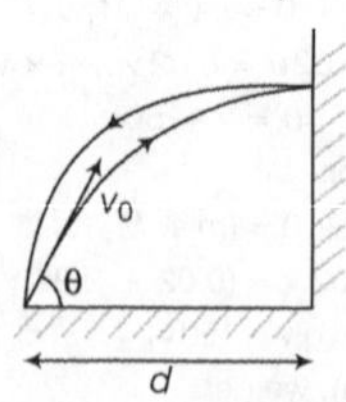

Time taken to reach the point of projection after impact

$$= \frac{d}{ev_0\cos\theta}$$

Note *Horizontal component after impact reverses and becomes equal to* $ev_0\cos\theta$

Total time of flight $= \dfrac{d}{v_0\cos\theta} + \dfrac{d}{ev_0\cos\theta}$

$$= \frac{d}{v_0\cos\theta}\left(\frac{e+1}{e}\right) \qquad \ldots\text{(i)}$$

There is no change in the vertical component of the velocity after impact, therefore, total time of flight is

$$T = \frac{2v_0\sin\theta}{g} \qquad \ldots\text{(ii)}$$

From Eqs. (i) and (ii), we get

$$\left(\frac{1+e}{e}\right)\frac{d}{v_0\cos\theta} = \frac{2v_0\sin\theta}{g}$$

$$\frac{1+e}{e} = \frac{2v_0^2\sin\theta\cos\theta}{gd}$$

$\Rightarrow \quad \dfrac{1}{e} = \dfrac{gh\sin 2\theta}{gd} - 1$

$$= \frac{h\sin 2\theta - d}{d}$$

$\therefore \quad e = \dfrac{d}{h\sin 2\theta - d}$

10. (d) Let u be the velocity of projection at the point A and velocity at B after collision can be formed as follows :

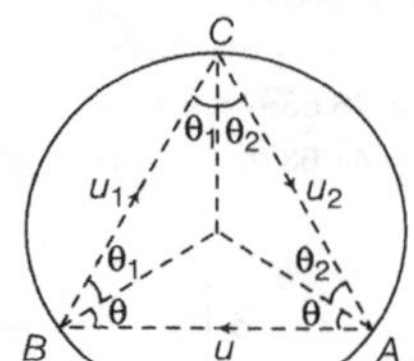

$$\tan\theta_1 = \frac{u\sin\theta}{eu\cos\theta} = \frac{\tan\theta}{e}$$

and $\quad u_1 = \sqrt{u^2\sin^2\theta + e^2u^2\cos^2\theta}$

Similarly, at $C \quad \tan\theta_2 = \dfrac{\tan\theta_1}{e} = \dfrac{\tan\theta}{e^2}$

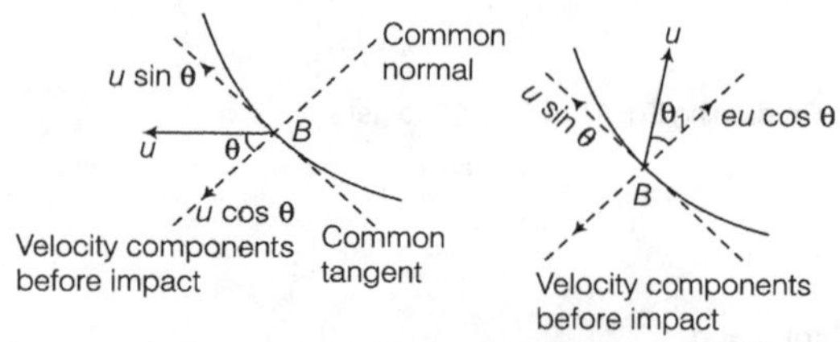

and $\quad u_2 = \sqrt{u_1^2\sin^2\theta_1 + e^2u_1^2\cos^2\theta_1}$

Since, the particle returns to the point of projection

$$\theta + \theta_1 + \theta_2 = \frac{\pi}{2}$$

or $\quad \tan(\theta + \theta_1 + \theta_2) = \infty$

or $\quad 1 - (\tan\theta\tan\theta_1 + \tan\theta_1\tan\theta_2 + \tan\theta_2\tan\theta) = 0$

$$1 - \tan^2\theta\left(\frac{\tan\theta_1}{\tan\theta} + \frac{\tan\theta_1\tan\theta_2}{\tan^2\theta} + \frac{\tan\theta_2}{\tan\theta}\right) = 0$$

$$1 - \tan^2\theta\left(\frac{1}{e} + \frac{1}{e^3} + \frac{1}{e^2}\right) = 0$$

$\Rightarrow \quad \tan^2\theta\left(\dfrac{1}{e} + \dfrac{1}{e^2} + \dfrac{1}{e^3}\right) = 1$

or $\quad \cot^2\theta = \dfrac{1}{e} + \dfrac{1}{e^2} + \dfrac{1}{e^3}$

$$= \frac{1}{0.5} + \frac{1}{(0.5)^2} + \frac{1}{(0.5)^3}$$

or $\quad \cot^2\theta = 14$

11. (a) $R = \dfrac{2u_x u_y}{g}$ and $T = \dfrac{2u_y}{g}$

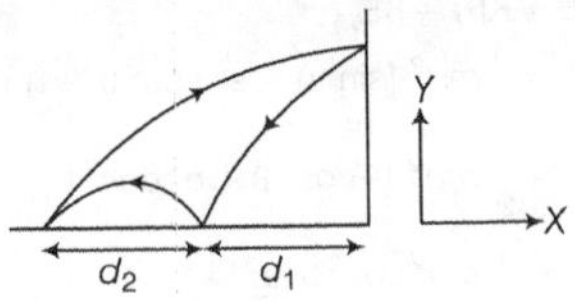

After 1st collision,

$$v_x = eu_x$$

Distance covered between 2nd and 1st collisions is

$$d_1 = v_x \frac{T}{2} = eu_x \frac{T}{2}$$

$$= \frac{eu_x}{2} \cdot \frac{2u_y}{g} = \frac{eu_x u_y}{g}$$

After 2nd collision,

$$v_x' = v_x = eu_x$$

and $\quad v_y' = eu_y$

$$\therefore \quad d_2 = \frac{2v_x' v_y'}{g} = \frac{2(eu_x)(eu_y)}{g} = \frac{2e^2 u_x u_y}{g}$$

Now, $\quad d_1 + d_2 = \dfrac{eu_x u_y}{g}(1 + 2e)$

But $\quad d_1 + d_2 = \dfrac{R}{2}$

$$\therefore \quad \frac{eu_x u_y}{g}(1 + 2e) = \frac{2u_x u_y}{2g}$$

$$e + 2e^2 = 1$$

$$\Rightarrow \quad 2e^2 + e - 1 = 0$$

By solving, we get

$$e = \frac{1}{2} = 0.5$$

12. (b) Let initial speed of big disc is u.

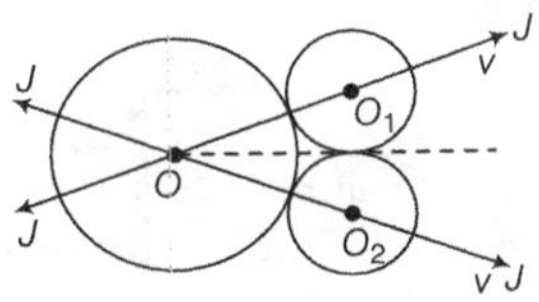

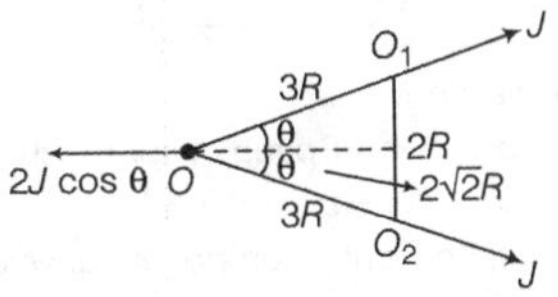

$$\cos\theta = \frac{2\sqrt{2}R}{3R} = \frac{2\sqrt{2}}{3}$$

$$2J\cos\theta = mu$$

$$\Rightarrow \quad J = \frac{mu}{2\cos\theta} = \frac{mu3}{2(2\sqrt{2})} = \frac{3mu}{4\sqrt{2}}$$

Also, $\quad J = mv$

$$\therefore \quad mv = \frac{3mu}{4\sqrt{2}}$$

$$\Rightarrow \quad v = \frac{3u}{4\sqrt{2}}$$

Along common normal direction, speed of approach $= u\cos\theta$ and speed of separation $= v$

$$\therefore \quad e = \frac{v}{u\cos\theta} = \frac{3u/4\sqrt{2}}{u(2\sqrt{2}/3)} = \frac{9}{16}$$

13. (i) $p_i = p_f$

$$\Rightarrow \quad m_1 u + 0 = -m_1 v + m_2 v$$

Before collision

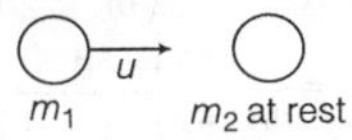

After collision

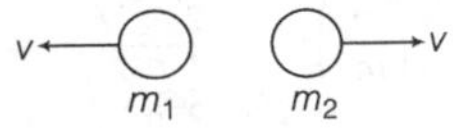

or $\quad m_1 u = m_2 v - m_1 v \quad \ldots$(i)

It is a perfectly elastic collision. So, $e = 1$,

Relative speed of approach = Relative speed of separation

$$u = v + v = 2v \quad \text{or} \quad v = \frac{u}{2}$$

Using this in Eq. (i), we get

$$m_1 u = m_2 \frac{u}{2} - m_1 \frac{u}{2}$$

$$\frac{m_1}{m_2} = \frac{1}{2} - \frac{m_1}{m_2} \times \frac{1}{2}$$

$$\Rightarrow \quad \frac{m_1}{m_2}\left(\frac{3}{2}\right) = \frac{1}{2}$$

$$\Rightarrow \quad \frac{m_1}{m_2} = \frac{1}{3}$$

(ii) $p_i = p_f$ along X-axis gives

$$m_1 u = m_1 v_1 \cos 30^\circ + m_2 v_2 \cos 30^\circ$$

or $\quad m_1 u = m_1 v_1 \dfrac{\sqrt{3}}{2} + m_2 v_2 \dfrac{\sqrt{3}}{2} \quad \ldots$(i)

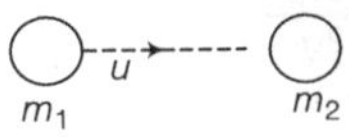

Before collision

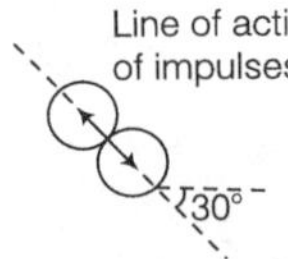

During collision

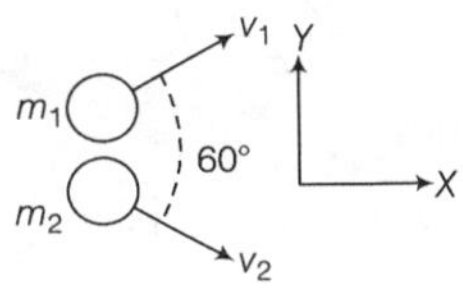

After collision

$p_i = p_f$ along Y-axis gives

$$0 = m_1 v_1 \sin 30^\circ - m_2 v_2 \sin 30^\circ$$

or $\quad m_1 v_1 = m_2 v_2$ or $v_2 = \dfrac{m_1}{m_2} v_1 \quad \ldots$(ii)

Along common normal (i.e. line of action of impulses),
Relative speed of approach = Relative speed of separation

$$u\cos 30^\circ = v_2 - v_1 \cos 60^\circ$$

$$\Rightarrow \quad \frac{u\sqrt{3}}{2} = v_2 - \frac{v_1}{2} \quad \ldots\text{(iii)}$$

From Eqs. (ii) and (iii), we get

$$\frac{u\sqrt{3}}{2} = \frac{m_1}{m_2} v_1 - \frac{v_1}{2}$$

$$\Rightarrow \quad v_1 = \frac{u\sqrt{3}/2}{\left(\dfrac{m_1}{m_2} - \dfrac{1}{2}\right)}$$

and $\quad v_2 = \dfrac{m_1}{m_2} \cdot \dfrac{u\sqrt{3}/2}{\left(\dfrac{m_1}{m_2} - \dfrac{1}{2}\right)} = \dfrac{u\sqrt{3}/2}{\left(1 - \dfrac{1}{2} \cdot \dfrac{m_2}{m_1}\right)}$

Substituting for v_1 and v_2 in Eq. (i), we get

$$m_1 u = m_1 \frac{\sqrt{3}}{2}\cdot\frac{u\sqrt{3}/2}{\left(\frac{m_1}{m_2}-\frac{1}{2}\right)} + m_2\frac{\sqrt{3}}{2}\cdot\frac{u\sqrt{3}/2}{\left(1-\frac{1}{2}\frac{m_2}{m_1}\right)}$$

$$\frac{m_1}{m_2} = \frac{m_1}{m_2}\left(\frac{3}{4}\right)\frac{1}{\left(\frac{m_1}{m_2}-\frac{1}{2}\right)} + \frac{3/4}{\left(1-\frac{1}{2}\frac{m_2}{m_1}\right)}$$

$$\frac{4}{3}\left(\frac{m_1}{m_2}\right) = \frac{2}{1-\frac{1}{2}\frac{m_2}{m_1}}$$

$$\Rightarrow \quad \frac{2}{3}\left(\frac{m_1}{m_2}\right)\left(1-\frac{1}{2}\frac{m_2}{m_1}\right) = 1$$

$$\frac{m_1}{m_2} - \frac{1}{2} = \frac{3}{2}$$

$$\Rightarrow \quad \frac{m_1}{m_2} = 2$$

14. The ball hits the box with a velocity $u\cos\theta$.

Let the velocity of M after the collision is u'.

Now, law of conservation of momentum gives

$$mu\cos\theta = Mu'$$

$$\Rightarrow \quad u' = mu\cos\theta/M$$

Law of conservation of energy for box M for its initial and final positions gives

$$\frac{1}{2}Mu'^2 = MgR$$

$$\Rightarrow \quad \frac{1}{2}\left(\frac{mu\cos\theta}{M}\right)^2 = gR$$

or $$u = \frac{M\sqrt{2gR}}{m\cos\theta}$$

Now, $$l = \frac{1}{2}\times \text{Range} = \frac{u^2\sin 2\theta}{2g}$$

$$= \frac{M^2 2gR}{m^2\cos^2\theta}\times\frac{2\sin\theta\cos\theta}{2g}$$

or $$l = \frac{2M^2R\tan\theta}{m^2}$$

15. $v\sin\phi = u\sin\theta$ ...(i)

$v\cos\phi = e\,u\cos\theta$ $\quad\left(\because e = \frac{\text{Speed of separation}}{\text{Speed of approach}}\right)$ ...(ii)

From Eqs. (i) and (ii), we get

$$\tan\phi = \left(\frac{\tan\theta}{e}\right)$$

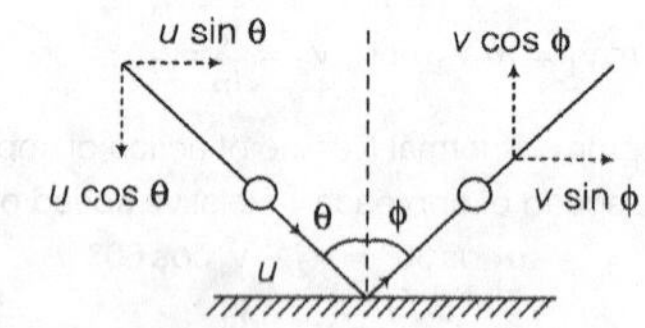

$$\therefore \quad \phi = \tan^{-1}\left(\frac{\tan\theta}{e}\right)$$

$$v = \sqrt{(v\sin\phi)^2 + (v\cos\phi)^2}$$

$$= \sqrt{u^2\sin^2\theta + e^2u^2\cos^2\theta}$$

or $$v = u\sqrt{\sin^2\theta + e^2\cos^2\theta}$$

Impulse is $|J| = mu\cos\theta + mv\cos\phi = mu\cos\theta + m\,eu\cos\theta$

$$J = mu\cos\theta(1+e)$$

$$\text{KE}_i = \frac{1}{2}mu^2$$

$$\text{KE}_f = \frac{1}{2}mu^2(\sin^2\theta + e^2\cos^2\theta)$$

$$\Delta\text{KE} = \text{KE}_f - \text{KE}_i$$

$$= \frac{1}{2}mu^2[\sin^2\theta + e^2\cos^2\theta - 1]$$

$$= \frac{1}{2}mu^2[-\cos^2\theta + e^2\cos^2\theta]$$

$$= \frac{1}{2}mu^2\cos^2\theta\,[e^2 - 1]$$

16. $u = \sqrt{2gh}$

Along common normal,

$$e = \frac{v\cos(90^\circ - \theta)}{u\cos\theta} \quad \text{or} \quad v = eu\cot\theta \qquad \text{...(i)}$$

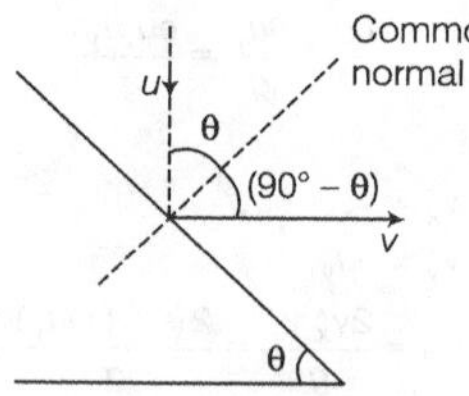

Along the common tangent,

$$u\sin\theta = v\cos\theta \quad \text{or} \quad v = u\tan\theta \qquad \text{...(ii)}$$

From Eqs. (i) and (ii), we get

$$eu\cot\theta = u\tan\theta$$

$$\Rightarrow \quad e = \tan^2\theta$$

If collision is elastic, magnitude of velocity along common normal remains unchanged besides the velocity along common tangent remains unchanged, impulse acts along common normal.

Along common normal,

$$|J| = m\Delta v = m(2u\cos\theta) = 2mu\cos\theta$$

or $$J = 2m\sqrt{2gh}\cos\theta$$

17. (2) Since the wall is smooth, the vertical component of velocity remains unaltered.

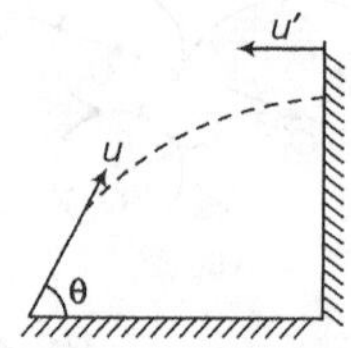

As collision is elastic, $e = 1$

$\therefore$ Speed of separation = Speed of approach

$$v' = u\cos\theta + u'$$

$\therefore$ After compact, horizontal component of velocity relative to the ground.

$$(u\cos\theta + u') + u' = u\cos\theta + 2u'$$

Time of flight depends on the vertical component of velocity, which is unchanged.

$$\therefore \quad T = \frac{2u\sin\theta}{g}$$

Distances covered in the horizontal direction before and after impact are same.

$$\therefore \quad (u\cos\theta)t = (u\cos\theta + 2u')(T - t)$$

$$t[u\cos\theta + u\cos\theta + 2u'] = T(u\cos\theta + 2u')$$

$$= \frac{2u\sin\theta}{g}(u\cos\theta + 2u')$$

$$t = \frac{1}{2(u\cos\theta + u')}\frac{2u\sin\theta}{g}\cdot(u\cos\theta + 2u')$$

or $$t = \frac{(u\cos\theta + 2u')u\sin\theta}{g(u\cos\theta + u')} \Rightarrow k = 2$$

18. (9) Using law of conservation of momentum between B and C

$$mu = mv_1 + 2mv_2 \quad \ldots(i)$$

Collision between B and C is elastic.

Note *Friction between A and B will be non-impulsive, therefore we ignore it.*

So, $e = 1$

$$\Rightarrow \quad u = v_2 - v_1 \quad \ldots(ii)$$

From Eqs. (i) and (ii), we get

$$v_2 = \frac{2u}{3}$$

For centre of mass of $(A + B)$ system,

$$(2m)\left(\frac{2u}{3}\right) = (3m)(v_f)$$

$$\therefore \quad v_f = \frac{4u}{9}$$

$$\Delta E = \frac{1}{2}(2m)\left(\frac{2u}{3}\right)^2 - \frac{1}{2}(3m)\left(\frac{4u}{9}\right)^2 = \frac{4}{27}mu^2 \Rightarrow \beta = 9$$

19. (a) While colliding with the wall x-component of velocity gets changed while y-component remains unaltered. Let speed in horizontal direction after collision with the wall is v_1.

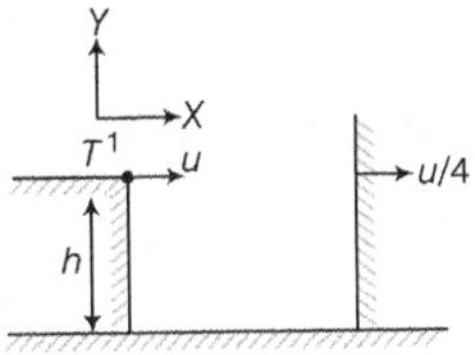

$$e = 1 = \frac{v_1 + \frac{u}{4}}{u - \frac{u}{4}} \Rightarrow v_1 = \frac{u}{2}$$

It should be noted that time of flight will be $2t$, where t is the time taken by the similar horizontal projectile to reach the ground. In this case, the collision must occur with the wall first and second with the ground. Whereas if the wall was moving towards the ball, then first collision must occur with the ground and second with the wall.

$$\therefore \quad \text{Time of flight} = 2t = 2\sqrt{\frac{2h}{g}}$$

20. (b) Let separation between point P and the wall is x_1, when the ball hits the wall. Thus,

$$\frac{x_1}{u} + \frac{x_1}{u/2} = T = 2\sqrt{\frac{2h}{g}}$$

$$\Rightarrow \quad 3\frac{x_1}{u} = 2\sqrt{\frac{2h}{g}}$$

Time taken to cover this distance x_1 is

$$t_1 = \frac{x_1}{u} = \frac{2}{3}\sqrt{\frac{2h}{g}}$$

$\therefore$ Initial separation is

$$x = x_1 - \frac{u}{4}\cdot t_1 = x_1 - \frac{u}{4}\cdot\frac{x_1}{u} = \frac{3}{4}x_1$$

or

$$x = \frac{3}{4}u\cdot\frac{2}{3}\sqrt{\frac{2h}{g}} = u\sqrt{\frac{h}{2g}}$$

21. (a)

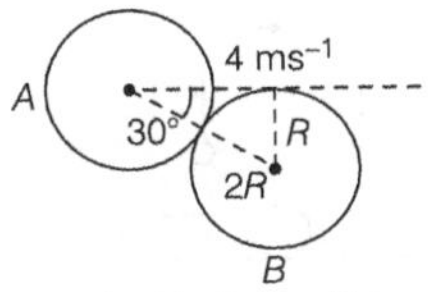

Just before collision

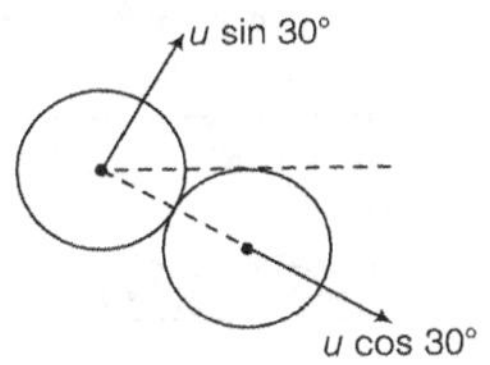

Just after collision

$$v_A = 4\sin 30° = 2\,\text{ms}^{-1}$$

$$\vec{v}_A = 2\cos 60°\,\hat{i} + 2\sin 60°\,\hat{j} = (\hat{i} + \sqrt{3}\,\hat{j})\,\text{ms}^{-1}$$

22. (c) $\vec{J}_{\text{on } B} = m(\vec{v}_B)_f - m(\vec{v}_B)_i$

$$= m\,[2\sqrt{3}\cos 30°\,\hat{i} - 2\sqrt{3}\sin 30°\,\hat{j}] - m(0)$$

$$(\because v_{B_f} = 2\sqrt{3}\,\text{ms}^{-1})$$

$$= (3m\hat{i} - \sqrt{3}m\hat{j})\,\text{kg-ms}^{-1}$$

23. (b) Along the normal, $e = \dfrac{\text{Speed of separation}}{\text{Speed of approach}}$

or

$$0.5 = \frac{(v_B - v_A)_{\text{normal}}}{4\cos 60° - 0}$$

$$\Rightarrow \quad (v_B - v_A)_{\text{normal}} = \sqrt{3} \quad \ldots(i)$$

Along common normal, $p_i = p_f$

$$\Rightarrow \quad m4\cos 30° = mv_A + mv_B$$

$$2\sqrt{3} = (v_A + v_B)_{\text{normal}} \quad \ldots(ii)$$

From Eqs. (i) and (ii), we get

$$(v_B)_{\text{normal}} = \frac{3\sqrt{3}}{2}\,\text{ms}^{-1}$$

and velocity component of B along perpendicular to normal is zero.

$$\therefore \quad \vec{v}_B = \frac{3\sqrt{3}}{2}\cos 30°\,\hat{i} - \frac{3\sqrt{3}}{2}\sin 30°\,\hat{j}$$

$$= \left(\frac{9}{4}\hat{i} - \frac{3\sqrt{3}}{4}\hat{j}\right)\text{ms}^{-1}$$

24. (a)

(i) Mass of striking ball and mass of (the box + C) are same. So, after the impact box with the ball will move with velocity v.

Let us assume the ball C moves with initial velocity v and later with velocity v_R.

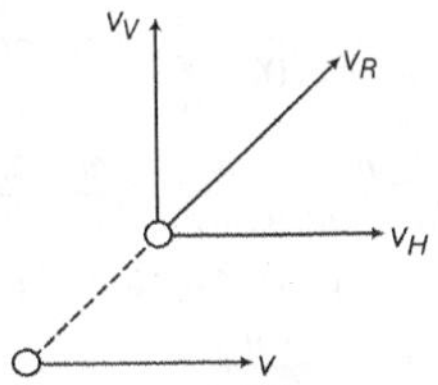

$$\frac{1}{2}(M + m)v^2 = \frac{1}{2}mv_R^2 + \frac{1}{2}Mv^2 + mgh$$

$$M = 2m \quad \text{(given)}$$

$$\frac{1}{2}3v^2 = \frac{1}{2}v_R^2 + \frac{1}{2}\frac{M}{m}v^2 + gh$$

$$3v^2 - \frac{M}{m}v^2 - 2gh = v_R^2$$

$$v_V^2 = v_R^2 - v_H^2 = v_R^2 - v^2 = 2v^2 - \frac{M}{m}v^2 - 2gh$$

At height h,

$$v_V = 0$$

$$\Rightarrow \quad h = \frac{v^2}{2g}\left(\frac{2m - M}{m}\right)$$

If $M = 2m$, $\quad h = 0$

(ii) By using similar arguments, we obtain if $M = 3m$

$$h = \frac{v^2}{2g}\left(\frac{3m - M}{m}\right)$$

So, for $M = 3m$, $h = 0$

(iii) If $M = k'm$, then $h = \dfrac{v^2}{2g}\left(\dfrac{k'm - M}{m}\right)$

So, for $M = k'm$, $h = 0$

Thus, for any value of M, ball C will not rise.

Hence, (i) → (p), (ii) → (p), (iii) → (s)

Revisal Problems (JEE Main)

1. (a) For horizontal part of length L,

$$x_1 = \frac{L}{2},\ y_1 = 0,\ m_1 = KL \qquad (\because m \propto L)$$

For vertical part of length $2L$,

$$x_2 = 0,\ y_2 = L,\ m_2 = K \times 2L$$

$$X_{COM} = \frac{m_1x_1 + m_2x_2}{m_1 + m_2} = \frac{KL\left(\frac{L}{2}\right) + 2KL\,(0)}{KL + 2KL} = \frac{L}{6} = \frac{1.2}{6} = 0.2\text{ m}$$

$$Y_{COM} = \frac{m_1y_1 + m_2y_2}{m_1 + m_2} = \frac{KL(0) + 2KL(L)}{KL + 2KL} = \frac{2}{3}L = \frac{2}{3}(1.2) = 0.8\text{ m}$$

$$\therefore\ \bar{Y}_{COM} = (0.2\,\hat{i} + 0.8\,\hat{j})\text{ m}$$

2. (b) A_1 = area of complete circle = πa^2

$$A_2 = \text{area of removed part} = \pi\left(\frac{a}{2}\right)^2 = \frac{\pi a^2}{4}$$

Coordinates of COM of original uncut lamina = $(x_1, y_1) = (0, 0)$

Coordinates of COM of removed portion = $(x_2, y_2) = \left(\frac{a}{2}, 0\right)$

For the residual part,

$$X_{COM} = \frac{A_1x_1 - A_2x_2}{A_1 - A_2} = \frac{\pi a^2(0) - \frac{\pi a^2}{4}\left(\frac{a}{2}\right)}{\pi a^2 - \frac{\pi a^2}{4}} = -\frac{a}{6}$$

Similarly, you will get $Y_{COM} = 0$

Hence, coordinates of COM of the uniform lamina = $\left(\frac{-a}{6}, 0\right)$

3. (d) Assuming origin at A, $(Y_{COM})_i = \frac{1(0) + 2(90)}{1+2} = 60$ m from A,

$$\bar{a}_{COM} = \frac{m_1\bar{a}_1 + m_2\bar{a}_2}{m_1 + m_2} = \frac{1(-10) + 2(-10)}{1+2}$$

$$= -10\text{ ms}^{-2} \text{ or } 10\text{ ms}^{-2} \text{ (downwards)}$$

$$\bar{u}_{COM} = \frac{m_1\bar{u}_1 + m_2\bar{u}_2}{m_1 + m_2}$$

$$= \frac{1\,(200) + 2(-55)}{1+2} = 30\text{ ms}^{-1}$$

$$v_{COM}^2 - u_{COM}^2 = 2a_{COM}\,s_{COM}$$

$$\Rightarrow \quad 0 - (30)^2 = 2(-10)s_{COM}$$

or $\quad s_{COM} = 45$ m

So, maximum height attained by COM from the ground is

$$h_{max} = 60 + 45 = 105\text{ m}$$

4. (d) Let $x = 0$ is located at the left end. As there is no external force acting on the system, position of COM does not change.

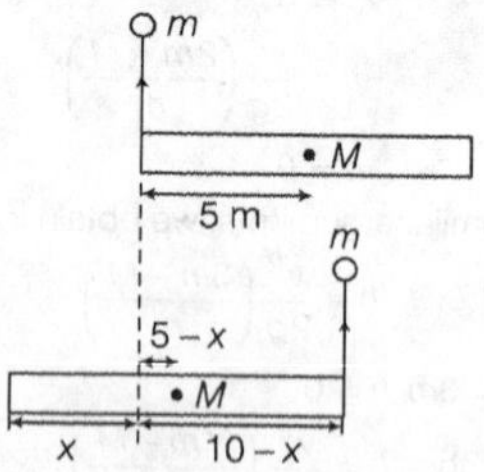

$$(X_i)_{COM} = (X_f)_{COM}$$

$$\frac{m(0) + M(5)}{m + M} = \frac{m(10 - x) + M(5 - x)}{m + M}$$

or $\quad 20(5) = 60(10 - x) + 20(5 - x)$

or $\quad x = 7.5$ m

5. (c) $X_{COM} = \dfrac{m\left(\frac{h}{2}\right) + (\alpha x)\frac{x}{2}}{m + \alpha x}$.

For X_{COM} to be minimum $\dfrac{dx_{COM}}{dx} = 0$

or $$(m + \alpha x)(\alpha x) - \left(m\frac{h}{2} + \frac{\alpha x^2}{2}\right)(\alpha) = 0$$

or $$m\alpha x + \alpha^2x^2 - \frac{mh\alpha}{2} - \frac{\alpha^2x^2}{2} = 0$$

$\Rightarrow$ $$\frac{\alpha^2x^2}{2} + m\alpha x - \frac{mh\alpha}{2} = 0$$

or $$\alpha x^2 + 2mx - mh = 0$$

$\Rightarrow$ $$x = \frac{-2m \pm \sqrt{4m^2 - 4(\alpha)(-mh)}}{2\alpha}$$

or $$x = \frac{\sqrt{m(m + h\alpha)} - m}{\alpha}$$

6. (a) From conservation of momentum,

$$mv_1 = Mv_2 \qquad \ldots\text{(i)}$$

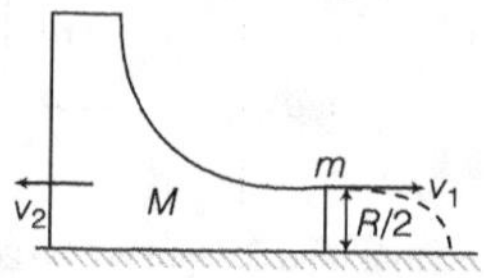

From conservation of energy,

$$mgR = \frac{1}{2}mv_1^2 + \frac{1}{2}Mv_2^2 \qquad \ldots\text{(ii)}$$

Using Eq. (i) in Eq. (ii), we get

$$mgR = \frac{1}{2}mv_1^2 + \frac{1}{2}M\left(\frac{m}{M}v_1\right)^2$$

$$2mgR = mv_1^2 + \frac{m^2}{M}v_1^2 \Rightarrow 2gR = v_1^2\left(1 + \frac{m}{M}\right)$$

$\Rightarrow$ $$v_1 = \sqrt{\frac{2MgR}{M + m}}$$

$\therefore$ $$v_2 = \frac{m}{M}\sqrt{\frac{2MgR}{M + m}}$$

Using concepts of projectile motion, $\dfrac{R}{2} = \dfrac{1}{2}gt^2$

$\Rightarrow$ $$t = \sqrt{\frac{R}{g}}$$

Desired distance, $s = (v_1 + v_2)t = \sqrt{\dfrac{2MgR}{M + m}}\left(1 + \dfrac{m}{M}\right)\sqrt{\dfrac{R}{g}}$

or $$s = R\sqrt{\frac{2(M + m)}{M}}$$

7. (d) From conservation of momentum, $mv_1 = Mv_2 \qquad \ldots$(i)

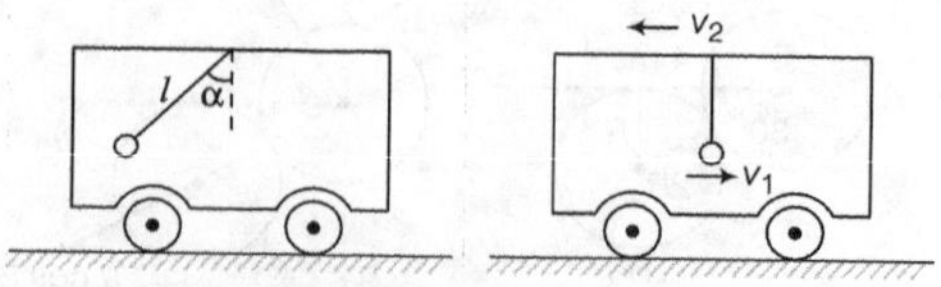

For conservation of mechanical energy,

$$mgl(1 - \cos\alpha) = \frac{1}{2}mv_1^2 + \frac{1}{2}Mv_2^2 \qquad \ldots\text{(ii)}$$

On solving Eqs. (i) and (ii), we get

$$v_2 = 2m\sin\left(\frac{\alpha}{2}\right)\sqrt{\frac{gl}{(M + m)M}}$$

8. (b) $e = \dfrac{v_1 + v_2}{v_0} \Rightarrow v_1 + v_2 = ev_0$

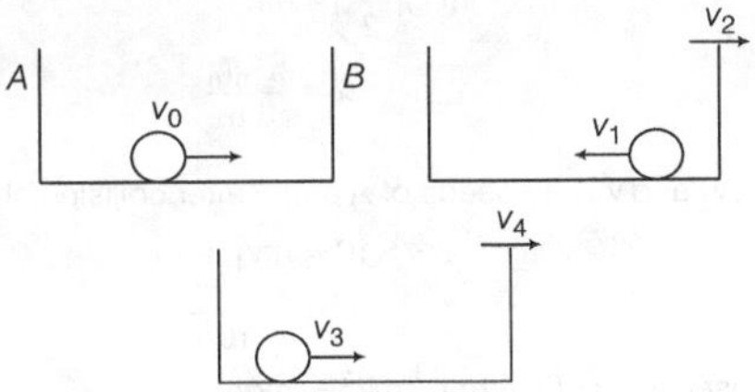

Also, $e = \dfrac{v_3 - v_4}{v_1 + v_2}$

$\Rightarrow \quad v_3 - v_4 = e(v_1 + v_2) = e^2 v_0$

The required time, $t = \dfrac{d}{v_0} + \dfrac{2d}{ev_0} + \dfrac{d}{e^2 v_0}$

$\Rightarrow \quad t = \dfrac{d}{v_0}\left(1 + \dfrac{2}{e} + \dfrac{1}{e^2}\right)$

9. (c) $v_1 = \sqrt{2gh}$

$v_2 = e\sqrt{2gh}$

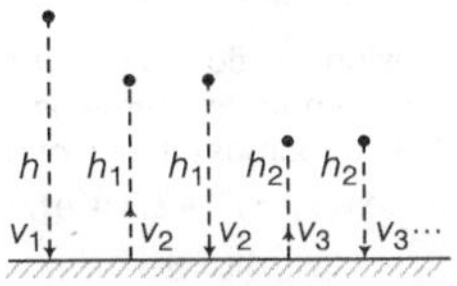

$v_3 = ev_2 = e^2\sqrt{2gh}$

$h_1 = \dfrac{v_2^2}{2g} = e^2 h$

$h_2 = \dfrac{v_3^2}{2g} = e^4 h$

Similarly, $h_3 = e^6 h$

Total distance travelled, $s = h + 2h_1 + 2h_2 + 2h_3 + \cdots$

$\therefore \quad s = h + 2e^2h + 2e^4h + 2e^6h + \cdots$

$= h + 2e^2h\,[1 + e^2 + e^4 + \cdots]$

or $s = h + 2e^2h\left[\dfrac{1}{1-e^2}\right] = h\left[1 + \dfrac{2e^2}{1-e^2}\right]$

$= h\left(\dfrac{1+e^2}{1-e^2}\right)$

10. (a) $t = \sqrt{\dfrac{2h}{g}}, t_1 = \sqrt{\dfrac{2h_1}{g}} = \sqrt{\dfrac{2e^2h}{g}} = e\sqrt{\dfrac{2h}{g}}$

$t_2 = \sqrt{\dfrac{2h_2}{g}} = e^2\sqrt{\dfrac{2h}{g}},$

$t_3 = \sqrt{\dfrac{2h_3}{g}} = e^3\sqrt{\dfrac{2h}{g}}, \ldots$

Total time before coming to rest, $T = t + 2t_1 + 2t_2 + 2t_3 + \cdots$

$T = \sqrt{\dfrac{2h}{g}} + 2e\sqrt{\dfrac{2h}{g}} + 2e^2\sqrt{\dfrac{2h}{g}} + 2e^3\sqrt{\dfrac{2h}{g}}$

$T = \sqrt{\dfrac{2h}{g}} + 2e\sqrt{\dfrac{2h}{g}}\,[1 + e + e^2 + \cdots]$

$= \sqrt{\dfrac{2h}{g}}\left[1 + 2e\left(\dfrac{1}{1-e}\right)\right]$

or $T = \sqrt{\dfrac{2h}{g}}\left(\dfrac{1+e}{1-e}\right)$

$\therefore$ Average speed $= \dfrac{\text{Total distance travelled}}{\text{Total time taken}}$

$= \dfrac{h\left(\dfrac{1+e^2}{1-e^2}\right)}{\sqrt{\dfrac{2h}{g}}\left(\dfrac{1+e}{1-e}\right)} = \sqrt{\dfrac{gh}{2}} \cdot \dfrac{(1+e^2)}{(1+e)^2}$

11. (c) Let us assume that the smaller ball is slightly separated from M just before M collides the floor.

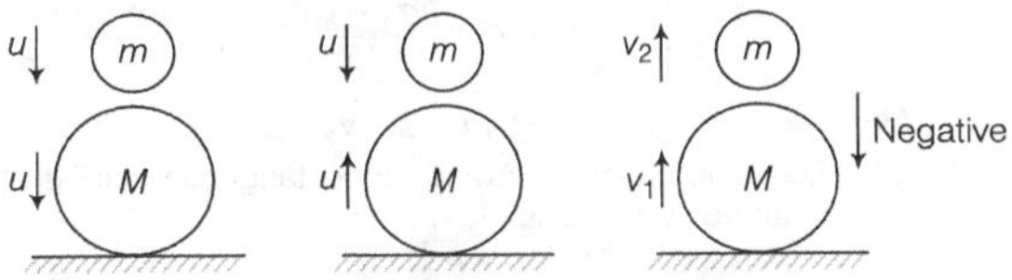

Just before collision of M with ground — Just after collision of M with ground — Just after the collision of M with m

$v_2 = \dfrac{m_2 - m_1}{m_1 + m_2} \cdot u_2 + \dfrac{2m_1}{m_1 + m_2} \cdot u_1 = \dfrac{m - M}{m + M}(-u) + \dfrac{2M}{m + M}(u)$

$\dfrac{m}{M} \approx 0$

$\therefore \quad v_2 \approx u + 2u = 3u$

But, $u = \sqrt{2gh}$, so $v_2 = 3\sqrt{2gh}$

$H = \dfrac{v_2^2}{2g} = \dfrac{9(2gh)}{2g} = 9h$

12. (d) When the particle hits the perfectly inelastic horizontal floor, the vertical component of velocity of the particle vanishes. Later, it moves with the horizontal velocity $u \cos \theta$ along the smooth horizontal floor. When the string becomes taut, the velocity component along the string vanishes due to impulse. At this instant speed of the particle is $v = (u \cos \theta) \cos \theta = u \cos^2 \theta$. The height raised by the particle later is

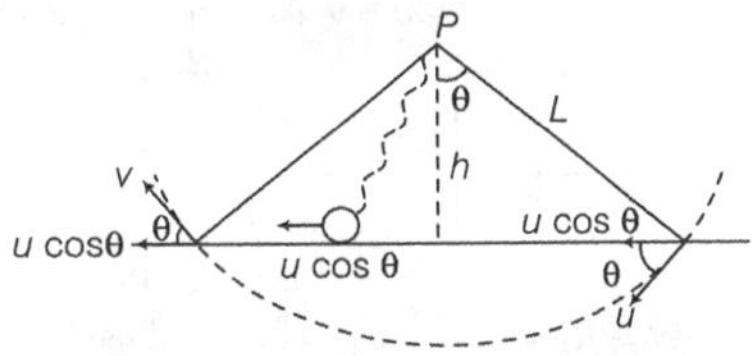

$h' = \dfrac{v^2}{2g} = \dfrac{u^2 \cos^4 \theta}{2g} = \dfrac{(2gh) \cos^4 \theta}{2g}$

or $h' = h\left(\dfrac{h}{L}\right)^4 = \dfrac{h^5}{L^4}$

13. (b) Impulse, $\mathbf{J} = m(\hat{i} + 3\hat{j}) - m(4\hat{i} - \hat{j}) = m(-3\hat{i} + 4\hat{j})$

To find the coefficient of restitution, we require the velocity components before and after impact, in the direction of impulse, i.e. along the line of impact.

The unit vector in the direction of $\mathbf{J}$ is $\dfrac{1}{5}(-3\hat{i} + 4\hat{j})$.

The magnitudes of the velocity components in this direction are

$(4\hat{i} - \hat{j}) \cdot \dfrac{1}{5}(-3\hat{i} + 4\hat{j}) = -\dfrac{16}{5}$ (just before impact)

$(\hat{i} + 3\hat{j}) \cdot \dfrac{1}{5}(-3\hat{i} + 4\hat{j}) = \dfrac{9}{5}$ (just after impact)

One is positive and the other negative because they are opposite to each other.

Now, $e = \dfrac{\text{Speed of separation}}{\text{Speed of approach}} = \dfrac{9/5}{16/5} = \dfrac{9}{16}$

14. (b)

	Mass m		Mass M	
	t-axis	n-axis	t-axis	n-axis
Before collision	v_1	v_2	0	0
After collision	v_1	v_3	0	v_4

$$v_3 = \left(\frac{m - eM}{m + M}\right)v_2 + \left(\frac{M + eM}{m + M}\right)(0)$$

$$\left[\text{Formula, } v_1 = \left(\frac{m_1 - em_2}{m_1 + m_2}\right)u_1 + \left(\frac{m_2 + em_2}{m_1 + m_2}\right)u_2\right]$$

As $\quad m = eM,\ v_3 = 0$

So, after collision m is moving along tangential axis and M is moving along normal axis.

$\therefore \quad x = \frac{\pi}{2}$

15. (a) $t = \frac{15000}{190} = 78.95$ s

$$v = u - gt + \left(\ln\frac{m_0}{m}\right)v_r$$

$$= 0 - (9.8)(78.95) + \left[\ln\left(\frac{21000}{6000}\right)\right](2800) \text{ or } v = 2735 \text{ ms}^{-1}$$

16. (b) Thrust force is

$$F_t = v_r\left(-\frac{dm}{dt}\right),\ v_r = 0 - v = -v$$

$$F_t = \frac{dm}{dt}\cdot v = \frac{dm}{dx}\cdot\frac{dx}{dt}\cdot v = \lambda v^2\ (\leftarrow)$$

$\therefore \quad F_{net} = F - \mu_k \lambda\, xg - \lambda v^2$

$$(\lambda x)\, a = F - \mu_k \lambda xg - \lambda v^2$$

$\Rightarrow \quad a = \frac{F}{\lambda x} - \mu_k g - \frac{v^2}{\lambda x}$

17. (b) $e = 1$

$\Rightarrow \quad v' - u = v + u \quad$ or $\quad v' = v + 2u$

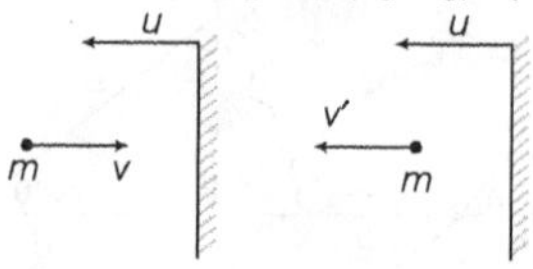

$$W = K_f - K_i = \frac{1}{2}m(v + 2u)^2 - \frac{1}{2}mv^2$$

$$= \frac{1}{2}m(v + 2u + v)(v + 2u - v) \text{ or } W = 2mu(u + v)$$

18. (c) $\int F dt = mu$

$\Rightarrow \quad \frac{1}{2}\pi(F_0)\left(\frac{T_0}{2}\right) = mu$

$\Rightarrow \quad u = \frac{\pi F_0 T_0}{4m}$

19. (b) Let v_1 and v_2 be speeds of A and B after collision, then

$$m \times 10 = mv_1 + mv_2 \quad \text{...(i)}$$

and $\quad e = \frac{v_2 - v_1}{10} \quad$ ...(ii)

By conservation of energy $\frac{1}{2}mv_2^2 = mgh$

$\Rightarrow \quad v_2^2 = 2 \times 10 \times 3.2$

$\therefore \quad v_2 = 8 \text{ ms}^{-1}$

Using this in Eq. (i), we get

$$v_1 = 2 \text{ ms}^{-1}$$

$$e \geq \frac{v_2 - v_1}{10}$$

$\Rightarrow \quad e \geq \frac{8-2}{10}$

$\Rightarrow \quad e \geq \frac{3}{5}$

20. (d) After each collision, vertical component becomes e times, whereas horizontal component remains same. So, after each collision time of flight becomes e times of previous one.

$\therefore$ Horizontal displacement $= R + eR + e^2R = (1 + e + e^2)R$

$$= \frac{u^2 \sin 2\theta}{g}(1 + e + e^2)$$

21. (a) N has no component along the surface of the wedge.

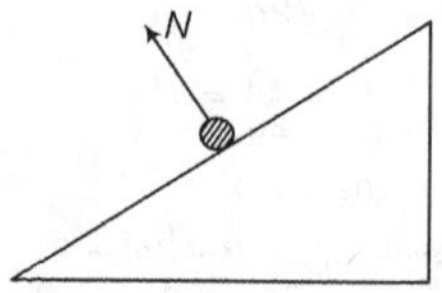

22. (c) $v_2 = \left(\frac{2m_1}{m_1 + m_2}\right)u_1 = \frac{2u_1}{\left(1 + \frac{m_2}{m_1}\right)}$

23. (d) Net external force is zero.

$\Rightarrow \quad \vec{a}_{COM} = 0$

So, COM may be at rest or may be moving with constant velocity.

Revisal Problems (JEE Advanced)

1. (a) All the surfaces are smooth. So, no net external force is acting on the system in horizontal direction. So, x-coordinate of COM of the system does not change.

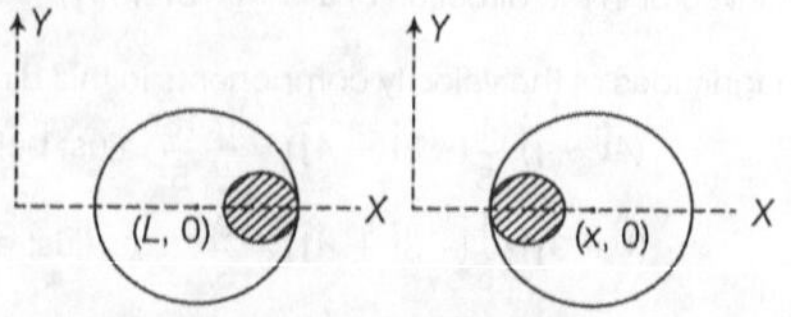

$$(X_{COM})_i = \frac{m_1x_1 + m_2x_2}{m_1 + m_2} = \frac{(4M)L + M(L + 5R)}{4M + M} = L + R \quad \text{...(i)}$$

$$(X_{COM})_f = \frac{(4M)(x) + M(x - 5R)}{4M + M} = x - R \quad \text{...(ii)}$$

Equating Eqs. (i) and (ii), we get

$$x = L + 2R$$

So, the coordinates of the centre of the large sphere $= (L + 2R, 0)$

2. (c)

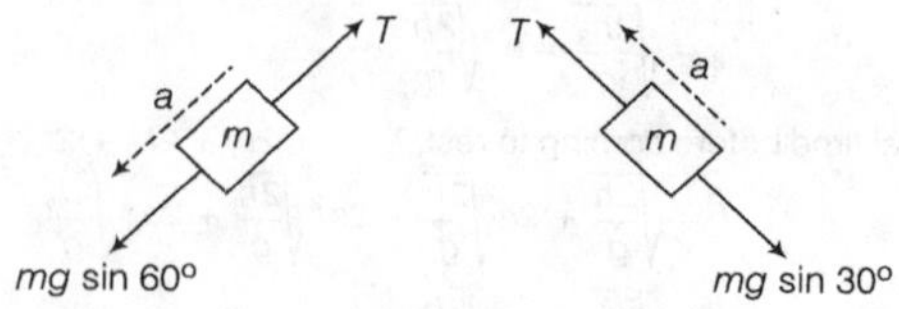

$$mg\frac{\sqrt{3}}{2} - T = ma \quad \text{...(i)}$$

$$T - \frac{mg}{2} = ma \quad \text{...(ii)}$$

On adding Eqs. (i) and (ii), we get

$$mg\left(\frac{\sqrt{3} - 1}{2}\right) = 2ma$$

$\therefore \quad a = g\left(\frac{\sqrt{3} - 1}{4}\right)$

$$\bar{a}_{COM} = \frac{m_1\bar{a}_1 + m_2\bar{a}_2}{m_1 + m_2}$$

$$= \frac{m(\bar{a}_1 + \bar{a}_2)}{2m} = \frac{\bar{a}_1 + \bar{a}_2}{2}$$

$$a_{COM} = \frac{1}{2}\sqrt{a^2 + a^2}$$

$$= \frac{a}{\sqrt{2}} = g\left(\frac{\sqrt{3}-1}{4\sqrt{2}}\right)$$

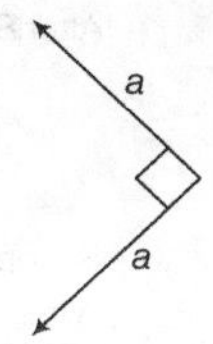

3. (b) $y = \sqrt{L^2 - x^2}$

$$\Rightarrow \quad \frac{dy}{dt} = -\frac{x}{\sqrt{L^2 - x^2}} \cdot \frac{dx}{dt}$$

or $$v_y = -\frac{3}{\sqrt{5^2 - 3^2}}(2) = -\frac{3}{2}\,\text{ms}^{-1}$$

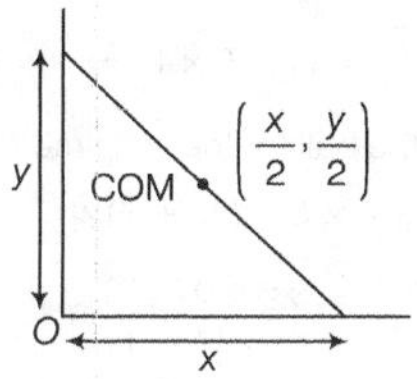

At any moment coordinates of COM are $\left(\frac{x}{2}, \frac{y}{2}\right)$.

Now, $$(v_{COM})_x = \frac{d}{dt}\left(\frac{x}{2}\right) = \frac{1}{2}\cdot\frac{dx}{dt} = \frac{1}{2}(2) = 1\,\text{ms}^{-1}$$

$$(v_{COM})_y = \frac{d}{dt}\left(\frac{y}{2}\right) = \frac{1}{2}\cdot\frac{dy}{dt} = \frac{1}{2}\left(\frac{3}{2}\right) = \frac{3}{4}\,\text{ms}^{-1}$$

$$v_{COM} = \sqrt{(v_{COM})_x^2 + (v_{COM})_y^2} = \sqrt{1 + \frac{9}{16}} = \frac{5}{4}$$

$$= 1.25\,\text{ms}^{-1}$$

4. (d) Acceleration of m_1 vertically downwards will be

$$a = g\sin\theta\,(\sin\theta) = g\sin^2\theta$$

$$\therefore \quad a_{COM} = \frac{m_1(g\sin^2\theta) + m_2(0)}{m_1 + m_2} = \frac{m_1 g\sin^2\theta}{m_1 + m_2}$$

For $(m_1 + m_2)$ system, from Newton's second law, we have

$(m_1 + m_2)g - N = (m_1 + m_2)\,a_{COM} = m_1 g\sin^2\theta$

$$\therefore \quad N = (m_1 + m_2)g - m_1 g\sin^2\theta$$

$$= m_1 g\cos^2\theta + m_2 g$$

or $$N = (m_1\cos^2\theta + m_2)g$$

5. (b) Let the velocities of the block and the plank, when the block leaves the spring be u and v, respectively. From energy conservation.

$$\frac{1}{2}kx^2 = \frac{1}{2}mu^2 + \frac{1}{2}Mv^2$$

$$\Rightarrow \quad \frac{1}{2}(100)(1)^2 = \frac{1}{2}(1)u^2 + \frac{1}{2}(5)v^2$$

$$100 = u^2 + 5v^2 \quad \ldots\text{(i)}$$

From momentum conservation,

$$mu + Mv = 0$$

$$\Rightarrow \quad u = -5v \quad \ldots\text{(ii)}$$

From Eqs. (i) and (ii), we get

$$v = \sqrt{\frac{10}{3}}\,\text{ms}^{-1}$$

From this moment, till the block falls, both plank and block keep their velocities constant.

Thus, when block falls, speed of plank $= \sqrt{\frac{10}{3}}\,\text{ms}^{-1}$

6. (a) $h = \frac{1}{2}gt^2$ (for horizontal projectile)

$$\therefore \quad t = \sqrt{\frac{2h}{g}}$$

Let v_1 be the horizontal velocity of the ball after collision.

$$\therefore \quad v_1 = \frac{d}{t} = d\sqrt{\frac{g}{2h}}$$

If v_2 is the velocity of wedge after collision, then $Mv_2 = mv_1$

$$v_2 = \frac{m}{M}v_1 = \frac{md}{M}\sqrt{\frac{g}{2h}}$$

From conservation of energy, $\frac{1}{2}kA^2 = \frac{1}{2}Mv_2^2$

So, $$\text{amplitude } A = v_2\sqrt{\frac{M}{k}} = \frac{md}{M}\sqrt{\frac{gM}{2hk}}$$

7. (d) As C moves, the string BC is slack and it gets taut and impulse acts along BC, when BC makes an angle 60° with AB as shown below.

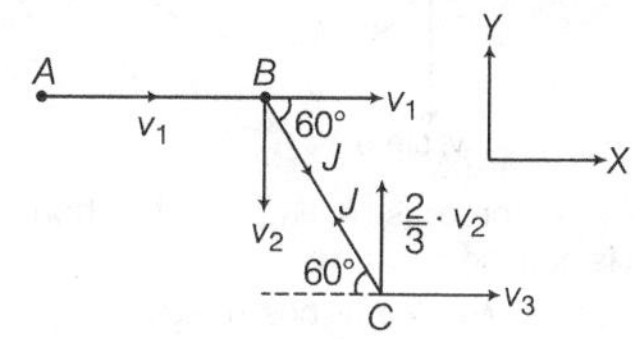

For BC system, from conservation of moment in Y-direction, we have

$$0 = 3m\,(v_C)_y + 2m(v_B)_y \Rightarrow (v_C)_y = -\frac{2}{3}(v_B)_y$$

If v_1 and $-v_2$ be x and y-components of final velocity of B, then

$$(v_B)_y = -v_2 \Rightarrow (v_C)_y = \frac{2}{3}v_2$$

For $(A + B + C)$ system, from conservation of momentum along X-direction, we have

$$3mv_0 = mv_1 + 2mv_1 + 3mv_3$$

$$\Rightarrow \quad v_0 = v_1 + v_3 \quad \ldots\text{(i)}$$

For particles B and C, the components of velocities along BC will be equal,

$$v_1\cos 60° + v_2\cos 30° = -\frac{2}{3}v_2\cos 30° + v_3\cos 60°$$

$$\Rightarrow \quad \frac{v_1}{2} + \frac{\sqrt{3}}{2}v_2 = \frac{v_3}{2} - \frac{v_2}{\sqrt{3}}$$

or $$v_1 - v_3 + \frac{5}{\sqrt{3}}v_2 = 0 \quad \ldots\text{(ii)}$$

Since, impulse, J = change in momentum, we can write

For C along X-axis, $J\cos 60° = 3\,m\,(v_0 - v_3)$

$$\Rightarrow \quad \frac{J}{2} = 3m\,(v_0 - v_3) \quad \ldots\text{(iii)}$$

For C along Y-axis, $J\cos 30° = 3\,m\left(\frac{2}{3}\cdot v_2\right) = 2mv_2$

or $$\frac{\sqrt{3}}{2}J = 2mv_2 \quad \ldots\text{(iv)}$$

On solving Eqs. (i), (ii), (iii) and (iv), we get

$$v_1 = \frac{2}{19}v_0$$

8. (a) Initial velocity is $u = 0$

Exhaust velocity is $v_r = u$

Initial mass is $m_0 = M_i$

The mass at the end of first stage is $m = \mu$

We know that $$v = u - gt + v_r \ln\frac{m_0}{m}$$

$$\therefore \quad v_1 = 0 - 0 + u\ln\frac{M_i}{\mu} \text{ or } v_1 = u\ln\frac{M_i}{\mu}$$

9. (c) Initial velocity for second stage is $u = v_1 = u \ln \frac{M_i}{\mu}$

Initial mass is $m_0 = \mu - m$

Exhaust velocity is $v_r = u$

Final mass is $m = M_f$

$$\therefore \quad v_2 = u \ln \frac{M_i}{\mu} - 0 + u \ln \frac{\mu - m}{M_f} = u \ln \frac{M_i(\mu - m)}{\mu M_f}$$

10. (b) Let v_r be the relative velocity of the washer relative to the centre of the hoop. In the horizontal direction, the component of velocity of washer is $(v_r \cos \phi - u)$.

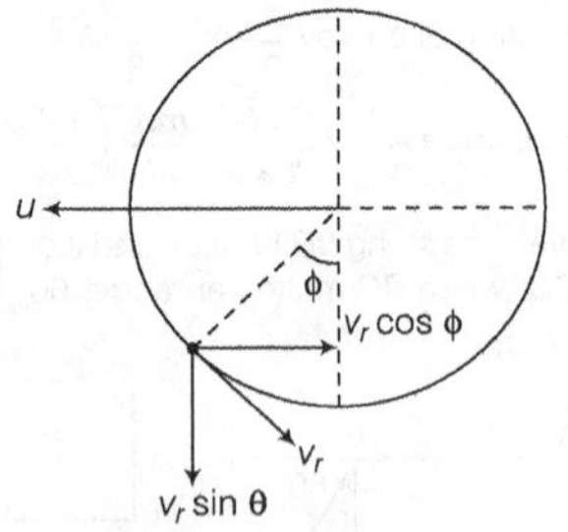

As no external force is acting in the horizontal direction, momentum is conserved.

$$Mu = m(v_r \cos \phi - u)$$

$$\Rightarrow \quad v_r = \frac{u}{\cos \phi}\left(\frac{M + m}{m}\right) \quad \ldots \text{(i)}$$

From conservation of mechanical energy

$$mgr\,(1 + \cos \phi) = \frac{1}{2}Mu^2 + \frac{1}{2}m\left[\sqrt{(v_r \cos \phi - u)^2 + (v_r \sin \phi)^2}\right]^2$$

Note *The vertical component of velocity of washer is $v_r \sin \phi$.*

$$\therefore \quad mgr(1 + \cos \phi) = \frac{1}{2}Mu^2 + \frac{1}{2}m(v_r^2 + u^2 - 2u\, v_r \cos \phi) \quad \ldots \text{(ii)}$$

Using Eq. (i) in Eq. (ii) will yield

$$u = m \cos \phi \sqrt{\frac{2gr\,(1 + \cos \phi)}{(M + m)\,(M + m \sin^2 \phi)}}$$

11. (b) In the figure $\mathbf{v}_{12}$ = velocity of ball w.r.t. wedge before collision and $\mathbf{v}'_{12}$ = velocity of ball w.r.t. wedge after collision.

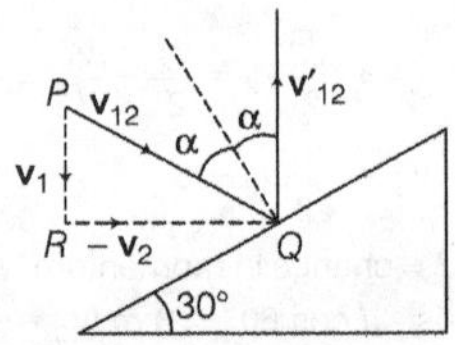

In elastic collision $\mathbf{v}_{12}$ and $\mathbf{v}'_{12}$ will make equal angle (say α) with the normal to the plane. We can show that $\alpha = 30°$

$$\therefore \quad \angle PQR = 30°$$

$$\Rightarrow \quad \tan 30° = \frac{v_1}{v_2} \Rightarrow \frac{v_1}{v_2} = \frac{1}{\sqrt{3}}$$

12. (b,c,d) Initially, when B is at the top of A, both are at rest. Finally when B reaches its highest point on A, both A and B will be moving with some common velocity. So, B cannot reach the top of A. So, option (a) is incorrect.

In the horizontal direction, no external force is acting on $(A + B)$ system. So, option (b) is correct.

After collision with the wall, the velocity of B reverses with no change in magnitude. Let the velocity of A is v_1 leftwards.

Before collision with the wall,

$$0 = mv + Mv_1$$

$$\text{or} \quad v_1 = -\frac{m}{M}v = \frac{m}{M}v \quad \text{(leftwards)}$$

After B collides with the wall, $v_{COM} = \frac{mv + Mv_1}{m + M}$

$$v_{COM} = \frac{mv + M\left(\frac{m}{M}\right)v}{M + m} = \frac{2\,mv}{M + m}$$

So, option (c) is correct.

When B reaches its highest part on A from momentum conservation.

$$mv + M\left(\frac{m}{M}\right)v = (m + M)v' \quad \text{or} \quad v' = \frac{2mv}{M + m}$$

So, option (d) is correct.

13. (a,c,d) When P throws the ball for the first time,

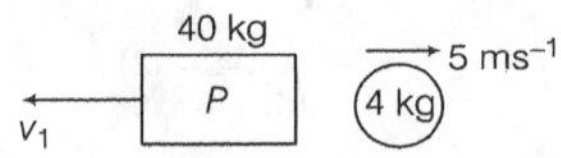

$$40v_1 = 5 \times 4 \Rightarrow v_1 = \frac{1}{2}\text{ ms}^{-1}$$

When P receives the ball for the first time,

$$40 \times \frac{1}{2} + 4 \times 5 = (40 + 4)v_1'$$

$$v_1' = \frac{40}{44} = \frac{10}{11}\text{ ms}^{-1}$$

When P throws the ball second time,

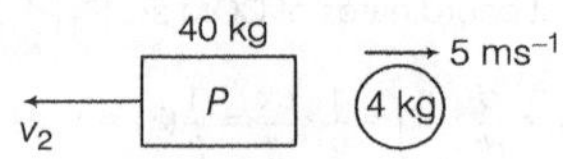

$$44 \times v_1' = 40 \times v_2 - 4 \times 5$$

$$44 \times \frac{10}{11} = 40v_2 - 20$$

$$\Rightarrow \quad v_2 = \frac{3}{2}\text{ ms}^{-1}$$

So, when P throws the ball for sixth time, his speed becomes

$$v_6 = \frac{2(6) - 1}{2} = 5.5\text{ ms}^{-1}$$

v_6 is greater than speed of ball. So, P can throw only for 6 times.

As no external force is acting on the system of $(P + Q + \text{ball})$, the COM remains stationary.

14. (b,c,d) In this collision, as masses of both are equal, the velocities along common normal direction get interchanged and along common tangent direction remains unaltered.

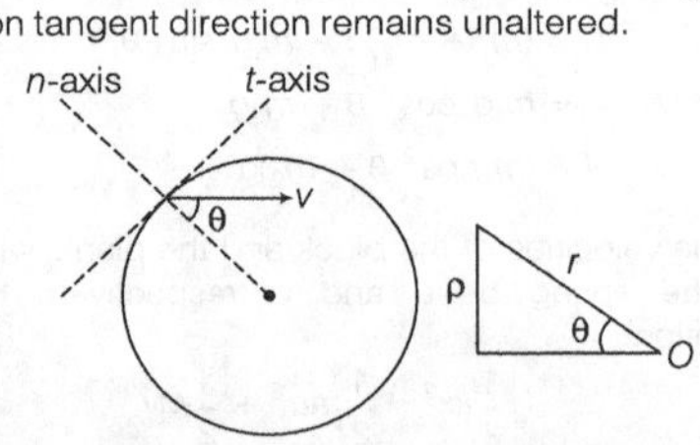

Speed of sphere will be $v \cos \theta = v\frac{\sqrt{r^2 - \rho^2}}{r}$

Speed of particle will be $v \sin \theta = \frac{v\rho}{r}$

Both these velocities will be perpendicular to each other.

15. (a,d) Here, mechanical energy is conserved.

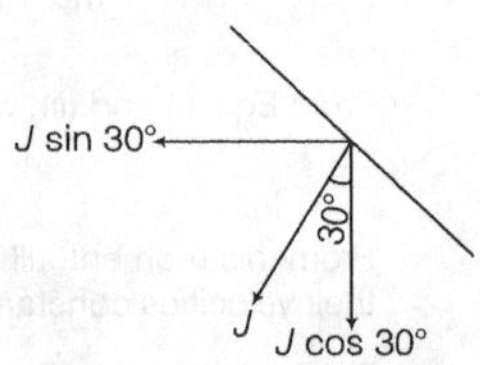

$$\therefore \quad \frac{1}{2}(1)(10)^2 = \frac{1}{2}(4)(4)^2 + \frac{1}{2}(1)v^2$$

$$\Rightarrow \quad v = 6\text{ ms}^{-1}$$

For wedge,

Impulse = Change in momentum

$$\therefore \quad J \sin 30° = 4(4) - 4(0)$$

$$J = 32\text{ N-s}$$

16. (c) Let the next collision happens after a time t', then

(as $a_{B_2} = 0.3 \times 10 = 3\ ms^{-2}$)

$$102t' - \frac{1}{2} 3t'^2 = v_t t' = 3t'$$

$$\therefore \quad t' = 6\ s$$

Say $v_{B_2} = 0$ after a time t, then

$$0 = 12 - 3t$$

$$\Rightarrow \quad t = 4\ s$$

As $t' > t$, B_2 comes to rest before second collision.

In $t = 4$ s, $\quad s_{B_2} = 12(4) - \frac{1}{2}(3)(4)^2 = 24\ m$

$$\therefore \quad s_{B_1} = 24\ m = 3 \times t''$$

$$\Rightarrow \quad t'' = 8\ s$$

17. (d) From previous discussion, velocity of blocks after collision will be v_{B_1} and $4v_{B_1}$, where v_{B_1} is the velocity of B_1 after collision.

So, time after which block B_2 comes to rest is obtained from

$$0 = 4v_{B_1} - 0.3 \times 10t$$

$$\Rightarrow \quad t = \frac{4v_{B_1}}{3}$$

Distance travelled by $B_2 = (4v_{B_1})\left(\frac{4v_{B_1}}{3}\right) - \frac{1}{2}(3)\left(\frac{4v_{B_1}}{3}\right)^2$

$$= \frac{16v_{B_1}^2}{3} - \frac{16v_{B_1}^2}{6} = \frac{16v_{B_1}^2}{6} = \frac{8v_{B_1}^2}{3}$$

Time taken by B_1 to collide with $B_2 = \frac{8v_{B_1}^2/3}{v_{B_1}} = \frac{8v_{B_1}}{3}$

Also, after each collision velocity of B_1 will be reduced to one-third of its velocity before collision.

Total time $= \frac{8}{3}(v_0/3) + \frac{8}{3}(v_0/9) + \frac{8}{3}(v_0/27) + \frac{8}{3}(v_0/81) + \ldots$ ...(i)

So, infinite collisions are possible.

18. (b) In the given interval,

$$s_{B_2} - s_{B_1} = x \text{ (say)}$$

Then, $\quad x = \left(12t - \frac{1}{2}3t^2\right) - 3t \quad$ or $\quad x = 9t - \frac{3}{2}t^2$

For x to be maximum, $\frac{dx}{dt} = 9 - 3t = 0$ or $t = \frac{9}{3} = 3\ s$

$$\therefore \quad x_{max} = 9(3) - \frac{3}{2}(3)^2 = 27 - \frac{27}{2} = \frac{27}{2} = 13.5\ m$$

Explanations for Q. Nos. (19-21)

According to the Bohr's model, the energy levels of the hydrogen atom are given by the formula

$$E_n = \frac{E_1}{n^2}$$

where, $n = 1, 2, 3, \ldots$. The ground state corresponds to $n = 1$, while the lowest excited state corresponds to $n = 2$. Thus, the smallest energy necessary for excitation of the hydrogen atom is

$$\Delta E = E_2 - E_1 = E_1\left(1 - \frac{1}{4}\right) = \frac{3}{4}E_1$$

During an inelastic collision a part of kinetic energy of the colliding particles is converted into their internal energy. The internal energy of the system of two hydrogen atoms considered in the problem cannot be changed by less than ΔE. It means that if the kinetic energy of the colliding atoms with respect to their centre of mass is less than ΔE, then the collision must be an elastic one. The value of v_0 can be found by considering the critical case, when the kinetic energy of the colliding atoms is equal to the smallest energy of excitation. With respect to the centre of mass, the atoms move in opposite direction with velocities $\frac{1}{2}v_0$.

Thus, $\quad \frac{1}{2}m_H\left(\frac{1}{2}v_0\right)^2 + \frac{1}{2}m_H\left(\frac{1}{2}v_0\right)^2 = \frac{3}{4}E_1$

and $\quad v_0 = \sqrt{\frac{3E_1}{m_H}} \quad (\approx 6.26 \times 10^4\ ms^{-1})$

Consider the case when $v = v_0$. The collision may be elastic or inelastic. When the collision is elastic, the atoms remain in their ground states and do not emit radiation. Radiation is possible only when the collision is inelastic. Of course, only the atom excited in the collision can emit the radiation.

19. (c) **20.** (a) **21.** (c)

22. (a) $v_{COM} = \frac{2(2) + 2(0)}{2 + 2} = 1\ ms^{-1}$

Total initial energy $= \frac{1}{2} \times 2 \times 2^2 = 4$ J

At maximum compression, $2v + 2v = 2(2) + 2(0)$

$$v = 1\ ms^{-1}$$

$$\frac{1}{2}kx_{max}^2 + \frac{1}{2}(2 + 2)v^2 = \frac{1}{2} \times 2 \times 2^2$$

$$\Rightarrow \quad \frac{1}{2} \times 4 \times x_{max}^2 = 2$$

$$\therefore \quad x_{max} = 1\ m$$

$$\frac{1}{2}k\,x_{max}^2 = \frac{1}{2} \times 4 \times (1)^2 = 2\ J$$

Hence, (i) → (q), (ii) → (q), (iii) → (q), (iv) → (p)

23. (8) $H = \frac{u^2 \sin^2\theta}{2g} = \frac{(20)^2 \times (3/4)}{2(10)} = 15\ m$

i.e. the shell strikes the ball at highest point of its trajectory. Velocity of (ball + shell) just after collision, using conservation of momentum is

$$v = \frac{u\cos 60^\circ}{2} = \frac{20(1/2)}{2} = 5\ ms^{-1}$$

Let v' be the velocity of the trolley, when combined reaches highest point. Then,

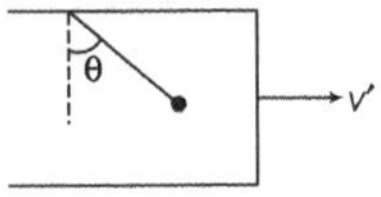

$$2 \times 5 = (2 + 18)v'$$

$$v' = \frac{1}{2}\ ms^{-1}$$

By conservation of mechanical energy, we get

$$\frac{1}{2}(2)(5)^2 - \frac{1}{2}(2 + 18)\left(\frac{1}{2}\right)^2 = (2)(10)(1)(1 - \cos\theta)$$

$$25 - 2.5 = 20(1 - \cos\theta)$$

$$1 - \cos\theta = 1.125$$

$$\cos\theta = 0.125 = \frac{1}{8}$$

$$\therefore \quad \alpha = 8$$

24. (6)

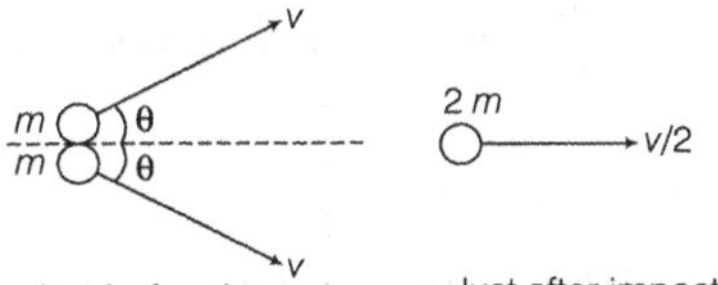

Just before impact Just after impact

$$mv\cos\theta + mv\cos\theta = (m + m)\frac{v}{2}$$

$$2mv\cos\theta = mv$$

$$\cos\theta = \frac{1}{2}$$

$$\therefore \quad \theta = 60^\circ$$

So, angle between the two before collision is

$$\phi = 2\theta = 120^\circ$$

$$\Rightarrow \quad \frac{\phi}{20} = 6^\circ$$

JEE Main & AIEEE Archive

1. (d) By conservation of energy,

Initial total energy = Final total energy

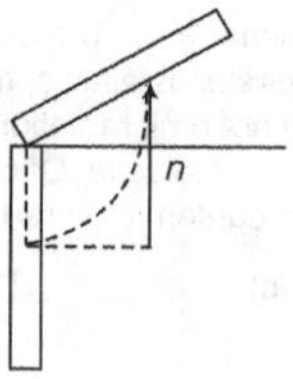

$$TE_i = TE_f$$

$$\frac{1}{2}I\omega^2 = mgh$$

$$\frac{1}{2} \times \frac{1}{3} ml^2\omega^2 = mgh$$

$$\Rightarrow \quad h = \frac{1}{6} \cdot \frac{l^2\omega^2}{g}$$

2. (a) $$x_{CM} = \frac{\int x dm}{\int dm}$$

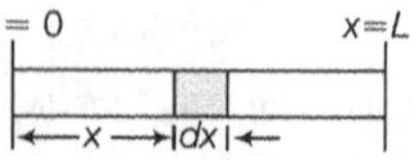

If $n = 0$,

Then, $$x_{CM} = \frac{L}{2}$$

As n increases, the centre of mass shift away from $x = \frac{L}{2}$ towards $x = L$ which only option (a) is satisfying.

Alternately, you can use basic concept.

$$x_{CM} = \frac{\int_0^L k\left(\frac{x}{L}\right)^n \times x dx}{\int_0^L k\left(\frac{x}{L}\right)^n dx}$$

$$= L\left[\frac{n+1}{n+2}\right]$$

3. (c) To keep the centre of mass at the same position, velocity of centre of mass is zero, so

$$\frac{m_1\mathbf{v}_1 + m_2\mathbf{v}_2}{m_1 + m_2} = 0$$

where, $\mathbf{v}_1$ and $\mathbf{v}_2$ are the velocities of particles 1 and 2, respectively.

$$\Rightarrow \quad m_1\frac{d\mathbf{r}_1}{dt} + m_2\frac{d\mathbf{r}_2}{dt} = 0 \quad \left(\because \mathbf{v}_1 = \frac{d\mathbf{r}_1}{dt} \text{ and } \mathbf{v}_2 = \frac{d\mathbf{r}_2}{dt}\right)$$

$$\Rightarrow \quad m_1 d\mathbf{r}_1 + m_2 d\mathbf{r}_2 = 0$$

where, $d\mathbf{r}_1$ and $d\mathbf{r}_2$ represent the change in displacement of particles.

Let 2nd particle has been displaced by distance x.

$$\Rightarrow \quad m_1(d) + m_2(x) = 0$$

or $$x = -\frac{m_1 d}{m_2}$$

Alternate Method

$$x_{CM} = \frac{m_1 x_1 + m_2 x_2}{m_1 + m_2} \quad \ldots(i)$$

After moving m_1 through a distance d towards centre of mass and to keep the position of centre of mass unchanged, let d' be the shift of m_2.

$$x_{CM} = \frac{m_1(x_1 - d) + m_2(x_2 + d')}{m_1 + m_2} \quad \ldots(ii)$$

From Eqs. (i) and (ii), we get

$$\frac{m_1x_1 + m_2x_2}{m_1 + m_2} = \frac{m_1(x_1 - d) + m_2(x_2 + d)}{m_1 + m_2}$$

$$\Rightarrow \quad -m_1 d + m_2 d' = 0$$

$$\therefore \quad d' = \frac{m_1}{m_2} d$$

4. (c) For pure translatory motion, net torque about centre of mass should be zero.

Thus, **F** is applied at centre of mass of system,

$$OP = \frac{m_1 \times 0 + m_2 \times l}{m_1 + m_2}$$

where, m_1 and m_2 are masses of horizontal and vertical section of the object. Assuming object is uniform,

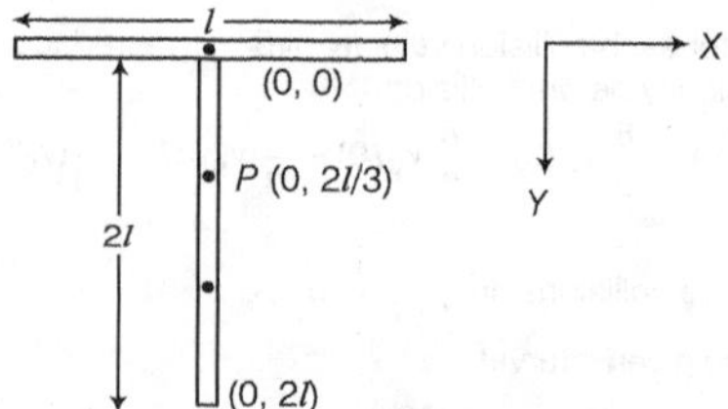

$$m_2 = 2m_1$$

$$\Rightarrow \quad OP = \frac{2l}{3}$$

$$PC = \left(l - \frac{2l}{3} + l\right)$$

$$= \left(2l - \frac{2l}{3}\right) = \frac{4l}{3}$$

5. (c) Let mass of each body be m. Their motion is represented as shown in figure.

From $$\mathbf{v}_{CM} = \frac{m_1\mathbf{v}_1 + m_2\mathbf{v}_2}{m_1 + m_2}$$

$$\mathbf{v}_{CM} = \frac{m \times 2v - mv}{m + m} = \frac{v}{2}$$

(the direction of motion of first particle is taken as positive.)

So, velocity of centre of mass of the system is $\frac{v}{2}$ in the direction of motion of particle having larger speed.

JEE Advanced & IIT JEE Archive

1. (4) After elastic collision,

$$v'_A = \left(\frac{m-2m}{m+2m}\right)(9) + \frac{2(2m)}{m+2m}(0)$$
$$= -3\ \text{ms}^{-1}$$

Now, from conservation of linear momentum after all collisions are complete,

$$m(+9\ \text{ms}^{-1}) = m(-3\ \text{ms}^{-1}) + 3m(v_C)$$

or $$v_C = 4\ \text{ms}^{-1}$$

2. (c) At first collision, one particle having speed $2v$ will rotate 240° $\left(\text{or } \frac{4\pi}{3}\right)$ while other particle having speed v will rotate 120° $\left(\text{or } \frac{2\pi}{3}\right)$. At first collision, they will exchange their velocities. Now as shown in figure, after two collisions they will again reach at point A.

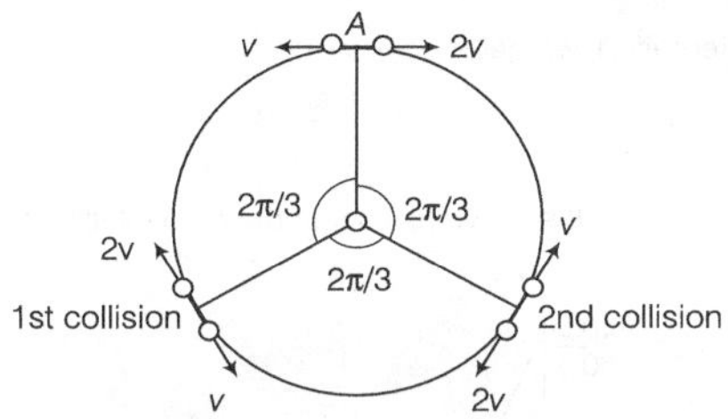

3. (a) $y_{CM} = \frac{m_1y_1 + m_2y_2 + m_3y_3 + m_4y_4 + m_5y_5}{m_1 + m_2 + m_3 + m_4 + m_5}$

$$= \frac{(6m)(0) + (m)(a) + m(a) + m(0) + m(-a)}{6m + m + m + m + m}$$
$$= \frac{a}{10}$$

4. (a) On a system of particles if

$$\Sigma \mathbf{F}_{\text{ext}} = 0$$

Then, $\mathbf{p}_{\text{system}}$ = constant

No other conclusions can be drawn.

5. (b) Between A and B, height fallen by block,

$$h_1 = \sqrt{3}\tan 60° = 3\ \text{m}$$

∴ Speed of block just before striking the second incline,

$$v_1 = \sqrt{2gh_1}$$
$$= \sqrt{2\times 10\times 3}$$
$$= \sqrt{60}\ \text{ms}^{-1}$$

In perfectly inelastic collision, component of v_1 perpendicular to BC will become zero, while component of v_1 parallel to BC will remain unchanged.

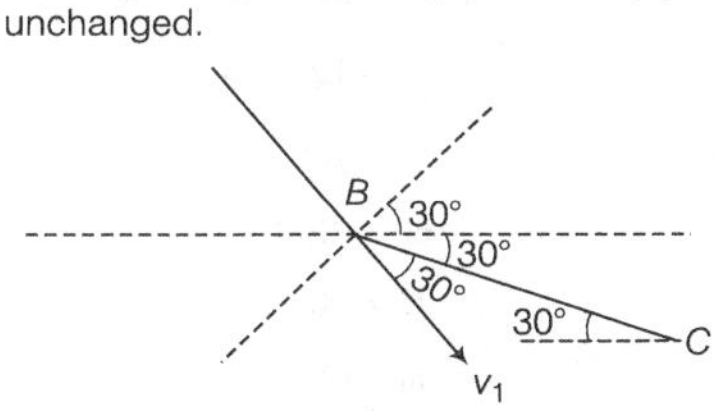

∴ Speed of block B immediately after it strikes the second incline is

$$v_2 = \text{component of } v_1 \text{ along } BC = v_1\cos 30°$$
$$= (\sqrt{60})\left(\frac{\sqrt{3}}{2}\right) = \sqrt{45}\ \text{ms}^{-1}$$

6. (b) Height fallen by the block from B to C

$$h_2 = 3\sqrt{3}\tan 30° = 3\ \text{m}$$

Let v_3 be the speed of block, at point C, just before it leaves the second incline, then

$$v_3 = \sqrt{v_2^2 + 2gh_2} = \sqrt{45 + 2\times 10\times 3} = \sqrt{105}\ \text{ms}^{-1}$$

7. (c) In elastic collision, compound of v_1 parallel to BC will remain unchanged, while component perpendicular to BC will remain unchanged in magnitude but its direction will be reversed.

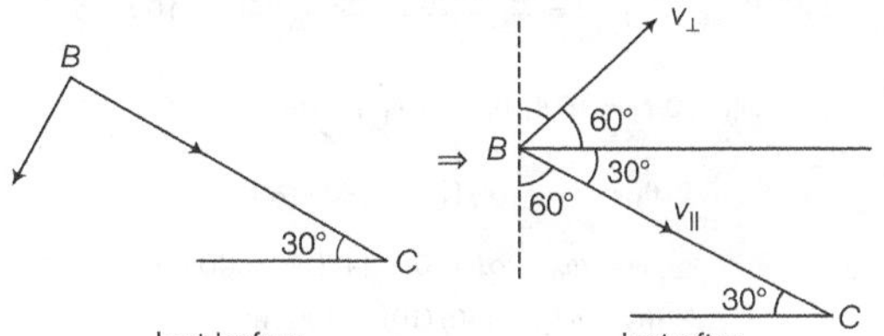

$$v_{||} = v_1\cos 30° = (\sqrt{60})\left(\frac{\sqrt{3}}{2}\right)$$
$$= \sqrt{45}\ \text{ms}^{-1}$$
$$v_\perp = v_1\sin 30° = (\sqrt{60})\left(\frac{1}{2}\right)$$
$$= \sqrt{15}\ \text{ms}^{-1}$$

Now, vertical component of velocity of block

$$v = v_\perp\cos 30° - v_{||}\cos 60°$$
$$= (\sqrt{15})\left(\frac{\sqrt{3}}{2}\right) - (\sqrt{45})\left(\frac{1}{2}\right) = 0$$

8. (a,d) Initial momentum of the system $\mathbf{p}_1 + \mathbf{p}_2 = 0$

∴ Final momentum $\mathbf{p}_1' + \mathbf{p}_2'$ should also be zero.

Option (b) is allowed because if we put $c_1 = -c_2 \neq 0$, $\mathbf{p}_1' + \mathbf{p}_2'$ will be zero. Similarly, we can check other options.

9. (d) $\mathbf{F} = \frac{d\mathbf{p}}{dt} = -kA\sin(kt)\,\hat{\mathbf{i}} - kA\cos(kt)\,\hat{\mathbf{j}}$

$$\mathbf{p} = A\cos(kt)\,\hat{\mathbf{i}} - A\sin(kt)\,\hat{\mathbf{j}}$$

Since, $\mathbf{F}\cdot\mathbf{p} = 0$

∴ Angle between $\mathbf{F}$ and $\mathbf{p}$ should be 90°.

10. (d) In case of elastic collision, coefficient of restitution $e = 1$.

Magnitude of relative velocity of approach

= Magnitude of relative velocity of separation.

But relative speed of approach ≠ Relative speed of separation

11. (d) If a force is applied at centre of mass of a rigid body, its torque about centre of mass will be zero, but acceleration will be non-zero. Hence, velocity will change.

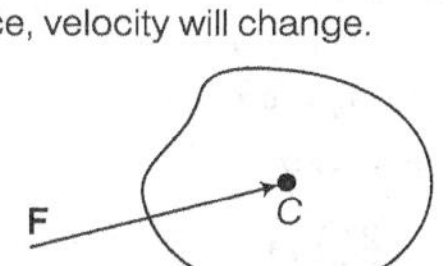

12. (10) $F = \frac{\Delta p}{\Delta t} = n \times \left(a \times \frac{b}{2}\right) \times (2mv)$

Equating the torque about hinge side, we have

$$n \times \left(a \times \frac{b}{2}\right) \times (2mv) \times \frac{3b}{4} = Mg \times \frac{b}{2}$$

Substituting the given values, we get

$$v = 10\ \text{ms}^{-1}$$

2. Rotational Motion

DPP-1 Rotational Kinematics

1. (d) Angular displacement θ in 60 s is 2π.

$$\therefore \quad \omega = \frac{\theta}{t} = \frac{2\pi}{60} = \frac{\pi}{30} \text{ rad s}^{-1}$$

2. (a) $\alpha = 1 \text{ rad s}^{-2}$, $\omega_0 = 5 \text{ rad s}^{-1}$, $\omega = 15 \text{ rad s}^{-1}$

$$\omega^2 - \omega_0^2 = 2\alpha\theta$$

$$\Rightarrow \quad \theta = \frac{\omega^2 - \omega_0^2}{2\alpha} = \frac{(15)^2 - (5)^2}{2(1)} = 100 \text{ rad}$$

3. (c) $\omega_0 = 0, t_1 = 10 \text{ s}, \theta_1 = \omega_0 t_1 + \frac{1}{2}\alpha t_1^2$

$$\theta_1 = 0 + \frac{1}{2}(4)(10)^2 = 200 \text{ rad}$$

$$\omega = \omega_0 + \alpha t = 0 + (4)(10) = 40 \text{ rad s}^{-1}$$

$$\theta_2 = \omega t_2 = (40)(10) = 400 \text{ rad}$$

$$\theta_3 = \omega t_3 - \frac{1}{2}\alpha t_3^2 = (40)(10) - \frac{1}{2}(4)(10)^2$$

$$= 200 \text{ rad}$$

$$\theta_{total} = 200 + 400 + 200 = 800 \text{ rad}$$

4. (a) $\omega_0 = 20 \text{ rad s}^{-1}$, $\alpha = -2 \text{ rad s}^{-2}$, $\omega = 0$

$$\omega = \omega_0 + \alpha t_1$$

$$\Rightarrow \quad t_1 = \frac{20}{2} = 10 \text{ s}$$

At $t_1 = 10$ s, body comes to a halt. If the torque continues to act, the body acquires initial angular speed in the opposite direction (i.e. initial kinetic energy) after another 10 s.

$\therefore$ Required time is $t = 10 + 10 = 20$ s

5. (c) $\mathbf{r} = 3\hat{i} + 4\hat{j}$ and $\mathbf{v} = 4\hat{i} - 3\hat{j}$

$\mathbf{v} = \omega \times \mathbf{r}$

As $\mathbf{r}$ and $\mathbf{v}$ are on XY-plane. ω should be along Z-axis.

Let $\omega = z\hat{\mathbf{k}}$

Then, $\omega \times \mathbf{r} = 3z\hat{j} - 4z\hat{i}$

$$\mathbf{v} = 4\hat{i} - 3\hat{j}$$

On comparing these two equations, we have

$$z = -1$$

$$\therefore \quad \omega = -\hat{\mathbf{k}}$$

6. (c) $\omega = \frac{d\theta}{dt} = \frac{d}{dt}\left(\frac{9t^3}{32}\right) = \frac{9}{32} \cdot 3t^2 = \frac{27}{32}t^2$

$$\alpha = \frac{d\omega}{dt} = \frac{27}{32} \cdot 2t = \frac{27}{16}t$$

Tangential acceleration, $a_t = r\alpha$

Normal acceleration, $a_n = \omega^2 r$

$$a_t = a_n \Rightarrow \alpha = \omega^2$$

$$\Rightarrow \quad \frac{27}{16}t = \left(\frac{27}{32}\right)^2 t^4$$

$$t^3 = \frac{27}{16} \times \frac{32}{27} \times \frac{32}{27} \quad \Rightarrow \quad t = \frac{4}{3} \text{ s}$$

$$a_{net} = \sqrt{a_n^2 + a_t^2} = \sqrt{\omega^2 r + r\alpha}$$

$$= \sqrt{\left(\frac{27}{32}\right)^2\left(\frac{4}{3}\right)^4\left(\frac{3}{4}\right) + \left(\frac{3}{4}\right)\left(\frac{27}{16}\right)\left(\frac{4}{3}\right)} = \frac{3}{4}\sqrt{6} \text{ ms}^{-2}$$

Linear velocity, $v = r\omega$

$$= (0.75)\left(\frac{27}{32}\right)\left(\frac{4}{3}\right)^2 = (0.75)\frac{3}{2} = \left(\frac{3}{4}\right)\left(\frac{3}{2}\right) = \frac{9}{8} \text{ ms}^{-1}$$

7. (b) Load L moves with constant acceleration, its velocity at time t is $v = at$.

Let ω_1 be the angular velocity of the pulley and the first gear wheel.

Peripheral velocity of the pulley $= \omega_1 r = at$

$$\therefore \quad \omega_1 = at/r$$

Peripheral velocity of first gear wheel is $\omega_1 r_1$.

Two gear wheels have same peripheral velocities. So,

$$\omega_2 r_2 = \omega_1 r_1$$

$$\Rightarrow \quad \omega_2 = \frac{\omega_1 r_1}{r_2} = \frac{atr_1}{rr_2}$$

Let θ be the angular displacement of the second year at time t.

Then, $\omega_2 = \frac{d\theta}{dt} = \frac{atr_1}{rr_2}$

On integrating, we get

$$\theta = \left(\frac{ar_1}{2rr_2}\right)t^2$$

8. (b) Area under the curve gives number of revolutions, n.

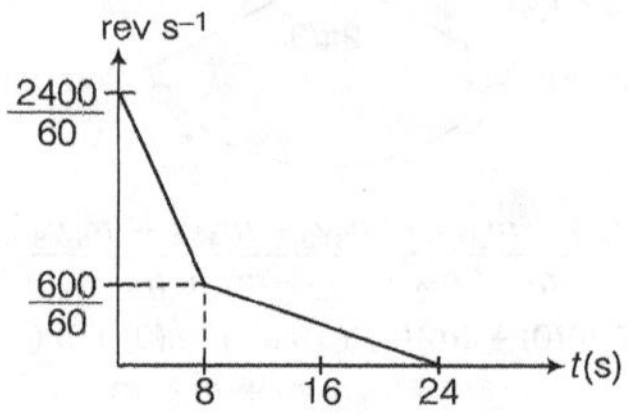

$$n = \frac{1}{2}(8)(40 - 10) + 8(10) + \frac{1}{2}(24 - 8)(10) = 280$$

9. (c) $\alpha = 12 - t$

$$\Rightarrow \quad \frac{d\omega}{dt} = 12 - t$$

$$\Rightarrow \quad d\omega = (12 - t)dt$$

$$\int_{60}^{\omega} d\omega = \int_4^6 (12 - t)dt$$

$$\Rightarrow \quad \omega - 60 = \left[12t - \frac{t^2}{2}\right]_4^6$$

$$\omega - 60 = 12(6 - 4) - \frac{1}{2}(6^2 - 4^2)$$

$$\Rightarrow \quad \omega = 74 \text{ rad s}^{-1}$$

$$\int d\omega = \int (12 - 1)dt$$

$$\Rightarrow \quad \omega = 12t - \frac{t^2}{2} + C$$

At $t = 4$ s, $\omega = 60$

$$\Rightarrow \quad 60 = 12(4) - \frac{(4)^2}{2} + C$$

$$\Rightarrow \quad C = 20$$

$$\therefore \quad \omega = 12t - \frac{t^2}{2} + 20$$

$$\Rightarrow \quad \frac{d\theta}{dt} = 12t - \frac{t^2}{2} + 20$$

$$\int_0^{\theta} d\theta = \int_0^6 \left(12t - \frac{t^2}{2} + 20\right)dt$$

$$\Rightarrow \quad \theta = \left[\frac{12t^2}{2} - \frac{t^3}{2 \times 3} + 20t\right]_0^6$$

$$\Rightarrow \quad \theta = 12\frac{(6)^2}{2} - \frac{(6)^3}{2(3)} + 20(6)$$

or $$\theta = 300 \text{ rad}$$

$$n = \frac{\theta}{2\pi} = 47.8$$

10. (b) $\alpha = -k\omega$ [given $\alpha \propto \omega$]

$$\omega\frac{d\omega}{d\theta} = -k\omega$$

$$\Rightarrow \quad d\omega = -k\,d\theta$$

$$\int_{\omega_0}^{\omega_0/2} d\omega = -k\int_0^{2\pi n} d\theta$$

$$\Rightarrow \quad \frac{\omega_0}{2} = k \times 2\pi n$$

$$\Rightarrow \quad n = \frac{\omega_0}{4\pi k} \quad \text{...(i)}$$

For $\frac{\omega_0}{2}$ to 0,

$$\int_{\omega_0/2}^{0} d\omega = -k\int_{2\pi n}^{2\pi n_1} d\theta$$

$$\Rightarrow \quad \frac{\omega_0}{2} = k \times 2\pi\,(n_1 - n)$$

$$n_1 - n = \frac{\omega_0}{4\pi k} \quad \text{...(ii)}$$

From Eqs. (i) and (ii), we get

$$n_1 - n = n$$

11. (c) $$\omega = \frac{d\theta}{dt} = \frac{d}{dt}(kt^2)$$

$$= 2kt = 2(1.5)t = 3t$$

Instantaneous angular acceleration, $\alpha = \frac{d\omega}{dt} = 3 \text{ rad s}^{-2}$.

Note that this is time independent.

$$(\omega)_{t=\frac{2}{3}\text{s}} = \alpha t = 3\left(\frac{2}{3}\right) = 2 \text{ rad s}^{-1}$$

$$v = r\omega$$

$$\Rightarrow \quad r = \frac{v}{\omega} = \frac{2}{2} = 1 \text{ m}$$

$$a_n = \frac{\omega^2}{r} = \frac{(2)^2}{1} = 4 \text{ ms}^{-2}$$

and $$a_t = r\alpha = (1)(3) = 3 \text{ ms}^{-2}$$

$$a_{net} = \sqrt{a_n^2 + a_t^2} = \sqrt{4^2 + 3^2} = 5 \text{ ms}^{-2}$$

12. (a,c,d) $\alpha = \pi \text{ rad s}^{-2}$, $\omega_0 = 0$, $t = 4$ s

(i) $$\omega = \omega_0 + \alpha t$$

$$\Rightarrow \quad \omega = 0 + \pi \times 4 = 4\pi \text{ rad s}^{-1}$$

(ii) $$\theta = \omega_0 t + \frac{1}{2}\alpha t^2$$

$$\Rightarrow \quad \theta = (0)(4) + \frac{1}{2}(\pi)(4)^2 = 8\pi \text{ rad}$$

(iii) Number of turns, $n = \frac{8\pi}{2\pi} = 4$

13. (a,d) $\alpha = 60 \text{ rad s}^{-2}$, $r = 0.2$ m, $\omega_0 = 0$

(i) $t = 0.15$ s,

$$\omega = \omega_0 + \alpha t = 0 + 60\,(0.15) = 9 \text{ rad s}^{-1}$$

$$a_n = \omega^2 r = (9)^2(0.2) = 16.2 \text{ ms}^{-2}$$

$$a_t = r\alpha = (0.2)(60) = 12 \text{ ms}^{-2}$$

$$a_{net} = \sqrt{a_n^2 + a_t^2} = \sqrt{(16.2)^2 + (12)^2} = 20.2 \text{ ms}^{-2}$$

(ii) $\theta = \frac{1}{2}\alpha t^2 = \frac{1}{2}(60)\,(0.25)^2 = 30 \times 0.0625$

Number of revolutions, $n = \frac{1.88}{2\pi} = 0.3$

14. (b,d)

(i) $\alpha = 4at^3 - 3bt^2$

$$\Rightarrow \quad \frac{d\omega}{dt} = 4at^3 - 3bt^2$$

$$d\omega = (4at^3 - 3bt^2)dt$$

$$\Rightarrow \quad \int_{\omega_0}^{\omega} d\omega = \int_0^t (4at^3 - 3bt^2)dt$$

$$\omega - \omega_0 = at^4 - bt^3$$

$$\Rightarrow \quad \omega = \omega_0 + at^4 - bt^3$$

(ii) $$\omega = \omega_0 + at^4 - bt^3$$

$$\Rightarrow \quad \frac{d\theta}{dt} = \omega_0 + at^4 - bt^3$$

$$d\theta = (\omega_0 + at^4 - bt^3)dt$$

$$\Rightarrow \quad \int_0^{\theta} d\theta = \int_0^t (\omega_0 + at^4 - bt^3)dt$$

$$\theta = \omega_0 t + \frac{1}{5}at^5 - \frac{1}{4}bt^4$$

15. (a,c) $\theta = 6t - 2t^2$, $\quad \omega = \frac{d\theta}{dt} = 6 - 4t$

Body comes to rest

When $\omega = 0 \Rightarrow 6 - ut = 0$

$$\Rightarrow \quad t = \frac{3}{2} = 1.5 \text{ s}$$

At $t = 0$, $\quad \omega = \omega_1 = 6 \text{ rad s}^{-1}$ and $\theta = \theta_1 = 0$

At $t = 1.5$ s, $\quad \omega = \omega_2 = 6 - 4 \times \frac{3}{2} = 0$

and $$\theta = \theta_2 = 6 \times \frac{3}{2} - 2 \times \frac{9}{4} = \frac{9}{2} \text{ rad}$$

So, $$\omega_{av} = \frac{\theta_2 - \theta_1}{t_2 - t_1} = \frac{\frac{9}{2} - 0}{\frac{3}{2} - 0} = 3 \text{ rad s}^{-1}$$

and $$\alpha_{av} = \frac{\omega_2 - \omega_1}{t_2 - t_1} = \frac{0 - 6}{\frac{3}{2} - 0} = -4 \text{ rad s}^{-2}$$

16. (b,c) $\omega_0 = 2\pi \times \frac{1200}{60} = 40\pi \text{ rad s}^{-1}$

$$\omega = 2\pi \times \frac{3120}{60} = 104\pi \text{ rad s}^{-1}$$

(i) $$\alpha = \frac{\omega - \omega_0}{t} = \frac{104\pi - 40\pi}{16} = 4\pi \text{ rad s}^{-2}$$

(ii) $$\theta = \omega_0 t + \frac{1}{2}\alpha t^2 = 40\pi(16) + \frac{1}{2}(4\pi)(16)^2 = 1152\,\pi \text{ rad}$$

$$n = \frac{\theta}{2\pi} = \frac{1152\,\pi}{2\pi} = 576$$

Explanations for Q. Nos. (17-19)

When A moves 20 m, then angular displacement θ is given by

$$s = r\theta$$

$$\Rightarrow \quad \theta = \frac{s}{r} = \frac{20}{1} = 20 \text{ rad}$$

$$\alpha = 2 \text{ rad s}^{-2}$$

and $$\omega_0 = 0 \quad \text{(given)}$$

$$\theta = \omega_0 t + \frac{1}{2}\alpha t^2 = 0 + \frac{1}{2}(2)t^2$$

$$\Rightarrow \quad t^2 = 20 \Rightarrow t = 4.47 \text{ s}$$

At this time angular velocity of pulley is $\omega = \omega_0 + \alpha t$

or $$\omega = 0 + (2)\,(4.47) = 8.94 \text{ ms}^{-1}$$

Now, velocity of A is $v_A = r_A\omega = (1)\,(8.94) = 8.94 \text{ ms}^{-1}$

and velocity of B is $v_B = r_B\omega = (0.75)\,(8.94) = 6.7 \text{ ms}^{-1}$

17. (a) **18.** (b) **19.** (d)

20. (a) $\omega = \omega_0 e^{-t/T}$

$$\Rightarrow \quad \ln\left(\frac{\omega}{\omega_0}\right) = -t/T$$

$$\Rightarrow \quad t = T\ln\left(\frac{\omega_0}{\omega}\right)$$

(i) $\omega = \frac{\omega_0}{2} \Rightarrow t = T\ln\left(\frac{\omega_0}{\omega_0/2}\right) = T\ln 2$

(ii) $\frac{d\omega}{dt} = \omega_0 e^{-t/T}\left(-\frac{1}{T}\right)$

or $\quad \alpha = -\frac{\omega_0}{T}e^{-t/T}$

Initially, $t = 0$

$$\therefore \quad \alpha_i = \frac{-\omega_0}{T}e^{\frac{-0}{T}} = \frac{-\omega_0}{T}$$

(iii) $\alpha_{av} = \frac{\omega_2 - \omega_1}{t_2 - t_1} = \frac{\frac{\omega_0}{2} - \omega_0}{T\ln 2 - 0} = \frac{-\omega_0}{2T\ln 2}$

(iv) $\omega = \omega_0 e^{-t/T}$

$$\Rightarrow \quad d\theta = \omega_0 e^{-t/T}dt$$

$$\int_0^\theta d\theta = \omega_0 \int_0^{T\ln 2} e^{-t/T}dt$$

$$\Rightarrow \quad \theta = \frac{\omega_0}{(-1/T)}(e^{-\ln 2} - 1)$$

or $\quad \theta = \omega_0 T\left(1 - \frac{1}{2}\right) = \omega_0 T/2$

Number of revolutions, $n = \frac{\omega_0 T}{2 \times 2\pi} = \frac{\omega_0 T}{4\pi}$

Hence, (i) → (s), (ii) → (r), (iii) → (q), (iv) → (p)

DPP-2 Moment of Inertia

1. (a) Mass of the elemental disc is

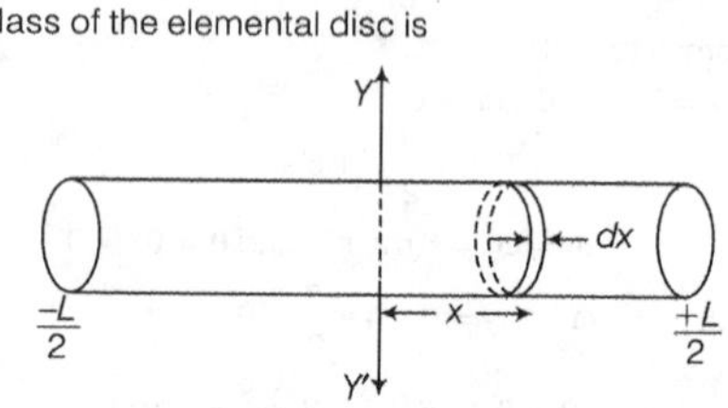

$$dm = \frac{M}{L}dx$$

$$(dI)_{YY'} = \left(\frac{M}{L}dx\right)\frac{R^2}{4} + \left(\frac{M}{L}dx\right)x^2$$

or $\quad (dI)_{YY'} = \frac{M}{L}\left[\frac{R^2}{4} + x^2\right]dx$

$$I_{YY'} = \frac{M}{L}\int_{-\frac{L}{2}}^{+\frac{L}{2}}\left(\frac{R^2}{4} + x^2\right)dx$$

$$= \frac{M}{L}\left[\frac{R^2}{4}\left(\frac{L}{2} + \frac{L}{2}\right) + \frac{1}{3}\left(\frac{L^3}{8} + \frac{L^3}{8}\right)\right]$$

$$I_{YY'} = \frac{MR^2}{4} + \frac{ML^2}{12}$$

2. (b)

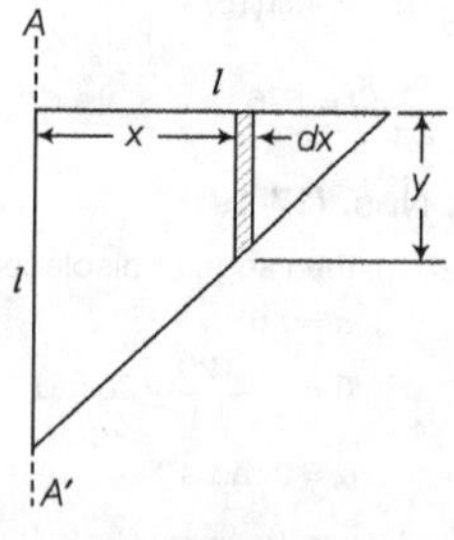

Consider an element of thickness dx at a distance x from AA'. Its height is $y = l - x$.

Mass of this element, $dm = \frac{M}{(l^2/2)}(l - x)dx = \frac{2M}{l^2}(l - x)dx$

$$(dI)_{AA'} = \frac{2M}{l^2}(l - x)dx \times x^2$$

$$I_{AA'} = \frac{2M}{l^2}\int_0^l (lx^2 - x^3)dx$$

$$= \frac{2M}{l^2}\left[\frac{l^4}{3} - \frac{l^4}{4}\right] = \frac{Ml^2}{6}$$

3. (d) We know

$$I_{BB'} = \frac{2}{5}(2M)R^2 \times \frac{1}{2} = \frac{2}{5}MR^2$$

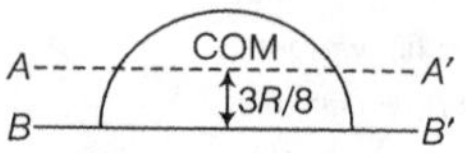

$$I_{BB'} = I_{COM} + M\left(\frac{3R}{8}\right)^2$$

$$\Rightarrow \quad I_{COM} = \frac{2}{5}MR^2 - \frac{9}{64}MR^2$$

$$I_{COM} = I_{AA'} = MR^2\left(\frac{128 - 45}{320}\right) = \frac{83}{320}MR^2$$

4. (c) $(I_{AA'})_{\text{removed disc}} = \frac{mr^2}{2}$

where, m is mass of disc of radius r.

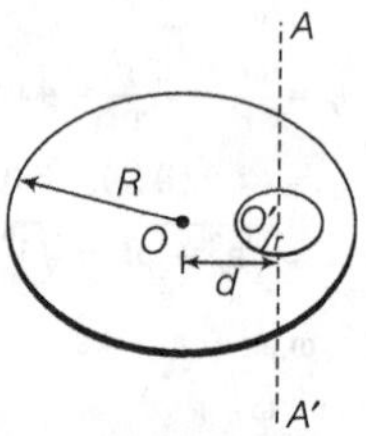

$$m = \frac{M}{R^2}r^2$$

$$\therefore \quad (I_{AA'})_{\text{removed disc}} = \frac{M}{R^2}\cdot\frac{r^4}{2}$$

$$(I_{AA'})_{\text{original uncut disc}} = \frac{MR^2}{2} + Md^2$$

$$(I_{AA'})_{\text{residual disc}} = (I_{AA'})_{\text{original uncut disc}} - (I_{AA'})_{\text{removed disc}}$$

$$= \frac{MR^2}{2} + Md^2 - \frac{Mr^4}{2R^2} = \frac{M}{2}\left[R^2 + 2d^2 - \frac{r^4}{R^2}\right]$$

5. (b) $(I_{AA'})_{\text{rod}} = \frac{ML^2}{12} + M\left(\frac{L}{2}\right)^2 = \frac{ML^2}{3}$

$$(I_{AA'})_{\text{sphere}} = \frac{2}{5}mr^2 + mL^2$$

$$(I_{AA'})_{\text{system}} = \frac{ML^2}{3} + \frac{2}{5}mr^2 + mL^2$$

$$= L^2\left(m + \frac{M}{3}\right) + \frac{2}{5}mr^2$$

6. (d) Consider an element of length dx at a distance x from the pivoted end. At the location of element $\lambda = ax^2 + b$.

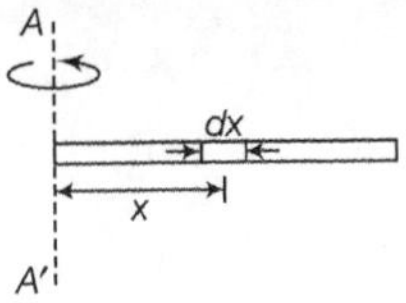

Mass of element is

$$dm = \lambda dx = (ax^2 + b)dx$$

$$(dI)_{AA'} = (dm)x^2 = (ax^2 + b)x^2 dx$$

$$I_{AA'} = \int_0^L (ax^4 + bx^2)dx$$

$$= \frac{aL^5}{5} + \frac{bL^3}{3}$$

7. (a) Mass of the removed disc is $m = M\dfrac{R_1^2}{R_2^2}$

$$(I_o)_{\text{original uncut disc}} = \frac{MR_2^2}{2} \quad \ldots\text{(i)}$$

$$(I_o)_{\text{removed disc}} = \left(\frac{MR_1^2}{R_2^2}\right)\frac{R_1^2}{2} = \frac{MR_1^4}{2R_2^2} \quad \ldots\text{(ii)}$$

On subtracting Eq. (ii) from Eq. (i), we get

$$(I_o)_{\text{residual disc}} = \frac{MR_2^2}{2} - \frac{MR_1^4}{2R_2^2}$$

$$I_o = \frac{M}{2}\left[\frac{R_2^4 - R_1^4}{R_2^2}\right] = \frac{M}{2R_2^2}(R_2^4 - R_1^4)$$

8. (b) Let L and m be the length and mass of the rod. Then, $I_1 = \dfrac{mL^2}{12}$

When it is made a ring, $2\pi R = L$

$$\Rightarrow \quad R = \frac{L}{2\pi}$$

$$I_2 = mR^2 + mR^2 = 2mR^2$$

$$= 2m\left(\frac{L}{2\pi}\right)^2 = \frac{mL^2}{2\pi^2}$$

Now, $$\frac{I_1}{I_2} = \frac{mL^2/12}{mL^2/2\pi^2} = \frac{\pi^2}{6}$$

9. (a) $I_2 = \dfrac{mL^2}{3}$, $I_1 = \dfrac{mL^2}{12} + m\left(\dfrac{L}{2}\right)^2 = \dfrac{mL^3}{3}$

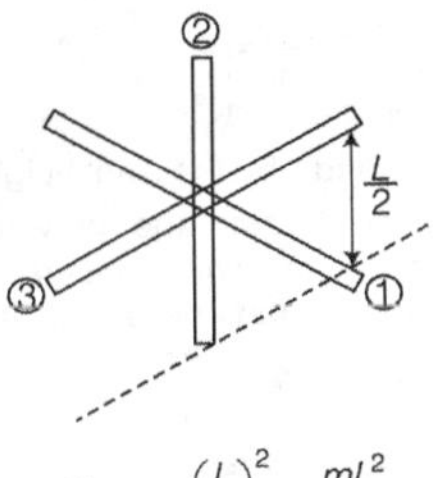

$$I_3 = m\left(\frac{L}{2}\right)^2 = \frac{mL^2}{4}$$

$$\therefore \quad I = mL^2\left(\frac{1}{3} + \frac{1}{3} + \frac{1}{4}\right) = \frac{11}{12}mL^2$$

10. (d)

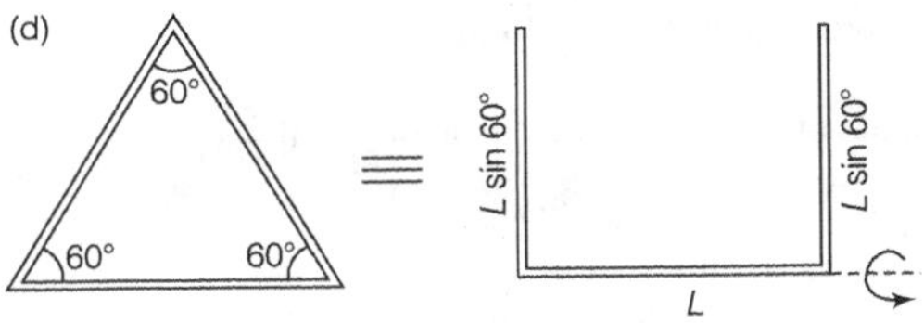

$$\therefore \quad I = [0] + \frac{m(L\sin 60°)^2}{3} + \frac{m(L\sin 60°)^2}{3}$$

$$= \frac{mL^2}{4} + \frac{mL^2}{4} = \frac{mL^2}{2}$$

11. (c) $I_{AB} = I_{COM} + Mr^2 = \dfrac{1.6Ma^2}{2} + M(2a)^2 = 4.8\,Ma^2$

Note *$I_{PQ} = I_{COM} = 1.6\,Ma^2/2$ from perpendicular axes theorem*

12. (b) The mass of the disc removed $= 9m\left[\dfrac{\pi(R/3)^2}{\pi R^2}\right] = m$

$$I_o = (I_{\text{whole disc}})_o - (I_{\text{removed disc}})_o$$

$$= \frac{9m(R)^2}{2} - \left[\frac{m(R/3)^2}{2} + m\left(\frac{2R}{3}\right)^2\right] = 4mR^2$$

13. (a) $$\frac{I_Y}{I_X} = \frac{m_2(4R)^2/2}{m_1 R^2/2} = \frac{\rho\pi(4R)^2\left(\frac{t}{4}\right)(4R)^2}{\rho(\pi R^2)(t)(R)^2} = 64$$

14. (a,d) $I_{22} = 1(0)^2 + 2(d_2)^2 + 3(0)^2 + 4(d_1)^2$

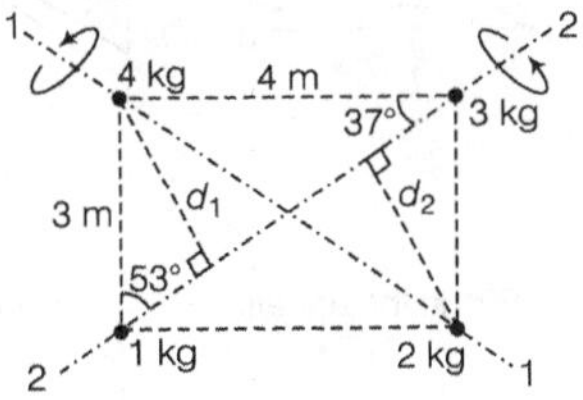

In the figure, $\quad d_1 = (\sin 53°) \times 3 = \dfrac{12}{5} = 2.4$ m

$\therefore \quad d_2 = 2.4$ m

So, $\quad I_{22} = 2(2.4)^2 + 4(2.4)^2 = 34.56$ kg-m^2

Similarly, $\quad I_{11} = 1(2.4)^2 + 3(2.4)^2 = 23.04$ kg-m^2

15. (b,d) (i) $I_1 = m(\sqrt{2}d)^2 + m(\sqrt{2}d)^2 + m(2d)^2 + m(0)^2 = 8\,md^2$

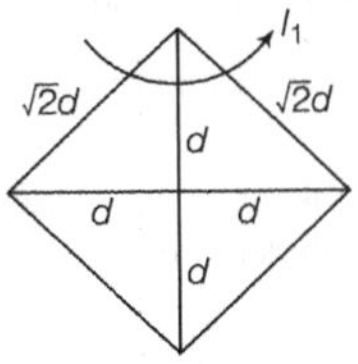

(ii) $\quad I_2 = m(d)^2 + m(d)^2 + m(\sqrt{2}d)^2 + m(0)^2 = 4\,md^2$

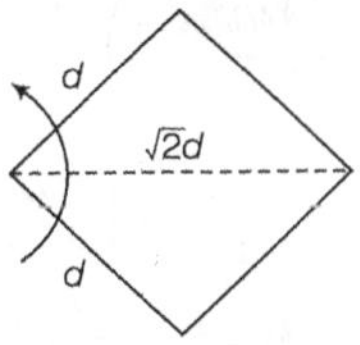

16. (a) (i) Consider an element of thickness dx at a distance x from the centre along its length.

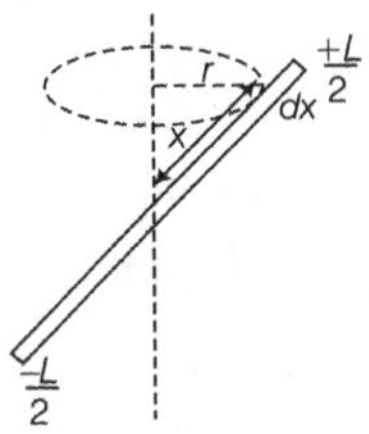

$$dm = \frac{m}{L}\cdot dx$$

$$r = x\sin\theta$$

$$dI = (dm)r^2 = \frac{m}{L}\sin^2\theta\, x^2 dx$$

$$I = \frac{m}{L}\sin^2\theta \int_{-\frac{L}{2}}^{+\frac{L}{2}} x^2 dx$$

$$= \frac{m}{L}\sin^2\theta\left[\frac{1}{3}\left(\frac{L^3}{8}+\frac{L^3}{8}\right)\right]$$

$$I = \frac{1}{12}mL^2\sin^2\theta$$

Note $I = \frac{1}{12}mL^2\sin^2\alpha = \frac{1}{12}mL_{\text{eff}}^2$

(ii) $I = \frac{1}{3}mL^2\sin^2\alpha = \frac{1}{3}mL_{\text{eff}}^2$

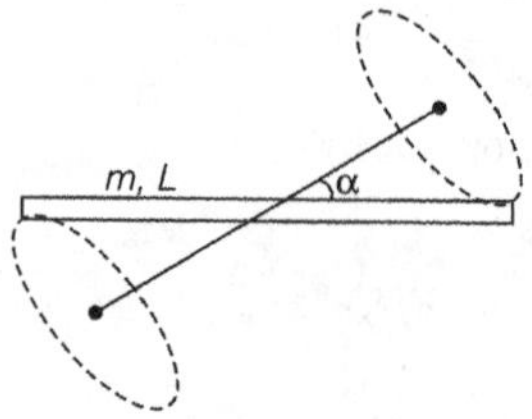

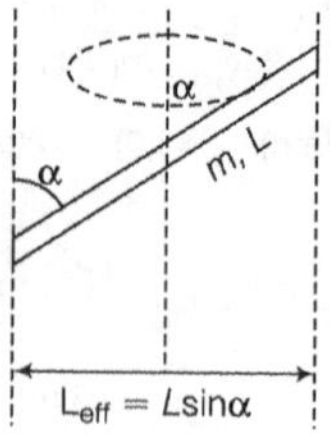

(iii) Put $\alpha = 90°$ in the equation $I = \frac{1}{3}mL^2\sin^2\alpha$

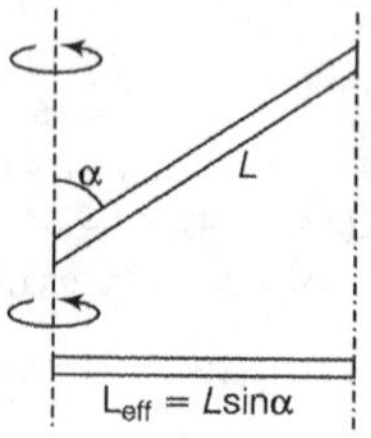

17. (d) Perpendicular axes theorem is applicable only to two dimensional bodies.

18. (b) Statement I is correct by parallel axes theorem $I = I_{COM} + mr^2$. Statement II is independently correct.

19.

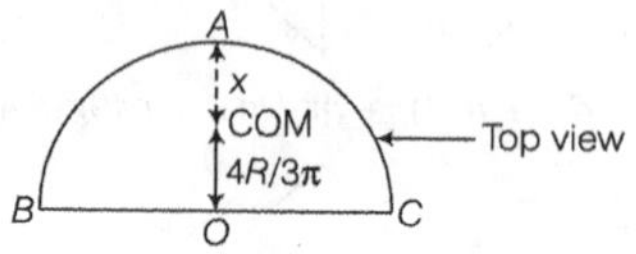

We know that

$$I_0 = \frac{(2M)R^2}{2}\times\frac{1}{2} = \frac{MR^2}{2}$$

$$I_0 = I_{COM} + M\left(\frac{4R}{3\pi}\right)^2$$

$$\Rightarrow \quad I_{COM} = \frac{MR^2}{2} - MR^2\frac{16}{9\pi^2}$$

$$\therefore \quad I_{COM} = MR^2\left(\frac{9\pi^2-32}{18\pi^2}\right)$$

Now, $I_A = I_{COM} + Mx^2$

where, $x = R - \frac{4R}{3\pi} = R\left(\frac{3\pi-4}{3\pi}\right)$

$$\therefore \quad I_A = MR^2\left(\frac{9\pi^2-32}{18\pi^2}\right) + MR^2\frac{(3\pi-4)^2}{9\pi^2}$$

$$= \frac{MR^2}{18\pi^2}[9\pi^2 - 32 + 18\pi^2 + 32 - 48\pi]$$

$$= \frac{MR^2}{18\pi^2}3\pi[9\pi - 16]$$

$$I_A = \left(\frac{9\pi-16}{6\pi}\right)MR^2$$

20. Let h is the height of the cone. Then, its mass per unit volume is

$$\rho = \frac{M}{\frac{1}{3}\pi R^2 h} = \frac{3M}{\pi R^2 h}$$

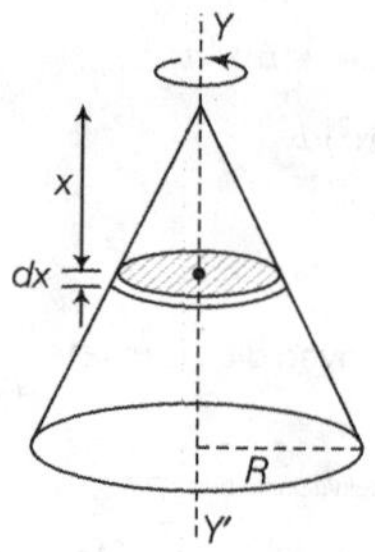

Consider a slice of thickness dx cut normal to the axis of the cone, at a distance x from the vertex measured along the altitude.

Mass of the element $= \rho\left[\pi\left(\frac{R}{h}x\right)^2 dx\right]$ ($\because$ mass = density × volume)

Moment of inertia of this element about the axis YY' is

$$(dI)_{YY'} = \frac{1}{2}\rho\left[\pi\left(\frac{R}{h}x\right)^2 dx\right]\left(\frac{R}{h}x\right)^2 = \frac{1}{2}\rho\pi\frac{R^4}{h^4}x^4 dx$$

$$\therefore \quad I_{YY'} = \frac{\rho\pi R^4}{2h^4}\int_0^h x^4 dx = \frac{\rho\pi R^4}{2h^4}\frac{h^5}{5} = \frac{\rho}{10}\pi R^4 h$$

$$I_{YY'} = \frac{3M}{\pi R^2 h}\cdot\frac{1}{10}\pi R^4 h = \frac{3}{10}MR^2$$

21.

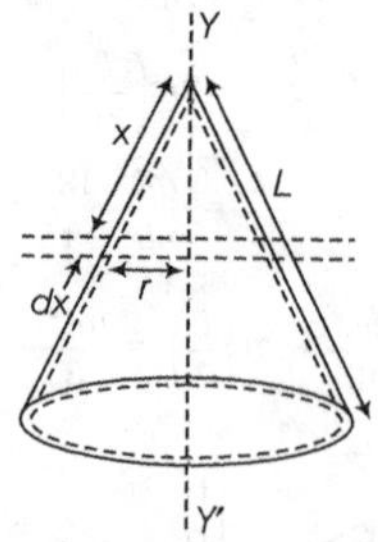

Let h be the altitude and L be the slant height of the cone.

[It should be noted that in the previous problem mass is distributed uniformly in the whole volume but here, unlike to that case, the mass is distributed only over its lateral surface area].

The mass per unit area is $\lambda = \frac{M}{\pi RL}$

Consider an element of the cone with its plane normal to the axis of the cone of width dx, at a distance x from the vertex measured along the slant height.

Radius of this elemental ring is $r = \frac{R}{L}x$

Its area $= (2\pi r)dx = 2\pi\left(\frac{R}{L}x\right)dx$

Mass of this element $= \lambda$ (area) $= \frac{M}{\pi RL}2\pi\left(\frac{R}{L}x\right)dx$

$$\therefore \quad dm = \frac{2m}{L^2}xdx$$

$$(dI)_{YY'} = (dm)(r^2)$$

$$= \left(\frac{2M}{L^2}xdx\right)\left(\frac{R^2}{L^2}x^2\right) = \frac{2MR^2}{L^4}x^3dx$$

$$\therefore \quad I_{YY'} = \frac{2MR^2}{L^4}\int_0^L x^3dx = \frac{2MR^2}{L^4}\cdot\frac{L^4}{4} = \frac{1}{2}MR^2$$

22. Consider an elemental shell of thickness dr at a distance r from the centre.

Volume of this shell is $dV = 4\pi r^2 dr$

Density of the sphere at the location of this element is $\rho = \rho_0\left(1 + \frac{r}{R}\right)$

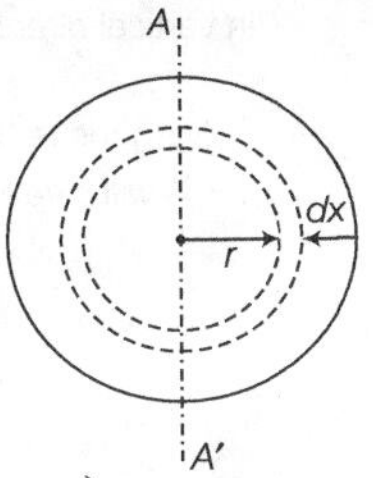

$\therefore$ Mass of element is

$$dm = \rho dV = \rho_0\left(1 + \frac{r}{R}\right)4\pi r^2 dr$$

or
$$dm = \left(4\pi\rho_0 r^2 + \frac{4\pi\rho_0}{R}r^3\right)dr$$

$$(dI)_{AA'} = \frac{2}{3}(dm)r^2 = \frac{2}{3}\left(4\pi\rho_0 r^2 + \frac{4\pi\rho_0}{R}r^3\right)r^2 dr$$

$$I_{AA'} = \frac{2}{3}\cdot 4\pi\rho_0\int_0^R\left(r^4 + \frac{r^5}{R}\right)dr = \frac{8\pi\rho_0}{3}\left(\frac{R^5}{5} + \frac{R^5}{6}\right)$$

$$I_{AA'} = \frac{8\pi\rho_0}{3}\frac{11R^5}{30} = \frac{44}{45}\pi\rho_0 R^5 \quad \ldots\text{(i)}$$

Mass of the sphere is

$$M = \int_0^R dm = \int_0^R\left(4\pi\rho_0 r^2 + \frac{4\pi\rho_0}{R}r^3\right)dr$$

or
$$M = 4\pi\rho_0\frac{R^3}{3} + 4\pi\rho_0\frac{R^3}{4} = \frac{7\pi\rho_0 R^3}{3}$$

$\therefore$
$$\pi\rho_0 R^3 = \frac{3M}{7}$$

Using this in Eq.(i), we get

$$I_{AA'} = \frac{44}{45}\frac{3M}{7}R^2 = \frac{44}{105}MR^2$$

Explanations for Q. Nos. (23-25)

(c) Angle between 1 and 2 is 90° and angle between 3 and 4 is 90°. From perpendicular axes theorem, $I_0 = I_1 + I_2$ and $I_0 = I_3 + I_4$

From symmetry, $I_1 = I_2$ and $I_3 = I_4$.

So, we can conclude that $I_1 = I_2 = I_3 = I_4$ in this case of square lamina.

23. (c) **24.** (c) **25.** (c)

DPP-3 Rotation about a Fixed Axis

1. (c) When impulse delivered by pivot is zero, the rod rotates smoothly without hindrance from pivot.

Linear impulse,
$$J = mv_c$$

Angular impulse is
$$Jx = I\omega = \frac{mL^2}{3}\omega$$

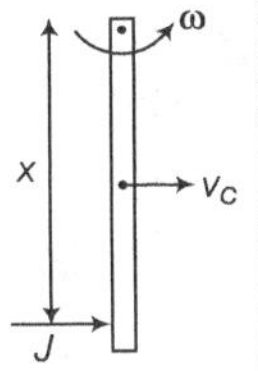

and
$$v_c = \frac{L}{2}\omega$$

$\therefore$
$$m\left(\frac{L}{2}\omega\right)x = \frac{mL^2}{3}\omega$$

$\Rightarrow$
$$x = \frac{2}{3}L$$

2. (c)
$$mgh = \frac{1}{2}mv^2 + \frac{1}{2}kh^2 + \frac{1}{2}I\omega^2$$

$$2mgh = mv^2 + kh^2 + \frac{MR^2}{2}\left(\frac{v}{R}\right)^2$$

$$2mgh = mv^2 + kh^2 + \frac{Mv^2}{2}$$

$$= v^2\left(m + \frac{M}{2}\right) + kh^2$$

$$v^2\left(m + \frac{M}{2}\right) = 2mgh - kh^2$$

$$v^2\left(4 + \frac{8}{2}\right) = 2(4)(9.8)(1) - 32(1)^2$$

$$v = 2.4 \text{ ms}^{-1}$$

3. (c) $I_{ZZ'} = \frac{ml^2}{3} + ml^2 = \frac{4}{3}ml^2$

From conservation of mechanical energy,

$$mg\left(\frac{l}{2}\right) + mgl = \frac{1}{2}I\omega^2$$

$$\frac{3mgl}{2} = \frac{1}{2}\cdot\frac{4}{3}ml^2\omega^2$$

$\Rightarrow$
$$\omega^2 = \frac{9}{4}\frac{g}{l}$$

$\Rightarrow$
$$\omega = \frac{3}{2}\sqrt{\frac{g}{l}}$$

4. (d) From the law of conservation of mechanical energy,

Decrease in GPE of block = Gain in KE of block + Gain in RKE of pulley + Gain in RKE of the shell

$$mgh = \frac{1}{2}mv^2 + \frac{1}{2}I\omega^2 + \frac{1}{2}\left[\frac{2}{3}MR^2\right]\omega'^2$$

where,
$$\omega = \frac{v}{r} \text{ and } \omega' = \frac{v}{R}$$

$\therefore$
$$2mgh = mv^2 + I\frac{v^2}{r^2} + \frac{2}{3}Mv^2$$

$\Rightarrow$
$$v^2 = \frac{2mgh}{m + \frac{I}{r^2} + \frac{2}{3}M}$$

or
$$v = \left[\frac{mgh}{\frac{m}{2} + \frac{I}{2r^2} + \frac{M}{3}}\right]^{\frac{1}{2}}$$

5. (a)
$$N_1 + \mu N_2 = mg \quad \ldots\text{(i)}$$
$$N_2 = \mu N_1 \quad \ldots\text{(ii)}$$

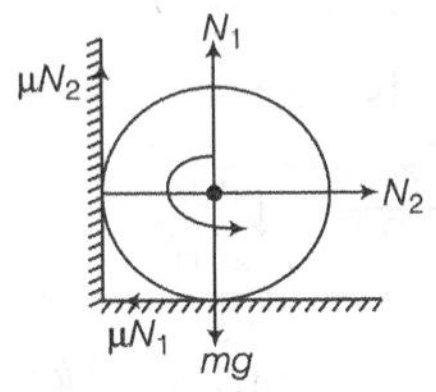

From Eqs. (i) and (ii), we get

$$N_1 = \frac{mg}{(1 + \mu^2)} \text{ and } N_2 = \frac{\mu mg}{(1 + \mu^2)}$$

$$T = (\mu N_1 + \mu N_2)R = I\alpha$$

$$\left(\frac{\mu mg}{1 + \mu^2} + \frac{\mu^2 mg}{1 + \mu^2}\right)R = \frac{mR^2}{2}\cdot\alpha$$

$\Rightarrow$
$$\frac{\mu mg}{1 + \mu^2}(\mu + 1) = \frac{mR}{2}\cdot\alpha$$

$$\alpha = \frac{2\mu g(\mu + 1)}{R(1 + \mu^2)}$$

$$\omega = 0, \omega_0 = \omega_0$$

$$\Rightarrow \quad 0 = \omega_0 - \alpha t$$

$$\Rightarrow \quad t = \frac{\omega_0 R(1+\mu^2)}{2\mu g(1+\mu)}$$

$$\theta = \omega_0 t - \frac{1}{2}\alpha t^2$$

$$= \frac{\omega_0^2 R(1+\mu^2)}{2\mu g(1+\mu)} - \frac{1}{2}\cdot\frac{2\mu g(\mu+1)}{R(1+\mu^2)}\cdot\frac{\omega_0^2 R^2(1+\mu^2)^2}{4\mu^2 g^2(1+\mu)^2}$$

$$\theta = \frac{\omega_0^2 R(1+\mu^2)}{4\mu g(1+\mu)}$$

Number of turns, $n = \frac{\theta}{2\pi}$

or $\quad n = \frac{\omega_0^2 R(1+\mu^2)}{8\pi\mu g\,(1+\mu)}$

6. (b) The forces can be drawn as shown,

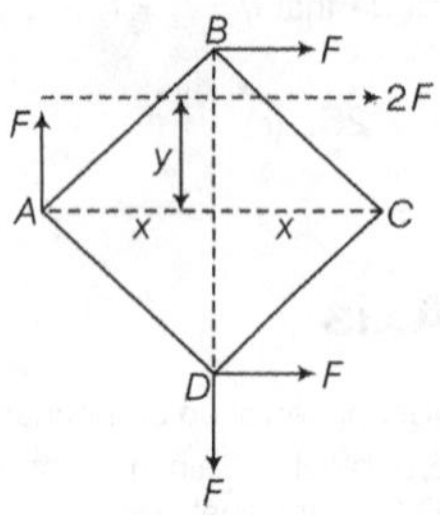

$$F_x = F + F = 2F$$

$$F_y = F - F = 0$$

Resultant of these forces is

$$R = \sqrt{F_x^2 + F_y^2}$$

or $\quad R = 2F$

Suppose the resultant force R passes from a point, a distance y from A, then taking moments about centre of lamina

$$Fx - 2Fy = 0$$

$$\Rightarrow \quad y = \frac{x}{2}$$

Thus, the resultant is applied at the mid-point of side BC.

Note *In the given situation only F at A can cause spinning about the centre of the square.*

7. (b) Just after the thread is burnt the forces acting on the rod are as shown in the figure.

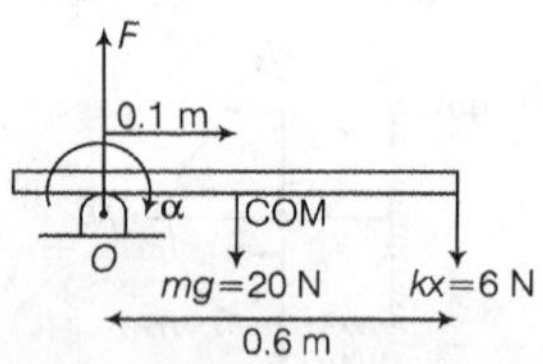

$$\tau_{\text{about } O} = (20)(0.1) + (6)(0.6) = 5.6 \text{ N-m}$$

$$\alpha = \frac{T}{I} = \frac{5.6}{\left\{\frac{(2)(1)^2}{12} + (2)(0.1)^2\right\}}$$

$$\Rightarrow \quad \alpha = 30 \text{ rad s}^{-2}$$

$$a_{COM} = (0.1)\alpha = (0.1)(30) = 3 \text{ ms}^{-2} (\downarrow)$$

$$F_{net} = ma$$

$$\Rightarrow \quad kx + mg - F = m\,a_{COM}$$

$$\Rightarrow \quad 6 + 20 - F = 2(3)$$

$$F = 20 \text{ N}$$

8. (b) **For equilibrium** In horizontal direction,

$$N_1 = \mu N_2 \quad \text{...(i)}$$

In vertical direction,

$$N_2 + \mu N_1 = F + mg \quad \text{...(ii)}$$

Solving these two equations and putting $F = 40$ N, $m = 2$ kg and $\mu = \frac{1}{3}$ will give $N_1 = 18$ N and $N_2 = 54$ N

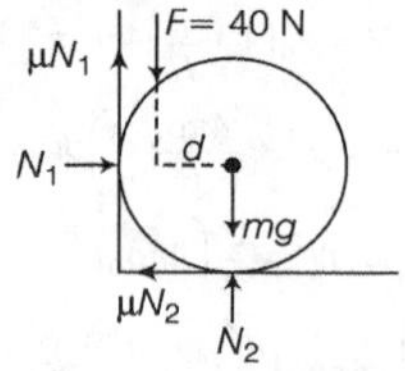

$$Fd = (\mu N_1 + \mu N_2)r$$

$$\Rightarrow \quad d = \frac{\mu(N_1 + N_2)r}{F}$$

$$= \frac{\frac{1}{3}(18+54)(0.1)}{40}$$

$$d = 0.06 \text{ m} = 6 \text{ cm}$$

9. (a,b,c) $\alpha = \frac{\tau}{I} = \frac{mg\,L/2}{mL^2/3} = \frac{3g}{2L}$

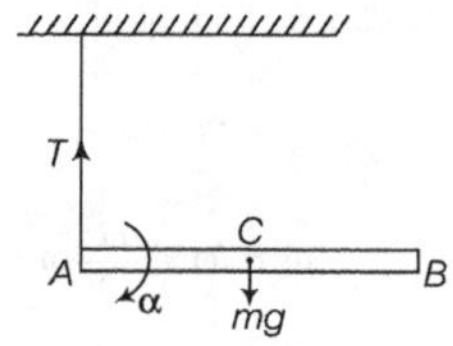

(i) $a_B = L\alpha = \frac{3}{2}g$

(ii) $a_C = \frac{L}{2}\alpha = \frac{3}{4}g$

(iii) $mg - T = ma_C$

$$T = mg - ma_C$$

or $\quad T = mg - m\cdot\frac{3}{4}g = \frac{mg}{4}$

10. (a,d) $\quad TR = I\alpha \quad$ ...(i)

$$mg - T = ma \quad \text{...(ii)}$$

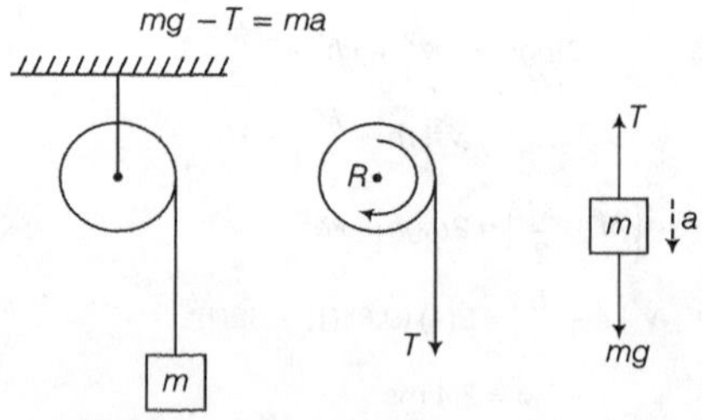

For no slipping the contact points of thread and pulley should have same acceleration.

$$\Rightarrow \quad a = R\alpha \quad \text{...(iii)}$$

Solving the above equations, we get

$$T = \frac{mg}{1 + \frac{mR^2}{I}}$$

and $\quad a = \frac{g}{1 + \frac{I}{mR^2}}$

11. (b,c,d)

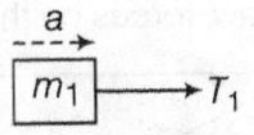

$$T_1 = m_1 a \quad \ldots \text{(i)}$$

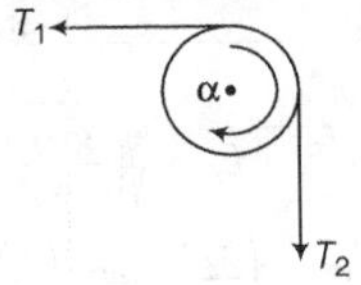

$$(T_2 - T_1)R = I\alpha \quad \ldots \text{(ii)}$$

$$m_2 g - T_2 = m_2 a \quad \ldots \text{(iii)}$$

For no slipping, $a = R\alpha$ ...(iv)

Solving the above equations, we get

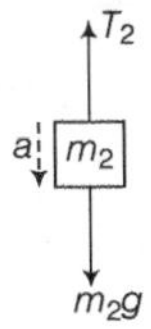

$$a = \frac{m_2 g}{m_1 + m_2 + \frac{I}{R^2}}$$

$$T_1 = \frac{m_1 m_2 g}{m_1 + m_2 + \frac{I}{R^2}}$$

$$T_2 = \frac{(m_1 + (I/R^2))m_2 g}{m_1 + m_2 + \frac{I}{R^2}}$$

12. (a,b,c,d)

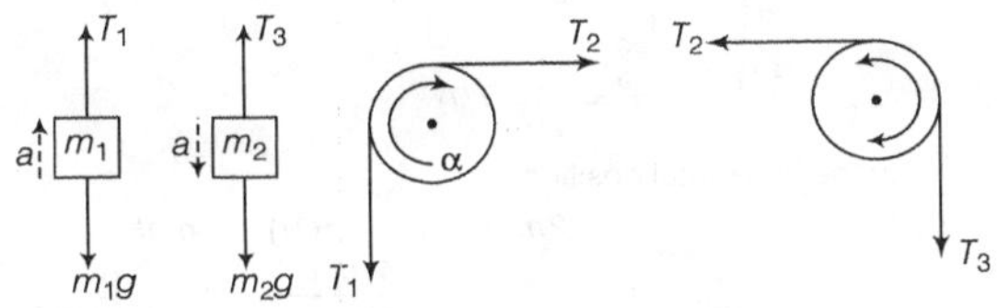

Equations of motion are

$$T_1 - m_1 g = m_1 a$$

$$m_2 g - T_3 = m_2 a$$

$$(T_2 - T_1)R = I\alpha$$

$$(T_3 - T_2)R = I\alpha$$

For no slipping on the pulley, $a = R\alpha$

Solving the above equations, we get

$$a = \frac{(m_2 - m_1)g}{\left(m_1 + m_2 + \frac{2I}{R^2}\right)}$$

$$T_1 = \frac{2m_1 g\left[m_2 + \frac{I}{R^2}\right]}{\left(m_1 + m_2 + \frac{2I}{R^2}\right)}$$

$$T_2 = \left[\frac{(m_1 + m_2)\frac{I}{R^2} + 2m_1 m_2}{m_1 + m_2 + \frac{2I}{R^2}}\right] g$$

$$T_3 = \frac{\left(m_1 + \frac{I}{R^2}\right)2m_2 g}{\left(m_1 + m_2 + \frac{2I}{R^2}\right)}$$

13. (c,d)

(i) $W = \Delta KE$

$$\Rightarrow \quad \tau\theta = \frac{1}{2} I\omega^2$$

$$\therefore \quad (40 \times 0.25)\left(\frac{5}{0.25}\right) = \frac{1}{2}(4)\,\omega^2$$

$$\Rightarrow \quad \omega = \sqrt{200/3} = 8.16 \text{ rad s}^{-1}$$

(ii) Mass of the block is $m = \frac{40}{10} = 4$ kg

For conservation of energy,

$$\begin{bmatrix}\text{Loss in GPE}\\ \text{of 4 kg mass}\end{bmatrix} = \begin{bmatrix}\text{Gain in KE of}\\ \text{4 kg mass}\end{bmatrix} + \begin{bmatrix}\text{Gain in RKE}\\ \text{of cylinder}\end{bmatrix}$$

$$mgh = \frac{1}{2}mv^2 + \frac{1}{2}I\omega^2$$

As $\quad v = R\omega,$

$$mgh = \frac{1}{2}m(R\omega)^2 + \frac{1}{2}I\omega^2$$

$$(4)(10)(5) = \frac{1}{2}(4)\left(\frac{1}{4}\right)^2 \omega^2 + \frac{1}{2}(4)\omega^2$$

$$\Rightarrow \quad \omega = 8 \text{ rad s}^{-1}$$

Note *Though the downward force is same in both the cases, this force has to accelerate only the cylinder in case (i) and it has to accelerate both cylinder and block in case (ii).*

14. About O, $\quad \tau_0 = I_0\alpha$

$$\Rightarrow \quad mgx = \left(\frac{mL^2}{12} + mx^2\right)\alpha$$

$$\alpha = \frac{gx}{\left(\frac{L^2}{12} + x^2\right)}$$

For α to be maximum, $\frac{d\alpha}{dx} = 0$

$$\therefore \quad \left(\frac{L^2}{12} + x^2\right)g - gx \times 2x = 0$$

$$\Rightarrow \quad 2x^2 = \frac{L^2}{12} + x^2$$

$$x^2 = \frac{L^2}{12}$$

$$\Rightarrow \quad x = \frac{L}{2\sqrt{3}}$$

$$\alpha_{max} = (\alpha)_{\text{for } x = \frac{L}{2\sqrt{3}}}$$

$$= \frac{g\left(\frac{L}{2\sqrt{3}}\right)}{\frac{L^2}{12} + \left(\frac{L}{2\sqrt{3}}\right)^2} = \frac{g\sqrt{3}}{L}$$

15. Let C be the centre of mass of the rod and ω be the angular speed of the rod about point O, at an angle θ. From conservation of energy,

$$mg\frac{L}{2}(\cos\theta - \cos\theta_0) = \frac{1}{2}\left(\frac{mL^2}{3}\right)\omega^2$$

$$\omega^2 = \frac{3g}{L}(\cos\theta - \cos\theta_0) \quad \ldots \text{(i)}$$

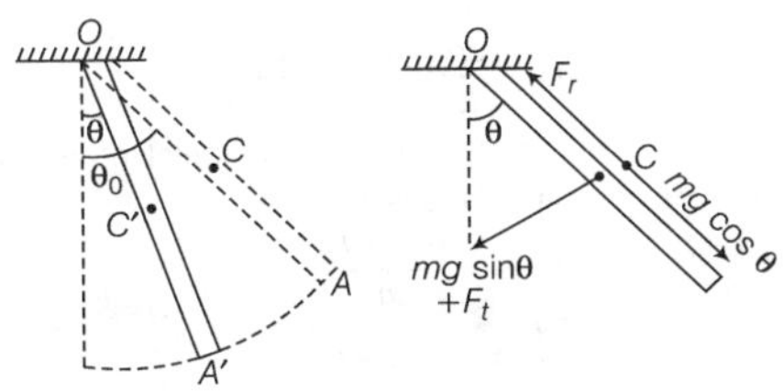

Now, $F_r - mg\cos\theta = m\left(\frac{L}{2}\right)\omega^2$ (This is centripetal force) ...(ii)

From Eqs. (i) and (ii), we get

$$F_r = \frac{mg}{2}(5\cos\theta - 3\cos\theta_0)$$

Angular acceleration of the rod at this instant is

$$\alpha = \frac{\tau}{I} = \frac{mg\left(\frac{L}{2}\sin\theta\right)}{(mL^2/3)} = \frac{3g\sin\theta}{2L}$$

$$a_{COM} = a_c = a_t = \left(\frac{L}{2}\right)\alpha = \frac{3}{4}g\sin\theta$$

(This is tangential acceleration of C)

$$F_t + mg\sin\theta = ma_t = \frac{3\,mg\sin\theta}{4L}$$

$\therefore F_t = -\frac{mg}{4}\sin\theta$, here negative sign indicates that the direction of F_t is opposite the component $mg\sin\theta$.

16. Let ω be the angular velocity of rod in horizontal position. From conservation of energy,

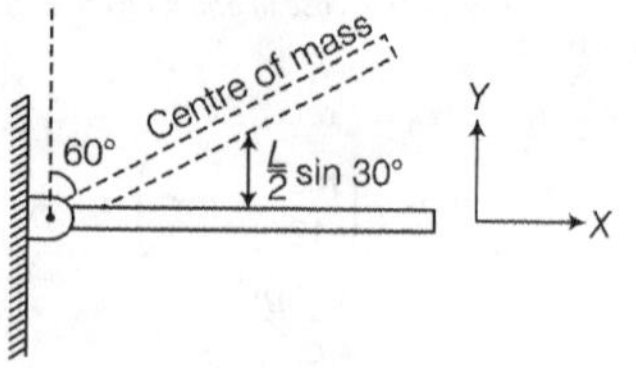

Loss in GPE of centre of mass of the rod

= Gain in RKE of the rod centre of mass of the rod

$$mg\left(\frac{L}{2}\sin 30°\right) = \frac{1}{2}I\omega^2$$

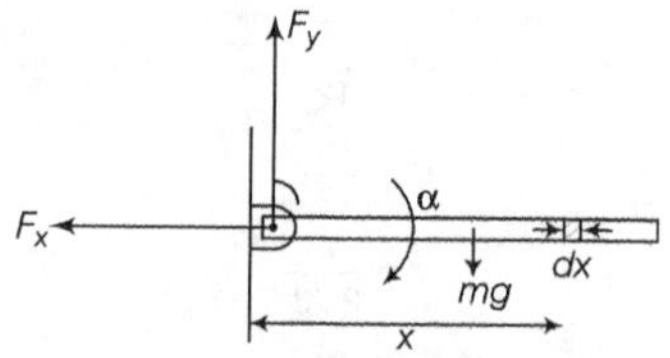

$$mg\,\frac{L}{4} = \frac{1}{2}\,\frac{mL^2}{3}\,\omega^2$$

$$\Rightarrow \quad \omega = \sqrt{\frac{3g}{2L}}$$

$$\alpha = \frac{\tau}{I} = \frac{mg\left(\frac{L}{2}\right)}{\left(\frac{mL^2}{3}\right)} = \frac{3g}{2L}$$

Mass of the element is $dm = \frac{m}{L}dx$

Centripetal force on it is $dF_x = (dm)\omega^2 x$

or $\quad dF_x = \left(\frac{m}{L}dx\right)\left(\frac{3g}{2L}\right)x = \frac{3mg}{2L^2}\,xdx$

$$F_x = \frac{3mg}{2L^2}\int_0^L x\,dx = \frac{3mg}{4},$$

This F_x is 'ON' the rod and supplies centripetal force on it.

Acceleration of centre of mass is

$$a_{COM} = \frac{mg - F_y}{m}$$

$$\Rightarrow \quad \frac{L}{2}\alpha = (mg - F_y)/m$$

$$F_y = mg - \frac{mL}{2}\cdot\alpha$$

$$= mg - \frac{mL}{2}\cdot\frac{3g}{2L}$$

$$= mg - \frac{3mg}{4} = \frac{mg}{4}$$

This, F_4 is also 'ON' the rod.

From Newton's third law, forces on the pivot are

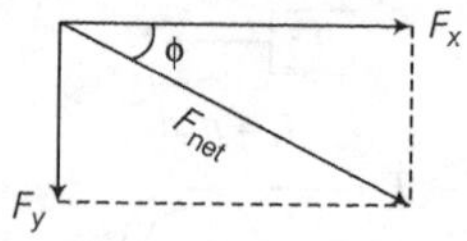

$$F_{net} = \sqrt{F_x^2 + F_y^2} = \sqrt{\left(\frac{3mg}{4}\right)^2 + \left(\frac{mg}{4}\right)^2} = \frac{\sqrt{10}}{4}mg$$

$$\phi = \tan^{-1}\left(\frac{F_y}{F_x}\right) = \tan^{-1}\left(\frac{1}{3}\right)$$

17. When the rod becomes horizontal from conservation of mechanical energy,

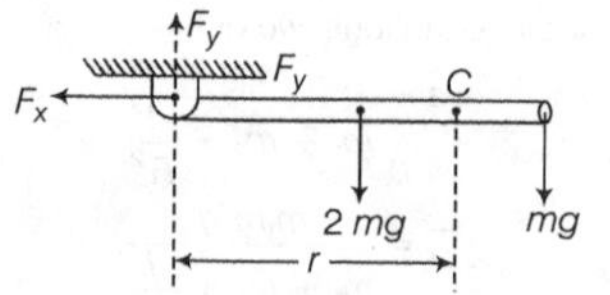

$$\begin{bmatrix}\text{Loss in RKE of}\\ \text{particle + Rod}\end{bmatrix} = \begin{bmatrix}\text{Gain in GPE}\\ \text{of centre of mass of rod}\end{bmatrix} + \begin{bmatrix}\text{Gain in GPE}\\ \text{of particle}\end{bmatrix}$$

$$\therefore \quad \frac{1}{2}I_A(\omega_0^2 - \omega^2) = (2m)(g)(l) + (m)(g)(2l)$$

$$\Rightarrow \quad \frac{1}{2}\left[\frac{(2m)(2l)^2}{3} + m(2l)^2\right]\left(\frac{49}{4}\frac{g}{l} - \omega^2\right) = 4\,mgl$$

$$\therefore \quad \omega^2 = \frac{221}{20}(g/l)$$

In the horizontal position,

$$\tau = (2m)(g)(l) + mg(2l) = 4\,mgl$$

$$\alpha = \frac{\tau}{I} = \frac{4\,mgl}{\left[\frac{2m(2l)^2}{3} + m(2l)^2\right]} = \frac{3}{5}\frac{g}{l}$$

Let the centre of mass of (rod + particle) be at point C. Then,

$$r = \frac{(2m)(l) + m(2l)}{2m + m} = \frac{4l}{3}$$

$$\therefore \quad a_t = r\alpha = \left(\frac{4l}{3}\right)\left(\frac{3g}{5l}\right) = 0.8\,g$$

(= acceleration of COM (a_c) in downward direction)

$$3mg - F_y = (3m)a_c$$

$$\Rightarrow \quad F_y = 3mg - (3m)(0.8)g = 0.6\,mg$$

Further, $\quad a_x = a_c(\leftarrow) = r\omega^2 = \left(\frac{4l}{3}\right)\left(\frac{221}{20}\cdot\frac{g}{l}\right) = 14.73\,g$

$$\therefore \quad F_x = (3m)(a_x) = 44.2\,mg$$

$\therefore$ Total hinge force is

$$F = \sqrt{F_x^2 + F_y^2} = 44.2\text{ mg} = 442\text{ mN}$$

18. (2) From the law of conservation of energy,

$$KE_i + GPE_i = KE_f + GPE_f$$

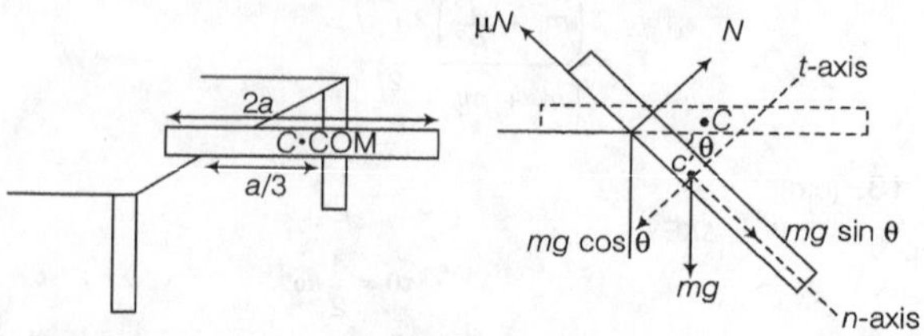

$$\therefore \quad 0 + 0 = \frac{1}{2}I\omega^2 + \left(-mg\frac{a}{3}\sin\theta\right) \quad \ldots(i)$$

From parallel axes theorem, $I = \frac{m(2a)^2}{12} + m\left(\frac{a}{3}\right)^2$.

Using this in Eq. (i) will get

$$\omega^2 = \frac{3g \sin \theta}{2a}$$

Differentiating w.r.t. θ, we get

$$2\omega \frac{d\omega}{d\theta} = \frac{3g}{2a} \cos \theta$$

As, $\alpha = \omega \frac{d\omega}{d\theta}$, $\alpha = \frac{3g}{4a} \cos \theta$

Along t-axis,

$$\Sigma F_t = mg \cos \theta - N = m\frac{a}{3}\alpha$$

or $$mg \cos\theta - N = m \cdot \frac{a}{3} \cdot \frac{3g}{4a} \cos\theta$$

$$\Rightarrow \quad N = \frac{3mg}{4} \cos\theta$$

Along n-axis,

$$\Sigma F_n = \mu N - mg \sin \theta = m\left(\frac{a}{3}\right)\omega^2$$

(this is centripetal force)

$$\therefore \mu\left(\frac{3mg}{4} \cos \theta\right) - mg \sin \theta = m\left(\frac{a}{3}\right)\left(\frac{3g \sin\theta}{2a}\right)$$

$$\mu \frac{3mg}{4} \cos \theta = \frac{3}{2} mg \sin \theta$$

$$\Rightarrow \quad \mu = 2 \tan \theta$$

$$\Rightarrow \quad \beta = 2$$

Explanations for Q. Nos. (19-21)

$$(\tau_O)_{net} = mg \cdot \frac{L}{2} \sin \theta$$

Also, $\tau = I\alpha$

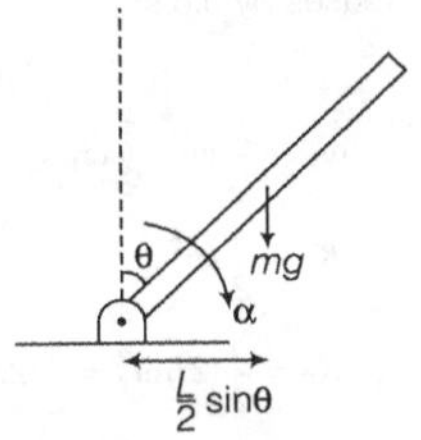

$$\therefore \quad \frac{mgL}{2} \sin \theta = \frac{mL^2}{3} \alpha$$

$$\Rightarrow \quad \alpha = \frac{3g \sin \theta}{2L}$$

We have

$$\alpha = \frac{3g \sin \theta}{2L}$$

$$\Rightarrow \quad \omega \frac{d\omega}{d\theta} = \frac{3g \sin \theta}{2L}$$

$$\int_0^{\omega} \omega \, d\omega = \frac{3g}{2L} \int_0^{\theta} \sin\theta \, d\theta$$

$$\Rightarrow \quad \frac{\omega^2}{2} = \frac{3g}{2L} (1 - \cos \theta)$$

or $$\omega = \sqrt{\frac{3g}{L} (1 - \cos \theta)}$$

Alternate Method From the conservation of mechanical energy,

Loss in GPE of centre of mass of rod = Gain in RKE of rod

$$mg \frac{L}{2}(1 - \cos \theta) = \frac{1}{2} I\omega^2 = \frac{1}{2}\left(\frac{mL^2}{3}\right)\omega^2$$

$$\Rightarrow \quad \omega = \sqrt{\frac{3g}{L}(1 - \cos \theta)}$$

When the rod is horizontal, $\theta = \frac{\pi}{2}$, so, $\alpha = \frac{3g}{2L}$. So, tangential acceleration of free end is $a_t = L\alpha = \frac{3g}{2}$

19. (d) **20.** (c) **21.** (b)

Explanations for Q. Nos. (22-23)

(i) $a_c = \omega^2 r$

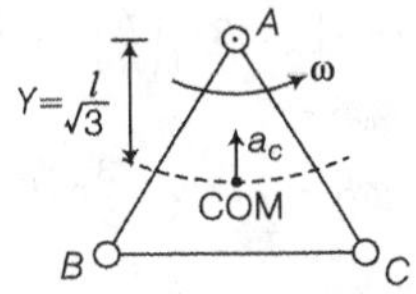

The centre of mass moves along a circular path with constant angular velocity. So, there must be a horizontal centripetal force directed towards the axis at the hinge.

$$F_c = 3m\left(\frac{l}{\sqrt{3}}\right)\omega^2 = \sqrt{3}\, ml\omega^2$$

(ii) Let F_x and F_y be the forces applied by the hinge along X-axis and Y-axis, respectively.

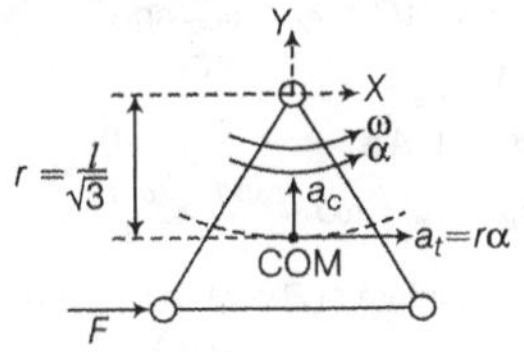

$$\Sigma F_x = F_x + F = (3m)\, a_t = 3m\left(\frac{l}{\sqrt{3}}\alpha\right) \quad \text{...(i)}$$

$$\Sigma F_y = F_y = 3m\left(\frac{l}{\sqrt{3}}\right)\omega^2 \quad \text{...(ii)}$$

$$\Sigma\tau = F\left(\frac{\sqrt{3}}{2} l\right) = (ml^2 + ml^2)\, \alpha \quad \text{...(iii)}$$

From Eq. (iii), we get

$$\alpha = \frac{\sqrt{3}F}{4ml}$$

From Eq. (i), we get

$$F_x + F = 3m\left(\frac{l}{\sqrt{3}} \cdot \frac{\sqrt{3}F}{4ml}\right) = \frac{F}{4}$$

$$\therefore \quad F_x = -\frac{3F}{4}$$

From Eq. (ii), we get

$$F_y = \sqrt{3}\, ml\, \omega^2$$

22. (c) **23.** (a)

DPP-4 Angular Momentum and Its Conservation

1. (c) The body is in combined translation and rotation. So, its angular momentum about O is

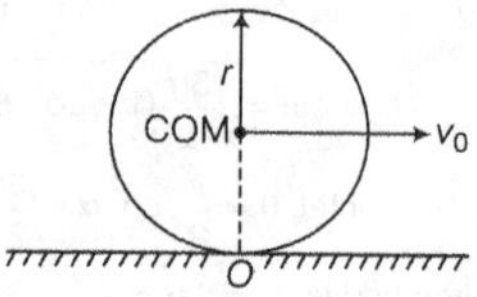

$$\mathbf{L} = \mathbf{L}_{COM} + m(\mathbf{r}_0 \times \mathbf{v}_0)$$

where, $\mathbf{L}_{COM}$ = angular momentum in centre of mass reference frame

and $m(\mathbf{r}_0 \times \mathbf{v}_0)$ = angular momentum of centre of mass w.r.t. point O

$$\therefore \quad \mathbf{L}_{COM} = I\omega = \left(\frac{1}{2}mR^2\right)\left(\frac{v}{R}\right) = \frac{1}{2}mvR \otimes$$

and $m(\mathbf{r}_0 \times \mathbf{v}_0) = mvR \otimes$

$$\therefore \quad |\mathbf{L}| = mvR + \frac{1}{2}mvR = \frac{3}{2}mvR$$

2. (a) In this new situation,

$$\mathbf{L}_{COM} = \frac{1}{2}mvR\odot$$

and $\quad m(\mathbf{r}_0 \times \mathbf{v}_0) = mvR\otimes$

$$\therefore \quad |\mathbf{L}| = -\frac{1}{2}mvR + mvR = \frac{1}{2}mvR$$

3. (b) The situation is shown in the figure.

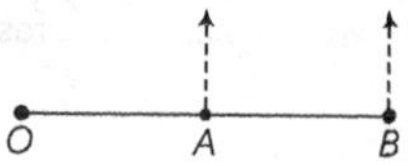

The velocity of A w.r.t O is $v_A = \omega\frac{d}{2}$ and that of B w.r.t. O is $v_B = \omega d$

$\therefore$ Velocity of B w.r.t. A is

$$v_B - v_A = \omega d - \frac{\omega d}{2} = \frac{\omega d}{2}$$

The angular momentum of B w.r.t. A is, therefore,

$$L = mvr = m\left(\frac{\omega d}{2}\right)\left(\frac{d}{2}\right) = \frac{1}{4}m\omega d^2$$

4. (b) Let the moment of inertia of the wheel about the axis be I. Initially, the first wheel is rotating at an angular speed ω about the axle and the second wheel is at rest. Take both the wheels together as the system. The total angular momentum of the system before the coupling is $I\omega + 0 = I\omega$.

When the second wheel is dropped into the axle, the two wheels ship on each other and exert forces of friction. The forces of friction have torques about the axis of rotation, but these are torques of internal forces. No external torque is applied on the two wheel system and hence the angular momentum of the system remains unchanged. If the common angular speed is ω', after coupling the total angular momentum of the two wheel system will be $(2I)\omega'$. Thus, $I\omega = 2I\omega' \Rightarrow \omega' = \frac{\omega}{2}$

5. (b) Since, there is no external torque in this process, we can apply the principle of conservation of angular momentum

$$I\omega = I'\omega'$$

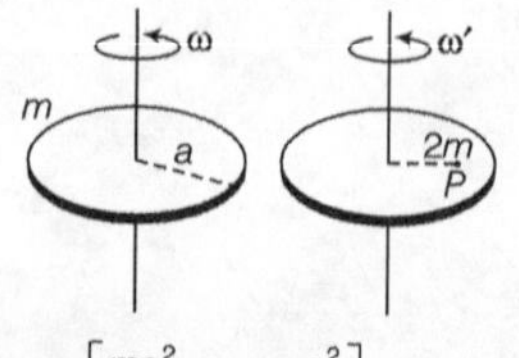

$$\Rightarrow \quad \frac{ma^2}{2}\cdot\omega = \left[\frac{ma^2}{2} + 2m\cdot\frac{a^2}{4}\right]\omega' \Rightarrow \omega' = \frac{\omega}{2}$$

6. (c) If we take ice caps + earth as the system, the external torque on the system, in the process of ice caps melting and water distributing uniformly over the surface of the earth is zero.

$\therefore$ Angular momentum is conserved, i.e. $I_1\omega_1 = I_2\omega_2$

$$I_1\frac{2\pi}{T_1} = I_2\frac{2\pi}{T_2}$$

$$\Rightarrow \quad \frac{I_1}{I_2} = \frac{T_1}{T_2}$$

Let I_1 be the moment of the inertia before the ice melts and I_2 be the moment of inertia after the ice melts.

$$T_1 = 24\text{ h and } T_2 = ?$$

Let M be the mass of the earth and m be the mass of ice. Thus, $I_1 = \frac{2}{5}MR^2$ and $I_2 = \frac{2}{5}MR^2 + \frac{2}{3}mR^2$ ($\because$ Water of total mass m is distributed on the surface of the earth. We can assume it as a hollow sphere of radius R and mass m).

$$I_1\omega_1 = I_2\omega_2$$

$$\Rightarrow \quad I_1\cdot\frac{2\pi}{T_1} = (I_1 + \Delta I)\left(\frac{2\pi}{T_2}\right)$$

$$\text{or} \quad \frac{T_2}{T_1} = \frac{I_1 + \Delta I}{I_1} = 1 + \frac{\Delta I}{I_1}$$

$$\Rightarrow \quad \Delta T = T_2 - T_1 = T_1\left(\frac{\Delta I}{I_1}\right)$$

$$\text{or} \quad \Delta T = (24 \times 3600)\left[\frac{\frac{2}{3}\times 2.3\times 10^{19}\, R^2}{\frac{2}{5}\times 5.94\times 10^{24}\, R^2}\right] \approx 0.6\text{ s}$$

So, time period increases by 0.6 s.

7. (d) $I_1\omega_1 = I_2\omega_2$

$$\Rightarrow \quad \omega_2 = \frac{I_1}{I_2}\omega_1 = \frac{1}{2}\omega_1$$

$$K_1 = \frac{1}{2}I\omega^2$$

$$K_2 = \frac{1}{2}(2I)\omega_2^2 = \frac{1}{2}(2I)\left(\frac{\omega_1}{2}\right)^2 = \frac{K_1}{2} = \frac{K}{2}$$

8. (b) Let the boy rotated the umbrella in clockwise sense. Initial angular momentum of the system was zero. To make the final angular momentum zero, the platform should rotate in anti-clockwise sense. We take here anti-clockwise +ve and clockwise – ve.

$$\omega_{up} = \omega_u - \omega_p \Rightarrow \omega_u = \omega_{up} + \omega_p = -2 + \omega$$

$$L_i = 0,\ L_f = (3\times 10^{-3})\omega + (2\times 10^{-3})(\omega - 2)$$

$$= 5\times 10^{-3}\omega - 4\times 10^{-3}$$

$$L_f = L_i \Rightarrow 5\times 10^{-3}\omega - 4\times 10^{-3} = 0$$

$$\Rightarrow \quad \omega = 0.8\text{ rev s}^{-1}$$

9. (a) Take the kid + ball + platform as the system. In the process of boy catching the ball, there is no external torque on the system, about the axis of rotation. So, angular momentum is conserved.

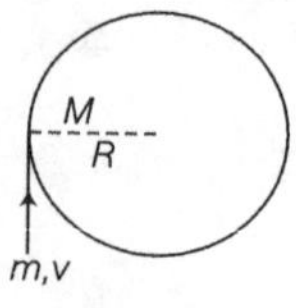

$$L_i = L_{ball} + L_{kid + platform} = mvR + 0 = mvR$$

$$L_f = [I_{kid} + I_{ball} + I_{platform}]\omega$$

$$= (MR^2 + mR^2 + I)\omega = [I + R^2(m + M)]\omega$$

$$L_i = L_f$$

$$\Rightarrow \quad mvR = [I + R^2(m + M)]\omega$$

$$\Rightarrow \quad \omega = \frac{mvR}{I + R^2(m + M)}$$

10. (c) Let ω be the angular velocity (boy + platform) just after he throws the ball. Just before throwing the ball, $L_i = 0$

$$L_f = mvR + (MR^2 + I)\omega$$

$$L_f = L_i$$

$$\Rightarrow \quad mvR + (MR^2 + I)\,\omega = 0$$

$$\Rightarrow \quad \omega = -\frac{mvR}{MR^2 + I}$$

Negative sign signifies that the sense of rotation is opposite to that of velocity of the ball.

11. (d) Let ω' be the new angular speed of the platform.

$$v_{bp} = v_b - v_p \Rightarrow v_b = v_{bp} + v_p = v + R\omega'$$

$$L_i = (MR^2 + I)\omega, L_f = L_{boy} + L_{platform}$$

$$= Mv_bR + I\omega'$$

or $\quad L_f = M(v + R\omega')R + I\omega' = MvR + (MR^2 + I)\omega'$

$$L_f = L_i$$

$$\Rightarrow \quad MvR + (MR^2 + I)\omega' = (I + MR^2)\omega$$

$$\therefore \quad \omega' = \omega - \frac{MvR}{I + MR^2}$$

12. (c) Since, there is no external torque, angular momentum will remain conserved. The moment of inertia will first decrease till the tortoise moves from A to C and then increase as it moves from C to D. Therefore, ω will initially increase and then decreases.

Moment of inertia when tortoise is at A is $I_1 = mR^2 + \frac{MR^2}{2}$ and moment of inertia when tortoise is at B is $I_2 = mr^2 + \frac{MR^2}{2}$

Here, $\quad r^2 = a^2 + [\sqrt{R^2 - a^2} - vt]^2$

$$L_i = L_f \Rightarrow I_1\omega_0 = I_2\omega(t)$$

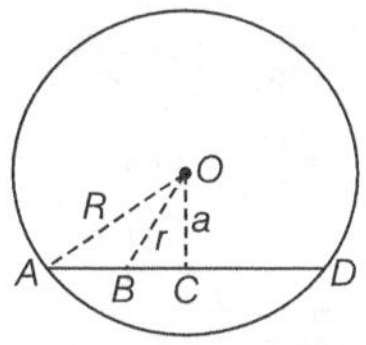

Substituting the values, we can see that variation of $\omega(t)$ is non-linear. So, option (c) is correct.

13. (b) Angular momentum of the system about point O will remain conserved.

$$L_i = L_f \Rightarrow mvL = I\omega \Rightarrow mvL = \left(mL^2 + \frac{ML^2}{3}\right)\omega$$

$$\Rightarrow \quad \omega = \frac{3mv}{L(3m + M)}$$

14. (a,c) Consider the situation shown in the figure. The structure rotates about the perpendicular bisector of the rod at an angular speed ω. The velocity of ball A w.r.t. centre O is $v = \omega\frac{d}{2}$.

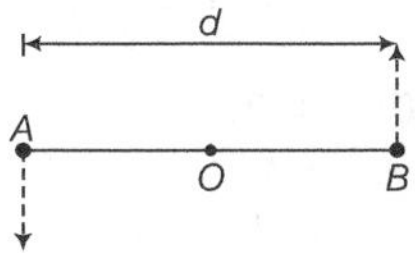

The angular momentum of ball A w.r.t. axis is

$$L_1 = mvR = m\left(\frac{\omega d}{2}\right)\left(\frac{d}{2}\right) = \frac{1}{4}m\omega d^2$$

The same is the angular momentum L_2 of the second ball. The angular momentum of the system is equal to the sum of these two angular momenta, i.e. $L = \frac{1}{2}m\omega d^2$.

15. (a,b)

(i) $\mathbf{L} = I\omega = M\left(\frac{R^2}{2}\right)\omega\hat{\mathbf{k}}$

(ii) $\mathbf{L} = I\omega = (I_{COM} + Md^2)\,\omega\hat{\mathbf{k}} = \left(M\frac{R^2}{2} + Md^2\right)\omega\hat{\mathbf{k}}$

$$\mathbf{L} = M\left(\frac{R^2}{2} + d^2\right)\omega\hat{\mathbf{k}}$$

16. (c,d)

(i) $\mathbf{L} = \mathbf{L}_{COM} + M(\mathbf{r}_0 \times \mathbf{v}_0)$

or $\quad \mathbf{L} = \frac{MR^2}{2}\omega(-\hat{\mathbf{k}}) + Mvd(-\hat{\mathbf{k}})$

$$= \left(\frac{MR^2}{2}\omega + Mvd\right)(-\hat{\mathbf{k}})$$

(ii) $\mathbf{L} = \mathbf{L}_{COM} + M(\mathbf{r}_0 \times \mathbf{v}_0)$

$$= \frac{MR^2}{2}\omega(\hat{\mathbf{k}}) + Mvd(-\hat{\mathbf{k}})$$

$$= \left(Mvd - \frac{MR^2}{2}\omega\right)(-\hat{\mathbf{k}})$$

17. (b,c) Here, the external force F is applied on the system along the axis of rotation. So, τ_{ext} is zero. Thus, I is conserved.

(i) $I_1 = \frac{MR^2}{2} + mR^2$

$$\omega_1 = \omega_0$$

$$I_2 = \frac{MR^2}{2}$$

$$\omega_2 = ?$$

$$I_1\omega_1 = I_2\omega_2$$

$$\Rightarrow \quad \omega_2 = \frac{I_1}{I_2}\omega_1$$

$$= \frac{\left(\frac{MR^2}{2} + mR^2\right)}{(MR^2/2)}\omega_0$$

$$= \left(1 + \frac{2m}{M}\right)\omega_0$$

(ii) The work performed by force,

F = change in KE of the system.

$$W = K_f - K_i = \frac{1}{2}I_2\omega_2^2 - \frac{1}{2}I_1\omega_1^2$$

$$= \frac{1}{2}\left(\frac{MR^2}{2}\right)\left(1 + \frac{2m}{M}\right)^2\omega_0^2 - \frac{1}{2}\left(\frac{MR^2}{2} + mR^2\right)\omega_0^2$$

$$= \frac{1}{2}mR^2\omega_0^2\left(1 + \frac{2m}{M}\right)$$

Explanations for Q. Nos. (18-20)

Since, there is no external torque in the entire process, $L_i = L_f$

(i) $I_1\omega_1 = I_2\omega_2$, $I_1 = 1.6 + 4(0.9)^2 \times 2 = 8.08$ kg-m^2

and $\quad \omega_1 = 2\pi \times 0.5 = \pi$ rad s^{-1}

$$I_2 = 1.6 + 4\,(0.15)^2 \times 2 = 1.78 \text{ kg-m}^2$$

$$\omega_2 = \frac{I_1}{I_2}\omega_1 = \frac{8.08}{1.78} \times \pi = 14.3 \text{ rad s}^{-1}$$

(ii) Initial KE is $K_i = \frac{1}{2}I_1\omega_1^2 = \frac{1}{2}(8.08)\,(\pi)^2 = 39.9$ J

Final KE is $K_f = \frac{1}{2}I_2\omega_2^2 = \frac{1}{2}(1.78)(14.3)^2 = 182$ J

(iii) Work done by the man $= K_f - K_i = 182 - 39.9 = 142.1$ J

18. (c) **19.** (a) **20.** (b)

21. System is free to rotate but not free to translate. During collision, net torque on the system (rod A + rod B + mass m) about point P is zero.

$\therefore$ Angular momentum of the system before collision = Angular momentum of the system after collision (about P).

Let ω be the angular velocity of the system just after collision, then

$$L_i = L_f$$

$$\Rightarrow \quad mv(2l) = I\omega$$

or $$mv(2l) = \left[m(2l)^2 + m_A\frac{l^2}{3} + m_B\left(\frac{l^2}{12} + \left(\frac{l}{2} + l\right)^2\right)\right]\omega$$

$$I = (0.05)\,(2 \times 0.6)^2 + (0.01)\frac{(0.6)^2}{3} + 0.02\left[\frac{0.6^2}{12} + \left(\frac{0.6}{2} + 0.6\right)^2\right]$$

$$= 0.09 \text{ kg-m}^2$$

$$\therefore \quad \omega = \frac{2mvl}{I} = \frac{2(0.05)(v)(0.06)}{0.09}$$

$$\Rightarrow \quad \omega = 0.067\, v$$

Now, after collision mechanical energy is conserved.

$\therefore$ Decrease in rotational KE = Increase in gravitational PE

or $$\frac{1}{2}I\omega^2 = mg(2l) + m_A g\left(\frac{l}{2}\right) + m_B g\left(l + \frac{l}{2}\right)$$

or $$\omega^2 = \frac{gl(4m + m_A + 3m_B)}{I} = \frac{(9.8)\,(0.6)\,(4 \times 0.05 + 0.01 + 3 \times 0.02)}{0.09}$$

$$\therefore \quad \omega = 4.2 \text{ rad s}^{-1}$$

$$v = \frac{\omega}{0.067} = \frac{4.2}{0.067} = 63 \text{ ms}^{-1}$$

22. About the centre of the disc, $\mathbf{L}_i = \mathbf{L}_f$

$$muR + 0 = \left(mR^2 + \frac{MR^2}{2}\right)\omega$$

$$\Rightarrow \quad (80)(4)(2) = (2)^2\left(80 + \frac{160}{2}\right)\omega$$

$$640 = 640\,\omega$$

$$\Rightarrow \quad \omega = 1 \text{ rad s}^{-1}$$

23. In this problem, we use L' for angular momentum

(i) $(L_i')_{\text{rod + insect about } O} = (L_f')_{\text{rod + insect about } O}$

$$0 + Mv\frac{L}{4} = \left[\frac{ML^2}{12} + M\left(\frac{L}{4}\right)^2\right]\omega \quad \Rightarrow \quad \omega = \frac{12v}{7L}$$

(ii) Due to the torque of weight of insect about O, angular momentum of the system will not remain conserved (although angular velocity ω is constant).

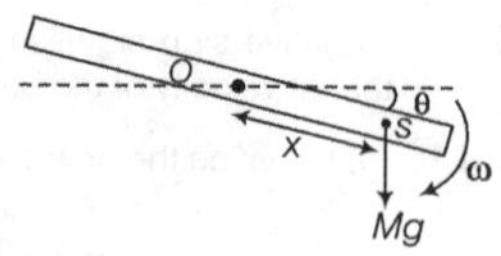

As the insect moves towards B, moment of inertia of the system increases, hence the angular momentum of the system increases. Let at time t, the insect be at a distance x from 0 and by then the rod has rotated through an angle θ. The angular momentum at that instant is

$$L' = \left(\frac{ML^2}{12} + Mx^2\right)\omega$$

$$\frac{dL'}{dt} = 2M\omega\, x\frac{dx}{dt} \qquad (\omega \text{ is constant})$$

$$\tau = 2M\omega\, x\frac{dx}{dt} \qquad \left(\because \frac{dL'}{dt} = \tau\right)$$

$$Mgx\cos\theta = 2M\omega\, x\frac{dx}{dt}$$

$$dx = \frac{g}{2\omega}\cos\omega t\, dt \qquad (\because \theta = \omega t)$$

At $t = 0$, $x = \frac{L}{4}$ and at time

$$t = \frac{T}{4} \text{ or } \frac{\pi}{2\omega},\ x = \frac{L}{2}$$

$$\int_{L/4}^{L/2} dx = \frac{g}{2\omega}\int_0^{\pi/2\omega}\cos\omega t\, dt$$

$$\Rightarrow \quad \frac{L}{2} - \frac{L}{4} = \frac{g}{2\omega^2}\left[\sin\frac{\pi}{2} - \sin 0^\circ\right]$$

$$\frac{L}{4} = \frac{g}{2\omega^2} \quad \text{or } \omega = \sqrt{\frac{2g}{L}}$$

$$\sqrt{\frac{2g}{L}} = \frac{12}{7}\cdot\frac{v}{L} \qquad \left(\because \omega = \frac{12v}{7L}\right)$$

$$v = \frac{7}{12}\sqrt{2gL} = \frac{7}{12}\sqrt{2 \times 10 \times 1.8} = 3.5 \text{ ms}^{-1}$$

DPP-5 Combined Translation and Rotation

1. (d) Applying law of conservation of linear momentum, we get

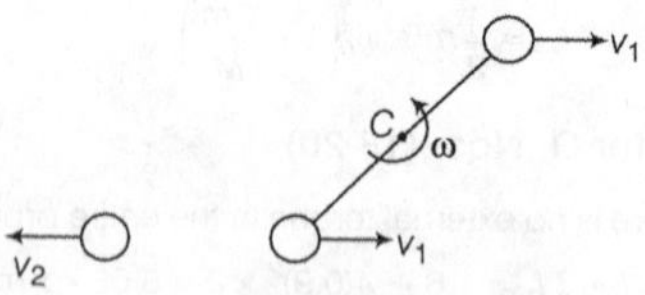

$$mv_0 = 2mv_1 - mv_2$$

$$\Rightarrow \quad 2v_1 - v_2 = v_0 \qquad \text{...(i)}$$

Applying law of conservation of angular momentum about centre of mass (of rod + two identical particles), we get

$$mv_0\left(\frac{L}{2}\sin 45^\circ\right) = 2m\left(\frac{L}{2}\right)^2\omega - mv_2\left(\frac{L}{2}\sin 45^\circ\right)$$

$$mv_0\frac{L}{2\sqrt{2}} = \frac{mL^2\omega}{2} - mv_2\frac{L}{2\sqrt{2}}$$

$$\Rightarrow \quad v_0 = \sqrt{2}L\omega - v_2 \qquad \text{...(ii)}$$

It is given that mechanical energy is conserved. So, the collision is elastic. Thus,

Relative speed of separation = Relative speed of approach

$$v_2 + v_1 + \left(\frac{L}{2}\omega\right)\left(\frac{1}{\sqrt{2}}\right) = v_0$$

$$\Rightarrow \quad v_0 = v_1 + v_2 + \frac{L\omega}{2\sqrt{2}} \qquad \text{...(iii)}$$

On solving the Eqs. (i), (ii) and (iii), we get

$$\omega = \frac{4\sqrt{2}}{7}\cdot\frac{v_0}{L}$$

2. (c) $N = mg = 50\,(10) = 500$ N

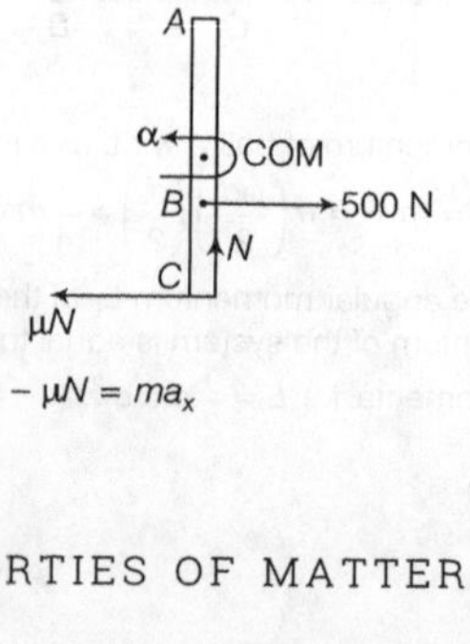

$$500 - \mu N = ma_x$$

$$\Rightarrow \quad a_x = \frac{500 - \mu N}{m}$$

or $$a_x = \frac{500 - (0.3)(500)}{50}$$

$$a_x = 7 \text{ ms}^{-2}$$

$$\tau_{COM} = 500(1) - (0.3)(500)(3)$$

$$\Rightarrow \quad \tau_{COM} = 50 \text{ N-m}$$

$$\tau_{COM} = \frac{mL^2}{12}\alpha = \frac{(50)(6)^2}{12}\alpha = 150\,\alpha$$

$$\therefore \quad 150\,\alpha = 50 \text{ N-m}$$

$$\Rightarrow \quad \alpha = \frac{1}{3} \text{ ms}^{-2}$$

$$\therefore \quad a_A = a_x - 3(\alpha) = 7 - 3\left(\frac{1}{3}\right) = 6 \text{ ms}^{-2} \text{ towards right.}$$

3. (a) $J = mv_c$ and $Jh = \frac{2}{5}mr^2\omega$

$$v_c = \frac{J}{m} \text{ and } \omega = \frac{5h(0.4r)}{2\,mr^2} = \frac{J}{mr}$$

$$(E_{\text{total}})_i = \frac{1}{2}mv_c^2 + \frac{2}{5}mr^2\omega^2$$

$$= \frac{1}{2}m\left(\frac{J}{m}\right)^2 + \frac{2}{5}mr^2\left(\frac{J}{mr}\right)^2$$

$$= \frac{J^2}{2m} + \frac{2J^2}{5m} = \frac{9}{10}\frac{J^2}{m}$$

$$(K_{\text{translational}})_i = \frac{1}{2}mv_c^2 = \frac{1}{2}m\left(\frac{J^2}{m^2}\right) = \frac{J^2}{2m}$$

$(K_{\text{rotational}})_i$ remains constant.

$(K_{\text{translational}})_i$ becomes

$(K_{\text{translational}})_f$ of $(m + 4m)$ and mgh'

From conservation of momentum,

$$mv_c = (m + 4m)v'$$

$$\Rightarrow \quad v' = \frac{v_c}{5}$$

$$(K_{\text{translational}})_f \text{ of } m + 4m = \frac{1}{2}(5m)(v')^2$$

$$= \frac{1}{2}5m\left(\frac{v_c}{5}\right)^2 = \frac{mv_c^2}{10}$$

$$= \frac{m}{10}\cdot\frac{J^2}{m^2} = \frac{J^2}{10\,m}$$

$$\therefore \quad \frac{J^2}{2m} = \frac{J^2}{10m} + mgh'$$

$$\Rightarrow \quad h' = \frac{1}{mg}\cdot\frac{J^2}{m}\left(\frac{1}{2} - \frac{1}{10}\right) = \frac{2}{5}\cdot\frac{J^2}{m^2 g}$$

4. (c,d) From geometry,

$$x = \frac{R}{\sin\theta}$$

$$\therefore \quad v = \frac{dx}{dt} = \frac{d}{dt}\left(\frac{R}{\sin\theta}\right) = \frac{R(d\theta/dt)(-\cos\theta)}{\sin^2\theta}$$

$$\omega = -\frac{d\theta}{dt}$$

so $$v = \frac{\omega R\cos\theta}{\sin^2\theta}$$

$$\because \quad \omega = \frac{v\sin^2\theta}{R\cos\theta} = \frac{v}{R}\cdot\frac{(R/x)^2}{\left(\sqrt{x^2 - R^2}/x\right)} = \frac{Rv}{x(x^2 - R^2)^{1/2}}$$

$$\alpha = \frac{d\omega}{dt} = \frac{d}{dt}\left[\frac{Rv}{x\sqrt{x^2 - R^2}}\right]$$

$$= -\frac{Rv^2(2x^2 - R^2)}{x^2(x^2 - R^2)^{3/2}}$$

5. (c) About C,

$$L_i = L_f$$

$$\Rightarrow \quad mv_0\frac{L}{2} + 0 = 0 + \frac{ML^2}{12}\cdot\omega$$

$$\omega = \frac{6v_0 m}{ML}$$

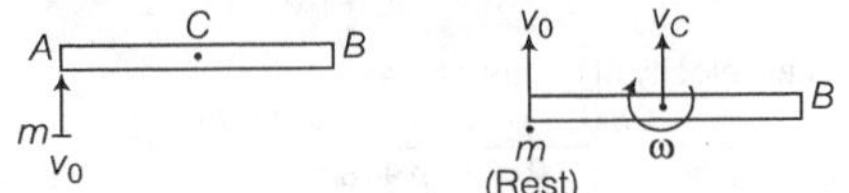

Before collision — (Rest) Immediately after collision, $t = 0$

$e = 1 \Rightarrow$ speed of separation = speed of approach

$$\Rightarrow \quad v_A = v_0 \text{ (at } t = 0)$$

$$\therefore \quad v_A = v_C + \frac{L}{2}\omega$$

$$\Rightarrow \quad v_0 = v_C + \frac{L}{2}\cdot\frac{6mv_0}{mL}$$

$$v_C = v_0\left[1 - 3\frac{m}{M}\right]$$

Conservation of linear momentum gives $mv_0 = Mv_C \Rightarrow v_C = \frac{m}{M}v_0$

$$v_0\left[1 - 3\frac{m}{M}\right] = \frac{m}{M}v_0 \Rightarrow 1 = 4\frac{m}{M}$$

$$\Rightarrow \quad \frac{m}{M} = \frac{1}{4}$$

6. (b) The point P has zero velocity if velocity due to rotation and velocity due to centre of mass (both possessed by point P) must be equal and opposite immediately after collision. This can happen only if point P is between C and B.

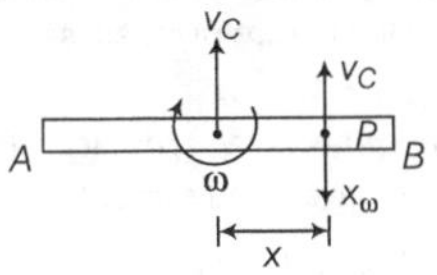

Let the point P is at a distance x from point C, between C and B.

Then, $$v_C = x\omega \quad \ldots\text{(i)}$$

As, $$\frac{m}{M} = \frac{1}{4},$$

$$\omega = \frac{6v_0}{L}\cdot\frac{1}{4} = \frac{3}{2}\frac{v_0}{L}$$

$$\Rightarrow \quad v_C = v_0\left[1 - 3\left(\frac{1}{4}\right)\right] = \frac{v_0}{4}$$

Thus, Eq. (i) can be written as $\frac{v_0}{4} = x\cdot\frac{3v_0}{2L}$

$$\Rightarrow \quad x = \frac{L}{6}$$

$$\therefore \quad AP = \frac{L}{2} + \frac{L}{6} = \frac{5L}{6}$$

7. (d) Time period, $T = \frac{2\pi}{\omega} = \frac{2\pi}{3v_0}\cdot 2L = \frac{4\pi L}{3v_0}$

Given, $$t = \frac{\pi L}{3v_0}$$

$$\Rightarrow \quad t = \frac{T}{4}$$

So, $$\theta = \omega t = \frac{2\pi}{T}\cdot\frac{T}{4} = \frac{\pi}{2}$$

The situation at $t = \frac{\pi L}{3v_0}\left(\text{i.e. } = \frac{T}{4}\right)$ is as shown in the figure.

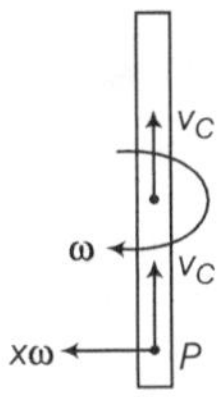

We know that $v_C = \frac{v_0}{4}$

and $x\omega = \frac{L}{6} \cdot \frac{3v_0}{2L} = \frac{v_0}{4}$

$\therefore$ Speed of P at this instant is

$$v_P = \sqrt{(v_C)^2 + (x\omega)^2} = \sqrt{\frac{v_0^2}{16} + \frac{v_0^2}{16}} = \frac{v_0}{2\sqrt{2}}$$

8. (a) Initial velocity of centre of mass

$$= \frac{2m(-v) + m(+2v) + 8m(0)}{2m + m + 8m} = 0$$

As, $F_{ext} = 0, (v_f)_{COM} = 0$

9. (b) $\tau_{ext} = 0$ [on two particles + bar system]

$\therefore$ $L_i = L_f$

$$\Rightarrow (2m)(v)(a) + (m)(2v)(2a) = \left[2m(a)^2 + m(2a)^2 + \frac{8m(6a)^2}{12}\right]\omega$$

By solving, we get $\omega = \frac{v}{5a}$

10. (d) $E = \frac{1}{2}I\omega^2 = \frac{1}{2}\left[2m(a)^2 + m(2a)^2 + \frac{8m(6a)^2}{12}\right]\omega^2$

or $E = \frac{1}{2}(30\,ma^2)\left(\frac{v^2}{25a^2}\right) = \frac{3}{5}mv^2$

11. (a) $\mathbf{v}_{COM} = \frac{m_1\mathbf{v}_1 + m_2\mathbf{v}_2}{m_1 + m_2} = \frac{m(v_0\hat{\mathbf{j}}) + m(0)}{m + m} = \frac{v_0}{2}\hat{\mathbf{j}}$

12. (c) $\mathbf{R}_{COM} = \frac{m_1\mathbf{r}_1 + m_2\mathbf{r}_2}{m_1 + m_2} = \frac{m\left(-\frac{L}{2}\hat{\mathbf{j}}\right) + m\left(\frac{L}{2}\hat{\mathbf{i}}\right)}{m + m} = \frac{1}{4}L(\hat{\mathbf{i}} - \hat{\mathbf{j}})$

13. (b) Before collision, in the centre of mass frame, the angular momentum about the centre of mass is

$$(\mathbf{L}')_{\text{initial, COM}} = \mathbf{L}'_{1\,COM} + \mathbf{L}'_{2\,COM}$$
$$= (\mathbf{r}_1 - \mathbf{R}_{COM}) \times m_1(\mathbf{v}_1 - \mathbf{v}_{COM}) + (\mathbf{r}_2 - \mathbf{R}_{COM}) \times m_2(\mathbf{v}_2 - \mathbf{v}_{COM})$$

(we have used symbol L' for angular momentum)

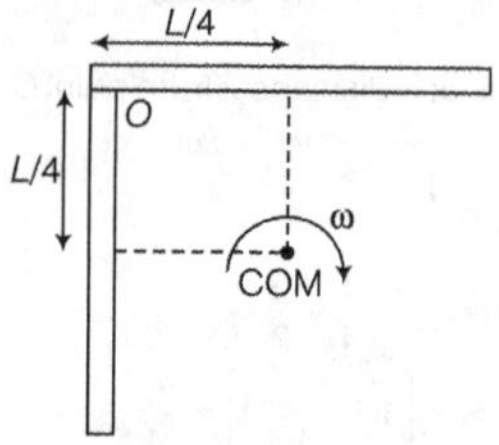

$$(\mathbf{L}')_{\text{initial, COM}} = \left[-\frac{L}{2}\hat{\mathbf{j}} - \frac{1}{4}(L\hat{\mathbf{i}} - L\hat{\mathbf{j}})\right] \times m\left[v_0\hat{\mathbf{j}} - \frac{v_0}{2}\hat{\mathbf{j}}\right]$$
$$+ \left[\frac{L}{2}\hat{\mathbf{i}} - \frac{1}{4}(L\hat{\mathbf{i}} - L\hat{\mathbf{j}})\right] \times m\left[0 - \frac{v_0}{2}\hat{\mathbf{j}}\right]$$
$$= -\frac{1}{4}mv_0L\,\hat{\mathbf{k}}$$

$I_0 = I_{COM} + \text{mass}\,(r^2)$ (from perpendicular axes theorem)

$I_{COM} = I_0 - \text{mass}\,(r^2)$

$$\therefore \quad I_{COM} = \left(\frac{mL^2}{3} + \frac{mL^2}{3}\right) - 2m\left(\sqrt{\left(\frac{L}{4}\right)^2 + \left(\frac{L}{4}\right)^2}\right)^2 = \frac{5}{12}mL^2$$

$$\mathbf{L}'_{\text{initial, COM}} = \mathbf{L}'_{\text{final, COM}}$$

$$\Rightarrow \quad -\frac{1}{4}mv_0L\,\hat{\mathbf{k}} = I_{COM}\,\omega_{COM}$$

$$\omega_{COM} = \frac{-\frac{1}{4}mv_0L}{\frac{5}{12}mL^2}\hat{\mathbf{k}} = \frac{-3}{5} \cdot \frac{v_0}{L}\hat{\mathbf{k}}$$

14. (d) The point A will have maximum velocity, when its velocity is in the direction of $\mathbf{v}_{COM}$ i.e. along $\hat{\mathbf{j}}$.

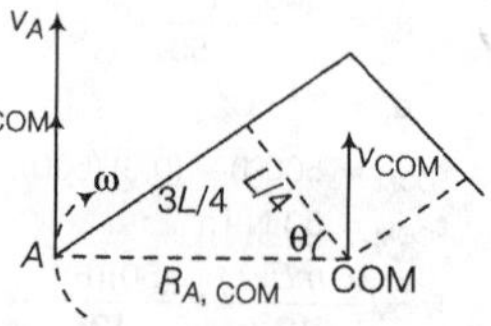

Now, $(v_A)_{max} = v_{COM} f\ \omega R_{A,\,COM}$

where $R_{A,\,COM} = \sqrt{\left(\frac{3L}{4}\right)^2 + \left(\frac{L}{4}\right)^2} = \frac{\sqrt{10}}{4}L$

$\therefore$ $(v_A)_{max} = \frac{v_0}{2} + \frac{3}{5}\frac{v_0}{L} \cdot \frac{\sqrt{10}}{4}L = \frac{1}{2}v_0\left[1 + \frac{3}{\sqrt{10}}\right]$

15. (a) Initially, finally (It has translated also)

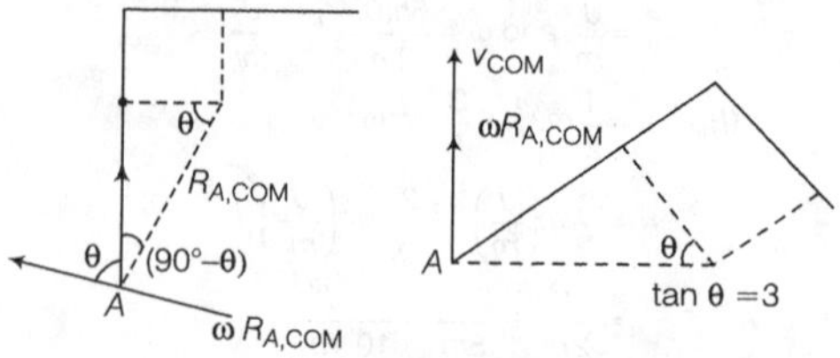

The time required for $R_{A,\,COM}$ to rotate by an angle $\tan^{-1}(3)$ is

$$t = \frac{\theta}{|\omega|} = \frac{\tan^{-1}(3)}{\frac{3}{5}\frac{v_0}{L}} = \frac{5L}{3v_0}\tan^{-1}(3)$$

16. (b) Initial angular momentum of the system about centre of mass of the system is conserved.

Thus, $\omega_0 = \frac{L_0}{I_0} = \frac{(mvL)^2}{2(mL^2)} = \frac{v}{L}$

17. (a) To hit m_3, the rod and two bullets must rotate at an angle of $\pi/2$. Therefore,

$$\omega_0 t = \frac{\pi}{2} \quad \Rightarrow \quad t = \frac{\pi}{2}\frac{L}{v} = \frac{\pi L}{2v}$$

18. (c) The new position $\mathbf{x}$ of centre of mass is (just before collision)

$$\mathbf{x} = \frac{m(-L) + m(L) + m(L)}{m + m + m} = \frac{L}{3}$$

The angular momentum about the centre of mass, now is conserved.

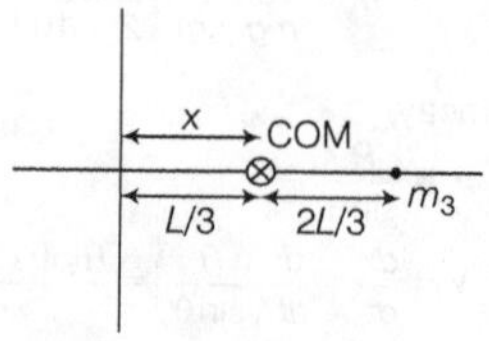

Just before collision,

$$L'_i = m(\omega_0 L)\left(L + \frac{L}{3}\right) + m(\omega_0 L)\left(L - \frac{L}{3}\right) + m(0)$$

or $L'_i = 2m\omega_0 L^2$

Just after collision $L'_f = I\omega = \left[2m\left(\frac{2L}{3}\right)^2 + m\left(\frac{4L}{3}\right)^2\right]\omega$

or $L'_f = \frac{24}{9}m\omega L^2$

$L'_f = L'_i$

$$\Rightarrow \quad \frac{24}{9}m\omega L^2 = 2m\omega_0 L^2 \quad \Rightarrow \quad \omega = \frac{3}{4}\omega_0 = \frac{3}{4}\cdot\frac{v}{L}$$

19. (a) The centre of mass is at rest all the times, because of conservation of linear momentum. So, linear velocity of the system must be zero.

20. (d) Let v be the centre of mass C of (rod + two particles) and ω be the angular velocity of the system and v its linear velocity, after collision.

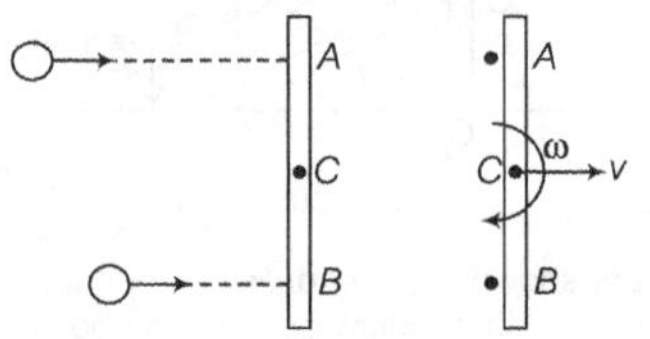

From law of conservation of linear momentum.

$$(0.08)(10) + (0.08)(6) = (0.08 + 0.08 + 0.16)\,v$$
$$v = 4\ \text{ms}^{-1}$$

21. (c) $AC = BC = 0.5\ \text{m}$

From conservation of angular momentum about C

$$(0.08)(10)(0.5) - (0.08)(6)(0.5) = I_{\text{system}}\,\omega$$

where, $I_{\text{system}} = I_{\text{rod}} + I_{\text{two particles}}$

$$= \frac{(0.16)(\sqrt{3})^2}{12} + 2(0.08)(0.5)^2$$
$$= 0.08\ \text{kg-m}^2$$

Using this in the equation above, we get $\omega = 2\ \text{rad s}^{-1}$

22. (a) Loss in KE $= -\Delta K = K_i - K_f$

$$-\Delta K = \frac{1}{2}(0.08)(10)^2 + \frac{1}{2}(0.08)(6)^2 - \frac{1}{2}(0.08 + 0.08 + 0.16)(4)^2 - \frac{1}{2}(0.08)(2)^2$$

or $\quad -\Delta K = 2.72\ \text{J}$

DPP-6 Instantaneous Axis of Rotation and Instantaneous Centre

1. (a) Let us consider the position of the disc at time t, then

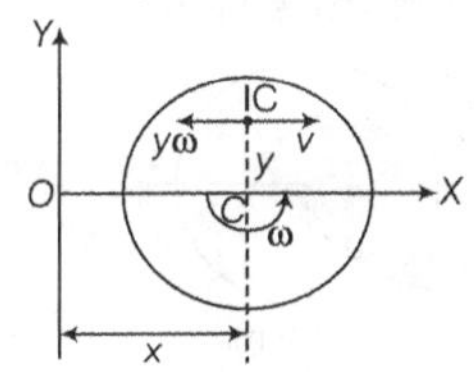

$$x = vt$$
$$\Rightarrow \quad t = \frac{x}{v}$$
$$\omega = \omega_0 + \alpha t = 0 + \alpha\left(\frac{x}{v}\right)$$
$$\omega = \frac{\alpha x}{v}$$

Let IC is at a distance y from centre of the disc.

$$v_{IC} = 0$$
$$\Rightarrow \quad v - y\omega = 0$$
$$\Rightarrow \quad y = \frac{v}{\omega}$$
$$\text{or} \quad y = \frac{v}{\left(\frac{\alpha x}{v}\right)}$$
$$\Rightarrow \quad xy = \frac{v^2}{\alpha} = \text{constant}$$

This is the equation of rectangular hyperbola.

2. (a) The coordinates of centre of mass C are

$$C = \left[\frac{L}{2}\cos\theta, \frac{L}{2}\sin\theta\right]$$
$$r = \frac{L}{2} = \text{Half of diagonal}$$

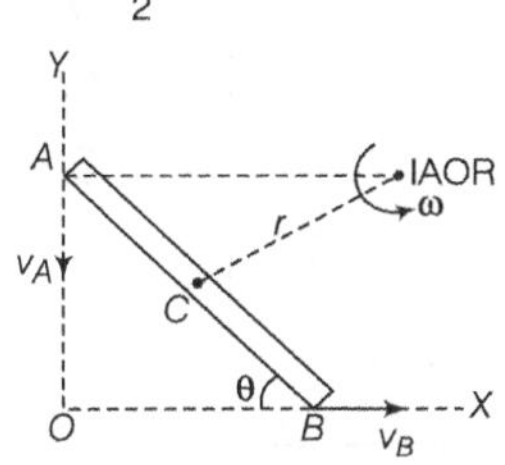

The position of instantaneous centre is shown in the diagram. As all the surfaces are smooth, mechanical energy is conserved.

$$\therefore \quad \left\{\begin{matrix}\text{Decrease in GPE}\\ \text{of COM of the rod}\end{matrix}\right\} = \left\{\begin{matrix}\text{Increase in RKE}\\ \text{of the rod about IAOR}\end{matrix}\right\}$$

$$mg\frac{L}{2}(1 - \sin\theta) = \frac{1}{2}I_{\text{IAOR}}\,\omega^2$$
$$I_{\text{IAOR}} = \frac{mL^2}{12} + mr^2$$
$$= \frac{mL^2}{12} + \frac{mL^2}{4} = \frac{mL^2}{3}$$
$$\therefore \quad mg\frac{L}{2}(1 - \sin\theta) = \frac{1}{2}\frac{mL^2}{3}\omega^2$$
$$\Rightarrow \quad \omega^2 = \frac{3g(1 - \sin\theta)}{L}$$
$$\therefore \quad \omega = \sqrt{\frac{3g(1 - \sin\theta)}{L}}$$

3. (c) $\omega^2 = \dfrac{3g(1 - \sin\theta)}{L}$

Differentiating w.r.t. time t, we get

$$2\omega\left(\frac{d\omega}{dt}\right) = \frac{3g}{L}\cos\theta\left(\frac{d\theta}{dt}\right)$$
$$2\omega\alpha = -\frac{3g}{L}\cos\theta(\omega)$$
$$\Rightarrow \quad \alpha = -\frac{3g}{2L}\cos\theta$$

The x-coordinate of centre of mass at an angle θ is

$$x = \frac{L}{2}\cos\theta$$

Differentiating w.r.t. t, we get

$$\frac{dx}{dt} = -\frac{L}{2}\sin\theta\cdot\frac{d\theta}{dt}$$

Differentiating again w.r.t. t, we get

$$\frac{d^2x}{dt^2} = -\frac{L}{2}\left[\sin\frac{d^2\theta}{dt^2} + \cos\theta\left(\frac{d\theta}{dt}\right)^2\right]$$

or $\quad a_x = -\dfrac{L}{2}[\alpha\sin\theta + \omega^2\cos\theta]$

or $\quad a_x = -\dfrac{\omega^2 L}{2}\cos\theta - \dfrac{\alpha L}{2}\sin\theta \quad \ldots\text{(i)}$

The rod will lose contact with the vertical wall at the instant when normal reaction by the vertical wall (which is along X-axis) becomes zero. So, $F_x = 0 \Rightarrow a_x = 0$.

Putting, $a_x = 0$ in Eq. (i), we have

$$-\omega^2 \cos\theta - \alpha \sin\theta = 0$$

or $$-\left[\frac{3g(1-\sin\theta)}{L}\cos\theta\right] - \left[-\frac{3g}{2L}\cos\theta\sin\theta\right] = 0$$

$$-(1-\sin\theta)\cos\theta + \frac{\cos\theta\sin\theta}{2} = 0$$

$$\Rightarrow \quad \frac{\cos\theta\sin\theta}{2} = \cos\theta - \cos\theta\sin\theta$$

$$\frac{3}{2}\cos\theta\sin\theta = \cos\theta$$

$$\Rightarrow \quad \sin\theta = \frac{2}{3} \Rightarrow \cos\theta = \frac{\sqrt{5}}{3}$$

By this moment, the lower has moved a distance d given by $d = L\cos\theta = \frac{\sqrt{5}}{3}L$

4. (d) In this case, instantaneous centre (i.e. point of zero velocity) is point of contact. We can assume that the hoop is in pure rotation (with same ω) about IAOR passing through instantaneous centre at this instant.

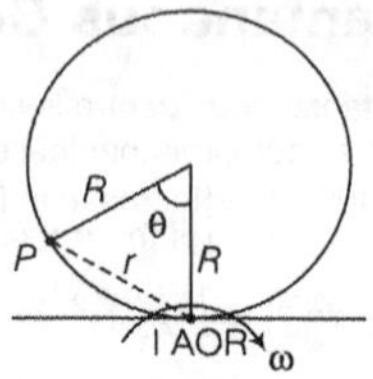

$$\frac{R}{\sin\left(90^\circ - \frac{\theta}{2}\right)} = \frac{r}{\sin\theta}$$

$$\therefore \quad r = 2R\sin\frac{\theta}{2}$$

$$\therefore \quad v_P = r\omega = 2R\sin\frac{\theta}{2}\frac{v_0}{R} \qquad \left(\because \omega = \frac{v_0}{R}\right)$$

$$\therefore \quad v_P = 2v_0\sin\frac{\theta}{2}$$

5. (a) These two velocities are anti-parallel. So, join their heads and their tails. The point of inter-section is IC.

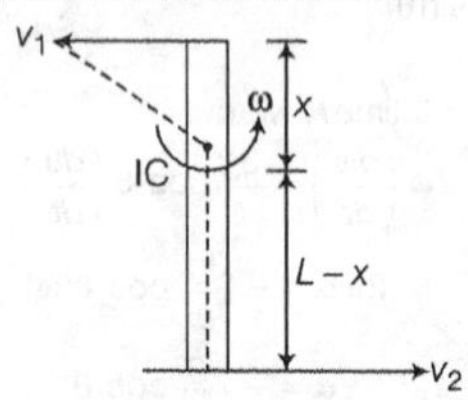

$$\therefore \quad \omega = \frac{v_1}{x} = \frac{v_2}{L-x} \Rightarrow v_1L - v_1x = v_2x$$

$$x = \left(\frac{v_1}{v_1+v_2}\right)L$$

6. (b) We can assume pure rotation about instantaneous centre at this instant. ω is same for all points and $v = r\omega$.

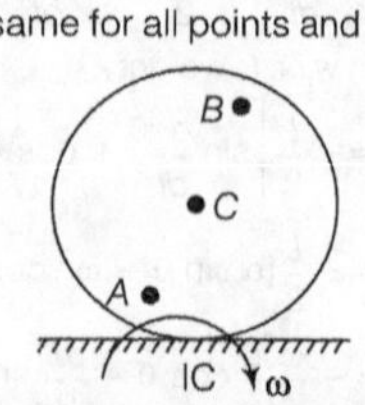

where, r is distance of the point under consideration from instantaneous cemtre, at that instant. From figure, it is clear that

$$r_A < r_C < r_B$$

$$\Rightarrow \quad v_A < v_C < v_B$$

7. (a) Let $OP = OP' = x$

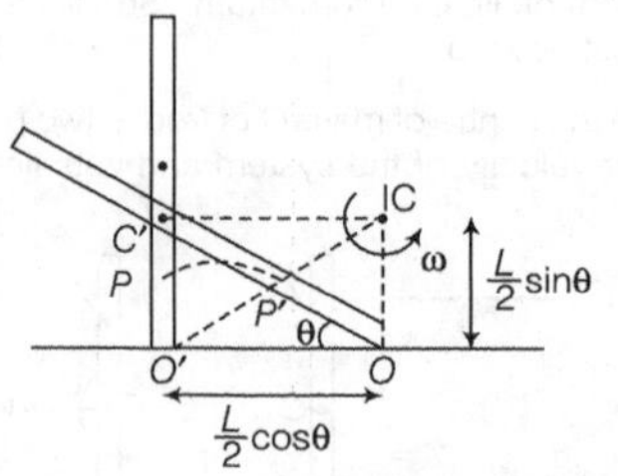

At the instant shown velocity of P' is perpendicular to OP' and is equal to $x\omega$. Simultaneously, point P' can be considered to be in pure rotation about instantaneous centre with same velocity, i.e. $x\omega$. And this velocity is perpendicular to line joining P' and instantaneous centre. So, distance between O and instantaneous centre is $2x$.

$$(2x)^2 = \left(\frac{L}{2}\cos\theta\right)^2 + \left(\frac{L}{2}\sin\theta\right)^2 = \frac{L^2}{4}$$

$$x^2 = \frac{L^2}{16} \Rightarrow x = \frac{L}{4}$$

8. (a) $v_0 = 4 - (10)(0.2) = 2\ \text{ms}^{-1}$

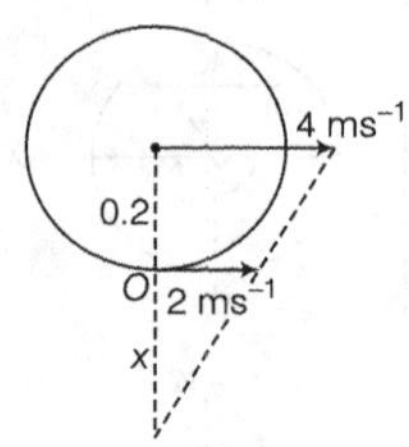

$$\frac{x+0.2}{4} = \frac{x}{2}$$

$$2x + 0.4 = 4x$$

$$2x = 0.4$$

$$x = 0.2\ \text{m}$$

9. (a)

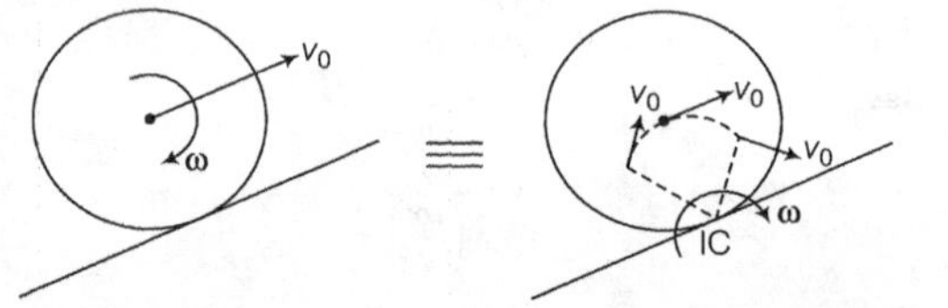

10. (2) The length of the rod is $L = \sqrt{3^2 + 3^2} = \sqrt{18}$ m. When the end A moves a distance $x = 1.5$ m, let the end B moves a distance y. From figure, $(3-x)^2 + (3+y)^2 = 18$

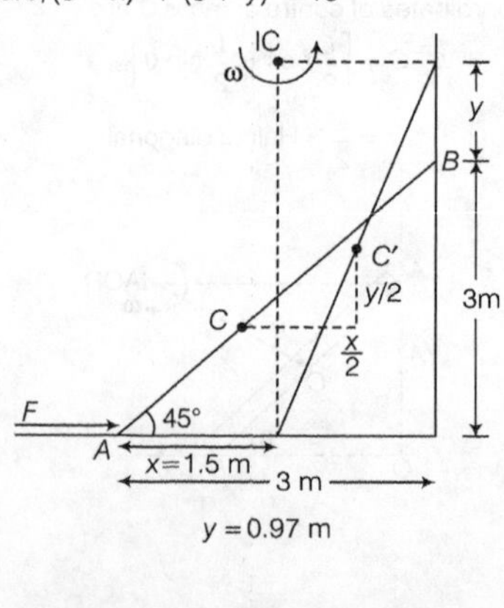

$$\Rightarrow \quad y = 0.97\ \text{m}$$

Now, let v_A and v_B are the velocities of A and B, respectively.

Then, $v_A = 3.97\omega$ and $v_B = 1.5\omega$

The velocity of centre C is

$$v_C^2 = \left(\frac{v_A}{2}\right)^2 + \left(\frac{v_B}{2}\right)^2$$

or $$v_C^2 = 4.52\,\omega^2$$

In this process, centre C will rise

$$\frac{y}{2} = \frac{0.97}{2} = 0.485 \text{ m}$$

From conservation of mechanical energy,

Work done by F = Increase in GPE of centre of mass + Increase in RKE of rod about instantaneous centre.

$$F \times 1.5 = mg(0.485) + \frac{1}{2} I_{\text{IAOR}}\omega^2$$

$$I_{\text{IAOR}} = \frac{M(\sqrt{18})^2}{12} + M\left(\frac{\sqrt{18}}{2}\right)^2 +$$

$$= 1.5\,M + 4.5\,M = 6\,M$$

$\therefore$ $$F \times 1.5 = 4.735\,M + \frac{1}{2}(6M)\omega^2$$

$$\frac{F}{M} \times 1.5 = 4.753 + 3\omega^2$$

$$1.5\frac{F}{M} - 4.753 = 3\omega^2 \quad \Rightarrow \quad \omega^2 = \frac{F}{2M} - 1.6$$

$$\omega = \left(\frac{F}{2M} - 1.6\right)^{1/2}$$

$\therefore$ $K = 2$

11. The floor is smooth, so the mechanical energy of the rod is conserved. Further, no horizontal force acts on the rod. So, the centre of mass of the rod moves vertically downwards.

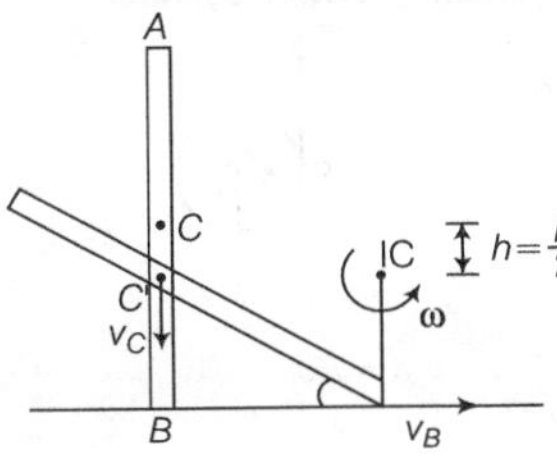

The velocities of centre of mass, C and point B are as shown. The location of instantaneous centre at this instant is found by drawing perpendiculars to $\mathbf{v}_C$ and $\mathbf{v}_B$ at respective points and instantaneous centre is located at the point of intersection of these two perpendiculars. So, the rod may now assumed to be in pure rotation about IAOR passing through instantaneous centre with angular speed ω.

Applying law of conservation of mechanical energy, we get

$$\left\{\begin{matrix}\text{Decrease in GPE} \\ \text{of COM of the rod}\end{matrix}\right\} = \left\{\begin{matrix}\text{Increase in RKE} \\ \text{or rod about IAOR}\end{matrix}\right\}$$

$\therefore$ $$mgh = \frac{1}{2} I_{\text{IAOR}}\omega^2$$

$\Rightarrow$ $$mg\frac{L}{2}(1 - \sin\theta) = \frac{1}{2}\left[\frac{mL^2}{12} + \frac{mL^2}{4}\cos^2\theta\right]\omega^2$$

Solving this equation, we get

$$\omega = \sqrt{\frac{12g(1 - \sin\theta)}{L(1 + 3\cos^2\theta)}}$$

$$|\mathbf{v}_C| = \left(\frac{L}{2}\cos\theta\right)\omega$$

$\Rightarrow$ $$v_C = \sqrt{\frac{3gL\,(1 - \sin\theta)\cos^2\theta}{(1 + 3\cos^2\theta)}}$$

12. $\sqrt{\dfrac{3gL\,(1 - \sin\theta)\sin^2\theta}{(1 + 3\cos^2\theta)}}$

$$|\mathbf{v}_B| = \left(\frac{L}{2}\sin\theta\right)\omega = \frac{L}{2}\sin\theta\sqrt{\frac{12g(1 - \sin\theta)}{L(1 + 3\cos^2\theta)}}$$

$$= \sqrt{\frac{3g\sin^2\theta\,(1 - \sin\theta)\,L}{(1 + 3\cos^2\theta)}}$$

Alternate Method The distance between two points of the rod should remain constant. So, the components of v_C and v_B along the rod should be same.

$\therefore$ $$v_C\sin\theta = v_B\cos\theta \quad \Rightarrow \quad v_B = v_C\tan\theta$$

or $$v_B = \sqrt{\frac{3gL\,(1 - \sin\theta)\cos^2\theta}{(1 + 3\cos^2\theta)}} \cdot \frac{\sin\theta}{\cos\theta}$$

$$= \sqrt{\frac{3gL(1 - \sin\theta)\sin^2\theta}{(1 + 3\cos^2\theta)}}$$

13. $v_C = \sqrt{\dfrac{3gL(1 - \sin\theta)\cos^2\theta}{(1 + 3\cos^2\theta)}}$

When the rod touches the ground, $\theta = 0°$

$\therefore$ $$v_C = \sqrt{\frac{3gL}{4}}$$

14. Decrease in GPE = Increase in RKE about instantaneous centre

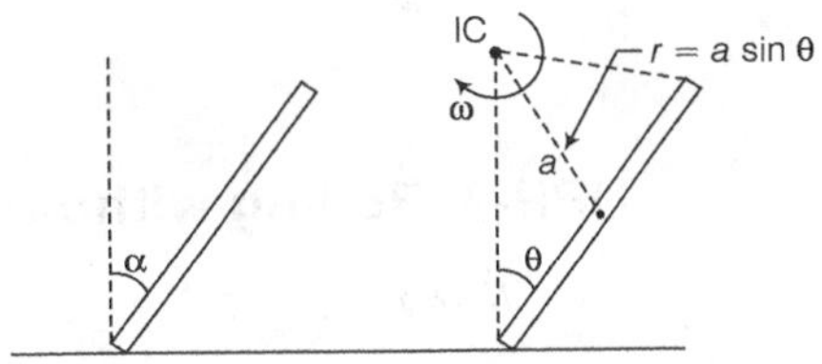

$$mga(\cos\alpha - \cos\theta) = \frac{1}{2}\left[\frac{m(2a)^2}{12} + m(a\sin\theta)^2\right]\omega^2$$

This gives $$\omega = \left[\frac{6g(\cos\alpha - \cos\theta)}{a(1 + 3\sin^2\theta)}\right]^{1/2}$$

15. (a) As there is no slipping,

$v_A = v(\leftarrow)$ and $v_B = 2v(\rightarrow)$

Two velocities are anti-parallel. The line joining tails of velocities intersects the line joining their heads at instantaneous centre.

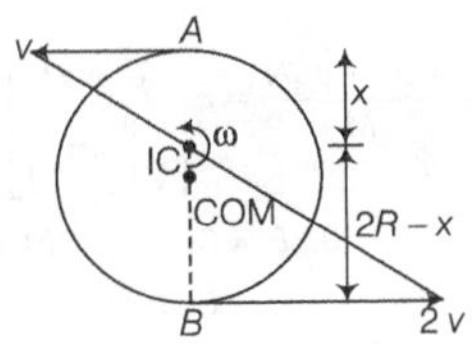

From similar triangles, $\dfrac{v}{x} = \dfrac{2v}{2R - x}$

$\Rightarrow$ $$2R - x = 2x$$

$\therefore$ $$x = \frac{2R}{3}$$

16. (b) We can assume that the disc is pinned at instantaneous centre and is in pure rotation about instantaneous centre at that instant with same ω.

$\therefore$ $$v_A = v = \frac{2R}{3}\omega \quad \Rightarrow \quad \omega = \frac{3v}{2R}$$

17. (c) $v_{\text{COM}} = r\omega$ $\qquad \left(\text{where, } r = R - \dfrac{2R}{3} = \dfrac{R}{3}\right)$

$$v_{\text{COM}} = \frac{R}{3} \cdot \frac{3v}{2R} = \frac{v}{2}$$

18. (d) $E_{total} = \frac{1}{2} I_{AOR}\omega^2 = \frac{1}{2}\left[\frac{mR^2}{2} + m\left(\frac{R}{3}\right)^2\right]\cdot\left(\frac{3v}{2R}\right)^2$

or $$E_{total} = \frac{1}{2}\left[\frac{11}{18}\right] mR^2 \left(\frac{9v^2}{4R^2}\right) = \frac{11}{16} mv^2$$

19. (b) $v = r\omega$

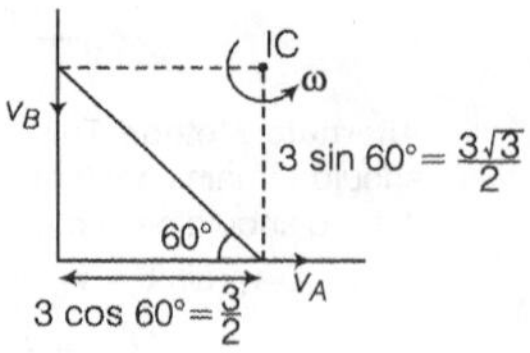

$\therefore \; v_A = \frac{3\sqrt{3}}{2}\omega$

or $\omega = 2 \times \frac{2}{3\sqrt{3}}$

$= \frac{4}{3\sqrt{3}}$ rad s^{-1}

$v_B = \frac{3}{2}\omega = \frac{3}{2}\cdot\frac{4}{3\sqrt{3}} = \frac{2}{\sqrt{3}} \approx 1.16$ ms^{-1}

20. (d) The motion of B can be looked as translation of end B with velocity v_A + rotation of B w.r.t. A with tangential velocity v_{BA}. Since, the velocity vector of end B is downwards, the resultant of $\bar{v}_A$ and $\bar{v}_{BA}$ must be equal to $\bar{v}_B$ (see figure).

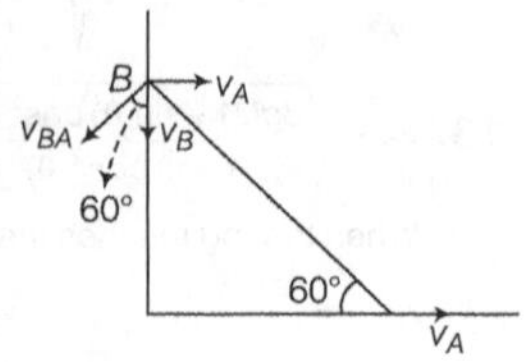

$$\sin 60° = \frac{v_A}{v_{BA}}$$

$\Rightarrow$ $$v_{BA} = \frac{v_A}{\sin 60°} = \frac{2\times 2}{\sqrt{3}}$$

or $$v_{BA} = 2.3 \text{ ms}^{-1}$$

The acceleration of end B has three components

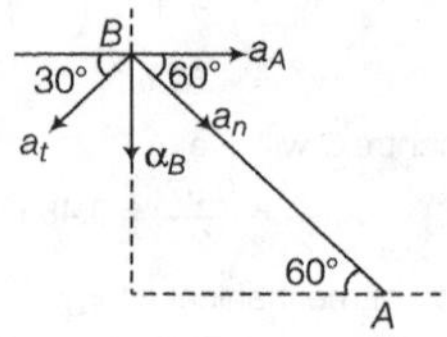

(i) Due to translation of end A, $a_A = 1.6$ ms^{-1}

(ii) Due to rotation of end B about A, i.e. tangential acceleration, $a_t = \alpha r_{AB} = \alpha(3) = 3\alpha$ ms^{-2}

(iii) Radial acceleration from B to A,

$$a_n = \frac{v_{BA}^2}{r_{AB}} = \frac{(2.3)^2}{3} = 1.76 \text{ ms}^{-1}$$

The net acceleration of end B is vertically downwards. So, net horizontal component must be zero.

$$a_x = 0$$

$\Rightarrow$ $$1.6 + 1.76\cos 60° - 3\alpha\cos 30° = 0$$

$\therefore$ $$\alpha = 0.95 \text{ rad s}^{-2}$$

$$a_B = a_y = 1.76\sin 60° + 3\alpha\sin 30° = 2.98 \text{ ms}^{-1}$$

DPP-7 **Rolling without Slipping and Rolling with Slipping**

1. (a)
$$N = Mg - \frac{3F}{5}$$
$$= Mg - 0.6F$$

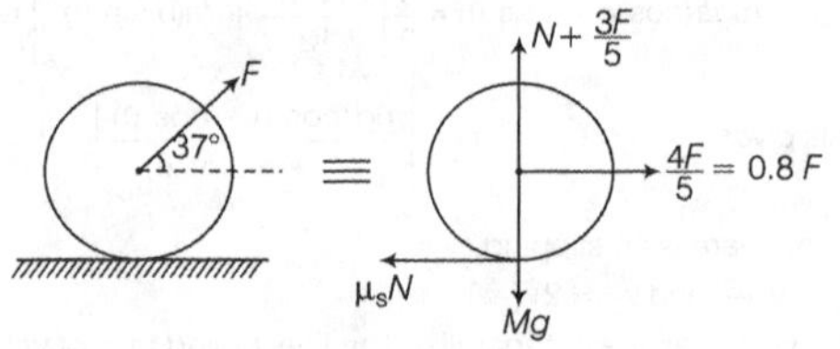

For pure rolling, $a = R\alpha$

$$\frac{0.8F - \mu_s N}{M} = R\left[\frac{(\mu_s N)R}{\left(\frac{1}{2}MR^2\right)}\right] \qquad \left(\because \alpha = \frac{T}{I}\right)$$

$$0.8F = 3\mu_s N = 3(0.4)(Mg - 0.6F)$$
$$F = 0.79\,Mg$$

$\therefore$ Maximum value of $F = 0.79\,Mg$

2. (b) For pure rolling, $a = R\alpha$

$$\frac{F - \mu Mg}{M} = R\left[\frac{\mu MgR - Fr}{MR^2/2}\right]$$

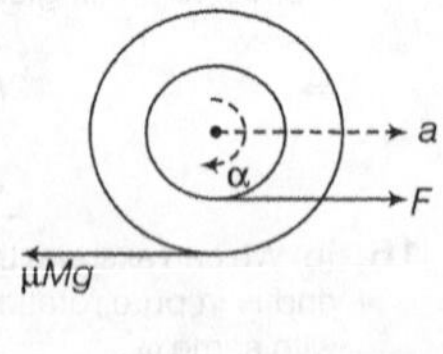

Solving this equation, we get

$$F = \frac{3\mu Mg R}{R + 2r}$$

This is the maximum value of F.

3. (a)
$$(2T)R = \frac{MR^2}{2}\alpha \qquad \text{...(i)}$$
$$Mg - 2T = Ma_{COM} \qquad \text{...(ii)}$$
$$2TR = \frac{MR^2}{2}\cdot\alpha' \qquad \text{...(iii)}$$

(For drawing FBD's, the system is viewed from left side)

Let acceleration of thread is a.

Upper Cylinder **Lower Cylinder**

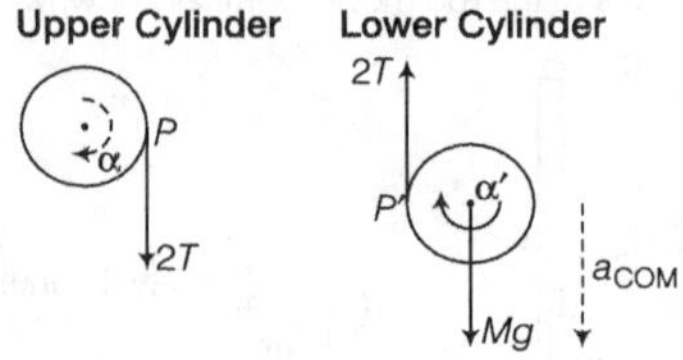

For no slipping at P, $\quad a = R\alpha$

For no slipping at P',

$$a = a_{COM} - R\alpha'$$

$\therefore$ $$a_{COM} = 2R\alpha \qquad \text{...(iv)}$$

Note $\alpha' = \alpha$ *from Eqs. (i) and (iii)*

Solving the above equations, we get $T = \frac{Mg}{10}$

4. (b) For the shell not to slide, it should start pure rolling immediately after being hit.

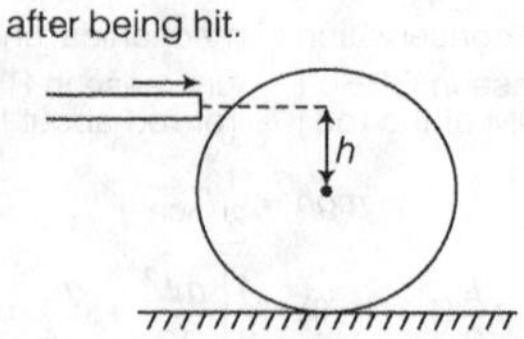

$$F dt = mv_0 \qquad \text{...(i)}$$
$$Fh dt = I\omega_0 = \frac{2}{3}mR^2\omega_0 \qquad \text{...(ii)}$$

and $$v_0 = R\omega_0 \qquad \text{...(iii)}$$

Dividing Eq.(ii) by Eq.(i), we get

$$h = \frac{2}{3}R^2\frac{\omega_0}{v_0}$$
$$= \frac{2}{3}R^2\frac{\omega_0}{R\omega_0} = \frac{2}{3}R$$

5. (a) $f = -ma \Rightarrow \mu mg = -ma$

$\Rightarrow \quad a = -\mu g$

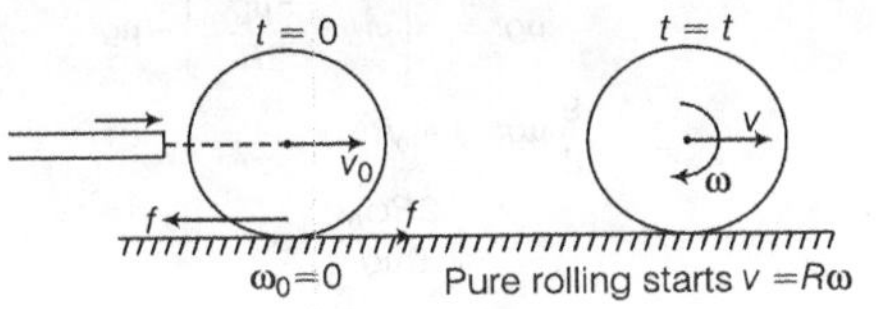

$$fR = I\alpha$$

$$\Rightarrow \quad \mu mgR = \frac{2}{5}mR^2\alpha \Rightarrow \alpha = \frac{5\mu g}{2R}$$

$$v = u + at$$

$$\Rightarrow \quad v = v_0 - \mu gt$$

$$\omega = \omega_0 + \alpha t$$

$$\Rightarrow \quad \omega = 0 + \frac{5\mu g}{2R}t = \frac{5\mu gt}{2R}$$

$$v = R\omega$$

$$\Rightarrow \quad (v_0 - \mu gt) = R\left(\frac{5\mu gt}{2R}\right) \Rightarrow \mu gt\left(1 + \frac{5}{2}\right) = v_0$$

$$t = \frac{2v_0}{7\mu g}$$

$$s = ut + \frac{1}{2}at^2$$

$$\Rightarrow \quad s = v_0 \cdot \frac{2v_0}{7\mu g} - \frac{1}{2}\mu g \cdot \frac{4v_0^2}{49\mu^2 g^2}$$

$$s = \frac{2v_0^2}{7\mu g} - \frac{2v_0^2}{49\mu g} = \frac{12v_0^2}{49\mu g}$$

6. (c) $Fhdt = \frac{2}{5}MR^2\omega_0$

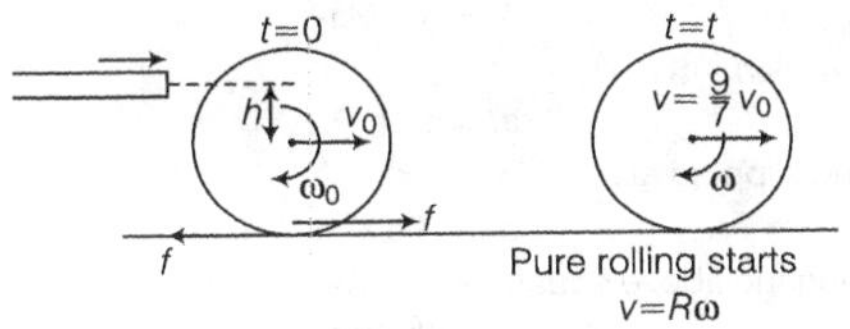

$$Fhdt = \frac{2}{5}mR^2\omega_0$$

$$Fdt = mv_0$$

$$\therefore \quad h = \frac{2}{5} \cdot \frac{R^2\omega_0}{v_0} \quad \text{...(i)}$$

$$f = ma \Rightarrow \mu mg = ma$$

$$\Rightarrow \quad a = \mu g$$

$$fR = -\frac{2}{5}mR^2\alpha$$

$$\Rightarrow \quad \alpha = -\frac{\mu mg}{\frac{2}{5}mR} = -\frac{5\mu g}{2R}$$

$u = v_0$, $v = \frac{9}{7}v_0$, $\quad v = u + at$

$$\Rightarrow \quad \frac{9}{7}v_0 = v_0 + \mu gt \Rightarrow t = \frac{2}{7} \cdot \frac{v_0}{\mu g}$$

$$\omega = \omega_0 + \alpha t$$

$$\Rightarrow \quad \omega = \omega_0 - \frac{5\mu g}{2R} \cdot t$$

$$v = R\omega$$

$$\Rightarrow \quad \frac{9}{7}v_0 = \left(\omega_0 - \frac{5\mu g}{2R} \cdot \frac{2}{7}\frac{v_0}{\mu g}\right)R$$

$$\Rightarrow \quad R\omega_0 = 2v_0$$

Using this in Eq. (i), we get $h = \frac{4}{5}R$

7. (a) Sphere

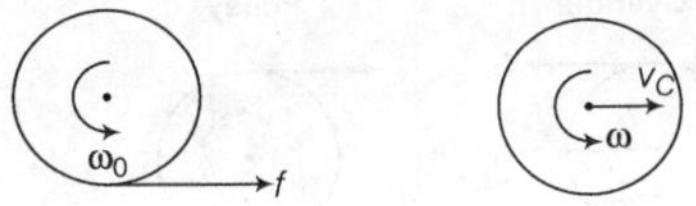

$t = 0$	$t = t$
$\mu = 0$	$v = v_C$
$\omega_0 = 0$	$\omega = \omega$

$$f = \mu mg$$

$$a = \mu g$$

$$\alpha = fR/I = \frac{5}{2}\frac{\mu g}{R}$$

$$v_C = u + at = 0 + \mu gt$$

$$\Rightarrow \quad v_C = \mu gt$$

$$\omega = \omega_0 + \alpha t = 0 + \frac{5}{2}\frac{\mu g}{R}t$$

$$\Rightarrow \quad \omega = \frac{5\mu gt}{2R}$$

At $t = t$, velocity of point of contact $= v_C + R\omega = \frac{7}{2}\mu gt$

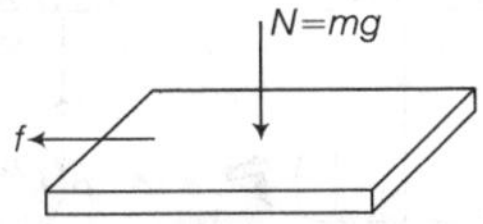

$$a = -\frac{f}{m} = -\frac{\mu mg}{M}$$

Instantaneous velocity, $v = v_0 - \frac{\mu mg}{M}t$

For pure rolling both points in contact should have same velocity

$$\therefore \quad \frac{7}{2}\mu gt = v_0 - \frac{\mu mg}{M}t$$

$$\mu gt\left[\frac{7}{2} + \frac{m}{M}\right] = v_0$$

$$\Rightarrow \quad \mu gt\left[\frac{7M + 2m}{2M}\right] = v_0$$

$$\text{or} \quad t = \frac{2Mv_0}{(7M + 2m)\mu g}$$

8. (a,c) For pure rolling to take place on the ground

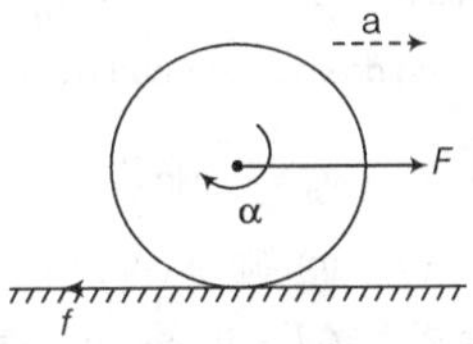

$$a = R\alpha \quad \text{...(i)}$$

$$F - f = Ma \quad \text{...(ii)}$$

$$I = MR^2$$

$$T = I\alpha = MR^2\alpha$$

$$fR = MR^2\alpha$$

$$f = MR\alpha \quad \text{...(iii)}$$

Using Eq. (i) in Eq. (iii), we get

$$f = Ma$$

Using this in Eq. (ii), we get

$$F - Ma = Ma$$

$$\Rightarrow \quad F = 2Ma$$

$$\therefore \quad f = \frac{F}{2}$$

$$\therefore \quad F - \frac{F}{2} = Ma \Rightarrow a = \frac{F}{2M}$$

9. (b,d)

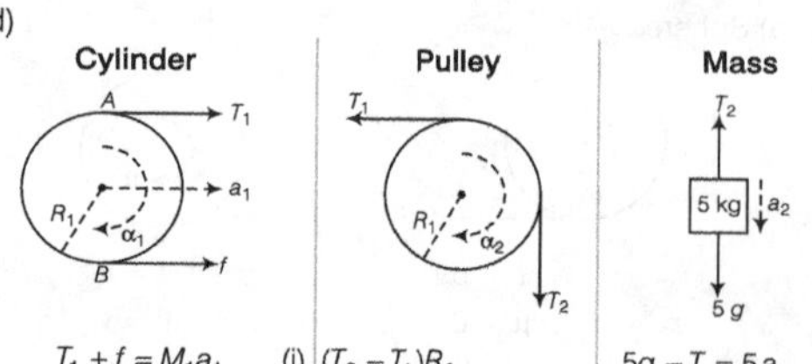

$T_1 + f = M_1 a_1$ ...(i)

$(T_1 - f)R_1 = \frac{M_1 R_1^2}{2} \cdot \alpha_1$ ...(ii)

$(T_2 - T_1)R_2 = \frac{M_2 R_2^2}{2} \cdot \alpha_2$ ...(iii)

$5g - T = 5a_2$ ...(iv)

For no slipping at A, $a_2 = a_1 + R_1\alpha_1$ ...(v)

For no slipping at B, $a_1 = R_1\alpha_1$ ...(vi)

Solving the above equations, we get

$$a_2 = \frac{4}{11} g, v = a_2 t = \frac{4gt}{11}$$

10. (a,d)

$$Mg - 2T = Ma_{COM} \quad ...(i)$$

$$2TR = \frac{MR^2}{2} \alpha \quad ...(ii)$$

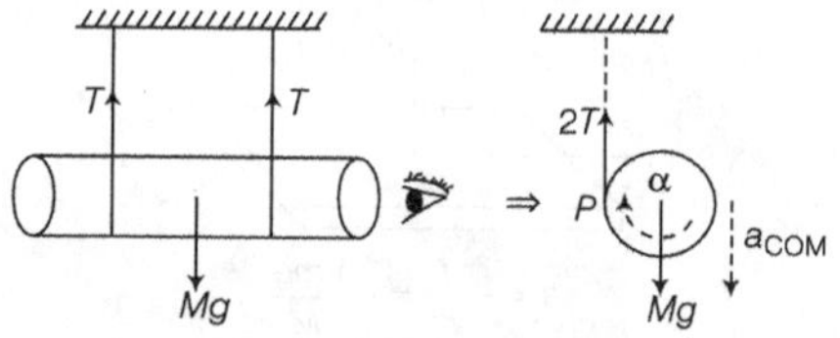

For no slipping at P, $a_{COM} = R\alpha$ ...(iii)

Solving the above equations, we get

$$a = a_{COM} = \frac{2g}{3} \text{ and } T = \frac{Mg}{6}$$

11. (d) Statement I is incorrect and Statement II is correct.

When a spherical body rolls without slipping on an inclined rough surfaces, frictional force is required to maintain pure rolling.

12. (b) Statement I is correct and Statement II is also correct and Statement II is not the correct explanation of Statement I.

If the surface is smooth, $f = 0$

Consider a sphere rolling down an inclined rough surface. Friction does work it increases KE_R. If the sphere is pure rolling on rough horizontal surface $w_f = 0$.

13. (b) For a disc rolling down an inclined plane, $f = \frac{1}{3} g \sin \alpha$

14. (a) $v_p = 2v_0 \sin \frac{\theta}{2}$

Hence, (i) → (q), (ii) → (p), (iii) → (s), (iv) → (r)

15. (a) $Fh \pm fR = I\alpha$, $F = ma_{COM}$, $a_{COM} = R\alpha$

Hence, (i) → (p,r,s), (ii) → (q,r,s), (iii) → (q,r,s), (iv) → (p,r,s)

16. (b) Let after time t (when pure rolling begins) v_1 and v_2 are the velocities of plank and the cylinder.

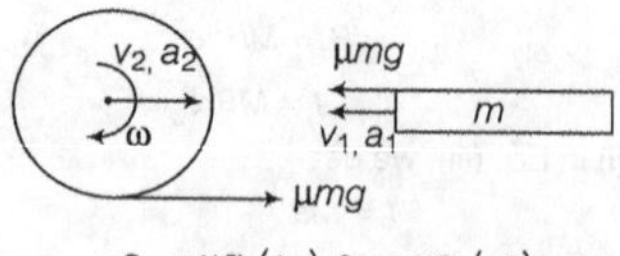

$$a_1 = \mu g \,(\leftarrow), a_2 = \mu g \,(\rightarrow)$$

$$v_1 = \mu g t \,(\leftarrow), v_2 = (\mu g t)\,(\rightarrow)$$

$$\alpha = -\frac{\mu mg \cdot R}{\frac{2}{5} mR^2} = -\frac{5 \mu g}{2R}$$

$$\omega = \omega_0 - \alpha t = \omega_0 - \left(\frac{5\mu g}{2R}\right) t$$

During pure rolling, velocities of contact points are same.

$$\therefore \quad v_1 = R\omega - v_1$$

$$\Rightarrow \quad \mu g t = R\left[\omega_0 - \frac{5\mu g}{2R} t\right] - \mu g t$$

$$\frac{9}{2} \mu g t = \omega_0 R$$

$$\Rightarrow \quad t = \frac{2R\omega_0}{9\mu g}$$

17. (a) $v_2 = \mu g t = \mu g \left[\frac{2}{9} \frac{R\omega_0}{\mu g}\right] = \frac{2}{9} R\omega_0$

18. (d)

$$s = ut + \frac{1}{2} at^2$$

$$\Rightarrow \quad s = 0 + \frac{1}{2}(\mu g)\left(\frac{2R\omega_0}{9\mu g}\right)^2$$

$$\text{or} \quad s = \frac{2}{81}\left[\frac{R^2\omega_0^2}{\mu g}\right]$$

19.

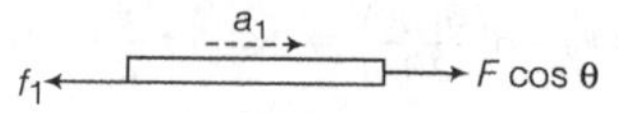

$$F \cos \theta - f_1 = ma_1 \quad ...(i)$$

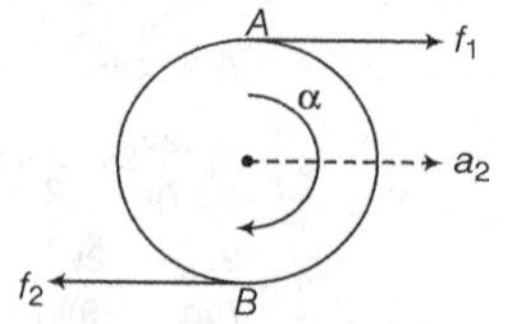

$$(f_2 + f_1) R = \frac{MR^2}{2} \cdot \alpha \quad ...(ii)$$

$$f_1 - f_2 = Ma_2 \quad ...(iii)$$

For no slipping at A,

$$a_1 = a_2 + R\alpha \quad ...(iv)$$

For no slipping at B,

$$a_2 - R\alpha = 0 \quad ...(v)$$

Solving the above equations, we get

$$a_2 = \frac{4F \cos \theta}{3M + 8m}$$

$$f_1 = \frac{3MF \cos \theta}{3M + 8m} \text{ and } f_2 = \frac{MF \cos \theta}{3M + 8m}$$

20. The hoop is rolling without supping. So, arc length AP' is equal to the displacement PA.

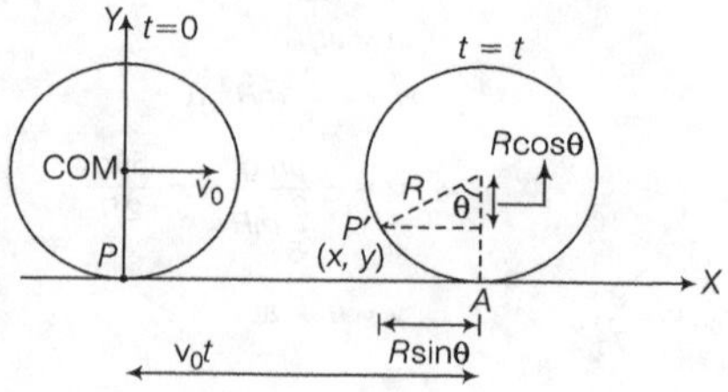

$$\therefore \quad AP' = v_0 t$$

$$\Rightarrow \quad R\theta = v_0 t$$

$$\Rightarrow \quad \theta = \frac{v_0 t}{R}$$

Coordinates of P' at time t are (x, y)

$$x = v_0 t - R \sin \theta = v_0 t - R \sin\left(\frac{v_0 t}{R}\right) \quad ...(i)$$

$$y = R - R\cos \theta = R - R \cos\left(\frac{v_0 t}{R}\right) \quad ...(ii)$$

Then, Eqs. (i) and (ii) combination of x and y will give rise to the trajectory called 'cycloid'.

$$v_x = \frac{dx}{dt} = v_0 - \frac{v_0}{R}\left[R\cos\left(\frac{v_0 t}{R}\right)\right]$$

and $$v_y = \frac{dy}{dt} = v_0 \sin\left(\frac{v_0 t}{R}\right)$$

$$\mathbf{v}_P = v_x\hat{\mathbf{i}} + v_y\hat{\mathbf{j}}$$

$$|\mathbf{v}_P| = \sqrt{v_x^2 + v_y^2}$$

$$= \sqrt{v_0^2 + v_0^2\cos^2\left(\frac{v_0 t}{R}\right) - 2v_0^2\cos\left(\frac{v_0 t}{R}\right) + v_0^2\sin^2\left(\frac{v_0 t}{R}\right)}$$

$$= v_0\sqrt{2 - 2\cos\left(\frac{v_0 t}{R}\right)} = v_0\sqrt{2}\sqrt{2\sin^2\left(\frac{v_0 t}{2R}\right)}$$

$$= 2v_0\sin\left(\frac{v_0 t}{2R}\right) = 2v_0\sin\left(\frac{\theta}{2}\right) \qquad \left(\because \theta = \frac{v_0 t}{R}\right)$$

$\therefore$ Velocity of point $P = v_0\left[1 - \cos\left(\frac{v_0 t}{R}\right)\right]\hat{\mathbf{i}} + v_0\sin\left(\frac{v_0 t}{R}\right)\hat{\mathbf{j}}$

Speed of point $P = 2v_0\sin\left(\frac{v_0 t}{2R}\right)$

Now, $$a_x = \frac{dv_x}{dt} = \frac{v_0^2}{R}\sin\left(\frac{v_0 t}{R}\right)$$

and $$a_y = \frac{dv_y}{dt} = \frac{v_0^2}{R}\cos\left(\frac{v_0 t}{R}\right)$$

$$\mathbf{a}_P = \frac{v_0^2}{R}\sin\left(\frac{v_0 t}{R}\right)\hat{\mathbf{i}} + \frac{v_0^2}{R}\cos\left(\frac{v_0 t}{R}\right)\hat{\mathbf{j}}$$

$$|\mathbf{a}_P| = \frac{v_0^2}{R}$$

21. $dr = v_P dt = 2v_0\sin\left(\frac{v_0 t}{2R}\right)dt$

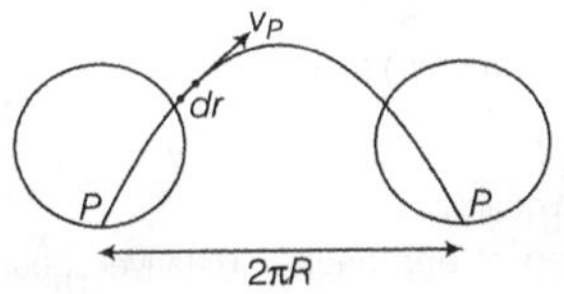

$$r = 2v_0\int_0^{2\pi R/v_0}\sin\left(\frac{v_0 t}{2R}\right)dt$$

$$r = 2v_0\left[\frac{-\cos\left(\frac{v_0 t}{2R}\right)}{(v_0 t/2R)}\right]_0^{2\pi R/v_0} = -4R\left[\cos\left(\frac{v_0}{2R}\times\frac{2\pi R}{v_0}\right) - \cos 0°\right]$$

$$= 8R$$

22. (i) Frictional force on each disc should be in forward direction. Let a_0 be the linear acceleration of the centre of mass of disc and α its angular acceleration about its centre of mass.

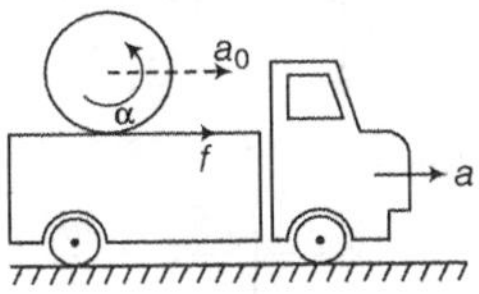

$$a_0 = \frac{f}{m} = \frac{f}{2} \qquad (\because m = 2\text{ kg})$$

$$\alpha = \frac{\tau}{I} = \frac{fR}{\left(\frac{mR^2}{2}\right)} = \frac{2f}{mR} = \frac{2f}{(2)(0.1)} = 10f$$

As, there is no slipping between disc and truck acceleration of point of contact = acceleration of truck

$$\therefore \quad a_0 + R\alpha = a \Rightarrow \frac{f}{2} + (0.1)(10f) = a$$

$$\frac{3}{2}f = a \Rightarrow f = \frac{2a}{3} = \frac{2(9)}{3} = 6\text{ N}$$

$$\mathbf{f} = (6\hat{\mathbf{i}})\text{ N}$$

(ii) $\mathbf{f} = (6\hat{\mathbf{i}})$ N (for both the discs)

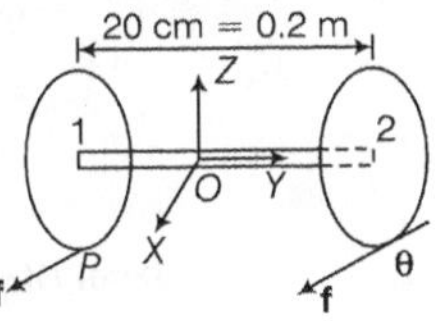

$$\mathbf{r}_P = \mathbf{r}_1 = -0.1\,\hat{\mathbf{j}} - 0.1\,\hat{\mathbf{k}}$$

and $$\mathbf{r}_Q = \mathbf{r}_2 = 0.1\,\hat{\mathbf{j}} - 0.1\,\hat{\mathbf{k}}$$

$\therefore$ Frictional torque on disc 1 about point O (i.e. centre of mass) is

$$\tau_1 = \mathbf{r}_1 \times \mathbf{f} = (-0.1\,\hat{\mathbf{j}} - 0.1\,\hat{\mathbf{k}}) \times (6\hat{\mathbf{i}})$$

$$= 0.6(\hat{\mathbf{k}} - \hat{\mathbf{j}})\text{ N-m}$$

Similarly, $$\tau_2 = 0.6(-\hat{\mathbf{j}} - \hat{\mathbf{k}})\text{ N-m}$$

$$|\tau_1| = |\tau_2| = 0.85\text{ N-m}$$

23. (i) $f = ma \Rightarrow a = \frac{f}{m} = \frac{\mu mg}{m} = \mu g$

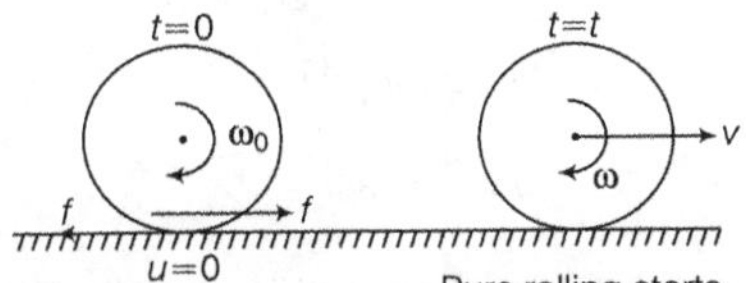

Pure rolling starts
$\therefore v = R\omega$

$$fR = -I\alpha$$

$$\Rightarrow \quad \alpha = -\frac{\mu mgR}{mR^2/2} = -\frac{2\mu g}{R}$$

$$v = u + at$$

$$\Rightarrow \quad v = 0 + \mu g t$$

$$\Rightarrow \quad t = \frac{v}{\mu g}$$

$$\omega = \omega_0 + \alpha t$$

$$\Rightarrow \quad \omega = \omega_0 - \frac{2\mu g}{R}.t$$

$$\therefore \quad \omega = \omega_0 - \frac{25\mu g}{R}\cdot\frac{v}{\mu g} = \omega_0 - \frac{2v}{R}$$

$$\therefore \quad v = R\omega$$

$$\Rightarrow \quad v = R\left[\omega_0 - \frac{2v}{R}\right]$$

$$\Rightarrow \quad v[1+2] = \omega_0 R$$

Note *a and α are constant till pure rolling starts. Later, they are zero, theoretically. In this problem, it is true because the body is rigid and only one point is in contact with the ground.*

$$v = \frac{\omega_0 R}{3}$$

$$\therefore \quad t = \frac{v}{\mu g} = \frac{\omega_0 R}{3\mu g}$$

(ii) $K_i = \frac{1}{2}I\omega_0^2 = \frac{1}{2}\frac{mR^2}{2}\omega_0^2 = \frac{mR^2\omega_0^2}{4}$

$$K_f = \frac{1}{2}mv^2 + \frac{1}{2}I\omega^2 = \frac{1}{2}mv^2 + \frac{1}{2}\frac{mR^2}{2}\left(\frac{v}{R}\right)^2$$

$$= \frac{3}{4}mv^2 = \frac{3}{4}m\left(\frac{\omega_0 R}{3}\right)^2 = \frac{mR^2\omega_0^2}{12}$$

$$W_{\text{friction}} = K_f - K_0 = \frac{mR^2\omega_0^2}{12} - \frac{mR^2\omega_0^2}{4} = -\frac{1}{6}m\omega_0^2 R^2$$

DPP-8 Angular Impulse

1. (d) From conservation of mechanical energy,

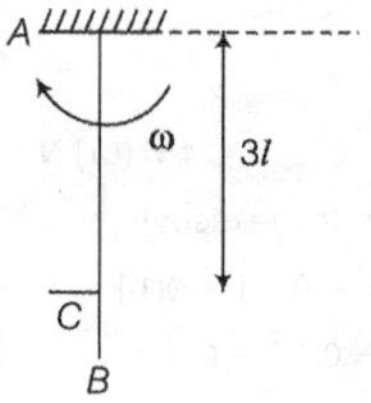

$$(3m)(g)(2l) = \frac{1}{2}I\omega^2 = \frac{1}{2}\left[\frac{(3m)(4l)^2}{3}\right]\omega^2$$

$$= 8ml^2\omega^2$$

$$\therefore \quad \omega = \frac{1}{2}\sqrt{\frac{3g}{l}}$$

Angular impulse = Change in angular momentum

$$J(3l) = I\omega$$

$$\Rightarrow \quad 3Jl = (16ml^2)\left(\frac{1}{2}\sqrt{\frac{3g}{l}}\right)$$

$$J = \frac{8}{3}ml\sqrt{\frac{3g}{l}}$$

$$\Rightarrow \quad J = \frac{8}{3}m\sqrt{3gl}$$

2. (a) Let ω' be the angular speed in opposite direction. Again applying conservation of mechanical energy.

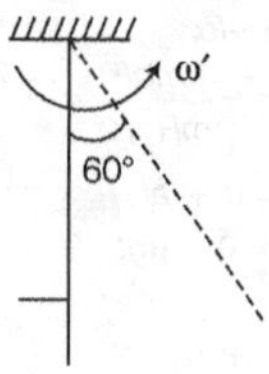

$$3m(g)(l) = \frac{1}{2}I(\omega')^2 = 8ml^2(\omega')^2$$

$$\therefore \quad \omega' = \frac{1}{2\sqrt{2}}\sqrt{\frac{3g}{l}}$$

Angular impulse = Change in angular momentum

$$J(3l) = I(\omega + \omega')$$

$$= 16ml^2\,\frac{1}{2}\sqrt{\frac{3g}{l}}\left(1 + \frac{1}{\sqrt{2}}\right)$$

$$\therefore \quad J = \frac{4}{3}m\sqrt{6gl}\,(\sqrt{2} + 1)$$

3. (b) Angular impulse = Change in angular momentum

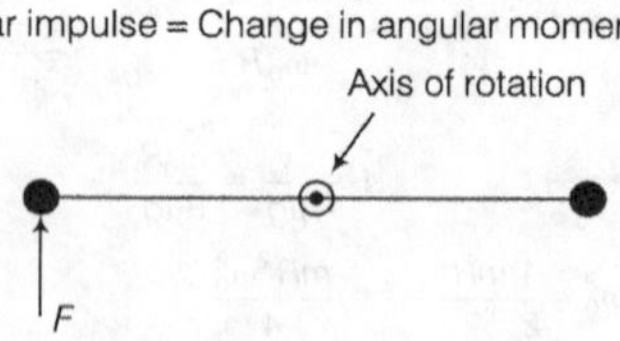

$$\tau dt = I(\omega_f - \omega_i)$$

$$F \times \frac{l}{2}dt = I(\omega_f - \omega_i)$$

$$\Rightarrow \quad 5\left(\frac{50 \times 10^{-2}}{2}\right)0.1 = 2\left[\left(\frac{1}{2}\right)\left(\frac{50 \times 10^{-2}}{2}\right)^2\right](\omega_f - 10)$$

$$\therefore \quad \omega_f = 12 \text{ rad s}^{-1}$$

4. (c) Let ω be the angular velocity of the rod.

Angular impulse = Change in angular momentum about centre of mass of system

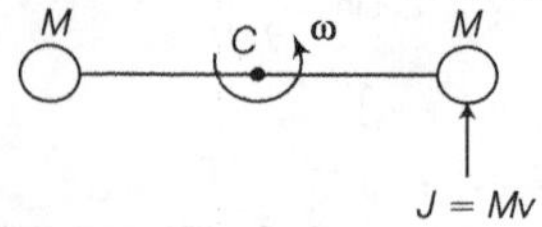

$$J \cdot \frac{L}{2} = I_C\omega$$

$$Mv\left(\frac{L}{2}\right) = 2\left[M\left(\frac{L}{2}\right)^2\right]\omega$$

$$\therefore \quad \omega = \frac{v}{L}$$

5. (a) Torque $= f \cdot R = I\alpha$

$$f = \frac{mg\sin\theta}{1 + \frac{MR^2}{I}} = \frac{2}{7}mg\sin\theta$$

Hence, $\Delta L = \frac{2}{7}mg\sin\theta\, Rt$

6. (b) Since, there is no external torque about A on the system (ball + rod), angular momentum is conserved about this point.

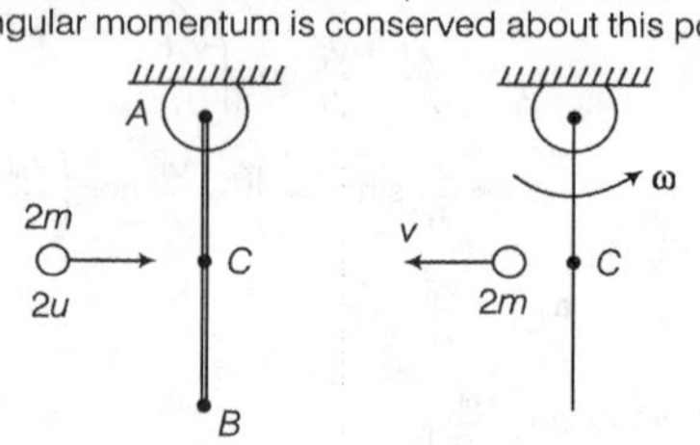

$$\therefore \quad L_i = L_f$$

$$(2m)(2u)(l) = \frac{(3m)(2l)^2}{3}\omega - (2m)(v)(l)$$

$$\therefore \quad 2u = 2\omega l - v \qquad \ldots\text{(i)}$$

For an elastic collision,

Relative speed of approach = Relative speed of separation at point C

$$\therefore \quad 2u = v + \omega l \qquad \ldots\text{(ii)}$$

Solving Eqs. (i) and (ii), we get $v = \frac{2u}{3}$

7. (a) In this case, component of velocity along AC will remain unaltered. Proceeding in the similar manner, we can replace.

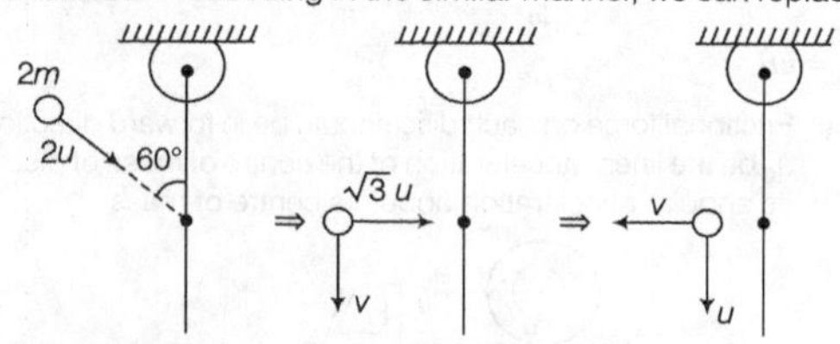

$2u$ by $\sqrt{3}u$ and find v

$$v = \frac{2}{3}\left(\frac{\sqrt{3}}{2}u\right) = \frac{u}{\sqrt{3}}$$

$\therefore$ Speed of the particle after impact $= \sqrt{v^2 + u^2}$

$$= \sqrt{\frac{u^2}{3} + u^2} = \frac{2}{\sqrt{3}}u$$

8. (a)

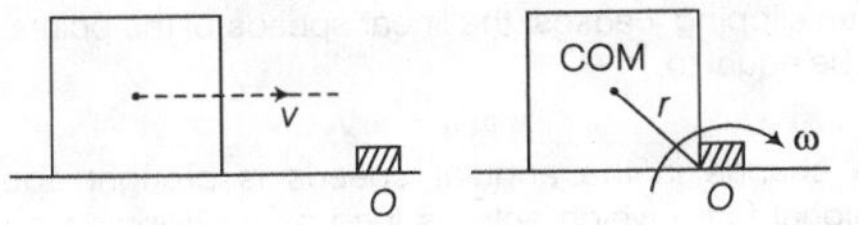

$$r = \sqrt{2}\,\frac{a}{2},\quad r^2 = \frac{a^2}{2}$$

Net torque about O is zero. So, angular momentum about O will be conserved or $L_i = L_f$.

$\therefore$ $$Mv\left(\frac{a}{2}\right) = I_0\omega = (I_{COM} + Mr^2)\omega$$

or $$\frac{Mva}{2} = \left[\frac{Ma^2}{6} + M\left(\frac{a}{2}\right)^2\right]\omega = \frac{2}{3}ma^2\omega$$

$\Rightarrow$ $$\omega = \frac{3v}{4a}$$

9. (c) $Mv_0\left(\frac{a}{2}\right) = I_0\omega = \left[\frac{Ma^2}{6} + M\left(\frac{a}{\sqrt{2}}\right)^2\right]\omega$

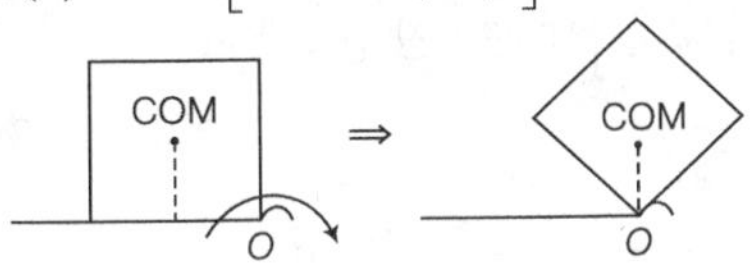

$$= \frac{2}{3}Ma^2\omega$$

$\therefore$ $$\omega = \frac{3v_0}{4a}$$

Now, from conservation of mechanical energy

$$\frac{1}{2}I\omega^2 = Mg(\Delta h) \Rightarrow \frac{1}{2}\left[\frac{2}{3}Ma^2\right]\left[\frac{3v_0}{4a}\right]^2 = Mg\left[\frac{a}{\sqrt{2}} - \frac{a}{2}\right]$$

$\Rightarrow$ $$v_0 = \sqrt{1.1ag}$$

10. (d) Angular impulse $= \Delta L$

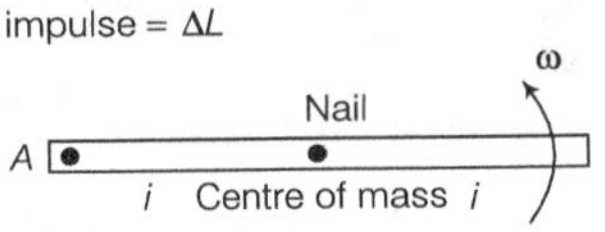

$$Jl = \frac{m(2l)^2}{3}\omega - 0$$

or $$J = \frac{4}{3}ml\omega$$

11. (a,c) Angular momentum is conserved about axis of rotation.

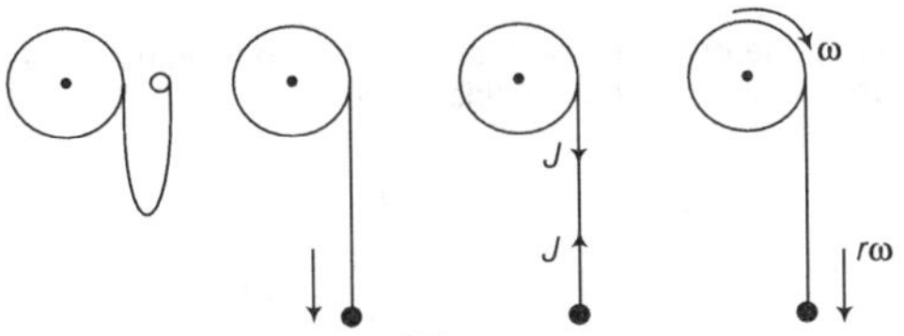

$$u = 4\sqrt{gr}$$

$\therefore$ $$mur = (mr\omega)r + I\omega$$

Here, $$I = \frac{1}{2}(2m)r^2 = mr^2$$

$\therefore$ $$m(4\sqrt{gr})r = mr^2\omega + mr^2\omega$$

$\Rightarrow$ $$\omega = 2\sqrt{\frac{g}{r}}$$

Considering the sudden change in the linear momentum of the particle caused by the impulsive tension, we have $J = -mr\omega - (4m\sqrt{gr})$.

or $$J = 2m\sqrt{gr}$$

12. (b,d) About A, angular momentum of the system (rod + bullet) is conserved.

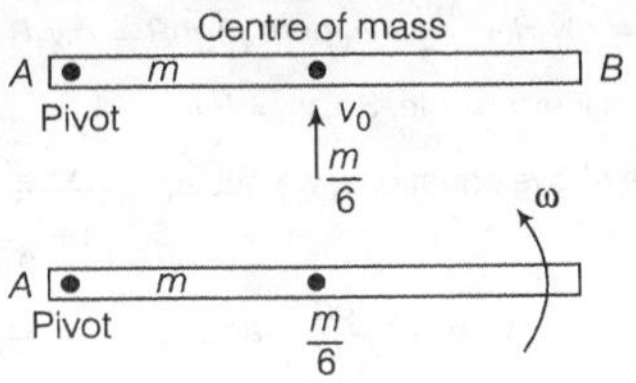

$$L_i = L_f$$

$$\frac{m}{6}(v_0)\left(\frac{L}{2}\right) = \left[\frac{mL^2}{3} + \frac{m}{6}\left(\frac{L}{2}\right)^2\right]\omega \quad \Rightarrow \quad \omega = \frac{2v_0}{9L}$$

$$K_i = \frac{1}{2}\left(\frac{m}{6}\right)v_0^2 = \frac{mv_0^2}{12}$$

$$K_f = \frac{1}{2}I_A\omega^2 = \frac{1}{2}\left[\frac{mL^2}{3} + \frac{mL^2}{24}\right]\omega^2$$

$$= \frac{1}{2}\left(\frac{9}{24}mL^2\right)\left(\frac{4v_0^2}{81L^2}\right) = \frac{mv_0^2}{108}$$

$\therefore$ $$\frac{K_f}{K_i} = \frac{mv_0^2/108}{mv_0^2/12} = \frac{1}{9}$$

13. (a,d) We can assume the rolling of cylinder as pure rotation about an axis passing through O (see figure). Using conservation of angular momentum about O.

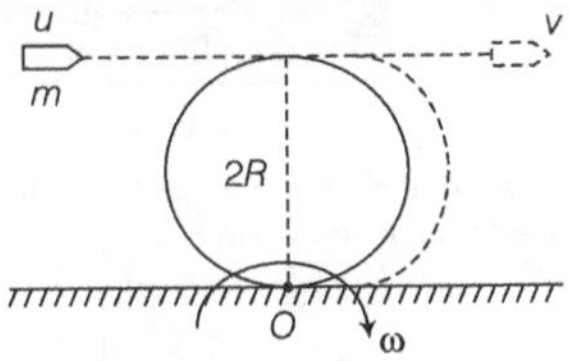

$$(m)(u)(2R) = (m)(v)(2R) + I_0\omega \quad \ldots\text{(i)}$$

Here, $$v = (2R)\omega \text{ and } I_0 = \frac{MR^2}{2} + MR^2 = \frac{3}{2}MR^2$$

Substituting these in Eq. (i), we get

$$\omega = \frac{4mu}{(8m + 3M)R} \text{ and } v = \frac{8mu}{(8m + 3M)}$$

14. (a,d) Point P has net velocity in rightward direction. So, friction is opposite and angular velocity decreases.

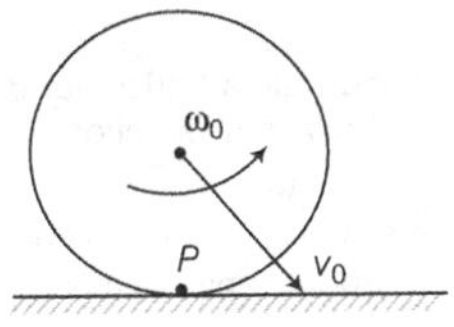

Case (i)

Point P' may have net velocity in leftward or rightward direction depending on values of v_0 and ω_0.

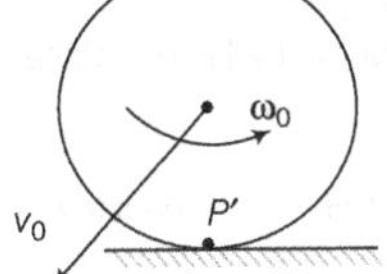

Case (ii)

15. (a,b,c,d) Just before the string is taut, cylinder is not rotating. So, $E_0 = mgh$ when the string gets taut, $\mathbf{L}_i = \mathbf{L}_f \Rightarrow$
$mv_0R + 0 = mv_1R + \frac{MR^2}{2}\omega_1$ or $m\sqrt{2ghR} = mv_1R + \frac{MR^2}{2}\omega_1$.
The, sring is inextensible. So, $v_1 = R\omega_1$

Solving the above equations, we get $\omega_1 = \frac{\sqrt{2gh}}{\left(1+\frac{M}{2m}\right)}$

$$v_1 = \frac{\sqrt{2gh}}{\left(1+\frac{M}{2m}\right)} \text{ and } E_1 = \frac{E_0}{\left(1+\frac{M}{2m}\right)}$$

16. (a) $L = I\omega_0 = \frac{1}{2}MR^2\omega_0$

17. (c) In the process, the mechanical energy of the chip remains constant.

$$\therefore \quad \frac{1}{2}mv^2 = mgh \Rightarrow h = \frac{v^2}{2g} = \frac{(R\,\omega_0)^2}{2g} = \frac{R^2\omega_0^2}{2g}$$

18. (d) $L_f = I'\omega_0 = \left(\frac{MR^2}{2} - mR^2\right)\omega_0 = \left(\frac{M}{2} - m\right)R^2\omega_0$

$$K_f = \frac{1}{2}I'\omega_0^2 = \frac{1}{2}\left(\frac{M}{2} - m\right)R^2\omega_0^2$$

19. (5) The forces during impact are as shown in Fig. (i).

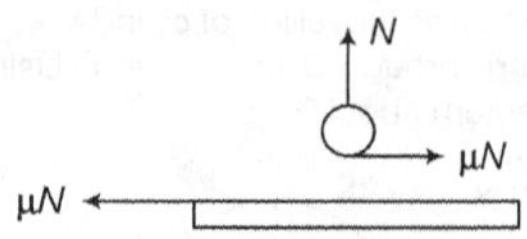

Fig (i)

Let the horizontal velocities of the plank and the ball be v_2 and v_1 in opposite directions as shown in Fig. (ii)

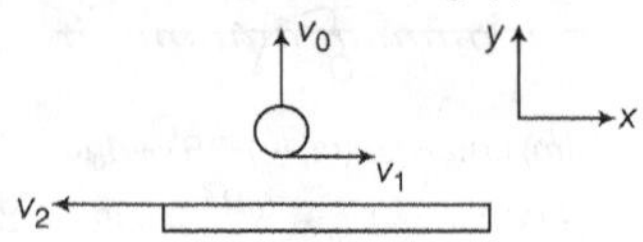

Fig (ii)

From conservation of linear momentum

$$mv_1 = Mv_2 \qquad \text{...(i)}$$

Linear impulse of the ball in vertical direction = Change in linear momentum in vertical direction.

$$\therefore \quad J = 2mv_0 \qquad \text{...(ii)}$$

$$[\because J = Ndt = \Delta p]$$

Linear impulse on the ball in horizontal direction. = Change in linear momentum in horizontal direction.

$$\therefore \quad \mu J = mv_1 \qquad \text{...(iii)}$$

Angular impulse on the ball about centre of mass = Change in angular momentum about centre of mass

$$\therefore \quad \mu JR = I\omega_0 = \frac{2}{5}mR^2\omega_0 \qquad \text{...(iv)}$$

Solving Eqs. (i), (ii), (iii) and (iv), we get

$$v_1 = \frac{2}{5}R\omega_0 \text{ and } v_2 = \frac{m}{M}\left(\frac{2}{5}R\omega_0\right)$$

Now, actual path of the ball is projectile, whose time of flight will be $T = \frac{2v_y}{g} = \frac{2v_0}{g}$

Relative velocity of ball with respect to plank in horizontal direction is

$$v_r = v_1 + v_2 = \frac{2}{5}\left(1 + \frac{m}{M}\right)R\omega_0$$

So, the desired distance is

$$s = v_rT = \frac{4}{5}\cdot\frac{v_0R\omega_0}{g}\left(1 + \frac{m}{M}\right)$$

20. When slipping ceases, the linear speeds of the points of contact will be equal to

$$\therefore \quad \omega_1'r_1 = \omega_2'r_2 \qquad \text{...(i)}$$

The change in the angular speeds is brought about by the frictional force which acts as long as the slipping exists. If this force f acts for a short time t, then torque on the first cylinder is fr_1 and that on the second is fr_2. The corresponding angular impulses are $-fr_1dt$ and fr_2dt.

$$\therefore \quad -fr_1dt = I_1(\omega_1' - \omega_2) \text{ and } fr_2\,dt = I_2(\omega_2' - \omega_2)$$

$$\Rightarrow \quad -\frac{I_1}{r_1}(\omega_1' - \omega_1) = \frac{I_2}{r_2}(\omega_2' - \omega_2) \qquad \text{...(ii)}$$

Solving Eqs. (i) and (ii), we get

$$\omega_1' = \frac{I_1\omega_1r_2 + I_2\omega_2r_1}{I_2r_1^2 + I_1r_2^2}\cdot r_2$$

and

$$\omega_2' = \frac{I_1\omega_1r_2 + I_2\omega_2r_1}{I_2r_1^2 + I_1r_2^2}\cdot r_1$$

21. Mass of the rod is

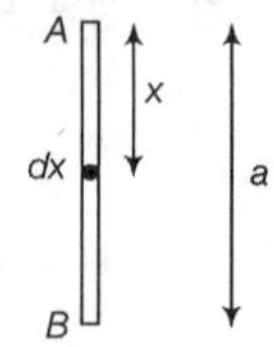

$$M = \int_0^a dm = \int_0^a \frac{\rho_0(a + x)}{a}\cdot dx = \frac{3a\rho_0}{2}$$

$$\therefore \quad M = \frac{3a\rho_0}{2}$$

$$X_{COM} = \frac{\int_0^a xdm}{\int dm} = \frac{\int_0^a \frac{\rho_0(a + x)}{a}\cdot xdx}{\frac{3a\rho_0}{2}} = \frac{5}{9}a$$

$$I_A = \int_0^a x^2dm = \int_0^a x^2\frac{\rho_0(a + x)}{9}dx = \frac{7\rho a^3}{12}$$

Angular impulse

= Change in angular momentum

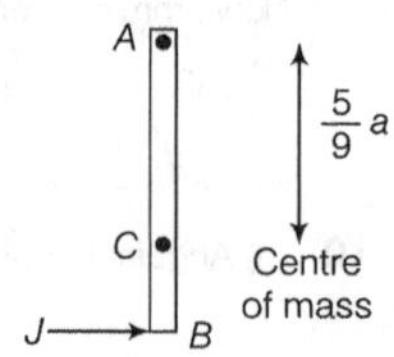

$$J(a) = I_A\omega = \frac{7\rho a^3}{12}\,\omega$$

$$\therefore \quad \omega = \frac{12J}{7\rho a^2}$$

B will pass through a point vertically above A, if

$$\frac{1}{2}I_A\,\omega^2 > Mg\,(2 \times AC)$$

$$\frac{1}{2}\left(\frac{7\rho a^3}{12}\right)\left(\frac{144\,J^2}{49\rho^2a^4}\right) > \left(\frac{3a\rho_0}{2}\right)g\,(2)\left(\frac{5}{9}a\right)$$

$$\Rightarrow \quad J > \frac{\sqrt{70a^3\rho_0^2g}}{6}$$

22. During the collision, the total angular momentum of the system about axis of rotation remains conserved.

$$\therefore \quad L_i = L_f$$

$$(I_{rods})\,\omega_i + (L_i)_{mud} = (I_{rods} + I_{mud})\,\omega_f$$

$$I_{rods} + I_{mud} = 4\left(\frac{Md^2}{3}\right) + md^2$$

$$= \frac{4(3m)d^2}{3} + md^2 = 5md^2$$

$$\therefore \quad \omega_f = (4md^2\,\omega_i + (L_i)_{mud})/5md^2 \qquad \text{...(i)}$$

Angular momentum is taken positive for counter clockwise rotation and is taken negative for clockwise rotation.

$(L_i)_{mud}$: Path 1 : $L_1 = -mvd$

Path 2 : $L_2 = 0$

Path 3 : $L_3 = mvd$ and

Path 4 : $L_4 = m(v\cos 60°)\,d = \frac{mvd}{2}$

Here, $v = 12\ \text{ms}^{-1}$

Substituting these values in Eq. (i),

We get corresponding ω_f. The values of ω_f, we would obtain

Path 1 : $\omega_f = -4 \text{ rad s}^{-1}$

Path 2 : $\omega_f = -1.6 \text{ rad s}^{-1}$

Path 3 : $\omega_f = 0.8 \text{ rad s}^{-1}$

Path 4 : $\omega_f = -0.4 \text{ rad s}^{-1}$

23. (i) As, the total initial linear momentum of the system (skaters + light pole) is zero, the centre of mass of the system will be at rest before and after collision.

After they are connected by the pole, the skaters will rotate around the centre of mass (at the mid-point of the pole). Free rotation of system always takes place about centre of mass.

Applying the conservation of angular momentum about an axis through centre of mass and perpendicular to the plane of the figure.

$$mvl + mvl = I\omega, \quad \text{where } I = (2m)(l)^2$$

$$\therefore \quad \omega = \frac{2mvl}{2ml^2} = \frac{v}{l} = \frac{10}{3/2} = \frac{20}{3} \text{ rad s}^{-1}$$

(ii) When the skaters reduce their separation, no external torque acts on the system. Hence, we can apply conservation of angular momentum about centre of mass. Due to redistribution of mass about axis of rotation, then moment of inertia changes.

The separation reduces to $2l' = 1 \text{ m}$

$$I\omega = I'\omega' \text{ (conservation of angular momentum)}$$

$$\omega' = \frac{I\omega}{I'} = \frac{(2m)l^2\omega}{2m(l')^2} = 9\omega$$

or $$\omega' = 9\left(\frac{20}{3}\right) = 60 \text{ rads}^{-1}$$

(iii) $$\frac{KE_f}{KE_i} = \frac{\frac{1}{2}I'\omega'^2}{\frac{1}{2}I\omega^2} = \frac{(I'\omega')\omega'}{(I\omega)\omega} = \frac{\omega'}{\omega} = 9$$

The kinetic energy increases because the skaters do positive work in pulling themselves towards the centre of the pole.

DPP-9 Toppling and Rolling on an Inclined Plane

1. (a) For toppling before slipping,

$$\mu \ge \frac{l}{2h} \text{ or } h \ge \frac{l}{2\mu} \text{ or } h \ge \frac{30}{2 \times 0.6} \text{ or } h \ge 25 \text{ cm}$$

2. (c) Let m be the mass of the block, l be length of each side and μ be the coefficient of friction, then the block will not slip if

$$F < \mu mg$$

The block will start toppling when torque of F about $S \ge$ torque of mg about S.

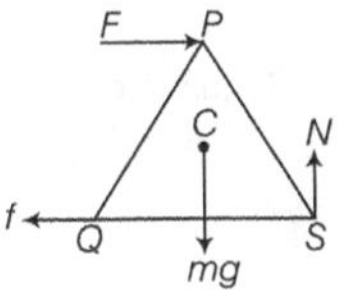

or $$F\frac{\sqrt{3}}{2}l \ge mg \cdot \frac{l}{2}$$

$$\Rightarrow \quad F \ge \frac{1}{\sqrt{3}}mg$$

Hence, the block will topple without slipping.

If $\mu mg \ge \frac{1}{\sqrt{3}}mg$ or $\mu \ge \frac{1}{\sqrt{3}}$

3. (a) $Mg \sin\theta - T = Ma$

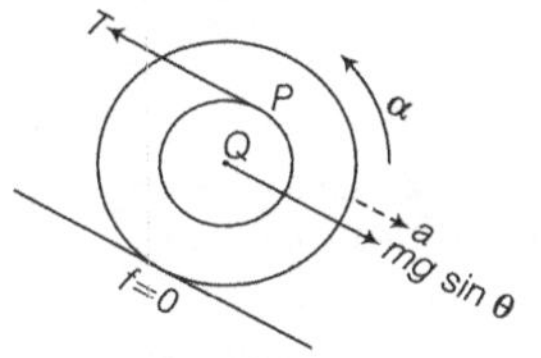

$$\therefore \quad a = \frac{Mg\sin\theta - T}{M} \quad \text{...(i)}$$

$$Tr = I\alpha \Rightarrow \alpha = \frac{Tr}{I} \quad \text{...(ii)}$$

$$a = r\alpha \quad \text{...(iii)}$$

Since, the points of contact P (on the thread) and Q on the spool should have same acceleration. $a_P = 0, a_Q = a - r\alpha$ and $a_P = a_Q$ $\Rightarrow a = r\alpha$.

Solving the Eqs. (i), (ii) and (iii), we get $a = \dfrac{g\sin\theta}{1 + \dfrac{I}{Mr^2}}$

4. (d) As, there is no slipping, then mechanical energy is conserved.

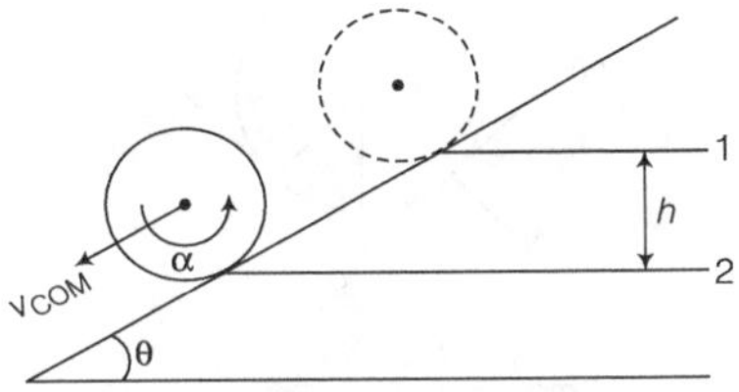

Total energy at position 1 is

$$E_1 = mgh$$

Total energy at position 2 is

$$E_2 = \frac{1}{2}mv_{COM}^2 + \frac{1}{2}I_{COM}\omega^2$$

Also at position 2, $v_{COM} = r\omega$

We know $I_{COM} = \dfrac{mR^2}{2}$

$$\therefore \quad E_2 = \frac{3}{4}mv_{COM}^2$$

$$E_1 = E_2$$

$$\Rightarrow \quad mgh = \frac{3}{4}mv_{COM}^2$$

or $$v_{COM} = \sqrt{\frac{4gh}{3}}$$

5. (c) If the block and cylinder move independently on the incline, their accelerations are

$$a_{block} = g\sin\theta - g\mu\cos\theta$$

$$= \frac{g}{\sqrt{2}} - \frac{g}{2\sqrt{2}} = \frac{g}{2\sqrt{2}}$$

$$a_{cylinder} = \frac{g\sin\theta}{1 + \dfrac{I}{2mR^2}}$$

$$= \frac{g/\sqrt{2}}{1 + \dfrac{2mR^2/2}{2mR^2}} = \frac{\sqrt{2}g}{3}$$

$a_{block} < a_{cylinder}$. So, both the bodies will move in contact with each other with common acceleration a which we have to calculate.

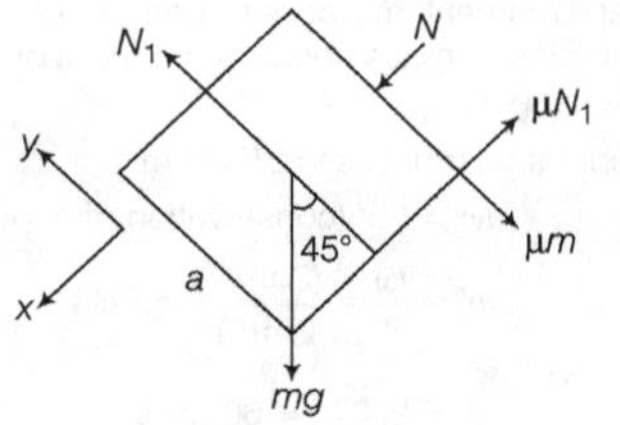

Free body diagram of block is as shown.

$$\Sigma F_x = ma$$

$$\therefore \quad \frac{mg}{\sqrt{2}} + N - \mu N_1 = ma \quad \ldots(i)$$

$$\Sigma F_y = 0$$

$$\Rightarrow \quad N_1 - \mu N = \frac{mg}{\sqrt{2}} \quad \ldots(ii)$$

From Eqs. (i) and (ii), we get

$$\frac{mg}{\sqrt{2}} + N - \mu^2 N - \frac{\mu mg}{\sqrt{2}} = ma \quad \ldots(iii)$$

The free body diagram of cylinder is as shown.

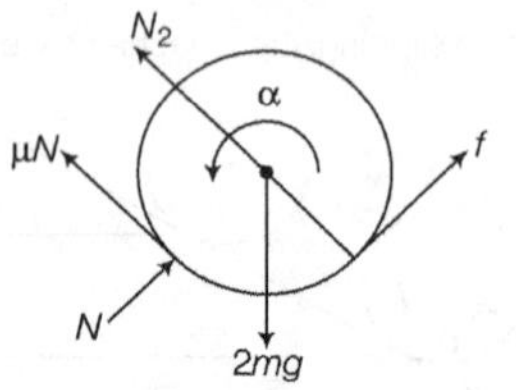

$$\Sigma F_x = 2ma$$

$$\Rightarrow \quad \frac{2mg}{\sqrt{2}} - N - f = 2ma \quad \ldots(iv)$$

and $$\Sigma\tau = fR - \mu NR = \frac{1}{2}(2m)(R^2)\left(\frac{a}{R}\right)$$

or $$f - \mu N = ma \quad \ldots(v)$$

From Eqs. (iv) and (v), we get

$$N = \frac{m(\sqrt{2}g - 3a)}{1+\mu}$$

Using this in Eq. (iii), we get $a = \dfrac{3g}{5\sqrt{2}}$

6. (b) $$\Sigma F_x = ma$$

$$\Rightarrow \quad 3mg \sin\theta - f = 3ma \quad \ldots(i)$$

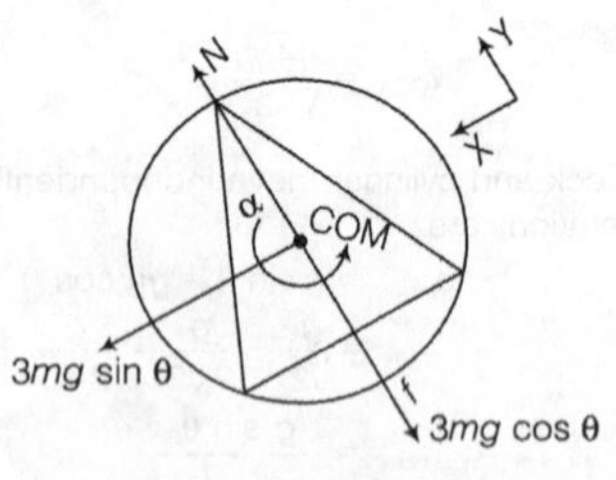

$$\Sigma F_y = 0$$

$$\Rightarrow \quad N - 3mg\cos\theta = 0 \quad \ldots(ii)$$

$$\Sigma\tau = fR = I\alpha \quad \ldots(iii)$$

For no slipping $a = R\alpha$ ...(iv)

The moment of inertia of the assembly about its centre of mass is

$$I = \frac{3}{2}mR^2$$

Now, on solving Eqs. (i), (ii), (iii) and (iv) simultaneously, we obtain

$$f = mg\sin\theta$$

If μ is the coefficient of fricti on at the contact surface, then

$$f \le \mu N$$

or $$f \le \mu \times 3mg\cos\theta$$

or $$mg\sin\theta \le 3\mu mg\cos\theta$$

or $$\mu \ge \frac{1}{3}\tan\theta$$

$$\therefore \quad \mu_{min} = \frac{1}{3}\tan\theta$$

7. (d) The net force along the incline is zero. So, cylinder will remain in position till it stops rotating. After that, it will start rolling downwards. $[mg\sin\theta = \mu mg\cos\theta]$

8. (c,d) The cylinder will slide if,

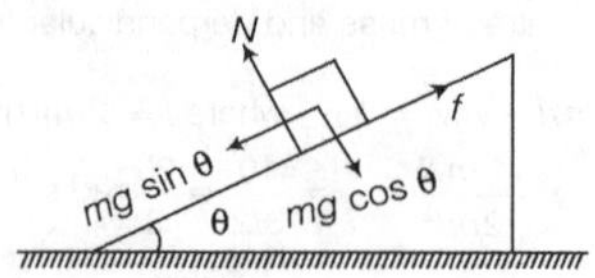

$$mg\sin\theta > \mu\, mg\cos\theta$$

or $$\tan\theta > \mu \quad \ldots(i)$$

The cylinder will topple if

$$(mg\sin\theta)\frac{h}{2} > (mg\cos\theta)r \text{ or } \tan\theta > \frac{2r}{h} \quad \ldots(ii)$$

Thus, the condition for sliding is $\tan\theta > \mu$ and the condition for toppling is $\tan\theta > \dfrac{2r}{h}$.

Hence, the cylinder will slide before toppling, if $\mu < \dfrac{2r}{h}$ and the cylinder will topple before sliding, if $\mu > \dfrac{2r}{h}$.

9. (a,b,c,d)

$$mg\sin\theta + f - T = ma_1 \quad \ldots(i)$$

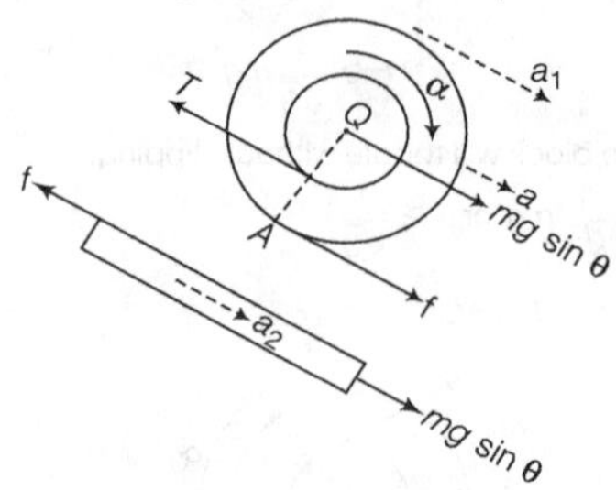

$$Mg\sin\theta - f = Ma_2 \quad \ldots(ii)$$

$$T(R) - f(2R) = I\alpha \quad \ldots(iii)$$

For no slipping at B,

$$a_1 - R\alpha = 0 \quad \ldots(iv)$$

For no slipping at A,

$$a_1 - 2R\alpha = a_2 \quad \ldots(v)$$

Solving Eqs. (i), (ii), (iii), (iv) and (v), we get

$$\alpha = \left[\frac{(m-M)g\sin\theta}{mR^2 + MR^2 + I}\right]R$$

For $m > M$, i.e. $\dfrac{m}{M} > 1$, α is positive. Thus, a_2 is negative. So, the plank will ascend.

For $m < M$, i.e. $\dfrac{m}{M} < 1$, α is negative. Thus, a_2 is positive. So, the plank will descend.

From $m = M$, i.e. $\dfrac{m}{M} = 1$, α is zero. Thus, a_2 is zero. So, the plank will remain stationary.

10. (a,c)

$m_A = m = 0.5$ kg, $m_P = M = 2$ kg, $R = 0.2$ m

Let a be the linear retardation of the block A and α the angular retardation of drum P. Then,

$$a = R\alpha \quad \text{...(i)}$$

$$mg \sin 30° - T = ma \Rightarrow \frac{mg}{2} - T = ma \quad \text{...(ii)}$$

Also, $$\alpha = \frac{\tau}{I} = \frac{TR}{\frac{1}{2}MR^2}$$

$$\Rightarrow \quad \alpha = \frac{2T}{MR} \quad \text{...(iii)}$$

Solving Eqs. (i), (ii) and (iii), we get

$$T = \frac{1}{2}\left[\frac{Mmg}{M+2m}\right]$$

$$\therefore \quad T = \frac{1}{2}\left[\frac{(2)(0.5)(9.8)}{2+2(0.5)}\right] = 1.63 \text{ N}$$

From Eq. (iii), angular retardation is $\alpha = \frac{2T}{MR} = \frac{(2)(1.63)}{(2)(0.2)}$

or $\alpha = 8.15$ rad s^{-2}

Linear retardation of the block is

$$a = R\alpha = (0.2)(8.15) = 1.63 \text{ ms}^{-2}$$

When $\omega_0 = 10$ rad s^{-1}, velocity of block will be

$$v_0 = \omega_0 R = 2 \text{ ms}^{-1}$$

$$\therefore \quad s = \frac{v_0^2}{2a} = \frac{(2)^2}{2(1.63)} = 1.227 \text{ m}$$

11. (a) Maximum frictional force can be

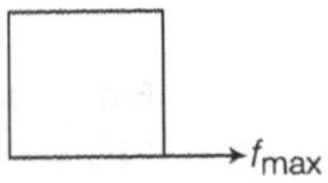

$$f_{max} = \mu mg$$

(m = mass of the block)

$\therefore$ The maximum acceleration of block can be

$$a_{max} = \frac{f_{max}}{m} = \mu g$$

or $a_{max} = (0.5)(9.8) = 4.9$ ms^{-2}

12. (b) In the critical case, when the block is about to topple about A, normal reaction will pass through A.

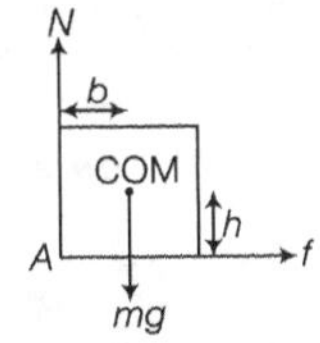

$$f = ma \quad \text{...(i)}$$

$$N = mg \quad \text{...(ii)}$$

The block will topple when torque of f about centre of mass > torque of N about centre of mass. Taking the limiting value,

$$\tau_f = \tau_N \text{ or } (f)(h) = (N)(b)$$

$$(ma)h = (mg)b$$

$$\Rightarrow \quad a = \left(\frac{b}{h}\right)g = \left(\frac{0.6}{0.9}\right)(9.8) = 6.53 \text{ ms}^{-2}$$

So, if the acceleration of car exceeds this value, the block will topple about A.

13. (c) The maximum acceleration/retardation at which the block is not disturbed, is the smaller of the two values obtained above, i.e. 4.9 ms^{-2}

Hence, the maximum retardation can be 4.9 ms^{-2}.

$$u = 72 \text{ km h}^{-1} = 72 \times \frac{5}{18} = 20 \text{ ms}^{-1}$$

$$v^2 - u^2 = 2as \; (v = 0)$$

$$s = \frac{u^2}{2a} = \frac{(20)^2}{2(4.9)^2} \text{ or } s = 40.82 \text{ m}$$

14. (d) In the limiting case, N passes through left edge. Taking moments about S,

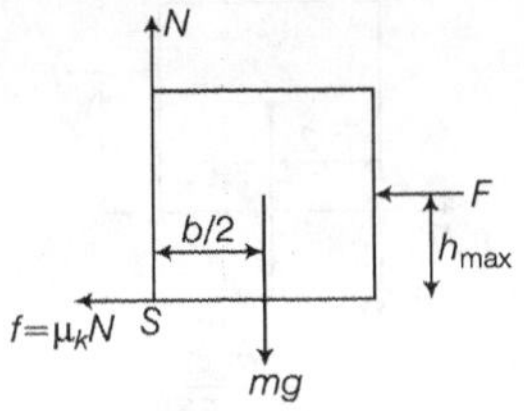

$$Fh_{max} - mg \cdot \frac{b}{2} = 0$$

or $$h_{max} = \frac{mgb}{2F}$$

Also, $N = mg$, $f = \mu_k N$ and $f = F$. So, $F = \mu_k mg$

$$\therefore \quad h_{max} = \frac{mgb}{2(\mu_k mg)} = \frac{b}{2\mu_k}$$

Alternate Method As, three forces F resultant of $\mu_k N$ and N, mg keep the body in equilibrium (dynamic), the resultant of $\mu_k N$ and N must pass through a point through which F acts, i.e. three forces F resultant of $\mu_k N$ and N and mg are concurrent.

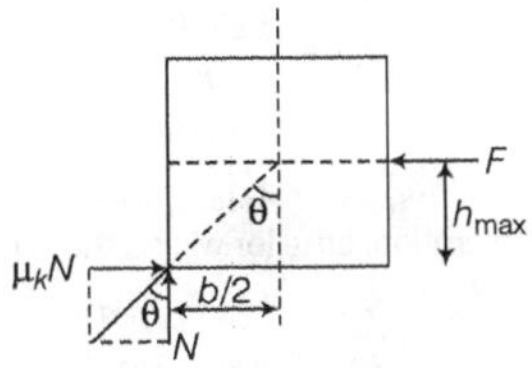

$$\tan\theta = \frac{\mu_k N}{N} = \frac{b/2}{h_{max}}$$

$$\Rightarrow \quad h_{max} = \frac{b}{2\mu_k}$$

15. (b) About S', $mgx = F \cdot \frac{H}{2}$

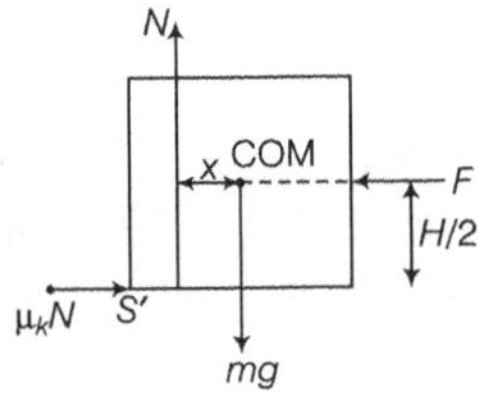

Also, $$F = \mu_k mg$$

$$\therefore \quad mgx = \mu_k mg \cdot \frac{H}{2}$$

$$x = \frac{\mu_k H}{2}$$

$\therefore$ Required distance is

$$\frac{b}{2} - \frac{\mu_k H}{2} = \frac{b - \mu_k H}{2}$$

Alternate Method

$$\tan\theta = \frac{\mu_k N}{N}$$

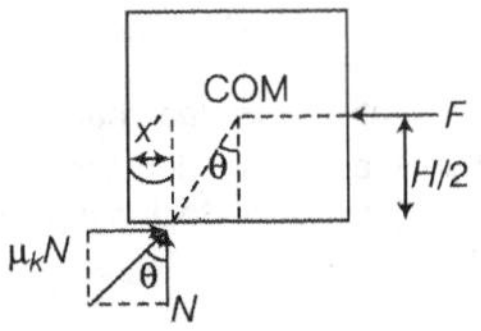

Also, $$\tan\theta = \left(\frac{b}{2} - x'\right)/H/2$$

$$\frac{2\left(\frac{b}{2} - x'\right)}{H} = \mu_k$$

$$\frac{b}{2} - x' = \frac{\mu_k H}{2}$$

$$x' = \frac{b - \mu_k H}{2}$$

16. (a) About S, $Fh = mg\dfrac{b}{2}$

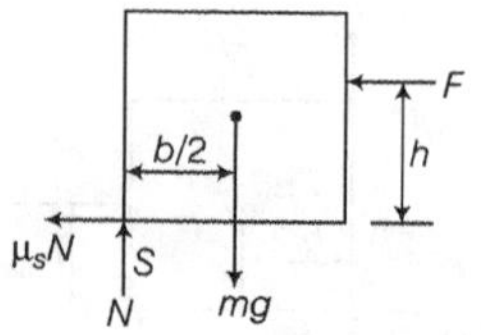

$$h = \frac{mgb}{2F}$$

Also, $F = \mu_s N,\ N = mg$

$\therefore$ $F = \mu_s mg$

$\therefore$ $h = \dfrac{mgb}{2\mu_s mg} = \dfrac{b}{2\mu_s}$

17. (a) Energy is conserved for all the three bodies. So, final KE for all will be same.

$$\frac{K^2}{R^2} = 1 \text{ for ring}, \frac{K^2}{R^2} = \frac{1}{2} \text{ for disc},$$

$$\frac{K^2}{R^2} = \frac{2}{5} = 0.4 \text{ for sphere}$$

$$a = \frac{g\sin\theta}{1 + \dfrac{K^2}{R^2}}$$

$\therefore$ $a_{sphere} > a_{disc} > a_{ring}$

Higher the acceleration, smaller will be the time taken.

Also, $v_{sphere} > v_{disc} > v_{ring}$

and $t_{sphere} < t_{disc} < t_{ring}$

Hence, (i) → (r), (ii) → (p), (iii) → (q), (iv) → (q)

18. Figure shows the rear view of the car that is turning towards left. Frictional force on the tyres provides centripetal force while turning.

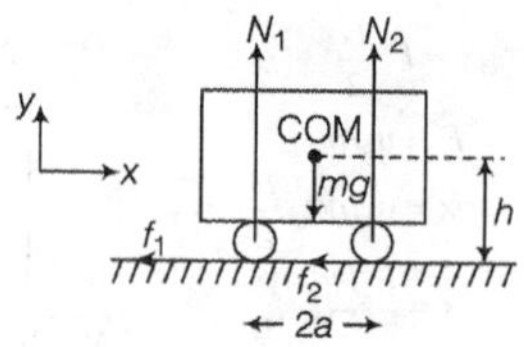

$$\Sigma F_x = f_1 + f_2 = \frac{mv^2}{r} \quad \text{...(i)}$$

$$\Sigma F_y = N_1 + N_2 - mg = 0 \quad \text{...(ii)}$$

From condition of rotational equilibrium, sum of moments about centre of mass must be zero.

$$\Sigma\tau = (f_1 + f_2)h + N_1 a - N_2 a = 0$$

or $$f_1 + f_2 = (N_2 - N_1)\frac{a}{h} \quad \text{...(iii)}$$

From the above equation, it is clear that $N_1 < N_2$.

So, as speed increases, N_1 decreases and at the time of overturning N_1 becomes zero. So, inner wheels lose their contact with the ground.

From Eqs. (i) and (iii), we get

$$N_2 - N_1 = \frac{mv^2 h}{ar} \quad \text{...(iv)}$$

By solving Eqs. (ii) and (iv), we get $2N_1 = mg - \dfrac{hmv^2}{ar}$ and $2N_2 = mg + \dfrac{hmv^2}{ar}$

At the time of overturning, $N_1 = 0$

$\Rightarrow$ $\dfrac{hmv^2}{ar} = mg \Rightarrow v = \sqrt{\dfrac{gra}{h}}$

Alternate Method We can solve this problem from the reference frame of car. At the instant, the car begins to overturn the forces N_1 and f_1 are zero. Taking moments about centre of mass for rotational equilibrium.

$$\Sigma\tau = \frac{mv^2}{r} \times h - mg \times a = 0$$

or $$v = \sqrt{\frac{gra}{h}}$$

Note *Overturning happens only when μ is sufficient to prevent slipping before overturning. If μ is not sufficient enough to prevent skidding, the vehicle skids first before toppling. In such case $(f_1 + f_2) \le \mu(N_1 + N_2)$ (note that $N_1 + N_2 = mg$)*

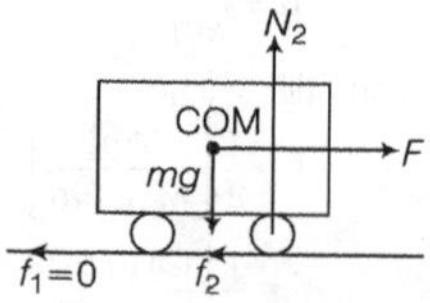

Equality sign holds good when skidding starts.

Also, $f_1 + f_2 = \dfrac{mv^2}{r}$

When skidding starts, $\dfrac{mv^2}{r} = \mu mg \Rightarrow v = \sqrt{\mu gr}$

Therefore, the limiting speed at which skidding uses is given by $v = \sqrt{\mu gr}$

So, if the car is to skid rather than overturn, then

$$\mu gr < \frac{gra}{h} \text{ or } \mu < \frac{a}{h}$$

19. (i) **From C to A**

$$v_A = \sqrt{2gh} = \sqrt{2 \times 10 \times 49} = 31.3 \text{ ms}^{-1}$$

From A to B

Let pure rolling starts after time t, then

$$v = R\omega \Rightarrow v_A - at = R(\alpha t)$$

$$v_A - \mu gt = R\left(\frac{\mu mgR}{\frac{2}{5}mR^2}\right)t = 2.5\,\mu gt$$

$\therefore$ $$t = \frac{v_A}{3.5\,\mu g} = \frac{31.3}{(3.5) \times \frac{2}{3} \times 10} = 1.34 \text{ s}$$

Now, $s = v_A t - \dfrac{1}{2}(\mu g)t^2 = (31.30)(1.34) - \dfrac{1}{2}\left(\dfrac{2}{3}\right)(10)(1.34)^2$

$\Rightarrow$ $s = 36$ m

$$v = v_A - at = v_A - \mu gt = 31.3 - \frac{2}{3}(10)(1.34)$$

$$= 22.36 \text{ ms}^{-1}$$

The remaining distance of $49 - 36 = 13$ m, the velocity is maintained at 22.36 ms^{-1}, because of pure rolling. So, the height attained above the horizontal on BD is

$$h = \frac{v^2}{2g} = \frac{(22.36)^2}{2 \times 10} = 25 \text{ m}$$

Note *From height h to point B, mechanical energy is constant.*

(ii) **From B to A**

Let pure rolling starts at time t'.

So, $v' = R\omega'$

$\Rightarrow$ $(v - \mu gt') = R(-\omega + \alpha t')$

$v - \mu gt' = -\omega R + R\alpha t'$

$\Rightarrow$ $$t' = \frac{v + R\omega}{3.5\,\mu g} = \frac{2v}{3.5\,\mu g}$$

$$t' = \frac{2 \times 22.36}{(3.5)\left(\frac{2}{3}\right)(10)} = 1.91 \text{s}$$

So, $v' = v - \mu g t' = 22.36 - \left(\frac{2}{3}\right)(10)(1.91) = 9.62 \text{ ms}^{-1}$

$$s' = (22.36)(1.91) - \frac{1}{2}\left(\frac{2}{3}\right)(10)(1.91)^2 = 30.35 \text{ m}$$

$s' < 49$ m $\Rightarrow$ velocity of sphere at A during the remaining journey is 9.62 ms^{-1}.

20. Given, $\mu > \tan\theta$

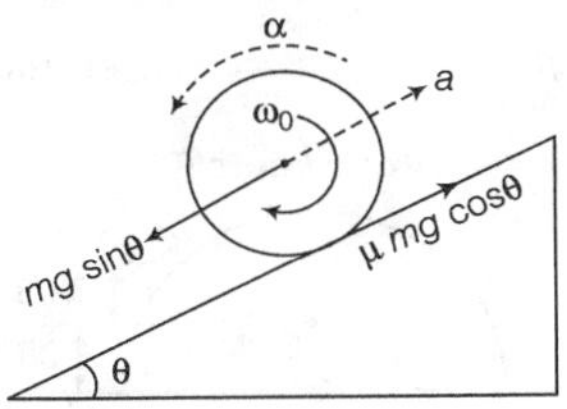

$\therefore \quad \mu mg \cos\theta > mg \sin\theta$

$$a = \mu g \cos\theta - g \sin\theta$$

$$\alpha = \frac{\tau}{I} = \frac{\mu mg \cos\theta R}{\frac{1}{2}mR^2} = \frac{2\mu g \cos\theta}{R}$$

Slipping will stop say at time t, then

$$v = R\omega$$

$\Rightarrow \quad at = R(\omega_0 - \alpha t)$

$\Rightarrow \quad t = \dfrac{R\omega_0}{(a + R\alpha)} = \dfrac{R\omega_0}{3\mu g \cos\theta - g\sin\theta}$

If s' is the distance travelled during this time t, then

$$s' = \frac{1}{2}at^2 = \frac{1}{2}(\mu g \cos\theta - g\sin\theta)\left(\frac{R\omega_0}{3\mu g\cos\theta - g\sin\theta}\right)^2$$

$$s' = \frac{R^2\omega_0^2(\mu\cos\theta - \sin\theta)}{2g\,(3\mu\cos\theta - \sin\theta)^2}$$

Now, $\quad v = at = (\mu g\cos\theta - g\sin\theta)\left(\dfrac{R\omega_0}{3\mu g\cos\theta - g\sin\theta}\right)$

$$= \frac{R\omega_0(\mu\cos\theta - \sin\theta)}{(3\mu\cos\theta - \sin\theta)}$$

Once slipping stops, pure rolling continues, if $\mu > \dfrac{\tan\theta}{1 + \dfrac{mR^2}{I}}$

$\Rightarrow \quad \mu > \dfrac{\tan\theta}{1+2}$

$\Rightarrow \quad \mu > \dfrac{\tan\theta}{3}$

In the question, it is given that $\mu > \tan\theta$. So, pure rolling continues.

$\therefore \quad a' = \dfrac{g\sin\theta}{1 + \dfrac{I}{mR^2}} = \dfrac{g\sin\theta}{1 + \dfrac{1}{2}} = \dfrac{2}{3}g\sin\theta$

If it travels a further distance s_2 under retardation a', then

$$s'' = \frac{v^2}{2a'} = \frac{3R^2\omega_0^2(\mu\cos\theta - \sin\theta)^2}{(3\mu\cos\theta - \sin\theta)^2(4g\sin\theta)}$$

$\therefore \quad s_{max} = s' + s'' = \dfrac{R^2\omega_0^2(\mu\cos\theta - \sin\theta)}{2g(3\mu\cos\theta - \sin\theta)^2}\left[1 + \dfrac{3(\mu\cos\theta - \sin\theta)}{2\sin\theta}\right]$

or $\quad s_{max} = \dfrac{R^2\omega_0^2(\mu\cos\theta - \sin\theta)}{4g\sin\theta\,(3\mu\cos\theta - \sin\theta)}$

Revisal Problems (JEE Main)

1. (b) $J = mv_0 \Rightarrow v_0 = \dfrac{J}{m}$

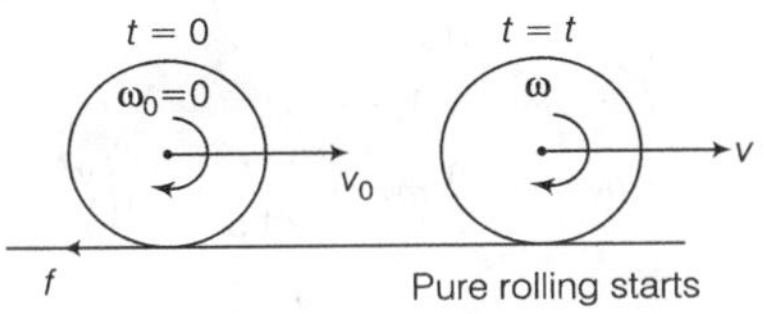

Pure rolling starts

$$v = R\omega$$

Angular momentum is conserved about bottom most point.

$$L_i = L_f$$

$\Rightarrow \quad mv_0R = mvR + I\omega$

$$mv_0R = mvR + (mR^2)\left(\frac{v}{R}\right)$$

$$= 2mvR$$

$$v = \frac{v_0}{2} = \frac{J}{2m}$$

2. (d) Components of velocity along the rod should be equal.

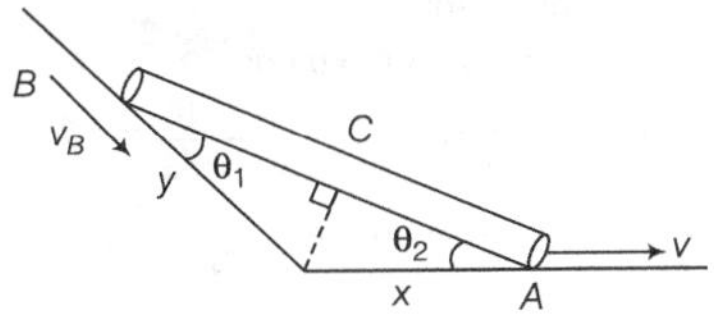

$\therefore \quad v_B\cos\theta_1 = v\cos\theta_2$

$\Rightarrow \quad v_B = v\dfrac{\cos\theta_2}{\cos\theta_1}$

3. (b) $I = \dfrac{1}{3}ml^2\sin^2\theta$ (in the figure)

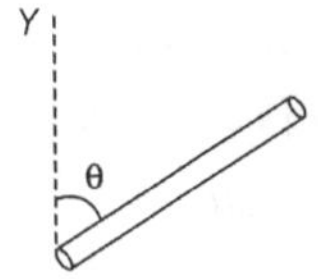

$\because \quad I_x = I_y = 2\left[\dfrac{ml^2}{3}\sin^2 45°\right] = \dfrac{ml^2}{3}$

$$I_z = 2\left(\frac{ml^2}{3}\right) = \frac{2}{3}ml^2$$

$\therefore \quad I_x = I_y < I_z$

4. (c) I_0 (about an axis passing through the centre and perpendicular to the plane of rods) is given by

$$I_0 = \frac{mL^2}{12} + \frac{mL^2}{12} = \frac{mL^2}{6}$$

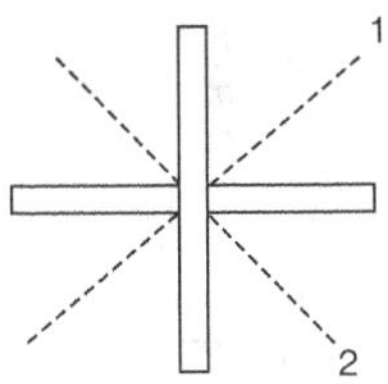

$I_0 = I_1 + I_2 = 2I_1$ $\quad (\because I_1 = I_2$ from symmetry)

$\therefore \quad I_1 = \dfrac{mL^2}{12}$

5. (c) Mass of each disc removed is

$$m = \frac{M}{16R^2} \cdot \pi R^2 = \frac{\pi}{16} M$$

$$(I_x)_{\text{of square}} = \frac{1}{12} M (4R)^2 = \frac{4}{3} MR^2$$

$$(I_x)_{\text{of each disc}} = \left(\frac{\pi}{16} M\right) \frac{R^2}{4} + \left(\frac{\pi}{16} M\right) R^2$$

$$= \frac{\pi}{16} MR^2 \cdot \frac{5}{4} = \frac{5\pi MR^2}{64}$$

$$(I_x)_{\text{of remaining portion}} = \frac{4}{3} MR^2 - 4\left(\frac{5\pi MR^2}{64}\right)$$

$$= \frac{4}{3} MR^2 - \frac{5\pi MR^2}{16}$$

$I_x = I_y$ From symmetry $I_z = I_x + I_y \Rightarrow I_z = 2I_x$

$$\therefore \quad I_z = 2\left[\frac{4}{3} MR^2 - \frac{5\pi MR^2}{16}\right]$$

$$= MR^2 \left[\frac{8}{3} - 5\frac{\pi}{8}\right]$$

or $$I_z = MR^2 \left[\frac{8}{3} - \frac{10\pi}{16}\right]$$

6. (b) $v = v_0 = 2v_0 \sin\left(\frac{\pi + \theta}{2}\right) = 2v_0 \cos\frac{\theta}{2}$

$$v^2 = 4v_0^2 \cos^2\frac{\theta}{2}$$

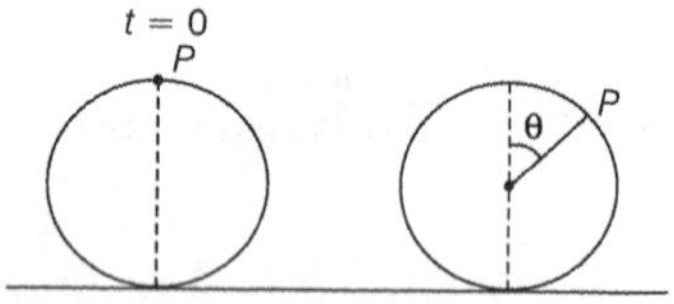

When $\theta = \pi$, $v^2 = 0$ (i.e. at the bottom most point)

At $t = \frac{T}{2} = \frac{\pi R}{v_0}$ (at the bottom most point)

$\therefore$ Correct graph is (b).

7. (c)

$$\tau = Fr_\perp = 2tR$$

$$\Delta L = \int \tau \, dt$$

$$\Rightarrow \quad L = \int_0^t 2tR \, dt$$

$$= Rt^2$$

8. (d) $\mathbf{v}_p = R\omega \, \hat{\mathbf{i}} + r\omega \cos(90° - \theta) \, \hat{\mathbf{i}} - r\omega \sin(90° - \theta) \, \hat{\mathbf{j}}$

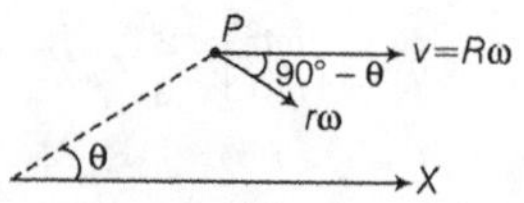

$$\mathbf{v}_p = \omega \, [(R + r\sin\theta) \, \hat{\mathbf{i}} - r\cos\theta \, \hat{\mathbf{j}}]$$

9. (c) $\alpha = \dfrac{F(h - R)}{\frac{2}{5} mR^2} = \dfrac{5F(h - R)}{2mR^2}$

$$a = \frac{F}{m}$$

For pure rolling, $\quad a = R\alpha$

$$\Rightarrow \quad \frac{F}{m} = \frac{5F(h - R)}{2mR}$$

$$2FR = 5Fh - 5FR$$

$$\Rightarrow \quad 5h = 7R$$

$$\Rightarrow \quad h = \frac{7}{5} R = 1.4\,R$$

10. (b) $\quad v_{2m} = \sqrt{2} v_0, \; v_m = 2v_0, \; v_m = \sqrt{2} \, v_0$

$$K_{\text{ring}} = \frac{1}{2} mv_0^2 + \frac{1}{2} \cdot mR^2 \cdot \left(\frac{v_0}{R}\right)^2 = mv_0^2$$

$$K_{2m+m+m} = \frac{1}{2} 2m (\sqrt{2} v_0)^2 + \frac{1}{2} m (2v_0)^2 + \frac{1}{2} m (\sqrt{2} \, v_0)^2$$

$$= 5mv_0^2$$

$$\therefore \quad K_{\text{system}} = mv_0^2 + 5mv_0^2 = 6mv_0^2$$

11. (d) $I_{XX'} = \frac{2}{5} mR^2$ (mass of the hemisphere = m)

$$I_{XX'} = I_{COM} + mr^2$$

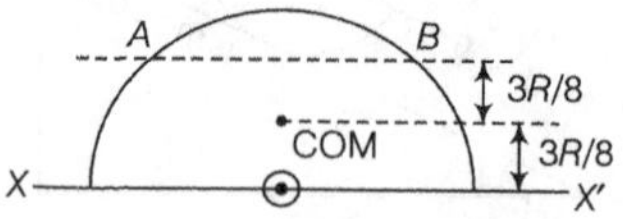

$$\frac{2}{5} mR^2 = I_{COM} + m\left(\frac{3R}{8}\right)^2$$

$$\Rightarrow \quad I_{COM} = \frac{83}{320} mR^2$$

$$I_{AB} = I_{COM} + mr^2$$

$$= \frac{83}{320} mR^2 + m\left(\frac{3}{8} R\right)^2 = \frac{128}{320 mR^2} = \frac{2}{5} mR^2$$

$$I_{AB} = mk^2 = \frac{2}{5} mR^2 \Rightarrow k = \sqrt{\frac{2}{5}} \, R$$

12. (b) In case (i) no energy is lost, whereas in case (ii) when the ball reaches B, sliding happens for a certain distance and friction does negative work and thus energy is lost in case (ii).

13. (b) $K_T = \frac{1}{2} mv^2$. When the disc climbs the smooth inclined plane, the rotational kinetic energy is conserved.

$$\therefore \quad K_T = mgh$$

$$\Rightarrow \quad \frac{1}{2} mv^2 = mgh \Rightarrow h = \frac{v^2}{2g}$$

or $$h = \frac{(6)^2}{2(10)} = 1.8 \text{ m}$$

14. (b) $E_1 = E_2$

$$\Rightarrow \quad mg(R - r) = \frac{1}{2} mv^2 + \frac{1}{2}\left(\frac{2}{5} mv^2\right)\left(\frac{v^2}{r^2}\right)$$

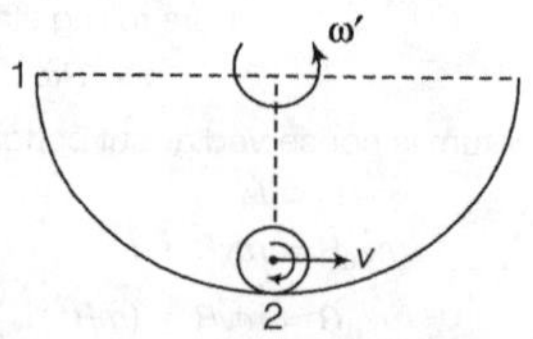

$$\therefore \quad v = \sqrt{\frac{10}{7} g (R - r)}$$

$$\omega' = \frac{v}{R - r} = \sqrt{\frac{10\,g}{7(R - r)}}$$

15. (d) As d remains constant,

$$v \cos 30° = u \cos 60°$$

$$v = 10\left(\frac{1}{2}\right)\left(\frac{2}{\sqrt{3}}\right) = \frac{10}{\sqrt{3}} \text{ ms}^{-1}$$

$$(v_\perp)_{\text{rel}} = 10 \sin 60° - \frac{10}{\sqrt{3}} \sin 30°$$

$$= 5\sqrt{3} - \frac{5}{\sqrt{3}} = \frac{10}{\sqrt{3}} \text{ ms}^{-1}$$

$$\omega_{BA} = \frac{(v_\perp)_{\text{rel}}}{d} = \frac{10/\sqrt{3}}{2} = \frac{5}{\sqrt{3}} \text{ rad s}^{-1}$$

16. (b) For no slipping at B, $v_0 = R\omega_0$

For no slipping at A, $v = v_0 + R\omega_0$

$\therefore \quad v = 2v_0$

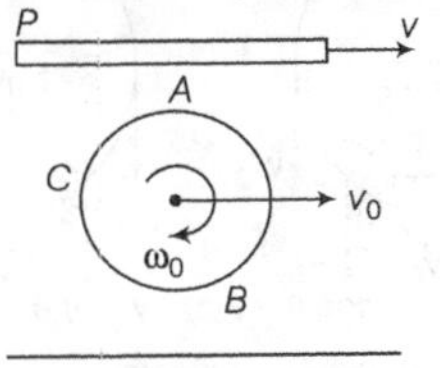

$$K_C = \frac{1}{2}Mv_0^2 + \frac{1}{2}(MR^2)\left(\frac{v_0}{R}\right)^2 = Mv_0^2$$

$$K_P = \frac{1}{2}Mv^2 = \frac{1}{2}M(2v_0)^2 = 2Mv_0^2$$

$\therefore \quad K_P : K_C = 2 : 1$

17. (d) Initially $Mg \sin\theta > kx$, so f is upwards.

When $kx > Mg \sin\theta$, f acts downwards.

18. (d) Let ω be the angular velocity of the table after the bultet is fired and m be the mass of bullet.

$$L_i = (I_0 + mR^2)\,\omega_0$$

Absolute velocity of bullet $= v - R\omega$

$$L_f = I_0\omega - m(v - R\omega)R$$

$\because \quad L_f = L_i$

$\Rightarrow \quad I_0\omega - mvR + mR^2\omega = I_0\omega_0 + mR^2\omega_0$

$\omega(I_0 + mR^2) - \omega_0(I_0 + mR^2) = mvR$

$$\omega - \omega_0 = \frac{mvR}{I_0 + mR^2}$$

19. (d) $T\cos\theta = mg$, $T\sin\theta = m\omega^2 \cdot r = m\omega^2 l \sin\theta$

$\therefore \quad T = m\omega^2 l$

For M to be stationary, $T = Mg$

$\therefore \quad Mg = m\omega^2 l$

$\Rightarrow \quad \omega = \sqrt{\dfrac{Mg}{ml}}, \quad f = \dfrac{1}{2\pi}\sqrt{\dfrac{Mg}{ml}}$

20. (c) Let the mass of cylinder is m and its radius is R. If C has to be at a fixed position, $f = mg \sin\theta$

For pure rolling,

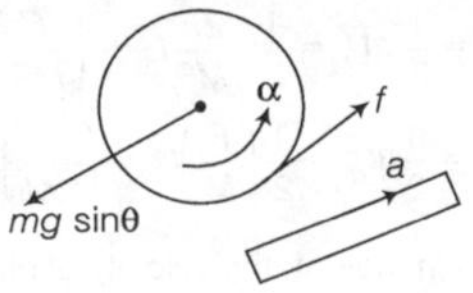

$$a = R\alpha$$

$$\alpha = \frac{f \cdot R}{mR^2/2} = \frac{2f}{mR}$$

$\because \quad a = \dfrac{2f}{m}$ or $a = \dfrac{2}{m}(mg\sin\theta) = 2g\sin\theta$

21. (d) Statement I is correct only when the two axes in comparison are parallel to each other.

22. (d)

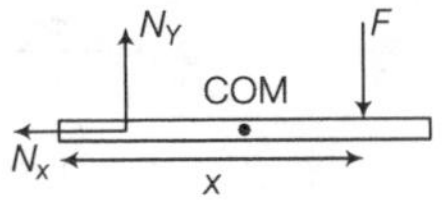

$$\tau = Fx$$

$$\alpha = \frac{Fx}{mL^2/3} \quad \text{...(i)}$$

$$a_{COM} = \frac{L}{2}\alpha = \frac{3Fx}{2mL}$$

$$F - N_y = m \cdot a_{COM} = \frac{3Fx}{2L}$$

$$N_y = F\left(1 - \frac{3x}{2L}\right) \quad \text{...(ii)}$$

From Eqs. (i) and (ii), it is clear that as x increases, N_y decreases and α increases.

23. (d) As, P is stationary w.r.t. ground. So, P has no linear acceleration but its centripital acceleration is non-zero.

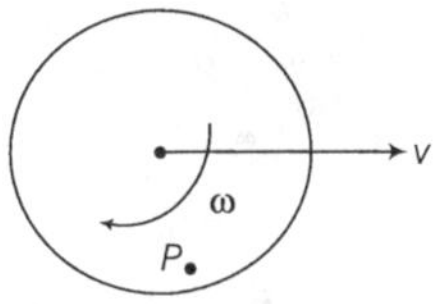

Revisal Problems (JEE Advanced)

1. (d) From law of conservation of mechanical energy,

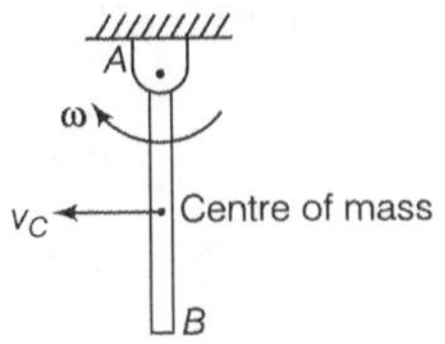

$$mg\left(\frac{l}{2}\right) = \frac{1}{2}I_A\omega^2 = \frac{1}{2}\cdot\frac{ml^2}{3}\omega^2$$

$$\omega^2 = \frac{3g}{l} \quad \text{...(i)}$$

Centripetal force on the rod at this instant is

$$F_c = m\left(\frac{l}{2}\right)\omega^2$$

or $\quad F_c = \left(\dfrac{l}{2}\right)\dfrac{3g}{l} = \dfrac{3}{2}mg$

Let F be the force exerted on the rod by the hinge then

$$F - mg = \frac{3}{2}mg$$

$\Rightarrow \quad F = \dfrac{5}{2}mg$

$|\text{Action}| = |\text{Reaction}|$

$\therefore$ Required force is $\dfrac{5}{2}mg$.

2. (c) Let speed of block is v and speed of COM of spherical body is v_{COM}.

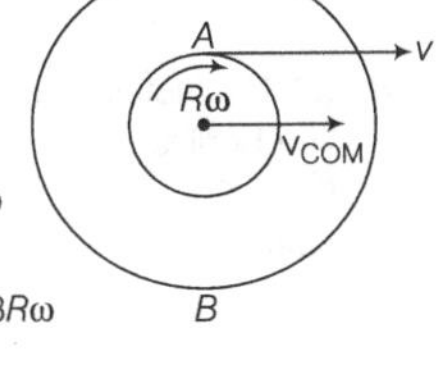

For no slipping at B,

$$v_{COM} - 2R\omega = 0$$

$\therefore \quad v_{COM} = 2R\omega$

For no slipping at A,

$$v = v_{COM} + R\omega = 3R\omega$$

$$K_b = \frac{1}{2}m(3R\omega)^2$$

$$= \frac{9}{2}mR^2\omega^2$$

$$K_s = \frac{1}{2}m(2R\omega)^2 + \frac{1}{2}(2mR^2)\,\omega^2$$

$$= 2mR^2\omega^2 + mR^2\omega^2$$

$$= 3mR^2\omega^2$$

$\therefore \quad \dfrac{K_s}{K_b} = \dfrac{3}{9/2} = \dfrac{2}{3}$

3. (c) $\tau = FR = I\alpha$

$$\Rightarrow \quad \alpha = \frac{FR}{MR^2/2} = \frac{2F}{MR}$$

$$\theta = \frac{1}{2}\alpha t^2 = \frac{1}{2}\cdot\frac{2F}{MR}t^2 = \left(\frac{F}{MR}\right)t^2$$

$$s = \frac{1}{2}at^2 = \frac{1}{2}\left(\frac{F}{M}\right)t^2 = \left(\frac{F}{2M}\right)t^2$$

4. (a) At the maximum height, the velocity of block will be same in all cases.

At maximum height, $N + mg = \frac{mv'^2}{R}$

$$N = \frac{mv'^2}{R} - mg$$

R is least in case (i) so, N is maximum there.

5. (b) $a = \frac{f}{m} = \mu g$

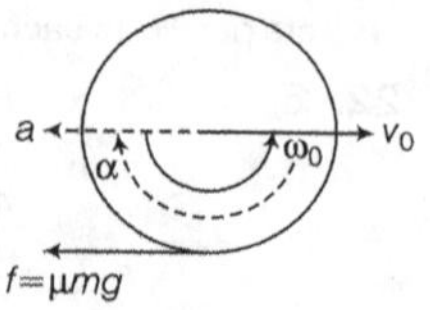

$$\alpha = \frac{\tau}{I} = \frac{f\cdot R}{\frac{2}{5}mR^2} = \frac{\mu mgR}{\frac{2}{5}mR^2} = \frac{5}{2}\cdot\frac{\mu g}{R}$$

$$0 = v_0 - at$$

$$\Rightarrow \quad t = \frac{v_0}{a}$$

Also, $0 = \omega_0 - \alpha t$

$$\Rightarrow \quad t = \frac{\omega_0}{\alpha}$$

$$\therefore \quad \frac{v_0}{a} = \frac{\omega_0}{\alpha}$$

$$\Rightarrow \quad \frac{v_0}{\omega_0} = \frac{a}{\alpha}$$

$$\Rightarrow \quad \frac{v_0}{\omega_0} = \frac{2R}{5}$$

$$\Rightarrow \quad 5v_0 = 2R\omega_0$$

6. (a)

$$\tau_1 = \mathbf{r}_1 \times \mathbf{F}_1 = F_1 r_1 \sin\theta_1 \odot$$

$$\tau_2 = \mathbf{r}_2 \times \mathbf{F}_2 = F_2 r_2 \sin\theta_2 \otimes$$

$$\therefore \quad \tau_{net} = (F_2 r_2 \sin\theta_2 - F_1 r_1 \sin\theta_1)\otimes$$

7. (a) $ml\omega^2 = Mg \quad \Rightarrow \quad \omega = \sqrt{\frac{Mg}{ml}}$

$$f = \frac{1}{2\pi}\sqrt{\frac{Mg}{ml}}$$

8. (b) The frictional force μmg is only horizontal force acting on the two bodies. So, each body has an acceleration μg in opposite direction.

Note *m is same.*

$$\therefore \quad a_r = 2\mu g$$

9. (a) $mg\sin\theta - f = ma_1$

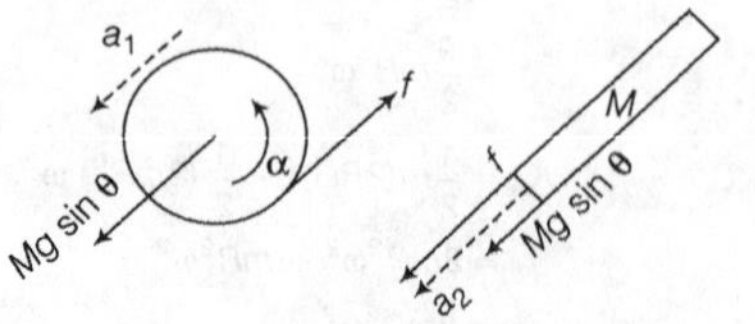

$$f\cdot R = \frac{2}{5}mR^2\alpha$$

For no slipping, $a_1 - R\alpha = a_2$

$$Mg\sin\theta + f = Ma_2$$

By solving above equation, we get $f = 0$

10. (a) $v_r = |2v\sin\theta| = |2v\sin\omega t|$

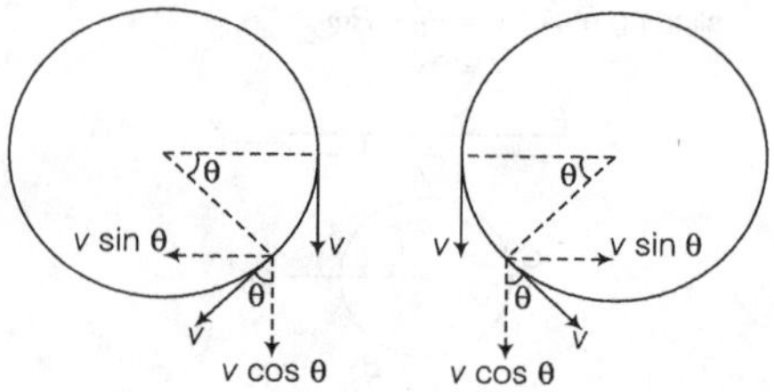

11. (a,b,d) In the horizontal direction, no force is acting. So, option (a) is correct. Initially, $N - mg = 0 \Rightarrow N = mg$.

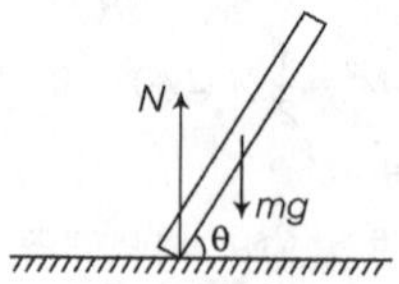

$$mg - N = ma \quad \ldots\text{(i)}$$

$$mg\left(\frac{L}{2}\cos\theta\right) = \frac{mL^2}{3}\cdot\alpha$$

$$\Rightarrow \quad \alpha = \frac{3}{2}\cdot\frac{g}{L}\cos\theta \quad \ldots\text{(ii)}$$

$$a = \frac{L}{2}\cdot\alpha \quad \ldots\text{(iii)}$$

From Eqs. (ii) and (iii), $a = \frac{L}{2}\cdot\frac{3}{2}\cdot\frac{g}{L}\cos\theta = \frac{3g}{4}\cos\theta$

From Eq. (i), $mg - N = m\cdot\frac{3g}{4}\cos\theta$

When the rod just becomes horizontal, $\theta = 0°$

$$\therefore \quad mg - N = \frac{3mg}{4} \quad \Rightarrow \quad N = \frac{mg}{4}$$

12. (b,d) Point of contact A is at rest. Momentarily, we can imagine pure rotation about point A as instantaneous centre. So, if only F_1 is applied, spool rotates clockwise.

If only F_2 is applied, spool does not rotate.

If only F_3 is applied, spool rotates anti-clockwise.

If only F_4 is applied, spool rotates anti-clockwise.

13. (c,d) Relative velocity between A and B along the rod should be zero.

$$\therefore \quad v_0\cos\theta = v_A\sin\theta \quad \Rightarrow \quad v_A = v_0\cot\theta$$

When $\theta = 37°$,

$$v_A = \frac{4}{3}v_0$$

$$\omega = \frac{\text{Component of relative velocity perpendicular to } AB}{AB}$$

$$= \frac{v_A\cos\theta - (-v_0\sin\theta)}{L} = \frac{v_A\cos\theta + v_0\sin\theta}{L}$$

when $\theta = 37°$,

$$\omega = \frac{\frac{4}{3}v_0\cdot\frac{4}{5} + v_0\frac{3}{5}}{L} = \frac{25}{15}\cdot\frac{v_0}{L} = \frac{5v_0}{3L}$$

14. (b,c,d) Here the cylinder has to rotate clockwise. So, friction acts backwards.

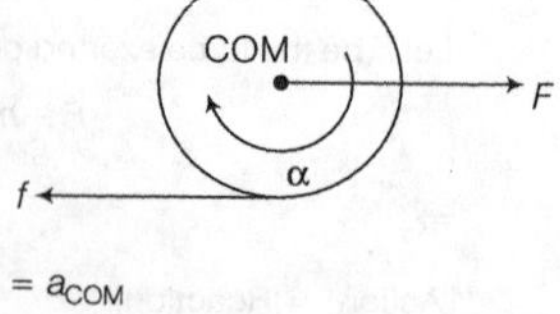

$$F - f = ma_{COM}$$

$$f\cdot R = \frac{mR^2}{2}\cdot\alpha$$

$a = R\alpha$, where $a = a_{COM}$

Solving these equations, we get

$$a = \frac{2F}{3m} \text{ and } f = \frac{F}{3}$$

15. (a,c) When the net force on a rigid body is zero, then the summation of moments about any point is constant.

Explanations for Q. Nos. (16-18)

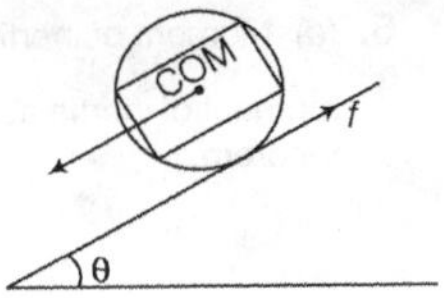

$$(4M + m)g\sin\theta - f = (4M + m)a$$

$$\tau = fR = I\left(\frac{a}{R}\right)$$

Solving the above, we get $a = \frac{7g}{24}$

$$I = 4\left[\frac{M(R\sqrt{2})^2}{12} + M\left(\frac{R}{\sqrt{2}}\right)^2\right] + mR^2 = 20 \text{ kg-m}^2$$

From the above, we get $\quad f = 20a$

$$\mu_{min} m'g\cos\theta = 20a$$

$\Rightarrow$ $$\mu_{min}(6\times 4 + 4)g\cos 30^\circ = 20\left(\frac{7g}{24}\right)$$

$$\mu_{min} = \frac{20\times 7}{24}\times\frac{2}{28\sqrt{3}} = \frac{5}{12\sqrt{3}}$$

16. (b) **17.** (c) **18.** (d)

19. (a)

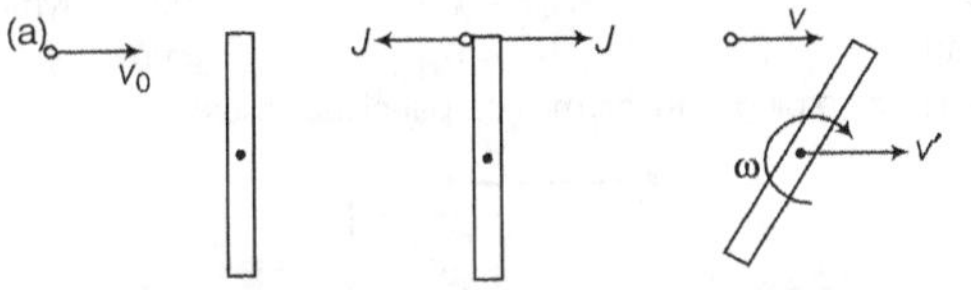

Just before impact — During impact — Just after impact

$$mv - mv_0 = -J \quad \ldots\text{(i)}$$

$$mv' = J \quad \ldots\text{(ii)}$$

$$mv_0 = mv + mv'$$

$\Rightarrow$ $$v_0 = v + v' \quad \ldots\text{(iii)}$$

$$e = 1 = \frac{v' + \frac{I\omega}{2} - v}{v_0}$$

$\Rightarrow$ $$v' + \frac{I\omega}{2} - v = v_0$$

$\Rightarrow$ $$v + v_0 = v' + \frac{I\omega}{2} \quad \ldots\text{(iv)}$$

Also, $$mv_0\left(\frac{I}{2}\right) = mv\,\frac{I}{2} + \left(\frac{mI^2}{12}\right)\omega$$

$\Rightarrow$ $$v_0 = v + \frac{I\omega}{6} \quad \ldots\text{(v)}$$

From Eqs. (iii) and (v), we get

$$v' = \frac{I\omega}{6}$$

So, from Eq. (iv), we get $\quad v + v_0 = \frac{2}{3}I\omega \quad \ldots\text{(vi)}$

Solving Eqs. (v) and (vi), we get $\omega = \frac{12v_0}{5I}$

$\therefore$ $$v = \frac{3v_0}{5},\ v' = \frac{2v_0}{5}$$

$$\frac{(K_f)_{ball}}{(K_i)_{ball}} = \frac{\frac{1}{2}mv^2}{\frac{1}{2}mv_0^2} = \frac{9}{25}$$

$$\frac{(J)_{rod}}{(J)_{ball}} = \frac{mv_0 - mv}{mv_0} = \frac{v_0 - \frac{3}{5}v_0}{v_0} = \frac{2}{5}$$

Note $mv_0 - mv = mv'$

$$\frac{(L_f)_{rod}}{(L_i)_{ball}} = \frac{\frac{mI^2}{12}\cdot\omega}{mv_0\cdot\frac{I}{2}} = \frac{I}{6}\cdot\frac{\omega}{v_0} = \frac{I}{6}\cdot\frac{12v_0}{5I}\cdot\frac{1}{v_0} = \frac{2}{5}$$

$$\frac{(K_{rod})_{rotation}}{(K_{rod})_{translation}} = \frac{\frac{1}{2}\cdot\frac{1}{12}mI^2\omega^2}{\frac{1}{2}mv'^2} = \frac{1}{12}\cdot I^2\left(\frac{12v_0}{5I}\right)^2\cdot\left(\frac{5}{2v_0}\right)^2 = 3$$

Hence, (i) → (r) (ii) → (p), (iii) → (p), (iv) → (s)

20. (a) $a_A + R\alpha = a_B$

$\Rightarrow$ $$a + 4\alpha = 6 \quad \ldots\text{(i)}$$

$$R\alpha - a_A = a_C$$

$\Rightarrow$ $$4\alpha - a = 2 \quad \ldots\text{(ii)}$$

From Eqs. (i) and (ii),

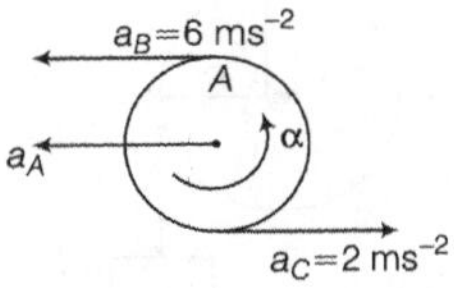

$$a = 2\text{ms}^{-2},\ \alpha = 1\text{ rads}^{-2}$$

$$l = l_0 + \frac{1}{2}(R\alpha)t^2 = 20 + \frac{1}{2}(4\times 1)(2)^2 = 28\text{ m}$$

$\therefore$ $$l - l_0 = 8\text{ m}$$

$$m_Bg - T = m_Ba_B$$

$\Rightarrow$ $$10m_B - T = 6m_B$$

$$T = m_Aa_A$$

$\Rightarrow$ $$T = 2m_A$$

$$\frac{m_A}{m_B} = 2$$

Hence, (i) → (s), (ii) → (r), (iii) → (q), (iv) → (p)

21. (2) At point P,

Decrease in GPH = Increase in $(K)_T$ + Increase in $(K)_R$

$$mg\cdot 7R = \frac{1}{2}mv^2 + \frac{1}{2}I\omega^2 = \frac{1}{2}mv^2 + \frac{1}{2}\left(\frac{2}{5}mr^2\right)\cdot\frac{v^2}{r^2}$$

$$= \frac{7}{10}mv^2 \quad (r \text{ is radius of ball})$$

$\therefore$ $$v^2 = 10gR$$

Centripetal force is $\frac{mv^2}{R} = \frac{m(10gR)}{R} = 10mg = 10(0.1)(10)$

$$= 10\text{ N} \Rightarrow x = 2$$

22. (6) **About the axis given**

$$I_A = \frac{mL^2}{12} + m(2R)^2 = \frac{m(2R)^2}{12} + m(2R)^2$$

$$= \left(\frac{1}{3} + 4\right)mR^2 = \frac{13}{3}mR^2$$

$$I_B = mR^2$$

$$I_{hoop} = \frac{mR^2}{2}$$

$$I_{net} = \left(\frac{13}{3} + 1 + \frac{1}{2}\right)mR^2 = \frac{26+6+3}{6}mR^2$$

$$= \frac{35}{6}mR^2 \simeq 6mR^2 \Rightarrow K = 6$$

23. (3) From conservation of angular momentum,

$$Mv_1R = I\omega + Mv_2R$$

For no slipping, $\quad \omega = \frac{v_2}{R}$

$\therefore$ $$Mv_1R = \frac{MR^2}{2}\cdot\frac{v_2}{R} + Mv_2R$$

$$v_1 = \frac{1}{2}v_2 + v_2$$

$\Rightarrow$ $$v_2 = \frac{2}{3}v_1 = \frac{2}{3}v_0$$

$\therefore$ $$\beta = 3$$

JEE Main & AIEEE Archive

1. (b) For the mass m,

$$mg - T = ma$$

As we know, $a = R\alpha$

So, $mg - T = mR\alpha$ ...(i)

As torque $= T \times R = mR^2\alpha$ ...(ii)

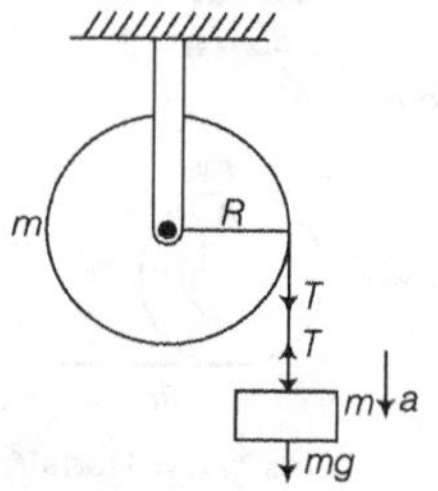

From Eqs. (i) and (ii), we get

$$a = g/2$$

Hence, the acceleration with the mass of a body fall is $\frac{g}{2}$.

2. (c) From the principle of conservation of angular momentum, we have $\Sigma I\omega$ = constant

where, I is moment of inertia ($I = mr^2$).

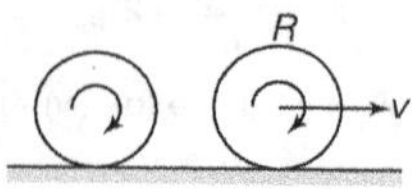

From law of conservation of angular momentum,

$$mr^2\omega_0 = mvr + mr^2 \times \frac{v}{r} \Rightarrow v = \frac{\omega_0 r}{2}$$

3. (c) Moment of inertia, $I = \frac{1}{2}MR^2 + mx^2$

where, m = mass of insect,

and x = distance of insect from centre.

Clearly, as the insect moves along the diameter of the disc, moment of inertia first decreases, then increases.

By conservation of angular momentum, angular speed first increases, then decreases.

4. (a) Given force, $F = 20t - 5t^2$

$$\alpha = \frac{FR}{I}$$

$$= \frac{(20t - 5t^2)2}{10} = 4t - t^2$$

$$\Rightarrow \frac{d\omega}{dt} = 4t - t^2 \Rightarrow \int_0^{\omega} d\omega = \int_0^t (4t - t^2)\, dt$$

$$\Rightarrow \omega = 2t^2 - \frac{t^3}{3}$$

When direction is reversed, $\omega = 0$, i.e.

$t = 0, 6$ s

Now, $d\theta = \omega dt$

$$\Rightarrow \int_0^{\theta} d\theta = \int_0^6 \left(2t^2 - \frac{t^3}{3}\right) dt$$

$$\Rightarrow \theta = \left[\frac{2t^3}{3} - \frac{t^4}{12}\right]_0^6$$

$$\Rightarrow \theta = 144 - 108 = 36 \text{ rad}$$

$\therefore$ Number of rotations,

$$n = \frac{\theta}{2\pi} = \frac{36}{2\pi} < 6$$

5. (d) Moment of inertia of square plate about xx' is $\frac{ma^2}{6}$, therefore moment of inertia about zz' can be computed using parallel axes theorem,

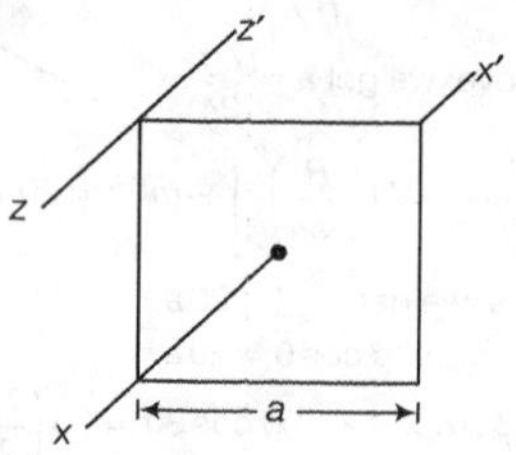

$$I_{zz'} = I_{xx'} + m\left(\frac{a}{\sqrt{2}}\right)^2 = \frac{ma^2}{6} + \frac{ma^2}{2} = \frac{2ma^2}{3}$$

6. (c) Let each side of square lamina be d.

So, $I_{EF} = I_{GH}$ (due to symmetry)

and $I_{AC} = I_{BD}$ (due to symmetry)

Now, according to theorem of perpendicular axes,

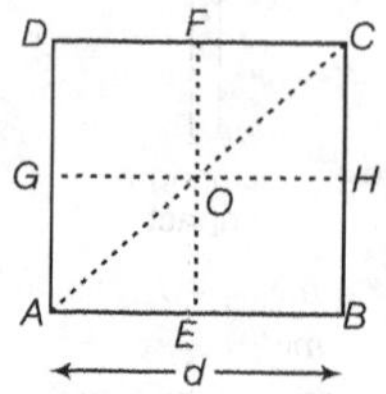

$$I_{AC} + I_{BD} = I_0$$

$$\Rightarrow 2I_{AC} = I_0 \quad \text{...(i)}$$

and $$I_{EF} + I_{GH} = I_0$$

$$\Rightarrow 2I_{EF} = I_0 \quad \text{...(ii)}$$

From Eqs.(i) and (ii), we get

$$I_{AC} = I_{EF}$$

$$\therefore I_{AD} = I_{EF} + \frac{md^2}{4}$$

$$= \frac{md^2}{12} + \frac{md^2}{4} \quad \left(\text{as } I_{EF} = \frac{md^2}{12}\right)$$

So, $$I_{AD} = \frac{md^2}{3} = 4I_{EF}$$

7. (a) Distance of centre of mass of new disc from the centre of mass of remaining disc is αR.

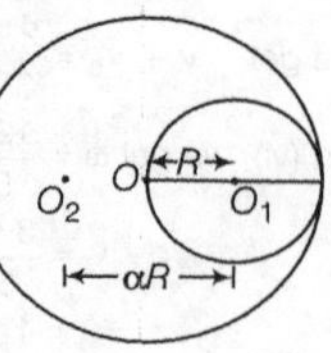

Mass of original disc $= M$

Mass of disc removed $= \frac{M}{\pi(2R)^2} \times \pi R^2 = \frac{M}{4}$

Mass of remaining disc $= M - \frac{M}{4} = \frac{3M}{4}$

$$\therefore -\frac{3M}{4}\alpha R + \frac{M}{4}R = 0 \Rightarrow \alpha = \frac{1}{3}$$

Note *In this question, the given distance must be αR for real approach to the solution.*

8. (a) Assuming that no energy is used up against friction, the total loss in potential energy is equal to the total gain in the kinetic energy.

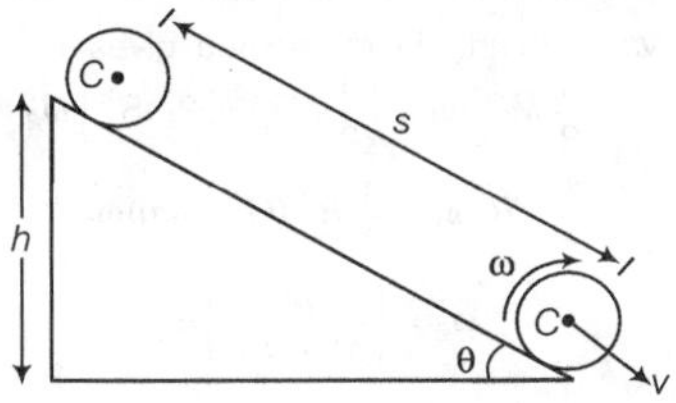

$$Mgh = \frac{1}{2} I\omega^2 + \frac{1}{2} Mv^2 \quad \text{(as } v = R\omega\text{)}$$

i.e. $$Mgh = \frac{1}{2} I\left(\frac{v^2}{R^2}\right) + \frac{1}{2} Mv^2$$

or $$\frac{1}{2} v^2 \left(M + \frac{I}{R^2}\right) = Mgh$$

or $$v^2 = \frac{2\,Mgh}{M + I/R^2} = \frac{2\,gh}{1 + I/MR^2}$$

If s is the distance covered along the plane, then

$$h = s\sin\theta$$

$\therefore$ $$v^2 = \frac{2gs\sin\theta}{1 + I/MR^2}$$

Now, $$v^2 = 2\,as$$

$\therefore$ $$2\,as = \frac{2\,g\,s\,\sin\theta}{1 + I/MR^2}$$

or $$a = \frac{g\sin\theta}{1 + I/MR^2}$$

Alternate Method

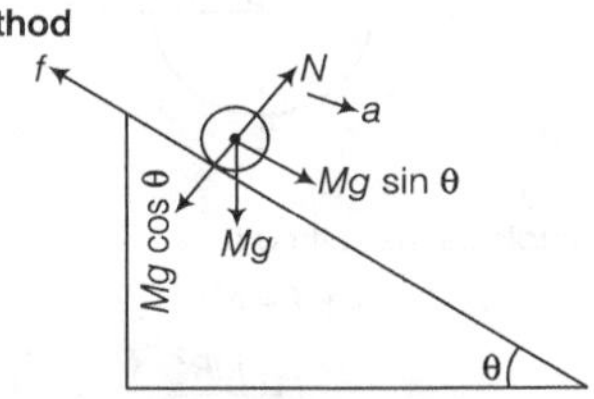

$$Mg\sin\theta - f = Ma \quad \ldots\text{(i)}$$

As torque, $$\tau = I\alpha = I\frac{a}{R} \quad \text{(as } a = R\alpha\text{)}$$

Torque, $$\tau = f \cdot R$$

So, $$f \cdot R = \frac{Ia}{R}$$

$\Rightarrow$ $$f = \frac{Ia}{R^2}$$

From Eq. (i),

$$Mg\sin\theta - \frac{Ia}{R^2} = Ma$$

$$a\left[M + \frac{I}{R^2}\right] = Mg\sin\theta$$

$$a = \frac{Mg\sin\theta}{M + \frac{I}{R^2}}$$

$\therefore$ $$a = \frac{g\sin\theta}{1 + \frac{I}{MR^2}}$$

9. (c) According to the principle of conservation of angular momentum, in the absence of external torque, the total angular momentum of the system is constant, i.e. zero torque $\left(\tau = \frac{dL}{dt}\right)$.

10. (c) The situation is shown in the figure.

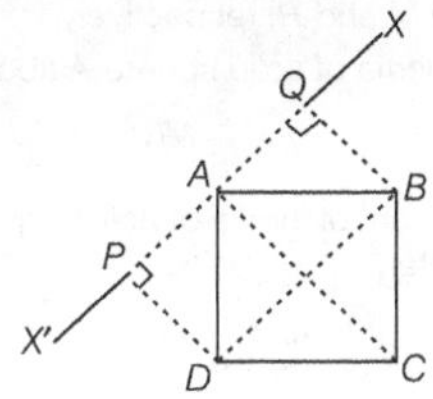

$$I_{XX'} = m \times DP^2 + m \times BQ^2 + m \times CA^2$$

$$= m \times 2 \times \left(\frac{\sqrt{2}\,l}{2}\right)^2 + m \times (\sqrt{2}\,l)^2 = 3ml^2$$

11. (c) The torque about the given position, $\tau = \mathbf{r} \times \mathbf{F}$

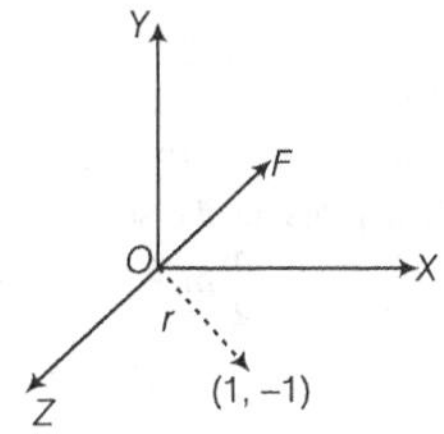

Here, $\mathbf{r} = \hat{\mathbf{i}} - \hat{\mathbf{j}}$ and $\mathbf{F} = -F\hat{\mathbf{k}}$

$\therefore$ $$\tau = (\hat{\mathbf{i}} - \hat{\mathbf{j}}) \times (-F\hat{\mathbf{k}}) = F[(-\hat{\mathbf{i}} \times \hat{\mathbf{k}}) + (\hat{\mathbf{j}} \times \hat{\mathbf{k}})]$$

$$= F(\hat{\mathbf{j}} + \hat{\mathbf{i}}) = F(\hat{\mathbf{i}} + \hat{\mathbf{j}})$$

12. (d) As no external torque is acting on the system, angular momentum of system remains conserved.

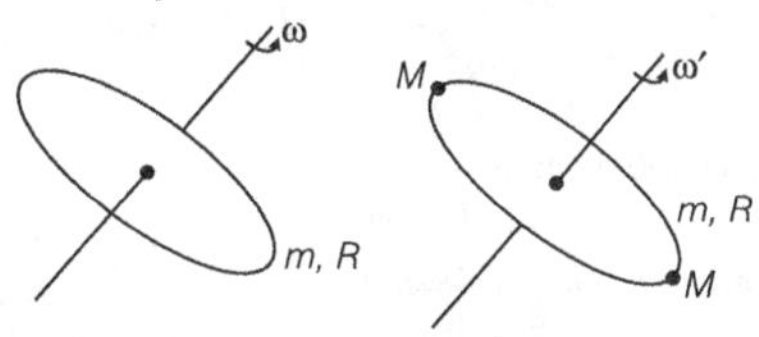

So, by conservation of angular momentum,

$$I_1\omega = I_2\omega'$$

$\Rightarrow$ $$mR^2\omega = (mR^2 + 2MR^2)\omega'$$

$\Rightarrow$ $$\omega' = \left(\frac{m}{m + 2M}\right)\omega$$

13. (d) The mass of complete (circular) disc is

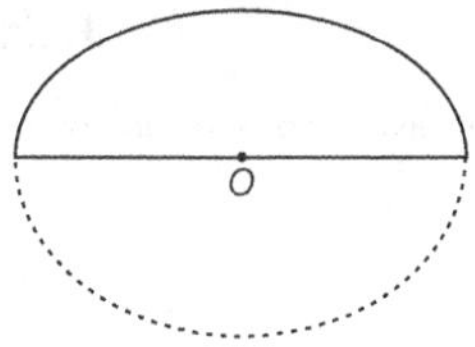

The moment of inertia of disc about the given axis is

$$I = \frac{2\,Mr^2}{2} = Mr^2$$

Let the moment of inertia of semi-circular disc be I_1.

The disc may be assumed as combination of two semi-circular parts.

Thus, $$I_1 = I - I_1$$

or $$I_1 = \frac{I}{2} = \frac{Mr^2}{2}$$

14. (c) Let same mass and same outer radii of solid sphere and hollow sphere be M and R, respectively.

The moment of inertia of solid sphere A about its diameter,

$$I_A = \frac{2}{5} MR^2 \quad \ldots \text{(i)}$$

Similarly, the moment of inertia of hollow sphere (spherical shell) B about its diameter,

$$I_B = \frac{2}{3} MR^2 \quad \ldots \text{(ii)}$$

It is clear from Eqs. (i) and (ii), $I_A < I_B$.

15. (b) In free space, neither acceleration due to gravity nor external torque act on the rotating solid sphere. Therefore, taking the same mass of sphere, if radius is increased, then moment of inertia, rotational kinetic energy and angular velocity will change but according to law of conservation of momentum, angular momentum will not change.

16. (d) Mass of disc (X),

$$m_X = \pi R^2 t \rho$$

where, ρ = density of material of disc

$$\therefore \quad I_X = \frac{1}{2} m_X R^2 = \frac{1}{2} \pi R^2 t \rho R^2$$

or $$I_X = \frac{1}{2} \pi \rho t R^4 \quad \ldots \text{(i)}$$

Mass of disc (Y),

$$m_Y = \pi (4R)^2 \frac{t}{4} \rho = 4\pi R^2 t \rho$$

and $$I_Y = \frac{1}{2} m_Y (4R)^2 = \frac{1}{2} 4\pi R^2 t \rho \cdot 16R^2$$

or $$I_Y = 32 \pi t \rho R^4 \quad \ldots \text{(ii)}$$

$$\therefore \quad \frac{I_Y}{I_X} = \frac{32 \pi t \rho R^4}{\frac{1}{2} \pi \rho t R^4} = 64 \quad \text{or} \quad I_Y = 64 I_X$$

17. (a) Angular momentum,

$$L = I\omega \quad \ldots \text{(i)}$$

Kinetic energy, $K = \frac{1}{2} I\omega^2 = \frac{1}{2} L\omega$ [from Eq. (i)]

$$\therefore \quad L = 2\frac{K}{\omega}$$

Now, $$L' = \frac{2\left(\frac{K}{2}\right)}{2\omega} \Rightarrow L' = \frac{L}{4}$$

18. (d) As, $\tau = \mathbf{r} \times \mathbf{F}$

τ is perpendicular to both $\mathbf{r}$ and $\mathbf{F}$, so $\mathbf{r} \cdot \tau$ as well as $\mathbf{F} \cdot \tau$ has to be zero.

19. (c) Conservation of angular momentum gives

$$\frac{1}{2} MR^2 \omega_1 = \left(\frac{1}{2} MR^2 + 2mR^2\right) \omega_2$$

$$\Rightarrow \quad \frac{1}{2} MR^2 \omega_1 = \frac{1}{2} R^2 (M + 4m) \omega_2$$

or $$\omega_2 = \left(\frac{M}{M + 4m}\right) \omega_1$$

20. (b) Angular momentum of particle about O,

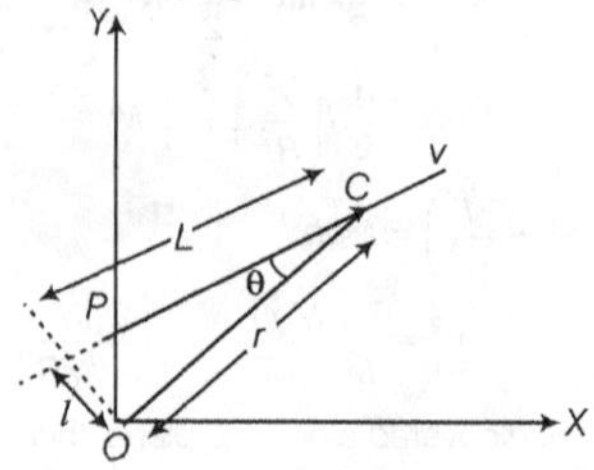

$$\mathbf{L} = m(\mathbf{r} \times \mathbf{v})$$

$$|\mathbf{L}| = mrv \sin\theta = mv(r \sin\theta) = mvl$$

21. (a) Moment of inertia of circular wire about its axis is MR^2. Consider two diameters XX' and YY'.

Moment of inertia about of these two diameters is same, say I.

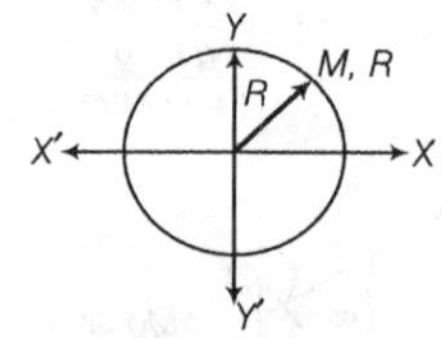

From perpendicular axes theorem,

$$I + I = MR^2$$

$$\Rightarrow \quad I = \frac{MR^2}{2}$$

22. (d) Since, the inclined plane is frictionless, then there will be no rolling and the mass will only slide down.

Hence, acceleration is same for all the given bodies.

JEE Advanced & IIT JEE Archive

1. (4) Applying conservation of angular momentum

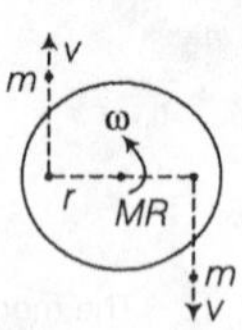

$$2mvr - \frac{MR^2}{2} \omega = 0$$

$$\omega = \frac{4mvr}{MR^2}$$

Substituting the values, we get

$$\omega = \frac{(4)(5 \times 10^{-2})(9)\left(\frac{1}{4}\right)}{45 \times 10^{-2} \times \frac{1}{4}}$$

$$\omega = 4 \text{ rad s}^{-1}$$

2. (8) $I_1\omega_1 = I_2\omega_2$

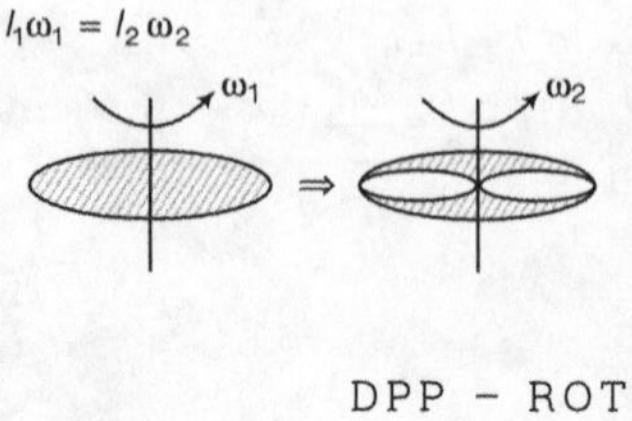

Disc → M, R
Ring → m, r

$$\therefore \quad \omega_2 = \left(\frac{I_1}{I_2}\right)\omega_1 = \left[\frac{\frac{1}{2}MR^2}{\frac{1}{2}MR^2 + 2(mr^2)}\right]\omega_1$$

$$= \left[\frac{50(0.4)^2}{50(0.4)^2 + 8 \times (6.25) \times (0.2)^2}\right](10)$$

$$= 8 \text{ rad/s}$$

3. (b,d) $I_P > I_Q$

In case of pure rolling,

$$a = \frac{g \sin\theta}{1 + I/mR^2}$$

$a_Q > a_P$ as its moment of inertia is less. Therefore, Q reaches first with more linear speed and more translational kinetic energy.

Further, $\omega = \frac{v}{R}$ or $\omega \propto v$

$$\therefore \quad \omega_Q > \omega_P \text{ as } v_P > v_Q$$

4. (a,b) Velocity of point O is

$$v_O = (3R\omega)\,\hat{\mathbf{i}}$$

$\mathbf{v}_{PO}$ is $\frac{R\omega}{2}$ in the direction shown in figure. In vector form,

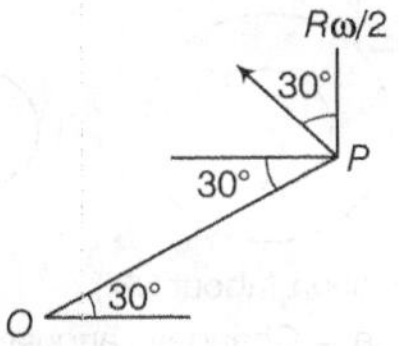

$$\mathbf{v}_{PO} = -\frac{R\omega}{2}\sin 30^\circ\,\hat{\mathbf{i}} + \frac{R\omega}{2}\cos 30^\circ\,\hat{\mathbf{k}}$$

$$= -\frac{R\omega}{4}\hat{\mathbf{i}} + \frac{\sqrt{3}R\omega}{4}\hat{\mathbf{k}}$$

But $\quad \mathbf{v}_{PO} = \mathbf{v}_P - \mathbf{v}_O$

$$\therefore \quad \mathbf{v}_P = \mathbf{v}_{PO} + \mathbf{v}_O = \left(-\frac{R\omega}{4}\hat{\mathbf{i}} + \frac{\sqrt{3}R\omega}{4}\hat{\mathbf{k}}\right) + 3R\omega\,\hat{\mathbf{i}}$$

$$= \frac{11}{4}R\omega\hat{\mathbf{i}} + \frac{\sqrt{3}}{4}R\omega\,\hat{\mathbf{k}}$$

5. (b) $|\mathbf{L}|$ or $L = I\omega$ (about axis of rod)

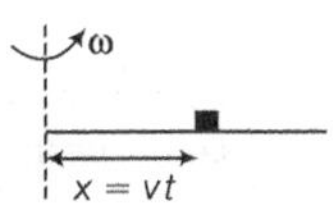

$$I = I_{rod} + mx^2 = I_{rod} + mv^2t^2$$

Here, $\quad m$ = mass of insect

$$\therefore \quad L = (I_{rod} + mv^2t^2)\,\omega$$

Now, $\quad |\tau| = \frac{dL}{dt} = (2mv^2t\omega)$

or $\quad |\tau| \propto t$

i.e. the graph is straight line passing through origin.

After time T, L = constant

$$\therefore \quad |\tau| \text{ or } \frac{dL}{dt} = 0$$

6. (a) Language of question is wrong because relative speed is not the correct word. Relative speed between two is always zero. The correct word is magnitude of relative velocity.

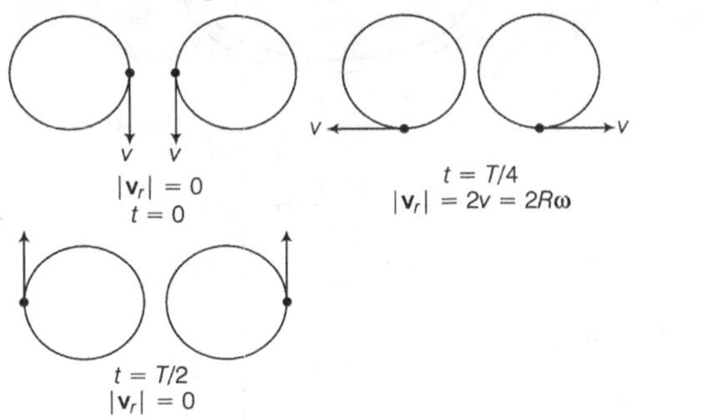

Corresponding to above values, the correct graph is (a).

7. (c) Angular momentum of a particle about a point is given by

$\mathbf{L} = \mathbf{r} \times \mathbf{p} = m\,(\mathbf{r} \times \mathbf{v})$

For $\mathbf{L}_O$

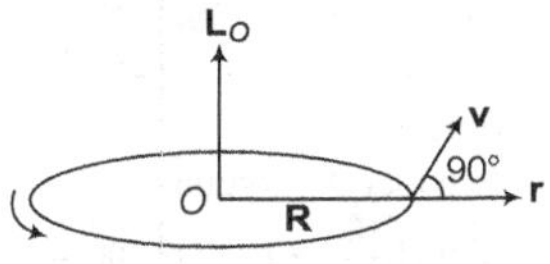

$|\mathbf{L}| = (mvr \sin\theta) \;= m\,(R\omega)\,(R)\sin 90^\circ$ = constant

Direction of $\mathbf{L}_O$ is always upwards. Therefore, complete $\mathbf{L}_O$ is constant, both in magnitude as well as direction.

For $\mathbf{L}_P$

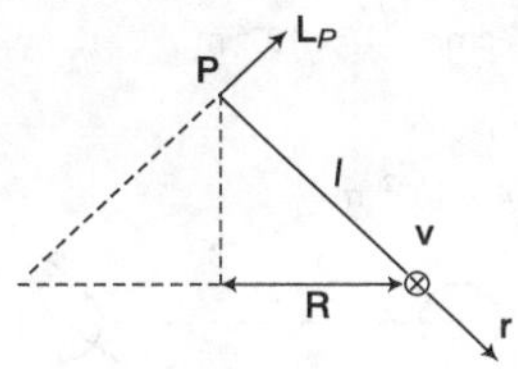

$|\mathbf{L}_P| = (mvr\sin\theta) = (m)\,(R\omega)\,(l)\sin 90^\circ = (mRl\omega)$

Magnitude of $\mathbf{L}_P$ will remain constant but direction of $\mathbf{L}_P$ keeps on changing.

8. (a) Language of question is not very clear, e.g. disc is rotating. Its different points have different velocities. Relative velocity of pebble with respect to which point, it is not clear. Further, actual initial positions of P and Q are also not given.

Explanations for Q. Nos. (9-10)

(i) Every particle of the disc is rotating in a horizontal circle.

(ii) Actual velocity of any particle is horizontal.

(iii) Magnitude of velocity of any particle is $v = r\omega$

where, r is the perpendicular distance of that particle from actual axis of rotation (Z-axis).

(iv) When it is broken into two parts, then actual velocity of any particle is resultant of two velocities.

$$v_1 = r_1\omega_1 \text{ and } v_2 = r_2\omega_2$$

Here,

r_1 = perpendicular distance of centre of mass from Z-axis

ω_1 = angular speed of rotation of centre of mass from Z-axis

r_2 = distance of particle from centre of mass and

ω_2 = angular speed of rotation of the disc about the axis passing through centre of mass

(v) Net v will be horizontal, if v_1 and v_2 both are horizontal. Further, v_1 is already horizontal, because centre of mass is rotating about a vertical Z-axis. To make v_2 also horizontal, second axis should also be vertical.

9. (d) **10.** (a)

11. (3) T = Total portion, R = Remaining portion and

C = Cavity and let, $\quad \sigma$ = mass per unit area.

Then,
$$m_T = \pi(2R)^2\sigma = 4\pi R^2\sigma$$
$$m_C = \pi(R)^2\sigma = \pi R^2\sigma$$

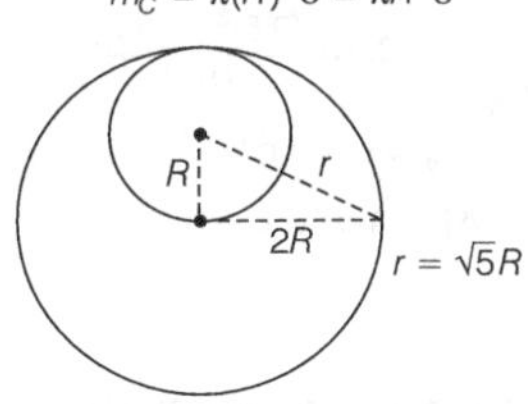

For I_P $\quad I_R = I_T - I_C$

$$= \frac{3}{2}m_T\,(2R)^2 - \left[\frac{1}{2}m_C R^2 + m_C r^2\right]$$

$$= \frac{3}{2}(4\pi R^2\sigma)\,(4R^2) - \left[\frac{1}{2}(\pi R^2\sigma) + (\pi R^2\sigma)\,(5R^2)\right]$$

$$= (18.5\,\pi R^4\sigma)$$

For I_O

$$I_R = I_T - I_C = \frac{1}{2}m_T\,(2R)^2 - \frac{3}{2}m_C\,R^2$$

$$= \frac{1}{2}(4\pi R^2\sigma)\,(4R^2) - \frac{3}{2}(\pi R^2\sigma)\,(R^2) = 6.5\,\pi R^4\sigma$$

$$\therefore \quad \frac{I_P}{I_O} = \frac{18.5\pi R^4\sigma}{6.5\pi R^4\sigma} = 2.846$$

Therefore, the nearest integer is 3.

12. (9) $r = \frac{d}{2} = \frac{\sqrt{5}}{2}\text{ cm} = \frac{\sqrt{5}}{2} \times 10^{-2}\text{ m}$

$m = 0.5\text{ kg}$

$a = 4\text{ cm} = 4 \times 10^{-2}\text{ m}$

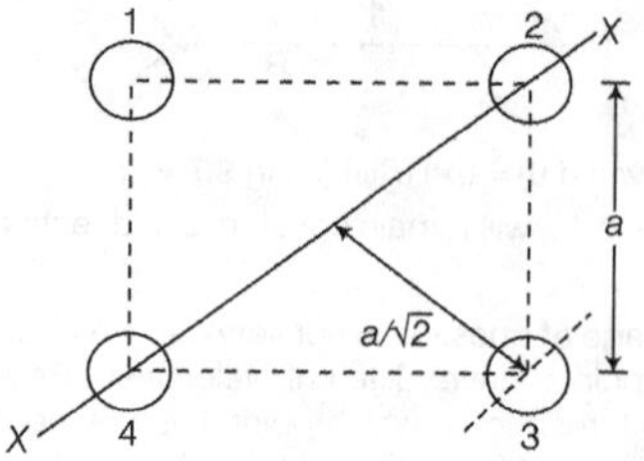

$$I_{XX} = I_1 + I_2 + I_3 + I_4$$

$$= \left[\frac{2}{5}mr^2 + m\left(\frac{a}{\sqrt{2}}\right)^2\right] + \frac{2}{5}mr^2 + \left[\frac{2}{5}mr^2 + m\left(\frac{a}{\sqrt{2}}\right)^2\right] + \frac{2}{5}mr^2$$

Substituting the values, we get

$$I_{XX} = 9 \times 10^{-4}\text{ kg-m}^2$$

$\therefore \quad N = 9$

13. (4) There is no slipping between ring and ground. Hence, f_2 is not maximum. But, there is slipping between ring and stick. Therefore, f_1 is maximum. Now, let us write the equations.

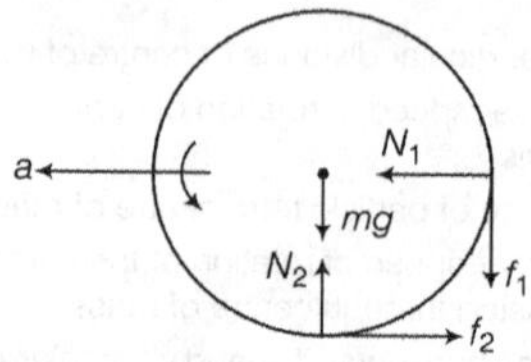

$$I = mR^2 = (2)\,(0.5)^2 = \frac{1}{2}\text{ kg-m}^2$$

$$N_1 - f_2 = ma$$

or $\quad N_1 - f_2 = (2)\,(0.3) = 0.6\text{ N} \quad \ldots\text{(i)}$

$$a = R\alpha = \frac{R\tau}{I}$$

$$= \frac{R(f_2 - f_1)R}{I} = \frac{R^2(f_2 - f_1)}{I}$$

$\therefore \quad 0.3 = \frac{(0.5)^2\,(f_2 - f_1)}{(1/2)}$

or $\quad f_2 - f_1 = 0.6\text{ N} \quad \ldots\text{(ii)}$

$$N_1^2 + f_1^2 = (2)^2 = 4 \quad \ldots\text{(iii)}$$

Further $\quad f_1 = \mu N_1 = \left(\frac{P}{10}\right)N_1 \quad \ldots\text{(iv)}$

Solving above four equations, we get $P \simeq 3.6$

Therefore, the correct answer should be 4.

14. (a,c) The data is incomplete. Let us assume that friction from ground on ring is not impulsive during impact.

From linear momentum conservation in horizontal direction, we have

$$(-2 \times 1) + (0.1 \times 20) = (0.1 \times 0) + (2 \times v) \quad \overset{-\text{ve} \qquad +\text{ve}}{\longleftrightarrow}$$

Here, v is the velocity of CM of ring after impact.

Solving the above equation, we have $v = 0$

Thus, CM becomes stationary.

$\therefore$ Correct option is (a).

Linear impulse during impact

(i) In horizontal direction,

$$J_1 = \Delta p = 0.1 \times 20 = 2\text{ N-s}$$

(ii) In vertical direction,

$$J_2 = \Delta p = 0.1 \times 10 = 1\text{ N-s}$$

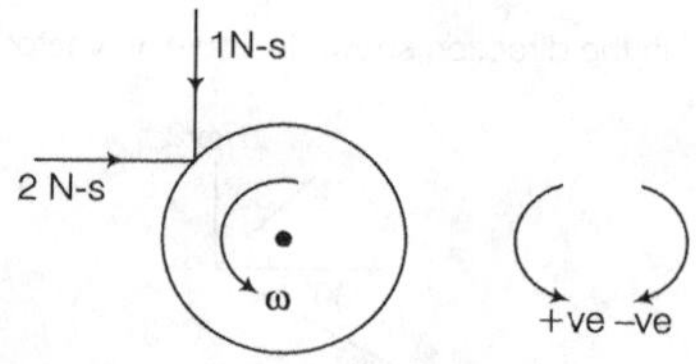

Writing the equation (about CM)

Angular impulse = Change in angular momentum

We have

$$1 \times \left(\frac{\sqrt{3}}{2} \times \frac{1}{2}\right) - 2 \times 0.5 \times \frac{1}{2} = 2 \times (0.5)^2\left[\omega - \frac{1}{0.5}\right]$$

Solving this equation, ω comes out to be positive or ω anti-clockwise. So, just after collision rightwards slipping is taking place.

Hence, friction is leftwards.

Therefore, option (c) is also correct.

15. (a) $W = \Delta U = U_f - U_i = U_\infty - U_P = -U_P = -mV_P$

$$= -V_P \qquad (\text{as } m = 1)$$

Potential at point P will be obtained by integration as given below.

Let dM be the mass of small ring as shown below:

$$dM = \frac{M}{\pi(4R)^2 - \pi\,(3R)^2}(2\pi r)\,dr = \frac{2Mr\,dr}{7R^2}$$

$$dV_P = -\frac{G\,dM}{\sqrt{16R^2 + r^2}}$$

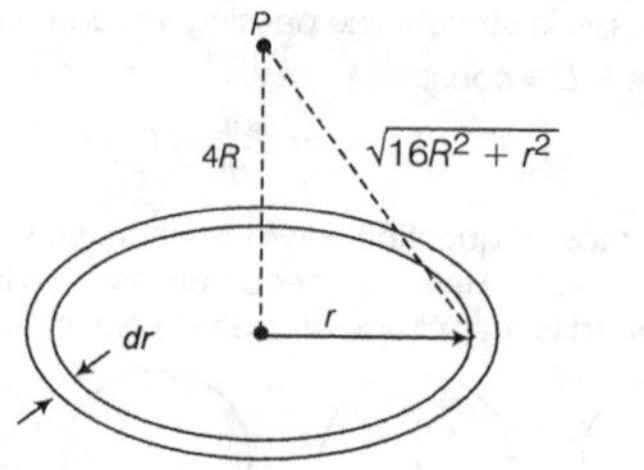

$$= -\frac{2GM}{7R^2}\int_{3R}^{4R}\frac{r}{\sqrt{16R^2 + r^2}}\,dr$$

$$= -\frac{2GM}{7R}(4\sqrt{2} - 5)$$

$\therefore \quad W = +\frac{2GM}{7R}(4\sqrt{2} - 5)$

16. (b) Condition of sliding is

$$mg\sin\theta > \mu mg\cos\theta$$

or $\quad \tan\theta > \mu$

or $\quad \tan\theta > \sqrt{3} \quad \ldots\text{(i)}$

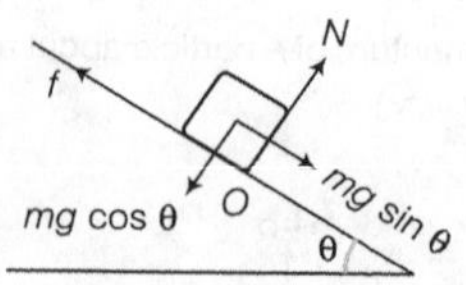

Condition of toppling is

Torque of $mg\sin\theta$ about O > torque of $mg\cos\theta$ about

$\therefore \quad (mg\sin\theta)\left(\frac{15}{2}\right) > (mg\cos\theta)\left(\frac{10}{2}\right)$

or $\quad \tan\theta > \frac{2}{3} \quad \ldots\text{(ii)}$

With increase in value of θ, condition of sliding is satisfied first.

17. (b,c) $v_A = 0$, $v_B = v$ and $v_C = 2v$

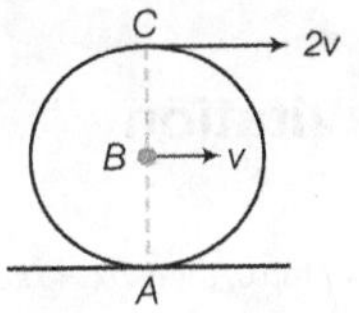

18. (d) In case of pure rolling on inclined plane,

$$a = \frac{g \sin\theta}{1 + I/mR^2}$$

$$I_{solid} < I_{hollow}$$

$$\therefore \quad a_{solid} > a_{hollow}$$

∴ Solid cylinder will reach the bottom first. Further, in case of pure rolling on stationary ground, work done by friction is zero. Therefore, mechanical energy of both the cylinders will remain constant.

∴ $(KE)_{hollow} = (KE)_{solid}$ = decrease in PE = mgh

19. (d) $\quad a = R\alpha$

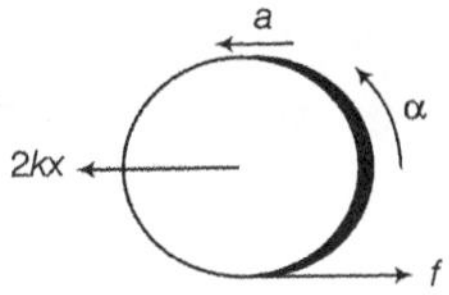

$$\therefore \quad \frac{2kx - f}{M} = R\left[\frac{fR}{\frac{1}{2}MR^2}\right]$$

Solving this equation, we get

$$f = \frac{2kx}{3}$$

$$\therefore \quad |F_{net}| = 2kx - f = 2kx - \frac{2kx}{3} = \frac{4kx}{3}$$

This is opposite to displacement.

$$\therefore \quad F_{net} = -\frac{4kx}{3}$$

20. (d) $\quad F_{net} = -\left(\frac{4kx}{3}\right)x$

$$\therefore \quad a = \frac{F_{net}}{M} = -\left(\frac{4k}{3M}\right)x = -\omega^2 x$$

$$\therefore \quad \omega = \sqrt{\frac{4k}{3M}}$$

21. (c) In case of pure rolling, mechanical energy will remain conserved.

$$\therefore \quad \frac{1}{2}Mv_0^2 + \frac{1}{2}\left(\frac{1}{2}MR^2\right)\left(\frac{v_0}{R}\right)^2 = 2\left[\frac{1}{2}kx^2_{max}\right]$$

$$\therefore \quad x_{max} = \sqrt{\frac{3M}{4k}}v_0$$

As, $\quad f = \frac{2kx}{3}$

$$\therefore \quad F_{max} = \mu Mg = \frac{2kx_{max}}{3} = \frac{2k}{3}\sqrt{\frac{3M}{4k}}v_0$$

$$\therefore \quad v_0 = \mu g\sqrt{\frac{3M}{k}}$$

22. (c) $\frac{1}{2}I(2\omega)^2 = \frac{1}{2}kx_1^2 \quad \ldots(i)$

$$\frac{1}{2}(2I)(\omega)^2 = \frac{1}{2}kx_2^2 \quad \ldots(ii)$$

From Eqs. (i) and (ii), we have

$$\frac{x_1}{x_2} = \sqrt{2}$$

23. (a) Let ω' be the common velocity. Then, from conservation of angular momentum, we have

$$(I + 2I)\omega' = I(2\omega) + 2I(\omega)$$

$$\omega' = \frac{4}{3}\omega$$

From the equation,

Angular impulse = Change in angular momentum

For any of the disc, we have

$$\tau t = I(2\omega) - I\left(\frac{4}{3}\omega\right) = \frac{2I\omega}{3}$$

$$\therefore \quad \tau = \frac{2I\omega}{3t}$$

24. (b) Loss of kinetic energy = $K_i - K_f$

$$= \left\{\frac{1}{2}I(2\omega)^2 + \frac{1}{2}(2I)(\omega)^2\right\} - \frac{1}{2}(3I)\left(\frac{4}{3}\omega\right)^2 = \frac{1}{3}I\omega^2$$

25. (d) $\frac{1}{2}mv^2 + \frac{1}{2}I\left(\frac{v}{R}\right)^2 = mg\left(\frac{3v^2}{4g}\right) \quad \therefore \quad I = \frac{1}{2}mR^2$

∴ Body is a disc.

26. (a) On smooth part BC, due to zero torque, angular velocity and hence the rotational kinetic energy remains constant. While moving from B to C translational kinetic energy converts into gravitational potential energy.

27. (a) $\frac{2}{5}MR^2 = \frac{1}{2}Mr^2 + Mr^2$ or $\frac{2}{5}MR^2 = \frac{3}{2}Mr^2$

$$\therefore \quad r = \frac{2}{\sqrt{15}}R$$

28. (c,d) In case of pure rolling,

$$f = \frac{mg\sin\theta}{1 + \frac{mR^2}{I}} \quad \text{(upwards)}$$

$$\therefore \quad f \propto \sin\theta$$

Therefore, as θ decreases force of friction will also decrease.

29. (a) $I_{remaining} = I_{whole} - I_{removed}$

or $\quad I = \frac{1}{2}(9M)(R)^2 - \left[\frac{1}{2}m\left(\frac{R}{3}\right)^2 + \frac{1}{2}m\left(\frac{2R}{3}\right)^2\right] \quad \ldots(i)$

Here, $\quad m = \frac{9M}{\pi R^2} \times \pi\left(\frac{R}{3}\right)^2 = M$

Substituting in Eq. (i), we have $I = 4MR^2$

30. (d) $\mathbf{L} = m(\mathbf{r} \times \mathbf{v})$

Direction of $(\mathbf{r} \times \mathbf{v})$, hence the direction of angular momentum remains the same.

31. Angular momentum of the system about an axis perpendicular to plane of paper and passing through O will remain conserved.

$$L_i = L_f$$

$$\therefore \quad mvL = I\omega = \left(mL^2 + \frac{ML^2}{3}\right)\omega$$

$$\therefore \quad \omega = \frac{3mv}{L(3m + M)}$$

32. For rolling without slipping, we have

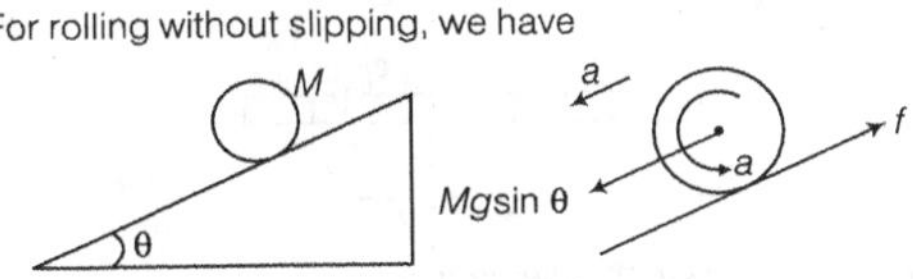

$$a = R\alpha$$

or $\quad \frac{Mg\sin\theta - f}{M} = R\left(\frac{fR}{\frac{1}{2}MR^2}\right)$

or $\quad \frac{Mg\sin\theta - f}{M} = \frac{2f}{M}$

$$\therefore \quad f = \frac{Mg\sin\theta}{3}$$

Therefore, linear acceleration of cylinder,

$$a = \frac{Mg\sin\theta - f}{M} = \frac{2}{3}g\sin\theta$$

3. Gravitation

DPP-1 Newton's Universal Law of Gravitation and Gravitational Field

1. (c)

$$r = 2R$$

$$M_1 = M_2 = \frac{4}{3}\pi R^3 \rho \qquad (\rho = \text{density})$$

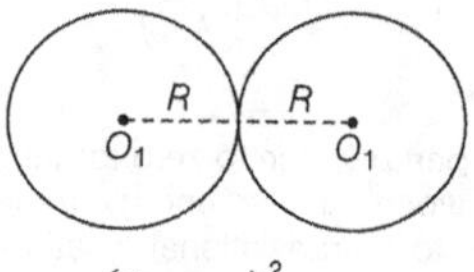

$$\therefore \quad F = G\frac{M_1M_2}{r^2} = \frac{G\left(\frac{4}{3}\pi R^3 \rho\right)^2}{(2R)^2} = \text{constant} \times R^4$$

$$\therefore \quad F \propto R^4$$

2. (a) Mass of the element is $dm = \frac{m}{L}dx$

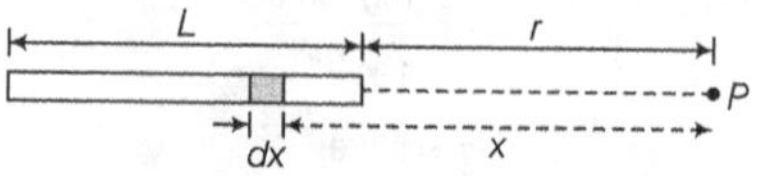

Gravitational field at point P due to this element is

$$dE_g = G\frac{\frac{m}{L}dx}{x^2}$$

Net gravitational field at point P due to the complete rod, is

$$E_g = G\frac{m}{L}\int_r^{r+L}\frac{1}{x^2}dx = G\frac{m}{L}\left[-\frac{1}{x}\right]_r^{r+L} \text{(taking } P \text{ as origin)}$$

or

$$E_g = G\frac{m}{L}\left[-\frac{1}{r+L} + \frac{1}{r}\right] = G\frac{m}{L}\left[\frac{-r+r+L}{r(r+L)}\right]$$

or

$$E_g = \frac{Gm}{r(L+r)}$$

3. (b) Mass of the element is

$$dm = \frac{m}{2L}dx$$

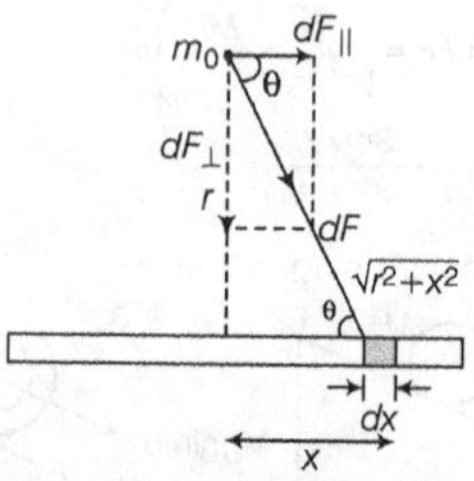

Force on m_0 due to this element is

$$dF = \frac{G m_0 \frac{m}{2L}dx}{r^2 + x^2}$$

From the it is clear that the components similar to $dF_{||}$ get cancelled. All $dF_\perp$ components get added.

$$dF_\perp = dF\sin\theta = \frac{Gmm_0}{2L(r^2+x^2)}dx \cdot \frac{r}{\sqrt{r^2+x^2}}$$

$$\therefore \quad F_{\text{net}} = \int_{-L}^{+L}\frac{Gmm_0r}{2L}\frac{1}{(r^2+x^2)^{3/2}}dx$$

$$= \frac{Gmm_0r}{2L}\int_{-L}^{+L}\frac{1}{(r^2+x^2)^{3/2}}dx$$

Let $\quad x = r\tan\theta$, then $dx = r\sec^2\theta\, d\theta$

When, $\quad x = -L,\ \theta_1 = \tan^{-1}\left(-\frac{L}{r}\right)$

and when $\quad x = +L,\ \theta_2 = \tan^{-1}\left(\frac{L}{r}\right)$

$$\therefore \quad \sin\theta_1 = -\frac{L}{\sqrt{r^2+L^2}}$$

and $\quad \sin\theta_2 = \frac{L}{\sqrt{r^2+L^2}}$ 4t6

$$\therefore \quad F_{\text{net}} = \frac{Gmm_0r}{2L}\int_{\theta_1}^{\theta_2}\frac{1}{r^3(1+\tan^2\theta)^{3/2}}r\sec^2\theta\, d\theta$$

$$= \frac{Gmm_0r}{2L}\cdot\frac{1}{r^2}\int_{\theta_1}^{\theta_2}\cos\theta\, d\theta$$

$$= \frac{Gmm_0}{2Lr}[\sin\theta]_{\theta_1}^{\theta_2}$$

$$= \frac{Gmm_0}{2Lr}[\sin\theta_2 - \sin\theta_1]$$

$$= \frac{Gmm_0}{2Lr}\left[\frac{L}{\sqrt{r^2+L^2}} + \frac{L}{\sqrt{r^2+L^2}}\right]$$

$$F_{\text{net}} = -\frac{Gmm_0}{r\sqrt{r^2+L^2}}\hat{j}$$

4. (d) Mass of one element is $dm = \lambda R d\theta$

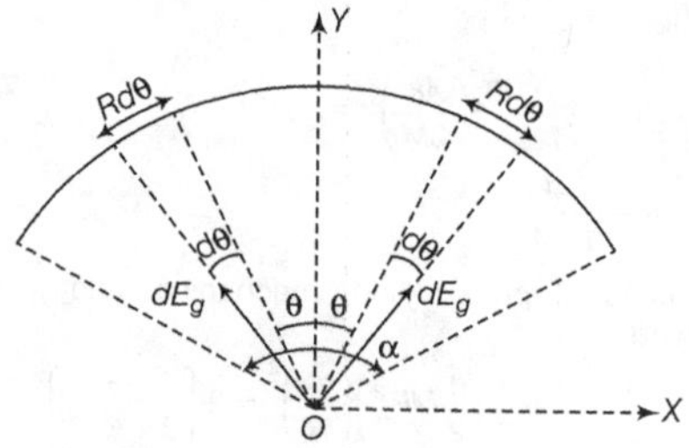

Field due to this element at the centre is

$$dE_g = G\frac{dm}{R^2}$$

or

$$dE_g = G\frac{\lambda R d\theta}{R^2} = \frac{G\lambda}{R}d\theta$$

From the given figure, it is clear that the components similar to $dE_g \sin\theta$ get cancelled and $dE_g\cos\theta$ get added.

$$\therefore \quad dE_g\cos\theta = \frac{G\lambda}{R}\cos\theta\, d\theta$$

$$\therefore \quad (E_g)_{\text{net}} = \frac{G\lambda}{R}\int_{-\alpha/2}^{+\alpha/2}\cos\theta\, d\theta$$

$$= \frac{G\lambda}{R}[\sin\theta]_{-\frac{\alpha}{2}}^{+\frac{\alpha}{2}}$$

$$= \frac{2G\lambda}{R}\sin\left(\frac{\alpha}{2}\right)$$

5. (a) Net force on M due to M and M, $2M$ and $2M$, $4M$ and $4M$, $5M$ and $5M$ and $7M$ and $7M$ is zero. The force due to $3M$ is

$$F = G\frac{M\,3M}{d^2}$$

$$= \frac{3GM^2}{d^2} \text{ along } X\text{-axis.}$$

6. (a) At point P, $|E_{g_1} - E_{g_2}| = E_g$

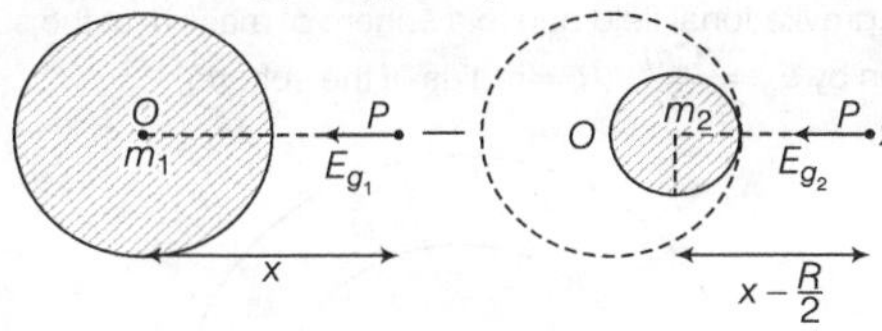

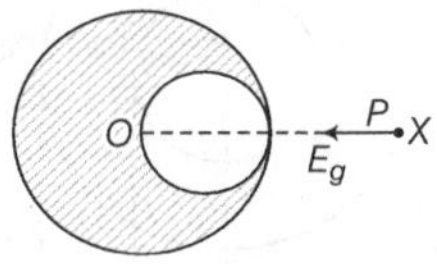

$$m_1 = \frac{4}{3}\pi R^3 \rho_0$$

$$m_2 = \frac{4}{3}\pi\left(\frac{R}{2}\right)^3 \rho_0$$

$$= \frac{4}{3}\pi R^3 \rho_0 \frac{1}{8}$$

$$E_{g_1} = \frac{Gm_1}{x^2} = \frac{G\frac{4}{3}\pi R^3 \rho_0}{x^2}$$

$$E_{g_2} = \frac{Gm_2}{\left(x - \frac{R}{2}\right)^2} = \frac{G\frac{4}{3}\pi R^3 \rho_0 \frac{1}{8}}{\left(x - \frac{R}{2}\right)^2}$$

$$\therefore \quad E_g = G\frac{4}{3}\pi R^3 \rho_0 \left[\frac{1}{x^2} - \frac{1}{8\left(x - \frac{R}{2}\right)^2}\right]$$

7. (b) Initially, field due to both is along positive X-axis. Due to both of them field increases till the observer reaches the centre of the ring. Here, field due to the ring is zero (not the net field). Later, the field due to ring is negative and due to sphere is positive. Somewhere, in the middle field becomes zero and later the influence of sphere will be greater than that of ring, field is positive and increases till the observer reaches the surface of sphere.

8. (a) Work done will be zero, when E_g is perpendicular to displacement.

E_g makes an angle $\theta_1 = \tan^{-1}\left(\frac{1}{4}\right)$ with positive X-axis.

So, line perpendicular to this have a slope of $m = \tan\theta = -4$

So, option (a) is correct.

9. (d) $L = \sqrt{3}R$

$$F = \frac{Gm^2}{L^2} = \frac{Gm^2}{3R^2}$$

$$F' = \sqrt{F^2 + F^2 + 2F^2\cos 60^\circ} = \sqrt{3}\,F$$

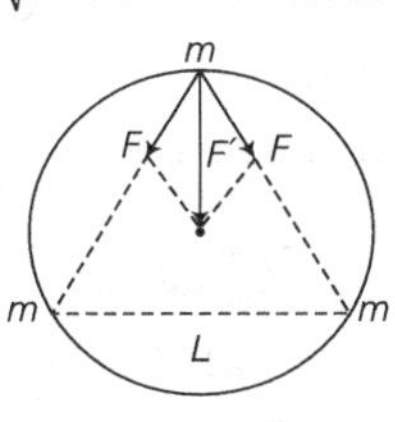

$$\therefore \quad F' = \frac{Gm^2}{3R^2}\cdot\sqrt{3} = \frac{Gm^2}{\sqrt{3}R^2}$$

This force F' supplies centripetal force, i.e.

$$\frac{mv^2}{R} = \frac{Gm^2}{\sqrt{3}R^2}$$

$$\Rightarrow \quad v = \sqrt{\frac{Gm}{\sqrt{3}R}}$$

10. (a,d) $|\mathbf{F}_1| = |\mathbf{F}_2| = |\mathbf{F}_3|$

Angles between $\mathbf{F}_1$ and $\mathbf{F}_2$, $\mathbf{F}_2$ and $\mathbf{F}_3$ and $\mathbf{F}_3$ and $\mathbf{F}_1$, each is equal to 120°.

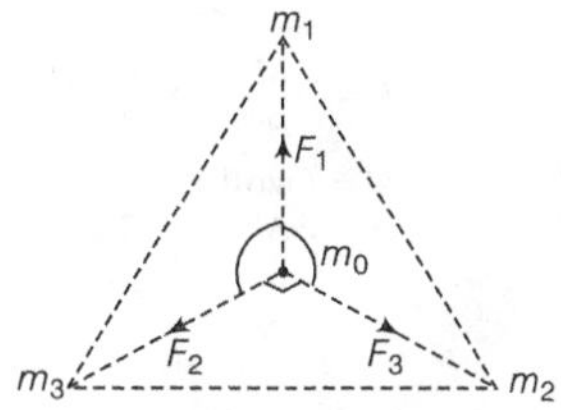

$$\therefore \quad F_{net} \text{ on } m_0 = 0$$

When m_0 is at the mid-point of a side, forces on m_0 due to masses at 2 and 3 cancel out each other. So, force due to m at 1, on m_0 is

$$F_{net} = G\frac{mm_0}{(\sqrt{3}L/2)^2} = \frac{4Gmm_0}{3L^2}$$

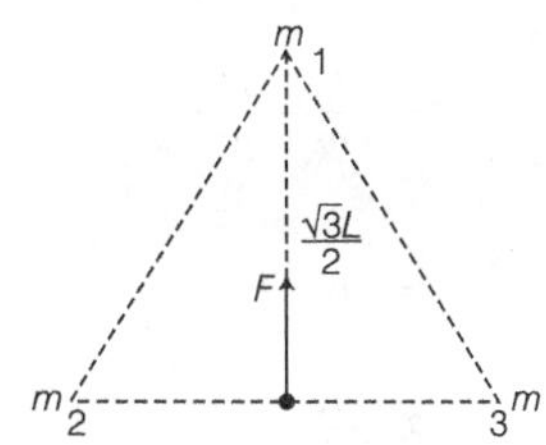

11. (a,b,c) (a) $r > R_2$: At point P_1

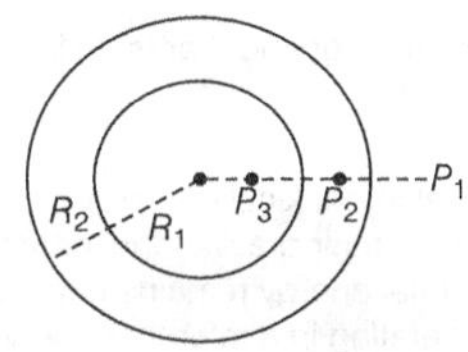

$$E_g = \frac{G(m_1 + m_2)}{r^2}$$

$$\therefore \quad F = E_g m_0 = \frac{G(m_1 + m_2)m_0}{r^2}$$

(b) $R_1 < r < R_2$: At point P_2,

$$E_g = \frac{Gm_1}{r^2}$$

$$\therefore \quad F = \frac{Gm_1 m_0}{r^2}$$

(c) $r < R_1$: At point P_3, $E_g = 0$ $\therefore$ $F = 0$

12. (a,b) $E_g = \frac{Gm}{r^2}$ if $r \geq R$ (i.e. outside)

$$\therefore \quad \frac{E_{g_1}}{E_{g_2}} = \frac{r_2^2}{r_1^2} \text{ when } r_1 > R \text{ and } r_2 > R$$

$$E_g = \frac{Gm}{R^3}r \quad \text{if } r < R \text{ (i.e. inside)}$$

$$\therefore \quad \frac{E_{g_1}}{E_{g_2}} = \frac{r_1}{r_2} \text{ when } r_1 < R \text{ and } r_2 < R$$

13. (b,c) When the distance of the particle from the centre is x, $E_g = \frac{GM}{R^3} \cdot x$

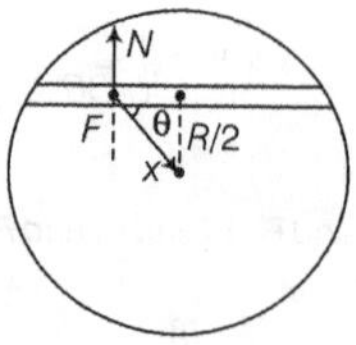

$$\therefore \quad F = \frac{GMm}{R^3} \cdot x$$

$$\therefore \quad N = F\sin\theta = \frac{GMm}{R^3} \cdot x \cdot \frac{R/2}{x} = \frac{GMm}{2R^2}$$

$\therefore$ N is constant and independent of x.

Tangential force is

$$F_t = F\cos\theta = \frac{GMm}{R^3} x \frac{\sqrt{x^2 - \frac{R^2}{4}}}{x}$$

$\therefore$ Acceleration is

$$a = \frac{F_t}{m} = \frac{GM}{R^3} \cdot \frac{1}{2}\sqrt{4x^2 - R^2}$$

At $x = \frac{R}{2}$,

$$a = 0$$

and at $x = R$,

$$a = \frac{\sqrt{3}}{2} \cdot \frac{GM}{R^2}$$

As, x increases, then a increases.

14. (a,d) Same force (in magnitude) acts on both the masses

$$\therefore \quad a \propto \frac{1}{m} \qquad (\because F = ma)$$

Due to gravitational force, kinetic energy of masses increases as they move closer but their gravitational potential energy decreases. Also, total energy remains constant. Centre of mass of system is accelerating in a reference frame attached to any of mass.

15. (d) We can take sphere as a point mass lying at its centre. But rod cannot be taken as a point mass lying at its centre.

16. (1) $\frac{G \times 10^3}{x^2} = \frac{G \times 10^5}{(1-x)^2}$

$$\Rightarrow \quad \frac{1-x}{x} = 10$$

$$\Rightarrow \quad 11x = 1$$

$$x = \frac{1}{11} \text{ km}$$

$$\therefore \quad p = 1$$

17. (4) m is attracted by only horizontal component of gravitational force due to a part of ring, which changes by a factor of

$$\frac{2r}{(4r^2 + r^2)^{3/2}} = \frac{4\sqrt{2}}{5\sqrt{5}} = \frac{4\sqrt{10}}{25}$$

So, $\quad k = 4$

Explanations for Q. Nos. (18-21)

The gravitational field due to a sphere of mass m at the surface is given by $E_g = \frac{Gm}{R^2}$, (R = radius of the sphere)

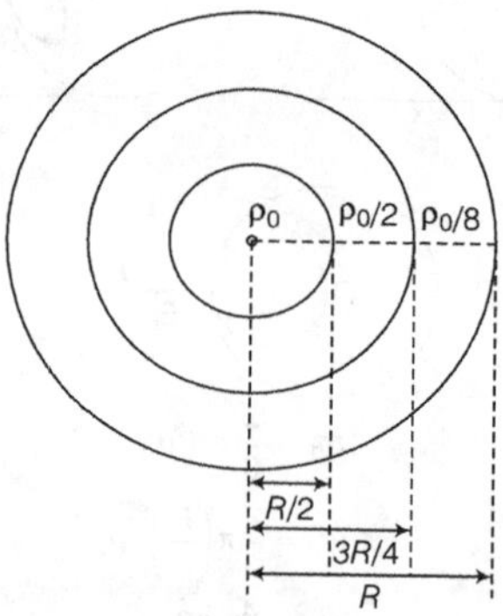

18. (a) For $r = \frac{R}{4}$, density $= \rho_0$

The mass of the sphere of radius $R/4$ is

$$m = \frac{4}{3}\pi\left(\frac{R}{4}\right)^3 \rho_0 = \pi R^3 \rho_0/48$$

$\therefore$ Gravitational field at a distance $\frac{R}{4}$ from the centre of sphere is

$$E_g = \frac{Gm}{(R/4)^2} = \frac{G(\pi R^3 \rho_0/48)}{(R^2/16)} = \frac{3\pi G R \rho_0}{9}$$

19. (d) For $r = \frac{R}{2}$, up to $\frac{R}{3}$, $\rho = \rho_0$ and between $\frac{R}{3}$ and $\frac{R}{2}$, $\rho = \frac{\rho_0}{2}$

$$\therefore \quad m = \frac{4}{3}\pi\left(\frac{R}{3}\right)^3 \rho_0 + \left[\frac{4}{3}\pi\left(\frac{R}{2}\right)^3 - \frac{4}{3}\pi\left(\frac{R}{3}\right)^3\right]\frac{\rho_0}{2}$$

$$= 0.108\pi R^3 \rho_0$$

$\therefore$ Gravitational field at a distance $r = R/2$ is

$$E_g = \frac{Gm}{(R/2)^2} = \frac{G\,\pi R^3 \rho_0 (0.018)}{R^2} \times 4 = \frac{35}{81}\pi G R \rho_0$$

20. (b) For $r = \frac{5R}{6}$, up to $\frac{R}{3}$, $\rho = \rho_0$, between $\frac{R}{3}$ and $\frac{3R}{4}$,

$\rho = \frac{\rho_0}{2}$ and between $\frac{3R}{4}$ and $\frac{5R}{6}$, $\rho = \frac{\rho_0}{8}$

$\therefore$ The mass of the sphere of radius $5R/6$ is

$$m = \frac{4}{3}\pi\left(\frac{R}{3}\right)^3 \rho_0 + \left[\frac{4}{3}\pi\left(\frac{3R}{4}\right)^3 - \frac{4}{3}\pi\left(\frac{R}{3}\right)^3\right]\rho_0/2$$

$$+ \left[\frac{4}{3}\pi\left(\frac{5R}{6}\right)^3 - \frac{4}{3}\pi\left(\frac{3R}{4}\right)^3\right]\rho_0/8$$

$$= 0.332\,\pi R^3 \rho_0$$

$\therefore$ Gravitational field at a distance, $r = \frac{5R}{6}$ is

$$E_g = \frac{Gm}{(5R/6)^2} = \frac{G \times 0.332\,\pi R^3 \rho_0}{25R^2} \times 36 = 0.48\,\pi G R \rho_0$$

DPP-2 Acceleration due to Gravity

1. (b) Initially, $W_i = 600$ N, finally $W_f = 240$ N. During the flight in between somewhere its weight will be zero because at that point gravitational pull of earth and mars will be equal.

2. (a) $g\left(1 - \frac{d}{R}\right) = g - \omega^2 R$

$$\Rightarrow \quad \omega^2 R = \frac{gd}{R} \Rightarrow d = \frac{\omega^2 R^2}{g}$$

3. (b) On the equator, $g_e = g - R\omega^2$

When $g_e = 0$, $\quad g - R\omega^2 = 0$

$$\Rightarrow \quad \omega = \sqrt{\frac{R}{g}}$$

$$\frac{2\pi}{T} = \sqrt{\frac{R}{g}}$$

$$\Rightarrow \quad T = 2\pi\sqrt{\frac{R}{g}} = 2\pi\sqrt{\frac{6400 \times 10^3}{10}} \times \frac{1}{3600} \text{ h}$$

or $\quad T = \frac{4\pi}{9}$

4. (c) $g = \frac{GM}{r_1^2} + \frac{GM}{r_2^2} + \cdots$

$$= GM\left(\frac{1}{1^2} + \frac{1}{2^2} + \frac{1}{4^2} + \cdots\right) = GM\left(\frac{1}{1 - \frac{1}{4}}\right) = GM \times \frac{4}{3} = 4G$$

5. (b) $g_h = g\left(1 - \frac{2h}{R}\right)$

$$w = mg \quad \Rightarrow \quad \frac{dw_h}{dh} = m\frac{dg_h}{dh}$$

$$\therefore \quad \left|\frac{dw_h}{dh}\right| = m\left|\frac{d}{dh}\left(g - \frac{2g_h}{R}\right)\right| = m\frac{2g}{R} = \frac{2mg}{R}$$

$$\Rightarrow \quad \left|\frac{dw_h}{dh}\right| = \frac{2mg}{R}h^0$$

$$\Rightarrow \quad \frac{dw_h}{dh} \propto h^0$$

6. (d) $g = \frac{GM}{R^2} = \frac{G}{R^2} \cdot \frac{4}{3}\pi R^3 \rho = \frac{4}{3}\pi G R \rho$

$$\therefore \quad g \propto R\rho$$

If R is increased by a factor of 2 to keep the value of g same, then value of ρ has to be changed by a factor of $\frac{1}{2}$.

7. (a) $g = \frac{GM}{R^2} = \frac{4}{3}\pi G R \rho$

$$\therefore \quad \frac{g_m}{g_e} = \frac{R_m}{R_e} \quad \Rightarrow \quad \frac{R_m}{R_e} = \frac{1}{6}$$

8. (c) At the equator, $g_e = g - R\omega^2$

For weightlessness at equator, $g_e = 0$

$$\omega = \sqrt{g/R} = \sqrt{\frac{10}{64 \times 10^5}} = \frac{1}{800} \text{ rad s}^{-1}$$

9. (a) For no acceleration, $mg' - mr\omega^2 = 0$

$$\therefore \quad g' = r\omega^2$$

$$\frac{GM}{R^3} r = r\omega^2$$

or $\quad \omega^2 = \frac{G}{R^3} \frac{4}{3}\pi R^3 \rho_0$

or $\quad \omega = \sqrt{\frac{4}{3}\pi G \rho_0}$

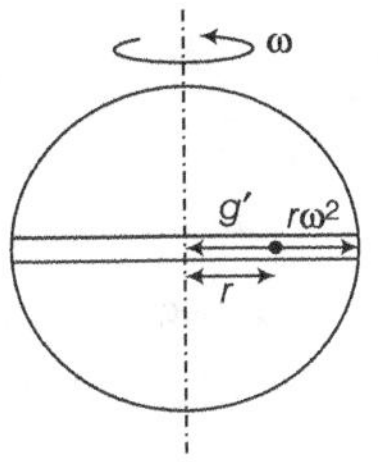

10. (b) Mass of the planet is M

$$M = \frac{4}{3}\pi R^3 \rho_1 + \frac{4}{3}\pi\{(2R)^3 - (R)^3\}\rho_2$$

$$= \frac{4}{3}\pi R^3 [\rho_1 + 7\rho_2]$$

$\therefore$ At the surface g_1

$$= \frac{GM}{(2R)^2} = \frac{G}{4R^2} \cdot \frac{4}{3}\pi R^3 (\rho_1 + 7\rho_2)$$

$$= \frac{\pi G R}{3}(\rho_1 + 7\rho_2)$$

Mass of the core is

$$m = \frac{4}{3}\pi R^3 \rho_1$$

$\therefore$ At a depth R,

$$g_2 = \frac{Gm}{(R)^2} = \frac{G}{R^2} \cdot \frac{4}{3}\pi R^3 \rho_1 = \frac{4\pi G R}{3}\rho_1$$

Given, $\quad g_1 = g_2$

$$\Rightarrow \quad \frac{\pi G R}{3}(\rho_1 + 7\rho_2) = \frac{4\pi G R}{3}\rho_1$$

$$\rho_1 + 7\rho_2 = 4\rho_1$$

$$\Rightarrow \quad 3\rho_1 = 7\rho_2$$

$$\Rightarrow \quad \frac{\rho_1}{\rho_2} = \frac{7}{3}$$

11. (d) During total eclipse, total attraction due to the sun and the moon is

$$F_1 = \frac{GM_s M_e}{r_1^2} + \frac{GM_m M_e}{r_2^2}$$

When the moon goes to the opposite side of the earth, then effective force of attraction is

$$F_2 = \frac{GM_s M_e}{r_1^2} - \frac{GM_m M_e}{r_2^2}$$

$$\Delta F = F_1 - F_2 = \frac{2GM_m M_e}{r_2^2}$$

Change in acceleration of earth,

$$\Delta a = \frac{\Delta F}{M_e} = \frac{2GM_m}{r_2^2}$$

Average force on earth $= F_{av} = \frac{F_1 + F_2}{2} = \frac{GM_s M_e}{r_1^2}$

Average acceleration is

$$a_{av} = \frac{F_{av}}{M_e} = \frac{GM_s}{r_1^2}$$

%change in acceleration $= \frac{\Delta a}{a_{av}} \times 100$

$$= \frac{2GM_m}{r_2^2} \cdot \frac{r_1^2}{GM_s} \times 100$$

$$= 2\left(\frac{r_1}{r_2}\right)^2 \frac{M_m}{M_s} \times 100$$

12. (c) Let $v = A\omega\cos\omega t$ and $a = -A\omega^2 \sin\omega t$ for the simple pendulum. Then,

$$v^2 = A^2\omega^2\cos^2\omega t$$

$$\Rightarrow \quad \frac{dv^2}{dt} = A^2\omega^2 \times 2\cos\omega t\,(-\sin\omega t) \cdot \omega$$

$$a^2 = A^2\omega^4 \sin^2\omega t$$

$$\Rightarrow \quad \frac{da^2}{dt} = A^2\omega^4 \times 2\sin\omega t\,(\cos\omega t)\omega$$

Now, $\quad \frac{dv^2}{da^2} = -\frac{1}{\omega^2}$

$\therefore$ |Slope of v^2-a^2 graph| $= \frac{1}{\omega^2}$

$$\tan\theta = \frac{1}{\omega^2}$$

$$\Rightarrow \quad \omega = \frac{1}{\sqrt{\tan\theta}}$$

$$T = \frac{2\pi}{\omega} = 2\pi\sqrt{\tan\theta}$$

Also, $\quad T = 2\pi\sqrt{\frac{l}{g}}$

$$\therefore \quad \tan\theta = \frac{l}{g}$$

$$\Rightarrow \quad g = \frac{l}{\tan\theta}$$

$$\therefore \quad \frac{g_{\text{on ground}}}{g_h} = \frac{\tan 60^\circ}{\tan 30^\circ} = 3$$

$$g_h = \frac{g_{\text{on ground}}}{3}$$

$$\frac{g}{\left(1+\frac{h}{R}\right)^2} = \frac{g}{3}$$

$$\Rightarrow \quad \left(1+\frac{h}{R}\right)^2 = 3$$

$$\Rightarrow \quad \frac{h}{R} = 0.73$$

$$h = 0.73R$$

13. (d) Value of g also depends on shape of earth. Statement II is correct.

14. (c) The length of the day is slowly increasing not due to viscous force between the earth and the atmosphere around it but due to expansion of universe. Distance of earth from the sun is increasing and $T^2 \propto \varepsilon^3$.

15. (d) Mass of the sphere of radius 1 m = M (given)

$\therefore$ Mass of the whole sphere of radius 4 m is

$$M' = \frac{M\left(\frac{4}{3}\pi(4)^3\right)}{\frac{4}{3}\pi(1)^3} = 64M$$

$\therefore$ The gravitational field at B is

E_B = field due to whole sphere – field due to sphere A

$$= \frac{GM'r}{R^3} - \frac{GM}{(AB)^2}$$

or $\quad E_B = \frac{G(64M)\times 2}{4^3} - \frac{GM}{4^2} = \frac{31}{16}GM$

$\therefore$ Acceleration of B is $a_B = \frac{31}{16}GM$

16. (a) The gravitational field at O is zero due to symmetry.

17. (4) $g = \frac{GM}{R^2}$ (if R increases, g decreases)

Taking logarithms on both the sides, we get

$$\log g = \log G + \log M - 2\log R$$

Differentiating, we get

$$\frac{dg}{g} = 0 + 0 - \frac{2dR}{R}$$

$$\frac{dg}{g} = -2\left(\frac{-2}{100}\right) = \frac{4}{100}$$

$$\% \text{ increase in } g = \frac{dg}{g} \times 100$$

$$= \frac{4}{100} \times 100 = 4\%$$

18. (5) $x^2 = 500y$ (given), differentiating w.r.t. time t, we get

$$2x\frac{dx}{dt} = 500\frac{dy}{dt}$$

Again differentiating w.r.t. time t, we get

$$2\left[x\frac{d^2x}{dt^2} + \left(\frac{dx}{dt}\right)\left(\frac{dx}{dt}\right)\right] = 500\frac{d^2y}{dt^2}$$

Given, $\frac{dx}{dt} = 360 \times \frac{5}{18}\ \text{ms}^{-1} = 100\ \text{ms}^{-1}$ and is constant.

So, $\quad \frac{d^2x}{dt^2} = 0$

$$\therefore \quad 2\left(\frac{dx}{dt}\right)^2 = 500\frac{d^2y}{dt^2}$$

$$\Rightarrow \quad \frac{d^2y}{dt^2} = \frac{2(100)^2}{500} = 40 = 4g = ay$$

$\therefore$ Effective value of g is

$$g_{\text{eff}} = 4g + g = 5g$$

$$\therefore \quad \beta = 5$$

19. (4) We know that, $g_\phi = g - R\omega^2\cos^2\phi$

At poles $\phi = 90^\circ$, $\therefore g_p = g$

At equator $\phi = 0^\circ$, $\therefore g_e = g - R\omega^2$

$$g_p - g_e = R\omega^2 = (6.4 \times 10^6)\left(\frac{2\pi}{24 \times 60 \times 60}\right)^2$$

$$= 34 \times 10^{-3}\ \text{ms}^{-2}$$

$$= (4 + 30) \times 10^{-3}\ \text{ms}^{-2}$$

$$\therefore \quad p = 4$$

20. (2) $a = -\frac{GM}{R^3}y$

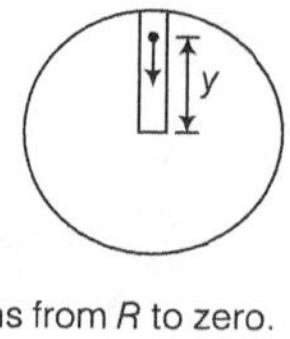

$$\therefore \quad v\frac{dv}{dy} = -\frac{GM}{R^3}y$$

$$vdv = -\frac{GM}{R^3}ydy$$

During downward journey, $u = 0$, $v = v$ and y runs from R to zero.

$$\therefore \quad \int_0^v vdv = -\frac{GM}{R^3}\int_R^0 ydy$$

$$\Rightarrow \quad \frac{v^2}{2} = -\frac{GM}{R^3}\left(0 - \frac{R^2}{2}\right)$$

$$\Rightarrow \quad v = \sqrt{\frac{GM}{R}}$$

After collision with the centre, velocity is v' (say).

Given, $\quad e = \frac{1}{2}$

$$\Rightarrow \quad \frac{v' - 0}{v - 0} = \frac{1}{2}$$

$$\Rightarrow \quad v' = \frac{1}{2}\sqrt{\frac{GM}{R}}$$

In the upward journey, let the ball reaches to a height h.

Now, $\quad \int_{v'}^0 vdv = -\frac{GM}{R^3}\int_0^h ydy$

$$\Rightarrow \quad \frac{v'^2}{2} = \frac{GM}{R^3}\cdot\frac{h^2}{2}$$

$$\frac{1}{4}\cdot\frac{GM}{R} = \frac{GM}{R^3}\cdot h^2$$

$$\Rightarrow \quad h = \frac{R}{2}$$

$\therefore$ Required distance is

$$R + \frac{R}{2} + \frac{R}{2} = 2R$$

$$\therefore \quad \beta = 2$$

DPP-3 Potential, Potential Energy and Self Energy

1. (d) Let the gravitational field is zero at a distance x from m_1.

$$\therefore \quad \frac{Gm_1}{x^2} = \frac{Gm_2}{(d-x)^2}$$

$$\Rightarrow \quad \frac{x}{d-x} = \sqrt{\frac{m_1}{m_2}}$$

$$x\sqrt{m_2} = d\sqrt{m_1} - x\sqrt{m_1}$$

$$\Rightarrow \quad x = \frac{d\sqrt{m_1}}{\sqrt{m_1} + \sqrt{m_2}}$$

$$\therefore \quad d - x = \frac{d\sqrt{m_2}}{\sqrt{m_1} + \sqrt{m_2}}$$

Net potential at this point is

$$V = -\frac{Gm_1}{x} - \frac{Gm_2}{(d-x)}$$

$$V = -G\left[\frac{m_1(\sqrt{m_1} + \sqrt{m_2})}{d\sqrt{m_1}} + \frac{m_2(\sqrt{m_1} + \sqrt{m_2})}{d\sqrt{m_2}}\right]$$

$$= -\frac{G}{d}(\sqrt{m_1} + \sqrt{m_2})^2$$

$$= -\frac{G}{d}(m_1 + m_2 + 2\sqrt{m_1 m_2})$$

2. (b) $m_A = \sigma 4\pi R_A^2$, $m_B = \sigma 4\pi R_B^2$, where σ = surface density

$$V_A = -\frac{Gm_A}{R_A}, V_B = -\frac{Gm_B}{R_B}$$

$$\frac{V_A}{V_B} = \frac{m_A}{m_B} \cdot \frac{R_B}{R_A} = \frac{\sigma 4\pi R_A^2}{\sigma 4\pi R_B^2} \cdot \frac{R_B}{R_A} = \frac{R_A}{R_B}$$

Given, $\quad \frac{V_A}{V_B} = \frac{3}{4}$

$$\therefore \quad \frac{R_A}{R_B} = \frac{3}{4} \Rightarrow R_B = \frac{4}{3}R_A$$

For new shell of mass m and radius R,

$$m = m_A + m_B = \sigma 4\pi R_A^2 + \sigma 4\pi R_B^2$$

$$\sigma 4\pi R^2 = \sigma 4\pi (R_A^2 + R_B^2)$$

$$\Rightarrow \quad R^2 = R_A^2 + \frac{16}{9}R_A^2$$

$$\therefore \quad R = \frac{5}{3}R_A; \frac{m}{m_A} = \frac{\sigma 4\pi R^2}{\sigma 4\pi R_A^2} = \frac{25}{9}$$

$$\therefore \quad \frac{V}{V_A} = \frac{m}{m_A} \cdot \frac{R_A}{R} = \frac{25}{9} \times \frac{3}{5} = \frac{5}{3}$$

3. (a) $W_{\text{external force}} + (W_{a \to b})_{\text{gravitational force}} = \Delta K$

$$-10 + (W_{a \to b})_{\text{gravitational force}} = 4$$

$$(W_{a \to b})_{\text{gravitational force}} = 14\,\text{J}$$

$$\therefore \quad m(V_A - V_B) = 14 \Rightarrow V_A - V_B = \frac{14}{2} = 7\,\text{Jkg}^{-1}$$

$$\therefore \quad V_B - V_A = -7\,\text{Jkg}^{-1}$$

4. (b) To assemble this, we assume that, we bring several mass elements dm step by step from infinity to a spherical surface of radius R with centre O. Every dm is uniformly distributed on this surface.

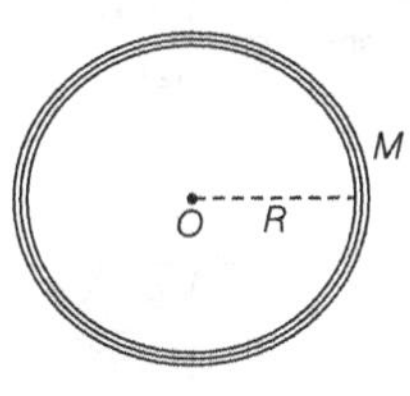

Now, we consider an intermediate situation when mass of surface becomes m and a further dm is brought from infinity to the surface. Thus, we can write the work done in bringing this elemental mass dm from infinity to a distance R from O is

$$dW = \frac{GM}{R}dM$$

Here, we can assume that a hollow sphere of mass m is behaving as it is concentrated at O for outer points.

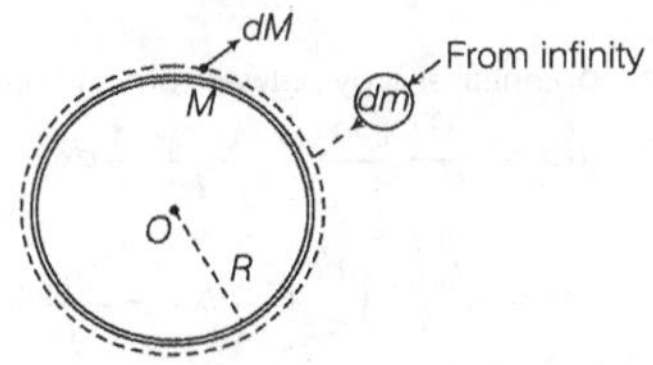

Now, total work done in increasing the mass from 0 to M can be obtained by integrating the above expression within proper limits as

$$W = \int dW = \int_0^M \frac{Gmdm}{R} = \frac{GM^2}{2R}$$

This work is done by the gravitational attractive forces of the body. This work is done by itself in assembling. Thus, gravitational self energy of a hollow sphere of mass M and radius R is given by

$$U_{\text{self}} = -\frac{GM^2}{2R}$$

5. (a) Here, radius is not constant. We assume that density of the sphere is ρ.

$$\therefore \quad \rho = \frac{M}{\frac{4}{3}\pi R^3}$$

Consider the intermediate situation.

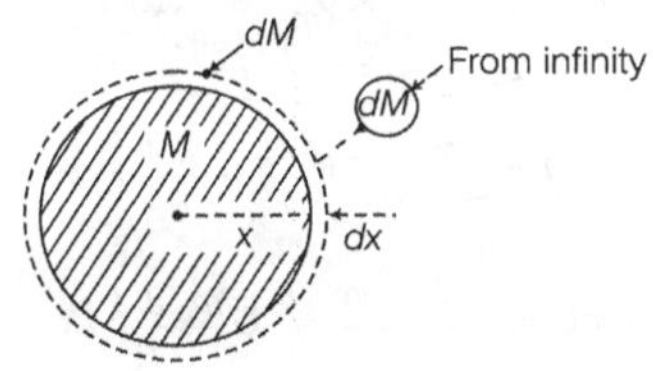

$$dM = (4\pi x^2 dx)\rho$$

$$M = \rho \cdot \frac{4}{3}\pi x^3$$

$$dW = \frac{Gm}{x}dM$$

$$= \frac{G}{x}\left(\rho \cdot \frac{4}{3}\pi x^3\right)(4\pi x^2 dx \cdot \rho)$$

$$= \frac{16\pi^2}{3}G\rho^2 x^4 dx$$

$$W = \frac{16}{3}\pi^2 G\rho^2 \cdot \int_0^R x^4 dx$$

$$= \frac{16\pi^2 G\rho^2}{3} \cdot \frac{R^5}{5}$$

$$W = \frac{16\pi^2 G}{3} \cdot \frac{R^5}{5} \cdot \left(\frac{M}{\frac{4}{3}\pi R^3}\right)^2$$

$$= \frac{3}{5} \cdot \frac{GM^2}{R}$$

$$\therefore \quad U_{\text{self}} = -\frac{3}{5} \cdot \frac{GM^2}{R}$$

6. (c) Consider an element as shown in the figure.

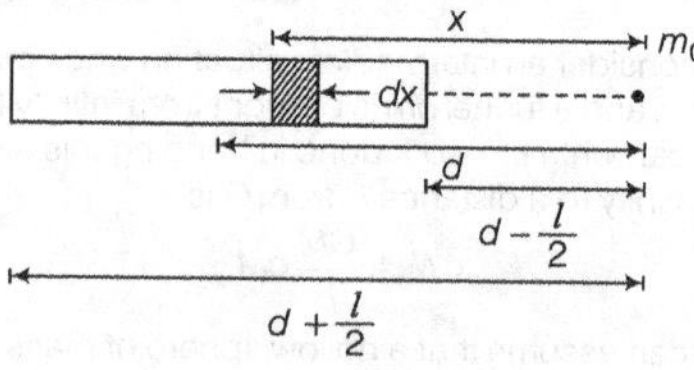

$$dm = \frac{m_0}{l}dx$$

Gravitational potential energy between dm and point mass is

$$dU = -\frac{G m_0 dm}{x} = -\frac{Gm_0^2}{l}\cdot\frac{1}{x}dx$$

Now, $$U = -\frac{Gm_0^2}{l}\int_{d-\frac{l}{2}}^{d+\frac{l}{2}}\frac{1}{x}dx = -\frac{Gm_0^2}{l}\ln\left(\frac{d+\frac{l}{2}}{d-\frac{l}{2}}\right)$$

or $$U = -\frac{Gm_0^2}{l}\ln\left(\frac{2d+l}{2d-l}\right)$$

7. (b) $U = -\frac{G m m_0}{R}$

$\therefore$ Binding energy $= |U| = \frac{Gmm_0}{R}$

This much energy is required to displace the particle from the centre of the ring to infinity.

8. (a) Consider an elementary hemispherical shell of radius x and thickness dx. Let dM is the mass of this element.

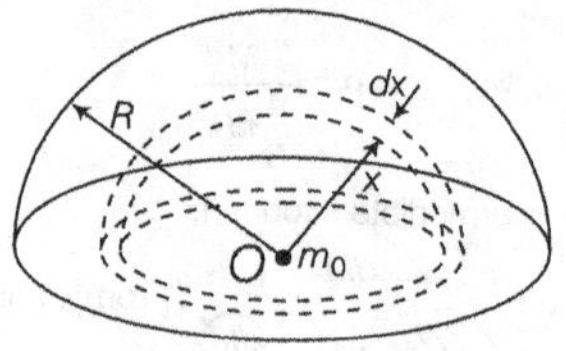

$$dM = \frac{M}{\frac{4}{6}\pi R^3}\cdot 2\pi x^2 dx = \frac{3M}{R^3}x^2 dx$$

$$dU = -\frac{G m_0 dm}{x} = -G\frac{m_0}{x}\frac{3M}{R^3}x^2 dx$$

$\therefore$ $$U = -\frac{3G Mm_0}{R^3}\int_0^R x dx = -\frac{3}{2}\frac{GM m_0}{R}$$

$\therefore$ $$U_i = -\frac{3}{2}\cdot\frac{GMm_0}{R},\ U_f = 0$$

$$W = (W_{i\to f})_{\text{gravitational force}} = -(U_f - U_i)$$

$$= U_i = -\frac{3}{2}\cdot\frac{GMm_0}{R}$$

9. (d) $U_i + K_i = U_f + K_f$

$\Rightarrow$ $$-\frac{3Gm^2}{d} + 0 = -\frac{3Gm^2}{2R} + 3\left(\frac{1}{2}mv^2\right)$$

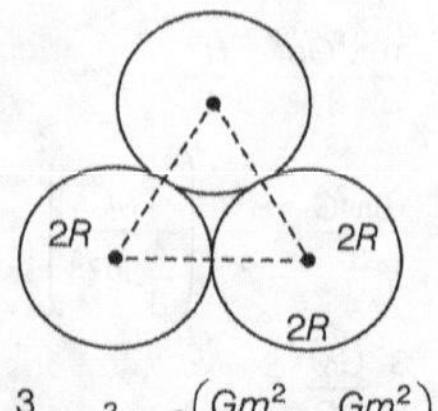

$$\frac{3}{2}mv^2 = 3\left(\frac{Gm^2}{2R} - \frac{Gm^2}{d}\right)$$

$\Rightarrow$ $$v = \sqrt{Gm\left(\frac{1}{R} - \frac{2}{d}\right)}$$

10. (b,d) Potential at A is

V_A = Potential at A due to 1st ring + Potential at A due to 2nd ring

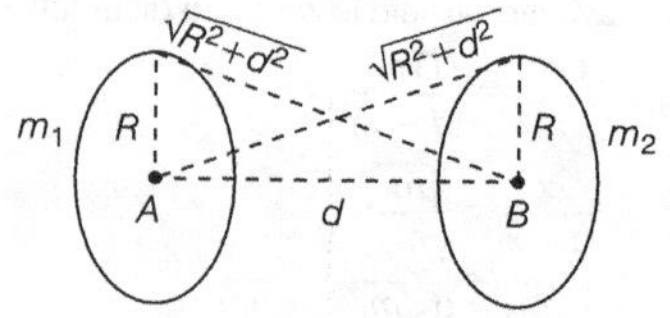

or $$V_A = -G\left(\frac{m_1}{R} + \frac{m_2}{\sqrt{R^2+d^2}}\right)$$

Similarly, $$V_B = -G\left(\frac{m_2}{R} + \frac{m_1}{\sqrt{R^2+d^2}}\right)$$

Now, $$\Delta V = V_B - V_A = G(m_1 - m_2)\left(\frac{1}{R} - \frac{1}{\sqrt{R^2+d^2}}\right)$$

Also, $$W = (\Delta V)m_0 = Gm_0(m_1 - m_2)\left(\frac{1}{R} - \frac{1}{\sqrt{R^2+d^2}}\right)$$

11. (a) Because gravitational force is always attractive in nature, the gravitational potential is negative.

12. (a) The radius of the given circle is 6 m. Obviously, the point will lie outside the sphere in YZ-plane . Consider point P such that $OP = 6$ m. Now, $AP = BP = \sqrt{6^2 + 2^2} = \sqrt{40}$ m

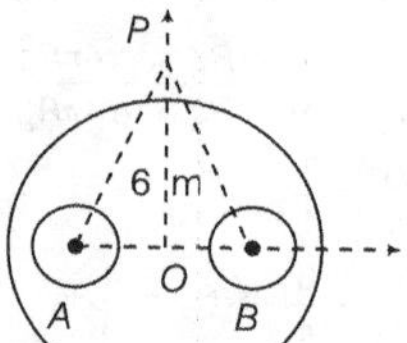

Potential at point P = Potential due to whole sphere – 2 (potential due to each cavity)

$\because$ Mass of the whole sphere is

$$m_0 = \frac{M}{\frac{4}{3}\pi(1)^2}\cdot\frac{4}{3}\pi(4)^3 \text{ or } m_0 = 64M$$

$\therefore$ $$V_P = -\frac{Gm_0}{OP} + 2\left[\frac{GM}{AP}\right]$$

$$= -\frac{G(64M)}{6} - \left(-2\frac{GM}{\sqrt{40}}\right)$$

$$= \left(\frac{\sqrt{10}}{100} - \frac{32}{3}\right)GM$$

13. (a) Potential at P = Potential of complete sphere

– Potential due to removed masses

$$= -22GM - 2\left(-\frac{GM}{2\sqrt{2}}\right)$$

$$= \left(\frac{\sqrt{2}}{2} - 22\right)GM$$

14. (a) At the surface of shell, $g = \frac{G(2M)}{(2R)^2}$ $(\because h << R)$

or $$g = \frac{GM}{2R^2}$$

$$t = \sqrt{\frac{2h}{g}}$$

$$= 2\sqrt{\frac{hR^2}{GM}}$$

15. (b) $v_A = \sqrt{2gh}$

$$= \sqrt{2\left(\frac{GM}{2R^2}\right)h} = \sqrt{\frac{GMh}{R^2}}$$

From A to B, field due to shell is zero, but field due to sphere is non-zero.

Hence, $t_{AB} < \frac{R}{v_A}$

$\Rightarrow$ $t_{AB} < \frac{R^2}{\sqrt{GMh}}$

16. (d) $h << R$. So, we can take $K_A \simeq 0$

Applying conservation of energy between A and B, we have

$$K_A + U_A = K_B + U_B$$
$$K_B = U_A - U_B$$
$$= m(V_A - V_B)$$
$$\frac{1}{2}mv_B^2 = m(V_A - V_B)$$

$\Rightarrow$ $v_B = \sqrt{2(V_A - V_B)}$

or $v_B = \sqrt{2\left[-\frac{GM}{2R} + \frac{GM}{R}\right]} = \sqrt{\frac{GM}{R}}$

17. (c) **Potential Sphere** $V = -\frac{GM}{2R^3}(3R^2 - r^2)$ for $r \le R$

$$V = -\frac{GM}{r} \text{ for } r > R$$

Shell $V = -\frac{GM}{R}$ for $r \le R$

$$V = -\frac{GM}{r} \text{ for } r > R$$

Field Sphere $E_g = \frac{GM}{R^3} \cdot r$ for $r \le R$

$$E_g = \frac{GM}{r^2} \text{ for } r > R$$

Shell $E_g =$ zero for $r < R$

$$E_g = \frac{GM}{r^2} \text{ for } r \ge R$$

Hence, (i) $\rightarrow$ (q), (ii) $\rightarrow$ (r), (iii) $\rightarrow$ (s), (iv) $\rightarrow$ (p)

18. (a) Inside a shell,

$$V = -\frac{GM}{R} \text{ and } E = 0$$

Outside the shell,

$$V = -\frac{GM}{r} \text{ and } E = \frac{GM}{r^2}$$

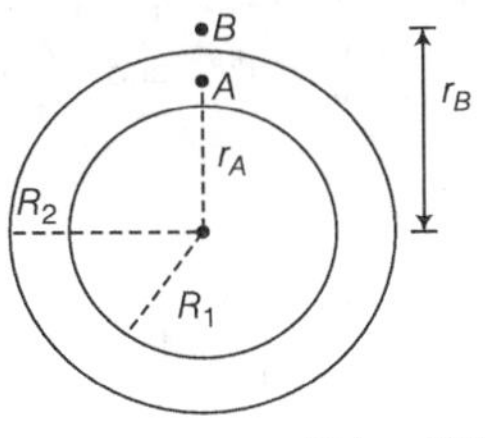

$$V_A = -\frac{GM_1}{R_2} - \frac{GM_2}{r_A}$$
$$V_B = -\frac{GM_1}{r_B} - \frac{GM_2}{r_B}$$

$\therefore$ $V_B > V_A$

Note *V is negative.*

$$E_B = \frac{GM_1}{r_B^2} + \frac{GM_2}{r_B^2}$$
$$E_A = \frac{GM_1}{r_A^2}$$

Hence, (i) $\rightarrow$ (q), (ii) $\rightarrow$ (t), (iii) $\rightarrow$ (r), (iv) $\rightarrow$ (s)

19. (c) $E_h = \frac{GM}{(R+h)^2} = \frac{GM}{4R^2} = \frac{E}{4}$

$$V_h = -\frac{GM}{2R} = -\frac{V}{2}$$
$$E_d = E\left(1 - \frac{d}{R}\right) = \frac{E}{2}$$
$$V_d = -\frac{GM}{2R^3}(3R^2 - r^2)$$
$$= -\frac{GM}{2R^3}\left(3R^2 - \frac{R^2}{4}\right)$$
$$= -\frac{11GM}{8R}$$

Note *V_d is more negative and V_h is less negative when compared to V.*

$\therefore$ $V_d < V$

and $V_h > V$

Hence, (i) $\rightarrow$ (p), (ii) $\rightarrow$ (q), (iii) $\rightarrow$ (s), (iv) $\rightarrow$ (t)

20. (3) $V_A = \begin{bmatrix}\text{Potential due to the}\\ \text{complete sphere at } A\end{bmatrix} - \begin{bmatrix}\text{Potential due to the}\\ \text{spherical cavity at } A\end{bmatrix}$

$$= -\frac{3}{2}\frac{GM}{R} - \left(-\frac{GM'}{r}\right)$$
$$M = \frac{4}{3}\pi R^3\rho$$
$$M' = \frac{4}{3}\pi\left(\frac{R}{2}\right)^3\rho$$
$$= \frac{1}{8}\left(\frac{4}{3}\pi R^3\rho\right), \; r = \frac{R}{2}$$

$\therefore$ $V_A = \frac{G}{R}\left[\frac{\pi\rho R^3}{3} - 2\pi\rho R^3\right]$

$$= -\frac{5}{3}\pi G\rho R^2$$

Now, $V_B = \begin{bmatrix}\text{Potential due to the}\\ \text{complete sphere at } B\end{bmatrix} - \begin{bmatrix}\text{Potential due to the}\\ \text{spherical cavity at } B\end{bmatrix}$

$$= -\frac{GM}{2R^3}(3R^2 - r^2) - \left[-\frac{3}{2}\frac{GM'}{r}\right]$$

or $V_B = -\frac{11}{8}\frac{GM}{R} + \frac{3GM'}{R}$

$$= \frac{G}{R}\left[\pi\rho R^3 - \frac{11\pi\rho R^3}{6}\right]$$
$$= -\frac{4}{3}\pi G\rho R^2$$

Now, $V_B - V_A = \frac{1}{3}\pi G\rho R^2$

$$\frac{1}{2}m_0 v^2 = U_B - U_A = m_0(V_B - V_A)$$

$\Rightarrow$ $v = \sqrt{2(V_B - V_A)}$

or $v = \frac{\sqrt{2\pi G\rho R^2}}{\sqrt{3}}$

$\Rightarrow$ $\beta = 3$

DPP-4 Escape Velocity and Binding Energy

1. (a)

$$U_i + K_i = U_f + K_f$$

$$-\frac{GMm}{R} + \frac{1}{2}mv_i^2 = -\frac{GMm}{r} + 0$$

$$\frac{1}{2}v_i^2 = \frac{GM}{R}\left(1-\frac{R}{r}\right) = gR\left(1-\frac{R}{r}\right)$$

$\therefore$ $$1-\frac{R}{r} = \frac{v_i^2}{2gR} \Rightarrow r = \frac{R}{1-\frac{v_i^2}{2gR}}$$

2. (c) Energy is conserved for the falling object $(K+U)_r = (K+U)_h$

$$\frac{1}{2}mv^2 - \frac{GM_Em}{r} = 0 - \frac{GM_Em}{R_E+h}$$

$\Rightarrow$ $$v = \frac{-dr}{dt} = \left[2\,GM_E\left(\frac{1}{r} - \frac{1}{R_E+h}\right)\right]^{1/2}$$

3. (b) Potential energy of m, when it is mid-way between M_1 and M_2 is

$$U = m(V_1 + V_2) = m\left[-\frac{GM_1}{d/2} - \frac{GM_2}{d/2}\right]$$

or $$U = -\frac{2Gm}{d}(M_1 + M_2)$$

Its kinetic energy is $K = 0$

Total energy is $E_T = U + K = -\frac{2Gm}{d}(M_1 + M_2)$

$\therefore$ Binding energy $= |E_T| = \frac{2Gm}{d}(M_1 + M_2)$

$\therefore$ To escape infinity KE to be supplied

$$= \frac{2Gm}{d}(M_1 + M_2)$$

or $$\frac{1}{2}mv_e^2 = \frac{2Gm}{d}(M_1 + M_2)$$

$\Rightarrow$ $$v_e = 2\sqrt{\frac{G(M_1 + M_2)}{d}}$$

4. (b) $mg_{pole} = \frac{GMm}{R^2} \Rightarrow GM = gR^2$ $\quad (\because g_{pole} = g)$

At the equator,

$$\frac{GMm}{R^2} - mg' = \frac{mv^2}{R}$$

$\Rightarrow$ $$mg - \frac{mg}{2} = \frac{mv^2}{R} \text{ or } g = \frac{2v^2}{R}$$

Now at the pole, total energy of a particle is

$$E_T = U + K$$

or $$E_T = -\frac{GMm}{R} + 0$$

$\therefore$ $$\text{Binding energy} = |E_T| = \frac{GMm}{R} = mgR$$

$\therefore$ $$\frac{1}{2}mv_e^2 = mgR = m\frac{2v^2}{R}R$$

$\Rightarrow$ $$v_e = 2v$$

5. (a) Let at a distance r from the centre of the earth, the body has a velocity v. From conservation of mechanical energy,

$$\frac{1}{2}mv^2 - \frac{GMm}{r} = \frac{1}{2}mv_e^2 - \frac{GMm}{R}$$

$$v^2 = v_e^2 + \frac{2GM}{R}\left[\frac{R}{r} - 1\right]$$

$$= 2gR + 2gR\left[\frac{R}{r} - 1\right]$$

$$\left(\because v_e = \sqrt{2gR} \text{ and } g = \frac{GM}{R^2}\right)$$

or $$v = \sqrt{\frac{2gR^2}{r}} \text{ or } \frac{dr}{dt} = R\frac{\sqrt{2g}}{\sqrt{r}}$$

or $$dt = \frac{1}{R\sqrt{2g}}r^{1/2}dr$$

$\Rightarrow$ $$\int_0^t dt = \frac{1}{R\sqrt{2g}}\int_R^{R+h} r^{1/2}dr$$

or $$t = \frac{2}{3}\cdot\frac{1}{R\sqrt{2g}}[(R+h)^{3/2} - R^{3/2}]$$

$$= \frac{1}{3}\sqrt{\frac{2R}{g}}\left[\left(1+\frac{h}{R}\right)^{3/2} - 1\right]$$

6. (b,c) We know that orbital speed of a satellite is $v_o = \sqrt{\frac{GM}{r}}$

$\therefore$ Its total energy is given by

$$E_T = U + K = -\frac{GMm}{r} + \frac{1}{2}mv_o^2$$

or $$E_T = -\frac{GMm}{r} + \frac{GMm}{2r} = -\frac{GMm}{2r}$$

$\therefore$ Binding energy $= |E_T| = \frac{GMm}{2r}$

So, extra KE to be given the satellite to escape the infinity is $\frac{GMm}{2r}$.

$\therefore$ $$\frac{1}{2}mv_e^2 = \frac{1}{2}mv_o^2 + \frac{GMm}{2r} = \frac{GMm}{r}$$

$\therefore$ $$v_e = \sqrt{2}\,v_o = 1.41v_o$$

7. (b,d) $v_e = \sqrt{\frac{2GM}{R}} = \sqrt{\frac{2G\left(\frac{4}{3}\pi R^3\right)\rho}{R}} = \sqrt{\frac{4G\rho}{3}}R$

$$v_e \propto R$$

$$A = 4\pi R_1^2$$

and $$4A = 4\pi R_2^2$$

$\Rightarrow$ $$R_2 = 2R_1$$

Mass of P_3 is $$M_3 = M_1 + M_2$$

$$\left(\frac{4}{3}\pi R_3^3\right)\rho = \left(\frac{4}{3}\pi R_1^3\right)\rho + \left(\frac{4}{3}\pi R_2^3\right)\rho$$

$$R_3^3 = R_1^3 + R_2^3$$

$\Rightarrow$ $$R_3^3 = 9R_1^3$$

$$R_3 = 9^{1/3}R_1$$

$\therefore$ $$R_3 > R_2 > R_1$$

So, $$v_3 > v_2 > v_1$$

Also, $$\frac{v_3}{v_1} = 9^{1/3} \text{ and } \frac{v_1}{v_2} = \frac{1}{2}$$

8. (b,d) $\rho_1 : \rho_2 = 1 : 8$

$\Rightarrow$ $$v_1 : v_2 = 8 : 1$$

$\Rightarrow$ $$R_1 : R_2 = 2 : 1$$

$$g \propto \frac{1}{R^2}$$

$\Rightarrow$ $$\frac{g_1}{g_2} = \frac{R_2^2}{R_1^2} = \frac{1}{4}$$

$$v_e \propto \sqrt{gR}$$

$\Rightarrow$ $$\frac{v_{e_1}}{v_{e_2}} = \sqrt{\frac{g_1}{g_2}\cdot\frac{R_1}{R_2}}$$

$$= \sqrt{\frac{1}{4}\cdot\frac{2}{1}} = 1 : \sqrt{2}$$

9. (d) The value of escape velocity is derived from the law of conservation of mechanical energy and energy is independent of direction.

10. (b) Inside the tunnel, $U = -\dfrac{GMm}{2R^2}(3R - r)$

$\therefore$ Total energy, $E_T = U + K = -\dfrac{GMm}{2R^2}(3R - r)$

Binding energy, $|E_T| = \dfrac{GMm}{2R^2}(3R - r)$

$$\therefore \quad \frac{1}{2}mv_e^2 = \frac{GMm}{2R^2}(3R - r) \text{ or } v_e = \sqrt{\frac{GM(3R - r)}{R^2}}$$

As r increases, then v_e decreases. At the surface of planet,

$$r = R$$

$$\therefore \quad v_e = \sqrt{\frac{2GM}{R}}$$

11. (a) $v_e = \sqrt{\dfrac{2GM}{R}} = \sqrt{2gR}$. This equation clearly says that v_e is independent of mass of the body.

12. (b) $v_0 = \sqrt{\dfrac{GM}{R}}$ and $v_e = \sqrt{\dfrac{2GM}{R}}$

If $v < v_0$, trajectory is an ellipse.

If $v = v_0$, trajectory is a circle.

If $v_0 < v < v_e$, trajectory is an ellipse.

If $v = v_e$, trajectory is a parabola.

Hence, (i) $\rightarrow$ (q), (ii) $\rightarrow$ (s), (iii) $\rightarrow$ (p), (iv) $\rightarrow$ (r)

13. (c) $g_d = g\left(1 - \dfrac{d}{R}\right) = g\left(1 - \dfrac{R - y}{R}\right) = \dfrac{g}{R}y$

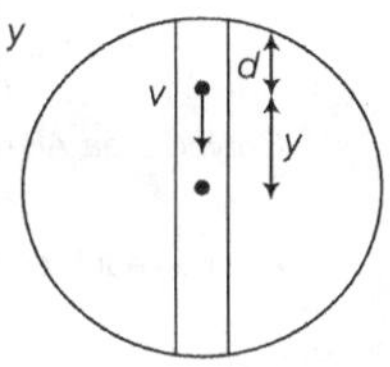

$$\therefore \quad a = -\frac{g}{R}y$$

$$v\frac{dv}{dy} = -\frac{g}{R}y$$

$$\Rightarrow \quad vdv = -\frac{g}{R}ydy$$

$$\int_{v_e}^{v} vdv = -\frac{g}{R}\int_R^y ydy$$

$$\Rightarrow \quad \frac{v^2}{2} - \frac{v_e^2}{2} = -\frac{g}{R}\left[\frac{y^2}{2} - \frac{R^2}{2}\right]$$

$$v^2 = -\frac{g}{R}y^2 + gR + v_e^2 = -\frac{g}{R}y^2 + 3gR \qquad (\because v_e = \sqrt{2gR})$$

$$v = \sqrt{3gR - \frac{g}{R}y^2} = \sqrt{g/R}\,[\sqrt{3R^2 - y^2}]$$

$$-\frac{dy}{dt} = \sqrt{\frac{g}{R}}\,[\sqrt{3R^2 - y^2}]$$

$$\Rightarrow \quad \frac{-1}{\sqrt{3R^2 - y^2}}dy = \sqrt{\frac{g}{R}}\,dt$$

$$\int_R^0 \frac{-1}{\sqrt{3R^2 - y^2}}dy = \sqrt{\frac{g}{R}}\int_0^t dt$$

$$\Rightarrow \quad -\left[\sin^{-1}\left(\frac{y}{\sqrt{3}R}\right)\right]_R^0 = \sqrt{\frac{g}{R}}\cdot t$$

$$-\left[\sin^{-1}(0) - \sin^{-1}\left(\frac{1}{\sqrt{3}}\right)\right] = \sqrt{\frac{g}{R}}\,t$$

$$\therefore \quad t = \sqrt{\frac{R}{g}}\sin^{-1}\left(\frac{1}{\sqrt{3}}\right)$$

14. (d) At a height h, acceleration is not proportional to the distance from the centre. So, motion will not be SHM but periodic in nature.

15. (a) $h = \dfrac{1}{2}gt^2 \Rightarrow t = \sqrt{\dfrac{2h}{g}}$

$$\therefore \quad \text{Time period} = 2\pi\sqrt{\frac{R}{g}} + 4t$$

$$= 2\pi\sqrt{\frac{R}{g}} + 4\sqrt{\frac{2h}{g}}$$

16. (2) From the law of conservation of energy,

$$(U + K)_{surface} = (U + K)_{at\ \infty}$$

$$-\frac{GMm}{R} + \frac{1}{2}mu^2 = 0 + \frac{1}{2}mv^2$$

$$-\frac{2GM}{R} + u^2 = v^2$$

$$-v_e^2 + u^2 = v^2 \qquad \left(\because v_e = \sqrt{\frac{2GM}{R}}\right)$$

$$v^2 = -(11.2)^2 + (15)^2 = -125 + 225 = 100$$

$$\therefore \quad v = 10\text{ kms}^{-1} \Rightarrow \alpha = 2$$

17. (3) We know that, $g = \dfrac{GM}{R^2} = \dfrac{4}{3}\pi GR\rho$

$$\therefore \quad \frac{g_p}{g_e} = \frac{R_p\rho_p}{R_e\rho_e}, \text{ also } v_e = \sqrt{2gR}$$

$$\therefore \quad \frac{(v_e)_p}{(v_e)_e} = \sqrt{\frac{g_pR_p}{g_eR_e}} = \sqrt{\frac{g_p}{g_e}\cdot\frac{g_p\rho_e}{g_e\rho_p}} = \frac{g_p}{g_e}\sqrt{\frac{\rho_e}{\rho_p}}$$

$$= \frac{\sqrt{6}}{11}\sqrt{\frac{3}{2}} = \frac{3}{11}$$

$$\therefore \quad (v_e)_p = \frac{3}{11}(v_e)_e = \frac{3}{11}(11) = 3\text{ km s}^{-1}$$

18. (2) Let v be the speed of projectile at the highest point. By conservation of angular momentum and conservation of mechanical energy, we have

$$muR\sin\alpha = mvr_{max}$$

and $$\frac{1}{2}mu^2 - \frac{GMm}{R} = \frac{1}{2}mv^2 - \frac{GMm}{r_{max}}$$

Solving these two equations, we give $r_{max} = \dfrac{3R}{2}$

$$\therefore \quad h_{max} = r_{max} - R = \frac{R}{2} \Rightarrow \psi = 2$$

19. (1) For a satellite in circular orbit,

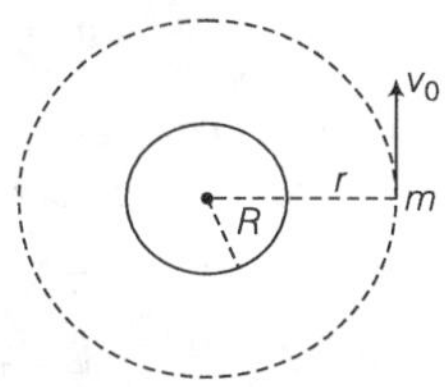

$$\frac{mv_0^2}{r} = \frac{GMm}{r^2} \text{ or } v_0^2 = \frac{GM}{R + h}$$

Given, $$v_0 = \frac{v_e}{2}$$

$$\Rightarrow \quad \sqrt{\frac{GM}{R + h}} = \frac{1}{2}\sqrt{\frac{2GM}{R}}$$

$$\frac{GM}{R + h} = \frac{GM}{2R} \Rightarrow h = R$$

When the satellite stops, $K = 0$

$$\therefore \quad U_i - U_f = K_f$$

$$\Rightarrow \quad -\frac{GMm}{2R} - \left(-\frac{GMm}{R}\right) = \frac{1}{2}mv^2$$

$$v = \sqrt{\frac{GM}{R}} = \sqrt{gR}$$

$$\therefore \quad k = 1$$

20. (3) $\dfrac{1}{2}mv_{esc}^2 = \dfrac{GM_mm}{R_m} \Rightarrow v_{lanuch} = 2\sqrt{\dfrac{2GM_m}{R_m}}$

For flight of stone from moon to earth, energy is conserved.

$$\therefore \quad (K + U)_{initial} = (K + U)_{final}$$

$$\frac{1}{2}mv_{lanuch}^2 - \frac{GmM_E}{R_m} - \frac{GmM_E}{R_{mE}} = \frac{1}{2}mv_{impact}^2 - \frac{GM_mm}{R_{mE}} - \frac{GmM_E}{R_E}$$

So, required velocity is

$$v_{impact} = \sqrt{2G\left[\frac{3M_m}{R_m} + \frac{M_m}{R_{mE}} + \frac{M_E}{R_E} - \frac{M_E}{R_{ME}}\right]}$$

So, $$2\times\frac{q}{p} = 2\times\frac{3}{2} = 3$$

DPP-5 Motion of Satellites and Planets

1. (a) $T \propto r^{3/2}$

$$E = -\frac{GMm}{2r}, \quad v = \sqrt{\frac{GM}{r}}$$

$$L = mvr = m\sqrt{GMr}$$

2. (c) $U = -\frac{GMm}{r}$

$$K = \frac{1}{2}mv_0^2 = \frac{GMm}{2r}$$

$$E_T = U + K = -\frac{GMm}{2r}$$

3. (c) Angular speed of earth = Angular speed of geostationary satellite

$$T \propto \frac{1}{\omega}$$

$$\Rightarrow \quad \frac{T_2}{T_1} = \frac{\omega_1}{\omega_2} = \frac{1}{2} \Rightarrow T_2 = \frac{T_1}{2}$$

Also, $\quad T \propto r^{3/2}$

$$\therefore \quad \left(\frac{r_2}{r_1}\right)^{3/2} = \frac{T_2}{T_1} = \frac{1}{2}$$

$$\frac{r_2}{r_1} = \left(\frac{1}{2}\right)^{2/3} = \frac{1}{4^{1/3}}$$

$$r_2 = \frac{r_1}{4^{1/3}}$$

4. (d) $mr\omega^2 = \frac{GMm}{r^2}$

$$\Rightarrow \quad GM = r^3\omega^2$$

$$g = \frac{r^3\omega^2}{R^2}$$

$$\therefore \quad g' = \frac{(4nR)^3\omega^2}{(nR)^2} = 64\,nR\omega^2$$

5. (b) Let M is mass of the planet.

Total energy of satellite at an orbital radius r is

$$E = -\frac{GMm}{2r}$$

$$\Rightarrow \quad dE = +\frac{GMm}{2r^2}dr \qquad \text{...(i)}$$

Also, $\quad \frac{dE}{dt} = -F \times v = -(kv^2)v = -kv^3$

$$= -k\left(\frac{GM}{r}\right)^{3/2}$$

or $\quad dE = -k\left(\frac{GM}{r}\right)^{3/2} dt \qquad \text{...(ii)}$

From Eqs. (i) and (ii), we get

$$\frac{GMm}{2r^2}dr = -k\left(\frac{GM}{r}\right)^{3/2} dt$$

$$dt = -\frac{GMm}{2r^2}\cdot\left(\frac{r}{GM}\right)^{3/2}\cdot\frac{1}{k}dr$$

$$= -\frac{m}{2\sqrt{GM}}\cdot\frac{1}{k}\cdot\frac{1}{\sqrt{r}}dr$$

$$\int_0^t dt = -\frac{m}{2k\sqrt{GM}}\int_{7R}^{R}\frac{1}{\sqrt{r}}dr$$

$$t = -\frac{m}{k\sqrt{GM}}[\sqrt{r}]_{7R}^{R}$$

$$t = \frac{mR}{k\sqrt{GM}}(\sqrt{7}-1)$$

$$\therefore \quad t = \frac{m}{k\sqrt{gR}}(\sqrt{7}-1) \qquad \left(\because g = \frac{GM}{R^2}\right)$$

6. (b) Let d = distance between stars and d_1 and d_2 be the distances of stars from centre of mass of system.

$$\therefore \quad d_1 = \left(\frac{d}{m_1+m_2}\right)m_2$$

and $\quad d_2 = \left(\frac{d}{m_1+m_2}\right)m_1$

$$m_1 + m_2 = M \qquad \text{(given)}$$

$$\therefore \quad d_1 = \frac{d}{M}m_2 \text{ and } d_2 = \frac{d}{M}m_1$$

Also, $\quad \frac{Gm_1m_2}{d^2} = m_1\omega^2 d_1 = m_2\omega^2 d_2$

$$\therefore \quad \frac{Gm_1(M-m_1)}{d^2} = m_1\omega^2\left[\frac{d}{M}(M-m_1)\right]$$

So, $\quad GM = \omega^2 d^3$

$$\Rightarrow \quad GM = \left(\frac{2\pi}{T}\right)^2 d^3$$

$$\therefore \quad d = \left[\frac{GMT^2}{4\pi^2}\right]^{1/3}$$

7. (a,c) Minimum colatitude is given by

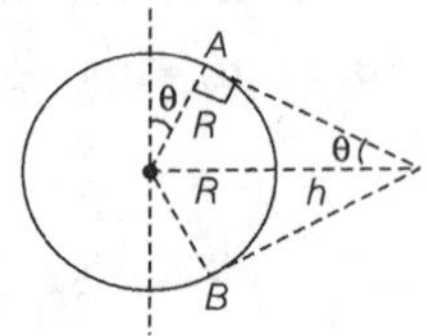

$$\sin\theta = \frac{R}{R+h}$$

Curved area AB on earth

$$= 2\pi R^2(1-\sin\theta)$$

Area on earth escaped from satellite

$$= 4\pi R^2 - 2\pi R^2(1-\sin\theta)$$

$$= 2\pi R^2(1+\sin\theta)$$

8. (b,c) A satellite while orbiting around the earth, it must be above the equator at sometime. The period of revolution of satellite is

$$T = \frac{2\pi}{R}\sqrt{\frac{(R+h)^3}{g}} > 2\pi\sqrt{\frac{R}{g}}$$

9. (a,b) $r_A = R + R = 2R$, $r_B = R + 3R = 4R$

$$K = \frac{GMm}{2r} \Rightarrow K \propto \frac{1}{r} \Rightarrow \frac{K_A}{K_B} = \frac{r_B}{r_A} = 2$$

$$U = -\frac{GMm}{r} \Rightarrow U \propto \frac{1}{r} \Rightarrow \frac{U_A}{U_B} = \frac{r_B}{r_A} = 2$$

10. (b,d) Just before collision, the orbital velocity is $v_o = \sqrt{\frac{GM}{R}}$

Now, the total energy of satellite is negative. Even after expelling Δm, if the satellite has to be in the gravitational field of planet, its total energy should be negative again.

$$\therefore \quad \frac{1}{2}(m-\Delta m)(v_o+v_r)^2 - \frac{GM(m-\Delta m)}{R} \le 0 \qquad \text{...(i)}$$

$$(v_o+v_r)^2 \le \frac{2GM}{R}$$

$$\Rightarrow \quad v_o + v_r \le \sqrt{\frac{2GM}{R}}$$

$$\sqrt{\frac{GM}{R}} + v_r \le \sqrt{\frac{2GM}{R}}$$

$$\Rightarrow \quad v_r \le (\sqrt{2}-1)\sqrt{\frac{GM}{R}}$$

Maximum value of v_r is

$$\therefore \quad x = (\sqrt{2}-1)\sqrt{\frac{GM}{R}}$$

In Eq. (i), $(m - \Delta m)$ gets cancelled out. So, no modification is required.

11. (a,d)

$$r_1 = R + \frac{R}{4} = \frac{5R}{4}$$

$$r_2 = R + \frac{R}{6} = \frac{7R}{6}$$

$$T \propto r^{3/2}$$

$$\Rightarrow \quad \frac{T_1}{T_2} = \left(\frac{r_1}{r_2}\right)^{3/2} = \left(\frac{\frac{5R}{4}}{\frac{7R}{6}}\right)^{3/2} = \left(\frac{15}{14}\right)^{3/2}$$

From the law of conservation of energy,

$$(U + K)_{\text{surface}} = (U + K)_{\text{orbit}}$$

$$-\frac{GMm_1}{R} + \frac{1}{2}m_1v_1^2 = -\frac{GMm_1}{2\left(R + \frac{R}{4}\right)}$$

$$\Rightarrow \quad v_1^2 = \frac{6GM}{5R}$$

Similarly, $-\frac{GMm_2}{R} + \frac{1}{2}m_2v_2^2 = -\frac{GMm_2}{2\left(R + \frac{R}{6}\right)}$

$$\therefore \quad v_2^2 = \frac{8GM}{7R}$$

$$\frac{v_1^2}{v_2^2} = \frac{\frac{6}{5}}{\frac{8}{7}} = \frac{21}{20} \Rightarrow \frac{v_1}{v_2} = \sqrt{\frac{21}{20}}$$

12. (d) $v_e = \sqrt{2}v_0 \Rightarrow v_e > v_0$

13. (b) $L_{\text{particle}} = m(v_0 + v)x = \sqrt{\frac{5}{4}}\, mv_0x$

$$v_0 = \sqrt{\frac{GM}{x}}$$

Total energy of particle

$$= \frac{1}{2}m(v_0 + v)^2 - \frac{GMm}{x} = \frac{1}{2}m\frac{5}{4}v_0^2 - \frac{GMm}{x}$$

$$= \frac{5}{8}\cdot\frac{GMm}{x} - \frac{GMm}{x}$$

$$= -\frac{3}{8}\cdot\frac{GMm}{x}$$

At any distance r, total energy of particle

$$= \frac{1}{2}mu^2 - \frac{GMm}{r}$$

But, angular momentum is conserved.

$$\therefore \quad mur = \sqrt{\frac{5}{4}}\, mv_0x = \sqrt{\frac{5}{4}}\, m\sqrt{\frac{GM}{x}}\,x$$

$$\Rightarrow \quad u = \sqrt{\frac{5}{4}\cdot\frac{GMx}{r^2}}$$

Total energy at a distance, $r = \frac{1}{2}m\cdot\frac{5}{4}\cdot\frac{GMx}{r^2} - \frac{GMm}{r}$

$$\therefore \quad \frac{1}{2}m\frac{5}{4}\frac{GMx}{r^2} - \frac{GMm}{r} = -\frac{3}{8}\cdot\frac{GMm}{x}$$

$$\frac{5}{8}\cdot\frac{x}{r^2} - \frac{1}{r} = -\frac{3}{8x}$$

$$\Rightarrow \quad \frac{5x - 8r}{8r^2} = -\frac{3}{8x}$$

$$(5x - 8r)x = -3r^2$$

$$3r^2 - (8x)r + 5x^2 = 0$$

$$\Rightarrow \quad (r - x)(3r - 5x) = 0$$

$$r = x \quad \text{and} \quad r = \frac{5}{3}x$$

Hence, (i) → (p), (ii) → (s), (iii) → (r), (iv) → (q)

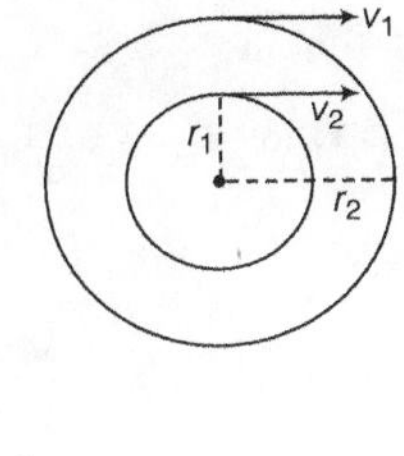

Explanations for Q. Nos. (14-15)

For circular orbits, $T^2 \propto r^3$

$$\therefore \quad \frac{T_1^2}{T_2^2} = \left(\frac{r_1}{r_2}\right)^3 \Rightarrow \frac{1}{64} = \frac{r_1^3}{r_2^3}$$

$$\therefore \quad r_2 = 4r_1 \Rightarrow r_2 = 4 \times 10^4 \text{ km}$$

$$v_1 = \frac{2\pi r_1}{T_1} = \frac{2\pi(10^4)}{1} \text{ km h}^{-1}$$

and $v_2 = \frac{2\pi(4 \times 10^4)}{8} = \pi \times 10^4 \text{ km h}^{-1}$

$$(v_2 - v_1) = \pi \times 10^4 \text{ kmh}^{-1}$$

$$\omega_r = \frac{v_2 - v_1}{r_2 - r_1} = \frac{\pi \times 10^4}{(4 \times 10^4 - 10^4)} = \frac{\pi}{3} \text{ rad h}^{-1}$$

14. (a) **15.** (b)

Explanations for Q. Nos. (16-17)

$$v_o = \frac{v_e}{2} = \frac{\sqrt{2gR}}{2} = \sqrt{\frac{gR}{2}}$$

$$\therefore \quad \sqrt{\frac{GM}{r}} = \sqrt{\frac{gR}{2}} \Rightarrow r = \frac{2GM}{gR} = \frac{2GM}{\frac{GM}{R^2}\cdot R} = 2R$$

$$\therefore \quad h = r - R = 2R - R = R$$

$$(U + K)_{\text{orbit}} = (U + K)_{\text{surface}}$$

$$-\frac{GMm}{(2R)} = -\frac{GMm}{R} + \frac{1}{2}mv^2$$

$$\frac{1}{2}v^2 = \frac{GM}{2R} = \frac{gR}{2} \Rightarrow v = \sqrt{gR}$$

16. (c) **17.** (a)

18. (b) $E_1 = (U + K)_{\text{surface}}$, $E_2 = (U + K)_{\text{1st orbit}}$

$$E_1 = -\frac{GMm}{R} + 0 = -\frac{GMm}{R}$$

$$E_2 = -\frac{GMm}{2(2R)} = -\frac{GMm}{4R}$$

$$W = \Delta E = -\frac{GMm}{4R} - \left(-\frac{GMm}{R}\right) = \frac{3GMm}{4R}$$

19. (d) E_3 = energy in 2nd orbit

$$= -\frac{GMm}{2(3R)} = -\frac{GMm}{6R}$$

$$W = E_3 - E_2 = -\frac{GMm}{6R} - \left(-\frac{GMm}{4R}\right)$$

$$= \frac{GMm}{12R} = \frac{mgR}{12}$$

20. (3) The fastest possible rate of rotation of a planet is that for which the gravitational force on a body at the equator just barely provides the centripetal force needed for the rotation. Let M be the mass and R be the radius of the planet and m is the mass of a body on its surface. Then,

$$mR\,\omega_{\max}^2 = \frac{GMm}{R^2}$$

$$\Rightarrow \quad \omega_{\max} = \sqrt{\frac{GM}{R^3}}$$

$$\text{or} \quad \omega_{\max} = \sqrt{\frac{G}{R^3}\cdot\frac{4}{3}\pi R^3\rho} = \sqrt{\frac{4G\rho\pi}{3}} = 2\sqrt{\frac{\pi G\rho}{3}}$$

$$T_{\min} = \frac{2\pi}{\omega_{\max}} = \sqrt{\frac{3\pi}{G\rho}}$$

$$\Rightarrow \quad \beta = 3$$

DPP-6 Kepler's Laws

1. (d) $\frac{dA}{dt} = \frac{L}{2m}$ (L = angular momentum)

$$= \frac{mv_{max} r_{min}}{2m}$$

$$\therefore \quad v_{max} = \frac{2}{r_{min}} \cdot \frac{dA}{dt}$$

$$= \frac{2}{2 \times 10^{12}} \times 4 \times 10^{16}$$

$$= 4 \times 10^4 \text{ ms}^{-1}$$

$$= 40 \text{ kms}^{-1}$$

2. (a) The time period T_e of the earth around the sun of mass M_s is given by

$T_e^2 = \frac{4\pi^2}{GM_s}(r_e)^3$, where r_e is the radius of the earth's orbit.

Similarly, time period T_m of the moon around the earth is given by

$$T_m^2 = \frac{4\pi^2}{GM_e} r_m^3$$

$$\therefore \quad \frac{T_e^2}{T_m^2} = \left(\frac{M_e}{M_s}\right)\left(\frac{r_e}{r_m}\right)^2$$

$$\frac{M_s}{M_e} = \left(\frac{T_m}{T_e}\right)^2 \left(\frac{r_e}{r_m}\right)^3 = \left(\frac{\frac{1}{13}}{1}\right)^2 (392)^3$$

3. (d) From the conservation of angular momentum,

$$mv_1 r_1 \sin 90° = mv_2 r_2 \sin 90°$$

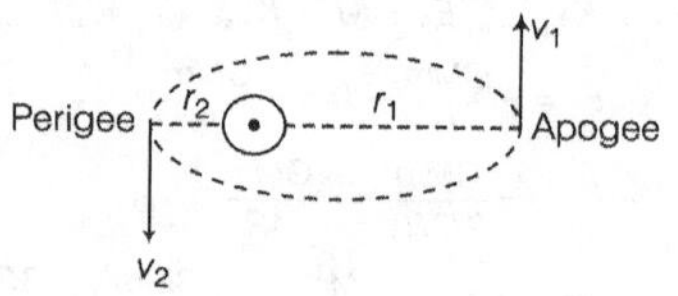

$$\frac{v_1}{v_2} = \frac{r_2}{r_1}$$

$$\Rightarrow \quad v_2 = \left(\frac{r_1}{r_2}\right) v_1 = \left(\frac{1.5 \times 10^7}{0.5 \times 10^7}\right)(5 \times 10^3)$$

$$\text{or} \quad v_2 = 1.5 \times 10^4 \text{ ms}^{-1}$$

4. (c) $\frac{mv^2}{r} = \frac{k}{r^{5/2}}$

$$\Rightarrow \quad v^2 = \frac{k}{mr^{3/2}}$$

$$T = \frac{2\pi r}{v} = 2\pi r\sqrt{\frac{mr^{3/2}}{k}}$$

$$T^2 = \frac{4\pi^2 m}{k} r^{2 + 3/2}$$

$$= \frac{4\pi^2 m}{k} r^{7/2}$$

$$\therefore \quad T^2 \propto r^{7/2}$$

5. (b) $T^2 = \frac{4\pi^2}{GM_s} a^3$, where a is semi-major axis.

$$a = \left[\frac{T^2 GM}{4\pi^2}\right]^{1/3}$$

In case of ellipse, we have

$$2a = r_{max} + r_{min}$$

$$r_{max} = 2a - r_{min}$$

6. (a,b)

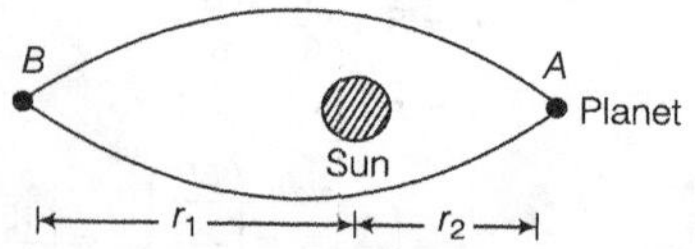

Equating total energy of planet at A and B

$$\frac{1}{2}mr_2^2\,\omega_2^2 - \frac{GMm}{r_2} = \frac{1}{2}m_1 r_1^2 \omega^2 - \frac{GMm}{r_1}$$

$$\Rightarrow \quad \frac{1}{2}\left(\frac{m^2 r_1^4 \omega_1^2}{mr_1^2} - \frac{m^2 r_2^4\,\omega_2^2}{mr_2^2}\right) = GMm\left(\frac{1}{r_1} - \frac{1}{r_2}\right)$$

$$\Rightarrow \quad \frac{1}{2}\left(\frac{I_1\,\omega_1^2}{mr_1^2} - \frac{I_2\,\omega_2^2}{mr_2^2}\right) = GMm\left(\frac{1}{r_1} - \frac{1}{r_2}\right)$$

As, $I_1\,\omega_1 = I_2\,\omega_2 = J$,

We have, $\frac{J^2}{2m}\left(\frac{1}{r_1^2} - \frac{1}{r_2^2}\right) = GMm\left(\frac{1}{r_1} - \frac{1}{r_2}\right)$

$$\Rightarrow \quad J = \sqrt{\frac{2GM_s\, m^2 r_1 r_2}{r_1 + r_2}}$$

Also, total energy of satellite is

$$E = \frac{-GMm}{2a}$$

where, a = semi-major axis

$\therefore$ Options (a) and (b) are correct.

7. (c,d) Orbital speed of satellite is

$$v_0 = \sqrt{\frac{GM}{a}} \quad \text{...(i)}$$

From conservation of angular momentum at P and Q, we have $mav_0 = mvr$

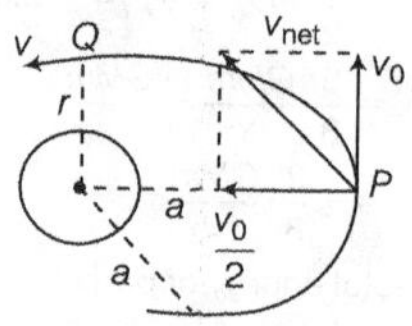

$$v = \frac{av_0}{r} \quad \text{...(ii)}$$

In the given figure, $v_{net} = \sqrt{v_0^2 + \frac{v_0^2}{4}} = \frac{\sqrt{5}}{2}v_0$

From conservation of mechanical energy at P and Q, we have

$$\frac{1}{2}m\left(\frac{\sqrt{5}}{2}v_0\right)^2 - \frac{GMm}{a} = \frac{1}{2}mv^2 - \frac{GMm}{r}$$

$$\frac{5}{8}v_0^2 - \frac{GM}{a} = \frac{v^2}{2} - \frac{GM}{r}$$

Substituting values of v and v_0 from Eqs. (i) and (ii), we get

$$\frac{5}{8} \cdot \frac{GM}{a} - \frac{GM}{a} = \frac{a^2}{r^2} \cdot \frac{1}{2}\left(\frac{GM}{a}\right) - \frac{GM}{r}$$

$$-\frac{3}{8a} = \frac{a}{2r^2} - \frac{1}{r}$$

$$\text{or} \quad -3r^2 = 4a^2 - 8ar$$

$$3r^2 - (8a)r + 4a^2 = 0$$

$$\Rightarrow \quad r = \frac{8a \pm \sqrt{64a^2 - 48a^2}}{6}$$

$$r = \frac{8a \pm 4a}{6}$$

$$\text{or} \quad r = 2a \quad \text{and} \quad r = \frac{2a}{3}$$

8. (b,c) At minimum and maximum distances velocity vector **v** makes an angle of 90° with radius vector. Hence, from conservation of angular momentum $mv_0 r_0 \sin\phi = mvr$ ...(i)

(m is mass of planet)

From energy conservation law, it follows that

$$\frac{1}{2}mv_0^2 - \frac{GMm}{r_0} = \frac{1}{2}mv^2 - \frac{GMm}{r} \quad \text{...(ii)}$$

Solving Eqs. (i) and (ii) for r, we get two values of r, one is r_{max} and another r_{min}, as

$$r_{max} = \frac{r_0}{2-K}\left(1 + \sqrt{1 - K(2-K)\sin^2\phi}\right)$$

and $$r_{min} = \frac{r_0}{2-K}\left(1 - \sqrt{1 - K(2-K)\sin^2\phi}\right)$$

where, $$K = \frac{r_0^2 v_0^2}{GM}$$

9. (a) Statement I and II are correct and Statement II is the correct explanation of Statement I. **F** is on planet and its **r** are anti-parallel to each other.

So, $\tau = 0$. Thus, its angular momentum is conserved.

10. (a) When the planet approaches the farthest point, the angle between force vector and displacement vector is obtuse so work done is negative and thus KE decreases. So, speed decreases.

11. (a) $K = \frac{1}{2}mv_0^2 = \frac{1}{2}m\left(\sqrt{\frac{GM}{r}}\right)^2 = \frac{GMm}{2r}$

$$U = -\frac{GMm}{r}$$

$$E = U + K = -\frac{GMm}{r} + \frac{GMm}{2r} = -\frac{GMm}{2r}$$

$$B = |E| = \frac{GMm}{2r}$$

Hence, (i) → (s), (ii) → (r), (iii) → (p), (iv) → (p)

12. (d) According to Kepler's third law, $T^2 \propto a^3$

Note $T^2 \propto r^3$, *here r is not defined.*

According to Kepler's second law, areal velocity is constant, i.e.

$$\frac{\Delta A}{\Delta t} = \frac{L}{2m} = \text{constant.}$$

In one time period, $\Delta t = T$ and $\Delta A = \pi ab$

$$\therefore \quad \frac{\pi ab}{T} = \frac{L}{2m}$$

$$\Rightarrow \quad L = \frac{2\pi mab}{T} = \text{constant} \quad (\because L \text{ is constant})$$

Explanations for Q. Nos. (13-15)

$$T^2 \propto a^3$$

$$T_e^2 = ka_e^3,\ T_m^2 = ka_m^3,\ T^2 = ka^3$$

$$a = \frac{a_e + a_m}{2}$$

$$a_e = \frac{T_e^{2/3}}{k^{1/3}},\ a_m = \frac{T_m^{2/3}}{k^{1/3}}$$

$$\therefore \quad T^2 = k\left[\frac{1}{2}\left\{\frac{T_e^{2/3}}{k^{1/3}} + \frac{T_m^{2/3}}{k^{1/3}}\right\}\right]^3$$

So, $$T = \left[\frac{T_e^{2/3} + T_m^{2/3}}{2}\right]^{3/2}$$

$$E_e = -\frac{GM_sM_e}{2a_e}$$

$$\Rightarrow \quad M_s = -\frac{2E_e a_e}{GM_e}$$

$$E_{\text{artificial planet}} = E_{ap} = -\frac{GM_sM}{2a} = -\frac{Gm}{2\left(\frac{a_e + a_m}{2}\right)}\left(-\frac{2E_e a_e}{GM_e}\right)$$

$$= \frac{2m}{M_e}\left[\frac{a_e E_e}{a_e + a_m}\right]$$

Areal velocity of the artificial planet around the sun will be more than that of earth.

13. (d) **14.** (a) **15.** (c)

Explanations for Q. Nos. (16-18)

$$\frac{GMm_2}{r_2^2} = m_2\omega_2^2 r_2$$

m_2 = mass of moon II

$$\therefore \quad M = \frac{\omega_2^2 r_2^3}{G} = \left(\frac{2\pi}{T_2}\right)^2 r_2^3 \cdot \frac{1}{G}$$

or $$M = \left(\frac{2\pi}{3\times10^5}\right)^2 (9\times10^7)^3 \left(\frac{1}{6.67\times10^{-11}}\right)$$

$$\therefore \quad M = 4.8\times10^{24}\ \text{kg}$$

$$g_3 = \frac{Gm_3}{R_3^2}$$

$$\Rightarrow \quad 0.2 = \frac{6.67\times10^{-11}\times m_3}{(2\times10^5)^2}$$

or m_3 = mass of moon III $= 1.2\times10^{20}$ kg

Centripetal acceleration of moon II is

$$a_c = \omega_2^2 r_2 = \left(\frac{2\pi}{3\times10^5}\right)^2 (9\times10^7)$$

$$= 0.04\ \text{ms}^{-2}$$

16. (d) **17.** (c) **18.** (b)

19. (2) When the earth's motion is suddenly stopped, it would fall into the sun and (suppose) it comes back. If the effect of temperature of sun is ignored, we can say that the earth would continue to move along a strongly extended flat ellipse whose extreme points are located at the earth's orbit and at the centre of the sun.

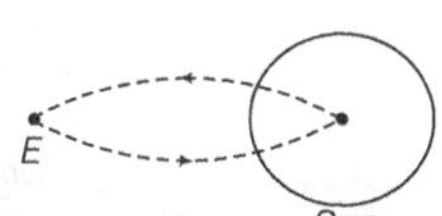

Sun

The semi-major axis of such an ellipse is $\frac{R}{2}$.

$\frac{T'^2}{T^2} = \frac{(R/2)^3}{R^3}$, where T is the time period of normal orbit of the earth.

$$\therefore \quad T'^2 = \frac{T^2}{8} \Rightarrow T' = \frac{T}{2\sqrt{2}}$$

Now, required time is

$$t = \frac{T'}{2} = \frac{T}{4\sqrt{2}} = \frac{365}{4\sqrt{2}} \simeq 65\ \text{days}$$

$$\therefore \quad \frac{130}{k} = 65$$

$$\Rightarrow \quad k = 2$$

20. (3) In a parking orbit, T = 1 day

From Kepler's law, $T^2 \propto R^3$

$$\therefore \quad \frac{T^2}{(1)^2} = \frac{(3R)^3}{(R)^3} \Rightarrow T^2 = 27$$

or $$T = \sqrt{27} = 3\sqrt{3}\ \text{days}$$

$$\Rightarrow \quad x = 3$$

Revisal Problems (JEE Main)

1. (c) The acceleration due to gravity is the measure of the gravitational field. Let the gravitational field is $\frac{1}{4}$th of the field at the surface, at a height h above the surface of the earth.

$$\therefore \quad g_h = \frac{g}{4} = \frac{gR^2}{(R+h)^2} \quad \Rightarrow \quad \frac{1}{4} = \left(\frac{R}{R+h}\right)^2$$

$$\frac{R}{R+h} = \frac{1}{2} \quad \Rightarrow \quad 2R = R + h$$

$$\Rightarrow \quad h = R$$

2. (b) $mg = 100$ N, $\quad g_d = g\left(1 - \frac{d}{R}\right), \frac{R}{2} = d \quad$ (given)

$$\therefore \quad g_d = g\left(1 - \frac{1}{2}\right) = \frac{g}{2}$$

$$\therefore \quad mg_d = \frac{mg}{2} = \frac{100}{2} = 50 \text{ N}$$

3. (a) Field strength on the axis of a ring at a distance x from the centre of the ring is

$$E = \frac{Gmx}{(R^2 + x^2)^{3/2}}$$

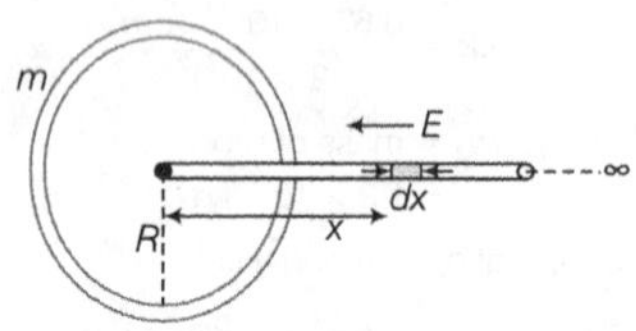

Mass of the element shown is $dm = \lambda dx$

Force on this element will be

$$dF = Edm = \frac{Gmx}{(R^2 + x^2)^{3/2}} \lambda dx$$

$$F = Gm\lambda \int_0^\infty \frac{x}{(R^2 + x^2)^{3/2}} dx = Gm\lambda \left[-\frac{1}{\sqrt{R^2 + x^2}}\right]_0^\infty$$

$$\therefore \quad F = \frac{Gm\lambda}{R}$$

4. (d) The mass of the shell from R_1 to r is

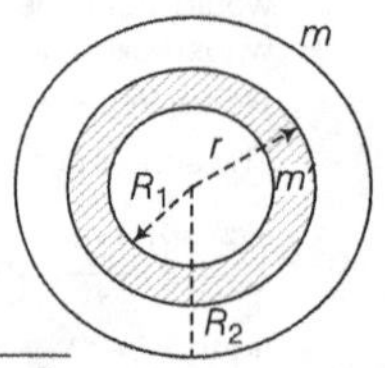

$$m' = \frac{4}{3}\pi(r^3 - R_1^3)\rho$$

where, $\quad \rho = \dfrac{m}{\frac{4}{3}\pi(R_2^3 - R_1^3)}$

$$E_g = \frac{Gm'}{r^2} = \frac{G}{r^2} \cdot \frac{4}{3}\pi(r^3 - R_1^3) \cdot \frac{m}{\frac{4}{3}\pi(R_2^3 - R_1^3)}$$

$$\text{or } E_g = \frac{Gm}{r^2}\left[\frac{r^3 - R_1^3}{R_2^3 - R_1^3}\right]$$

5. (a) Total number of interactions is

$$\frac{n(n-1)}{2} = \frac{8(8-1)}{2} = 28$$

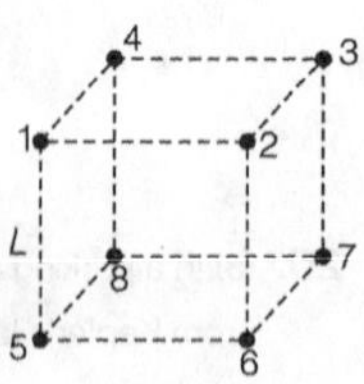

Out of these, 12 interactions are due to masses of separation L, 12 interactions are due to masses of separation $\sqrt{2}L$ (face diagonal) and 4 interactions are due to masses of separation $\sqrt{3}L$ (body diagonal).

$$\therefore U = -12\left(\frac{Gm^2}{L}\right) - 12\left(\frac{Gm^2}{\sqrt{2}L}\right) - 4\left(\frac{Gm^2}{\sqrt{3}L}\right)$$

$$= -\frac{Gm^2}{L}\left[12 + \frac{12}{\sqrt{2}} + \frac{4}{\sqrt{3}}\right] = -\frac{Gm^2}{L}\left[12 + 6\sqrt{2} + \frac{4}{\sqrt{3}}\right]$$

6. (b) Mass of the element is $dm = \frac{m}{L}dx$.

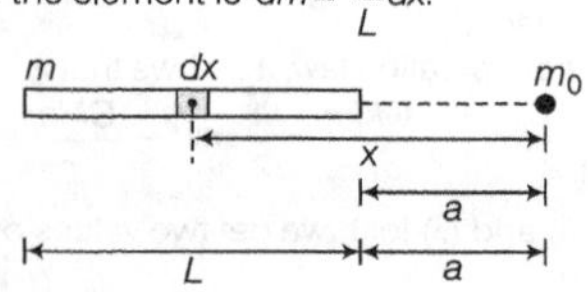

The gravitational potential energy of the element dm and the point mass m_0 is

$$dU = -\frac{Gm_0 dm}{x} = \frac{Gm_0\left(\frac{m}{L}dx\right)}{x}$$

$$\therefore \quad U = \int dU = -\frac{Gmm_0}{L}\int_a^{a+L} \frac{1}{x}dx = -\frac{Gmm_0}{L}\ln\left(\frac{a+L}{a}\right)$$

7. (d) V_P = Potential due to the shell

+ Potential due to the point mass

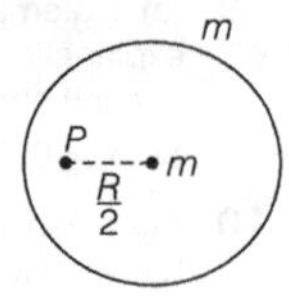

$$= -\frac{GM}{R} + \left(-\frac{GM}{R/2}\right) = -\frac{3GM}{R}$$

8. (c) $V = 2x + 3y - z$ (given)

$$\mathbf{E} = -\left[\frac{\partial V}{\partial x}\hat{\mathbf{i}} + \frac{\partial V}{\partial y}\hat{\mathbf{j}} + \frac{\partial V}{\partial z}\hat{\mathbf{k}}\right]$$

$$\frac{\partial V}{\partial x} = 2, \frac{\partial V}{\partial y} = 3, \frac{\partial V}{\partial z} = -1$$

$$\therefore \quad \mathbf{E} = -(2\hat{\mathbf{i}} + 3\hat{\mathbf{j}} - \hat{\mathbf{k}}) = -2\hat{\mathbf{i}} - 3\hat{\mathbf{j}} + \hat{\mathbf{k}}$$

9. (b) $-\frac{GMm}{R} + \frac{1}{2}mv^2 = -\frac{GMm}{R+h} + 0$ and $g = \frac{GM}{R^2}$

From the above equations, we get

$$h = \frac{v^2}{2g - \frac{v^2}{R}}$$

$$\therefore \quad h = \frac{(4000)^2}{2 \times 9.8 - \frac{(4000)^2}{6400 \times 10^3}} = 9.35 \times 10^5 = 935 \text{ km}$$

10. (a) Gravitational potential of the system is

$$U_i = -\frac{2Gm_1m_2}{R^2}(\sqrt{R^2 + L^2} - L) \quad \text{...(i)}$$

$$\text{or } U_i = -\frac{2Gm_1m_2L}{R^2}\left[\left(1 + \frac{R^2}{L^2}\right)^{1/2} - 1\right]$$

$$\frac{R}{L} << 1 \Rightarrow \left(1 + \frac{R^2}{L^2}\right)^{1/2} = 1 + \frac{R^2}{2L^2}$$

$$\therefore \quad U_i = -\frac{2Gm_1m_2L}{R^2}\left[1 + \frac{R^2}{2L^2} - 1\right] = -\frac{2Gm_1m_2L}{R^2} \cdot \frac{R^2}{2L^2}$$

$$\therefore \quad U_i = -\frac{Gm_1m_2}{L}$$

When the particle collides with the disc at its centre (i.e. $L = 0$) then from Eq. (i), we get

$$U_f = -\frac{2Gm_1m_2}{R}$$

From conservation of energy, $U_i + K_i = U_f + U_f$

$$-\frac{Gm_1m_2}{L} + 0 = -\frac{2Gm_1m_2}{R} + \frac{1}{2}\mu v_r^2$$

where, μ = reduced mass $= \dfrac{m_1m_2}{m_1 + m_2}$

$$\therefore \quad Gm_1m_2\left(\frac{2}{R} - \frac{1}{L}\right) = \frac{1}{2} \cdot \frac{m_1m_2}{m_1 + m_2}v_r^2$$

$$\therefore \quad v_r = \sqrt{2G(m_1 + m_2)\left(\frac{2}{R} - \frac{1}{L}\right)}$$

11. (b) On the line joining two stars, let at point P net gravitational field is zero.

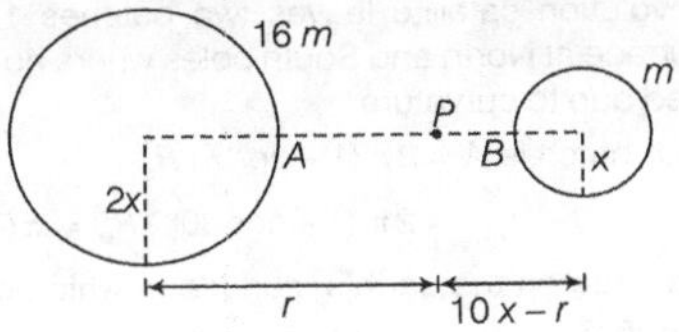

$$\frac{G(16m)}{r^2} = \frac{G(m)}{(10x - r)^2}$$

$$\Rightarrow \quad 4(10x - r) = r$$

$$5r = 40x \Rightarrow r = 8x$$

Once the body reaches at point P, the gravitational pull of attraction due to bigger star vanishes and the gravitational pull due to smaller star takes the lead to make m_0 move towards it automatically.

So, a minimum KE has to be imparted to m_0 from the surface of bigger star such that it is just able to reach point P.

Applying law of conservation of energy between A and P, we get

$$\frac{1}{2} m_0 v_{min}^2 + \left[-\frac{G(16m)m_0}{2x} - \frac{G(m)\,m_0}{8x}\right] = 0 + \left[-\frac{G\,m m_0}{2x} - \frac{G(16m)m_0}{8x}\right]$$

$$\therefore \quad \frac{1}{2} m_0 v_{min}^2 = \frac{G m m_0}{8x}(45) \Rightarrow v_{min} = \frac{3\sqrt{5}}{2}\sqrt{\frac{Gm}{x}}$$

12. (c) $M v_1 r_1 \sin 90^\circ = M v_2 r_2 \sin 90^\circ$

$$v_1 = v_2\left(\frac{r_2}{r_1}\right)$$

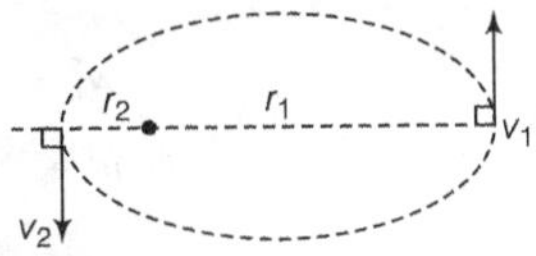

Also,

$$\frac{1}{2} M v_1^2 - \frac{GM_s M}{r_1} = \frac{1}{2} M v_2^2 - \frac{GM_s M}{r_2}$$

$$\frac{1}{2} v_2^2\left(\frac{r_2}{r_1}\right)^2 - \frac{1}{2} v_2^2 = -GM_s\left(\frac{-1}{r_1} + \frac{1}{r_2}\right)$$

$$v_2^2\left[\frac{r_2^2 - r_1^2}{r_1^2}\right] = 2GM_s\left(\frac{1}{r_1} - \frac{1}{r_2}\right)$$

$$v_2^2 \frac{(r_2 + r_1)(r_2 - r_1)}{r_1^2} = 2GM_s \frac{r_2 - r_1}{r_1 r_2}$$

$$v_2^2 = 2GM_s \frac{r_1}{r_2} \cdot \frac{1}{r_1 + r_2} \Rightarrow v_2 = \sqrt{\frac{2GM_s r_1}{(r_1 + r_2) r_2}}$$

$$L = M v_2 r_2 = M\sqrt{\frac{2GM_s r_1 r_2}{r_1 + r_2}}$$

13. (d) $r^2 = b^2 + a^2 e^2$

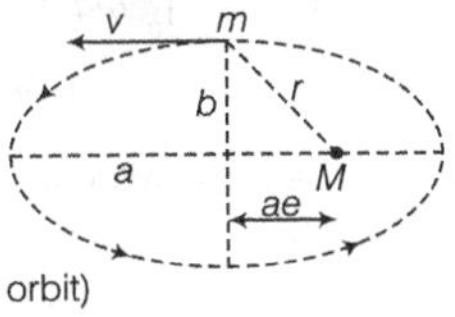

But, $\quad b^2 = a^2(1 - e^2)$

$$\therefore \quad r^2 = a^2e^2 + a^2 - a^2e^2$$

$$\Rightarrow \quad r = a$$

$$E = -\frac{GMm}{2a} \text{ (in elliptical orbit)}$$

At the given point, $E = \frac{1}{2}mv^2 + \left(-\frac{GMm}{r}\right) = \frac{1}{2}mv^2 - \frac{GMm}{a}$

$$\therefore \quad \frac{1}{2}mv^2 - \frac{GMm}{a} = -\frac{G\,Mm}{2a}$$

$$\Rightarrow \quad \frac{1}{2}mv^2 = \frac{1}{2} \cdot \frac{GMm}{a}$$

$$v^2 = \frac{GM}{R^2} \cdot \frac{R^2}{a} = \frac{gR^2}{a}$$

$$\Rightarrow \quad v = R\sqrt{\frac{g}{a}}$$

14. (a) The outer shell has no contribution in the gravitational field at point P.

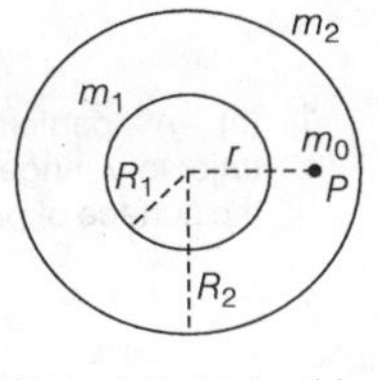

$$\therefore \quad E_P = \frac{Gm_1}{r^2}$$

$\therefore$ Force on m_0 is

$$F = \frac{Gm_1 m_0}{r^2}$$

15. (d) The speed of spaceship in a circular orbit close to the earth's surface is given by $\quad v_o = \sqrt{gR}$

We know that, $\quad v_e = \sqrt{2gR}$

Additional velocity required to escape

$$= v_e - v_o = (\sqrt{2} - 1)\sqrt{gR}$$

16. (c) Applying conservation of angular momentum between points P and A.

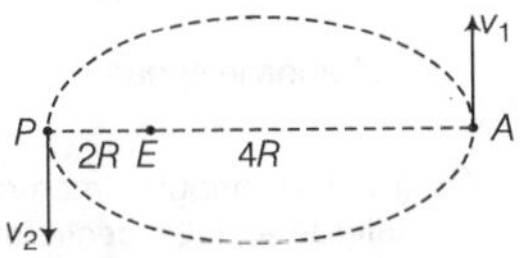

$$mv_2(2R) = mv_1(4R)$$

$$\therefore \quad v_2 = 2v_1 \qquad \ldots\text{(i)}$$

From conservation of energy,

$$\frac{1}{2}mv_2^2 - \frac{GMm}{2R} = \frac{1}{2}mv_1^2 - \frac{GMm}{4R} \qquad \ldots\text{(ii)}$$

From Eqs. (i) and (ii), we get

$$v_1 = \sqrt{\frac{GM}{6R}}, \; v_2 = \sqrt{\frac{2GM}{3R}}$$

Let r be radius of curvature at point P, then

$$\frac{mv_2^2}{r} = \frac{GMm}{(2R)^2}$$

$$\therefore \quad r = \frac{4v_2^2 R^2}{GM} = \frac{4R^2}{GM} \cdot \frac{2GM}{3R} = \frac{8R}{3}$$

17. (c) On m_0, force due to outer shell is zero.

So, F_{net} on m_0 is

$$\frac{G\,m_0 m_1}{x^2} + \frac{G\,m_0 m_2}{x^2} = \frac{Gm_0}{x^2}(m_1 + m_2)$$

18. (d) Angular momentum is conserved. So,

$$\frac{2}{5}MR^2\omega = \frac{2}{5}M\left(\frac{R}{2}\right)^2\omega'$$

$$\omega' = 4\omega$$

$$\Rightarrow \quad \frac{2\pi}{T'} = 4\left(\frac{2\pi}{T}\right)$$

$$\Rightarrow \quad T' = \frac{T}{4} = \frac{24}{4} = 6\text{ h}$$

19. (b) $E_{net} = G(1)\left[\frac{1}{1^2} + \frac{1}{2^2} + \frac{1}{4^2} + \frac{1}{8^2} + \cdots\right]$

$$\left(S_\infty \text{ in a GP} = \frac{a}{1 - r}, \text{ here } a = 1, r = \frac{1}{4}\right)$$

$$\therefore \quad E_{net} = G\left[\frac{1}{1 - \frac{1}{4}}\right] = \frac{4G}{3}$$

20. (d) Potential energy of particle at a height $R = -\frac{G\,Mm}{2R}$

$$= -\frac{mgR}{2} \qquad (\because g = GM/R^2)$$

21. (d) $g = \frac{GM}{R^2} = \frac{G}{R^2} \cdot \frac{4}{3}\pi R^3 \rho = \frac{4}{3}G\pi R\rho$

Greater the value of ρR, greater will be the value of g.

22. (a) On the earth atmosphere is present. So, the balloon due to less density moves up on the earth whereas on the moon there is no atmosphere, so it falls down with $a = \frac{g}{6}$.

Revisal Problems (JEE Advanced)

1. (c) A particle follows a parabolic trajectory tangential to a planet, when at the surface of planet it has a velocity.

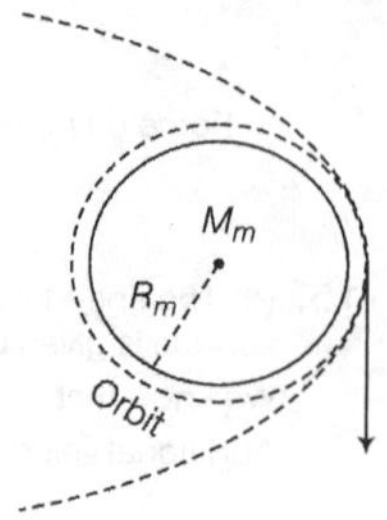

$$\therefore \quad v = \sqrt{\frac{2GM_m}{R_m}}$$

Now, to transform it into a circular orbit, its speed should be decreased to orbital speed.

$$v_o = \sqrt{\frac{GM_m}{R_m}}$$

$$\therefore \text{Change in speed} = v - v_o = \sqrt{\frac{GM_m}{R_m}}(\sqrt{2} - 1)$$

2. (b) The angular momentum of the satellite w.r.t. the centre of orbit is

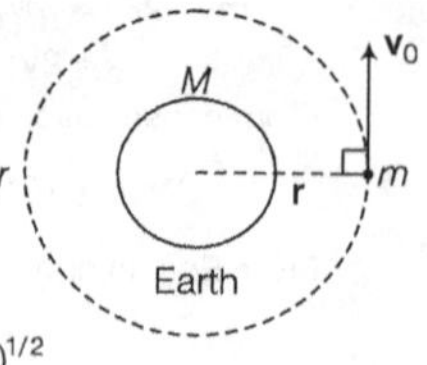

$$L = \mathbf{r} \times m\mathbf{v}_o$$

$$L = mv_o r \sin 90° = mv_o r$$

Orbital speed is $v_o = \sqrt{\frac{GM}{r}}$

$$\therefore \quad L = m\sqrt{\frac{GM}{r}}r = (GMm^2r)^{1/2}$$

3. (a) In this process during displacement, the size of sphere is also changing. So, we have to account for the self energy of sphere also.

The total initial energy of the system is

U_i = Self energy of m_0 + Self energy of m + Interaction energy of m_0 and m

$$= 0 + \left(-\frac{3}{5}\frac{Gm^2}{r}\right) + \left(-\frac{Gmm_0}{5r}\right)$$

The new situation is shown in figure.

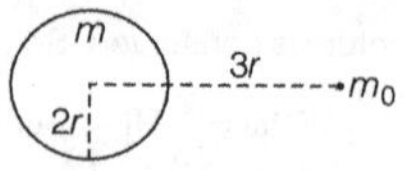

The total final energy of the system is

$$U_f = 0 + \left(-\frac{3}{5}\frac{Gm^2}{2r}\right) + \left(-\frac{Gmm_0}{3r}\right)$$

$$W = U_f - U_i = \left[-\frac{3}{5}\frac{Gm^2}{r} - \frac{Gmm_0}{5r}\right] - \left[-\frac{3}{5}\frac{Gm^2}{2r} - \frac{Gmm_0}{3r}\right]$$

$$= \frac{Gm}{30r}(9m - 4m_0)$$

4. (d) For the figure shown, when the satellite revolves, it covers a complete circular belt on the earth's surface for communication. If the colatitude of the farthest point (point F) on surface up to which signals can be received is λ, then we have

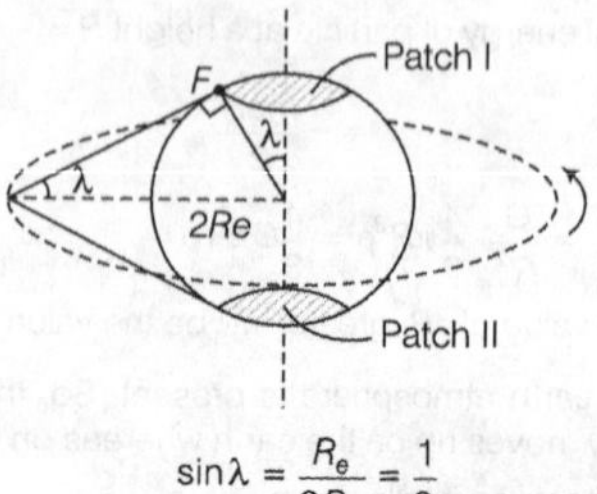

$$\sin\lambda = \frac{R_e}{2R_e} = \frac{1}{2}$$

$$\therefore \quad \lambda = 30°$$

During revolution satellite leaves two patches 1 and 2 on the earth's surface at North and South poles where no signals can be transmitted due to curvature.

Area of one patch is $A = 2\pi(1 - \cos\lambda)R_e^2$

$$= 2\pi(1 - \cos 30°)R_e^2 = \pi(2 - \sqrt{3})R_e^2$$

Thus, total area on the earth's surface to which communication can be made is

$$A_c = 4\pi R_e^2 - 2A$$

$$= 4\pi R_e^2 - 2\pi(2 - \sqrt{3})R_e^2 = 2\pi R_e^2[2 - 2 + \sqrt{3}]$$

$$= 2\sqrt{3}R_e^2$$

5. (c) $\omega_{\text{satellite}} = \omega_{\text{earth}}$

Time period of satellite,

$$T = 24\text{ h}$$

Required time is

$$t = \frac{T}{4} = 6\text{ h}$$

6. (a) From conservation of angular momentum,

$$mv_0l_0 = mvr \Rightarrow v = \frac{v_0l_0}{r}$$

From conservation of mechanical energy

$$\frac{1}{2}mv_0^2 = -\frac{GM_sm}{r} + \frac{1}{2}m\left(\frac{v_0l_0}{r}\right)^2$$

$$v_0^2 = \frac{-2GM_sr + v_0^2l_0^2}{r^2}$$

$$\Rightarrow \quad (v_0^2)r^2 + (2GM_s)r - (v_0^2l_0^2) = 0$$

$$r = \frac{-2GM_s \pm \sqrt{4G^2M_s^2 + 4(v_0^2)(v_0^2l_0^2)}}{2v_0^2}$$

Neglecting negative sign.

$$r_{\min} = \frac{GM_s}{v_0^2}\left[-1 + \sqrt{1 + \left(\frac{v_0^2l_0}{GM_s}\right)^2}\right]$$

or

$$r_{\min} = \frac{GM_s}{v_0^2}\left[\sqrt{1 + \left(\frac{v_0^2l_0}{GM_s}\right)^2} - 1\right]$$

7. (c) $v_i = 0$ (given). At maximum height, $v_f = 0$

$\therefore$ Work done by all forces = 0 (from work-energy theorem)

So, $\int_0^h (\mathbf{F} + mg)\,dy = 0$

$$\Rightarrow \quad \int_0^h mg(1 - 2ky)\,dy = 0 \Rightarrow mg(h - kh^2) = 0$$

$$mgh(1 - kh) = 0 \Rightarrow h = \frac{1}{k}$$

The work performed by force **F** over the first half of the ascent is

$$W = \int_0^{h/2} F\,dy = \int_0^{1/2k} 2mg(1 - ky)\,dy = 2mg\left(\frac{1}{2k} - \frac{1}{8k}\right)$$

or $\quad W = \frac{3mg}{4k}$

8. (b) Centre point is the unstable equilibrium position, where potential energy is maximum.

9. (a,d) Time period in both the cases is

$$T_1 = T_2 = 2\pi\sqrt{\frac{R}{g}} = 84.6\text{ min}$$

However, $v_1 > v_2$, because the difference in potential energy between the extreme position and mean position will be more in the first case.

10. (a,c) $E_g = -\dfrac{dV}{dr}$

$E_g = 0 \Rightarrow V$ is constant and this constant may be zero.

11. (a,b,c)

Spherical shell   Solid sphere

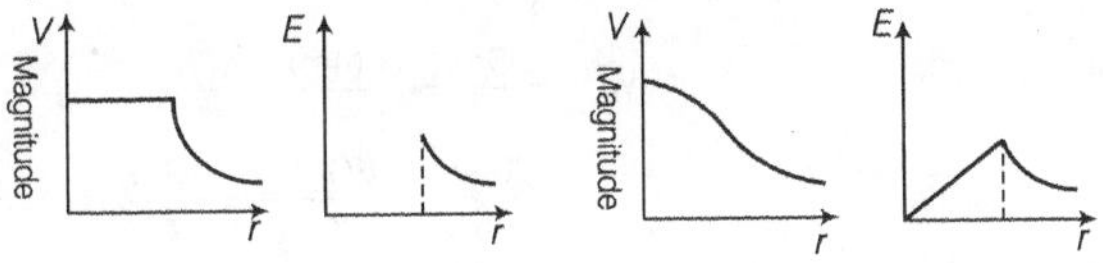

12. (a,c)

$$t = nT \Rightarrow T = \frac{t}{n} = \frac{40}{20} = 2\,\text{s}$$

$$\frac{\Delta t}{t} = \frac{\Delta T}{T} \Rightarrow \frac{1}{40} = \frac{\Delta T}{2}$$

$$\Rightarrow \quad \Delta T = \frac{2}{40} = \frac{1}{20} = 0.05\,\text{s}$$

$$T = 2\pi\sqrt{\frac{l}{g}} \Rightarrow \quad \frac{\Delta T}{T} = -\frac{1}{2}\cdot\frac{\Delta g}{g}$$

$$-\frac{\Delta g}{g} = 2\cdot\frac{\Delta T}{T} = \frac{2\,(0.05)}{2} = 0.05$$

$$-\frac{\Delta g}{g} \times 100 = 5$$

$\therefore$ Percentage error in $g = 5\%$

13. (b,c) At all the places, potential will be increased by $\dfrac{GM}{R}$.

So,
$$V_\infty = 0 + \frac{GM}{R} = \frac{GM}{R}$$

$$V_{\text{centre}} = -\frac{3GM}{2R} + \frac{GM}{R} = -\frac{GM}{2R}$$

14. (d) $T^2 \propto r^3 \Rightarrow \left(\dfrac{T_1}{T_2}\right)^2 = \left(\dfrac{r_1}{r_2}\right)^3$

$$\frac{T_1}{T_2} = \left(\frac{r_1}{r_2}\right)^{3/2}$$

$$\Rightarrow \quad T_2 = T_1\left(\frac{r_2}{r_1}\right)^{3/2} = 1 \times \left(\frac{6}{15}\right)^{3/2} \quad \text{or} \quad T_2 = \left(\frac{2}{5}\right)^{3/2}\text{yr}$$

15. (a) $v_o \propto \dfrac{1}{\sqrt{r}} \Rightarrow \dfrac{(v_o)_1}{(v_o)_2} = \sqrt{\dfrac{r_2}{r_1}}$

$$\frac{(v_o)_2}{(v_o)_1} = \sqrt{\frac{r_1}{r_2}} = \sqrt{\frac{15}{6}} = \sqrt{\frac{5}{2}} = \sqrt{\frac{10}{4}}$$

16. (c) If we consider origin at the heavier mass, the centre of mass is at a distance $\dfrac{L}{3}$ from the origin. So,the heavier star revolves in an orbit of radius $\dfrac{L}{3}$.

17. (c) For lighter star, $\dfrac{G\,(m)(2m)}{L^2} = m\left(\dfrac{2L}{3}\right)\omega^2$ ...(i)

For heavier star, $\dfrac{G\,(m)(2m)}{L^2} = (2m)\left(\dfrac{L}{3}\right)\omega^2$ ...(ii)

On adding Eqs. (i) and (ii), we get

$$\frac{4Gm^2}{L^2} = \frac{4}{3}mL\omega^2 \quad \Rightarrow \quad \omega = \sqrt{\frac{3Gm}{L^3}}$$

18. (a) $K_L = \dfrac{1}{2}m\left(\dfrac{2L}{3}\omega\right)^2 = \dfrac{2}{9}mL^2\omega^2$

$$K_H = \frac{1}{2}(2m)\left(\frac{L}{3}\omega\right)^2 = \frac{1}{9}mL^2\omega^2$$

$$K_L + K_H = \left(\frac{2}{9} + \frac{1}{9}\right)mL^2\omega^2 = \frac{mL^2\omega^2}{3}$$

19. (4) Consider only left rod (in the absence of right rod). Consider a hypothetical point P as shown.

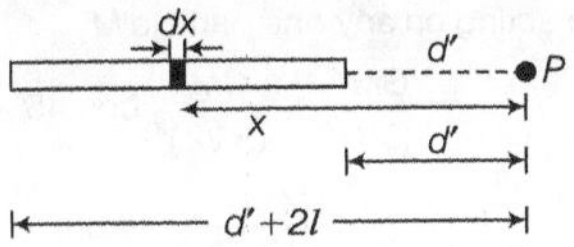

Gravitational field due to element, at P is

$$dE_g = \frac{G\,dm}{x^2} = \frac{G\dfrac{m}{2l}dx}{x^2}$$

At P, $\quad E_g = \dfrac{Gm}{2l}\displaystyle\int_{d'}^{d'+2l}\frac{1}{x^2}dx = -\frac{Gm}{2l}\left[\frac{1}{x}\right]_{d'}^{d'+2l}$

$$\therefore \quad E_g = \frac{Gm}{2l}\left[\frac{1}{d'} - \frac{1}{d'+2l}\right] = \frac{Gm}{2l}\cdot\frac{d'+2l-d'}{d'(d'+2l)}$$

or $\quad E_g = \dfrac{Gm}{d'(d'+2l)}$ ...(i)

Now, let us consider two rods.

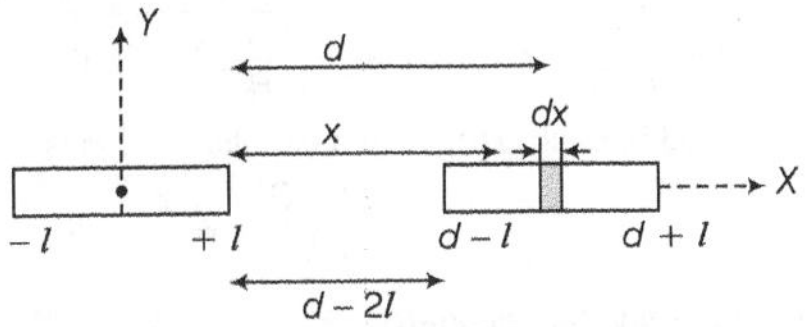

Consider an element of length dx on right side rod. This element is at a distance x from right end of left rod. Field due to left rod at the location of this element is $\dfrac{Gm}{x(x+2l)}$ by using Eq. (i),

Force on this element is

$$dF = E_g dm = \frac{Gm}{x(x+2l)}\cdot\frac{m}{2l}dx$$

or $\quad dF = \dfrac{Gm^2}{2l}\cdot\dfrac{1}{x(x+2l)}dx$

$$F_{\text{net}} = \frac{Gm^2}{2l}\int_{d-2l}^{d}\frac{1}{x(x+2l)}dx = \frac{Gm^2}{2l}\left[-\frac{1}{2l}\ln\left(\frac{2l+x}{x}\right)\right]_{d-2l}^{d}$$

$$F_{\text{net}} = \frac{Gm^2}{4l^2}\ln\left(\frac{d^2}{d^2-4l^2}\right)$$

$\Rightarrow \beta = 4$

20. (9) When the particle is at the centre, let their respective speeds are v_1 and v_2.

From conservation of linear momentum,

$$0 = 5.4 \times 10^9 v_1 - 6 \times 10^8 v_2$$

$\therefore \quad v_2 = 9v_1$

From conservation of energy,

$$(U+K)_i = (U+K)_f$$

$$-\frac{Gm_1m_2}{\sqrt{4^2+3^2}} + 0 = -\frac{Gm_1m_2}{4} + \frac{1}{2}m_1v_1^2 + \frac{1}{2}m_2v_2^2$$

$$Gm_1m_2\left(\frac{1}{4} - \frac{1}{5}\right) = \frac{1}{2}m_1v_1^2 + \frac{1}{2}m_2\,(81v_1^2)$$

$$G\,\frac{(9m_2^2)}{20} = \frac{1}{2}\,(90m_2)\,v_1^2$$

Note $m_1 = 9m_2$ (given)

$$v_1 = \sqrt{\frac{Gm_2}{100}} = \sqrt{\frac{6.67 \times 10^{-11} \times 6 \times 10^8}{100}} = 0.02\,\text{ms}^{-1}$$

$$\therefore \quad v_2 = 9v_1 = 0.18\,\text{ms}^{-1} = 18\,\text{cms}^{-1}$$

$$= 2 \times 9\,\text{cms}^{-1} \Rightarrow \alpha = 9$$

JEE Main & AIEEE Archive

1. (d) Net force acting on any one particle M

$$= \frac{GM^2}{(2R)^2} + \frac{GM^2}{(R\sqrt{2})^2}\cos 45^\circ + \frac{GM^2}{(R\sqrt{2})^2}\cos 45^\circ$$

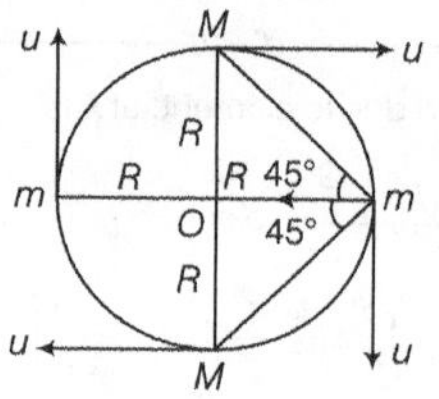

$$= \frac{GM^2}{R^2}\left(\frac{1}{4} + \frac{1}{\sqrt{2}}\right)$$

This force will equal to centripetal force.

So, $\frac{Mu^2}{R} = \frac{GM^2}{R^2}\left(\frac{1}{4} + \frac{1}{\sqrt{2}}\right)$

$\Rightarrow \quad u = \sqrt{\frac{GM}{4R}(1 + 2\sqrt{2})} = \frac{1}{2}\sqrt{\frac{GM}{R}(2\sqrt{2} + 1)}$

Hence, speed of each particle in a circular motion is

$$= \frac{1}{2}\sqrt{\frac{GM}{R}(2\sqrt{2} + 1)}$$

2. (a) From conservation of energy,

Total energy at the planet = Total energy at altitude

$$-\frac{GMm}{R} + (KE)_{surface} = -\frac{GMm}{3R} + \frac{1}{2}mv_A^2 \quad \ldots(i)$$

In the orbital of planet, the necessary centripetal force is obtained by gravitational force.

$\therefore \quad \frac{mv_A^2}{R + 2R} = \frac{GMm}{(R + 2R)^2}$

$\Rightarrow \quad \frac{v_A}{2} = \frac{GM}{3R} \quad \ldots(ii)$

From Eqs. (i) and (ii), we get

$$(KE)_{surface} = \frac{5}{6}\cdot\frac{GMm}{R}$$

3. (d) Potential energy on the earth's surface is $-mgR$ while in free space, it is zero. So, to free the spaceship, minimum required energy is

$$K = mgR = 10^3 \times 10 \times 6400 \times 10^3 \text{ J} = 6.4 \times 10^{10} \text{ J}$$

4. (b) As, gravitational force provides necessary centripetal force.

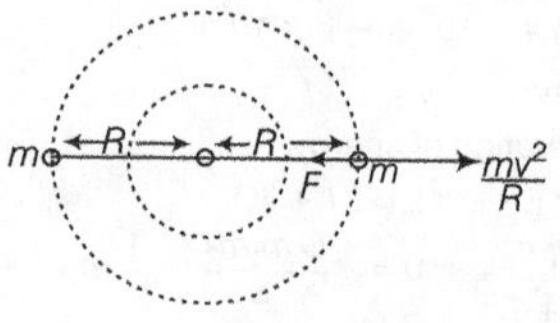

i.e. $\quad F = \frac{Gm^2}{(2R)^2} = \frac{mv^2}{R}$

$\Rightarrow \quad v = \sqrt{\frac{Gm}{4R}}$

5. (c) Let gravitational field be zero at point P as shown in the figure.

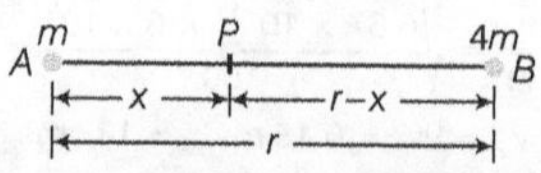

$\therefore \quad \frac{Gm}{x^2} = \frac{G(4m)}{(r - x)^2}$

$\Rightarrow \quad 4x^2 = (r - x)^2$

$\Rightarrow \quad 2x = r - x$

$\Rightarrow \quad x = \frac{r}{3}$

$\therefore \quad V_P = -\frac{Gm}{x} - \frac{G(4m)}{r - x}$

$$= -\frac{3Gm}{r} - \frac{6GM}{r} = -\frac{9Gm}{r}$$

6. (a) As, the value of acceleration due to gravity above the earth's surface is g.

$g' = \frac{GM}{(R + h)^2}$, acceleration due to gravity at height h

$\Rightarrow \quad \frac{g}{9} = \frac{GM}{R^2}\cdot\frac{R^2}{(R + h)^2} = g\left(\frac{R}{R + h}\right)^2$

$\Rightarrow \quad \frac{1}{9} = \left(\frac{R}{R + h}\right)^2$

$\Rightarrow \quad \frac{R}{R + h} = \frac{1}{3}$

$\Rightarrow \quad 3R = R + h$

$\Rightarrow \quad 2R = h$

7. (c) Mass of planet, $M_p = 10M_e$, where M_e is mass of earth.

Radius of planet, $R_p = \frac{R_e}{10}$, where R_e is radius of earth.

Escape velocity is given by $v_e = \sqrt{\frac{2GM}{R}}$

For planet, $\quad v_p = \sqrt{\frac{2G \times M_p}{R_p}} = \sqrt{\frac{100 \times 2GM_e}{R_e}}$

$$= 10 \times v_e = 10 \times 11 = 110 \text{ kms}^{-1}$$

8. (a) Correct option is (a), you can make an analogy with Gauss law in electrostatics.

9. (a) According to Millikan's oil drop experiment, electronic charge is given by

$$q = \frac{6\pi\eta r(v_1 + v_2)}{E}$$

which is independent of g.

So, $\quad \frac{\text{electronic charge on the moon}}{\text{electronic charge on the earth}} = 1$

10. (c) Mass of earth = Volume × Density

As, $\quad g = \frac{GM}{R^2}, \; M = \left(\frac{4}{3}\pi R^3\right)\rho$

$\therefore \; g = \frac{4G}{3}\cdot\frac{\pi R^3}{R^2}\rho$ or $g = \left(\frac{4G\pi R}{3}\right)\rho \quad (\because \rho = \text{average density})$

$\Rightarrow \quad g \propto \rho$ or $\rho \propto g$

11. (c) Acceleration due to gravity at height h.

$$g_h = g\left(1 - \frac{2h}{R}\right) \quad (\text{if } h << R) \ldots(i)$$

Acceleration due to gravity at depth d.

$$g_d = g\left(1 - \frac{d}{R}\right) \quad \ldots(ii)$$

As per statement of the problem, $g_h = g_d$

i.e. $\quad g\left(1 - \frac{2h}{R}\right) = g\left(1 - \frac{d}{R}\right)$

$\Rightarrow \quad 2h = d$

12. (d) The initial potential energy of the particle = Work done

$$U_i = -\frac{GMm}{r}$$

$$U_i = -\frac{6.67 \times 10^{-11} \times 100 \times 10^{-2}}{0.1}$$

$$U_i = -\frac{6.67 \times 10^{-11}}{0.1} = -6.67 \times 10^{-10}\ \text{J}$$

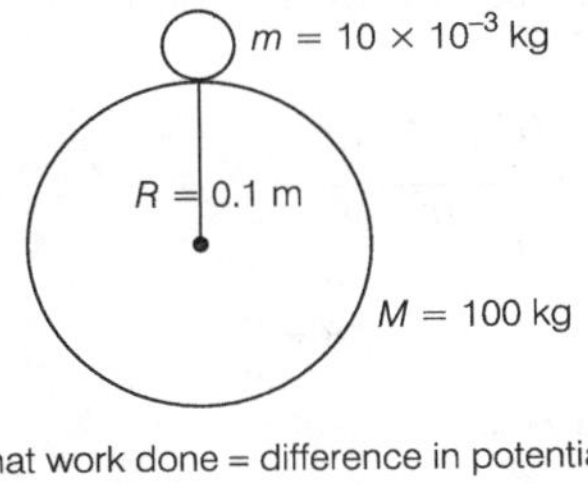

We know that work done = difference in potential energy

$\therefore \quad W = \Delta U = U_f - U_i$

$\Rightarrow \quad W = -U_i \qquad (\because U_f = 0)$

$= 6.67 \times 10^{-10}$ J

13. (d) The gravitational force exerted on satellite at a height x is

$$F_G = \frac{GM_e m}{(R+x)^2}$$

where, M_e = mass of the earth

Since, gravitational force provides the necessary centripetal force, so

$$\frac{GM_e m}{(R+x)^2} = \frac{mv_o^2}{(R+x)}$$

where, v_o is orbital speed of satellite.

$$\Rightarrow \quad \frac{GM_e m}{(R+x)} = mv_o^2$$

$$\text{or} \quad \frac{gR^2 m}{(R+x)} = mv_o^2 \qquad \left(\because g = \frac{GM_e}{R^2}\right)$$

$$\text{or} \quad v_o = \sqrt{\left[\frac{gR^2}{(R+x)}\right]} = \left[\frac{gR^2}{(R+x)}\right]^{1/2}$$

14. (a) Time period of satellite,

$$T = 2\pi\sqrt{\frac{(R+h)^3}{GM_e}}$$

where, $R + h$ = orbital radius of satellite

and M_e = mass of earth

Thus, time period does not depend on the mass of satellite.

15. (b) Gravitational potential energy of body on the earth's surface is

$$U = -\frac{GM_e m}{R}$$

At a height h from earth's surface, its value is

$$U_h = -\frac{GM_e m}{(R+h)} = -\frac{GM_e m}{2R} \qquad (\text{as } h = R)$$

where, M_e = mass of earth

m = mass of body

and R = radius of earth

$$\therefore \text{ Gain in potential energy} = U_h - U = -\frac{GM_e m}{2R} - \left(-\frac{GM_e m}{R}\right)$$

$$= -\frac{GM_e m}{2R} + \frac{GM_e m}{R} = \frac{GM_e m}{2R} = \frac{gR^2 m}{2R} \qquad \left(\text{as } g = \frac{GM_e}{R^2}\right)$$

$$= \frac{1}{2} mgR$$

16. (a) The necessary centripetal force required for a planet to move round the sun = Gravitational force exerted on it

$$\text{i.e.} \quad \frac{mv^2}{R} = \frac{GM_e m}{R^n} \quad \text{or} \quad v = \left(\frac{GM_e}{R^{n-1}}\right)^{1/2}$$

$$\text{Now,} \quad T = \frac{2\pi R}{v} = 2\pi R \times \left(\frac{R^{n-1}}{GM_e}\right)^{1/2}$$

$$= 2\pi\left(\frac{R^2 \times R^{n-1}}{GM_e}\right)^{1/2} = 2\pi\left(\frac{R^{(n+1)/2}}{(GM_e)^{1/2}}\right)$$

$$\text{or} \quad T \propto R^{(n+1)/2}$$

17. (c) According to Kepler's law,

$$T^2 \propto r^3$$

$$\Rightarrow \quad 5^2 \propto r^3 \qquad \text{...(i)}$$

$$\text{and} \quad (T')^2 \propto (4r)^3 \qquad \text{...(ii)}$$

From Eqs. (i) and (ii), we have

$$\frac{25}{(T')^2} = \frac{r^3}{64r^3} \quad \text{or} \quad T' = \sqrt{1600}$$

$$\text{or} \quad T' = 40 \text{ h}$$

18. (c) Just before collision, total distance travelled by both

$$= 12R - R - 2R = 9R$$

If x is the distance travelled by block of mass M and $9R - x$, the distance travelled by block of mass $5M$.

As centre of mass of system not change, then

$$Mx = 5M(9R - x)$$

$$\Rightarrow \quad x = 7.5R$$

19. (c) The escape velocity is independent of angle of projection, hence it will remain same, i.e. 11 kms^{-1}.

20. (d) Gravitational potential energy of body will be

$$E = -\frac{GMm}{r}$$

where, M = mass of earth

m = mass of the body

and R = radius of earth

At $r = 2R$,

$$E_1 = -\frac{GMm}{(2R)}$$

At $r = 3R$,

$$E_2 = -\frac{GMm}{(3R)}$$

Energy required to move a body of mass m from an orbit of radius $2R$ to $3R$ is

$$\Delta E = \frac{GMm}{R}\left[\frac{1}{2} - \frac{1}{3}\right] = \frac{GMm}{6R}$$

21. (c) The minimum kinetic energy required to project a body of mass m from earth's surface to infinity is known as escape energy. Therefore,

$$\text{KE} = \frac{GM_e m}{R} = mgR \qquad \left(\because gR = \frac{GM_e}{R}\right)$$

22. (a) Escape velocity $= \sqrt{2gR_e}$

So, escape velocity is independent of m. So, it depends upon the mass of m as m^0.

23. (b) When gravitational force becomes zero, then centripetal force required cannot be provided. So, the satellite will move with the velocity as it has at the instant when gravitational force becomes zero, i.e. it moves tangentially to the original orbit.

1. (b) Given, $R_{planet} = \frac{R_{earth}}{10}$

and density, $\rho = \frac{M_{earth}}{\frac{4}{3}\pi R^3_{earth}} = \frac{M_{planet}}{\frac{4}{3}\pi R^3_{planet}}$

$\Rightarrow \quad M_{planet} = \frac{M_{earth}}{10^3}$

$$g_{\text{surface of planet}} = \frac{GM_{planet}}{R^2_{planet}} = \frac{GM_e \cdot 10^2}{10^3 \cdot R_e^2} = \frac{GM_e}{10R_e^2} = \frac{g_{\text{surface of earth}}}{10}$$

$$g_{\text{depth of planet}} = g_{\text{surface of planet}}\left(\frac{x}{R}\right)$$

where, x = distance from centre of planet

$\therefore$ Total force on wire,

$$F = \int_{4R/5}^{R} \lambda\, dx\, g\left(\frac{x}{R}\right) = \frac{\lambda g}{R}\left[\frac{x^2}{2}\right]_{4R/5}^{R}$$

Here, $g = g_{\text{surface of planet}}$

$R = R_{planet}$

Substituting the given values, we get

$$F = 108 \text{ N}$$

2. (b,d)

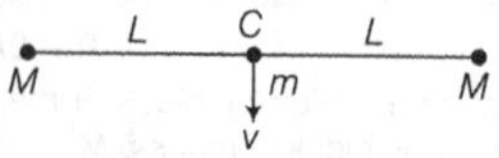

Let v is the minimum velocity. From energy conservation,

$$U_C + K_C = U_\infty + K_\infty$$

$\therefore \quad mv_C + \frac{1}{2}mv^2 = 0 + 0$

$\therefore \quad v = \sqrt{-2v_C}$

$$= \sqrt{(-2)\left(\frac{-2GM}{L}\right)} = 2\sqrt{\frac{GM}{L}}$$

The energy of the mass m remains constant.

3. (b,d) Surface area of Q is four times. Therefore, radius of Q is two times and volume is eight times. Therefore, mass of Q is also eight times.

So, let $M_P = M$ and $R_P = r$

Then, $M_Q = 8\,M$ and $R_Q = 2r$

Now, mass of R is $(M_P + M_Q)$ or $9\,M$. Therefore, radius of R is $(9)^{1/3}r$. Now, escape velocity from the surface of a planet is given by

$$v = \sqrt{\frac{2GM}{r}} \quad (r = \text{radius of that planet})$$

$\therefore \quad v_P = \sqrt{\frac{2GM}{r}} \Rightarrow v_Q = \sqrt{\frac{2G(8M)}{(2r)}}$

$$v_R = \sqrt{\frac{2G(9M)}{(9)^{1/3}r}}$$

From here, we can see that

$$\frac{v_P}{v_Q} = \frac{1}{2}$$

and $\quad v_R > v_Q > v_P$

4. (b) In circular orbit of a satellite, potential energy

$$= -2 \times (\text{kinetic energy})$$
$$= -2 \times \frac{1}{2}mv^2 = -mv^2$$

Just to escape from the gravitational pull, its total mechanical energy should be zero. Therefore, its kinetic energy should be positive mv^2.

5. (3) $g = \frac{GM}{R^2} = \frac{G\left(\frac{4}{3}\pi R^3\right)\rho}{R^2}$

or $\quad g \propto \rho R$

or $\quad R \propto \frac{g}{\rho}$

Now, escape velocity, $v_e = \sqrt{2gR}$

or $\quad v_e \propto \sqrt{gR}$

or $\quad v_e \propto \sqrt{g \times \frac{g}{\rho}} \propto \sqrt{\frac{g^2}{\rho}}$

$\therefore \quad (v_e)_{planet} = (11 \text{ kms}^{-1})\sqrt{\frac{6}{121} \times \frac{3}{2}}$

$$= 3 \text{ kms}^{-1}$$

6. (6) $\frac{L_{Total}}{L_B} = \frac{(I_A + I_B)}{I_B \cdot \omega}$ (as ω will be same in both cases)

$= \frac{I_A}{I_B} + 1 = \frac{m_A r_A^2}{m_B r_B^2} + 1 = \frac{r_A}{r_B} + 1$ (as, $m_A r_A = m_B r_B$)

$= \frac{11}{2.2} + 1$ $\left(\text{as, } r \propto \frac{1}{m}\right)$

$= 6$

7. (c) For $r \le R$,

$$\frac{mv^2}{r} = \frac{GmM}{r^2} \quad \ldots(i)$$

Here, $\quad M = \left(\frac{4}{3}\pi r^3\right)\rho_0$

Substituting the value in Eq. (i), we get

$$v \propto r$$

i.e. v-r graph is a straight line passing through the origin.

For $r > R$,

$$\frac{mv^2}{r} = \frac{Gm\left(\frac{4}{3}\pi R^3\right)\rho_0}{r^2}$$

or $\quad v \propto \frac{1}{\sqrt{r}}$

The corresponding v-r graph will be as shown in option (c).

8. (a) Force acting on astronaut is utilised in providing necessary centripetal force, thus he feels weightlessness, as he is in a state of free fall.

9. (d) In case of binary star system, angular velocity and hence the time period of both the stars are equal.

4. Elasticity

DPP-1 Stresses and Strains

1. (b) L = Original length = 2 m

$$\Delta L = \text{Increase in length}$$
$$= 1\,\text{cm} = 1 \times 10^{-2}\,\text{m}$$

Longitudinal strain is

$$\varepsilon = \frac{\Delta L}{L} = \frac{1 \times 10^{-2}}{2}$$
$$= 5 \times 10^{-3} = 0.005$$

2. (a)

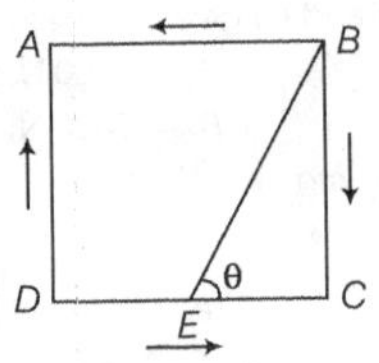

Consider a wedge BEC, which is shown. Its free body diagram is drawn.

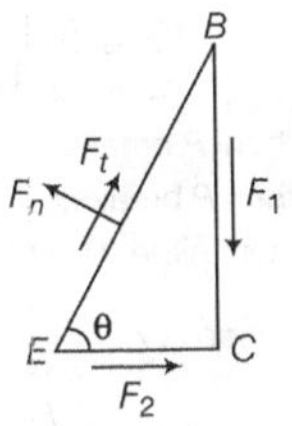

The wedge BEC is acted upon by the following forces :

(i) $F_1 = \sigma\,(BC \times 1)$ along the face BC.

(ii) $F_2 = \sigma\,(EC \times 1)$ along the face EC.

(iii) F_n normal to face EB.

(iv) F_t tangential to face EB.

Resolving these forces normal and along the plane EB,

$$F_n = \sigma(BC)\cos\theta + \sigma(EC)\sin\theta$$
$$F_t = \sigma(EC)\cos\theta - \sigma(BC)\sin\theta$$

Area of the section $BE = BE \times 1$

$$\therefore \text{ Normal stress} = \frac{F_n}{BE}$$
$$= \frac{\sigma(BC)\cos\theta + \sigma(EC)\sin\theta}{BE}$$
$$= \sigma\sin\theta\cos\theta + \sigma\cos\theta\sin\theta$$
$$= \sigma\sin 2\theta$$

For the plane to carry maximum normal stress, $\sin 2\theta$ should be maximum, i.e. 1.

$\therefore$ Maximum normal stress = σ

Note *This plane (with maximum normal stress) makes $\theta = 45°$ with DC.*

3. (d) The net pressure at a depth of 1 km in the ocean is

$$p = \rho g h + p_{atm}$$
$$= (10^3)(9.8)(10^3) + 10^5$$
$$= 99 \times 10^5\,\text{Pa} = 9.9 \times 10^6\,\text{Pa}$$

This pressure acts uniformly on all sides of the balloon (which is in equilibrium) and the restoring forces with in the balloon are equal to external forces. So, normal stress will be same as external pressure.

4. (a) Let the length of unstretched wire is L_0.

$$T\sin\theta = m\omega^2 r = m(4\pi)^2 r = 16\pi^2 mr \quad \ldots\text{(i)}$$
$$T\cos\theta = mg \quad \ldots\text{(ii)}$$

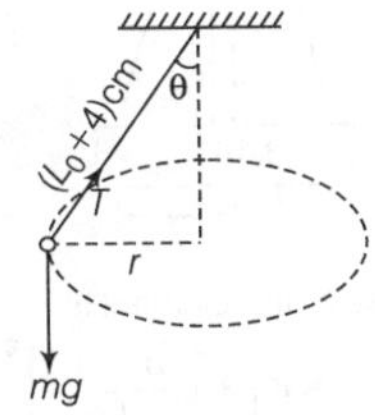

On dividing Eq. (i) by Eq. (ii), we get

$$\tan\theta = 16\pi^2 \frac{r}{g}$$

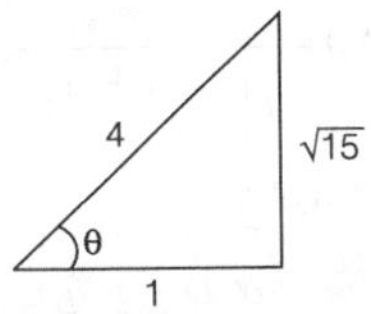

The wire is elastic. So, it obeys Hooke's law.

$\therefore$ Initially, $mg = k(1)$

$\Rightarrow$ $k = mg$

When the mass rotates,

$T = 4k = 4mg$ and from Eq. (ii), we have

$\therefore$ $4mg\cos\theta = mg$

$\Rightarrow$ $\cos\theta = \dfrac{1}{4}$

$\therefore$ $\tan\theta = \sqrt{15}$

So, $\sqrt{15} = 16\pi^2 \dfrac{r}{g}$

$$\Rightarrow \quad r = \frac{g\sqrt{15}}{16\pi^2} = \frac{(981)(\sqrt{15})}{16(3.14)^2}$$
$$= 24\,\text{cm}$$

Also, from geometry, $\sin\theta = \dfrac{24}{L_0 + 4}$

$$\therefore \quad \frac{\sqrt{15}}{4} = \frac{24}{L_0 + 4}$$

$$\Rightarrow \quad L_0 = \frac{24 \times 4}{\sqrt{15}} - 4 = \frac{(24 - \sqrt{15}) \times 4}{\sqrt{15}}\,\text{cm}$$

5. (a) Let T be the tension in the rope. Then,

$$2T = 10\,\text{kN}$$
$$\Rightarrow \quad T = 5000\,\text{N}$$

Longitudinal stress in the rope is

$$\sigma = \frac{T}{A} = \frac{5000\,\text{N}}{10^3\,\text{mm}^2} = 5\,\text{N mm}^{-2}$$

$$\text{Extension in the rope} = \frac{\text{Stress}}{Y} \times L$$
$$= \frac{5}{10^3} \times (600 + 900)$$
$$= 7.5\,\text{mm}$$

$\therefore$ Downward deflection of the load $= \dfrac{7.5}{2} = 3.75\,\text{mm}$

6. (c) Change in length of element is

$$d(\Delta L) = \frac{F\,dx}{AY}$$

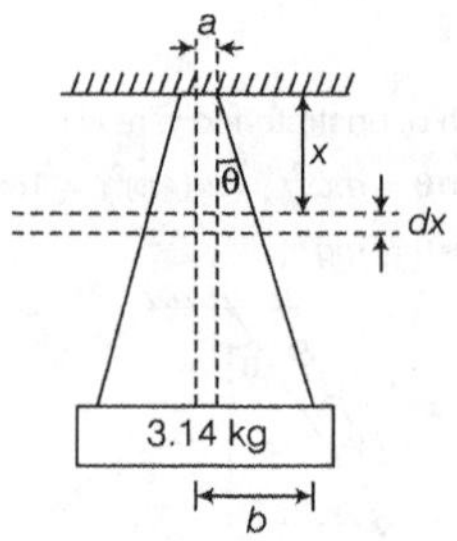

Cross-sectional area of the element is

$$A = \pi\left[a + \frac{b-a}{L}x\right]^2$$

where, $a = \frac{b-a}{L}x$ is radius of the wire at the location of element.

$$d(\Delta L) = \frac{F}{\pi\left[a + \frac{b-a}{L}x\right]^2 Y}dx$$

Total change in length of wire is

$$\Delta L = \frac{F}{\pi Y}\int_0^L \frac{1}{\left(a + \frac{b-a}{L}x\right)^2}dx$$

Let $\quad a + \frac{b-a}{L}x = t$

$\Rightarrow \quad \frac{b-a}{L}dx = dt$

$\Rightarrow \quad dx = \frac{L}{b-a}dt$

When $x = 0, t = a$ and when $x = L, t = b$

$$\therefore \quad \Delta L = \frac{F}{\pi Y}\int_a^b t^{-2}\cdot\frac{L}{b-a}dt$$

$$= \frac{FL}{\pi Y(b-a)}(-1)\left(\frac{1}{b} - \frac{1}{a}\right)$$

$$= \frac{FL}{\pi Y(b-a)}\cdot\frac{(b-a)}{ab} = \frac{FL}{\pi Y ab}$$

$$= \frac{(3.14)(9.8)(10)}{(3.14)(2\times10^{11})(5\times10^{-4})(9.8\times10^{-4})}$$

$$= 10^{-3}\text{m}$$

$$\therefore \quad \Delta L = 10^{-3}\text{m}$$

7. (a,b) FBD's of two portions are shown below :

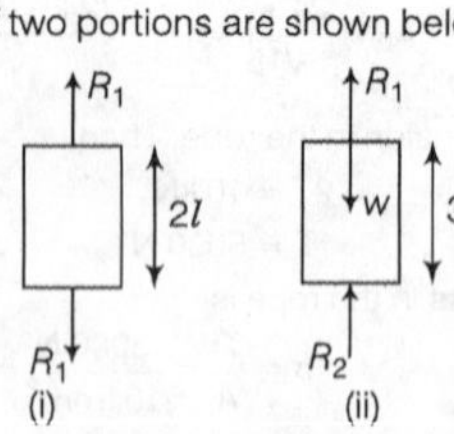

(i) (ii)

For the equilibrium of second part,

$$R_1 + R_2 = w = 5000 \text{ N} \quad \ldots\text{(i)}$$

Stresses in two parts are

$$\sigma_1 = \frac{R_1}{A} \quad\text{and}\quad \sigma_2 = \frac{R_2}{A}$$

Elongation in the portion $2l$ is

$$\delta l_1 = \frac{\sigma_1}{Y}(2l) = \frac{R_1}{AY}(2l)$$

Compression in the portion $3l$ is

$$\delta l_2 = \frac{\sigma_2}{Y}(3l) = \frac{R_2}{AY}(3l)$$

Elongation in the upper portion = Compression in the lower portion

(∵ There is no net change in length)

$$\therefore \quad \frac{R_1}{AY}(2l) = \frac{R_2}{AY}(3l)$$

$$\Rightarrow \quad 2R_1 = 3R_2 \quad \ldots\text{(ii)}$$

On solving Eqs. (i) and (ii) gives,

$$R_1 = 3000 \text{ N}$$
$$R_2 = 2000 \text{ N}$$

8. (a,b,c) When, $T_Q = \frac{mg}{3}$

and $\quad T_P = mg + \frac{mg}{3} = \frac{4mg}{3}$

Stress, $\quad \sigma = \frac{T}{\pi r^2}$

$$\frac{\sigma_P}{\sigma_Q} = \frac{T_P}{T_Q}\cdot\left(\frac{r_Q}{r_P}\right)^2$$

If $r_P = r_Q$, $\sigma_P = 4\sigma_Q$, then P breaks.
If $r_P < 2r_Q$, $\sigma_P > \sigma_Q$, then P breaks.
If $r_P = 2r_Q$, $\sigma_A = \sigma_B$, then either P or Q may break.

9. (a,c) Area of cross-section is $A = \frac{\pi d^2}{4}$

$$\Delta L = \frac{F}{\pi d^2/4}\cdot\frac{L}{Y}$$

$\Rightarrow \quad \Delta L \propto \frac{1}{d^2}$

$\therefore \quad \frac{\Delta L_B}{\Delta L_A} = \frac{(2d)^2}{(d)^2}$

$\Rightarrow \quad \Delta L_B = 4\Delta L_A$

$\quad \text{Strain} = \frac{\Delta L}{L} = \frac{4F}{\pi d^2 Y}$

$\Rightarrow \quad \text{Strain} \propto \frac{1}{d^2}$

$\therefore \quad (\text{Strain})_B = 4(\text{Strain})_A$

10. (c,d) If stress in steel = stress in brass,

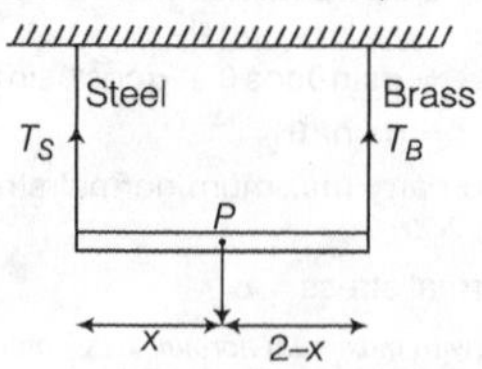

then $\quad \frac{T_S}{A_S} = \frac{T_B}{A_B}$

$$\frac{T_S}{T_B} = \frac{A_S}{A_B} = \frac{10^{-3}}{2\times10^{-3}} = \frac{1}{2} \quad \ldots\text{(i)}$$

System is in equilibrium. So, taking moments about P

$$T_S \cdot x = T_B(2 - x)$$

$$\Rightarrow \quad \frac{T_S}{T_B} = \frac{2 - x}{x} \quad \ldots(ii)$$

From Eqs. (i) and (ii), we get

$$x = 1.33 \text{ m}$$

$$\text{Strain} = \frac{\text{Stress}}{Y}$$

If strain in steel = strain in brass,

$$\text{then} \quad \frac{T_S / A_S}{Y_S} = \frac{T_B / A_B}{Y_B}$$

$$\therefore \quad \frac{T_S}{T_B} = \frac{A_S\, Y_S}{A_B\, Y_B} = \frac{(10^{-3})(2 \times 10^{11})}{(2 \times 10^{-3})(10^{11})} = 1 \quad \ldots(iii)$$

From Eqs. (ii) and (iii), we get

$$x = 1 \text{ m}$$

11. (a,b) Area of steel rod,

$$A_S = 16 \text{ cm}^2$$

Area of two brass rods,

$$A_B = 2 \times 10 = 20 \text{ cm}^2$$

$$F = 5000 \text{ kg}$$

$$\sigma_S = \text{Stress in steel}$$

and $\quad \sigma_B = \text{Stress in brass}$

Decrease in length of steel rod = Decrease in length of brass rod

$$\frac{\sigma_S}{Y_S} \cdot L_S = \frac{\sigma_B}{Y_B} \cdot L_B$$

$$\Rightarrow \quad \sigma_S = \frac{Y_S}{Y_B} \cdot \frac{L_B}{L_S} \cdot \sigma_B$$

$$\text{or} \quad \sigma_S = \left(\frac{2 \times 10^6}{10^6}\right)\left(\frac{20}{30}\right) \sigma_B$$

$$\Rightarrow \quad \sigma_S = \frac{4}{3} \sigma_B \quad \ldots(i)$$

Now, $\quad F = \sigma_S A_S + \sigma_B A_B$

or $\quad 5000 = \sigma_S \times 16 + \sigma_B \times 20 \quad \ldots(ii)$

From Eqs. (i) and (ii), we get

$$\sigma_B = 120.9 \text{ kg cm}^{-2}$$

and $\quad \sigma_S = 161.2 \text{ kg cm}^{-2}$

12. (a,d) Let PQ is part 1, QR part 2 and RS part 3.

For equilibrium,

$$F_2 + F_4 = F_1 + F_3$$

$$F_2 = F_1 + F_3 - F_4$$

$$= 4500 + 45000 - 13000$$

$$= 36500 \text{ kg}$$

For the sake of simplicity, the force of 36500 kg (acting at Q) may be split up into two forces of 4500 kg and 32000 kg. The force of 45000 kg acting at R may be split into two forces 32000 kg and 13000 kg. Now, it can be seen that the part PQ of the member is subjected to a tension of 4500 kg, part QR is subjected to a compression of 32000 kg and part RS is subjected to a tension of 13000 kg.

$$\text{Now,} \quad \Delta l = \frac{1}{Y}\left[\frac{F_1 L_1}{A_1} - \frac{F_2' L_2}{A_2} + \frac{F_3' L_3}{A_3}\right]$$

$$= \frac{1}{2.1 \times 10^6}\left[\frac{4500 \times 120}{6.25} - \frac{32000 \times 60}{25} + \frac{13000 \times 90}{12.5}\right] \text{cm}$$

$$= 0.049 \text{ cm} = 0.49 \text{ mm}$$

13. (c) ***Case I*** $\quad F = ma$

$$\Rightarrow \quad a = \frac{F}{m}$$

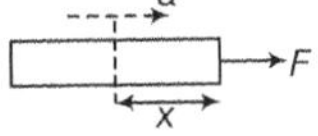

$$F - T = m'a = \frac{m}{L} xa$$

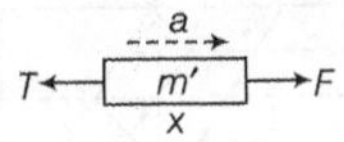

$$T = F - \frac{m}{L} x \cdot \frac{F}{m} = F\left(1 - \frac{x}{L}\right) \quad \ldots(i)$$

Case II $\quad F - mg = ma'$

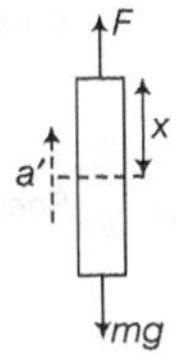

$$a' = \frac{F}{m} - g$$

$$F - T - m'g = m'a'$$

$$F - T - m'g = m'\left(\frac{F}{m} - g\right)$$

$$F - T = m' \cdot \frac{F}{m}$$

$$F - T = \frac{m}{L} x \cdot \frac{F}{m} = \frac{x}{L} F$$

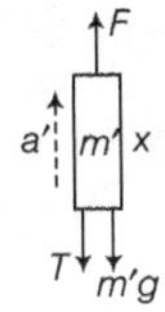

$$\Rightarrow \quad T = F\left(1 - \frac{x}{L}\right) \quad \ldots(ii)$$

Tension at a point on the rod (of length L) at a distance x from point of application of force is same in both cases. [from Eqs. (i) and (ii)]

Hence, weight has no effect on tension in case (ii).

Note *You can appreciate the extension of rod in first case, by comparing it with a case of many identical blocks connected by ideal springs.*

Extension in rod occurs due to force acting at any point on the rod. In certain cases, when net force acts at the centre of rod like weight, extension may not occur like the given case (ii).

14. (b) Initially, $\Delta L = \dfrac{FL}{AY}$

$$= \frac{10 \times 1}{10^{-3} \times 2 \times 10^5}$$

$$= 0.05 \text{ m} = 5 \text{ cm}$$

15. (c) Force constant of string is

$$k = \frac{\text{Force}}{\text{Elongation}} = \frac{F}{FL/AY}$$

$$\text{or} \quad k = \frac{YA}{L} = \frac{1 \times 10^5 \times 10^{-3}}{1} = 200 \text{ Nm}^{-1}$$

Initial elastic potential energy of string $= \dfrac{1}{2} k\, (0.05)^2$

$$= \frac{1}{2}(200)(25 \times 10^{-4}) = 25 \times 10^{-2} \text{ J}$$

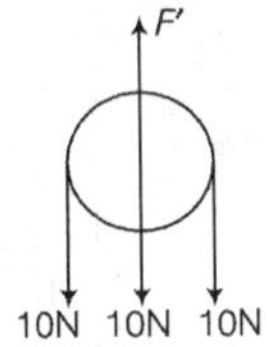

Let after force $F = 10$ N is applied extra elongation is x, then

$$25 \times 10^{-2} + 30(x) = \frac{1}{2} k (x + 0.05)^2$$

$$0.25 + 30x = \frac{1}{2}(200)(x^2 + 25 \times 10^{-4} + 0.1x)$$

$$0.25 + 30x = 100x^2 + 0.25 + 10x$$

$$\therefore \quad 100x^2 = 20x$$

$$\Rightarrow \quad x = \frac{20}{100} = \frac{1}{5} \text{ m} = 0.2 \text{ m} = 20 \text{ cm}$$

$$\therefore \quad x_{max} = 20 + 5 = 25 \text{ cm}$$

16. (d) When the displacement of the pulley is 25 cm, the string gets loosened on both the sides. So, point A moves down by 50 cm.

17. (a)

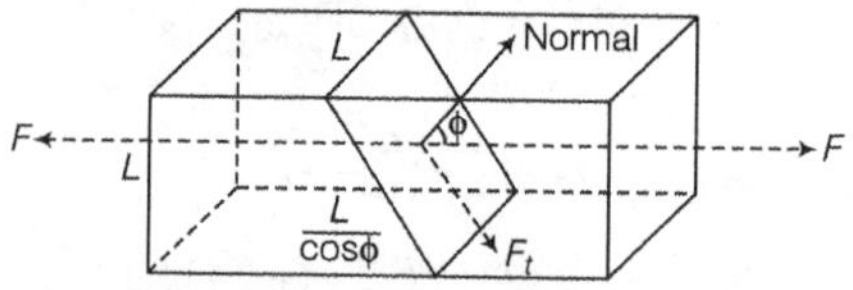

Tangential force is $F_t = F\sin\phi$

Area of section considered is

$$A' = L\left(\frac{L}{\cos\phi}\right) = \frac{L^2}{\cos\phi} = \frac{A}{\cos\phi}$$

$$\text{Shearing stress} = \frac{F_t}{A'} = \frac{F\sin\phi}{A/\cos\phi} = \frac{F}{A}\sin\phi\cos\phi \quad \ldots\text{(i)}$$

Normal force on plane section considered is

$$F_N = F\cos\phi$$

$$\therefore \text{Tensile stress} = \frac{F_N}{A'} = \frac{F\cos\phi}{A/\cos\phi} = \frac{F}{A}\cos^2\phi \quad \ldots\text{(ii)}$$

Shearing stress is maximum when

$\phi = 45°$ [from Eq. (i)]

Tensile stress is maximum when

$\phi = 0°$ [from Eq. (ii)]

Hence, (i) → (r), (ii) → (p), (iii) → (q), (iv) → (t)

18. (1) Consider an element AB of length dl. Let T be the tension and a be the area of cross-section of the wire.

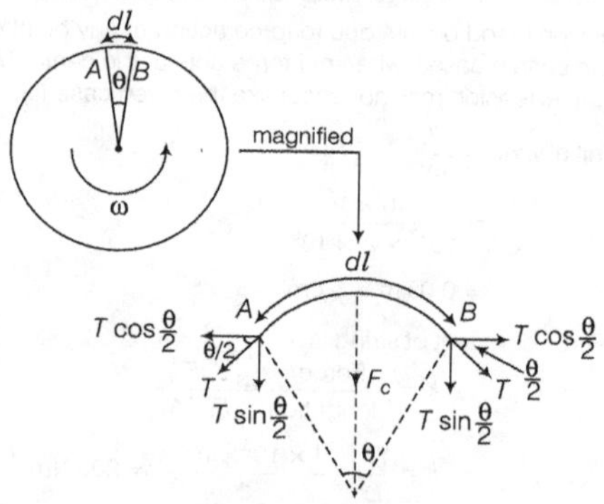

Mass of the element is dm = (Volume) (Density) = $(a\,dl)\rho$

The component of T, i.e. $\left(T\sin\frac{\theta}{2} + T\sin\frac{\theta}{2}\right)$ provides required centripetal force to the element considered.

$$\therefore \quad F_C = 2T\sin\frac{\theta}{2}$$

$$(dm)R\,\omega^2 = 2T\left(\frac{\theta}{2}\right) \quad \left(\because \theta \text{ is small, } \sin\frac{\theta}{2} \simeq \frac{\theta}{2}\right)$$

$$T = \frac{(a\,dl)\rho R\omega^2}{\theta}$$

$$\theta = \frac{dl}{R}$$

$$\therefore \quad T = aR^2\rho\omega^2 \quad \ldots\text{(i)}$$

Let ΔR be the increase in radius of the ring.

$$\text{Longitudinal strain} = \frac{\Delta l}{l} = \frac{\Delta(2\pi R)}{2\pi R} = \frac{\Delta R}{R}$$

$$Y = \frac{T/a}{\Delta R/R} \Rightarrow \Delta R = \frac{TR}{aY}$$

From Eq. (i), $\quad \Delta R = \frac{R}{aY}(aR^2\rho\omega^2)$

$$\therefore \quad \Delta R = \frac{\rho R^3\omega^2}{Y}$$

$$\Rightarrow \quad x = 1$$

19. (5) Let ΔL be the extension of wire when the sphere is at mean position. Then,

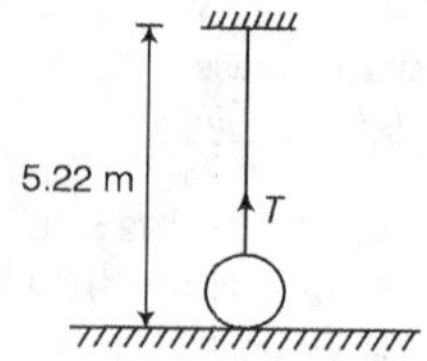

$$L + \Delta L + 2r = 5.22$$

$$\Delta L = 5.22 - L - 2r$$

$$= 5.22 - 5 - 2(0.1)$$

$$= 0.02 \text{ m}$$

Let T be the tension in the wire at mean position during oscillations. Then,

$$Y = \frac{T/A}{\Delta L/L}$$

$$\Rightarrow \quad T = \frac{YA\,\Delta L}{L} = \frac{Y(\pi r^2)\Delta L}{L}$$

$$\text{or} \quad T = \frac{(2\times10^{11})(\pi)(0.5\times10^{-3})^2\times0.02}{5}$$

$$= 200\pi$$

Also, at mean position

$$T - mg = \frac{mv^2}{R}$$

(here, $R = 5.22 - r = 5.22 - 0.1 = 5.12$ m)

$$\therefore \quad (200\pi) - (8\pi)(9.8) = \frac{(8\pi)v^2}{5.12}$$

$$\Rightarrow \quad v^2 = 5\times20.48$$

$$\therefore \quad N = 5$$

20. (5) If L is the maximum length of the wire that can be suspended. Then $F = mg$, where $m = \rho LA$ (where, A = cross-sectional area)

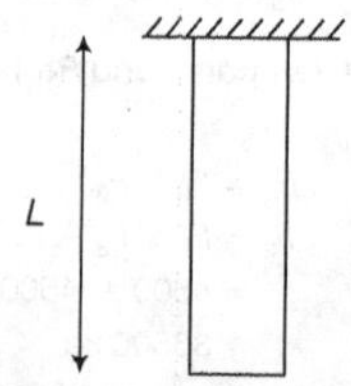

$$\therefore \text{Maximum stress} = \frac{mg}{A} = g\cdot\frac{\rho LA}{A} = \rho Lg$$

Given, breaking stress = $\sigma_b = 7.8\times10^9$ Nm^{-2}

$$\therefore \quad \rho Lg = 7.8\times10^9$$

$$\Rightarrow \quad L = 7.8\times10^9\times\frac{1}{7800\times10}$$

$$\text{or} \quad L = 10^5 \text{ m}$$

$$\Rightarrow \quad k = 5$$

DPP-2 Modulii of Elasticity

1. (a) Length of the wire is $l = 20\,\text{m}$

Radius of the wire is $r = 2 \times 10^{-3}\,\text{m}$

Increase in length is $\Delta l = 0.031 \times 10^{-3}\,\text{m}$

$$g = 3.1 \times \pi\ \text{ms}^{-2}$$

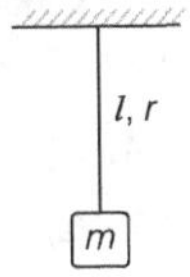

Young's modulus is

$$Y = \frac{Fl}{A\,\Delta l} = \frac{(mg)l}{\pi r^2 (\Delta l)}$$

$$\therefore \quad Y = \frac{4 \times 3.1 \times \pi \times 20}{\pi (2 \times 10^{-3})^2 \times 0.031 \times 10^{-3}}$$

$$= 2 \times 10^{12}\ \text{Nm}^{-2}$$

2. (b) Volume of wire is $V = L \times \pi r^2 = 1 \times \pi (10^{-3})^2 = \pi \times 10^{-6}\,\text{m}^3$

Area of square cross-section $= (2 \times 10^{-3})^2 = 4 \times 10^{-6}\,\text{m}^2$

Length of new wire $= \dfrac{\text{Volume}}{\text{Area}} = \dfrac{\pi \times 10^{-6}}{4 \times 10^{-6}} = \dfrac{\pi}{4}\,\text{m}$

Initially extension is $x = \dfrac{FL}{AY} = \dfrac{F}{Y}\left(\dfrac{1}{\pi \times 10^{-6}}\right)$

$$\therefore \quad \frac{F}{Y} = (\pi \times 10^{-6})x$$

Finally extension is $x' = \dfrac{FL'}{A'Y} = \dfrac{F}{Y} \cdot \dfrac{L'}{A'}$

or $$x' = (\pi \times 10^{-6})\,x \cdot \frac{\pi/4}{4 \times 10^{-6}} = \frac{\pi^2}{16}x$$

3. (c) Extension in the first case is

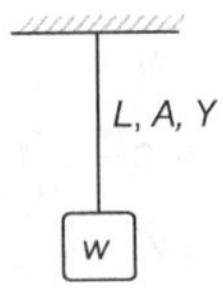

$$l = \frac{FL}{AY} = \frac{wL}{AY}$$

Extension in the second case is

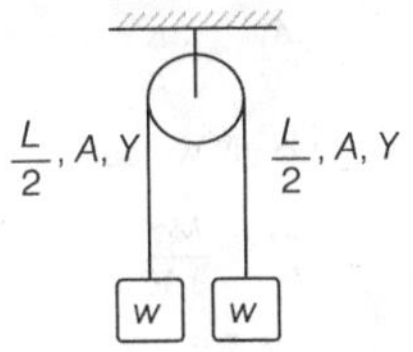

$$l' = \frac{w(L/2)}{AY} + \frac{w(L/2)}{AY} = \frac{wL}{AY}$$

It is clear that $\quad l' = l$

Note *In the above problem, even if lengths of wire are unequal on two sides of the pulley, the elongation will still be l.*

4. (b) $Y = \dfrac{Fl}{A\,\Delta l}$

$$\therefore \quad Y = \frac{4l}{A(l_1 - l)} = \frac{5l}{A(l_2 - l)}$$

$$\Rightarrow \quad 4l_2 - 4l = 5l_1 - 5l$$

$$\therefore \quad l = 5l_1 - 4l_2$$

5. (c) $Y = \dfrac{\text{Stress}}{\text{Strain}}$

If Y = stress, then strain = 1

$$\Rightarrow \quad \frac{\Delta l}{l} = 1 \Rightarrow \Delta l = l$$

6. (d) Force on wire = Load carried

$$= m(g + a) = m\left(g + \frac{g}{2}\right) = \frac{3mg}{2}$$

Stress $\sigma = \dfrac{F}{A} \quad \Rightarrow \quad A = \dfrac{F}{\sigma}$

Hence, $\quad A_{\min} = \dfrac{F}{\sigma_{\max}}$

$$\frac{\pi d_{\min}^2}{4} = \frac{\frac{3}{2}m \cdot g}{\sigma} \Rightarrow d_{\min}^2 = \frac{\frac{3}{2}mg \times 4}{\sigma \pi}$$

$$d_{\min} = \sqrt{\frac{6}{\pi} \cdot \frac{mg}{\sigma}}$$

7. (b) Consider a very small (differential) disc element at a distance y from the upper base.

Let r and $r + dr$ denote the radii of upper and lower surfaces, respectively and A and $A + dA$ be the corresponding areas. Let w denotes the weight of the body above the elemental disc and dw that of the elemental disc. Normal compressive stress at the top surface of the whole body is $\sigma = \dfrac{P}{\pi r_0^2}$ $\quad$ ...(i)

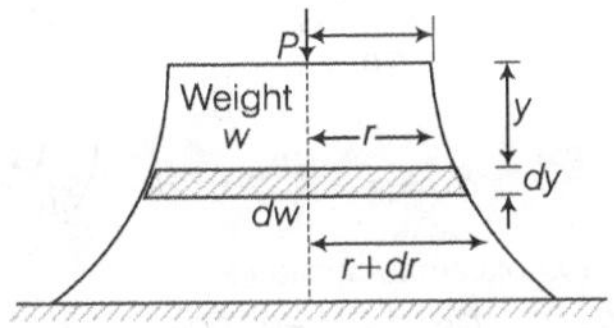

Now, consider the normal compressive stress acting over both the surfaces of the element. We can write

$$\frac{P + w}{A} = \frac{P + w + dw}{A + dA} = \sigma = \text{constant}$$

$$PA + PdA + wA + wdA = PA + wA + Adw$$

$$dA(P + w) = Adw$$

$$\Rightarrow \quad \frac{dA}{dw} = \frac{A}{P + w} = \frac{1}{\sigma} \quad \text{...(ii)}$$

The increment in area between upper and lower surfaces of the differential disc is

$$dA = \pi(r + dr)^2 - \pi r^2 \simeq 2\pi r dr$$

The increment in weight is

$$dw = \pi r^2 \rho (dy)$$

Using these in Eq. (ii),

$$\frac{2\pi r dr}{\pi r^2 \rho dy} = \frac{1}{\sigma}$$

$$2\frac{dr}{r} = \frac{\rho}{\sigma}dy$$

$$2\int_{r_0}^{r} \frac{1}{r}dr = \frac{\rho}{\sigma}\int_0^y dy$$

$$\Rightarrow \quad 2\ln\frac{r}{r_0} = \frac{\rho}{\sigma}y$$

From Eq. (i), $$2\ln\frac{r}{r_0} = \frac{\rho}{(P/\pi r_0^2)}y$$

$$\Rightarrow \quad \ln\frac{r}{r_0} = \frac{\pi r_0^2 \rho}{2P}y$$

$$r = r_0 e^{(\pi r_0^2 \rho y/2P)}$$

8. (a) In the figure, the reciprocal of slope of stress-strain (X and Y-axes) curve, upto proportionality limit, gives Young's modulus. The measure of ductility is obtained as the length of the stress-strain curve between yield point and ultimate load.

9. (a) Change in volume,

$$\Delta V = \frac{-\Delta p \cdot V_i}{B}$$

Hence, density at depth of about 11 km is

$$= \frac{\text{Mass}}{\text{Volume}} = \frac{\rho_0 \times V_i}{V_i - \frac{\Delta p V_i}{B}} = \frac{\rho_0 B}{(B - \Delta p)}$$

$$= \frac{\rho_0}{1 - \frac{\Delta p}{B}} = \frac{\rho_0}{1 - \frac{1 \times 10^8}{2 \times 10^9}} = \frac{\rho_0}{1 - \frac{1}{20}} = \frac{\rho_0}{0.95}$$

$\Rightarrow$ $\rho = \frac{\rho_0}{0.95} \Rightarrow \rho_0 = 0.95\rho$

10. (a,b,d) $P = Fv = F \times 2\pi R \times N = 2\pi NT$

$$P = \omega T$$

Also, for a linear twist r and radial twist θ,

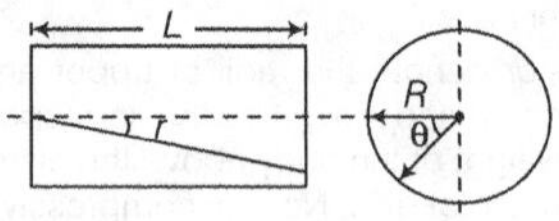

$$Lr = R\theta$$

and $G = \frac{\tau}{r}$

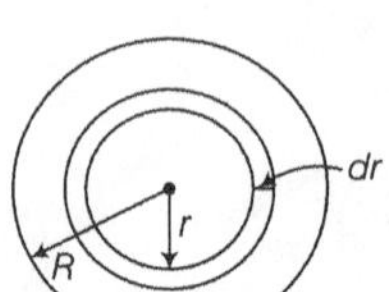

$\Rightarrow$ $\frac{G\theta}{L} = \frac{\tau}{r} = \frac{T}{R}$

Also, $dT = \tau = \frac{G\theta}{L} 2\pi r^3\, dr$

$\therefore$ Torque on whole cross-section is

$$T = \frac{G\theta}{L} 2\pi \int_0^R r^3 dr = \frac{G\theta R^4 \pi}{2L} = \frac{\pi R^3}{2}\left(\frac{R\theta}{L} G\right) = \frac{\pi R^3}{2} \cdot \sigma_{max}$$

Note σ_{max} is not a constant

$\therefore$ $T \propto R^4$

Also, $\frac{T}{j} = \frac{G\theta}{L} = \frac{\tau}{r}$

11. (d) Young's modulus is a material constant.

12. (b) On crossing the yield region, the material will experience the breaking stress and further elongation causes reduction in stress and breaking of the wire.

13. (a) Stress-strain curve flattens on crossing elastic region.

Explanations for Q. Nos. (14-16)

For vertical equilibrium of L-shaped structure,

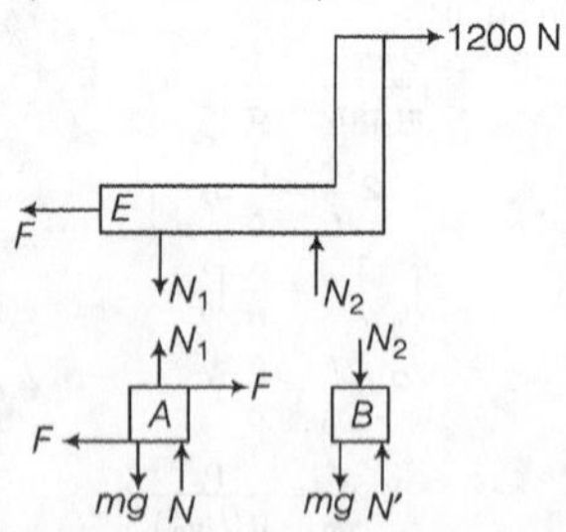

$$N_1 = N_2$$

For horizontal equilibrium, $F = 1200$ N

For rotational equilibrium, torque about E (or any other point) = 0

$\therefore$ $N_2 \times 4 - 1200 \times 3 = 0$

$\Rightarrow$ $N_2 = 900$ N

Shearing stress in $A = \frac{F}{A} = \frac{1200}{5}\,\text{Ncm}^{-2} = 240\,\text{Ncm}^{-2}$

Shearing stress in $B = 0$

Longitudinal stress in $A = \frac{N_1}{A} = \frac{900}{5} = 180\,\text{Ncm}^{-2}$

(tensile in nature)

Longitudinal stress in $B = \frac{N_2}{A} = 180\,\text{Ncm}^{-2}$

(compressive in nature)

14. (b) 15. (a) 16. (c)

17. (a) $\Delta l_{copper} = \frac{Fl}{AY} = \frac{500 \times 8}{0.5 \times 10^{-4} \times 10^{11}} = 0.8$ mm

$$\Delta l_{steel} = \frac{Fl}{AY} = \frac{500 \times 4}{0.5 \times 10^{-4} \times 2 \times 10^{11}} = 0.2 \text{ mm}$$

$$(\Delta l)_{net} = \Delta l_{copper} + \Delta l_{steel} = 0.8 + 0.2 = 1.0 \text{ mm}$$

Hence, (i) → (r), (ii) → (p), (iii) → (q)

18. (a) For A and B, the rod is in equilibrium and hence internal restoring force developed per unit area across any cross-section is same and thus stress is uniform. But in C and D, the case is opposite.

Hence, (i) → (p,r), (ii) → (p,s), (iii) → (q,s), (iv) → (q,r)

19. (2) The increase in pressure is $\Delta p = \rho g h$

or $\Delta p = 10^3 \times 9.8 \times 3000 = 29.4 \times 10^6\,\text{Nm}^{-2}$

The volumetric strain $= \frac{\Delta V}{V} = \frac{\Delta p}{B}$

$$= \frac{29.4 \times 10^6}{2.2 \times 10^9} = 1.36$$

So, percentage compression = 1.36% = 2 (0.68)%

$\Rightarrow$ $\beta = 2$

20. (3) Increase in pressure is $\Delta p = \frac{Mg}{A}$

Bulk modulus is $B = \frac{\Delta p}{(\Delta V / V)}$

$\therefore$ $\frac{\Delta V}{V} = \frac{\Delta p}{B} = \frac{Mg}{AB}$ ...(i)

Also, the volume of the sphere is

$$V = \frac{4}{3}\pi R^3$$

$\Rightarrow$ $\frac{\Delta V}{V} = \frac{3\Delta R}{R}$

or $\frac{\Delta R}{R} = \frac{1}{3} \cdot \frac{\Delta V}{V}$, using Eq. (i), we get

$$\frac{\Delta R}{R} = \frac{Mg}{3AB}$$

$\therefore$ $\alpha = 3$

21. (4) Force constant of the wire is

$$k = \frac{F}{\Delta L} = \frac{F}{FL/AY} = \frac{YA}{L}$$

$$\omega = \sqrt{\frac{k}{m}} = \sqrt{\frac{YA}{Lm}}$$

$\therefore$ $\sqrt{\frac{YA}{Lm}} = 140$

$\Rightarrow$ $\frac{Y \times 4.9 \times 10^{-7}}{1 \times 0.1} = 140 \times 140$

$$Y = \frac{140 \times 140}{49 \times 10^{-7}} = 4 \times 10^9$$

$\Rightarrow$ $p = 4$

DPP-3 Elastic Potential Energy

1. (d) Elastic forces are conservative when loading and unloading curves coinciders even when they are not linear as in case of rubber.
All other options are incorrect.

2. (b) $U = \frac{1}{2}\sigma\varepsilon = \frac{1}{2}(Y\varepsilon)\varepsilon$

$\Rightarrow \quad U = \frac{1}{2}Y\varepsilon^2$ is similar to $x = ky^2$

which is a parabola passing through origin and symmetric about X-axis.

3. (c) $\Delta U = \frac{1}{2} \times \text{Stress} \times \text{Strain} \times \text{Volume}$

$$= \frac{1}{2} \times \frac{mg}{A} \times \frac{h}{L} \times A \times L$$

$$= \frac{1}{2}\,mgh$$

4. (a) Work done $= \frac{1}{2}$(Stress) (Strain) (Volume)

$$= \frac{1}{2}(Y)\,(\varepsilon)^2\,(\text{Volume})$$

$$\frac{1}{2}mv^2 = \frac{1}{2}(Y)\left(\frac{\Delta l}{l}\right)^2 (Al)$$

$$v^2 = \frac{Y\,A\,l(\Delta l)^2}{ml^2}$$

$$\Rightarrow \quad v = \sqrt{\frac{YA(\Delta l)^2}{lm}}$$

5. (c) $U = \frac{1}{2}$(Stress) (Strain) (Volume)

$$= \frac{1}{2}\frac{F}{A} \cdot \frac{\Delta l}{l} \cdot Al = \frac{1}{2}F(\Delta l) = \frac{1}{2}Mg\,(\Delta l)$$

$\therefore \quad \frac{1}{2}Mg\,(\Delta l) = mC\,\Delta T$

So, rise in temperature of the wire is

$$\Delta T = \frac{Mg\,(\Delta l)}{2\,mC}$$

As, load also do work of breaking so, $\Delta T > \frac{Mg\,\Delta l}{2\,mC}$

6. (b) $W = U = \frac{1}{2}F(\Delta l) = \frac{1}{2} \cdot \frac{YA(\Delta l)^2}{l}$

$\therefore \quad U \propto \frac{A}{l}$ $\quad$ ($\because$ In both the cases Δl is same)

$$\frac{U_2}{U_1} = \frac{A_2}{A_1} \cdot \frac{l_1}{l_2} = \left(\frac{r_2}{r_1}\right)^2 \cdot \frac{l_1}{l_2} = (4)(2)$$

$\therefore \quad U_2 = 8\,U_1 = 8(2) = 16\,\text{J}$

7. (d) Hystersis loss corresponding to elasticity per unit volume of a substance is given by the area of hystersis loop, i.e. stress-strain curve corresponding to one complete loading and deloading.

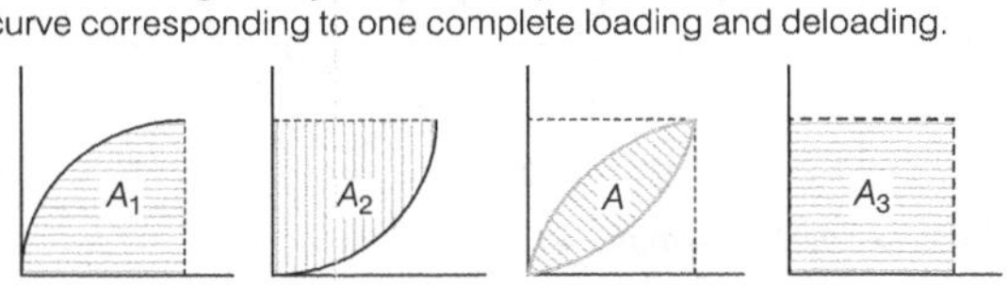

Area of an ellipse $= \pi \times$ Semi-major axis $\times$ Semi-minor axis

$$A_1 = \frac{1}{4}(\pi \times 8 \times 4 \times 10^2)$$

and $\quad A_2 = \frac{1}{4}(\pi \times 8 \times 4 \times 10^2)$

Also, $\quad A_3 = 8 \times 4 \times 10^2$

Area of hysteresis loop is

$$A = A_1 + A_2 - A_3$$

$$A = 2\left[\frac{\pi}{4} \times 8 \times 4 \times 10^2\right] - [8 \times 4 \times 10^2]$$

= Work done per cycle

= Energy lost per cycle per unit volume

8. (d) It is clear from the figures that area enclosed in hysteresis loop is larger for R_2 than for R_1. So, rubber R_2 will absorb more energy than R_1.
In a heavy machine, the energy of vibrations is quite large. So, rubber R_2 is suitable for this purpose.
In a car tyre, small energy is to be absorbed by tyre (more by air inside it). So, rubber R_1 is suitable for this purpose.

9. (d) Tension, $T = \frac{F}{L_0} \cdot x$

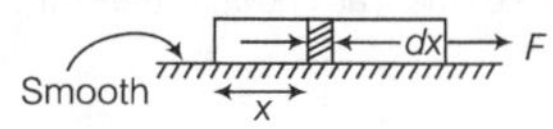

$$\text{Stress, } \sigma = \frac{T}{A} = \frac{F}{AL_0}x$$

$$dU = \frac{1}{2} \cdot \frac{\sigma^2}{Y}Adx = \frac{1}{2} \cdot \frac{F^2}{A^2L_0^2} \cdot x^2\,\frac{A}{Y}dx$$

or $\quad dU = \frac{F^2}{2AL_0^2Y}x^2dx$

$\Rightarrow \quad U = \frac{F^2}{2AY\,L_0^2}\int_0^{L_0} x^2dx$

$$U = \frac{F^2}{2AY\,L_0^2} \cdot \frac{L_0^3}{3} = \frac{F^2L_0}{6AY}$$

10. (a,d) $U = \frac{1}{2}k\,x^2$. It is a parabola symmetric about U-axis.

At $x = 0.2$ mm, $\quad U = 0.2$ J $\quad$ (from the figure)

$\therefore \quad 0.2 = \frac{1}{2}k(2 \times 10^{-4})^2$

$\Rightarrow \quad k = 10^7\ \text{Nm}^{-1}$

$$k = \frac{YA}{L}$$

$\Rightarrow \quad \frac{A}{L} = \frac{k}{Y} = \frac{10^7}{2 \times 10^{11}} = 5 \times 10^{-5}$ $\quad$ ...(i)

$AL = \text{Volume} = 200 \times 10^{-6}\ \text{m}^3$ $\quad$ ...(ii)

On solving Eqs. (i) and (ii), we get

$$A = 10^{-4}\ \text{m}^2 \text{ and } L = 2\,\text{m}$$

11. (a) $U = \frac{1}{2}$(Stress) (Strain) (Volume)

$$= \frac{1}{2}\left(\frac{F}{A}\right)\left(\frac{l}{L}\right)(AL) = \frac{1}{2}Fl = \frac{1}{2}Mgl$$

$\therefore$ Loss in energy to surroundings $= Mgl - \frac{1}{2}Mgl = \frac{Mgl}{2}$

12. (a) $Y_{steel} > Y_{copper}$. So, for the same strain, stress to be produced in steel is more.

Also, $\quad$ Work done $= \frac{1}{2}$Stress $\times$ Strain

$$\text{Volume} = \frac{1}{2}y \times \varepsilon^2$$

13. (b) Bulk modulus is defined as $B = \frac{\Delta p}{(\Delta V/V_0)}$

$\therefore \quad$ Decrease in volume is $-\Delta V = \frac{(\Delta p)V_0}{B} = \frac{(p - p_0)V_0}{B}$

14. (a) For any small change of pressure dp, there will be a change of volume dV and $dp = -B\frac{dV}{V}$. In this change, work is done on the system and the energy stored in the material is

$$dW = -pdV = \left(\frac{V}{B}\right)pdp$$

In the change mentioned in the question, the total work done is

$$W = -\int_{V_1}^{V_2} pdV = \int_{p_1}^{p_2} \frac{V}{B}pdp$$

The change in volume is negligible and volume can be treated as constant.

$$W = \frac{V_0}{B}\int_{p_0}^{p} pdp = \frac{V_0}{B}\left(\frac{p^2 - p_0^2}{2}\right)$$

$\therefore$ Extra energy stored per unit volume is

$$\frac{W}{V_0} = \frac{1}{2B}(p^2 - p_0^2)$$

15. (6) Consider an element as shown in the figure.

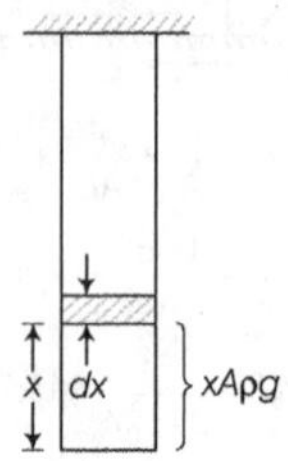

Stress in the element $= \frac{\text{Force}}{\text{Area}} = \frac{xA\rho g}{A} = x\rho g$

Now, elastic potential energy stored in the wire is

$$dU = \frac{1}{2}\text{(Stress) (Strain) (Volume)}$$

$$= \frac{1}{2}\cdot\frac{(\text{Stress})^2}{Y}(\text{Volume})$$

$$dU = \frac{1}{2}\cdot\frac{(x\rho g)^2}{Y}Adx = \frac{1}{2}\cdot\frac{\rho^2 g^2 A}{Y}x^2dx$$

Total elastic potential energy $= \frac{1}{2}\cdot\frac{\rho^2 g^2 A}{Y}\int_0^L x^2dx$

$$= \frac{\rho^2 g^2 AL^3}{6Y}$$

16. (8) Energy per unit volume

$$= \frac{1}{2}\text{(Stress) (Strain)}$$

$$u = \frac{U}{V} = \frac{1}{2}\text{(Stress)}\left(\frac{\text{Stress}}{\text{Bulk modulus}}\right)$$

or $\quad u = \frac{1}{2}(\text{Stress})^2(\text{Compressibility})$

$$= \frac{1}{2}(\rho gh)^2(\text{Compressibility})$$

$$= \frac{1}{2}(1000 \times 9.8 \times 1000)^2(5 \times 10^{-10})$$

$$= 2.4 \times 10^4 \text{ Jm}^{-3} = 24 \times 10^3 \text{ Jm}^{-3}$$

$$= 3(8) \times 10^3 \text{ Jm}^{-3}$$

$\therefore \quad x = 8$

DPP-4 Thermal Stresses and Strains

1. (a) As one end is free, it can expand freely and no thermal stress is induced.

2. (a) Young's modulus is

$$Y = \frac{mg\,L}{A\Delta L}$$

Also, $\quad \Delta L = L\alpha\Delta\theta$

$\therefore \quad Y = \frac{mgL}{A\cdot L\alpha\Delta\theta} = \frac{mg}{A\alpha\Delta\theta}$

$\Rightarrow \quad m = \frac{YA\alpha\Delta\theta}{g}$

$\therefore \quad m = \frac{(10^{11})(\pi \times (10^{-3})^2) \times 10^{-5}(40 - 30)}{10} = \pi$

$\therefore \quad m \simeq 3$ kg

3. (b) The thermal stress in the rod $= Y\alpha\Delta\theta$

Thus, for same rise of temperature ($\Delta\theta$), the thermal stress will be equal, if $Y_1\alpha_1 = Y_2\alpha_2$

or $\quad Y_1 : Y_2 = \alpha_2 : \alpha_1$

$\therefore \quad Y_1 : Y_2 = 3 : 2$

4. (c) Strain developed is

$$\varepsilon = \frac{\Delta L}{L} = \alpha\Delta\theta$$

or $\quad \varepsilon = (12 \times 10^{-6})(50)$

$$= 6 \times 10^{-4}$$

Strain will be negative as the rod is in compressed state.

5. (d) The increase in length of the rod, due to rise in temperature in the absence of walls is $\Delta L = L\alpha\Delta\theta$

$\therefore \quad \Delta L = (1000)(10^{-4})(20) = 2$ mm

But the rod can expand upto 1001 mm only.

$\therefore \quad$ Compression = 1mm

So, $\quad$ stress developed $= Y\frac{\Delta L}{L} = 10^{11}\left(\frac{1}{1000}\right) = 10^8$ Pa

6. (d) Thermal stress is $\sigma = Y\alpha\Delta\theta$

$$\Delta\theta = 70 - 35 = 35°\text{C}$$

$\therefore \quad \sigma = 2.0 \times 10^{11} \times 1.1 \times 10^{-5} \times 35$

$$= 7.7 \times 10^7 \text{ Nm}^{-2}$$

The increase in tension (increase because the temperature falls)

$$= \sigma\cdot A = (7.7 \times 10^7)(1 \times 10^{-6}) = 77 \text{ N}$$

7. (d) E = Decrease in potential energy of bob $= (mg \times \Delta L) = mg\Delta L$

$$U = \frac{1}{2}\times\sigma\times\varepsilon\times V = \frac{1}{2}\cdot\frac{mg}{A}\cdot\frac{\Delta L}{L}\cdot AL$$

$$U = \frac{1}{2}mg\Delta L$$

$$E - U = Q$$

$\Rightarrow \quad mg\Delta L - \frac{1}{2}mg\Delta L = Q$

$$Q = \frac{1}{2}mg\Delta L$$

So, $\quad Q = U$

Also,
$$Q = mC\Delta\theta$$
$$\therefore \quad \frac{Q_1}{Q_2} = \frac{mC_1\,\Delta\theta_1}{mC_2\,\Delta\theta_2} = \frac{U_1}{U_2}$$
$$\frac{C_1}{C_2} = \frac{\Delta\theta_2}{\Delta\theta_1}$$
$$\Rightarrow \quad U_1 = U_2$$
$$\therefore \quad \frac{U_1}{U_2} = \frac{\sigma_1 \times \varepsilon_1 \times V_1}{\sigma_2 \times \varepsilon_2 \times V_2}, \text{ but } \sigma_1 = \sigma_2$$
$\because mg$ and A are same in both the cases.

Also,
$$\frac{V_1}{V_2} = \frac{L_1}{L_2}$$
$$\therefore \quad \frac{U_1}{U_2} = \frac{\varepsilon_1}{\varepsilon_2} \cdot \frac{L_1}{L_2}$$
$$\therefore \quad \frac{\varepsilon_1}{\varepsilon_2} = \frac{U_1}{U_2} \cdot \frac{L_2}{L_1}$$
$$\Rightarrow \quad \frac{\varepsilon_1}{\varepsilon_2} = \frac{L_2}{L_1} \quad (\because U_1 = U_2)$$

8. (a) $\sigma = Y\alpha\Delta\theta = (1.2 \times 10^{11})(1.1 \times 10^{-5})(20 - 10)$
$$= 1.32 \times 10^7 \text{ Nm}^{-2}$$

9. (c) Strain produced in rods is
$$\varepsilon = \varepsilon_1 + \varepsilon_2 = \alpha_1\Delta\theta + \alpha_2\Delta\theta$$
or
$$\varepsilon = (\alpha_1 + \alpha_2)\,\Delta\theta$$
$$= (10^{-5} + 2 \times 10^{-5})\,100 = 3 \times 10^{-3}$$
If σ is the stress induced by preventing this strain, then
$$\varepsilon = \frac{\sigma}{Y_1} + \frac{\sigma}{Y_2} = \sigma\left(\frac{1}{Y_1} + \frac{1}{Y_2}\right)$$
$$\sigma = \frac{\varepsilon}{\frac{1}{Y_1} + \frac{1}{Y_2}} = \frac{3 \times 10^{-3}}{\left(\frac{1}{3 \times 10^{10}} + \frac{1}{10^{10}}\right)} = 2.25 \times 10^7 \text{ Nm}^{-2}$$

10. (a) $\Delta L = L\,\alpha\Delta\theta$, longitudinal strain $= \frac{\Delta L}{L} = \alpha\Delta\theta$
$$\therefore \quad \varepsilon = 12 \times 10^{-6} \times 20 = 24 \times 10^{-5}$$
Tensile stress in the wire $= Y \times$ strain $= 21 \times 10^{11} \times 24 \times 10^{-5}$
$$= 5.04 \times 10^8 \text{ dyne cm}^{-2}$$
$\therefore$ Tension = Tensile stress × Area of cross-sectional
$$= (5.04 \times 10^8)(0.5 \times 10^{-2}) = 25.2\text{ N}$$

11. (a) $(\Delta L)_{\text{copper}} = L\,\alpha_c\Delta\theta - \frac{2FL}{AY_c}$
$$(\Delta L)_{\text{steel}} = L\,\alpha_s\,\Delta\theta + \frac{FL}{AY_s}$$
As the assembly does not bend,
$$(\Delta L)_{\text{copper}} = (\Delta L)_{\text{steel}}$$
$$L\,\alpha_c\,\Delta\theta - \frac{2FL}{AY_c} = L\alpha_s\Delta\theta + \frac{FL}{AY_s}$$
$$\Rightarrow \quad F = \frac{AY_cY_s(\alpha_c - \alpha_s)(\Delta\theta)}{2Y_s + Y_c}$$

12. (b) Force on copper plate will be $2F$.
$$\therefore \quad F_c = 2F = \frac{2\,AY_cY_s(\alpha_c - \alpha_s)(\Delta\theta)}{2Y_s + Y_c}$$

13. (a,c,d) Force at the left end is same as force at right end. But area of cross-section at left end is minimum. So, stress at the left end is maximum.

14. (a,b) Copper expands more than invar. So, the strip will be concave at upper surface. So, signal lamp and bell will be switched ON.

15. (a)When the ends do not yield,
$$\Delta\theta = 80°\text{C} : \sigma = Y\alpha\Delta\theta = (2 \times 10^6)(12 \times 10^{-6})(80)$$
$$= 1920 \text{ kg cm}^{-2}$$
$$\Delta\theta = 100°\text{C} : \sigma = (2 \times 10^6)(12 \times 10^{-6})(100) = 2400 \text{ kg cm}^{-2}$$
When the ends yield,
$$\Delta\theta = 80°\text{C, yield is 1 mm}$$
$$(\Delta L)_{\text{net}} = L\alpha\Delta\theta - 0.1$$
$$\text{Strain, } \varepsilon = \frac{(\Delta L)_{\text{net}}}{L} = \alpha\Delta\theta - \frac{0.1}{L}$$
$$\sigma = Y\varepsilon = Y\left(\alpha\Delta\theta - \frac{0.1}{L}\right)$$
$$= 2 \times 10^6\left[12 \times 10^{-6} \times 80 - \frac{0.1}{600}\right]$$
$$= 1587 \text{ kg cm}^{-2}$$
$\Delta\theta = 100°$C and yield is 2 mm.
$$\sigma = Y\left(\alpha\Delta\theta - \frac{0.2}{L}\right) = 2 \times 10^6\left[12 \times 10^{-6} \times 100 - \frac{0.2}{600}\right]$$
$$= 1733 \text{ kg cm}^{-2}$$
Hence, (i) → (p), (ii) → (q), (iii) → (r), (iv) → (s)

16. (c) $\Delta\theta = 120°\text{C} - 20°\text{C} = 100°\text{C}$
$$\Delta L_c = L_c\,\alpha_c\,\Delta\theta = (0.5)(\alpha_c)(100) = 50\alpha_c$$
$$\Delta L_s = L_s\,\alpha_s\,\Delta\theta = (1.5)\,\alpha_s(100) = 150\alpha_s$$
$$\Delta L = \Delta L_c + \Delta L_s = 50\alpha_c + 150\alpha_s$$
But $\quad \Delta L = 0.002\text{ m} \quad$ (given)
$$\therefore \quad 50\alpha_c + 150\alpha_s = 0.002$$
$$\alpha_s = \frac{0.002 - 50\alpha_c}{150} = \frac{0.002 - 50 \times 1.6 \times 10^{-5}}{150}$$
$$= 0.8 \times 10^{-5}\,°\text{C}^{-1}$$

17. (d) If there is no change in the length of individual rod, then stress in both the rods must be equal. So,
$$\sigma_s = \sigma_c \Rightarrow Y_s\,\alpha_s\,\Delta\theta = Y_c\,\alpha_c\Delta\theta$$
$$Y_s = \frac{Y_c\alpha_c}{\alpha_s} = \frac{1.3 \times 10^{11} \times 1.6 \times 10^{-5}}{0.8 \times 10^{-5}}$$
$$= 2.6 \times 10^{11} \text{ Nm}^{-2}$$

18. (b) $\alpha_t > \alpha_b$. So, when the assembly is heated, tube tries to expand more than the bolt. But the tube is tightened by the bolt. Therefore, its expansion cannot be more than that of the bolt. Hence, a compressive stress is developed in the tube and a tensile stress is developed in the bolt.

Let the initial length of the assembly be L and let its elongation be ΔL.

Elongation of the tube if it would be free to expand
$$= L\alpha_t\Delta\theta$$
But its actual elongation is ΔL. So, elongation prevented in it
$$= L\alpha_t\Delta\theta - \Delta L$$
$$\text{So, compressive strain in the tube} = \frac{L\alpha_t\Delta\theta - \Delta L}{L}$$
$$= \frac{50 \times 2 \times 10^{-5} \times 10 - 0.0075}{50} = 5 \times 10^{-5}$$

19. (a) Compressive stress in the tube is
$$\sigma_t = \varepsilon_t\,Y_t$$
or
$$\sigma_t = (5 \times 10^{-5})10^{11} = 5 \times 10^6 \text{ Nm}^{-2}$$

20. (b) Tensile stress in bolt $= \sigma_b = \frac{Y_b(\Delta L - L\alpha_b\Delta\theta)}{L}$
$$\text{or} \quad \sigma_b = \frac{(2 \times 10^{11})[0.0075 - 50 \times 10^{-5} \times 10]}{50} = 10^7 \text{ Nm}^{-2}$$

Revisal Problems (JEE Main)

1. (c) $Y = \dfrac{FL}{A\Delta L} \Rightarrow \Delta L = \dfrac{FL}{YA} = \dfrac{4FL}{Y(\pi d^2)}$

For steel wire, $F_s = (6 + 4) \times 9.8 = 98$ N

For copper wire, $F_c = 6 \times 9.8 = 58.8$N

$$\therefore \quad \frac{(\Delta L)_s}{(\Delta L)_c} = \left(\frac{F_s}{F_c}\right)\left(\frac{L_s}{L_c}\right)\left(\frac{Y_c}{Y_s}\right)\left(\frac{d_c^2}{d_s^2}\right)$$

$$= \left(\frac{98}{58.8}\right)\left(\frac{2}{1}\right)\left(\frac{1}{2}\right)\left(\frac{2}{1}\right)^2$$

$$= \frac{98 \times 4}{58.8} = \frac{5 \times 4}{3} = \frac{20}{3}$$

2. (a) $B = -\dfrac{\Delta p}{\Delta V/V} = -\dfrac{10^7}{-0.015/100} = 6.67 \times 10^{10}$ Pa

3. (b) $V = L^3$

$$\Rightarrow \frac{\Delta V}{V} = 3 \cdot \frac{\Delta L}{L} = 3(-1\%) = -\frac{3}{100}$$

$$B = -\frac{p}{\Delta V/V} = -\frac{10^6}{-3/100} = 3.33 \times 10^7 \text{ Pa}$$

4. (d) Area of top surface is $A = 3\text{cm} \times 1\text{ cm} = 3\text{cm}^2$

$$= 3 \times 10^{-4} \text{ m}^2$$

Shearing stress, $\sigma = \dfrac{F}{A} = \dfrac{2.1}{3 \times 10^{-4}}$

$$= 0.7 \times 10^4 \text{ Nm}^{-2}$$

Angle of shear, $\theta = \dfrac{\text{Shearing stress}}{\text{Shear modulus}} = \dfrac{0.7 \times 10^4}{1.4 \times 10^5} = 0.05$ rad

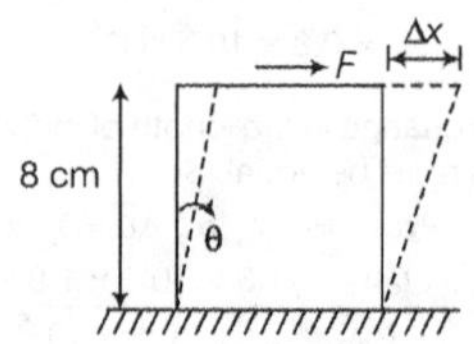

From given figure, $\tan\theta = \dfrac{\Delta x}{8\text{cm}} \Rightarrow \theta \approx \dfrac{\Delta x}{8\text{cm}}$

$$\therefore \quad \Delta x = (0.05)\, 8 = 0.4\text{cm} = 4\text{mm}$$

5. (a) Average force is

$$F_{av} = \frac{mg}{2} = \frac{AL\rho g}{2}$$

$$\Delta L = \frac{F_{av}L}{AY} = \frac{\frac{AL\rho g}{2} \cdot L}{A \cdot Y} = \frac{\rho g L^2}{2Y}$$

or $\quad \Delta L = \dfrac{(1.5 \times 10^{-3}) \times 10 \times (8)^2}{2\,(5 \times 10^6)}$

$$= 9.6 \times 10^{-8} \text{ m}$$

6. (b) $U = \dfrac{1}{2}(\sigma)\,(\varepsilon)\,(\text{Volume}) = \dfrac{1}{2} \cdot \dfrac{F}{A} \cdot \dfrac{l}{L} \cdot AL = \dfrac{1}{2} Fl$

7. (a) $U = \dfrac{1}{2} F \cdot \Delta L = \dfrac{1}{2}(20)\,(10^{-3}) = 0.01$J

8. (d) $\Delta V = \dfrac{pV}{B}$ (this is decrease in volume)

This is to be compensated by an increase in volume by heating through a temperature $\Delta\theta$.

$$\Delta V = V\gamma\Delta\theta$$

$$\therefore \quad V\gamma\Delta\theta = \frac{pV}{B}$$

$$\Rightarrow \quad \Delta\theta = \frac{p}{\gamma B}$$

9. (a) $Y = \dfrac{FL}{A\Delta L}$, Y and F are same for all wires.

So, $\quad \Delta L \propto \dfrac{L}{A}$, i.e. $\Delta L \propto \dfrac{L}{d^2}$

$\dfrac{L}{d^2}$ is highest for first wire.

So, the correct option is (a).

10. (b) $Y = \dfrac{FL}{A\Delta L} \Rightarrow \dfrac{F}{\Delta L} = \dfrac{YA}{L} \Rightarrow k = \dfrac{YA}{L}$

$$T = 2\pi\sqrt{\frac{m}{k}} = 2\pi\sqrt{\frac{m \cdot L}{YA}}$$

$$\Rightarrow \quad T \propto \frac{1}{\sqrt{Y}}$$

$$\frac{T_1}{T_2} = \sqrt{\frac{Y_2}{Y_1}} = \sqrt{\frac{2}{3}}$$

11. (d) Increase in pressure $(\Delta p) = np_0 - p_0 = (n-1)p_0$

Let V be the volume of a certain mass m of water at the surface, so that $m = \rho V$. The decrease in volume under the pressure Δp is

$$\Delta V = \frac{V\Delta p}{B}$$

$\therefore$ Volume of the same mass m of water at the given depth is

$$V' = V - \Delta V = V - \frac{V\Delta p}{B} = V\left(1 - \frac{\Delta p}{B}\right) = \frac{V}{B}(B - \Delta p)$$

Density of water at that depth is

$$\rho' = \frac{m}{V'} = \frac{\rho V}{V'} = \frac{\rho V}{\frac{V}{B}(B - \Delta p)}$$

$$\rho' = \frac{\rho B}{B - \Delta p} = \frac{\rho B}{B - (n-1)p_0}$$

12. (b) At the mid-point, $F = \dfrac{mg}{2} + mg = \dfrac{3mg}{2}$

Stress is $\sigma = \dfrac{F}{A} = \dfrac{3mg}{2A}$

13. (b) $F = -\dfrac{dU}{dx}$

$\therefore$ F is repulsive from A to B and attractive from B to C.

14. (d) $T = 2\pi\sqrt{\dfrac{m'}{k}} = 2\pi\sqrt{\dfrac{m + m_p}{k}}$

$$T^2 = \frac{4\pi^2}{k}(m + m_p)$$

where, m_p is the mass of the pan, m is the mass placed in the pan. Above equation shows that T^2 *versus* m graph will be a straight line, not passing through origin.

15. (b) Y and η depend on nature of the material.

16. (c) Breaking stress is same for both.

$$\sigma_b = \frac{F}{A}$$

$$\because F \propto A, \quad \frac{F_1}{F_2} = \frac{A_1}{A_2} = \frac{r_1^2}{r_2^2} = \left(\frac{d_1}{d_2}\right)^2 = \left(\frac{3}{1.5}\right)^2 = 4$$

$$\therefore \quad F_2 = \frac{1.5 \times 10^5}{4} = 0.375 \times 10^5 \text{ N}$$

17. (d) $\quad Y = \dfrac{FL}{A\Delta L} \Rightarrow A = \dfrac{FL}{Y\Delta L} = \left(\dfrac{10^4}{7 \times 10^9}\right) \cdot \dfrac{1}{0.2\%}$

$$A = \frac{10^4}{7 \times 10^9} \times \frac{100}{0.2} = \frac{1}{1400} = 7.1 \times 10^{-4} \text{ m}^2$$

18. (c) $$V = \frac{\pi D^2}{4} \times L$$

$\Rightarrow$ $$\Delta V = \frac{\pi L}{4} \cdot 2D\,\Delta D + \frac{\pi D^2}{4} \cdot \Delta L$$

or $$\Delta V = \frac{\pi L D}{2} \Delta D + \frac{\pi D^2}{4} \Delta L$$

$\because$ $\Delta V = 0$ (given)

$\therefore$ $$\frac{\pi L D}{2} \Delta D = -\frac{\pi D^2}{4} \Delta L$$

$\Rightarrow$ $$\left(\frac{\Delta D}{D}\right) \Big/ (\Delta L / L) = -\frac{1}{2} = -0.5$$

$\therefore$ Poisson's ratio $= -0.5$

19. (a) $$Y = \frac{FL}{A\Delta L}$$

$$\Delta L = \frac{FL}{\pi r^2 \cdot Y}$$

$$\Delta L = \frac{F \cdot L \cdot \pi r^2}{\pi^2 r^4 \cdot Y} = \frac{F\,(\text{Volume})}{\pi^2 r^4 \cdot Y}$$

$\therefore$ $$\Delta L \propto \frac{1}{r^4}; \frac{\Delta L_A}{\Delta L_B} = \left(\frac{r_B}{r_A}\right)^4 = 2^4 = 16$$

20. (a) $Y = \frac{FL}{A\Delta L}$

Here, $Y \propto \frac{1}{\Delta L}$ and wire A has least extension.

Revisal Problems (JEE Advanced)

1. (a) $k = \frac{YA}{L}$

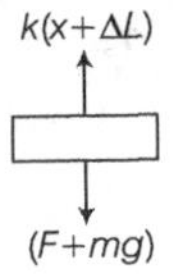

Work done by the force the wire exerts is

$$W = \int_0^{\Delta x} k\,(x + \Delta L)\,dx$$

$$W = \frac{YA}{L} \int_0^{\Delta x} (x + \Delta L)\,dx = \frac{YA}{L} \left[\frac{x^2}{2} + x\,\Delta L\right]_0^{\Delta x}$$

$$= \frac{YA}{L} \left[\frac{(\Delta x)^2}{2} + \Delta x\,\Delta L\right]$$

But $$\Delta L = \frac{mgL}{AY}$$

$\therefore$ $$W = \frac{YA}{L} \left[\frac{(\Delta x)^2}{2} + \Delta x \left(\frac{mgL}{AY}\right)\right]$$

$$W = \frac{YA}{L} \cdot \frac{(\Delta x)^2}{2} + mg\,(\Delta x)$$

2. (d) $$Y = \frac{\text{Stress}}{\text{Strain}} = \frac{\sigma_b L}{\Delta L}$$

where, σ_b = breaking stress and ΔL = elongation

$$\Delta L = \frac{\sigma_b L}{Y} = \frac{mg}{A} \cdot \frac{L}{Y} = \frac{AL\rho g}{A} \cdot \frac{L}{Y} = \frac{L^2 \rho g}{Y}$$

3. (c) $k = \frac{YA}{L}$,

$$k_{eq} = k_1 + k_2 = \frac{Y_1 A}{L} + \frac{Y_2 L}{L} = \frac{A}{L}(Y_1 + Y_2)$$

$\therefore$ $$\frac{Y(2A)}{L} = \frac{A}{L}(Y_1 + Y_2) \Rightarrow Y = \frac{Y_1 + Y_2}{2}$$

We can imagine this system to be equivalent to a wire of cross-sectional area $2A$ and length L and Young's modulus $\frac{Y_1 + Y_2}{2}$.

$\therefore$ $$U = \frac{1}{2} \times \text{Stress} \times \text{Strain} \times \text{Volume}$$

$$= \frac{1}{2} \frac{(\text{Stress})^2}{Y} \text{Volume} = \frac{1}{2} \left[\frac{Mg}{2A}\right]^2 \cdot \frac{1}{Y} \cdot (2A)(L)$$

$$= \frac{M^2 g^2 L}{4AY} = \frac{M^2 g^2 L}{4A\left(\frac{Y_1 + Y_2}{2}\right)} = \frac{M^2 g^2 L}{2A\,(Y_1 + Y_2)}$$

4. (b) $YA = L$ (given)

$$\frac{F.L}{A\Delta L} A = L \Rightarrow \frac{F}{\Delta L} = 1$$

From the given graph, slope = $\tan 45° = 1$

5. (c) $\frac{\varepsilon}{F} = 10^{-1}$ (given)

$\Rightarrow$ $$\frac{\Delta L}{LF} = 0.1$$

But $$\frac{F}{\Delta L} = 1$$

$\therefore$ $$\frac{1}{L} = 0.1 \Rightarrow L = 10\text{ m} = 1000\text{ cm}$$

6. (c) $$W = \frac{1}{2} kx^2 = \frac{1}{2} \cdot \frac{YA}{L} x^2 \qquad \left(\because k = \frac{YA}{L}\right)$$

$$Y = \frac{2LW}{Ax^2}$$

7. (d) Stress $= \frac{\text{Weight}}{\text{Area}} \Rightarrow \sigma_b = \frac{AL\rho g}{A} \Rightarrow L = \frac{\sigma_b}{\rho g}$

So, at length $\frac{\sigma_b}{\rho g}$ wire will break.

Given, length of wire $= \frac{5\sigma_b}{4\rho g}$

From O to D length $= \frac{4\sigma_b}{4\rho g} = \frac{\sigma_b}{\rho g}$

$\therefore$ At point D, the wire breaks.

8. (c) $$Y = \frac{(Mg/2)}{A'} \cdot \frac{L}{\Delta L} = \frac{gL}{2\,(\Delta L)} \cdot \frac{M}{A'} \left(\frac{L}{L}\right) = \frac{gL^2}{2\,(\Delta L)} \cdot \frac{(M/L)}{A'}$$

$A = \frac{M/L}{A'}$ = mass per unit length per unit area

9. (a) $$\text{Slope} = \frac{\Delta L}{F} = \frac{10^{-4}}{20}$$

$\Rightarrow$ $$\frac{F}{\Delta L} = 2 \times 10^5$$

$$Y = \frac{FL}{A\Delta L} = \frac{F}{\Delta L} \cdot \frac{L}{A}$$

$$= (2 \times 10^5)\left(\frac{1}{10^{-6}}\right) = 2 \times 10^{11}\text{ Nm}^{-2}$$

10. (b,c,d) $Y = \frac{TL}{A\Delta L}$, here A, ΔL and L are same for all the 3 wires.

$\therefore$ $Y \propto T$ ...(i)

For rotational equilibrium, $T_1\,(2x) = T_3(x)$

$\therefore$ $2T_1 = T_3$

So, $2Y_1 = Y_3$ [from Eq. (i)]

Given, $Y_2 = Y_3$

$\therefore$ $T_2 = T_3$ [from Eq. (i)]

$T_1 + T_2 + T_3 = Mg$ (for translational equilibrium)

$T_1 + 2T_1 + 2T_1 = Mg \Rightarrow T_1 = Mg/5$

11. (a,b,c,d) $f_0 = \frac{1}{2L}\sqrt{\frac{F}{\mu}} = \frac{1}{2L}\sqrt{\frac{F}{\rho A}}$

where, F = tension

μ = mass per unit length

$$\text{Stress} = \frac{F}{A} = Y\,(\text{Strain}) = Y\left(\frac{\Delta L}{L}\right) = Y\left(\frac{L\alpha\theta}{L}\right) = Y\alpha\theta$$

$\therefore$ $$f_0 = \frac{1}{2L}\sqrt{\frac{Y\alpha\theta}{\rho}}$$

$\therefore$ $$f_0 \propto \frac{1}{L} \propto \sqrt{Y} \propto \sqrt{\theta} \propto \sqrt{\frac{\alpha}{\rho}}$$

12. (b,c) The more the modulus of elasticity, the more is the resistance offered to external deforming forces.

13. (a,b) $\Delta L = \frac{FL}{AY}$, ΔL can be increased by making F or L two times.

14. (a) In tyre, air is compressed, the elasticity involved is bulk. Automobile shaft is sheared, the elasticity involved is shear, i.e. rigidity. The deformation of wire of spring is twisting strain. The elasticity involved is shear, i.e. rigidity.

Hence, (i) $\rightarrow$ (p), (ii) $\rightarrow$ (r), (iii) $\rightarrow$ (r), (iv) $\rightarrow$ (q)

15. (b) In case of shear stress, shape changes but volume of the body does not change.

In case of longitudinal (tensile/compressive) stress, shape does not change but volume changes.

Hence, (i) $\rightarrow$ (p,s), (ii) $\rightarrow$ (q,r), (iii) $\rightarrow$ (q,r), (iv) $\rightarrow$ (q,r)

16. (9) $mg = k(9) \Rightarrow k = \frac{mg}{9}$

$k \propto \frac{1}{L} \Rightarrow$ Each of the two parts has spring constant $2k$ when they are in parallel.

$$k_{eq} = 2k + 2k = 4k$$

Now, $$mg = 4kx'$$

$\Rightarrow$ $$x' = \frac{mg}{4k} = \frac{mg \cdot 9}{4 \cdot mg} = \frac{9}{4}\text{ cm}$$

$\therefore$ $$\beta = 9$$

17. (4) Tension, $T = \sigma_b \cdot A = 4.8 \times 10^7 \times 10^{-6} = 48$ N

$$T = mr\omega^2$$

$\Rightarrow$ $$\omega = \sqrt{\frac{T}{mr}} = \sqrt{\frac{48}{10\,(0.3)}} = 4\text{ rad s}^{-1}$$

18. (b) Among the given materials, ultimate strength of bone (or human femur) is maximum.

19. (b) Strain $= \frac{\text{Stress}}{Y}$

$$\left(\frac{\Delta L}{L}\right)_{max} \times 100 = \frac{\text{Maximum stress}}{Y} \times 100$$

$$= \frac{400 \times 10^6}{200 \times 10^9} \times 100 = 0.2\%$$

20. (d) Smaller the value of Young's modulus, greater the flexibility.

JEE Main & AIEEE Archive

1. (a) By Hooke's law,

$$\text{Young's modulus } (Y) = \frac{\text{Tensile stress}}{\text{Tensile strain}}$$

So, $$Y = \frac{F/A}{\Delta L/L} = \frac{FL}{A\Delta L}$$

If the rod is compressed, then compressive stress and strain appear. Their ratio Y is same as that for tensile case.

Given, length of a steel wire (ΔL) = 10 cm, Temperture (t) = 100°C

As length is constant.

$\therefore$ Strain $= \frac{\Delta L}{L} = \alpha\,\Delta\theta$ [as $\Delta L = (L\Delta\theta)$]

Now, Pressure = Stress = $Y \times$ Strain

$$= 2 \times 10^{11} \times 1.1 \times 10^{-5} \times 100 = 2.2 \times 10^8\text{ Pa}$$

2. (b) Frequency, $f = \frac{v}{2l} = \frac{1}{2l}\sqrt{\frac{T}{\mu}} = \frac{1}{2l}\sqrt{\frac{T}{Ad}}$

where, T is tension in the wire and μ is the mass per unit length of wire.

Also, Young's modulus, $Y = \frac{Tl}{A\Delta l} \Rightarrow \frac{T}{A} = \frac{Y\Delta l}{l}$

Putting this value in expression of frequency, we have

$$f = \frac{1}{2l}\sqrt{\frac{Y\Delta l}{ld}}$$

Given, $l = 1.5$ m, $\frac{\Delta l}{l} = 0.01$, $d = 7.7 \times 10^3$ kg m^{-3},

$Y = 2.2 \times 10^{11}$ Nm^{-2}

Putting these values, we have

$$f = \frac{1}{2 \times 1.5}\sqrt{\frac{2.2 \times 10^{11} \times 0.01}{7.7 \times 10^3}}$$

$$f = \sqrt{\frac{2}{7}} \times \frac{10^3}{3}$$

$$f \approx 178.2\text{ Hz}$$

3. (d) As no relation between k_1 and k_2 is given in the question, that is why, nothing can be predicted about Statement I. But as in Statement II, $k_1 < k_2$.

Then, for same force,

$$W = F \cdot x = F \cdot \frac{F}{k} = \frac{F^2}{k}$$

$\Rightarrow$ $$W \propto \frac{1}{k}$$

i.e. $$W_1 > W_2$$

But for same displacement,

$$W = F \cdot x$$

$$= \frac{1}{2}kx \cdot x = \frac{1}{2}kx^2$$

$\Rightarrow$ $$W \propto k$$

i.e. $$W_1 < W_2$$

Thus, Statement II is correct and Statement I is incorrect.

4. (d) Elongation due to change in temperature,

$$\Delta l = L\alpha\Delta T$$

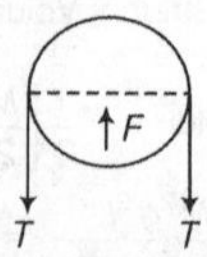

which is compensated by elastic strain, when temperature becomes normal, i.e.

$$\Delta l = \frac{TL}{YS}$$

Thus, $$\frac{TL}{YS} = L\alpha\Delta T$$

$\Rightarrow$ $$T = YS\alpha\Delta T$$

At equilibrium, force exerted by one half on other,

$$F = 2T = 2YS\alpha\Delta T$$

5. (c) As change in length,

$$\Delta L = \alpha L \Delta T = \frac{FL}{AY}$$

$\Rightarrow$ $$\text{Stress} = \frac{F}{A} = Y\alpha\Delta T$$

6. (c) From FBD of sphere, using Lami's theorem,

$$\frac{F}{mg} = \tan\theta \quad \ldots(\text{i})$$

When suspended in liquid, as θ remains same.

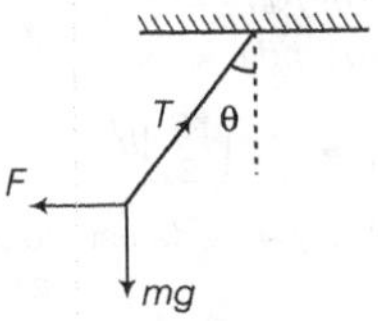

$\therefore$ $$\frac{F'}{mg\left(1-\frac{\rho}{d}\right)} = \tan\theta \quad \ldots(\text{ii})$$

From Eqs. (i) and (ii), we get

$$\frac{F}{mg} = \frac{F'}{mg\left(1-\frac{\rho}{d}\right)}, \text{ where } F' = \frac{F}{K}$$

$\therefore$ $$\frac{F}{mg} = \frac{F}{mg\,K\left(1-\frac{\rho}{d}\right)}$$

or $$K = \frac{1}{1-\frac{\rho}{d}} = \frac{1}{\left(1-\frac{0.8}{1.6}\right)} = 2$$

7. (d) As volume is same,

$\therefore$ $$A_1 l_1 = A_2 l_2$$

$\Rightarrow$ $$l_2 = \frac{A_1 l_1}{A_2} = \frac{A \times l_1}{3A} = \frac{l_1}{3}$$

$\Rightarrow$ $$\frac{l_1}{l_2} = 3$$

$\because$ $$\Delta x_1 = \frac{F_1}{AY} \times l_1 \quad \ldots(\text{i})$$

and $$\Delta x_2 = \frac{F_2}{3AY} \times l_2 \quad \ldots(\text{ii})$$

Here, $$\Delta x_1 = \Delta x_2$$

$\therefore$ $$\frac{F_2}{3AY} l_2 = \frac{F_1}{AY} l_1$$

$\Rightarrow$ $$F_2 = 3F_1 \times \frac{l_1}{l_2} = 3F_1 \times 3 = 9F$$

8. (b) Energy stored in wire $= \frac{1}{2} \times$ Stress $\times$ Strain $\times$ Volume

and Young's modulus $= \frac{\text{Stress}}{\text{Strain}}$

$\Rightarrow$ $$\text{Strain} = \frac{S}{Y}$$

$\therefore$ $$\frac{\text{Energy stored in wire}}{\text{Volume}} = \frac{1}{2} \times \text{Stress} \times \text{Strain} = \frac{1}{2} \times S \times \frac{S}{Y} = \frac{S^2}{2Y}$$

9. (d) As work done in stretching the wire

$$= \text{Potential energy stored}$$
$$= \frac{1}{2} \times \text{Stress} \times \text{Strain} \times \text{Volume}$$
$$= \frac{1}{2} \times \frac{F}{A} \times \frac{l}{L} \times AL = \frac{1}{2} Fl$$

10. (d) Elastic energy stored in the wire is

$$U = \frac{1}{2} \times \text{Stress} \times \text{Strain} \times \text{Volume}$$
$$= \frac{1}{2} \times \frac{F}{A} \times \frac{\Delta l}{L} \times AL = \frac{1}{2} F\Delta l$$
$$= \frac{1}{2} \times 200 \times 1 \times 10^{-3}$$
$$= 0.1 \text{ J}$$

JEE Advanced & IIT JEE Archive

1. (c) $$\Delta l = \frac{FL}{AY} = \frac{FL}{(\pi r^2)Y}$$

$\therefore$ $$\Delta l \propto \frac{L}{r^2}$$

$\therefore$ $$\frac{\Delta l_1}{\Delta l_2} = \frac{L/R^2}{2L/(2R)^2} = 2$$

2. (a,d) On small sphere,

$$\frac{4}{3}\pi R^3(\rho)g + kx = \frac{4}{3}\pi R^3(2\rho)g \quad \ldots(\text{i})$$

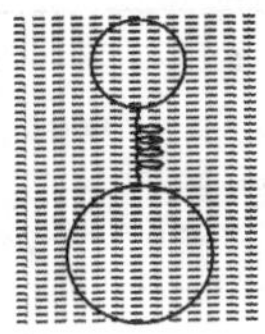

On second sphere (large),

$$\frac{4}{3}\pi R^3(3\rho)g = \frac{4}{3}\pi R^3(2\rho)g + kx \quad \ldots(\text{ii})$$

From Eqs. (i) and (ii), we get

$$x = \frac{4\pi R^3 \rho g}{3k}$$

The complete arrangement has high density than that of the liquid. So, the light sphere is completely submerged.

3. (d) From the definition of bulk modulus,

$$B = \frac{-dp}{(dV/V)}$$

Substituting the values, we have

$$B = \frac{(1.165 - 1.01) \times 10^5}{(10/100)}$$
$$= 1.55 \times 10^5 \text{ Pa}$$

5. Fluid Mechanics

DPP-1 Pressure in Static/Accelerated Fluids

1. (c) $p_1 = p_0 + 10$ (in cm of Hg) and $p_2 = p_0 + 5$ (in cm of Hg)

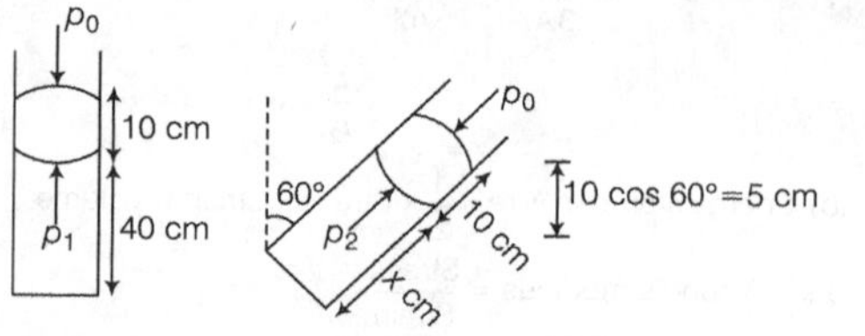

Let A be the area of cross-section of the tube. In this process, there is no temperature change. So, $p_1V_1 = p_2V_2$.

$\therefore \quad (p_0 + 10)40A = (p_0 + 5)xA$

$\Rightarrow \quad (76 + 10)40 = (76 + 5)x$

$$x = \frac{86}{81} \times 40 = 42.47\text{cm}$$

2. (d) $p_1 = p_2 \Rightarrow \dfrac{F_1}{A_1} = \dfrac{F_2}{A_2}$

or $\quad \dfrac{500 \times 9.8}{5} = \dfrac{F_2}{200 \times 10^{-4}}$

$\therefore \quad F_2 = 19.6\text{ N}$

3. (a) Force on the side walls is

F_1 = Average pressure × curved surface area

or $\quad F_1 = \left(\dfrac{\rho g h}{2}\right)(2\pi Rh) = \pi R\rho g h^2$ (ρ = density of liquid)

Force on the bottom is F_2 = Pressure at the bottom × area of the bottom

$$= (\rho g h)(\pi R^2) = \pi \rho g h R^2$$

Given, $\quad F_1 = F_2$

$\therefore \quad \pi R \rho g h^2 = \pi \rho g h R^2 \Rightarrow h = R$

4. (b) Pressure at point $Q_2 = \rho g(H - y)$

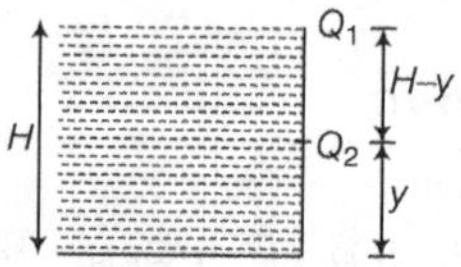

Area between Q_1 and $Q_2 = (H - y)(1) = H - y$

Average pressure between Q_1 and $Q_2 = \rho g\left(\dfrac{H-y}{2}\right)$

$\therefore$ Required force is

$$F = \rho g\left(\frac{H-y}{2}\right)(H-y)$$

$$= \frac{1}{2}\rho g(H-y)^2$$

So, $\quad (H-y)^2 = \dfrac{2}{\rho g}F$

At $y = H$, $\quad F = 0 \quad$ (here, depth = 0)

At $y = 0$, $\quad F = \dfrac{\rho g H^2}{2} \quad$ (here, depth = H)

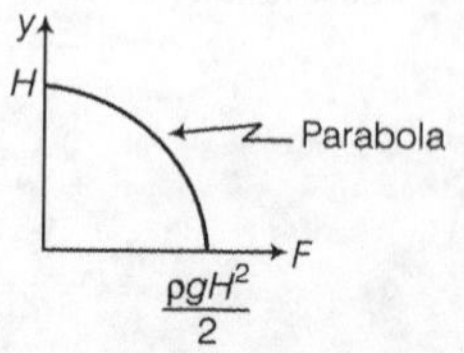

5. (b) Consider an element of liquid of thickness dh as shown in the figure. Density at this location is $\rho = kh$ (where, k is a constant). Pressure due to this element is

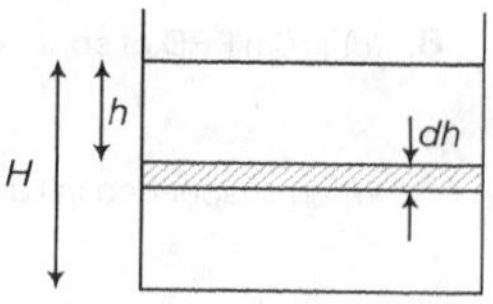

$$dp = (kh)g\,dh$$

$$\Rightarrow \quad p = kg\int_0^h h\,dh$$

$$= \frac{1}{2}kg\,h^2 = \left(\frac{kg}{2}\right)h^2$$

$p = \dfrac{kg}{2}h^2$ is similar to $y = kx^2$ (a parabola).

6. (a) Centripetal force on the element considered is

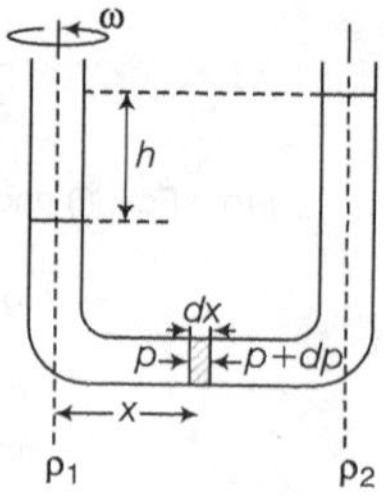

$$(dp)A = (dm)x\omega^2$$

$$= (dx\,A\rho)x\omega^2$$

$$dp = \rho\omega^2 x\,dx$$

$$\int_{p_1}^{p_2} dp = \rho\omega^2\int_0^L x\,dx$$

$$p_2 - p_1 = \frac{\rho\omega^2L^2}{2},$$

Also, $\quad p_2 - p_1 = \rho g h$

$\therefore \quad \rho g h = \dfrac{\rho\omega^2 L^2}{2}$

$\Rightarrow \quad h = \dfrac{\omega^2L^2}{2g}$

7. (c) $a_x = a_0\cos\alpha = 5\cos 37° = 4\text{ ms}^{-2}$

$a_y = a_0\sin\alpha = 5\sin 37° = 3\text{ ms}^{-2}$

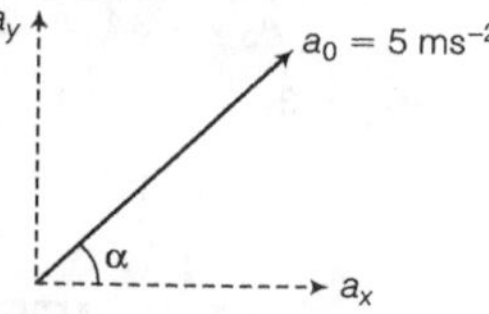

Now, $\theta = \tan^{-1}\left(\dfrac{a_x}{a_y + g}\right) = \tan^{-1}\left(\dfrac{4}{3+10}\right) = \tan^{-1}\left(\dfrac{4}{13}\right)$

8. (a) As, the volumes are equal, the distances are as shown in figure.

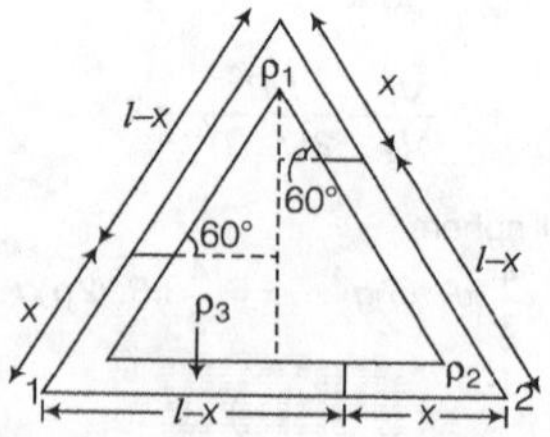

Let $\quad \rho_1 = \rho, \rho_2 = \rho + d$ and $\rho_3 = \rho + 2d$

Pressures at 1 and 2 are equal. So,

$\rho_1 g(l - x)\sin 60° + \rho_3\,gx\sin 60° = \rho_1 gx\sin 60° + \rho_2 g(l - x)\sin 60°$

$\therefore \quad \rho_1(l - x) + \rho_3 x = \rho_1 x + \rho_2(l - x)$

$$\rho_1 l - \rho_1 x + \rho_3 x = \rho_1 x + \rho_2 l - \rho_2 x$$

$$x(\rho_3 - \rho_1 - \rho_1 + \rho_2) = (\rho_2 - \rho_1)l$$

$$x(\rho + 2d - \rho - \rho + \rho + d) = (\rho + d - \rho)\,l$$

$$3xd = ld$$

$$\Rightarrow \quad x = \frac{l}{3}$$

9. (d) Let acceleration of tube is a' and radius of cylinder is R.

There is no slipping between tube and cylinders. So, points A and A' should have same acceleration.

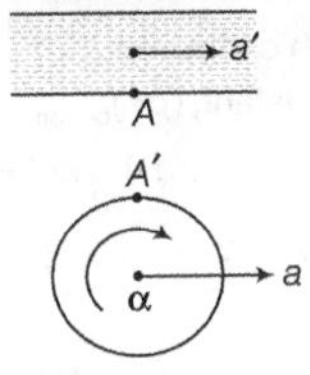

$\therefore \quad a' = a + R\alpha \quad \ldots(i)$

Similarly, there is no slipping between cylinder and the ground.

So, points B and B' should have same acceleration.

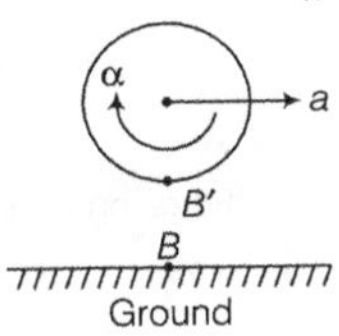

$\therefore \quad 0 = a - R\alpha \quad \ldots(ii)$

From Eqs. (i) and (ii), we get

$$a' = 2a$$

Pressure at A is

$$p_A = p_{atm} + \rho g H \quad \text{(from vertical limb)}$$

Also, pressure at A is

$$p'_A = p_{atm} + \rho a' L \quad \text{(from horizontal limb)}$$

$$= p_{atm} + \rho(2a)L$$

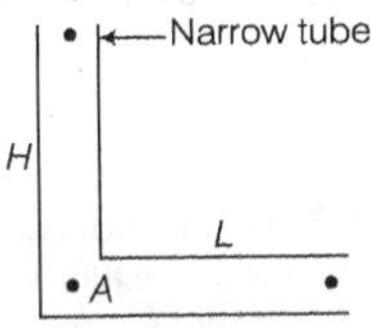

For no spilling from any opening,

$p_A = p'_A$

$\therefore \quad p_{atm} + \rho g H = p_{atm} + \rho(2a)L$

$\Rightarrow \quad a = \dfrac{gH}{2L}$

10. (a) Pressure at the location of element is

$$p = \rho g h$$

or $$p = \rho g y \sin\theta$$

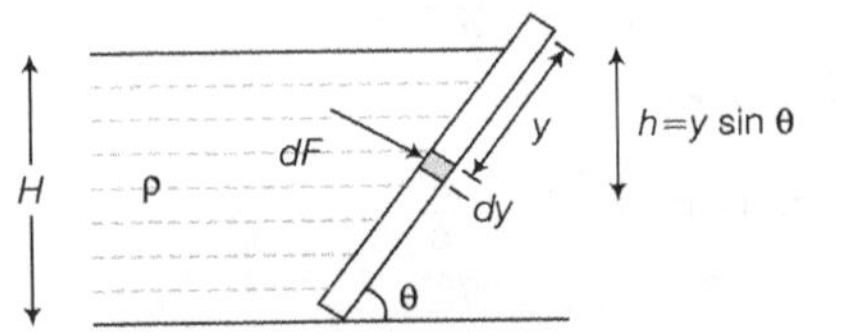

Force on the element is $dF = p(bdy)$, where b is the width of the wall.

$$dF = \rho g y \sin\theta (bdy) = \rho g b \sin\theta \; y dy$$

$\therefore$ Total force on the wall due to the liquid is

$$F = \rho g b \sin\theta \int_0^{H/\sin\theta} y dy$$

or $$F = \rho g \, b \sin\theta \frac{H^2}{2\sin^2\theta} = \frac{\rho g b H^2}{2\sin\theta}$$

or $$\frac{F}{b} = \frac{\rho g H^2}{2\sin\theta}$$

11. (d) $p_B - p_A = (\rho g h + \rho a_0 L + P_{atm}) - P_{atm} = (gh + a_0 L)\rho$

12. (a) Consider a fluid particle P of mass m at coordinates (x, y) as shown in the figure. From a non-inertial (rotating) frame of reference two forces are acting on it.

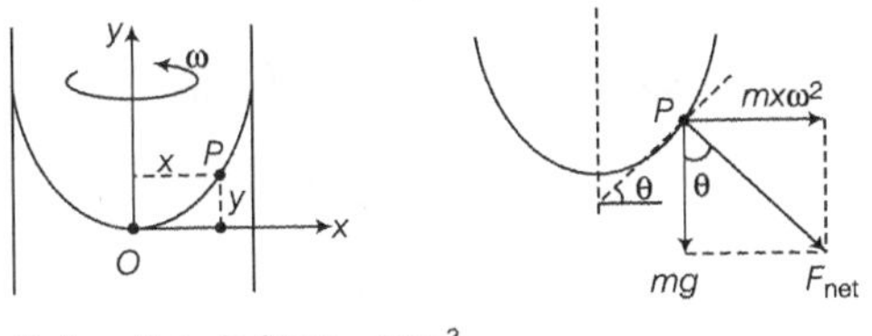

(i) F_p = Pseudo force = $mx\omega^2$

(ii) Weight = mg

Net force on the particle should be perpendicular to the free surface (in equilibrium).

Hence, $$\tan\theta = \frac{mx\omega^2}{mg} = \frac{x\omega^2}{g}$$

But $\dfrac{dy}{dx} = \tan\theta$ so, $\dfrac{dy}{dx} = \dfrac{x\omega^2}{g} \quad \Rightarrow \quad dy = \dfrac{\omega^2}{g} x dx$

$$\int_0^y dy = \frac{\omega^2}{g}\int_0^x x dx \quad \Rightarrow \quad y = \frac{\omega^2}{g} \cdot \frac{x^2}{2}$$

$\Rightarrow \quad y = \left(\dfrac{\omega^2}{2g}\right) x^2$

This is the equation of the free surface of liquid, which is a parabola.

At radial distance $x = r$, $y = \dfrac{\omega^2}{2g} r^2$

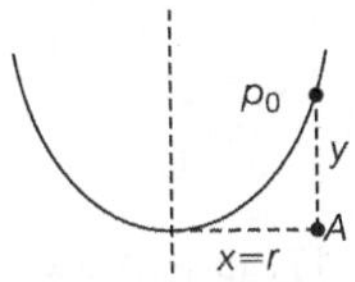

$\therefore$ Pressure at $A = p_0 + \rho g y$

$$= p_0 + \frac{\rho\omega^2 r^2}{2}$$

13. (c) We know that

$$\tan\theta = \frac{a_0}{g} = \frac{9}{10}$$

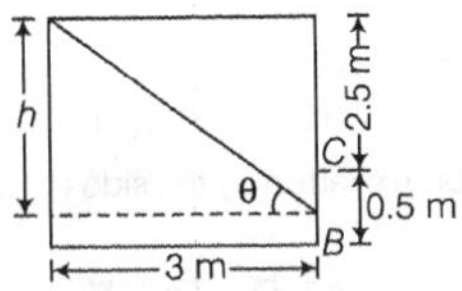

Also, $\quad \tan\theta = \dfrac{h}{3} \quad$ (from figure)

$\therefore \quad \dfrac{h}{3} = \dfrac{9}{10} \quad \Rightarrow \quad h = 2.7 \text{ m}$

So, when water level goes below C no water comes out through C.

Initial volume of water is

$$V_i = 3 \times 3 \times 2 = 18 \text{ m}^3$$

Final volume of water is

$$V_f = \frac{1}{2} \times 3 \times 2.7 \times 3 + 3 \times 0.3 \times 3 = 14.85 \text{ m}^3$$

$\therefore$ Volume of water spilled $= V_i - V_f = 18 - 14.85 = 3.15 \text{ m}^3$

$= 3150 \text{ L}$

14. (a,b,c) If a_0 is the horizontal acceleration of the vessel, then the water surface makes an angle $\theta = \tan^{-1}\left(\dfrac{a_0}{g}\right)$ with the horizontal.

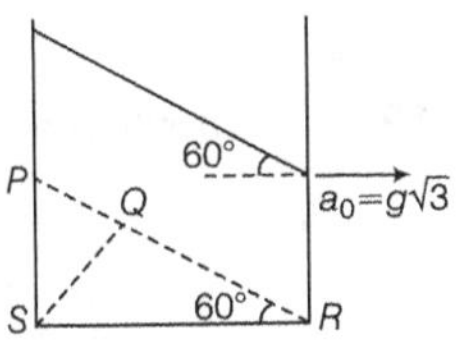

$\therefore \quad \theta = \tan^{-1}\left(\dfrac{g\sqrt{3}}{g}\right)$

$$= \tan^{-1}(\sqrt{3}) = 60^\circ$$

So, in the accelerating container, free surface of water is parallel to PQR. So, pressure at every point on the PQR will be same. And as one moves from Q to S, pressure increases.

15. (a,c) Let z_{min} is the minimum level of the dotted parabola.

$\therefore \quad z_1 = z_{min} + \dfrac{r_1^2\omega^2}{2g} = z_{min} + \dfrac{(0.1)^2(10)^2}{2 \times 10}$

and $\quad z_2 = z_{min} + \dfrac{r_2^2\omega^2}{2g} = z_{min} + \dfrac{(0.2)^2(10)^2}{2 \times 10}$

But $\quad z_1 + z_2 = 2 \times 0.4 = 0.8 \text{ m}$

Solving the above equations, we get

$$z_{min} = 0.275 \text{ m}$$

$$\therefore \quad z_1 = 0.275 + \frac{(0.1)^2(10)^2}{2 \times 10} = 0.325 \text{ m}$$

and $$z_2 = 0.275 + \frac{(0.2)^2(10)^2}{2 \times 10} = 0.475 \text{ m}$$

So, $$\frac{z_1}{z_2} = \frac{0.325}{0.475} = 0.684$$

16. (b) Force exerted by water on the bottom is

$$F_1 = (p_0 + \rho gh) A_1$$

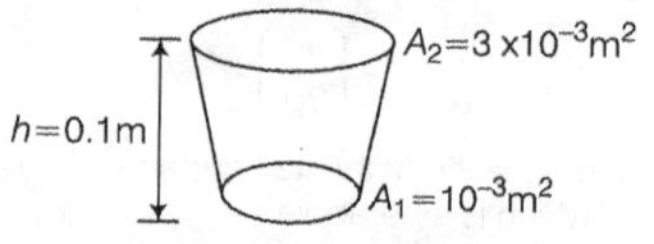

$$\therefore \quad F_1 = (1.01 \times 10^5 + 1000 \times 10 \times 0.1)\, 10^{-3}$$
$$= 102 \text{ N (downwards)}$$

17. (d) FBD of water only is shown in the figure,

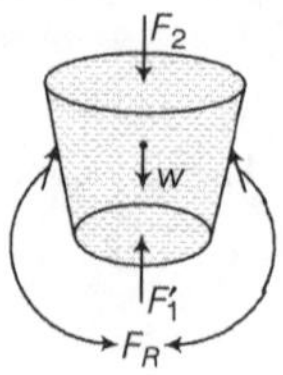

Forec exerted by atmosphere on water is

$$F_2 = p_0(A_2) = 1.01 \times 10^5 (3 \times 10^{-3})$$
$$= 303 \text{ (downwards)}$$

Force exerted by bottom on water is $F_1' = 102$ N (upwards)

Weight of water is

$$w = V\rho g = (10^{-3})(1000)(10) = 10 \text{ N (downwards)}$$

F_R = Resultant force exerted by the sides of the glass on the water. This is upwards.

For equilibrium, $$F_R + F_1' = F_2 + w$$
$$F_R + 102 = 303 + 10$$
$$\Rightarrow \quad F_R = 211 \text{ N (upwards)}$$

18. (b) Now, force on the bottom is

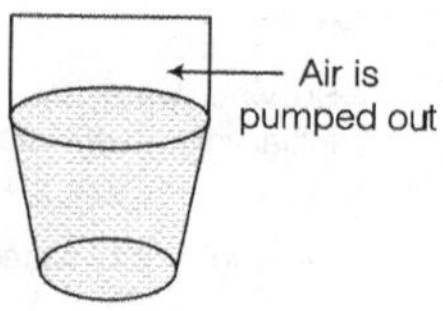

$$F_1 = (\rho g h) A_1$$
$$= (1000 \times 10 \times 0.1)\, 10^{-3}$$
$$= 1 \text{ N (downwards)}$$

Force on water by atmosphere on water is zero i.e. $F_2 = 0$

Weight is $w = 10$ N (downwards)

Again for equilibrium,

$$F_R + F_1' = F_2 + w$$
$$F_R + 1 = 0 + 10$$
$$\Rightarrow \quad F_R = 9 \text{ N (upwards)}$$

19. (6) At a depth h, density is $\rho = \rho_0(1 + h)$

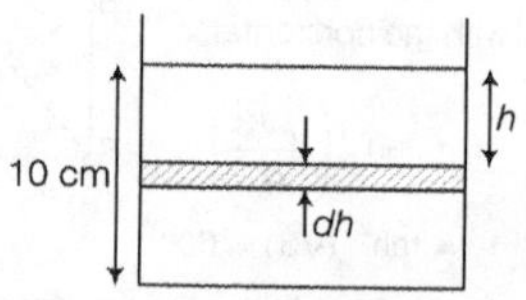

$\therefore$ Mass of the element is $dm = \rho_0(1 + h)Adh$

or $$dm = (0.8)(1 + h)(100)\, dh$$

or $$dm = 80\,(1 + h)\, dh$$

$\therefore$ Mass of the liquid is $$m = 80 \int_0^{10} (1 + h) dh$$

$$= 80 \left[10 + \frac{10^2}{2}\right] = 4800 \text{ g}$$

Volume is $V = 100 \times 10 = 1000 \text{ cm}^3$

$$\therefore \quad \rho_{av} = \frac{m}{V} = \frac{4800}{1000} = 4.8 \text{ gcm}^{-3} = \frac{8k}{10} \text{ gcm}^{-3}$$

$$\Rightarrow \quad k = 6$$

20. (2) Gauge pressure at the depth H is ρgH. The depth of point P $\frac{h}{2}$, $\therefore$ $p_{gauge} = \rho g\left(\frac{h}{2}\right)$.

21. (7) Let H be the depth of the lake.

As temperature is constant, $(pV)_{bottom} = (pV)_{surface}$

$$[\rho gh + H(1)(g)]\, \frac{4}{3}\pi r^3 = (\rho gh)\, \frac{4}{3}\pi (2r)^3$$

$$\therefore \quad H = 7h\rho \Rightarrow \beta = 7$$

22. Force on slant face is

$$F_1 = p_{av} \times \text{Area of slant face}$$
$$= \frac{\rho gh}{2}(5 \times 1)$$
$$= \frac{1000 \times 10 \times 4}{2} \times 5 = 10^5 \text{ N}$$

Force on vertical face is

$$F_2 = p_{av} \times \text{Area of vertical face}$$
$$= \frac{\rho gh}{2}(4 \times 1) = \frac{1000 \times 10 \times 4}{2}(4 \times 1)$$
$$= 8 \times 10^4 \text{ N}$$

Force on bottom face is

$$F_3 = p_{av} \times \text{Area of bottom face}$$
$$= \rho gh (3 \times 1)$$
$$= 1000 \times 10 \times 4 \times 3 = 12 \times 10^4 \text{ N}$$

23. Let the width of gate perpendicular to the plane of the figure is b. Force at a depth x on the element, is $(\rho gx) bdx$.

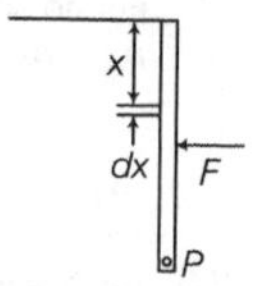

Torque of this force about pivot is

$$d\tau = \rho gxbdx.(h - x)$$

$$\therefore \quad \tau = \rho gb \int_0^h (xh - x^2)\, dx = \rho gb\left(\frac{h^3}{2} - \frac{h^3}{3}\right)$$

$$= \frac{\rho gbh^3}{6} = \frac{\rho gAh^2}{6} \qquad (\because A = bh)$$

$$\tau_{\text{due to } F} = F.\frac{2h}{3}$$

For equilibrium, $$F.\frac{2h}{3} = \frac{\rho gAh^2}{6}$$

$$\Rightarrow \quad F = \frac{\rho gAh}{4}$$

For translational equilibrium,

$$F_{net} = 0$$

Total force due to liquid is

$$F' = \rho gb \int_0^h x dx = \frac{\rho gbh^2}{2}$$

or $$F' = \frac{\rho gAh}{2}$$

$$F_{net} = 0$$
$$\Rightarrow \quad F + R_x = F'$$
$$\Rightarrow \quad R_x = F' - F = \frac{\rho gAh}{2} - \frac{\rho gAh}{4}$$

or $$R_x = \frac{\rho gAh}{4}$$

24. For horizontal portion of the liquid,

$$F_{net} = ma$$

$$\Rightarrow (p_0 + 2\rho gh)A - (p_0 + \rho gh)A = \left[2\rho\left(\frac{Al}{2}\right) + \rho\left(\frac{Al}{2}\right)\right] a_0$$

$$\rho ghA = \frac{3}{2}\rho Ala_0$$

$$h = \frac{3a_0 l}{2g}$$

DPP-2 Archimedes' Principle and Buoyant Force

1. (c) $$V_i \rho_L g = V_s \rho_s g$$

Where, V_i = immersed volume

V_s = volume of solid

ρ_L = density of liquid

ρ_s = density of solid

$$\therefore \quad \frac{V_i}{V_s} = \frac{\rho_s}{\rho_L} = \frac{900}{1000} = 0.9$$

$\Rightarrow$ Fraction of volume outside water = 1 – 0.9 = 0.1

2. (a) In floatation, weight of the floating body = buoyant force.

Let m be the mass of ice. Then, $mg = V_i \rho_w g$

$$\Rightarrow \quad V_i = \frac{m}{\rho_w} \quad \text{...(i)}$$

When ice of mass m melts, let the volume of water formed is V. Then,

$$V = \frac{m}{\rho_w} \quad \text{...(ii)}$$

From Eqs. (i) and (ii), $V_i = V$

Hence, level of water does not change.

3. (b) Let mass of copper = m

Then, mass of gold = 50 – m

Volume of copper is $V_c = \dfrac{m}{10}$

Volume of gold is $V_g = \dfrac{50 - m}{20}$ $\quad \left(\because V = \dfrac{m}{\rho}\right)$

When immersed in water, decrease in weight = upthrust

$$\therefore \quad (50 - 46)\, g = (V_c + V_g)\rho_w g$$

$$4 = \left[\frac{m}{10} + \frac{50 - m}{20}\right](1) \quad (\because \rho_w = 1 \text{ g cc}^{-1})$$

So, $80 = 2m + 50 - m$

$\Rightarrow$ $m = 30$ g

4. (d) When the system is at rest,

B_F = Buoyant force

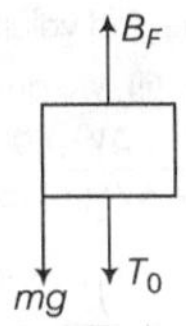

$$mg + T_0 = B_F$$

$$\Rightarrow \quad V\rho_S g + T_0 = V\rho_L g$$

$$T_0 = Vg\,(\rho_L - \rho_S) \quad \text{...(i)}$$

When the system is accelerating,

(from non-inertial frame)

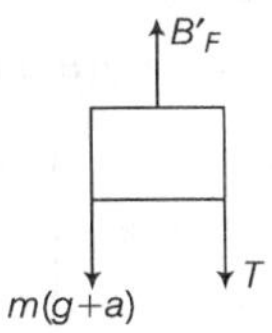

$$B'_F = m(g + a) + T$$

$$V\rho_L\,(g + a) = V\rho_s\,(g + a) + T$$

$$T = V(g + a)\,(\rho_L - \rho_s) \quad \text{...(ii)}$$

On dividing Eq. (ii) by Eq. (i), we get

$$\frac{T}{T_0} = \frac{g + a}{g}$$

$$\Rightarrow \quad T = T_0\left(1 + \frac{a}{g}\right)$$

5. (a) $B_{F_1} + B_{F_2} = mg$

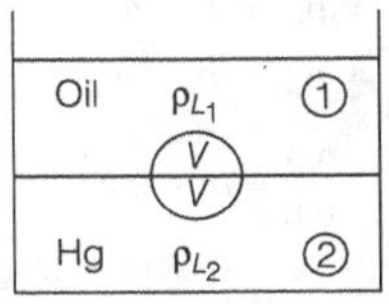

$$V\rho_{L_1} g + V\rho_{L_2} g = (2V)\rho_s g$$

$$\therefore \quad \rho_s = \frac{\rho_{L_1} + \rho_{L_2}}{2} = \frac{0.8 + 13.6}{2}$$

$$\rho_s = 7.2 \text{ g cc}^{-1}$$

6. (c) V = volume of amount of solid in the hollow sphere

$$= \frac{4}{3}\pi\,(R^3 - r^3)$$

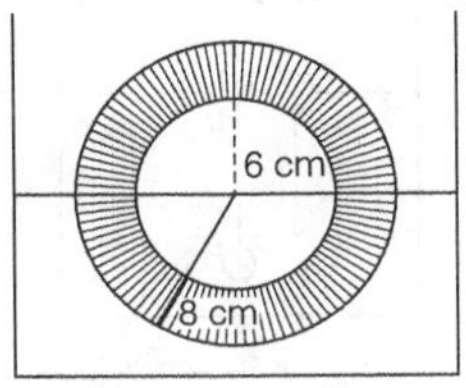

$$mg = B_F$$

$$V\rho_s g = V_i \rho_L g$$

$$\Rightarrow \quad \frac{4}{3}\pi\,(R^3 - r^3)\rho_s = \left(\frac{2}{3}\pi R^3\right)\rho_L$$

$$\rho_L = 1$$

$$\therefore \quad \rho_s = \frac{R^3}{2\,(R^3 - r^3)}$$

$$= \frac{1}{2\left[1 - \dfrac{r^3}{R^3}\right]} = \frac{1}{2\left[1 - \left(\dfrac{6}{8}\right)^3\right]}$$

$$\therefore \quad \rho_s = 0.865 \text{ g cc}^{-1}$$

$\therefore$ Specific gravity = 0.865

7. (b) $$B_{F_1} + B_{F_2} = (m_1 + m_2)\,g$$

$$V_1\rho_w g + V_2\rho_w g = (m_1 + m_2)\,g$$

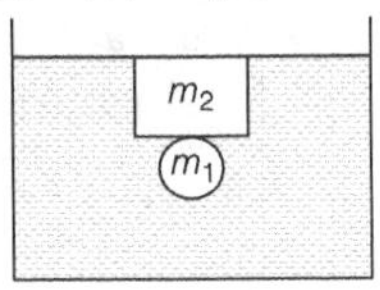

$$\frac{m_1}{\rho_1}\cdot\rho_w + \frac{m_2}{\rho_2}\cdot\rho_w = m_1 + m_2$$

$$\frac{m}{7.8} + \frac{0.2}{0.8} = m + 0.2$$

$$\therefore \quad m = 0.0573 \text{ kg} = 57.3 \text{ g}$$

8. (b)

$$B_{F_2} = (m_1 + m_2)\,g$$
$$V_2 \rho_w g = (m_1 + m_2)\,g$$

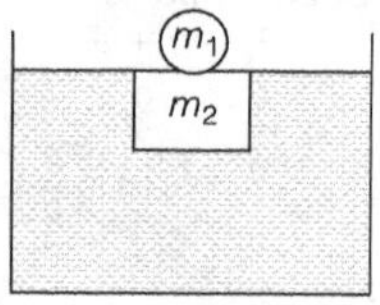

$$V_2 \rho_w = m_1 + m_2$$
$$\frac{m_2}{\rho_2} \cdot \rho_w = m_1 + m_2$$
$$\Rightarrow \quad \frac{0.2}{0.8} = m + 0.2$$
$$\Rightarrow \quad m = 0.05 \text{ kg} = 50 \text{ g}$$

9. (b) When the piston moves downwards, the density of air trapped will increase. So, the buoyancy force due to air on the block will become slightly more and the cube will move up slightly. Similar argument holds good when the piston moves upwards.

10. (b) Upto the instant, the block completely enters the water, tension increases as B_F increases. Therefore, T becomes constant. Hence, graph (b) is correct.

11. (a,b) $u = \sqrt{2gh} = \sqrt{2 \times 9.8 \times 96} = 19.6 \text{ ms}^{-1}$

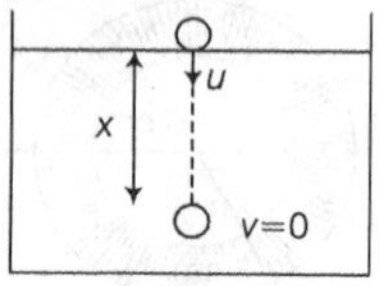

$$F_{net} = B_F - mg \text{ (upwards)}$$
$$\therefore \quad a = \frac{B_F - mg}{m}$$
$$= \frac{V\rho_L g - V\rho g}{V\rho} = \left(\frac{2\rho - \rho}{\rho}\right) g$$
$$\therefore \quad a = g \text{ (upwards)}$$

So, $0 - u^2 = 2(-g)\,x$

$$\Rightarrow \quad x = \frac{u^2}{2g} = \frac{19.6 \times 19.6}{2 \times 9.8} = 19.6 \text{ m}$$
$$v = u + at$$
$$\Rightarrow \quad 0 = 19.6 - 9.8\,t_1$$
$$\Rightarrow \quad t_1 = 2\text{s}$$

Required time $= t_1 + t_2 = 2 + 2 = 4$ s

12. (a,d) When the string is not cut.

From non-inertial frame (i.e. the tank),

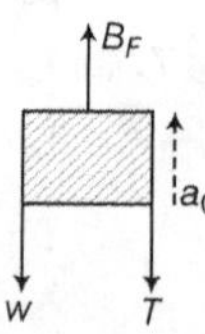

$$B_F = m(g + a_0) + T$$
$$\Rightarrow \quad V\rho_w (g + a_0) = V\rho\,(g + a_0) + T$$
$$\left(\frac{1}{800}\right)(1000)(11) = \left(\frac{1}{800}\right)(800)(11) + T$$
$$T = \frac{55}{4} - 11$$
$$= \frac{11}{4} = 2.75 \text{ N}$$

When the string is cut $T = 0$

From ground frame, $B_F - mg = ma'$

$$V\rho_w (g + a_0) - mg = ma'$$
$$a' = \frac{\left(\frac{1}{800}\right)(1000)(11) - 1(10)}{1} = \frac{55}{4} - 10$$
$$a' = \frac{15}{4} = 3.75 \text{ ms}^{-2}$$

13. (b) Vertical force is buoyant force (B_F) upwards

$$B_F = V_i \rho_L g = \left(\frac{\pi R^2}{2} L\right)\rho g = \frac{\pi R^2 L \rho g}{2}$$

14. (c)

$$p_1 = \rho g R$$
$$p_2 = \rho g \cdot 3R$$

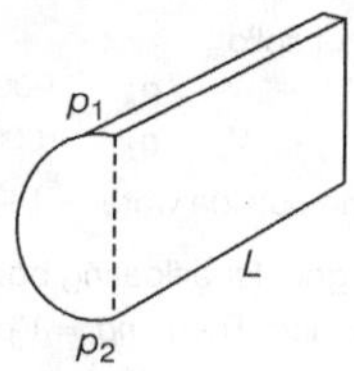

$$\Rightarrow \quad p_{av} = \frac{p_1 + p_2}{2} = 2\rho g R$$
$$\text{Area} = 2R \cdot L$$

Horizontal force is

$$F_H = (2\rho g R) \cdot 2RL = 4\rho g R^2 L$$

15. (d) At every point F_{net} is radial.

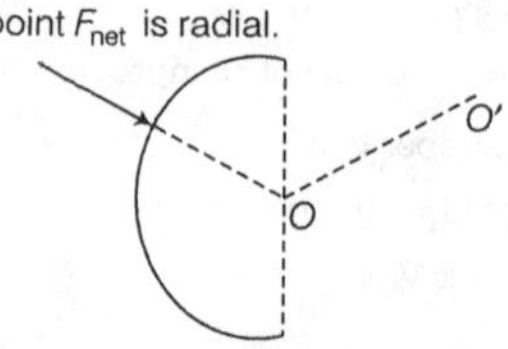

$$\therefore \quad \tau_{\text{about } OO'} = \text{zero}$$

16. (a) $\because$ Buoyant force = Weight of floating body

$$\therefore \quad V_1 \rho_w g = (V - \Delta V)\rho_i g + (\Delta V)(\rho_{m_1})(g) \qquad \ldots \text{(i)}$$

and $\quad V_2 \rho_w g = (V - \Delta V)\rho_i g + (\Delta V)(\rho_{m_2})(g) \qquad \ldots \text{(ii)}$

where, ρ_i = density of ice and

V = total volume of the cube

ΔV = volume of cavity in each cube

m_1 = mass of A, ρ_{m_1} = density of A

m_2 = mass of B, ρ_{m_2} = density of B

V_1 = immersed volume of cube 1 initially

V_2 = immersed volume of cube 2 initially

On dividing Eq. (i) by Eq. (ii), we get

$$\frac{V_1}{V_2} = \frac{(V - \Delta V)\rho_i g + (\Delta V)(\rho_{m_1})(g)}{(V - \Delta V)(\rho_i)(g) + (\Delta V)(\rho_{m_2})(g)}$$
$$= \frac{\left(\frac{V}{\Delta V} - 1\right)\rho_i + \rho_{m_1}}{\left(\frac{V}{\Delta V} - 1\right)\rho_i + \rho_{m_2}}$$

or

$$\frac{V_1}{V_2} = \frac{\left(\frac{1000}{20} - 1\right)(0.9) + 4.9}{\left(\frac{1000}{20} - 1\right)(0.9) + 1.4} = \frac{49}{46} \qquad (\text{so, } h_1 > h_2)$$

17. (c) Mass of cavity A is more than B. So, A sinks first.

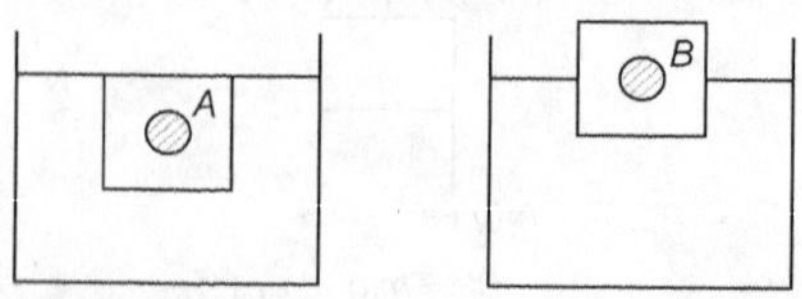

18. (d) As long as cubes are floating, respective water levels do not change. First A sinks, later B sinks. After sinking, water level decreases due to melting of ice. And also $\frac{dh_1}{dt} = \frac{dh_2}{dt}$. So, the best option is (d).

19. (b) When the plastic ball is in the container, it displaces water whose weight is equal to the weight of plastic ball. When the plastic ball is in water, it floats and displaces water of equal weight. So, d remains constant and h decreases.

When the stone is in the container, it displaces water whose weight is equal to the weight of stone. When the stone is in water, it sinks and displaces water whose volume is equal to the volume of stone. So, d and h decrease.

When ice cubes are in the container, before and after melting, displace same amount of water. So, d and h remain same.

When temperature rises, balloon expands. So, B_F on it increases. Thus, tension h increases and d also increases.

Hence, (i) → (p,t), (ii) → (p,s), (iii) → (q,t), (iv) → (r)

20. (5) $\rho_s < \rho_w$. So, block tends to move up and thus the spring has elongation.

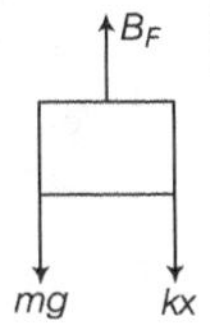

$$B_F = mg + kx$$
$$V\rho_w g = V\rho_s g + kx$$
$$\frac{0.5}{800} \times 1000 \times 10 = (0.5)(10) + 50x$$
$$x = 0.025 \text{ m} = 2.5 \text{ cm} = \frac{5}{2} \text{ cm}$$
$$\Rightarrow \quad \beta = 5$$

21. (9) $T + m_1 g = B_{F_1}$

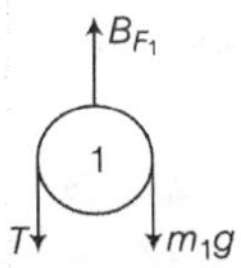

or $\quad T + (0.8)(250)g = (250)\rho_L g \quad$ ...(i)

$$T + B_{F_2} = m_2 g$$

or $\quad T + (250)(\rho_L)(g) = (1.2)(250)(g) \quad$ ...(ii)

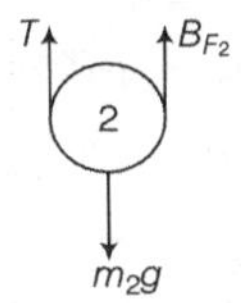

From Eqs. (i) and (ii), $\quad T = 0.5$ N

$\therefore \quad \frac{k}{18} = \frac{1}{2} \quad \Rightarrow \quad k = 9$

22. (1) $T_{max} = (7 \times 10^6)(10^{-6}) = 7$ N

Initial buoyant force is

$$B_{F_i} = (0.1)^3 (1000)(10) = 10 \text{ N}$$

Weight of the block is

$$w = (0.1)^3 (1500)(10) = 15 \text{ N}$$

$\therefore \quad$ Initial tension $T_i = 15 - 10 = 5$ N

At the time of breaking tension is $T_{max} = 7$ N

$\therefore \quad B_{F_{min}} = 15 - 7 = 8$ N

(at the time of breaking of wire)

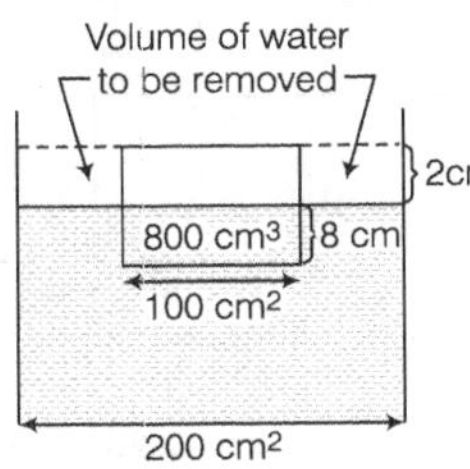

$\therefore \quad (V_i)(1000)(10) = 8$

$\Rightarrow \quad V_i$ = volume immersed $= 8 \times 10^{-4}$ m³

or $\quad V_i = 800$ cm³

Volume of water to be removed = (200 − 100) (2) = 200 cm³

$\therefore \quad 2t = 200$

$\Rightarrow \quad t = 100$ s

$\Rightarrow \quad p = 1$

23. $V_2 = V - V_1 = k(h^3 - d^3)$

V_1 = volume of the cone in top liquid

V_2 = volume of the cone in bottom liquid

$V = kh^3$ = total volume of the cone

$$\rho_s(V_1 + V_2) = \rho V_1 + 2\rho V_2$$
$$\rho_s h^3 = \rho d^3 + 2\rho(h^3 - d^3)$$
$$= 2\rho h^3 - \rho d^3$$
$$\rho d^3 = h^3(2\rho - \rho_s)$$
$$d^3 = h^3\left(2 - \frac{\rho_s}{\rho}\right)$$
$$\Rightarrow \quad d = \left(2 - \frac{\rho_s}{\rho}\right)^{1/3} \cdot h$$

24. Density of block $= \frac{1}{2} \times$ Density of water

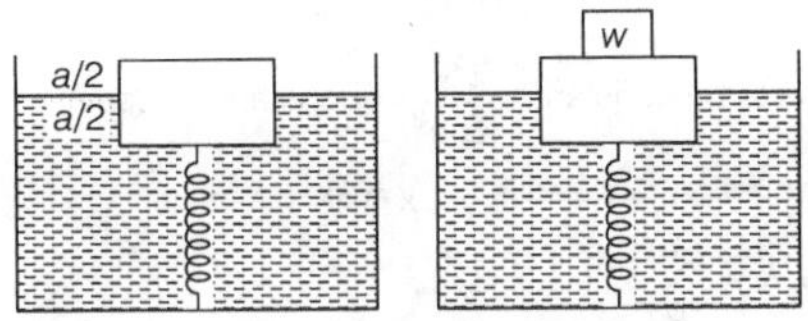

So, 50% of its volume is immersed in water. In second case, half-volume of the block is further immersed into water.

$\therefore \quad w$ = extra uptrust + spring force

$$= \left(a \times a \times \frac{a}{2}\right)(2\rho)(g) + k\left(\frac{a}{2}\right)$$

or $\quad w = a\left(a^2\rho g + \frac{k}{2}\right)$

25. Volume of rod $= \frac{m}{\rho} = \frac{1.5}{0.6 \times \rho_w} = \frac{2.5}{\rho_w}$

Upthrust (B_F) on the rod $= V_{rod} \times \rho_w \times g$

or $\quad B_F = \frac{2.5}{\rho_w} \times \rho_w \times g = 2.5g$

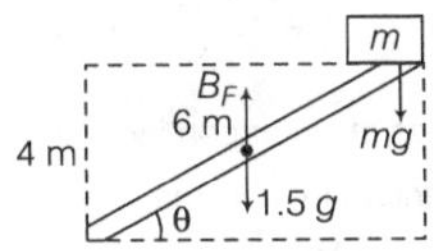

For equilibrium τ_{net} about hinge = 0

$\therefore \quad mg(6\cos\theta) + 1.5g(3\cos\theta) = 2.5g(3\cos\theta)$

$$6m + 4.5 = 7.5$$

$\Rightarrow \quad m = \frac{3}{6} = 0.5$ kg

26. For block,

$$T - mg\sin 30° = ma \quad \text{...(i)}$$

For sphere, Weight–Buoyant force $-T = ma$

$$mg - \frac{mg}{2} - T = ma \quad \text{...(ii)}$$

On solving Eqs. (i) and (ii), we get

$$a = 0$$

DPP-3 Bernoulli's Principle and Its Applications

1. (a) Initial velocity of water blade is $v = \sqrt{2gh} - \omega R$

So, the mass of water impinging on the blade in unit time is

$$m = \rho Av = \rho A(\sqrt{2gh} - \omega R)$$

After the impact, the velocity of water w.r.t. blade is zero (given).

$\therefore$ Change in momentum in unit time = mv

According to Newton's second law, the force required

$$= mv = \rho A(\sqrt{2gh} - \omega R)^2$$

2. (d) A snapshot of the system at an instant is shown in the figure. v = speed with which the water level lowers (this has to be constant).

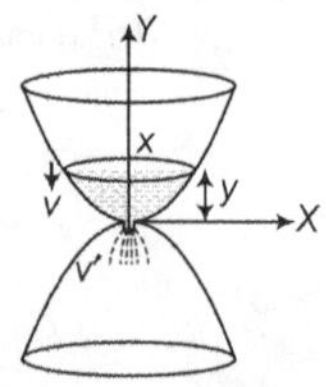

$A = \pi x^2$ = area of surface of water

$v' = \sqrt{2gy}$ = speed of water through orifice.

a = area of orifice (given)

From equation of continuity, at this instant $Av = av'$

$$\pi x^2 v = a\sqrt{2gy}$$

$$2gy = \frac{\pi^2 x^4 v^2}{a^2}$$

$$\Rightarrow \quad y = \frac{\pi^2 v^2}{2ga^2} \cdot x^4$$

If v has to be constant, $\frac{\pi^2 v^2}{2ga^2}$ = constant = k (say)

then equation required is $y = kx^4$

where, $k = \frac{\pi^2 v^2}{2ga^2}$

3. (b) Mass of water coming out in unit time is $m = (\rho A)v$

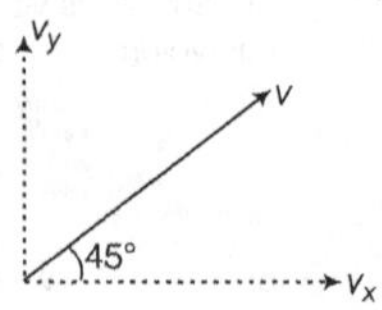

$$m = (1000)\frac{\pi(0.05)^2}{5}(5) = 9.82 \text{ kg s}^{-1}$$

$$v_x = v\cos 45° = 5\left(\frac{1}{\sqrt{2}}\right) = 3.54 \text{ ms}^{-1}$$

Applying momentum equation in X-direction

$$\text{Spring force magnitude} = \frac{mv_x}{1\text{ s}} = \frac{(9.82)(3.54)}{1}$$

$$F_{SF} = 34.76 \text{ N}$$

$$\therefore \quad kx = 34.76 \text{ N}$$

$$\Rightarrow \quad x = \frac{34.76}{20} = 1.74 \text{ cm}$$

4. (c) $v = \sqrt{2gh}$, mass of water coming out in unit time is $m = Av\rho$

Change in momentum in unit time,

$$\frac{|\Delta p|}{t} = mv = A\rho v^2$$

$$\therefore \quad |F| = A\rho \cdot 2gh$$

$$F_{net} = 2A\rho gh - \mu Mg$$

$$= 2A\rho gh - \mu V\rho g$$

$$= 2A\rho gh - \frac{Ah}{V}V\rho g = A\rho gh$$

$$\left[\because \mu = \frac{Ah}{V}, \text{ (given)}\right]$$

$$a = \frac{A\rho gh}{M} = \frac{A\rho gh}{V\rho} = \frac{Ah}{V}g$$

5. (c)

$$A_1v_1 = A_2v_2 + A_3v_3$$

$$A \times 3 = A \times 1.5 + 1.5A \times v$$

$$\Rightarrow \quad v = 1 \text{ ms}^{-1}$$

6. (c)

$$A_Av_A = A_Bv_B = A_cv_c$$

$$\Rightarrow \quad v_B > v_A > v_C$$

$$\Rightarrow \quad p_B < p_A < p_C$$

As A and D are at the same horizontal level and $p_D = p_0$ (i.e. atmospheric pressure)

If we apply $p + \frac{1}{2}\rho v^2$ = constant, we get

$$p_A = p_0$$

$\therefore$ $p_B < p_0$ and $p_C > p_0$

The liquid comes out only if inside pressure is greater than p_0. So, he will be successful if he drills hole at point C.

7. (b)

$$p_1 + \frac{1}{2}\rho v_1^2 = p_2 + \frac{1}{2}\rho v_2^2 \quad \text{(pipe is horizontal)}$$

$$p_1 - p_2 = \frac{1}{2}\rho(v_2^2 - v_1^2)$$

$$\Rightarrow \quad \frac{1}{2}(v_2^2 - v_1^2) = \frac{p_1 - p_2}{\rho}$$

$\therefore$ Change in KE per kg mass $= \frac{p_1 - p_2}{\rho} = \frac{8}{800} = 10^{-2}$ Jkg^{-1}

8. (b) From the non-inertial frame of tube, force on the liquid element is zero.

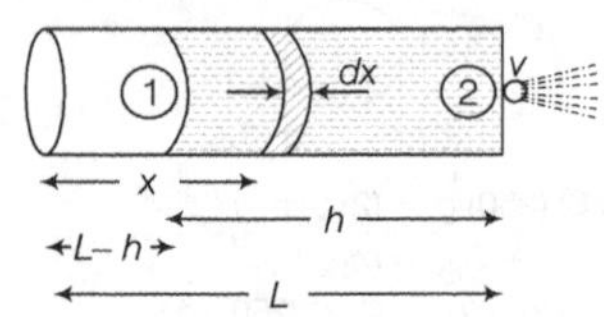

$$(dp)A + (dx\,A\rho)\omega^2 x = 0 \quad (dx\,A\rho = \text{mass of element})$$

$$dp = -\rho\omega^2 x dx$$

$$\Rightarrow \quad \int_{p_1}^{p_2} dp = -\rho\omega^2 \int_{L-h}^{L} x dx$$

$$p_2 - p_1 = -\frac{\rho\omega^2}{2}[L^2 - (L-h)^2]$$

$$= -\frac{\rho\omega^2}{2}[L^2 - L^2 - h^2 + 2Lh]$$

$$p_1 - p_2 = \frac{\rho\omega^2}{2}[2Lh - h^2] \quad \ldots(i)$$

On applying Bernoulli's theorem between points (1) and (2), we have

$$p_1 + \frac{1}{2}\rho(0)^2 = p_2 + \frac{1}{2}\rho v^2$$

$$p_1 - p_2 = \frac{1}{2}\rho v^2 \quad \ldots(ii)$$

From Eqs. (i) and (ii), we get

$$\frac{1}{2}\rho v^2 = \frac{\rho\omega^2}{2}h^2\left[\frac{2L}{h} - 1\right]$$

$$v = \omega h\sqrt{\frac{2L}{h} - 1}$$

9. (c)

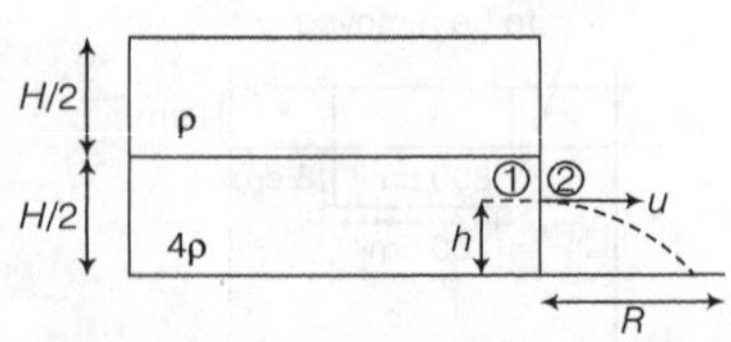

On applying Bernoulli's theorem between points (1) and (2), we have

$$p_0 + \rho g\frac{H}{2} + 4\rho g\left(\frac{H}{2} - h\right) = p_0 + \frac{1}{2}4\rho u^2$$

$$\rho g \frac{H}{2} + 2\rho gH - 4\rho gh = 2\rho u^2$$

$$\frac{5\rho gH}{2} - 4\rho gH = 2\rho u^2$$

$$\Rightarrow \quad u^2 = \left[\frac{5}{4}gH - 2gh\right]$$

$$u^2 = \frac{g}{4}[5H - 8h] \qquad \Rightarrow u = \frac{\sqrt{g}}{2}\sqrt{5H - 8h}$$

$$t = \sqrt{\frac{2h}{g}}$$

$$R = ut = \frac{\sqrt{g}}{2}\sqrt{5H - 8h} \cdot \sqrt{\frac{2h}{g}} = \frac{1}{\sqrt{2}}\sqrt{5Hh - 8h^2}$$

For R to be maximum,

$$\frac{dR}{dh} = 0$$

$$\frac{dR}{dh} = \frac{1}{\sqrt{2}} \cdot \frac{1}{2\sqrt{5Hh - 8h^2}}(5H - 16h) = 0$$

$$\Rightarrow \quad 5H - 16h = 0$$

$$\Rightarrow \quad h = \frac{5}{16}H$$

10. (a,d)

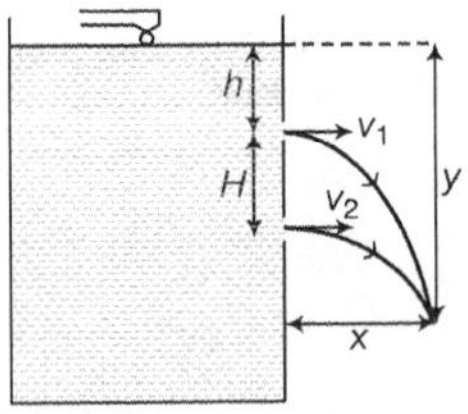

$$v_1 = \sqrt{2gh}$$

$$v_2 = \sqrt{2g(h + H)}$$

$$x = v_1 t_1 = v_2 t_2$$

and $$y = h + \frac{1}{2}gt_1^2$$

and $$y = h + H + \frac{1}{2}gt_2^2$$

The point of intersection will remain at the same place if the level of water in the vessel does not change. That means $Q = A_1v_1 + A_2v_2$

or $$Q = a(v_1 + v_2)$$

$$v_1 + v_2 = \frac{Q}{a}$$

$$\Rightarrow \sqrt{2gh} + \sqrt{2g(h + H)} = \frac{Q}{a}$$

$$\sqrt{2g(h + H)} = \frac{Q}{a} - \sqrt{2gh}$$

On squaring both sides, we get

$$2gh + 2gH = \frac{Q^2}{a^2} + 2gh - 2\frac{Q}{a}\sqrt{2gh}$$

$$2\frac{Q}{a}\sqrt{2gh} = \frac{Q^2}{a^2} - 2gH$$

$$2\left[\frac{140}{0.2}\right]\sqrt{2 \times 980 \times h} = \frac{140^2}{(0.2)^2} - 2 \times 980 \times 50$$

$$\Rightarrow \quad h = 40 \text{ cm}$$

$$\therefore \quad v_1 = \sqrt{2 \times 980 \times 40} = 280 \text{ cms}^{-1}$$

and $$v_2 = \sqrt{2 \times 980\,(40 + 50)} = 420 \text{ cms}^{-1}$$

$$v_1 t_1 = v_2 t_2$$

$$\Rightarrow \quad t_1 = \frac{v_2}{v_1}t_2 = \frac{420}{280}t_2 = \frac{3}{2}t_2$$

$$\therefore \quad y = 40 + \frac{1}{2} \times 980 \times \frac{9}{4}t_2^2$$

$$= 40 + 50 + \frac{1}{2} \times 980\,t_2^2$$

$$\frac{1}{2} \times 980\left(\frac{5}{4}t_2^2\right) = 50$$

$$\Rightarrow \quad t_2 = \left(\frac{4}{49}\right) = \frac{2}{7}\text{ s}$$

$$\therefore \quad t_1 = \frac{3}{7}\text{ s}$$

$$\therefore \quad x = v_1 t_1 = (280)\left(\frac{3}{7}\right) = 120 \text{ cm}$$

and $$y = h + \frac{1}{2}gt_1^2$$

$$= 40 + \frac{1}{2}(980)\left(\frac{9}{49}\right) = 130 \text{ cm}$$

11. (a,c) Applying Bernoulli's theorem between points (1) and (2), we get

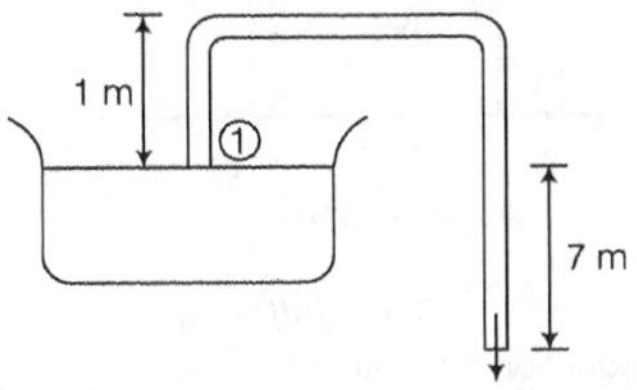

$$p_1 + \frac{1}{2}\rho v_1^2 + \rho gh_1 = p_2 + \frac{1}{2}\rho v_2^2 + \rho gh_2$$

$p_1 = p_2 = p_{atm}$, $h_1 = 0$, $h_2 = -7$ m and $v_1 \approx 0$,

$$\therefore \quad 1000 \times 9.8 \times 0 = \frac{1}{2}(1000)v_2^2 + 1000 \times 9.8(-7)$$

$$\therefore \quad v_2^2 = 2 \times 9.8 \times 7$$

$$\Rightarrow \quad v_2 \approx 11.7 \text{ ms}^{-1}$$

Minimum pressure in the bend will be at A. Again, applying Bernoulli's theorem between points (A) and (1)

$$p_A + \frac{1}{2}\rho v_A^2 + \rho gh_A = p_1 + \frac{1}{2}\rho v_1^2 + \rho gh_1$$

Again, $v_1 \approx 0$, $h_1 = 0$, $h_A = 1$ m, $p_1 = 1.01 \times 10^5 \text{ Nm}^{-2}$

$$p_A + \frac{1}{2}(1000)v_A^2 + (1000)(9.8)(1) = 1.01 \times 10^5 + 0 + 0$$

$$v_A = v_2$$

$$\therefore \quad p_A + 500\,(11.7)^2 = 101000 - 9800$$

$$\Rightarrow \quad p_A = 2.27 \times 10^4 \text{ Nm}^{-2}$$

12. (a) According to Bernoulli's theorem, the pressure decreases in a region, where the velocity of the wind is high. When the pressure difference is sufficiently high, the described phenomena happen.

13. (a) Statements I and II are correct and Statement II is the correct explanation of Statement I.

Explanations for Q. Nos. (14-16)

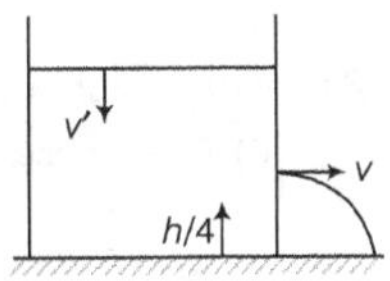

Time taken to reach the ground is

$$T = \sqrt{\frac{2(h/4)}{g}} = \sqrt{\frac{h}{2g}}$$

This is constant.

$$v = \sqrt{2gy}$$

as y decreases, v also decreases

$$x = vT = \sqrt{2gy}\,T$$

$\therefore$ As y decreases, x also decreases.

$\because$ T is constant.

At some instant, $Av' = av$ (from equation of continuity)

$$A\left(-\frac{dy}{dt}\right) = a\sqrt{2gy}$$

$$-\int_{\frac{3h}{4}}^{0} y^{-\frac{1}{2}}dy = \frac{a\sqrt{2g}}{A}\int_0^t dt$$

$$\Rightarrow \quad -2[\sqrt{y}]_{\frac{3h}{4}}^{0} = \frac{a\sqrt{2g}}{A} \cdot t$$

$$2\sqrt{\frac{3h}{4}} = \frac{a\sqrt{2g}}{A} \cdot t \quad \Rightarrow t = \frac{A}{a}\sqrt{\frac{3h}{2g}}$$

14. (a) **15.** (c) **16.** (a)

Explanations for Q. Nos. (17-19)

$\therefore \quad u = \sqrt{2gh}$

Time taken to reach the ground, $T = \sqrt{\dfrac{2(H-h)}{g}}$

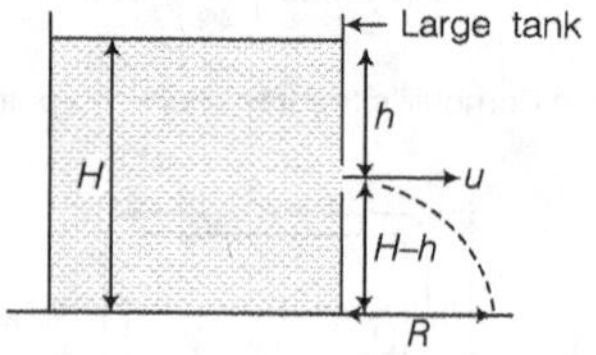

$$R = uT = \sqrt{2gh}\sqrt{\frac{2(H-h)}{g}}$$

or $\quad R = 2\sqrt{h(H-h)}$

From this equation it is clear that

$$R_h = R_{H-h}$$

For R_{max}, $\quad \dfrac{dR}{dh} = 0$

$\Rightarrow \quad h = \dfrac{H}{2}$

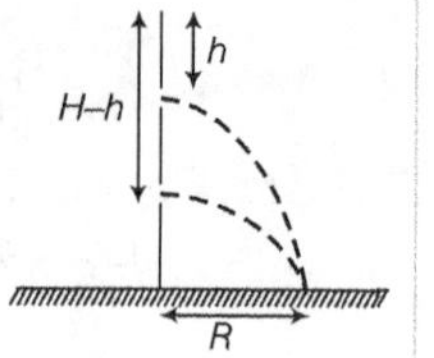

$$u = \sqrt{2gh}$$

$\Rightarrow \quad v_1 < v_2 < v_3$

$$x_1 = \sqrt{2gh}\cdot T$$

$$= \sqrt{2gh}\cdot\sqrt{\frac{2(3h)}{g}} = 2\sqrt{3}h$$

17. (b) **18.** (d) **19.** (c)

20. (6) Applying Bernoulli's theorem for above the plate (2) and below the plate (1), we have

$$p_1 = p_2 + \frac{1}{2}\rho v^2$$

$\Rightarrow \quad p_1 - p_2 = \dfrac{1}{2}\rho v^2$

Upward force is

$$(p_1 - p_2)A = \frac{1}{2}\rho v^2 A$$

$\therefore \quad \dfrac{1}{2}\rho v^2 A = mg$

$\Rightarrow \quad v = \sqrt{\dfrac{2mg}{\rho A}} = 6\sqrt{\dfrac{mg}{18\rho A}}$

$\Rightarrow \quad \gamma = 6$

21. (i) $u = \sqrt{2gh} = \sqrt{2 \times 10 \times 5} = 10\ \text{ms}^{-1}$

(ii) From conservation of energy, $v^2 = u^2 + 2gH$

$$v^2 = (10)^2 + 2(10)(5) = 200$$

$\Rightarrow \quad v = 14.1\ \text{ms}^{-1}$

(iii)

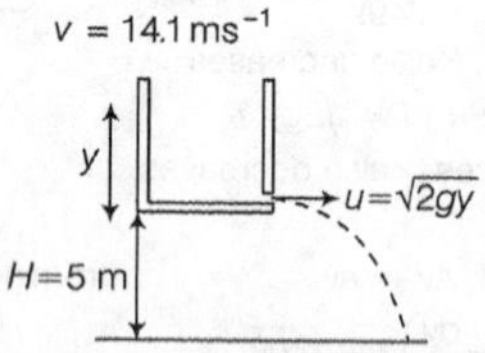

a = area of orifice and A = area of tank

$$au = A\left(\frac{-dy}{dt}\right)$$

$$a\sqrt{2gy} = -A\frac{dy}{dt}$$

$$y^{-1/2}dy = \frac{-a}{A}\sqrt{2g}\,dt$$

$$\int_5^{2.5} y^{-1/2}dy = \frac{-a}{A}\sqrt{2g}\int_0^t dt$$

$$2[y^{1/2}]_5^{2.5} = \frac{-a}{A}\sqrt{2g}\,t$$

$$2(\sqrt{5} - \sqrt{2.5}) = \frac{10^{-4}}{\pi(1)^2}\sqrt{2 \times 10}\,t$$

$$t = \frac{2\pi}{\sqrt{20}} \times 10^4(\sqrt{5} - \sqrt{2.5}) \approx 9200\ \text{s}$$

22. $\dfrac{1}{2}\rho u^2 = \rho gh + \dfrac{mg}{A}$

$$h = 1 - 0.5 = 0.5\ \text{m},\ A = 0.5\ \text{m}^2$$

$$u = \sqrt{2gh + \frac{2mg}{\rho A}} = \sqrt{2 \times 9.8 \times 0.5 + \frac{2 \times 20 \times 9.8}{1000 \times 0.5}}$$

$$= 3.25\ \text{ms}^{-1}$$

Speed with which it hits the surface is

$$v = \sqrt{u^2 + 2gh} = \sqrt{(3.25)^2 + 2(9.8)(0.5)} = 4.51\ \text{ms}^{-1}$$

23. $u = \sqrt{2g(10-h)}$

Component of this velocity parallel to the plane is $u\cos 30°$. Let the stream strikes the plane after time t.

Then, $\quad 0 = u\cos 30° - g\sin 30° t$

$$t = \frac{u}{g}\cot 30°$$

Further (x, y) are coordinates of the point where stream strikes the plane.

$$x = ut = \frac{u^2}{g}\cot 30°$$

Also, $\quad y = x\tan 30°$

$\Rightarrow \quad x = \sqrt{3}y$ and $y = h - \dfrac{1}{2}gt^2$

$\therefore \quad \dfrac{u^2\cot 30°}{g} = \sqrt{3}\left(h - \dfrac{1}{2}gt^2\right)$

$$\frac{\sqrt{3}u^2}{g} = \sqrt{3}\left(h - \frac{1}{2}g.\frac{u^2\cot^2 30°}{g^2}\right)$$

$$\frac{u^2}{g} = h - \frac{3}{2}\cdot\frac{u^2}{g}$$

$\Rightarrow \quad h = \dfrac{5}{2}\cdot\dfrac{u^2}{g} = \dfrac{5}{2g}[2g(10-h)]$

$$h = 5(10-h)$$

$\Rightarrow \quad 6h = 50$

$\Rightarrow \quad h = \dfrac{50}{6} = 8.33\ \text{m}$

24. $\quad \dfrac{F}{A} = \dfrac{1}{2}\rho u^2$

$\Rightarrow \quad F = \dfrac{1}{2}\rho A u^2 \quad \ldots(i)$

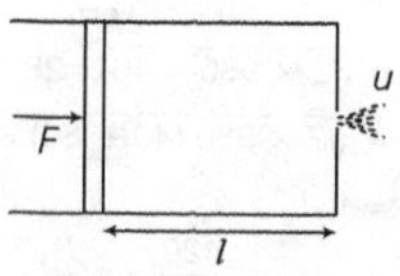

Here, u is the velocity of water, with which it comes out of the whole.

Also, $\quad V = Al \quad \ldots(ii)$

$$V = (au)t$$

$\Rightarrow \quad t = \dfrac{V}{au} \quad \ldots(iii)$

and $\quad W = F \cdot l \quad \ldots(iv)$

From Eqs. (iv) and (i), we get

$$W = \left(\frac{1}{2}\rho Au^2\right)\left(\frac{V}{A}\right)$$

or $\quad W = \dfrac{1}{2}\rho\left(\dfrac{V}{at}\right)^2 V = \dfrac{1}{2}\cdot\dfrac{\rho V^3}{a^2t^2}$

DPP-4 Viscosity

1. (b) Given, $R = 0.04$ m, $\eta = 0.001$ Pa-s, $L = 4000$ m

and $Q = 20\text{Ls}^{-1} = 20 \times 10^{-3}\,\text{m}^3\text{s}^{-1}$

$$Q = \frac{\pi}{8} \cdot \frac{R^4}{L} \cdot \frac{p_1 - p_2}{\eta}$$

$$\Rightarrow \quad \Delta p = p_1 - p_2 = \frac{8Q\eta L}{\pi R^4} = \frac{8(20 \times 10^{-3})(0.001)(4000)}{(3.14)(0.04)^4}$$

or $\Delta p = p_1 - p_2 = 7.96 \times 10^4$ Pa

$$h = \frac{\Delta p}{\rho g} = \frac{7.96 \times 10^4}{13600 \times 9.8} = 0.5972 \text{ m} = 59.72 \text{ cm}$$

2. (d) $$\frac{4}{3}\pi R^3 = 8\left(\frac{4}{3}\pi r^3\right)$$

$$\Rightarrow \quad R = 2r$$

Terminal velocity is given by

$$v_T = \frac{2}{9} \cdot \frac{r^2}{\eta}(\rho_S - \rho_L)g$$

or $v_T \propto r^2$

$$\Rightarrow \quad \frac{v_{T_2}}{v_{T_1}} = \frac{R^2}{r^2} = \frac{4r^2}{r^2} = 4$$

$$\Rightarrow \quad v_{T_2} = 4(v_{T_1}) = 4(0.1)$$

or $v_{T_2} = 0.4 \text{ ms}^{-1}$

3. (c) The sphere, as it falls from a large height, has a large velocity when it enters the viscous liquid. So, it experiences a large viscous force upwards. The velocity from then keeps on decreasing. Thus, the viscous force and the magnitude of (negative) acceleration go on decreasing non-linearly. At a particular point of time, acceleration becomes zero and velocity remains constant. So, the best curve is C.

4. (d) Buoyant force + Viscous force = mg

$$yg + 6\pi\eta r v_T = xg$$

$$\Rightarrow \quad v_T = \frac{(x-y)g}{r(6\pi\eta)}$$

$$\therefore \quad v_T \propto \frac{x-y}{r}$$

5. (c) Gravitational force is constant. As velocity increases, viscous force increases. Thus, net force decreases. So, the required sequence is P, Q, R.

6. (a) Terminal velocity (for the given depth and time) is

$$v_T = \frac{s}{t} = \frac{2 \times 10^{-2}\text{m}}{3600 \text{ s}} = 5.56 \times 10^{-6} \text{ ms}^{-1}$$

But $$v_T = \frac{2}{9} \cdot \frac{r^2}{\eta}(\rho - \rho_L)g$$

$$5.56 \times 10^{-6} = \frac{2}{9} \cdot \frac{r^2}{(10^{-2} \times 10^{-1})}(1.8 \times 10^3 - 1 \times 10^3)(10)$$

$$r = 1.77\,\mu\text{m}$$

$$\Rightarrow \quad \text{Diameter} = 3.54\,\mu\text{m}$$

As radius increases, terminal velocity increases, so they settle down fast. But the least value of v_T, in the given problem, we are interested in, is $5.56 \times 10^{-6}\text{ms}^{-1}$. Corresponding radius is 1.77 μm. The particles with radius less than this value are not settled in 1 h.

7. (a) $$v_T = \frac{2}{9} \cdot \frac{r^2(\rho - \rho_L)g}{\eta}$$

$$\Rightarrow \quad \sqrt{2gh} = \frac{2}{9} \cdot \frac{(3 \times 10^{-4})^2 (10^4 - 10^3)(9.8)}{9.8 \times 10^{-6}}$$

$$\therefore \quad h \simeq 1650 \text{ m}$$

8. (b) Thickness of annular space

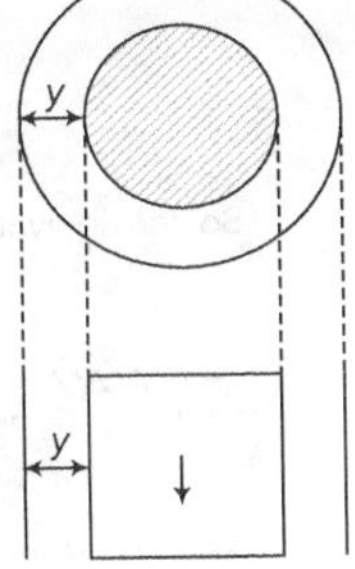

$$= \frac{20.0628 - 20}{2}$$

$$= 0.0314 \text{ cm} = 0.000314 \text{ m}$$

In steady state,

Gravitational force = Viscous force

or $$mg = \eta A \frac{\Delta v}{\Delta y} \qquad \ldots(\text{i})$$

But $A = 2\pi r l$

$$= 2 \times 3.14 \times (10 \times 10^{-2})(20 \times 10^{-2})$$

$$= 0.1256 \text{ m}^2$$

From Eq. (i),

$$\therefore \quad 1 \times 10 = (10 \times 10^{-1})(0.1256)\left(\frac{v - 0}{0.000314}\right)$$

$$\Rightarrow \quad v = 0.025 \text{ ms}^{-1} = 2.5 \text{ cms}^{-1}$$

9. (b) Relative to liquid, the velocity of sphere is $2v_0$ (upwards).

$\therefore$ Viscous force on sphere $= 6\pi\eta r(2v_0)$ (downwards)

$= 12\pi\eta r v_0$ (downwards)

10. (c) Volume flow rate in capillary tube is

$$Q = \frac{V}{t} = \frac{\pi r^4}{8\eta l}\Delta p \qquad \ldots(\text{i})$$

$$\Delta p = \rho g h \qquad \ldots(\text{ii})$$

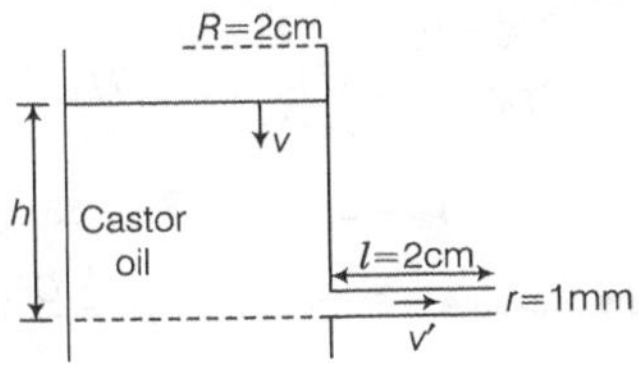

Also, volume flow rate in capillary tube

$$\frac{V}{t} = A'v' = \pi r^2 v' \qquad \ldots(\text{iii})$$

Using Eqs. (ii) and (iii) in Eq. (i), we get

$$\pi r^2 v' = \frac{\pi r^4}{8\eta l} \cdot \rho g h \quad \text{or} \quad v' = \frac{r^2 \rho g h}{8\eta l} \qquad \ldots(\text{iv})$$

From equation of continuity,

$$v'A' = vA$$

$$\pi r^2 v' = v \cdot \pi R^2$$

$$\Rightarrow \quad v = \frac{r^2}{R^2} \cdot v'$$

or $$v = \frac{r^2}{R^2} \cdot \frac{r^2 \rho g h}{8\eta l} = \frac{r^4 \rho g h}{8R^2 \eta l} \qquad \text{[from Eq. (iv)]}$$

11. (a,b,c,d) Poiseuille's equation can be written as,

$$Q = \frac{p_1 - p_2}{\left(\frac{8\eta L}{\pi R^4}\right)} = \frac{\Delta p}{X}$$

where, $$X = \frac{8\eta L}{\pi R^4}$$

This equation can be compared with the ohm's law equation, $i = \frac{\Delta V}{R}$. Problem of series and parallel combination of pipes can be solved in the similar manner as done in case of an electrical circuit.

$$X = \frac{8\eta L}{\pi R^4}$$

$$\Rightarrow \qquad X \propto \frac{L}{R^4}$$

$$X_{AB} : X_{CD} : X_{EF} : X_{GH}$$

$$= \frac{\frac{1}{2}}{\left(\frac{1}{2}\right)^4} : \frac{\frac{1}{2}}{\left(\frac{1}{2}\right)^4} : \frac{1}{\left(\frac{1}{2}\right)^4} : \frac{1}{(1)^4} = 8:8:16:1$$

So, the equivalent electrical circuit can be drawn as under

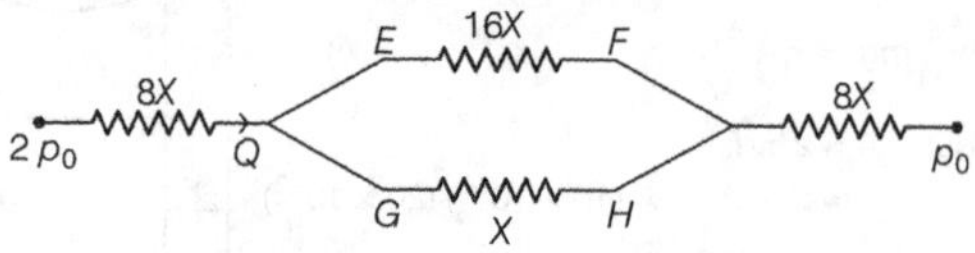

As, the current is distributed in the inverse ratio of the resistances in parallel. The Q will be distributed in the inverse ratio of X.

Thus, volume flow rate through EF will be $\frac{Q}{17}$ and that from GH will be $\frac{16}{17}Q$.

$$X_{net} = 8X + \frac{(16X)(X)}{16X + X} + 8X = \frac{288}{17}X$$

$$Q = \frac{\Delta p}{X_{net}} \qquad \left(\text{same as, } i = \frac{\Delta V}{R}\right)$$

$$Q = \frac{2p_0 - p_0}{\frac{288}{17}X} = \frac{17 p_0}{288X}$$

Let p_E be the pressure at E. Then,

$$2p_0 - p_E = Q(8X) \qquad (\Delta V = iR)$$

$$2p_0 - p_E = \frac{17p_0}{288X} \cdot 8X = \frac{17p_0}{36}$$

$$\Rightarrow \qquad p_E = 2p_0 - \frac{17p_0}{36} = 1.53\,p_0$$

Let p_F be the pressure at F. Then, $p_E - p_F = \frac{Q}{17}(16X)$

$$p_F = p_E - \frac{Q}{17}(16X) = p_E - \frac{16}{17}X\left(\frac{17p_0}{288X}\right)$$

$$= p_E - \frac{p_0}{18} = 1.53p_0 - 0.056\,p_0$$

$$\therefore \qquad p_F = 1.47\,p_0$$

12. (b,c,d) Viscous force $= mg\sin\theta$ $\qquad (\because F_{net} = 0)$

$$\eta a^2 \frac{V}{t} = a^3 \rho g \sin\theta \Rightarrow \eta = \frac{t a \rho g \sin\theta}{V}$$

13. (a) With the decrease in temperature, the viscosity of lubricants increases, due to which the machine parts are jammed in winter.

14. (d) Objects falling in a viscous medium attain terminal velocity (v_T). But this terminal velocity depends on radius of the raindrop falling and is not same for all sizes of raindrops.

15. (b) Volume of fluid crossing the section of the tube of length l and radius R in unit time is

$$V = \int_0^R 2\pi r v dr = 2\pi \int_0^R r v_0 \left(\frac{R^2 - r^2}{R^2}\right) dr$$

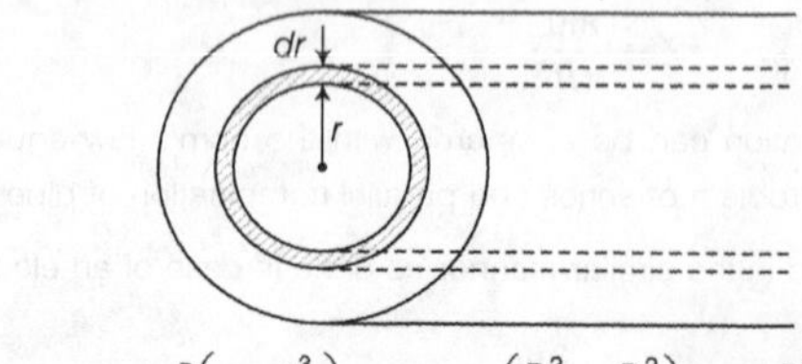

$$V = 2\pi v_0 \int_0^R \left(r - \frac{r^3}{R^2}\right) dr = 2\pi v_0 \left(\frac{R^2}{2} - \frac{R^2}{4}\right) = \frac{\pi v_0 R^2}{2}$$

16. (c) Kinetic energy of the fluid within the volume of the tube is

$$\text{KE} = \int_0^R \frac{1}{2}(dm)v^2 = \frac{1}{2}\int_0^R (2\pi r\, dr\, l\rho) v_0^2 \left[\frac{R^2 - r^2}{R^2}\right]^2$$

$$= \frac{1}{2} 2\pi l \rho v_0^2 \int_0^R r\left[1 + \frac{r^4}{R^4} - 2\frac{r^2}{R^2}\right] dr$$

$$= \pi l \rho v_0^2 \int_0^R \left(r + \frac{r^5}{R^4} - 2\frac{r^3}{R^2}\right) dr$$

$$= \pi l \rho v_0^2 \left[\frac{R^2}{2} + \frac{R^2}{6} - \frac{R^2}{2}\right] = \frac{\pi l \rho v_0^2 R^2}{6}$$

17. (a) $v = v_0\left[1 - \frac{r^2}{R^2}\right] \Rightarrow \frac{dv}{dr} = -\frac{v_0}{R^2} \cdot 2r$

$$\Rightarrow \qquad \left(\frac{dv}{dr}\right)_{r=R} = -\frac{2v_0}{R}$$

Friction force is due to viscosity.

$$|F_v| = \left|\eta A \frac{dv}{dr}\right| = \left|\eta(2\pi R l)\left(\frac{dv}{dr}\right)_{r=R}\right|$$

$$= \eta 2\pi R l \cdot \frac{2v_0}{R} = 4\pi \eta l v_0$$

18. (a)

(i) Stoke's law, $F = 6\pi\eta r v$

(ii) Terminal velocity, $v_T = \frac{2}{9} \cdot \frac{r^2 g(\rho_S - \rho_L)}{\eta}$

(iii) Excess pressure inside a mercury drop is $\Delta p = \frac{2T}{r}$

(iv) Viscous force is $F_V = -\eta A \frac{dv}{dy}$

Hence, (i) → (p,r), (ii) → (p,q,r), (iii) → (p,s), (iv) → (r,t)

19. (4) $D = 2r = 2 \times 2 \times 10^{-3}\text{m} = 4 \times 10^{-3}\text{m}$

For laminar flow with maximum velocity, Reynold's number is $N_R = 2000$

$$N_R = \frac{\rho V D}{\eta}$$

$$\Rightarrow \qquad V = \frac{N_R \eta}{\rho D} = \frac{(2000)(2.12 \times 10^{-3})}{(1.06 \times 10^3)(4 \times 10^{-3})} = 1\,\text{ms}^{-1}$$

Corresponding flow rate is $Q = Av = (\pi r^2)v$

$$Q = \pi(2 \times 10^{-3})^2 \times 1 = 4\pi \times 10^{-6}\,\text{m}^3\text{s}^{-1}$$

$$\Rightarrow \qquad x = 4$$

20. (4) Velocity gradient $= \frac{\Delta v}{\Delta y} = \frac{2-0}{1-0} = 2\,\text{ms}^{-1}\text{m}^{-1}$

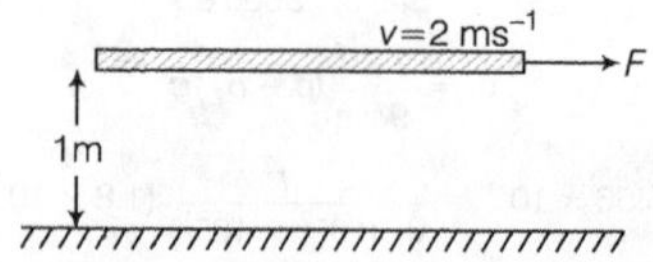

Viscous force, $|F| = \eta A \frac{\Delta v}{\Delta y} = (0.01 \times 10^{-1})(2)(2) = 0.004$ N

So, to keep the plate moving with constant velocity, a force of 4×10^{-3}N has to be applied. Thus, $\alpha = 4$.

21. (5) Velocity gradient $= \frac{\Delta v}{\Delta y} = \frac{0.5 - 0}{1.25 \times 10^{-2}} = 40\,\text{ms}^{-1}\text{m}^{-1}$

$$F = 2\left[\eta A \frac{\Delta v}{\Delta y}\right] \Rightarrow 1 = 2 \times \eta \times 0.5 \times 40$$

$$\Rightarrow \qquad \eta = \frac{1}{40}\,\text{kg-sm}^{-2}$$

or $\qquad \eta = 0.25 \times 10^{-1} = 25 \times 10^{-3}\,\text{kgm}^{-2}$

$$\Rightarrow \qquad \beta = 5$$

22. Let r be the radius of the bubble at any time t.

$\therefore$ At time t, excess pressure is $\Delta p = \frac{4T}{r}$

The rate at which the volume changes is

$$\frac{dV}{dt} = \frac{d}{dt}\left(\frac{4}{3}\pi r^3\right) = 4\pi r^2 \frac{dr}{dt}$$

From Poiseuillie's equation,

$$\left|\frac{dV}{dt}\right| = \frac{\pi(\Delta p)b^4}{8\eta l}$$

Hence, $$4\pi r^2 \frac{dr}{dt} = -\frac{\pi b^4}{8\eta l}\left(\frac{4T}{r}\right) \qquad \left(\because \frac{dr}{dt} \text{ is negative}\right)$$

$$dt = -\frac{8\eta l}{b^4 T} r^3 dr$$

If t_0 is the sought time, then

$$\int_0^{t_0} dt = -\frac{8\eta l}{b^4 T}\int_{r_1}^{r_2} r^3 dr = \frac{2\eta l}{b^4 T}(r_1^4 - r_2^4)$$

$$\therefore \qquad t_0 = \frac{2\eta l}{b^4 T}(r_1^4 - r_2^4)$$

23. (i) $F_V = 6\pi\eta r v = 6\pi(0.8)(10^{-3})(10^{-2}) = 1.5 \times 10^{-4}$ N

(ii) Hydrostatic force = Upthrust $= \frac{4}{3}\pi r^3 \rho g$

$$= \frac{4}{3}\pi(10^{-3})^3 \times 1260 \times 9.8 = 5.2 \times 10^{-5} \text{ N}$$

(iii) At terminal velocity, w = Upthrust + Viscous force

$$\therefore \quad 50 \times 10^{-3} \times 9.8 = 5.2 \times 10^{-5} + 6\pi(0.8)(10^{-3})v_T$$

$$v_T = 32.5 \text{ ms}^{-1}$$

DPP-5 Surface Tension

1. (b) Soap bubble has two surfaces. In increasing the diameter of soap bubble, the new surface area created $= 8\pi r_2^2 - 8\pi r_1^2$

$$= 8\pi(r_2^2 - r_1^2) = 8\pi[(2.5)^2 - (1)^2] \times 10^{-4}$$

$$\Delta A = 13.2 \times 10^{-3} \text{m}^2$$

Surface tension is numerically equal to surface energy per unit area. So, work done is

$$W = \text{Surface tension} \times \Delta A$$

$$= 3 \times 10^{-2} \times 13.2 \times 10^{-3} = 3.96 \times 10^{-4} \text{ J}$$

2. (d) When thousand drops coalesce, the area of free surface of water decreases. So, the corresponding surface energy (i.e. energy associated with the free surface) decreases. This decrease in surface energy is equal to the heat liberated.

Final volume = Initial volume $\Rightarrow \frac{4}{3}\pi R^3 = (1000)\left(\frac{4}{3}\pi r^3\right)$

$$R = 10r = (10)(10^{-7})\text{m} = 10^{-6}\text{m}$$

Water drop has only one free surface unlike a soap bubble.

$\therefore$ Decrease in surface area $= (4\pi r^2)\,1000 - 4\pi R^2$

$$= 4\pi[1000r^2 - R^2] = 4\pi[10^{-11} - 10^{-12}]$$

$$= 4\pi \times 10^{-11}[1 - 0.1] = 0.9 \times 4\pi \times 10^{-11}$$

$$= 1.13 \times 10^{-10} \text{ m}^2$$

Surface tension (T) is numerically equal to surface energy possessed by unit surface area.

So, heat liberated = (decrease in surface area) T

$$= (1.13 \times 10^{-10})(7 \times 10^{-2}) = 7.91 \times 10^{-12} \text{J}$$

3. (d) Mass of air is conserved

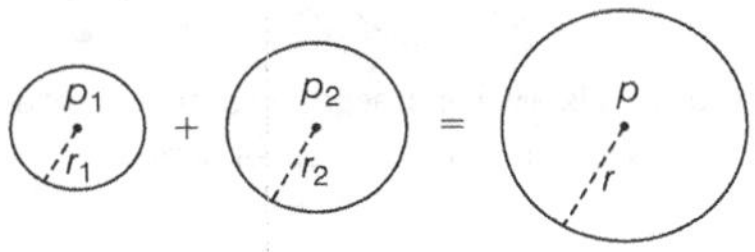

$\therefore \qquad n_1 + n_2 = n$

As $pV = nRT$, $$\frac{p_1V_1}{RT_1} + \frac{p_2V_2}{RT_2} = \frac{pV}{RT}$$

$$T_1 = T_2 = T \; (\because \text{temperature is constant})$$

$\therefore \qquad p_1V_1 + p_2V_2 = pV$

Let us denote surface tension by S. Then,

$$\left(p_0 + \frac{4S}{r_1}\right)\left(\frac{4}{3}\pi r_1^3\right) + \left(p_0 + \frac{4S}{r_2}\right)\left(\frac{4}{3}\pi r_2^3\right) = \left(p_0 + \frac{4S}{r}\right)\left(\frac{4}{3}\pi r^3\right)$$

Solving this, we get $S = \frac{p_0(r^3 - r_1^3 - r_2^3)}{4(r_1^2 + r_2^2 - r^2)}$

4. (b) F is the tension in the wire.

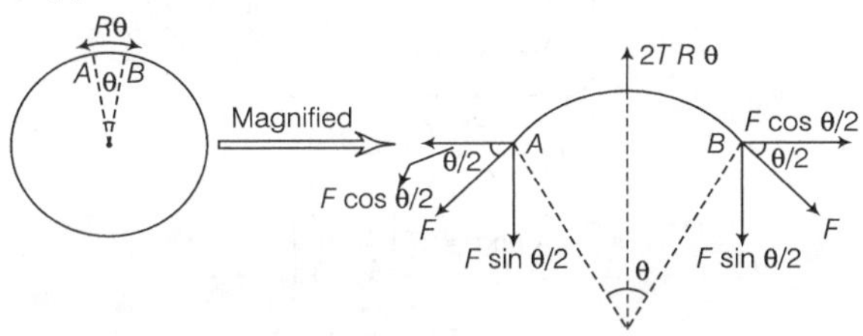

Force due to tension in the wire, on the element is

$$F_1 = 2F\sin\left(\frac{\theta}{2}\right)\downarrow$$

Force on the element due to surface tension of soap film is

$$F_2 = 2(TR\theta)\uparrow$$

$\because$ The wire is in contact with the soap film along upper length $R\theta$ and along lower length $R\theta$ both of the element AB.

For equilibrium, $$F_2 = F_1$$

$$2(TR\theta) = 2F\sin\left(\frac{\theta}{2}\right)$$

$$2(TR\theta) = 2F\left(\frac{\theta}{2}\right) \qquad \left(\because \theta \text{ is small, } \sin\frac{\theta}{2} \simeq \frac{\theta}{2}\right)$$

$$F = 2TR$$

$$Y = \frac{FL}{A\Delta L} = \frac{FR}{A\Delta R}$$

$$\Rightarrow \qquad \Delta R = \frac{FR}{AY} = (2TR)\frac{R}{AY}$$

$$\Delta R = \frac{2TR^2}{AY}$$

5. (c) By virtue of surface tension, the soap bubble tries to contract and the excess pressure inside soap bubble is $\Delta p_1 = \frac{4T}{r}$.

Due to the presence of charge on outer surface, the particle repel one another. If this pressure due to repulsion is equal to Δp_1 in magnitude, then inside pressure = outside pressure = atmospheric pressure. From electrostatics, electric pressure is

$$\Delta p_2 = \frac{\sigma^2}{2\varepsilon_0} \Rightarrow \Delta p_2 = \Delta p_1$$

$$\frac{\sigma^2}{2\varepsilon_0} = \frac{4T}{r}$$

$$\Rightarrow \qquad \frac{Q^2}{16\pi^2 r^4 \cdot (2\varepsilon_0)} = \frac{4T}{r} \qquad \left(\because \sigma = \frac{Q}{4\pi r^2}\right)$$

$$Q^2 = 64 \times 2\pi^2 r^3 T\varepsilon_0$$

$$\therefore \qquad Q = 8\pi r\sqrt{2rT\varepsilon_0}$$

6. (c) Weight of the drop = Force due to surface tension + Buoyant force

$$\therefore \quad \frac{4}{3}\pi r^3 \rho g = (2\pi r)T + \frac{1}{2}\left[\frac{4}{3}\pi r^3\right]\rho_L g$$

$$\frac{4}{3}\pi r^3 g\left[\rho - \frac{\rho_L}{2}\right] = 2\pi r \cdot T$$

$$\Rightarrow \quad r^2 \frac{4}{3} g\left[\frac{2\rho - \rho_L}{2}\right] = 2T$$

$$\Rightarrow \quad r^2 = \frac{3T}{g(2\rho - \rho_L)} \Rightarrow r = \sqrt{\frac{3T}{g(2\rho - \rho_L)}}$$

7. (d) Pressure at A = Pressure at A'

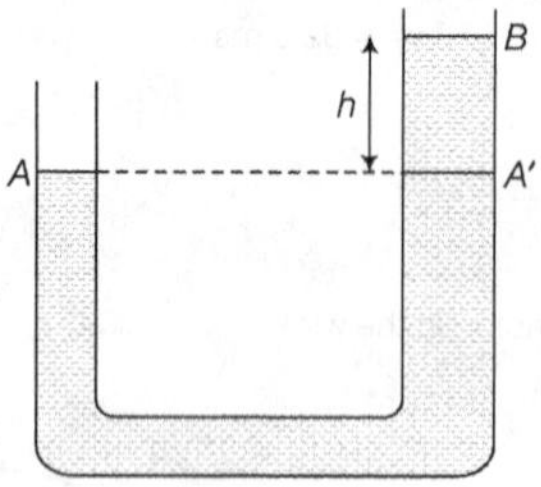

$$p_0 - \frac{2T}{r_1} = \left(p_0 - \frac{2T}{r_2}\right) + h\rho g$$

$$h\rho g = 2T\left[\frac{1}{r_2} - \frac{1}{r_1}\right]$$

$$h = \frac{2T}{\rho g}\left[\frac{1}{r_2} - \frac{1}{r_1}\right]$$

Here, $r_2 > r_1$, so h is negative. That means level of B must be below the level of A.

8. (a) The pressure of the water changes linearly with the increase in height. At the bottom of the meniscus it is equal to the external atmospheric pressure p_0 and at the top it is equal to $p_0 - \rho gh$. The average pressure exerted on the wall is $p_{av} = p_0 - \frac{\rho gh}{2}$. The force corresponding to this value, for an aquarium with side walls of length L is $F_1 = Lh\, p_{av}$

or $$F_1 = Lh\left(p_0 - \frac{\rho gh}{2}\right)$$

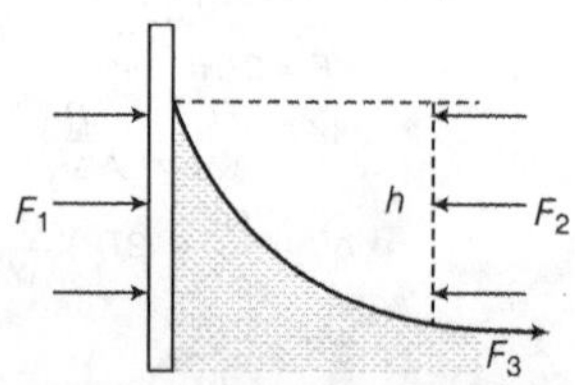

Consider the horizontal forces acting 'ON' the volume of water enclosed by the dashed lines in the figure. F_1 is towards right. (F_1 is by the wall). External air pushes it to the left with a force $F_2 = p_0 Lh$ and due to surface tension along a line of length L, a force $F_3 = LT$ acts towards right due to remaining water.

The volume itself is at rest.

$$\therefore \quad F_1 + F_3 - F_2 = 0$$

$$Lh\left(p_0 - \frac{\rho gh}{2}\right) + LT - p_0 Lh = 0$$

$$\frac{\rho gh^2 L}{2} = LT$$

$$\Rightarrow \quad h = \sqrt{\frac{2T}{\rho g}}$$

or $$h = \sqrt{\frac{2 \times 0.073}{1000 \times 9.8}} = 0.00386\text{m} \approx 4\text{mm}$$

9. (a) Insect is in equilibrium. So,

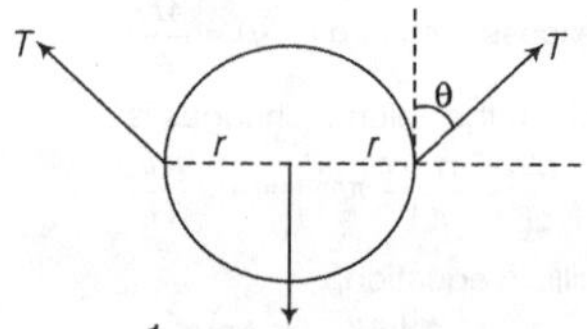

$\frac{1}{6}$(weight of insect)

$$2\pi r T\cos\theta = \frac{1}{6} \times \text{weight of insect}$$

$$2(3.14)(2 \times 10^{-5})(0.072)\cos\theta = \frac{1}{6}(0.003 \times 10^{-3} \times 9.8)$$

$$\therefore \quad \cos\theta \approx 0.54$$

$$\Rightarrow \quad \theta = \cos^{-1}(0.54)$$

10. (b) $$d = 2R\cos\theta$$

For water-glass, $$\theta = 0°$$

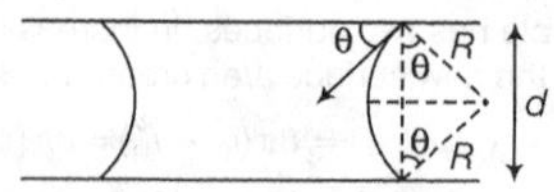

$$\therefore \quad d = 2R$$

$$\Rightarrow \quad R = \frac{d}{2}$$

Excess pressure is $\Delta p = T\left[\frac{1}{R_1} + \frac{1}{R_2}\right]$

Here, for one surface, $R_1 = R$

For other surface, $R_2 = \infty$

$$\therefore \quad \Delta p = \frac{T}{R} = \frac{2T}{d}$$

$$V \approx Ad$$

$$\Rightarrow \quad d = \frac{V}{A}$$

$$\therefore \quad \Delta p = \frac{2TA}{V}$$

$$\therefore \quad \text{Force, } F = A \cdot \Delta p = \frac{2TA^2}{V}$$

11. (a,c) $$r_1 < r_2$$

$$\Rightarrow \quad p_1 > p_2$$

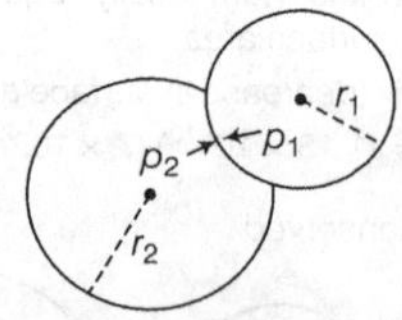

Excess pressure is always present on concave side.

So, interface is concave towards small bubble.

$$p_1 = p_0 + \frac{4T}{r_1}$$

$$p_2 = p_0 + \frac{4T}{r_2}$$

Excess pressure $= p_1 - p_2 = 4T\left[\frac{1}{r_1} - \frac{1}{r_2}\right]$

This excess pressure for the interface is also equal to $\frac{4T}{R}$.

$$\therefore \quad 4T\left(\frac{1}{r_1} - \frac{1}{r_2}\right) = \frac{4T}{R}$$

$$R = \frac{r_1 r_2}{r_2 - r_1} = \frac{(0.002)(0.004)}{0.004 - 0.002} = 0.004\text{m}$$

12. (a,b) $(QQ')T \times 2 = F$

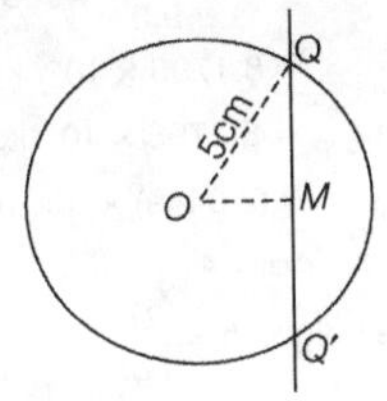

$\therefore \quad QQ' = \frac{F}{2T} = \frac{16 \times 10^{-3}}{2(0.1)} = 0.08 \text{ m} = 8 \text{ cm}$

$\therefore \quad QM = 4\text{cm}$

$\therefore \quad OM = \sqrt{5^2 - 4^2} = 3\text{cm}$

Required distance = (5+ 3) cm or (5 − 3) cm

i.e. 8 cm or 2 cm

13. (d) F_{net} (by water) = Buoyant force $= \left[\frac{1}{2} \times l \times l \frac{\sqrt{3}}{2}\right] L \times \rho_w g$

$$= \frac{\sqrt{3}}{4} L l^2 \rho_w g$$

14. (d) $p_{av} = \dfrac{\left[p_0 + \rho_w g \left(H - \frac{l\sqrt{3}}{2}\right)\right] + [p_0 + \rho g H]}{2}$

$$= p_0 + \rho_w g \left(H - \frac{l\sqrt{3}}{4}\right)$$

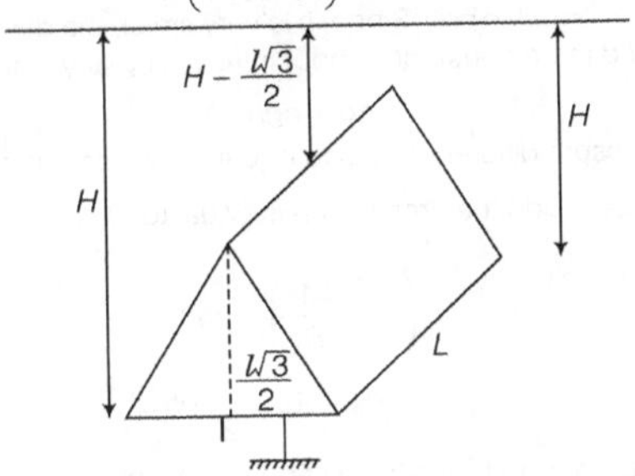

On slant face,

$$F_{net} = p_{av} \times \text{Area} = \left[p_0 + \rho_w g \left(H - \frac{l\sqrt{3}}{4}\right)\right] Ll$$

15. (a) $[M_{styrofoam} + m]g + F_{surface\ tension}$ = Buoyant force

$$\left[\frac{1}{2} \times l \times \frac{l\sqrt{3}}{2} \times L \times \rho_{sf} + m\right] g$$

$$+ \left[2L + 2 \cdot \frac{l}{2}\right] T = \left[\frac{1}{2} \times l \times \frac{l\sqrt{3}}{2} - \frac{1}{2} \times \frac{l}{2} \times \frac{l}{2} \frac{\sqrt{3}}{2}\right] \times L\rho_w g$$

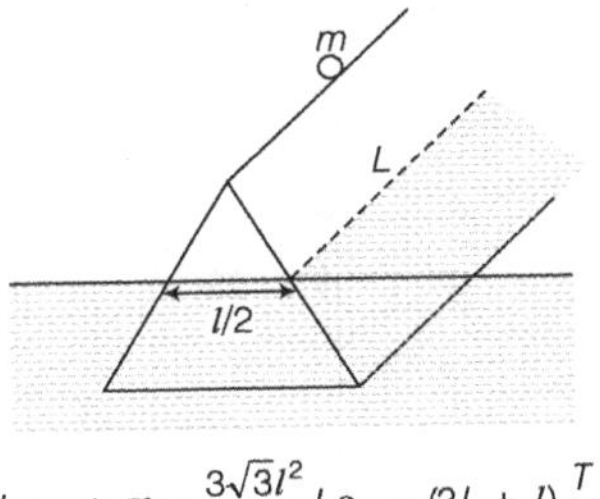

$$\frac{l^2\sqrt{3}}{4} L\rho_{sf} + m = \frac{3\sqrt{3}l^2}{16} L\rho_w - (2L + l)\frac{T}{g}$$

$$m = \frac{l^2 3\sqrt{3}}{16} L\rho_w - \frac{l^2\sqrt{3}}{4} L\rho_{sf} - (2L + l)\frac{T}{g}$$

$$= \frac{Ll^2\sqrt{3}}{4}\left[\frac{3}{4}\rho_w - \rho_{sf}\right] - (2L + l)\frac{T}{g}$$

16. (a) When capillary is in vertical position, the upper meniscus is concave and pressure due to surface tension is

$p_1 = \frac{2T}{R_u}$ and is directed vertically upwards.

Here, wetting is completed i.e. $\theta = 0°$, so

R_u = radius of capillary tube, $r = \frac{1}{2}$mm = 0.5 mm.

$\therefore$ In all cases, $\quad R_u = r = 0.5$ mm

The pressure above the drop and below the drop is same i.e. atmospheric. But pressure due to length of water drop i.e. hydrostatic pressure is always vertically downwards. To compensate for this the lower meniscus has to be concave or convex or flat depending on the situation.

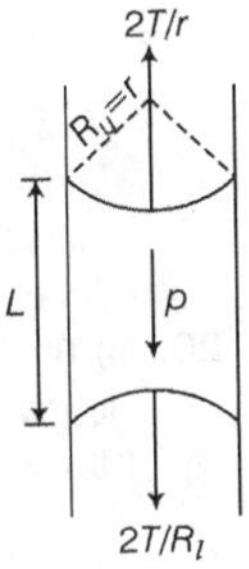

When $L = 2$cm,

$p_1 = \frac{2T}{r} = \frac{2(0.075)}{0.005} = 300 \text{ Nm}^{-2}$ (in all cases)

Hydrostatic pressure is $p = \rho g L$

$= 1000 \times 10 \times 0.02$

$= 200 \text{ Nm}^{-2}$

$p_1 > p$. So, the bottom meniscus has to be concave downwards for equilibrium.

$$p_2 = \frac{2T}{R_l} = 300 - 200$$

$$\Rightarrow \quad \frac{2T}{R_l} = 100$$

$$\Rightarrow \quad R_l = \frac{2 \times 0.075}{100}$$

$$R_l = 1.5 \times 10^{-3}\text{m} = 1.5 \text{ mm}$$

17. (b) When $L = 4$cm,

hydrostatic pressure is

$p = \rho g L$

$= 1000 \times 10 \times 0.04 = 400 \text{ Nm}^{-2}$

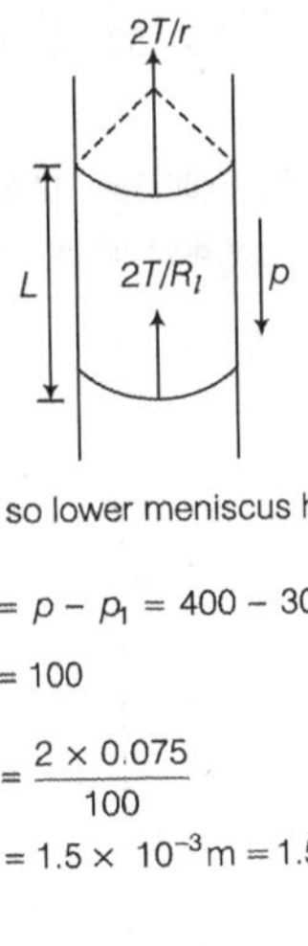

In this case $p_1 < p$, so lower meniscus has to be concave upwards.

$$p_2 = p - p_1 = 400 - 300 = 100 \text{ Nm}^{-2}$$

$$\frac{2T}{R_l} = 100$$

$$\Rightarrow \quad R_l = \frac{2 \times 0.075}{100}$$

$$R_l = 1.5 \times 10^{-3}\text{m} = 1.5 \text{ mm}$$

18. (c) When $L = 3$cm,

hydrostatic pressure is $\quad p = \rho g L$

$p = 1000 \times 10 \times 0.03$

$= 300 \text{ Nm}^{-2}$

Here, $p_1 = p$, so $p_2 = p - p_1 = 0$

$$\frac{2T}{R_l} = 0$$

$$\Rightarrow \quad R_l = \infty$$

i.e. lower meniscus is flat.

19. (5) $\cos 60° = \frac{1}{R} \Rightarrow R = 2\text{cm}$

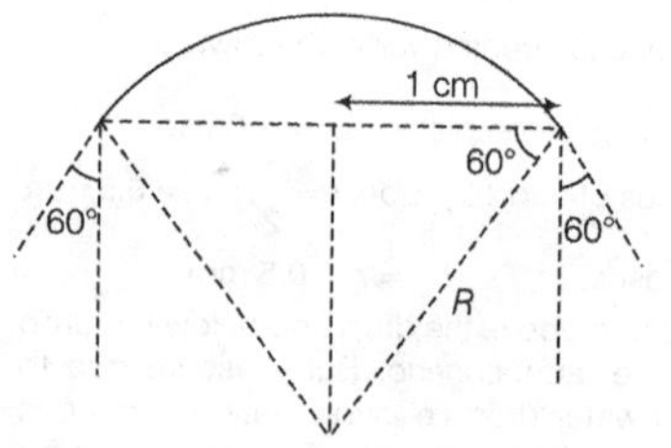

$$\Delta p = \frac{4T}{R} = \frac{4(0.05)}{0.02} = 10\text{Pa}$$

$\therefore \quad \beta = 5$

20. (4) The forces acting on the plate are

(i) $mg\,(\downarrow)$ (ii) Force due to surface tension $(F_T)(\downarrow)$

(iii) Buoyant force $(B_F)(\uparrow)$

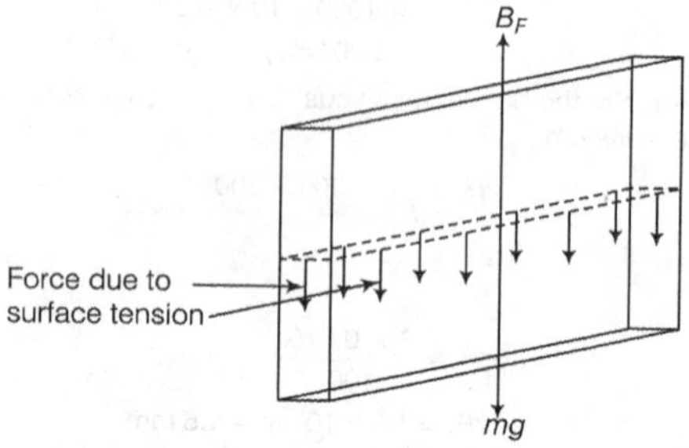

$$w = mg = (8.2 \times 10^{-3}) \times 9.8 = 80.36 \times 10^{-3}\text{N}$$

The water is in contact along the length,

$$L = 2(10 + 0.2) = 20.4\text{cm}$$

$\therefore \quad F_T = (20.4 \times 10^{-2})(7.3 \times 10^{-2}) = 148.92 \times 10^{-4}\text{N}$

$$B_F = \left[\frac{10 \times 0.2 \times 1.54}{2} \times 10^{-6}\right] \times 1000 \times 9.8 = 15.092 \times 10^{-3}\text{N}$$

$\therefore$ Apparent weight

$$w + F_T - B_F = (80.36 + 14.892 - 15.092) \times 10^{-3}$$
$$= 80.16 \times 10^{-3}\text{N}$$

$$= \frac{80.16 \times 10^{-3}}{9.8}\text{ kg-wt}$$
$$= 8.1796 \times 10^{-3}\text{ kg-wt}$$
$$w_{app} = 8.1796 \times 10^{-3}\text{kg-wt}$$
$$= (2.0449) \times 4 \times 10^{-3}\text{ kg-wt}$$

$\Rightarrow \quad \alpha = 4$

21. The water rises to a height, $h = \frac{2T}{\rho g r}$

The potential energy of water column is $U = \frac{mgh}{2} = \frac{2\pi T^2}{\rho g}$

The forces of surface tension perform work, $W = 2\pi r T h$

or $\quad W = \frac{4\pi T^2}{\rho g}$

From energy conservation, the amount of heat liberated is

$$Q = W - U = \frac{4\pi T^2}{\rho g} - \frac{2\pi T^2}{\rho g} = \frac{2\pi T^2}{\rho g}$$

22. Pressure inside the bubble $= p_0 + \rho g h + \frac{2T}{r}$

This pressure is greater than atmospheric pressure by an amount

$$\Delta p = \rho g h + \frac{2T}{r}$$
$$= (1000)(9.8)(0.1) + \frac{2 \times 0.075}{10^{-3}} = 1130\text{ Nm}^{-2}$$

23. The difference in pressure at the upper and lower surfaces of the bubble is equal to $\rho g h$. Since, h is very small, the density ρ can be assumed to be constant without introducing any significant error.

$\therefore \quad \Delta p = \rho g h \quad$ ...(i)

This pressure difference is also equal to excess pressure difference inside the drop which is equal to $2T\left(\frac{1}{R_1} - \frac{1}{R_2}\right)$.

$\therefore$
$$2T\left(\frac{1}{R_1} - \frac{1}{R_2}\right) = \rho g h$$
$$R_2 - R_1 = \frac{1}{2}\rho g h \frac{R_1 R_2}{T}$$

Now, as the drop is very small and nearly spherical $R_1 + R_2 \simeq h$ and $R_1 \simeq R_2$, by no means bad approximations.

Hence, $\quad R_2 - R_1 = \frac{\rho g h^3}{8T}$

Revisal Problems (JEE Main)

1. (c) $\tan 37° = \frac{r}{8}$

$\Rightarrow \quad \frac{r}{8} = \frac{3}{4}$

$\Rightarrow \quad r = 6\text{ m}$

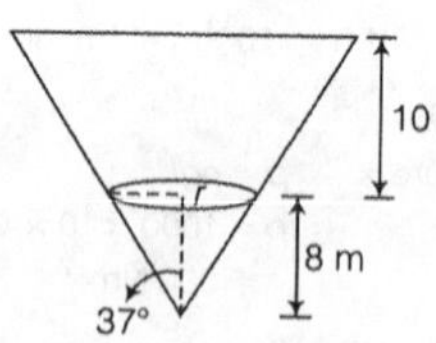

$$F = [p_0 + \rho g \times 10]\,\pi r^2$$
$$= [10^5 + 800 \times 10 \times 10]\,\pi \times 36$$
$$\simeq 2 \times 10^7\text{ N}$$

2. (d)

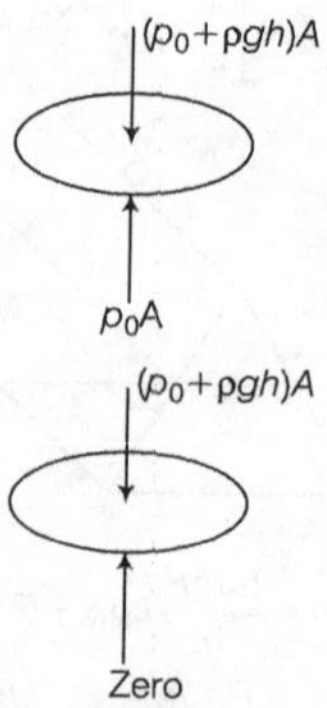

$\therefore \quad F_1 = \rho g h A$

$\therefore \quad F_2 = (p_0 + \rho g h)A$

$$F_2 - F_1 = p_0 A$$
$$= (101300)(1) = 101300\text{N}$$

3. (b) Downward force – Upward force = Force of buoyancy = mg

$$[p_0 + \rho g \cdot 2R]\,\pi R^2 - F_U = \left(\frac{1}{3}\pi R^2\, 2R\right)\rho g - \left(\frac{1}{3}\pi R^2 R\right)\rho g$$

$$= \frac{1}{3}\pi R^2 \cdot R\rho g = \frac{1}{3}\pi R^3 \rho g$$

$$F_U = \pi R^2\left[p_0 + \rho g \cdot 2R - \frac{1}{3}R\rho g\right]$$

$$= \pi R^2\left[p_0 + \frac{5R\rho g}{3}\right]$$

or $$F_U = \frac{\pi R^3 \rho g}{3}\left[5 + \frac{3p_0}{R\rho g}\right]$$

4. (a)
$$|F| = \eta A \frac{dv}{dy}$$

$$= \eta 2\pi R L \frac{v - 0}{R' - R}$$

$$\frac{|F|}{L} = 2\pi\eta R\left(\frac{v}{R' - R}\right)$$

5. (c) Consider a cylinder in place of the irregular body shown.

V_i = volume immersed

= volume of displaced liquid

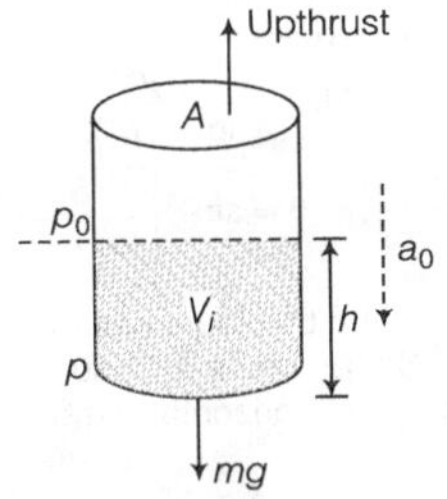

Pressure at depth h is,

$$p = p_0 + \rho(g - a_0)h$$

Upthrust = $(p - p_0)A$

$= [p_0 + \rho(g - a_0)h - p_0]A$

$= h\rho g A - \rho a_0 A h$

$\rho g h A$ = weight of the liquid displaced.

∴ Upthrust < weight of the liquid displaced. This is true in case of a body with irregular shape also.

6. (a) $\tan 53° = \frac{R_1}{x} = \frac{R_2}{x + 30} \Rightarrow R_2 - R_1 = 40$ m

$p_0(\pi R_2^2) = (p_0 + 1000 \times 10 \times 30)\,\pi R_1^2$ (given)

∴ $R_2^2 = 4R_1^2$

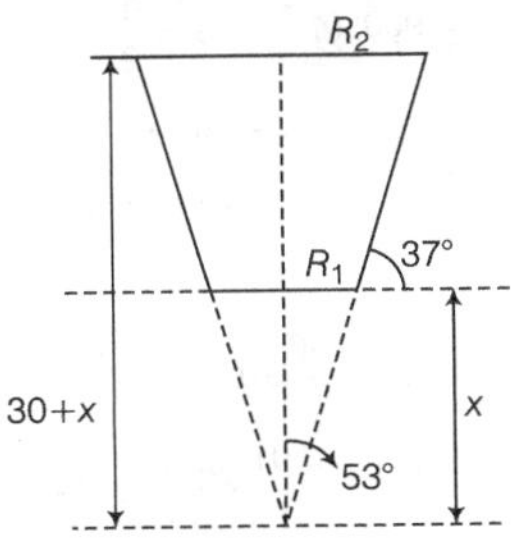

$R_2 = 2R_1$

∴ $2R_1 - R_1 = 40 \Rightarrow R_1 = 40$ m

and $R_2 = 80$ m

So, $R_1 + R_2 = 120$ m $\Rightarrow 20\alpha = 120$

$\Rightarrow \alpha = 6$

7. (d) In the figure shown, ranges will be equal for depths h and $H - h$ or for heights (from the bottom) $H - h$ and h.

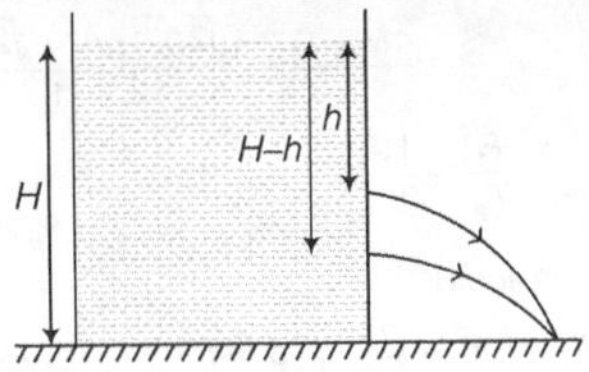

$$(H - h) + h = H$$

∴ $$h_1 + h_2 = H$$

We know that range will be maximum for a height $\frac{H}{2}$.

∴ Range will be maximum when height $= \frac{H}{2} = \frac{h_1 + h_2}{2}$

8. (d) Pressure inside the mercury at the level of S will be less than the atmospheric pressure. So, mercury will not come out of the hole at S.

9. (c) Liquids having acute angle of contact wet the solid surface in contact.

10. (a) $r_B < r_A$ (given). So, excess pressure in B will be larger than that in A. So, air flows from B to A till B be becomes a drop.

11. (c) Length of the rod inside the water

$$1 \sec\theta = \sec\theta$$

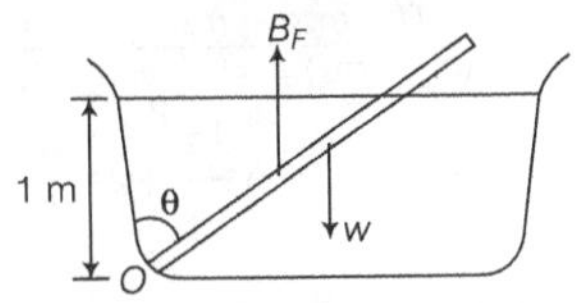

Upthrust or buoyant force is B_F.

$$B_F = V_i \rho_L g$$

$$= \left(\frac{\text{Toal volume}}{\text{Total length}}\right)(\text{length immersed})\,\rho_L g$$

$$= \left(\frac{2/500}{2}\right)(\sec\theta)(1000)(10)$$

or $$B_F = 20\sec\theta$$

Weight of rod is $w = 2 \times 10 = 20$N

For rotational equilibrium of the rod, net torque about point O should be zero.

$$(B_F)\left(\frac{\sec\theta}{2}\right)(\sin\theta) = w\left(\frac{\text{Length}}{2}\right)\sin\theta$$

$$20\sec\theta\left(\frac{\sec\theta}{2}\right) = 20\left(\frac{2}{2}\right)$$

$\Rightarrow$ $$\sec^2\theta = 2 \Rightarrow \sec\theta = \sqrt{2}$$

$\Rightarrow$ $$\theta = 45°$$

∴ $$B_F = 20\sec 45° = 20\sqrt{2}\text{ N}$$

For vertical equilibrium of the rod, force exerted by the hinge on the rod will be $(20\sqrt{2} - 20)$ N downwards or 8.28N ≈ 8.3 N downwards.

12. (b) Applying Bernoulli's equation between points 1 and 2, we have

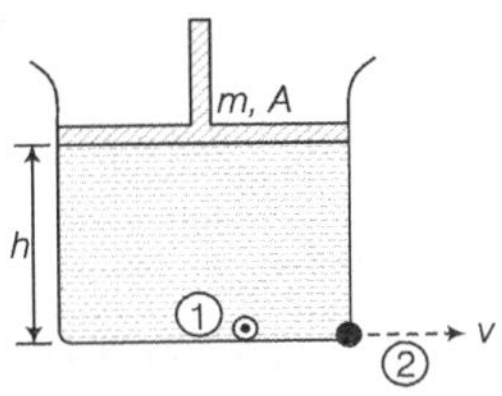

$$p_0 + \frac{mg}{A} + \rho g h = p_0 + \frac{1}{2}\rho v^2$$

$$v^2 = \frac{2mg}{\rho A} + 2gh$$

$\Rightarrow$ $$v = \sqrt{2\left(gh + \frac{mg}{\rho A}\right)}$$

13. (a) Let ρ be the density of liquid. Then, $F_1 = (\Delta p)A = \rho gh A$

In the second case F_2 = rate of change of momentum

$$= \frac{mv - 0}{1} = \frac{(\sqrt{2gh}\cdot A\rho)\sqrt{2gh} - 0}{1} = 2\rho gh A$$

$$\therefore \quad \frac{F_1}{F_2} = \frac{1}{2}$$

14. (a) When B is in liquid,

$$F_{net} \text{ on } B = T - m_B g + \frac{m_B}{\rho_B}\cdot\rho_L g = T - m_B g\left(1 - \frac{1}{2}\right) = T - \frac{1}{2}m_B g$$

$$F_{net} \text{ on } A = m_A g - T = \frac{2}{3}m_B g - T$$

$$\therefore \quad T - \frac{1}{2}m_B g = m_B a_1$$

$$\frac{2}{3}m_B g - T = m_A a_1$$

$$m_B g\left(\frac{2}{3} - \frac{1}{2}\right) = (m_A + m_B)a_1$$

$$\Rightarrow \quad a_1 = g\left(\frac{2}{3} - \frac{1}{2}\right)\cdot\frac{3}{5}$$

$$a_1 = +\frac{1}{6}\cdot\frac{3}{5}g = +\frac{g}{10}$$

$\therefore$ Acceleration of B is $a_1 = \frac{g}{10}$ upwards and this is constant.

When B is in air,

$$m_B g - T' = m_B a_2$$

$$T' - m_A g = m_A a_2$$

$$(m_B - m_A)g = (m_B + m_A)a_2$$

$$a_2 = \frac{1 - \frac{2}{3}}{1 + \frac{2}{3}}g = \frac{g}{5}$$

$\therefore$ Acceleration of B is $a_2 = \frac{g}{5}$ downwards and this is constant.

So, B oscillates periodically not simple harmonically.

15. (b) When the ball moves up, viscous force (F_v) acts downwards. So as the velocity increases, F_v becomes larger and the acceleration decreases.

16. (d) Range will become twice if velocity becomes twice for same h.

$$\left(\because R = v\sqrt{\frac{2h}{g}}\right)$$

$v = \sqrt{2gh}$. If $v = 2v$, then h well become 4 times or $h = 40$ m.

Thus, an extra pressure equivalent to 30 m of water should be applied.

$$1\text{atm} = 0.76 \times 13.6\text{m of water} \approx 10\text{m of water}$$

$$\therefore \quad 30 \text{ m of water} \approx 3 \text{ atm}$$

17. (a) $Q_1 = Q_2$

$$\Rightarrow \quad \sqrt{2gy}\cdot L^2 = \pi R^2\cdot\sqrt{2g\cdot 4y}$$

$$\pi R^2 = \frac{L^2}{2} \Rightarrow R = \frac{L}{\sqrt{2\pi}}$$

18. (c) $k_1 = \left(\frac{\rho_{Fe}}{\rho_{Hg}}\right)_{0°C}$ and $k_2 = \left(\frac{\rho_{Fe}}{\rho_{Hg}}\right)_{60°C}$

$$\frac{k_1}{k_2} = \frac{(\rho_{Fe})_{0°C}}{(\rho_{Fe})_{60°C}}\cdot\frac{(\rho_{Hg})_{60°C}}{(\rho_{Hg})_{0°C}} = \frac{1 + 60\,\gamma_{Fe}}{1 + 60\,\gamma_{Hg}} = \frac{1 + 60x}{1 + 60y}$$

19. (b) Pressure at A is $p_A = p_{atm} - \rho_1 g l\sin\theta$

Pressure at B is $p_B = p_{atm} + \rho_2 gh$

Also, $\quad p_B = p_A + \rho_3 g l\sin\theta$

$$\therefore \quad p_{atm} + \rho_2 gh = p_{atm} - \rho_1 g l\sin\theta + \rho_3 g l\sin\theta$$

$$g l\sin\theta(\rho_3 - \rho_1) = \rho_2 gh$$

$$\sin\theta = \frac{\rho_2 h}{(\rho_3 - \rho_1)l}$$

$$\Rightarrow \quad \theta = \sin^{-1}\left[\frac{\rho_2 h}{(\rho_3 - \rho_1)l}\right]$$

20. (c) With respect to tube the liquid has to be in equilibrium. So, pressure at C due to vertical limb and horizontal limb should be equal.

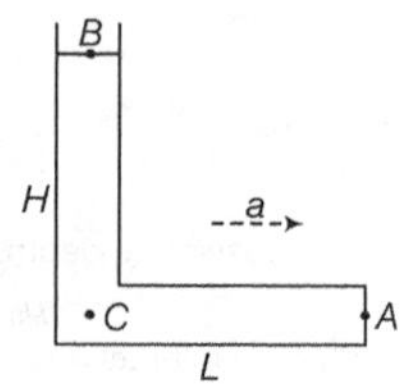

$$\therefore \quad p_B + \rho gH = p_A + \rho La$$

$$p_B = p_{atm}$$

If p_A has to be equal to p_{atm}, then the equation reduces to $\rho gH = \rho La$

$$\Rightarrow \quad a = \frac{gH}{L}$$

Revisal Problems (JEE Advanced)

1. (c)

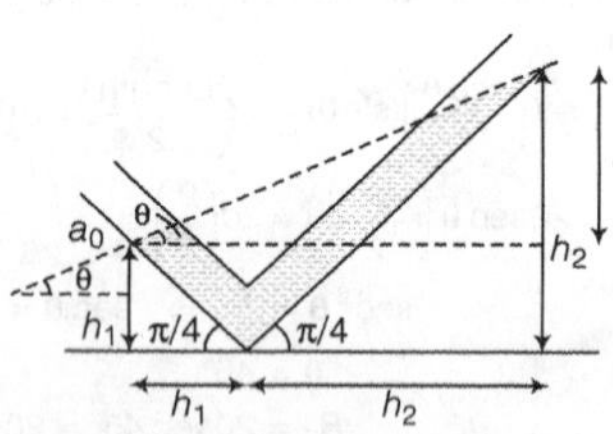

From the figure, $\quad \tan\theta = \frac{h_2 - h_1}{h_1 + h_2}$

Also, we know

$$\tan\theta = \frac{a_0}{g}$$

$$\therefore \quad \frac{a_0}{g} = \frac{h_2 - h_1}{h_1 + h_2}$$

$$\Rightarrow \quad a_0 = g\left(\frac{h_2 - h_1}{h_1 + h_2}\right)$$

2. (a) $A_1 v_1 = A_2 v_2$ (equation of continuity)

$$0.2 \times 4 = \frac{0.2}{4} \times v_2$$

$$\Rightarrow \quad v_2 = 16 \text{ ms}^{-1}$$

Applying Bernoulli's theorem between points P_1 and P_2, we have

$$P_1 + \frac{1}{2}\rho v_1^2 = P_2 + \frac{1}{2}\rho v_2^2$$

$$280 \times 10^3 + \frac{1}{2}(900)(4)^2 = P_2 + \frac{1}{2}(900)(16)^2$$

$$P_2 = 280 \times 10^3 + \frac{1}{2}(900)(16 - 256)$$

$$= 172 \text{ k Nm}^{-2}$$

3. (a) $Av = A_P v_P + A_Q v_Q$

$$\Rightarrow \quad 0.1 = \frac{A}{3}(20) + \left(\frac{2A}{3}\right)v_Q$$

$$0.1 = \frac{0.01}{3}(20) + \frac{2}{3}(0.01)v_Q$$

$$\Rightarrow \quad 0.1 = \frac{0.2}{3} + \frac{0.02}{3}v_Q$$

$$v_Q = \left(0.1 - \frac{0.2}{3}\right)\left(\frac{3}{0.02}\right)$$

$$= \frac{0.1}{3}\left(\frac{3}{0.02}\right)$$

$$= \frac{10}{2} = 5 \text{ ms}^{-1}$$

4. (b) Velocity of efflux of water, $v = \sqrt{2g\left(\frac{h}{2}\right)} = \sqrt{gh}$

Force on ejected water = Rate of change of momentum on ejected water

$$F = \rho a v^2$$

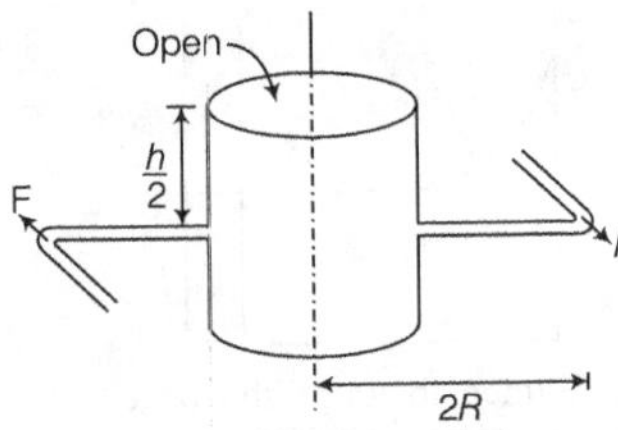

Torque of these forces about central line = $(\rho a v^2)(4R)$

$$= 4R\rho a v^2 = 4R\rho a\left(\sqrt{2g\frac{h}{2}}\right)^2 = 4\rho a g h R$$

5. (d) $v_0 = \sqrt{2gh}$

$$h' = \frac{h}{\sqrt{2}}$$

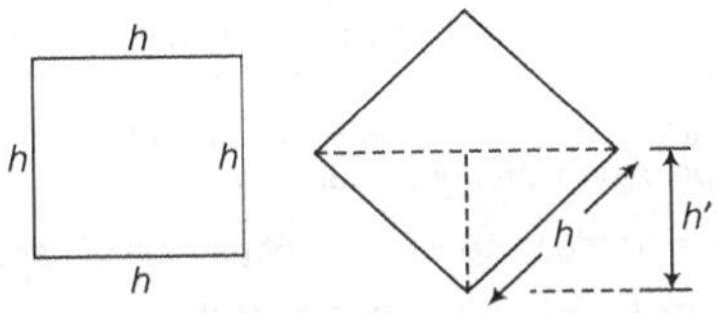

$$v = \sqrt{2gh'} = \sqrt{2g\cdot\frac{h}{\sqrt{2}}} = \frac{\sqrt{2gh}}{2^{1/4}} = \frac{v_0}{2^{1/4}}$$

6. (c) Volume of sphere $= \frac{4}{3}\pi r^3 = \frac{4}{3}\pi\left(\frac{3}{40\pi}\right) = 0.1\text{m}^3$

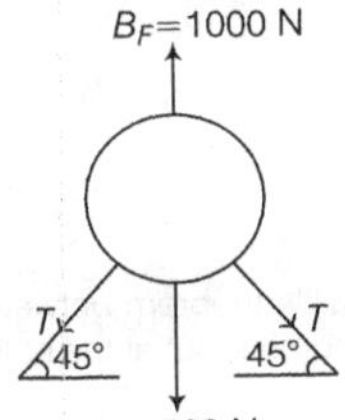

Weight of the sphere is $w = 50 \times 10 = 500\text{N}$

Buoyant force on sphere is

$$B_F = V_i \rho_L g = (0.1)(1000)(10) = 1000\text{N}$$

$$w + 2T\sin 45° = B_F$$

$$500 + 2T\frac{1}{\sqrt{2}} = 1000$$

$$T\sqrt{2} = 500$$

$$T = \frac{500}{\sqrt{2}} = 250\sqrt{2}\,\text{N}$$

7. (c) Maximum level is when,

rate of inflow = rate of outflow

$$A_1v_1 = A_2v_2$$

where, A_1, A_2 are cross-sectional areas of pipe and hole respectively and v_1, v_2 are flow speeds through pipe and hole, respectively.

$$10^{-4} \times 2 = 0.5 \times 10^{-4}\sqrt{2gh_{max}}$$

$\Rightarrow$ $$2g\,h_{max} = 16$$

$$h_{max} = \frac{16}{20} = 0.8\text{ m}$$

Now, at time t, let the level be at h. Then,

$$A\frac{dh}{dt} = A_1v_1 - A_2v_2$$

$$= 10^{-4} \times 2 - 0.5 \times 10^{-4}\sqrt{2gh}$$

$$= \frac{10^{-4}}{2}[4 - \sqrt{20h}]$$

$\therefore$ $$\frac{dh}{dt} = \frac{10^{-4}}{2 \times 0.4}[4 - \sqrt{20h}]$$

$$\int_0^t dt = \frac{8}{10^{-3}}\int_0^h \frac{1}{4 - \sqrt{20h}}dh \quad (\because h = 0 \text{ at } t = 0)$$

This implies that h is not linear in t. Thus, we can exclude options (b) and (d). Since, $h_{max} = 0.8$, correct option is (c).

8. (b) $$A_1v_1 = A_2v_2$$

$$v_2 = \frac{4 \times 10^{-3}}{8 \times 10^{-3}} \times 1 = 0.5\text{ ms}^{-1}$$

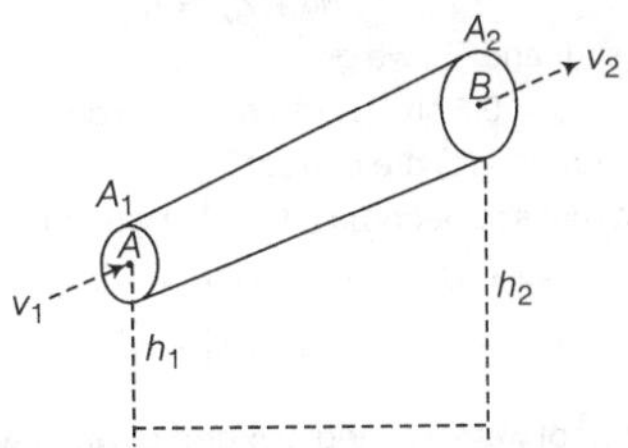

Applying Bernoulli's equation at sections A and B, we have

$$p_1 + \frac{1}{2}\rho v_1^2 + \rho g h_1 = p_2 + \frac{1}{2}\rho v_2^2 + \rho g h_2$$

$$p_1 - p_2 = \rho g(h_2 - h_1) + \frac{1}{2}\rho(v_2^2 - v_1^2)$$

Work done per unit volume by the pressure as the fluid flows from A to B is

$$W = p_1 - p_2$$

$$= \rho g(h_2 - h_1) + \frac{1}{2}\rho(v_2^2 - v_1^2)$$

or $$W = 1000 \times 9.8(5 - 2) + \frac{1}{2} \times 1000\left(\frac{1}{4} - 1\right)$$

$$= (29400 - 375)\text{Jm}^{-3}$$

$$= 29025$$

$$\simeq 2.9 \times 10^4\text{ Jm}^{-3}$$

9. (d) According to Archimedes' principle,

Upthrust = Weight of the liquid displaced

$$F_{bottom} - F_{top} = V\rho g$$

$$F_2 - F_1 = V\rho g$$

But $$F_1 = (\rho g h)\cdot \pi R^2$$

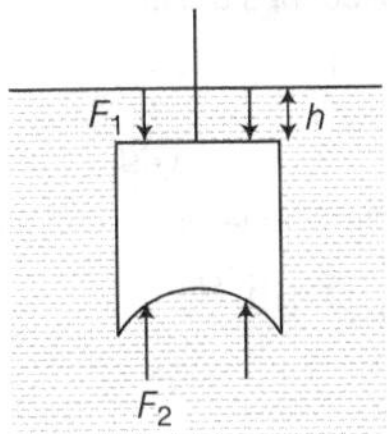

$\therefore$ $$F_2 = V\rho g + \rho g h \cdot \pi R^2$$

$$= \rho g[V + \pi R^2 h]$$

10. (a) $F = \eta A \frac{dv}{dy} \Rightarrow$ Shear stress $\frac{F}{A} = \eta \frac{dv}{dy}$

$$v = k_1 \left[\frac{2y^2}{k_2^2} - \frac{y^3}{k_2^3} \right] \quad \text{(given)}$$

$$\frac{dv}{dy} = k_1 \left[\frac{4y}{k_2^2} - \frac{3y^2}{k_2^3} \right]$$

$$\left(\frac{dv}{dy} \right)_{\text{at } y = k_2} = k_1 \left[\frac{4}{k_2} - \frac{3}{k_2} \right] = \frac{k_1}{k_2}$$

$\therefore$ Shear stress at $y = k_2$ is $\frac{\eta k_1}{k_2}$.

11. (a,b,c) According to condition of floating,

$$(V_i)_k \rho_k g + (V_i)_w \rho_w g = V \rho_{\text{ice}} g$$

$$(Ah_k)(0.8\rho_w) + (Ah_w)\rho_w = (Ah)(0.9\rho_w)$$

$$0.8h_k + h_w = 0.9h \quad \text{...(i)}$$

Also,
$$h_k + h_w = h \quad \text{...(ii)}$$

Solving Eqs. (i) and (ii), we get

$$h_k = 0.5 \text{ cm} \quad \text{and} \quad h_w = 0.5 \text{ cm}$$

So, options (a) and (b) are correct.

1 cm^3 of ice melts to become 0.9cm^3 of water.

After the ice melts, 0.5cm^3 of volume will be occupied by the kerosene. So, the fall in the level of kerosene is $\Delta h_k = \frac{0.5}{A}$.

Out of 0.9cm^3 of water formed, 0.5 cm^3 of water will be filled in the vacant part of ice and remaining (0.9 − 0.5) = 0.4cm^3 of water will rise the level of water.

$\therefore$ Rise in the level of water is

$$\Delta h_w = \frac{0.9 - 0.5}{A} = \frac{0.4}{A}$$

Hence, the net fall in the overall level is

$$\Delta h = \frac{0.5}{A} - \frac{0.4}{A} = \frac{0.1}{A}$$

or
$$\Delta h = \frac{0.1 \text{ cm}^3}{10 \text{ cm}^2} = 0.01 \text{ cm} = 0.1 \text{ mm}$$

So, option (c) is correct.

12. (a,b,c) Equation of free surface is

$$y = \frac{\omega^2 r^2}{2g}$$

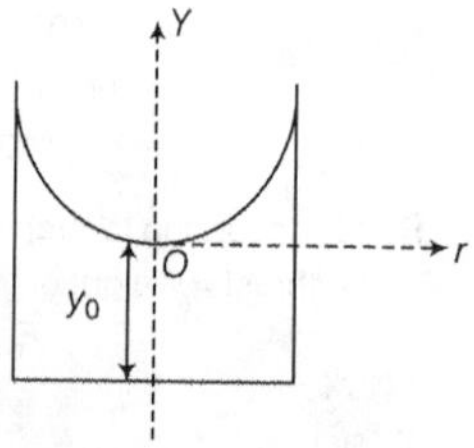

If a parabola $y = kx^2$ is rotated about Y-axis, the volume of the paraboloid formed is $V = \frac{\pi}{2k} y^2$.

Let us assume free surface first touches the brim.

When free surface touches the brim $r = R$

Volume of paraboloid is $V = \frac{\pi}{2\left(\frac{\omega^2}{2g}\right)} y^2 = \frac{g\pi}{\omega^2} \cdot y^2$

Initial volume of water = Final volume of water

$$\therefore \quad \pi R^2 \cdot \frac{2}{3} H = \pi R^2 H - \frac{g}{\omega^2} \cdot \pi y^2$$

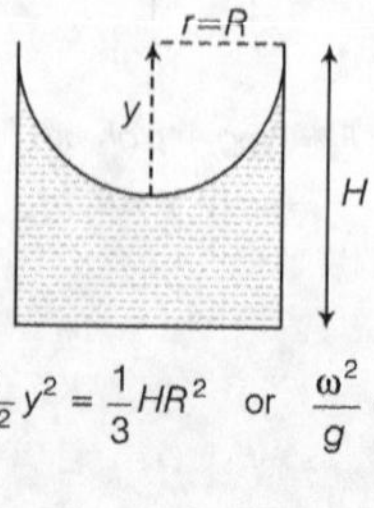

$$\Rightarrow \quad \frac{g}{\omega^2} y^2 = \frac{1}{3} H R^2 \quad \text{or} \quad \frac{\omega^2}{g} = \left(\frac{3}{HR^2} \right) y^2$$

Now,
$$y = \frac{\omega^2 r^2}{2g}$$

$$\Rightarrow \quad y = \frac{1}{2} \left(\frac{3}{HR^2} \right) y^2 \cdot R^2 \quad \text{or} \quad y = \frac{2H}{3}$$

So, the situation (where, $r = R$) becomes as shown in the figure.

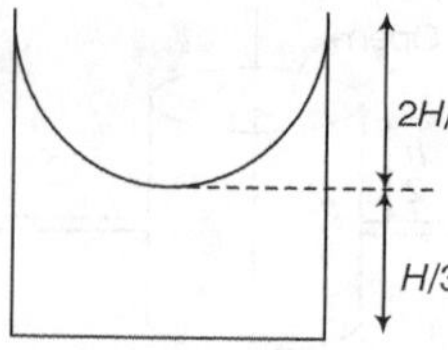

For the free surface to touch the base, ω has to be further increased and water has to spill out.

13. (a,b,d) As the liquid is non-viscous, there is no loss of energy, i.e. energy per unit volume of the liquid is the same at every point.

Liquid is flowing from A to B. So, $p_A > p_B$.

$$a_A v_A = a_B v_B$$

and
$$a_A < a_B$$

$\therefore$
$$v_A > v_B$$

$p_A > p_B$ and $v_A > v_B$ and energy per unit volume of the liquid is constant. So, potential energy at A has to be smaller than potential energy at B (in the light of Bernoulli's equation)

$\therefore$ $(PE)_A < (PE)_B$ i.e. section A has to be at lower level than section B. So, pipe cannot be horizontal.

14. (b,c) Force on the base = $(p_0 + \rho g h) \times$ area of base

h_3 is maximum. So, option (b) is correct.

Net force exerted by the liquid (on the vessel) is its weight, which is same in all the vessels. So, option (c) is correct.

15. (c) Water at the same level is at the same pressure.

$$\therefore \quad p_0 + \rho g h_2 + \rho g h_0 = 2p_0$$

$$\Rightarrow \quad \rho g (h_2 + h_0) = p_0$$

$$h_2 + h_0 = \frac{p_0}{\rho g}$$

$$\Rightarrow \quad h_2 = \frac{p_0}{\rho g} - h_0$$

16. (b) Applying Bernoulli's theorem between the top of water in the tank and the water coming out at the hole, we have

$$2p_0 + \rho g (h_1 - h_0) = p_0 + \frac{1}{2} \rho v^2$$

$$\frac{1}{2} \rho v^2 = p_0 + \rho g (h_1 - h_0)$$

$$\Rightarrow \quad v = \left\{ \frac{2}{\rho} [p_0 + \rho g (h_1 - h_0)] \right\}^{1/2}$$

17. (d) When the water stops coming out of the hole, pressure inside the tank at the level of hole should be p_0.

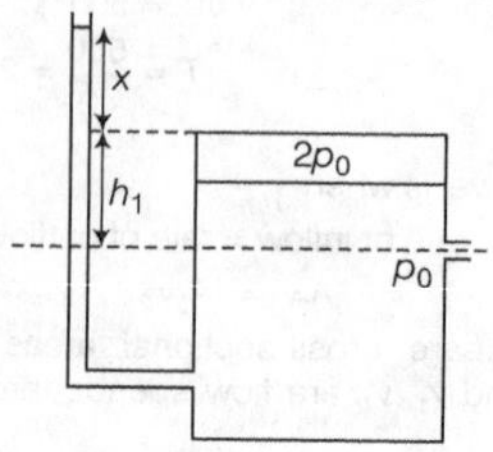

$$p_0 + \rho g x + \rho g h_1 = p_0 \quad (\because \text{same level})$$

$$x = -h_1$$

or in the long tube, the water should be h_1 metre below the top or $-h_1$ metre above the top.

18. (a)

(i) B_F = Buoyant force

$$= \left(\frac{m}{\rho_s}\cdot\rho\right)g_{\text{eff}}$$

$$= m\left(\frac{\rho}{\rho_s}\right)\sqrt{a^2+g^2}$$

$$= m\left(\frac{\rho}{\rho_s}\right)\sqrt{\frac{g^2}{4}+g^2}$$

$$= \frac{m}{2}\cdot\frac{\rho}{\rho_s}g\sqrt{5}$$

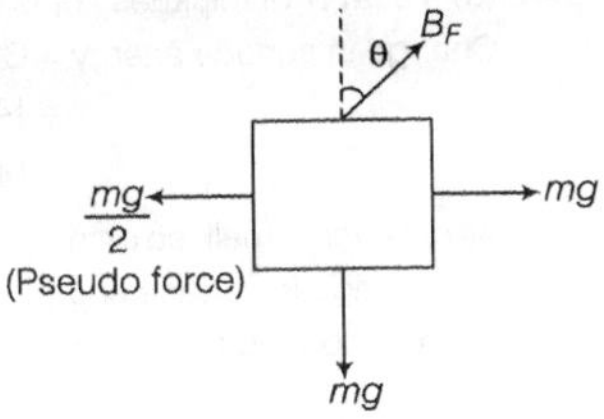

We know, $\tan\theta = \frac{a}{g} = \frac{1}{2}$

$\Rightarrow \quad \sin\theta = \frac{1}{\sqrt{5}}$ and $\cos\theta = \frac{2}{\sqrt{5}}$

Let acceleration of cube w.r.t. container in horizontal direction is a'_x.

Then, $\quad mg - \frac{mg}{2} + B_F\sin\theta = ma'_x$

$$\frac{mg}{2} + \frac{m}{2}\cdot\frac{\rho}{\rho_s}g\sqrt{5}\left(\frac{1}{\sqrt{5}}\right) = ma'_x$$

$$a'_x = \frac{g}{2}\left[1+\frac{\rho}{\rho_s}\right]$$

$$\therefore \quad a_x = a'_x + \frac{g}{2} = g + \frac{g}{2}\cdot\frac{\rho}{\rho_s} = g\left[\frac{2\rho_s+\rho}{2\rho_s}\right]$$

In the vertical direction,

$$B_F\cos\theta - mg = ma_y$$

$$\left(\frac{m}{2}\cdot\frac{\rho}{\rho_s}g\sqrt{5}\right)\left(\frac{2}{\sqrt{5}}\right) - mg = ma_y$$

$$a_y = g\frac{\rho}{\rho_s} - g = g\left[\frac{\rho-\rho_s}{\rho_s}\right]$$

$$\therefore \quad \frac{a_x}{a_y} = \frac{2\rho_s+\rho}{2(\rho-\rho_s)}$$

(ii) $mg = ma_x \Rightarrow a_x = g$

$$B_F = \frac{m}{\rho_s}\rho g$$

$$B_F - mg = ma_y$$

$$\frac{m}{\rho_s}\rho g - mg = ma_y$$

$$\Rightarrow \quad a_y = g\left(\frac{\rho}{\rho_s}-1\right)$$

$$\therefore \quad \frac{a_x}{a_y} = \frac{\rho_s}{\rho-\rho_s}$$

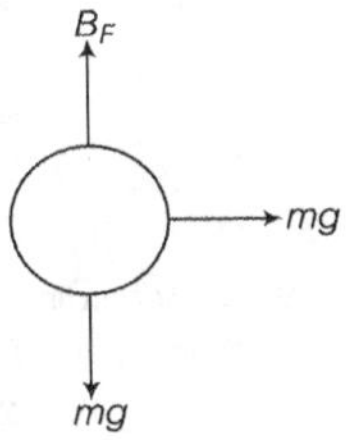

(iii) $(p_2A - p_1A) = ma_x$

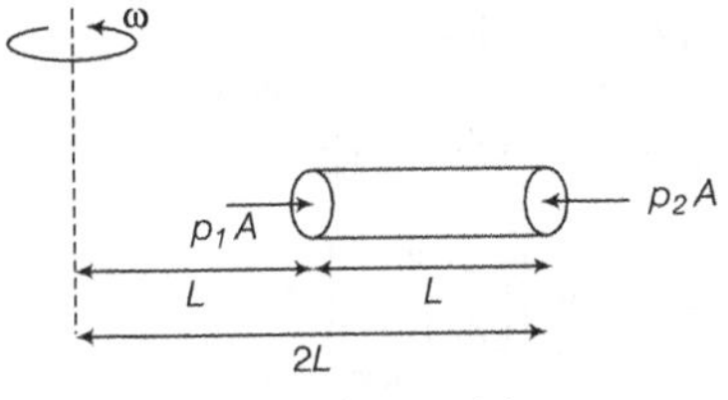

$$(p_2-p_1)A = ma_x$$

$$\left[\frac{\rho\omega^2}{2}(2L)^2 - \frac{\rho\omega^2}{2}(L)^2\right]A = ma_x$$

$$\frac{\rho\omega^2}{2}\cdot 3L^2A = (\rho_s AL)a_x$$

$$\frac{3}{2}L^2\rho\cdot\left(\frac{2g}{3L}\right) = \rho_s La_x$$

$$\therefore \quad \frac{a_x}{g} = \frac{\rho}{\rho_s}$$

(iv) When there is no liquid, $\tan\theta = \frac{a_0}{g}$ ...(i)

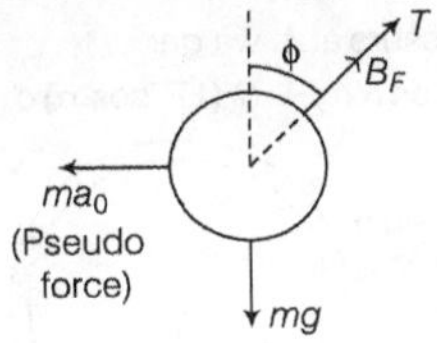

when there is liquid

$$(T+B_F)\sin\phi = ma_0$$

$$(T+B_F)\cos\phi = mg$$

$$\therefore \quad \tan\phi = \frac{a_0}{g} \quad \text{...(ii)}$$

From Eqs. (i) and (ii), we get

$$\frac{\tan\theta}{\tan\phi} = 1:1$$

Hence, (i) → (s), (ii) → (r), (iii) → (p), (iv) → (q)

19. (3) The vertical force due to surface at the hole

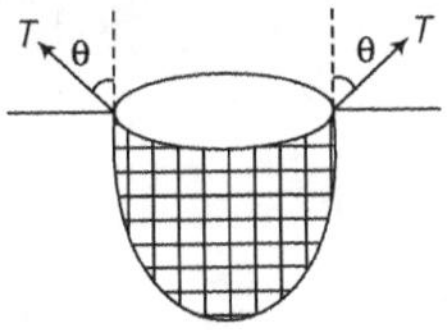

$= T\cos\theta \times L = T\cos\theta \times 2\pi r$

This will balance weight $mg\ (= \pi r^2 h\rho g)$.

$\therefore \quad T\cos\theta \times 2\pi r = \pi r^2 h\rho g$

$$\Rightarrow \quad h = \frac{2T\cos\theta}{r\rho g}$$

h is maximum, when $\cos\theta = 1$

$$\therefore \quad h_{\max} = \frac{2T}{r\rho g} = \frac{2\times 0.075}{0.0005\times 1000\times 10}$$

$$= 0.03 \text{ m} = 3 \text{ cm}$$

20. (4) Mass of air entering in 1 s is $m = \pi b^2 v\rho$

Change in momentum in 1 s $= mv - 0 = \pi b^2\rho v^2$

Thus, $\quad F = \pi b^2\rho v^2$

Pressure due to this force is

$$\Delta p = \frac{F}{A} = \frac{\pi b^2\rho v^2}{\pi b^2} = \rho v^2$$

When this excess pressure equals to $\frac{4T}{R}$, the bubble gets separated.

$$\therefore \quad \rho v^2 = \frac{4T}{R} \Rightarrow R = \frac{4T}{\rho v^2}$$

So, $\quad \beta = 4$

Alternate Method The soap bubble will separate from the ring, when the force of surface tension becomes equal to the force due to air.

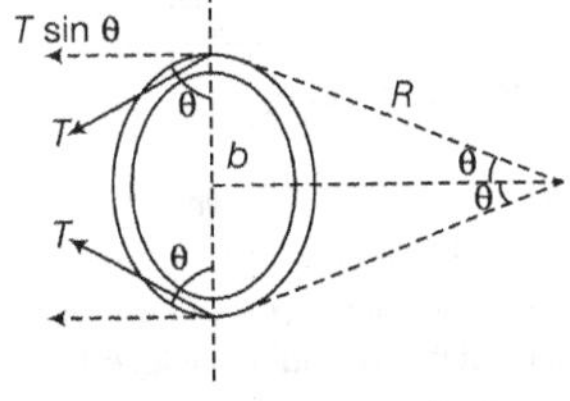

$$2\pi b\cdot(2T\sin\theta) = \pi b^2\rho v^2$$

Note *Circular frame has two edges.*

$$4bT\sin\theta = b^2\rho v^2$$

$$4bT\left(\frac{b}{R}\right) = b^2\rho v^2$$

$$R = \frac{4T}{\rho v^2} \Rightarrow \beta = 4$$

JEE Main & AIEEE Archive

1. (c) Equating pressure at A, we get

$$R \sin \alpha\, d_2 + R \cos\alpha\, d_2 + R\,(1 - \cos \alpha)\, d_1 = R\,(1 - \sin \alpha)\, d_1$$

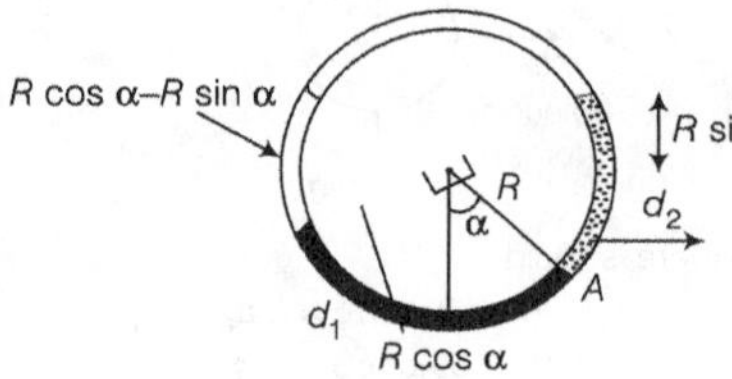

$$(\sin \alpha + \cos \alpha)\, d_2 = d_1\,(\cos \alpha - \sin \alpha)$$

$$\Rightarrow \quad \frac{d_1}{d_2} = \frac{1 + \tan \alpha}{1 - \tan \alpha}$$

2. (a) The bubble will detach, if

Buoyant force ≥ Surface tension force

i.e. $$\frac{4}{3}\pi R^3 \rho_w g \geq \int T \times dl \sin \theta$$

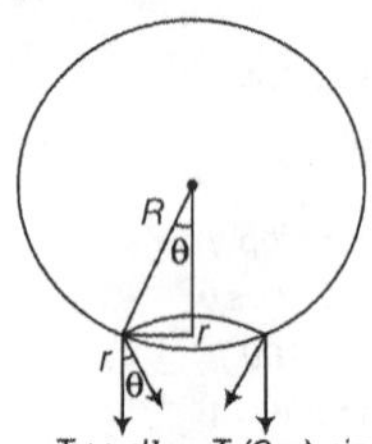

$$T \times dl = T\,(2\pi r) \sin \theta$$

$$(\rho_w)\left(\frac{4}{3}\pi R^3\right) g \geq (T)\,(2\pi r) \sin \theta$$

$$\sin \theta = \frac{r}{R}$$

On solving, we get

$$r = \sqrt{\frac{2\rho_w R^4 g}{3T}} = R^2\sqrt{\frac{2\rho_w g}{3T}}$$

3. (c) The various forces acting on the cylinder are shown in the figure.

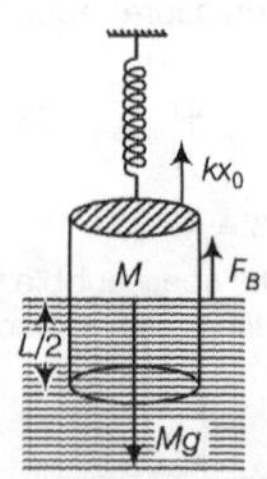

Let x_0 be the extension in spring.

At equilibrium,

Upward forces = Downward forces

$$kx_0 + F_B = Mg$$

where, k is spring constant and F_B is force due to buoyancy which is equal to weight of liquid displaced.

$$F_B = \sigma \cdot \frac{L}{2} \cdot A \cdot g$$

$$\therefore \quad kx_0 + \sigma \frac{L}{2} Ag = Mg$$

$$\Rightarrow \quad x_0 = \frac{Mg - \dfrac{\sigma LAg}{2}}{k} = \frac{Mg}{k}\left(1 - \frac{\sigma LA}{2M}\right)$$

4. (d) If a layer of thickness dr is evaporated, then

Change in surface energy = Change in surface area × T

$$= [4\pi\,(r + dr)^2 - 4\,\pi\, r^2]\, T$$

$$= [4\pi r^2 + 4\pi\, dr^2 + 8\pi r dr - 4\pi r^2]\, T$$

As dr is very small, so dr^2 neglected.

So, change in surface energy = $8\pi r dr\, T$

Energy required to evaporate layer of thickness dr is

$$(4\pi r^2 dr)\rho L$$

The process of evaporation just starts, if change in surface energy is just sufficient to evaporate the water layer.

$$\therefore \quad (4\pi r^2 dr)\rho L = (8\pi r dr)\, T$$

$$\Rightarrow \quad r = \frac{2T}{\rho L}$$

5. (d) At equilibrium, weight of the given block is balanced by force due to surface tension, i.e.

$$2L \cdot S = w$$

or $$S = \frac{w}{2L} = \frac{1.5 \times 10^{-2}\ \text{N}}{2 \times 0.3\ \text{m}} = 0.025\ \text{Nm}^{-1}$$

6. (d) Let R be the radius of the bigger drop, then

Volume of bigger drop = 2 × Volume of small drop

$$\frac{4}{3}\pi R^3 = 2 \times \frac{4}{3}\pi r^3$$

$$R = 2^{1/3} r$$

Surface energy of bigger drop,

$$E = 4\pi R^2 T$$

$$= 4 \times 2^{2/3}\, \pi r^2 T = 2^{8/3} \pi r^2 T$$

7. (b) As terminal velocity, $v = \dfrac{2}{9} r^2 \dfrac{(\rho - \rho_0)}{\eta}$

Then, $$v \propto \frac{\rho - \rho_0}{\eta}$$

$$\therefore \quad \frac{v_2}{v_1} = \frac{\rho - \rho_{02}}{\rho - \rho_{01}} \times \frac{\eta_1}{\eta_2} = \frac{7.8 - 1.2}{7.8 - 1} \times \frac{8.5 \times 10^{-4} \times 10}{13.2}$$

$$= 6.25 \times 10^{-4}\ \text{cms}^{-1}$$

8. (c) According to work-energy theorem,

Work done = Change in surface energy

$$\Rightarrow \quad W = 2T \times 4\pi\,(R_2^2 - R_1^2)$$

$$= 2 \times 0.03 \times 4\pi\,[(5)^2 - (3)^2] \times 10^{-4}\ \text{J} = 0.4\pi\ \text{mJ}$$

9. (c) From Bernoulli's theorem,

$$\rho g h = \frac{1}{2}\rho\,(v_2^2 - v_1^2)$$

$$\Rightarrow \quad gh = \frac{1}{2} v_1^2 \left[\left(\frac{v_2}{v_1}\right)^2 - 1\right]$$

$$\Rightarrow \quad gh = \frac{1}{2} v_1^2 \left[\left(\frac{A_1}{A_2}\right)^2 - 1\right] \qquad (\because A_1 v_1 = A_2 v_2)$$

$$\Rightarrow \quad \left(\frac{A_1}{A_2}\right)^2 = 1 + \frac{2hg}{v_1^2}$$

$$\Rightarrow \quad \left(\frac{D_1}{D_2}\right)^4 = 1 + \frac{2gh}{v_1^2}$$

$$\Rightarrow \quad D_2 = \frac{D_1}{\left(1 + \dfrac{2gh}{v_1^2}\right)^{1/4}} = \frac{8 \times 10^{-3}}{\left(1 + \dfrac{2 \times 10 \times 0.2}{(0.4)^2}\right)^{1/4}}$$

$$\approx 3.6 \times 10^{-3}\ \text{m}$$

10. (b) $\rho_{oil} < \rho < \rho_{water}$

Oil is the least dense of them, so it should settle at the top with water at the base. Now, the ball is less denser than oil but less denser than water. So, it will sink through oil but will not sink in water. So, it will stay at the oil-water interface.

11. (a) The forces acting on the ball are gravity force, buoyancy force and viscous force. When ball acquires terminal speed, it is in dynamic equilibrium, let terminal speed of ball be v_T. So,

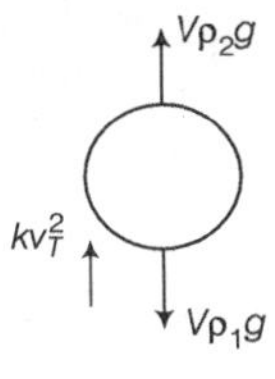

$$V\rho_2 g + kv_T^2 = V\rho_1 g$$

or $$v_T = \sqrt{\frac{V(\rho_1 - \rho_2)g}{k}}$$

12. (d) $\rho_1 < \rho_2$ as denser liquid acquires lowest position of vessel.

$\rho_3 > \rho_1$ as ball sinks in liquid 1 and $\rho_3 < \rho_2$ as ball does not sink in liquid 2, so

$$\rho_1 < \rho_3 < \rho_2$$

13. (c) Soap solution has lower surface tension T as compared to pure water and capillary rise $h = \dfrac{2T\cos\theta}{\rho r g}$, so h is less for soap solution.

14. (c) Terminal speed of spherical body in a viscous liquid is given by

$$v_T = \frac{2r^2(\rho - \sigma)g}{9\eta}$$

where, ρ = density of substance of a body

σ = density of liquid.

From given data,

$$\frac{v_T(\text{Ag})}{v_t(\text{Au})} = \frac{\rho_{Ag} - \sigma_l}{\rho_{Au} - \sigma_l}$$

$$\Rightarrow \quad v_T(\text{Ag}) = \frac{10.5 - 1.5}{19.5 - 1.5} \times 0.2$$

$$= \frac{9}{18} \times 0.2 = 0.1 \text{ ms}^{-1}$$

15. (d) Water fills the tube entirely in gravity less condition. Hence, length of water column in the capillary tube is 20 cm.

16. (b) Retarding force acting on a ball falling into a viscous fluid,

$$F = 6\pi\eta R v$$

where, R = radius of ball,

v = velocity of ball

and η = coefficient of viscosity.

$\therefore$ $F \propto R$ and $F \propto v$

or in other words, retarding force is directly proportional to both radius R and velocity v.

17. (c) The excess pressure inside the soap bubble is inversely proportional to radius of soap bubble i.e. $p \propto 1/r$, r being the radius of bubble. It follows that pressure inside a smaller bubble is greater than that inside a bigger bubble.

Thus, if these two bubbles are connected by a tube, air will flow from smaller bubble to bigger bubble and the bigger bubble grows at the expense of the smaller one.

18. (b) Applying Bernoulli's theorem just inside and outside the hole. Take reference line for gravitational potential energy at the bottom of the vessel.

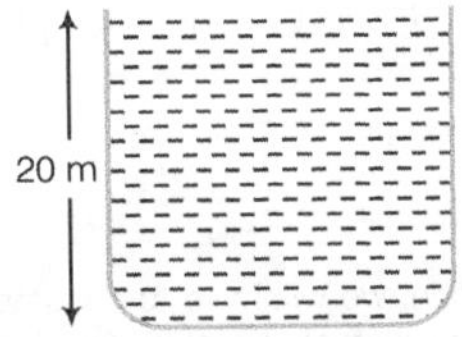

Let p_0 be the atmospheric pressure, ρ be the density of liquid and v be the velocity at which water is coming out.

$$p_{inside} + \rho gh + 0 = p_{outside} + \frac{\rho v^2}{2}$$

$$\Rightarrow \quad p_0 + \rho gh = p_0 + \frac{\rho v^2}{2}$$

or $$v = \sqrt{2gh} = \sqrt{2 \times 10 \times 20} = 20 \text{ ms}^{-1}$$

JEE Advanced & IIT JEE Archive

1. (d) Using geometry,

$$\frac{b}{R} = \cos\left(\theta + \frac{\alpha}{2}\right)$$

$$\Rightarrow \quad R = \frac{b}{\cos\left(\theta + \frac{\alpha}{2}\right)}$$

Using pressure equation along the path $MNTK$

$$p_0 - \frac{2S}{R} + h\rho g = p_0$$

Substituting the value of R, we get

$$h = \frac{2S}{R\rho g} = \frac{2S}{b\rho g}\cos\left(\theta + \frac{\alpha}{2}\right)$$

2. (a) Let V_1 = total material volume of shell

V_2 = total inside volume of shell and

x = fraction of V_2 volume filled with water.

In floating condition,

Total weight = Upthrust

$$\therefore V_1\rho_c g + (xV_2)(1)g = \left(\frac{V_1 + V_2}{2}\right)(1)g$$

or $$x = 0.5 + (0.5 - \rho_c)\frac{V_1}{V_2}$$

From here we can see that, $x > 0.5$ if $\rho_c < 0.5$

3. (a,b,d)

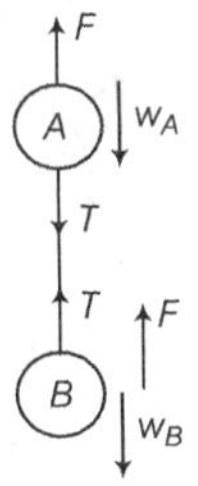

$$F = \text{Upthrust} = Vd_F g$$

Equilibrium of A, $Vd_F g = T + w_A = T + Vd_A g$ ...(i)

Equilibrium of B, $T + Vd_F g = Vd_B g$ ...(ii)

On adding Eqs. (i) and (ii), we get

$$2d_F = d_A + d_B$$

From Eq. (i), we can see that

$$d_F > d_A \quad (\text{as } T > 0)$$

From Eq. (ii), we can see that,

$$d_B > d_F$$

4. (c) Vertical force due to surface tension,

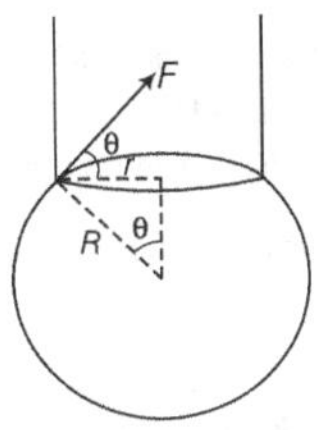

$$F_v = F\sin\theta = (T2\pi r)(r/R) = \frac{2\pi r^2 T}{R}$$

5. (a) $\frac{2\pi r^2 T}{R} = mg = \frac{4}{3}\pi R^3 \cdot \rho \cdot g$

$\therefore \quad R^4 = \frac{3r^2T}{2\rho g} = \frac{3 \times (5 \times 10^{-4})^2 (0.11)}{2 \times 10^3 \times 10}$

$= 4.125 \times 10^{-12}\ m^4$

$\therefore \quad R = 1.425 \times 10^{-3} m \approx 1.4 \times 10^{-3}\ m$

6. (b) Surface energy,

$E = (4\pi R^2)\, T$

$= (4\pi)\,(1.4 \times 10^{-3})^2\,(0.11)$

$= 2.7 \times 10^{-6}\ J$

7. (6) Although not given in the question, but we will have to assume that temperatures of A and B are same.

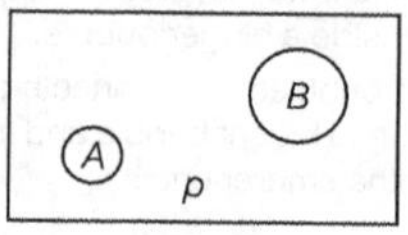

$p = 8\ Nm^{-2}$

$\frac{n_B}{n_A} = \frac{p_B V_B/RT}{p_A V_A/RT} = \frac{p_B V_B}{p_A V_A} = \frac{(p + 4S/r_A) \times 4/3\pi\,(r_A)^3}{(p + 4S/r_B) \times 4/3\pi\,(r_B)^3}$

(S = surface tension)

Substituting the values, we get

$\frac{n_B}{n_A} = 6$

8. (6) In this question, we will have to assume that temperature of enclosed air about water is constant (or pV = constant)

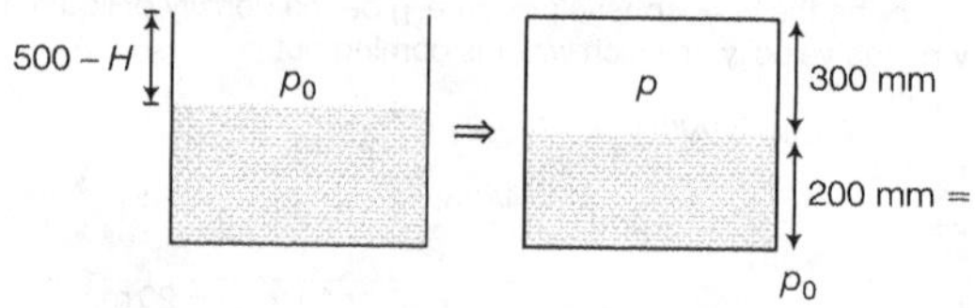

$p = p_0 - \rho g h \quad \ldots(i)$

$p_0[A(500 - H)] = p[A(300)] \quad \ldots(ii)$

On solving Eqs. (i) and (ii), we get

$H = 206$ mm

$\therefore$ Level fall = (206 – 200) mm = 6 mm

9. (b) $\Delta p_1 = \frac{4T}{r_1}$ and $\Delta p_2 = \frac{4T}{r_2}$

$r_1 < r_2$

$\therefore \quad \Delta p_1 > \Delta p_2$

$\therefore$ Air will flow from 1 to 2 and volume of bubble at end 1 will decrease.

10. (a) From continuity equation, Av = constant

or $\quad A \propto \frac{1}{v}$

At lower heights, speed will be more. Therefore, area of cross-section will be less.

11. (b) Force from right hand side liquid on left hand side liquid.

(i) Due to surface tension force

$= 2RT$ (towards right)

(ii) Due to liquid pressure force $= \int_{x=0}^{x=h} (p_0 + \rho g h)(2R \cdot x)\, dx$

$= (2p_0 Rh + R\rho g h^2)$ (towards left)

$\therefore$ Net force is $|2p_0 Rh + R\rho g h^2 - 2RT|$.

12. (c) Let A_1 = Area of cross-section of cylinder = $4\pi r^2$

A_2 = Area of base of cylinder in air = πr^2

and A_3 = Area of base of cylinder in water = $A_1 - A_2 = 3\pi r^2$

The free body diagram of cylinder

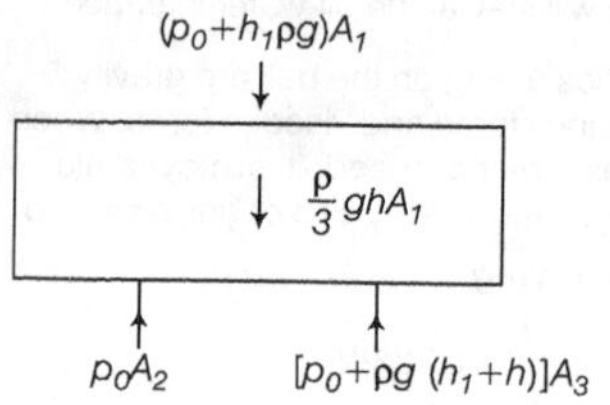

Equating the net downward forces and net upward forces, we get, $h_1 = \frac{5}{3}h$.

13. (a) Again equating the forces, we get

$h_2 = \frac{4h}{9}$

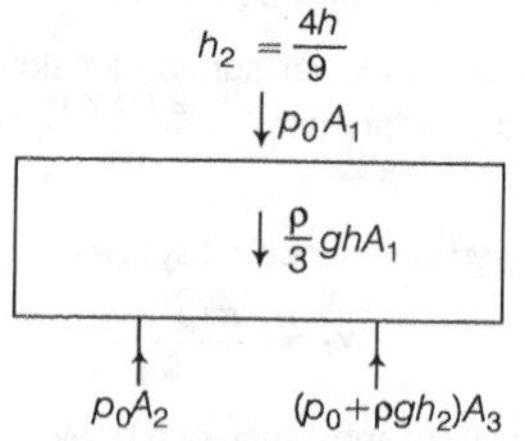

14. (a) For $h_2 < \frac{4h}{9}$, buoyant force will further decrease. Hence, the cylinder remains at its original position.

15. For circular motion of small element dx, we have

$dF = (dm) \times \omega^2$

$\therefore \quad (dp)\, A = (\rho A dx) \cdot \times \omega^2$

or $\quad dp = \rho\, \omega^2\, x \cdot dx$

$\therefore \quad \int_{p_1}^{p_2} dp = \rho\, \omega^2 \int_0^L x\, dx$

$\therefore \quad p_2 - p_1 = \frac{\rho\, \omega^2 L^2}{2}$

$\therefore \quad \rho g H = \frac{\rho\, \omega^2 L^2}{2}$

$\therefore \quad H = \frac{\omega^2 L^2}{2g}$

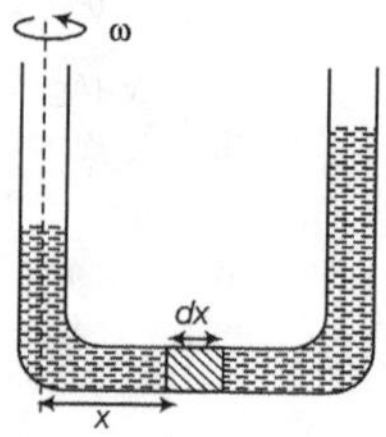

16. (a) Applying continuity equation at 1 and 2, we have

$A_1 v_1 = A_2 v_2 \quad \ldots(i)$

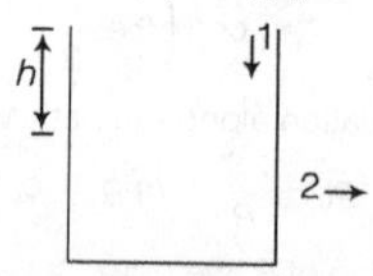

Further applying Bernoulli's equation at these two points, we have

$p_0 + \rho g h + \frac{1}{2}\rho v_1^2 = p_0 + 0 + \frac{1}{2}\rho v_2^2 \quad \ldots(ii)$

On solving Eqs. (i) and (ii), we have

$v_2^2 = \frac{2gh}{1 - \frac{A_2^2}{A_1^2}}$

On substituting the values, we have

$v_2^2 = \frac{2 \times 10 \times 2.475}{1 - (0.1)^2} = 50\ m^2 s^{-2}$

Last 3 Years' Questions
JEE Main & Advanced
(2016-2018)

JEE Main

2016

1. A particle of mass m is moving along the side of a square of side a, with a uniform speed v in the XY-plane as shown in the figure.

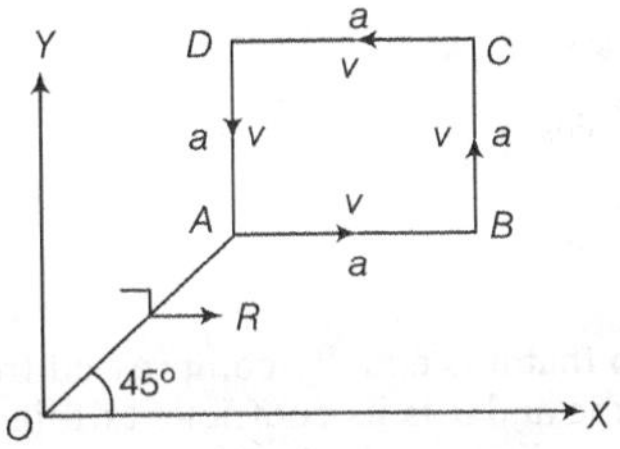

Which of the following statements is **false** for the angular momentum **L** about the origin? **[Rotational Motion]**

(a) $\mathbf{L} = \dfrac{-mv}{\sqrt{2}} R\hat{\mathbf{k}}$ when the particle is moving from A to B

(b) $\mathbf{L} = mv\left(\dfrac{R}{\sqrt{2}} - a\right)\hat{\mathbf{k}}$ when the particle is moving from C to D

(c) $\mathbf{L} = mv\left(\dfrac{R}{\sqrt{2}} + a\right)\hat{\mathbf{k}}$ when the particle is moving from B to C

(d) $\mathbf{L} = \dfrac{mv}{\sqrt{2}} R\hat{\mathbf{k}}$ when the particle is moving from D to A

2. A roller is made by joining together two corners at their vertices O. It is kept on two rails AB and CD which are placed asymmetrically (see the figure), with its axis perpendicular to CD and its centre O at the centre of line joining AB and CD (see the figure). It is given a light push, so that it starts rolling with its centre O moving parallel to CD in the direction shown. As it moves, the roller will tend to **[Gravitation]**

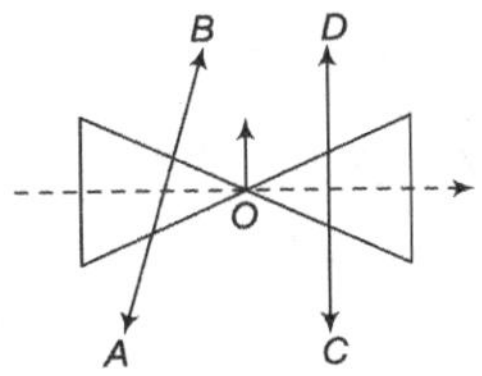

(a) turn left
(b) turn right
(c) go straight
(d) turn left and right alternately

3. A satellite is revolving in a circular orbit at a height h from the Earth's surface (radius of earth $R, h << R$). The minimum increase in its orbital velocity required, so that the satellite could escape from the earth's gravitational field, is close to (neglect the effect of atmosphere) **[Gravitation]**

(a) $\sqrt{2gR}$
(b) $\sqrt{gR}$
(c) $\sqrt{gR/2}$
(d) $\sqrt{gR}\,(\sqrt{2} - 1)$

2017

4. The moment of inertia of a uniform cylinder of length l and radius R about its perpendicular bisector is I. What is the ratio l/R such that the moment of inertia is minimum? **[Rotational Motion]**

(a) $\frac{\sqrt{3}}{2}$ (b) 1

(c) $\frac{3}{\sqrt{2}}$ (d) $\sqrt{\frac{3}{2}}$

5. A slender uniform rod of mass M and length l is pivoted at one end so that it can rotate in a vertical plane (see the figure). There is negligible friction at the pivot. The free end is held vertically above the pivot and then released. The angular acceleration of the rod when it makes an angle θ with the vertical, is **[Rotational Motion]**

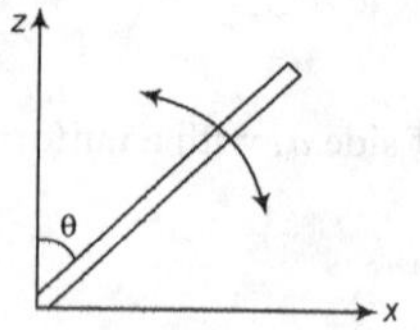

(a) $\frac{2g}{3l}\sin\theta$ (b) $\frac{3g}{2l}\cos\theta$

(c) $\frac{2g}{3l}\cos\theta$ (d) $\frac{3g}{2l}\sin\theta$

6. An external pressure P is applied on a cube at 0°C so that it is equally compressed from all sides. K is the bulk modulus of the material of the cube and α is its coefficient of linear expansion. Suppose we want to bring the cube to its original size by heating. The temperature should be raised by **[Elasticity]**

(a) $\frac{P}{\alpha K}$ (b) $\frac{3\alpha}{PK}$

(c) $3PK\alpha$ (d) $\frac{P}{3\alpha K}$

7. The variation of acceleration due to gravity g with distance d from centre of the Earth is best represented by
(R = Earth's radius) **[Gravitation]**

(a) 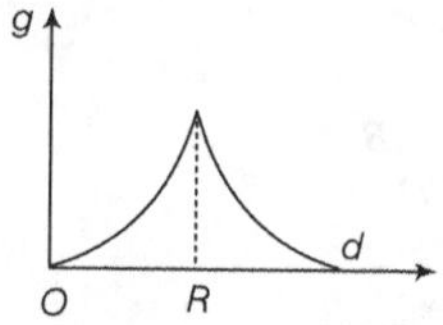

(b) 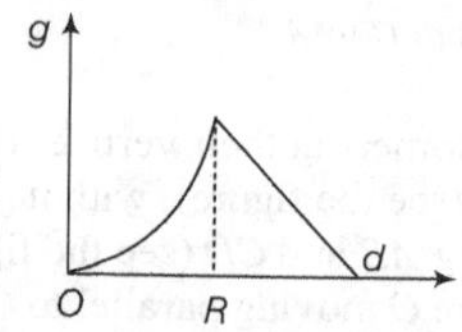

(c) 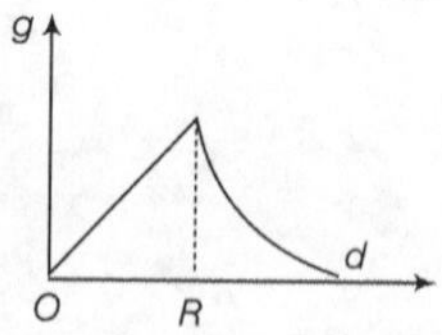

(d) 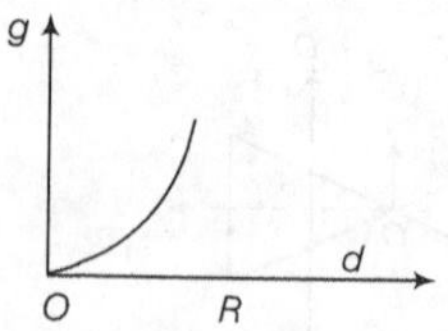

8. A man grows into a giant such that his linear dimensions increase by a factor of 9. Assuming that his density remains same, the stress in the leg will change by a factor of **[Elasticity]**

(a) $\frac{1}{9}$ (b) 81

(c) $\frac{1}{81}$ (d) 9

2018

9. Seven identical circular planar discs, each of mass M and radius R are welded symmetrically as shown in the figure. The moment of inertia of the arrangement about the axis normal to the plane and passing through the point P is **[Rotational Motion]**

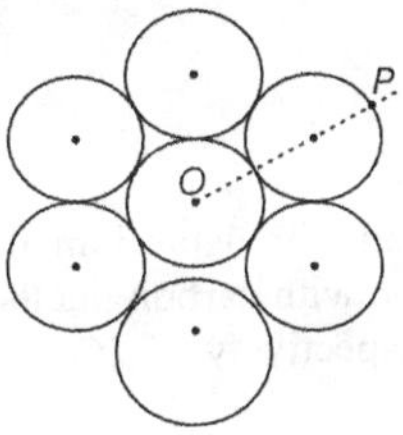

(a) $\frac{19}{2}MR^2$
(b) $\frac{55}{2}MR^2$
(c) $\frac{73}{2}MR^2$
(d) $\frac{181}{2}MR^2$

10. From a uniform circular disc of radius R and mass $9\,M$, a small disc of radius $\frac{R}{3}$ is removed as shown in the figure. The moment of inertia of the remaining disc about an axis perpendicular to the plane of the disc and passing through centre of disc is **[Rotational Motion]**

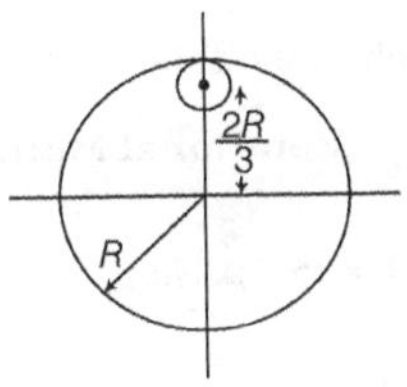

(a) $4MR^2$
(b) $\frac{40}{9}MR^2$
(c) $10MR^2$
(d) $\frac{37}{9}MR^2$

11. A particle is moving with a uniform speed in a circular orbit of radius R in a central force inversely proportional to the nth power of R. If the period of rotation of the particle is T, then : **[Gravitation]**

(a) $T \propto R^{3/2}$ for any n
(b) $T \propto R^{\frac{n}{2}+1}$
(c) $T \propto R^{(n+1)/2}$
(d) $T \propto R^{n/2}$

12. A solid sphere of radius r made of a soft material of bulk modulus K is surrounded by a liquid in a cylindrical container. A massless piston of area a floats on the surface of the liquid, covering entire cross-section of cylindrical container. When a mass m is placed on the surface of the piston to compress the liquid, the fractional decrement in the radius of the sphere, $\left(\frac{dr}{r}\right)$ is **[Elasticity]**

(a) $\frac{Ka}{mg}$
(b) $\frac{Ka}{3mg}$
(c) $\frac{mg}{3Ka}$
(d) $\frac{mg}{Ka}$

13. The mass of a hydrogen molecule is 3.32×10^{-27} kg. If 10^{23} hydrogen molecules strike per second, a fixed wall of area 2 cm^2 at an angle of 45° to the normal and rebound elastically with a speed of 10^3 m/s, then the pressure on the wall is nearly **[Centre of Mass]**

(a) $2.35 \times 10^3\ \text{N/m}^2$
(b) $4.70 \times 10^3\ \text{N/m}^2$
(c) $2.35 \times 10^2\ \text{N/m}^2$
(d) $4.70 \times 10^2\ \text{N/m}^2$

14. In a collinear collision, a particle with an initial speed v_0 strikes a stationary particle of the same mass. If the final total kinetic energy is 50% greater than the original kinetic energy, the magnitude of the relative velocity between the two particles after collision, is

[Centre of Mass]

(a) $\frac{v_0}{4}$ (b) $\sqrt{2}\, v_0$

(c) $\frac{v_0}{2}$ (d) $\frac{v_0}{\sqrt{2}}$

15. It is found that, if a neutron suffers an elastic collinear collision with deuterium at rest, fractional loss of its energy is P_d; while for its similar collision with carbon nucleus at rest, fractional loss of energy is P_c. The values of P_d and P_c are respectively **[Centre of Mass]**

(a) (0.89, 0.28) (b) (0.28, 0.89)

(c) (0, 0) (d) (0, 1)

JEE Advanced

2016

1. The position vector **r** of particle of mass m is given by the following equation

$\mathbf{r}(t) = \alpha t^3 \hat{\mathbf{i}} + \beta t^2 \hat{\mathbf{j}}$

where, $\alpha = \frac{10}{3}\,\text{ms}^{-3}$, $\beta = 5\,\text{ms}^{-2}$ and $m = 0.1$ kg. At $t = 1$s, which of the following statement(s) is (are) true about the particle? **[Rotational Motion, Type 2]**

(a) The velocity **v** is given by $\mathbf{v} = (10\hat{\mathbf{i}} + 10\hat{\mathbf{j}})\,\text{ms}^{-1}$

(b) The angular momentum **L** with respect to the origin is given by $\mathbf{L} = (5/3)\hat{\mathbf{k}}$ Nms

(c) The force **F** is given by $\mathbf{F} = (\hat{\mathbf{i}} + 2\hat{\mathbf{j}})$N

(d) The torque τ with respect to the origin is given by $\tau = -\frac{20}{3}\hat{\mathbf{k}}$ Nm

2. Consider two solid spheres P and Q each of density 8 gm cm^{-3} and diameters 1 cm and 0.5 cm, respectively. Sphere P is dropped into a liquid of density 0.8 gm cm^{-3} and viscosity $\eta = 3$ poiseulles.

Sphere Q is dropped into a liquid of density 1.6 gm cm^{-3} and viscosity $\eta = 2$ poiseulles. The ratio of the terminal velocities of P and Q is **[Fluid Mechanics, Type 3]**

3. Two thin circular discs of mass m and $4m$, having radii of a and $2a$, respectively, are rigidly fixed by a massless, rigid rod of length $l = \sqrt{24}\,a$ through their centers. This assembly is laid on a firm and flat surface and set rolling without slipping on the surface so that the angular speed about the axis of the rod is ω. The angular momentum of the entire assembly about the point 'O' is **L** (see the figure). Which of the following statement(s) is (are) true? **[Rotational Motion, Type 2]**

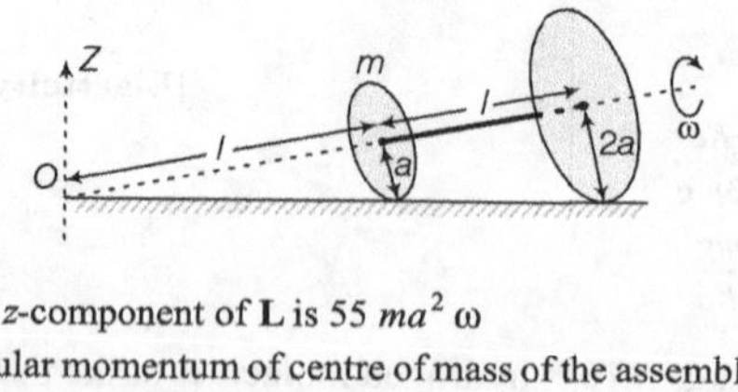

(a) The magnitude of the z-component of **L** is $55\,ma^2\omega$

(b) The magnitude of angular momentum of centre of mass of the assembly about the point O is $81\,ma^2\omega$

(c) The centre of mass of the assembly rotates about the Z-axis with an angular speed of $\frac{\omega}{5}$

(d) The magnitude of angular momentum of the assembly about its centre of mass is $17\,ma^2\frac{\omega}{2}$

Note ***Type-1*** *Only one correct option,* ***Type-2*** *More then one correct option,* ***Type-3*** *Single digit Answer,* ***Type-4*** *Numerical value based*

Paragraph 1 (Q.Nos. 4 & 5)

A frame of the reference that is accelerated with respect to an inertial frame of reference is called a non-inertial frame of reference. A coordinate system fixed on a circular disc rotating about a fixed axis with a constant angular velocity ω is an example of a non-inertial frame of reference. The relationship between the force $\vec{F}_{rot}$ experienced by a particle of mass m moving on the rotating disc and the force $\vec{F}_{in}$ experienced by the particle in an inertial frame of reference is,

$$\vec{F}_{rot} = \vec{F}_{in} + 2m(\vec{v}_{rot} \times \vec{\omega}) + m(\vec{\omega} \times \vec{r}) \times \vec{\omega},$$

where, $\mathbf{v}_{rot}$ is the velocity of the particle in the rotating frame of reference and $\mathbf{r}$ is the position vector of the particle with respect to the centre of the disc.

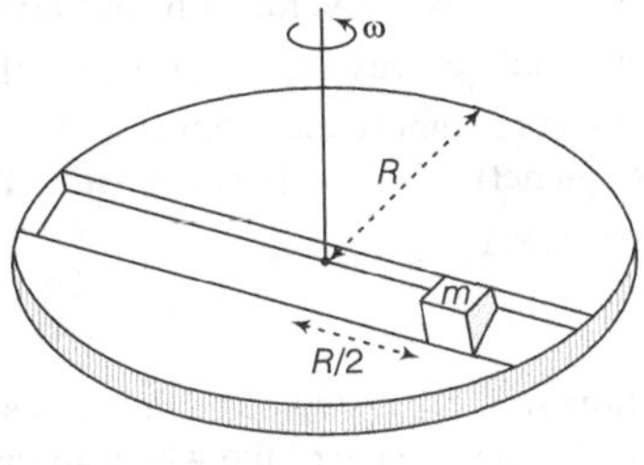

Now, consider a smooth slot along a diameter of a disc of radius R rotating counter-clockwise with a constant angular speed ω about its vertical axis through its centre. We assign a coordinate system with the origin at the centre of the disc, the X-axis along the slot, the Y-axis perpendicular to the slot and the Z-axis along the rotation axis ($\omega = \omega\hat{\mathbf{k}}$). A small block of mass m is gently placed in the slot at $\mathbf{r} = (R/2)\hat{\mathbf{i}}$ at $t = 0$ and is constrained to move only along the slot.

[Rotational Motion]

4. The distance r of the block at time t is

(a) $\frac{R}{2}\cos 2\omega t$ (b) $\frac{R}{2}\cos \omega t$

(c) $\frac{R}{4}(e^{\omega t} + e^{-\omega t})$ (d) $\frac{R}{4}(e^{2\omega t} + e^{-2\omega t})$

5. The net reaction of the disc on the block is

(a) $m\omega^2 R \sin \omega t\hat{\mathbf{j}} - mg\hat{\mathbf{k}}$ (b) $\frac{1}{2}m\omega^2 R\,(e^{\omega t} - e^{-\omega t})\hat{\mathbf{j}} + mg\hat{\mathbf{k}}$

(c) $\frac{1}{2}m\omega^2 R\,(e^{2\omega t} - e^{-2\omega t})\hat{\mathbf{j}} + mg\hat{\mathbf{k}}$ (d) $-m\omega^2 R \cos \omega t\hat{\mathbf{j}} - mg\hat{\mathbf{k}}$

2017

6. A drop of liquid of radius $R = 10^{-2}$ m having surface tension $S = \frac{0.1}{4\pi}\ \text{Nm}^{-1}$ divides itself into K identical drops. In this process the total change in the surface energy $\Delta U = 10^{-3}$ J. If $K = 10^{\alpha}$, then the value of α is **[Fluid Mechanics, Type 3]**

7. Consider regular polygons with number of sides $n = 3, 4, 5 \ldots\ldots$ as shown in the figure. The centre of mass of all the polygons is at height h from the ground. They roll on a horizontal surface about the leading vertex without slipping and sliding as depicted. The maximum increase in height of the locus of the centre of mass for each each polygon is Δ. Then, Δ depends on n and h as **[Centre of Mass, Type 1]**

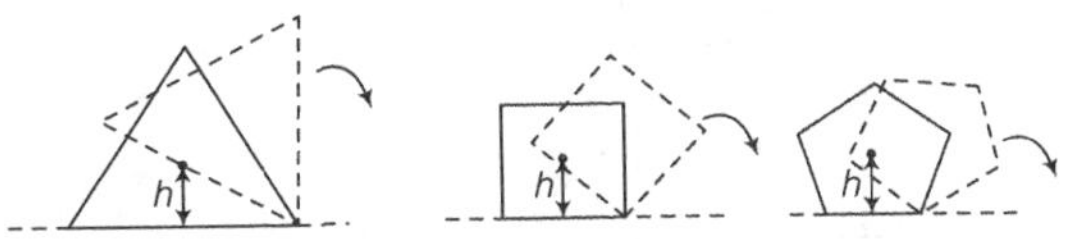

(a) $\Delta = h\sin^2\left(\frac{\pi}{n}\right)$ (b) $\Delta = h\sin\left(\frac{2\pi}{n}\right)$

(c) $\Delta = h\tan^2\left(\frac{\pi}{2n}\right)$ (d) $\Delta = h\left(\frac{1}{\cos\left(\frac{\pi}{n}\right)} - 1\right)$

8. Consider an expanding sphere of instantaneous radius R whose total mass remains constant. The expansion is such that the instantaneous density ρ remains uniform throughout the volume. The rate of fractional change in density $\left(\frac{1}{\rho}\frac{d\rho}{dt}\right)$ is constant. The velocity v of any point of the surface of the expanding sphere is proportional to **[Fluid Mechanics, Type 1]**

(a) R (b) $\frac{1}{R}$

(c) R^3 (d) $R^{\frac{2}{3}}$

9. A rocket is launched normal to the surface of the Earth, away from the Sun, along the line joining the Sun and the Earth. The Sun is 3×10^5 times heavier than the Earth and is at a distance 2.5×10^4 times larger than the radius of Earth. The escape velocity from Earth's gravitational field is $v_e = 11.2\,\text{km s}^{-1}$. The minimum initial velocity (v_s) required for the rocket to be able to leave the Sun-Earth system is closest to (Ignore the rotation and revolution of the Earth and the presence of any other planet) **[Gravitation, Type 1]**

(a) $v_s = 72\,\text{km s}^{-1}$ (b) $v_s = 22\,\text{km s}^{-1}$

(c) $v_s = 42\,\text{km s}^{-1}$ (d) $v_s = 62\,\text{km s}^{-1}$

10. A wheel of radius R and mass M is placed at the bottom of a fixed step of height R as shown in the figure. A constant force is continuously applied on the surface of the wheel so that it just climbs the step without slipping. Consider the torque τ about an axis normal to the plane of the paper passing through the point Q. Which of the following options is/are correct? **[Rotational Motion, Type 2]**

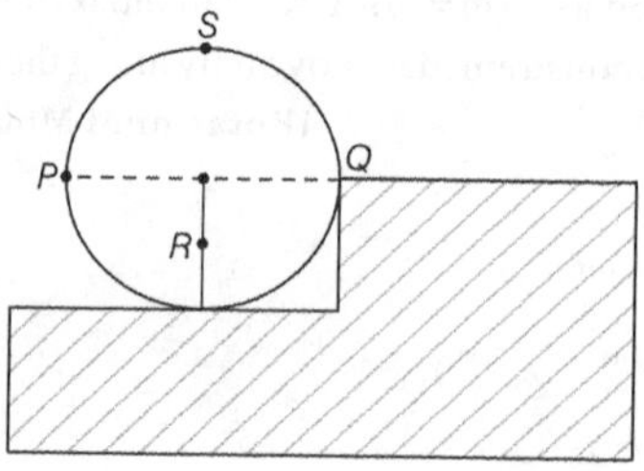

(a) If the force is applied normal to the circumference at point P, then τ is zero
(b) If the force is applied tangentially at point S, then $\tau \neq 0$ but the wheel never climbs the step
(c) If the force is applied at point P tangentially, then τ decreases continuously as the wheel climbs
(d) If the force is applied normal to the circumference at point X, then τ is constant

11. A rigid uniform bar AB of length L is slipping from its vertical position on a frictionless floor (as shown in the figure). At some instant of time, the angle made by the bar with the vertical is θ. Which of the following statements about its motion is/are correct? **[Rotational Motion, Type 2]**

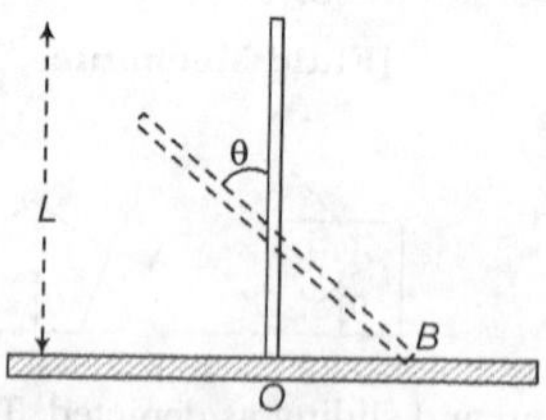

(a) Instantaneous torque about the point in contact with the floor is proportional to $\sin\theta$
(b) The trajectory of the point A is parabola
(c) The mid-point of the bar will fall vertically downward
(d) When the bar makes an angle θ with the vertical, the displacement of its mid-point from the initial position is proportional to $(1 - \cos\theta)$

Paragraph 2 (Q. No. 12 & 13)

One twirls a circular ring (of mass M and radius R) near the tip of one's finger as shown in Figure 1. In the process the finger never loses contact with the inner rim of the ring. The finger traces out the surface of a cone, shown by the dotted line. The radius of the path traced out by the point where the ring and the finger is in contact is r. The finger rotates with an angular velocity ω_0. The rotating ring rolls without slipping on the outside of a smaller circle described by the point where the ring and the finger is in contact (Figure 2). The coefficient of friction between the ring and the finger is μ and the acceleration due to gravity is g. **[Rotational Motion]**

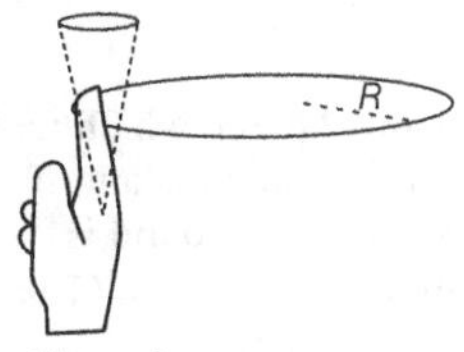

Figure 1

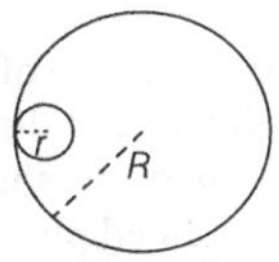

Figure 2

12. The total kinetic energy of the ring is

(a) $M\omega_0^2(R-r)^2$
(b) $\frac{1}{2}M\omega_0^2(R-r)^2$
(c) $M\omega_0^2R^2$
(d) $\frac{3}{2}M\omega_0^2(R-r)^2$

13. The minimum value of ω_0 below which the ring will drop down is

(a) $\sqrt{\frac{g}{2\mu(R-r)}}$
(b) $\sqrt{\frac{3g}{2\mu(R-r)}}$
(c) $\sqrt{\frac{g}{\mu(R-r)}}$
(d) $\sqrt{\frac{2g}{\mu(R-r)}}$

14. A block of mass M has a circular cut with a frictionless surface as shown. The block rests on the horizontal frictionless surfaced of a fixed table. Initially the right edge of the block is at $x = 0$, in a coordinate system fixed to the table. A point mass m is released from rest at the topmost point of the path as shown and it slides down. When the mass loses contact with the block, its position is x and the velocity is v. At that instant, which of the following option is/are correct? **[Rotational Motion, Type 2]**

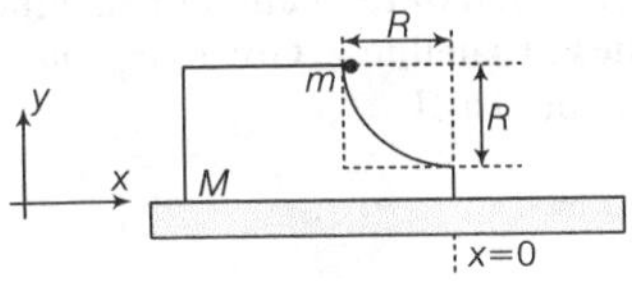

(a) The velocity of the point mass m is $v = \sqrt{\frac{2gR}{1+\frac{m}{M}}}$

(b) The x component of displacement of the centre of mass of the block M is $-\frac{mR}{M+m}$

(c) The position of the point mass is $x = -\sqrt{2}\frac{mR}{M+m}$

(d) The velocity of the block M is $V = -\frac{m}{M}\sqrt{2gR}$

2018

15. The potential energy of mass m at a distance r from a fixed point O is given by $V(r) = kr^2/2$, where k is a positive constant of appropriate dimensions. This particle is moving in a circular orbit of radius R about the point O. If v is the speed of the particle and L is the magnitude of its angular momentum about O, which of the following statements is (are) true ? **[Rotational Motion, Type 2]**

(a) $v = \sqrt{\frac{k}{2m}}R$
(b) $v = \sqrt{\frac{k}{m}}R$
(c) $L = \sqrt{mk}\,R^2$
(d) $L = \sqrt{\frac{mk}{2}}R^2$

16. A uniform capillary tube of inner radius r is dipped vertically into a beaker filled with water. The water rises to a height h in the capillary tube above the water surface in the beaker. The surface tension of water is σ. The angle of contact between water and the wall of the capillary tube is θ. Ignore the mass of water in the meniscus. Which of the following statements is (are) true ? **[Center of Mass, Type 2]**

(a) For a given material of the capillary tube, h decreases with increase in r
(b) For a given material of the capillary tube, h is independent of σ
(c) If this experiment is performed in a lift going up with a constant acceleration, then h decreases
(d) h is proportional to contact angle θ

17. A ring and a disc are initially at rest, side by side, at the top of an inclined plane which makes an angle 60° with the horizontal. They start to roll without slipping at the same instant of time along the shortest path. If the time difference between their reaching the ground is $(2-\sqrt{3})/\sqrt{10}$ s, then the height of the top of the inclined plane, in metres, is .... (Take, $g = 10\text{ms}^{-2}$) **[Rotational Motion, Type 4]**

18. Consider a thin square plate floating on a viscous liquid in a large tank. The height h of the liquid in the tank is much less than the width of the tank. The floating plate is pulled horizontally with a constant velocity u_0. Which of the following statements is (are) true? **[Elasticity, Type 2]**

(a) The resistive force of liquid on the plate is inversely proportional to h
(b) The resistive force of liquid on the plate is independent of the area of the plate
(c) The tangential (shear) stress on the floor of the tank increases with u_0
(d) The tangential (shear) stress on the plate varies linearly with the viscosity η of the liquid

19. A planet of mass M, has two natural satellites with masses m_1 and m_2. The radii of their circular orbits are R_1 and R_2, respectively. Ignore the gravitational force between the satellites. Define v_1, L_1, K_1 and T_1 to be respectively, the orbital speed, angular momentum, kinetic energy and time period of revolution of satellite 1; and v_2, L_2, K_2 and T_2 to be the corresponding quantities of satellite 2. Given, $m_1 / m_2 = 2$ and $R_1 / R_2 = 1/4$, match the ratios in List-I to the numbers in List-II. **[Gravitation]**

	List-I		List-II
P.	$\frac{v_1}{v_2}$	1.	$\frac{1}{8}$
Q.	$\frac{L_1}{L_2}$	2.	1
R.	$\frac{K_1}{K_2}$	3.	2
S.	$\frac{T_1}{T_2}$	4.	8

(a) P $\rightarrow$ 4; Q $\rightarrow$ 2; R $\rightarrow$ 1; S $\rightarrow$ 3
(b) P $\rightarrow$ 3; Q $\rightarrow$ 2; R $\rightarrow$ 4; S $\rightarrow$ 1
(c) P $\rightarrow$ 2; Q $\rightarrow$ 3; R $\rightarrow$ 1; S $\rightarrow$ 4
(d) P $\rightarrow$ 2; Q $\rightarrow$ 3; R $\rightarrow$ 4; S $\rightarrow$ 1

20. In the List-I below, four different paths of a particle are given as functions of time. In these functions, α and β are positive constants of appropriate dimensions and $\alpha \neq \beta$. In each case, the force acting on the particle is either zero or conservative. In List-II, five physical quantities of the particle are mentioned: **p** is the linear momentum, **L** is the angular momentum about the origin, K is the kinetic energy, U is the potential energy and E is the total energy. Match each path in List-I with those quantities in List-II, which are conserved for that path. **[Rotational Motion]**

	List-I		List-II
P.	$\mathbf{r}(t) = \alpha t\,\mathbf{i} + \beta t\,\mathbf{j}$	1.	$\mathbf{p}$
Q.	$\mathbf{r}(t) = \alpha \cos \omega t\,\mathbf{i} + \beta \sin \omega t\,\mathbf{j}$	2.	$\mathbf{L}$
R.	$\mathbf{r}(t) = \alpha (\cos \omega t\,\mathbf{i} + \sin \omega t\,\mathbf{j})$	3.	K
S.	$\mathbf{r}(t) = a t\,\mathbf{i} + \frac{\beta}{2} t^2\,\mathbf{j}$	4.	U
		5.	E

(a) P $\rightarrow$ 1, 2, 3, 4, 5; Q $\rightarrow$ 2, 5; R $\rightarrow$ 2, 3, 4, 5; S $\rightarrow$ 5
(b) P $\rightarrow$ 1, 2, 3, 4, 5; Q $\rightarrow$ 3, 5; R $\rightarrow$ 2, 3, 4, 5; S $\rightarrow$ 2, 5
(c) P $\rightarrow$ 2, 3, 4; Q $\rightarrow$ 5; R $\rightarrow$ 1, 2, 4; S $\rightarrow$ 2, 5
(d) P $\rightarrow$ 1, 2, 3, 5; Q $\rightarrow$ 2, 5; R $\rightarrow$ 2, 3, 4, 5; S $\rightarrow$ 2, 5

21. Consider a body of mass 1.0 kg at rest at the origin at time $t = 0$. A force $\mathbf{F} = (\alpha t\,\hat{\mathbf{i}} + \beta\,\hat{\mathbf{j}})$ is applied on the body, where $\alpha = 1.0\text{Ns}^{-1}$ and $\beta = 1.0\text{N}$. The torque acting on the body about the origin at time $t = 1.0\text{s}$ is τ. Which of the following statements is (are) true ?

[Rotational Motion, Type 2]

(a) $|\tau| = \frac{1}{3}\text{N-m}$

(b) The torque τ is in the direction of the unit vector $+\hat{\mathbf{k}}$

(c) The velocity of the body at $t = 1$ s is $\mathbf{v} = \frac{1}{2}(\hat{\mathbf{i}} + 2\hat{\mathbf{j}})\,\text{ms}^{-1}$

(d) The magnitude of displacement of the body at $t = 1$ s is $\frac{1}{6}$m

Answer *with* Explanations

JEE Main

1. (a) We can apply $\mathbf{L} = m(\mathbf{r} \times \mathbf{v})$ for different parts.

For example :

In part (a), coordinates of A are $\left(\frac{R}{\sqrt{2}}, \frac{R}{\sqrt{2}}\right)$

Therefore, $\mathbf{r} = \frac{R}{\sqrt{2}}\hat{\mathbf{i}} + \frac{R}{\sqrt{2}}\hat{\mathbf{j}}$ and $\mathbf{v} = v\hat{\mathbf{i}}$

So, substituting in $\mathbf{L} = m(\mathbf{r} \times \mathbf{v})$ we get,

$$\mathbf{L} = -\frac{mvR}{\sqrt{2}}\hat{\mathbf{k}}$$

Hence, option (a) is correct. Similarly, we can check other options also.

2. (a)

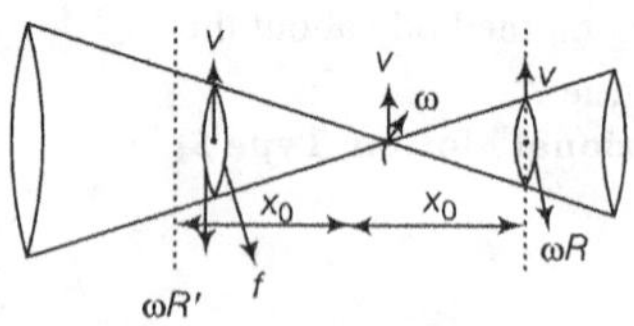

At distance x_0 from O, $v = \omega R$

distance less than x_0, $v > \omega R$

Initially, there is pure rolling at both the contacts. As the cone moves forward, slipping at AB will start in forward direction, as radius at left contact decreases.

Thus, the cone will start turning towards left. As it moves further slipping at CD will start in backward direction which will also turn the cone towards left.

3. (d) $v_{\text{orbital}} = \sqrt{\frac{GM}{R}} = \sqrt{gR}$

$$v_{\text{escape}} = \sqrt{2gR}$$

$\therefore$ Extra velocity required

$$= v_{\text{escape}} - v_{\text{orbital}}$$

$$= \sqrt{gR}\,(\sqrt{2} - 1)$$

4. (d) MI of a solid cylinder about its perpendicular bisector of length is

$$I = m\left(\frac{l^2}{12} + \frac{R^2}{4}\right)$$

$$\Rightarrow \quad I = \frac{mR^2}{4} + \frac{ml^2}{12} = \frac{m^2}{4\pi\rho l} + \frac{ml^2}{12} \qquad [\because \rho\pi R^2 l = m]$$

For I to be maximum,

$$\frac{dI}{dl} = -\frac{m^2}{4\pi\rho}\left(\frac{1}{l^2}\right) + \frac{ml}{6} = 0$$

$$\Rightarrow \quad \frac{m^2}{4\pi\rho} = \frac{ml^3}{6}$$

Now, putting $m = \rho\pi R^2 l$

$$\therefore \quad l^3 = \frac{3}{2\pi\rho} \cdot \rho\pi R^2 l \qquad \frac{l^2}{R^2} = \frac{3}{2}$$

$$\therefore \quad \frac{l}{R} = \sqrt{\frac{3}{2}}$$

5. (d) As the rod rotates in vertical plane so a torque is acting on it, which is due to the vertical component of weight of rod.

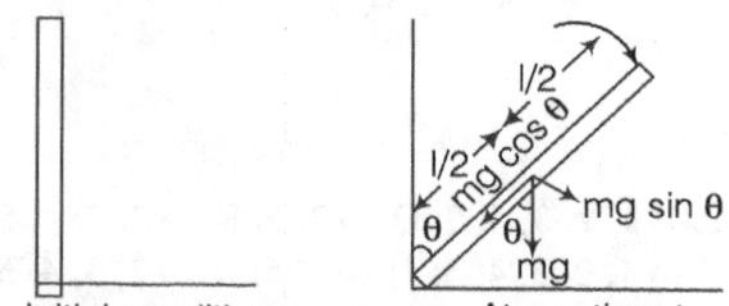

Initial condition — At any time t

Now, Torque τ = force × perpendicular distance of line of action of force from axis of rotation

$$= mg\sin\theta \times \frac{l}{2}$$

Again, Torque $\tau = I\alpha$

Where, I = moment of inertia $= \frac{ml^2}{3}$

[Force and Torque frequency along axis of rotation passing through in end]

α = angular acceleration

$$\therefore \quad mg\sin\theta \times \frac{l}{2} = \frac{ml^2}{3}\alpha$$

$$\therefore \quad \alpha = \frac{3g\sin\theta}{2l}$$

6. (d) $K = \frac{P}{(-\Delta V/V)}$

$$\Rightarrow \quad -\frac{\Delta V}{V} = \frac{P}{K}$$

$$\Rightarrow \quad -\Delta V = \frac{PV}{K}$$

Change in volume $\Delta V = \gamma v \Delta T$

Where r = coefficient of volume expansion

Again, $\gamma = 3\alpha$

α is coefficient of linear expansion

$$\therefore \quad \Delta V = V(3\alpha)\Delta T$$

$$\therefore \quad \frac{PV}{K} = V(3\alpha)\Delta T$$

$$\therefore \quad \Delta T = \frac{P}{3\alpha K}$$

7. (c) Inside the earth surface $g = \frac{GM}{R^3}r$

i.e. $\quad g \propto r$

Out the earth surface $g = \frac{Gm}{r^2}$ i.e. $g \propto \frac{1}{r^2}$

So, till earth surface 'g' increases linearly with distance r, shown only in graph (c).

8. (d) Stress $= \frac{\text{Weight}}{\text{Area}}$

Volume will become (9^3) times.

So weight = volume × density × g will also become $(9)^3$ times.

Area of cross section will become $(9)^2$ times.

$$= \frac{9^3 \times W_0}{9^2 \times A_0} = 9\left(\frac{W_0}{A_0}\right)$$

Hence, the stress increases by a factor of 9.

9. (d) **Key Idea** *First we found moment of inertia (MI) of system using parallel axis theorem about centre of mass, then we use it to find moment of inertia about given axis.*

Moment of inertia of an outer disc about the axis through centre is $= \frac{MR^2}{2} + M(2R)^2 = MR^2\left(4 + \frac{1}{2}\right) = \frac{9}{2}MR^2$

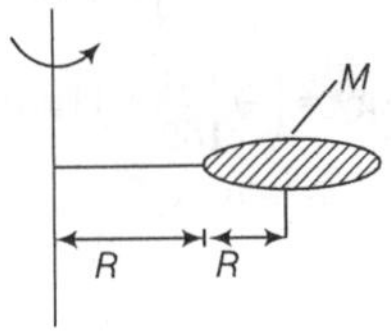

For 6 such discs,

moment of inertia $= 6 \times \frac{9}{2}MR^2 = 27MR^2$

So, moment of inertia of system

$$= \frac{MR^2}{2} + 27MR^2 = \frac{55}{2}MR^2$$

Hence, $\quad I_P = \frac{55}{2}MR^2 + (7M \times 9R^2)$

$\Rightarrow \quad I_P = \frac{181}{2}MR^2$

$$I_{system} = \frac{181}{2}MR^2$$

10. (a) Moment of inertia of remaining solid = Moment of intertia of complete solid − Moment of inertia of removed portion

$$\therefore \quad I = \frac{9MR^2}{2} - \left[\frac{M\left(\frac{R}{3}\right)^2}{2} + M\left(\frac{2R}{3}\right)^2\right]$$

$\Rightarrow \quad I = 4MR^2$

11. (c) $\therefore$ Force = Mass × Acceleration = $m\omega^2 R$

and given, $\quad F \propto \frac{1}{R^n}$

$\Rightarrow \quad F = \frac{k}{R^n}$

So, we have

$$\frac{k}{R^n} = m\left(\frac{2\pi}{T}\right)^2 R$$

$\Rightarrow \quad T^2 = \frac{4\pi^2 m}{k} \cdot R^{n+1}$

$\Rightarrow \quad T \propto R^{\frac{n+1}{2}}$

12. (c) $\therefore$ Bulk modulus,

$$K = \frac{\text{Volumetric stress}}{\text{Volumetric strain}} = \frac{\Delta p}{\frac{\Delta V}{V}}$$

$\Rightarrow \quad K = \dfrac{mg}{a\left(\frac{3\Delta r}{r}\right)} \qquad \left[\because V = \frac{4}{3}\pi r^3, \text{ so } \frac{\Delta V}{V} = \frac{3\Delta r}{r}\right]$

$\Rightarrow \quad \frac{\Delta r}{r} = \frac{mg}{3aK}$

13. (a)

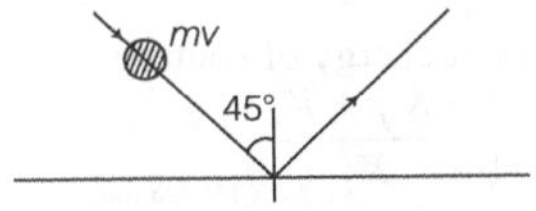

Momentum imparted due to first collision

$= 2mv\sin 45° = \sqrt{2}mv \qquad \left[\because \sin 45° = \frac{1}{\sqrt{2}}\right]$

$\therefore$ Pressure on surface

$$= \frac{n\sqrt{2}mv}{\text{Area}} = \frac{10^{23} \times \sqrt{2} \times 3.32 \times 10^{-27} \times 10^3}{(2 \times 10^{-2})^2}$$

$p = 2.35 \times 10^3 \text{ N/m}^2$

14. (b) **Key Idea** *Momentum is conserved in all type of collisions,*

Final kinetic energy is 50% more than initial kinetic energy

$\Rightarrow \quad \frac{1}{2}mv_2^2 + \frac{1}{2}mv_1^2 = \frac{150}{100} \times \frac{1}{2}mv_0^2 \qquad \ldots(i)$

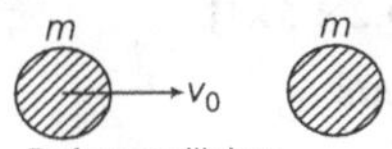

Before collision

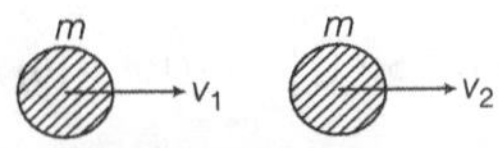

After collision

Conservation of momentum gives,

$$mv_0 = mv_1 + mv_2$$

$v_0 = v_2 + v_1 \qquad \ldots(ii)$

From Eqs. (i) and (ii), we have

$$v_1^2 + v_2^2 + 2v_1v_2 = v_0^2$$

$\Rightarrow \quad 2v_1v_2 = \frac{-v_0^2}{2}$

$\therefore \quad (v_1 - v_2)^2 = (v_1 + v_2)^2 - 4v_1v_2 = 2v_0^2$

or $\quad v_{rel} = \sqrt{2}v_0$

15. (a) Neutron-Deuterium collision;

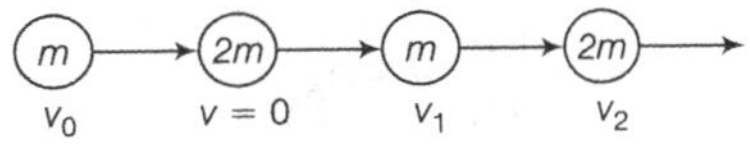

Momentum conservation gives;

$$mv_0 = mv_1 + 2mv_2$$

$\Rightarrow \quad v_0 = v_1 + 2v_2 \qquad \ldots(i)$

Collision given is elastic .

So, coefficient of restitution, $e = 1$

$$\therefore \qquad e = 1 = \frac{\text{Velocity of separation}}{\text{Velocity of approach}}$$

$$\Rightarrow \qquad 1 = \frac{v_2 - v_1}{v_0 - 0} \Rightarrow v_0 = v_2 - v_1 \qquad \ldots\text{(ii)}$$

On adding Eqs. (i) and (ii), we get

$$2v_0 = 3v_2 \Rightarrow \frac{2v_0}{3} = v_2$$

So, from Eq. (i), we get

$$v_1 = v_0 - 2v_2 = v_0 - \frac{4v_0}{3} \Rightarrow v_1 = -\frac{v_0}{3}$$

Fractional loss of energy of neutron

$$= \left(\frac{-K_f + K_i}{K_i}\right)_{\text{for neutron}}$$

$$= \frac{-\frac{1}{2}mv_1^2 + \frac{1}{2}mv_0^2}{\frac{1}{2}mv_0^2}$$

$$= \frac{-\frac{v_0^2}{9} + v_0^2}{v_0^2}$$

$$= \left(-\frac{1}{9} + 1\right) = \frac{8}{9} = 0.8\overline{8} = 0.89$$

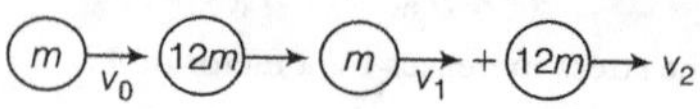

Similarly, for neutron-carbon atom collision;
Momentum conservation gives;

$$v_0 = v_1 + 12v_2 \text{ and } e = 1$$

$$\Rightarrow \qquad v_0 = v_2 - v_1$$

So, $\qquad v_1 = \frac{11}{13}v_0$

$$\therefore \text{Loss of energy} = \left(-\frac{121}{169} + 1\right) = 0.28$$

So, $P_d = 0.89$ and $P_c = 0.28$

JEE Advanced

1. (a,b,d) $\quad \mathbf{r} = \alpha t^3\hat{\mathbf{i}} + \beta t^2\hat{\mathbf{j}}$

$$\mathbf{v} = \frac{d\mathbf{r}}{dt} = 3\alpha t^2\hat{\mathbf{i}} + 2\beta t\hat{\mathbf{j}}$$

$$\mathbf{a} = \frac{d^2\mathbf{r}}{dt^2} = 6\alpha t\,\hat{\mathbf{i}} + 2\beta\hat{\mathbf{j}}$$

At $\qquad t = 1\,\text{s}$,

(a) $\mathbf{v} = 3 \times \frac{10}{3} \times 1\hat{\mathbf{i}} + 2 \times 5 \times 1\hat{\mathbf{j}}$

$$= (10\hat{\mathbf{i}} + 10\hat{\mathbf{j}})\ \text{m/s}$$

(b) $\hat{\mathbf{L}} = \hat{\mathbf{r}} \times \hat{\mathbf{p}}$

$$= \left(\frac{10}{3} \times 1\hat{\mathbf{i}} + 5 \times 1\hat{\mathbf{j}}\right) \times 0.1(10\hat{\mathbf{i}} + 10\hat{\mathbf{j}})$$

$$= \left(-\frac{5}{3}\hat{\mathbf{k}}\right)\text{N-ms}$$

(c) $\mathbf{F} = m\mathbf{a} = m \times \left(6 \times \frac{10}{3} \times 1\hat{\mathbf{i}} + 2 \times 5\hat{\mathbf{j}}\right)$

$$= (2\hat{\mathbf{i}} + \hat{\mathbf{j}})\text{N}$$

(d) $\tau = r \times \mathbf{F} = \left(\frac{10}{3}\hat{\mathbf{i}} + 5\hat{\mathbf{j}}\right) \times (2\hat{\mathbf{i}} + \hat{\mathbf{j}})$

$$= +\frac{10}{3}\hat{\mathbf{k}} + 10(-\hat{\mathbf{k}})$$

$$= \left(-\frac{20}{3}\hat{\mathbf{k}}\right)\text{N-m}$$

2. **(3)** Terminal velocity is given by

$$v_T = \frac{2}{9}\frac{r^2}{\eta}(d - \rho)g$$

$$\frac{v_P}{v_Q} = \frac{r_P^2}{r_Q^2} \times \frac{\eta_Q}{\eta_P} \times \frac{(d - \rho_P)}{(d - \rho_Q)}$$

$$= \left(\frac{1}{0.5}\right)^2 \times \left(\frac{2}{3}\right) \times \frac{(8 - 0.8)}{(8 - 1.6)}$$

$$= 4 \times \frac{2}{3} \times \frac{7.2}{6.4} = 3$$

3. (c,d)

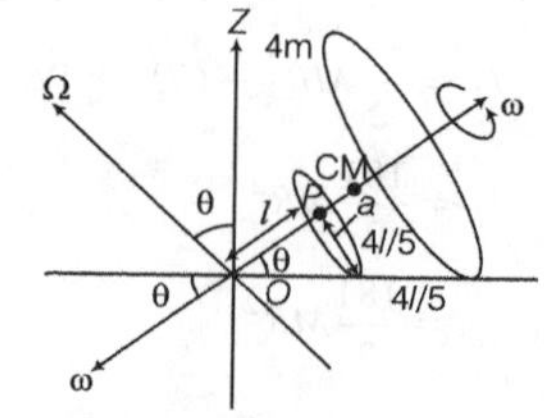

$$\cos\theta = \frac{1}{\sqrt{l^2 + a^2}} = \frac{\sqrt{24}}{5}$$

(c) Velocity of point P : $a\omega = 1\,\Omega$ then

$\Omega = \frac{a\omega}{1}$ = Angular velocity of CM w.r.t. point O. Angular velocity fo CM w.r.t. Z-axis = $\Omega\cos\theta$

$$\omega_{CM-z} = \frac{a\omega}{1}\frac{\sqrt{24}}{5}$$

$$= \frac{a\omega}{\sqrt{24}\,a}\frac{\sqrt{24}}{5}$$

$$\omega_{CM-z} = \frac{a\omega}{5}$$

(a) $L_z = L_{CM-O}\cos\theta - L_{D-CM}\sin\theta$

$$= \frac{81\sqrt{24}}{5}a^2m\omega \times \frac{\sqrt{24}}{5} - \frac{17\,ma^2\omega}{2} \times \frac{1}{\sqrt{24}}$$

$$= \frac{81 \times 24ma^2\omega}{25} - \frac{17ma^2\omega}{2\sqrt{24}}$$

(b) $L_{CM-O} = (5m)\left[\frac{9l}{5}\Omega\right]\frac{9l}{5} = \frac{81ml^2\omega}{5}$

$$= \frac{81ml^2}{5} \times \frac{a\omega}{l}$$

$$L_{CM-O} = \frac{81mla\omega}{5} = \frac{81\sqrt{24}\, a^2 m\omega}{5}$$

(d) $L_{D-CM} = \frac{ma^2}{2}\omega + \frac{4m(2a)^2}{2}\omega = \frac{17ma^2\omega}{2}$

4. (c) Force on block along slot $= m\omega^2 r$

$$= ma = m\left(\frac{vdv}{dr}\right)$$

$$\int_0^v vdv = \int_{R/2}^r \omega^2 r dr$$

$$\Rightarrow \quad \frac{v^2}{2} = \frac{\omega^2}{2}\left(r^2 - \frac{R^2}{4}\right)$$

$$\Rightarrow \quad v = \omega\sqrt{r^2 - \frac{R^2}{4}} = \frac{dr}{dt}$$

$$\Rightarrow \int_{R/4}^r \frac{dr}{\sqrt{r^2 - \frac{R^2}{4}}} = \int_0^t \omega dt$$

$$\ln\left(\frac{r + \sqrt{r^2 - \frac{R^2}{4}}}{\frac{R}{2}}\right) - \ln\left(\frac{\frac{R}{2} \text{ and } + \sqrt{\frac{R^2}{4} - \frac{R^2}{4}}}{\frac{R}{2}}\right) = \omega t$$

$$\Rightarrow \quad r + \sqrt{r^2 - \frac{R^2}{4}} = \frac{R}{2}e^{\omega t}$$

$$\Rightarrow \quad r^2 - \frac{R^2}{4} = \frac{R^2}{4}e^{2\omega t} + r^2 - 2r\frac{R}{2}e^{\omega t}$$

$$\Rightarrow \quad r = \frac{\frac{R^2}{4}e^{2\omega t} + \frac{R^2}{4}}{Re^{\omega t}} = \frac{R}{4}(e^{\omega t} + e^{-\omega t})$$

5. (b)

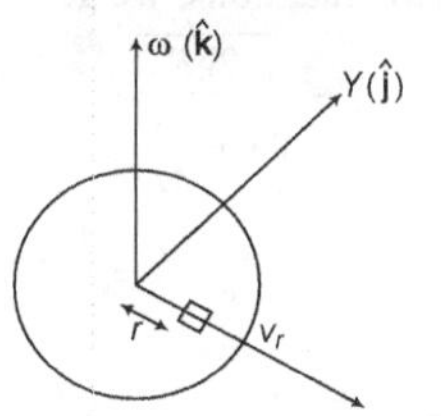

$$\mathbf{F}_{rot} = \mathbf{F}_{in} + 2m(v_{rot}\hat{\mathbf{i}}) \times \omega\hat{\mathbf{k}} + m(\omega\hat{\mathbf{k}} \times r\hat{\mathbf{i}}) \times \omega\hat{\mathbf{k}}$$

$$mr\omega^2\hat{\mathbf{i}} = \mathbf{F}_{in} + 2mv_{rot}\omega(-\hat{\mathbf{j}}) + m\omega^2 r\hat{\mathbf{j}}$$

$$\mathbf{F}_{in} = 2mv_r\omega\hat{\mathbf{j}},$$

$$r = \frac{R}{4}[e^{\omega t} + e^{-\omega t}]$$

$$\frac{dr}{dt} = v_r = \frac{R}{4}[\omega e^{\omega t} - \omega e^{-\omega t}]$$

$$\mathbf{F}_{in} = 2m\frac{R\omega}{4}[e^{\omega t} - e^{-\omega t}]\omega\hat{\mathbf{j}}$$

$$\mathbf{F}_{in} = \frac{mR\omega^2}{2}[e^{\omega t} - e^{-\omega t}]\hat{\mathbf{j}}$$

Also, reaction is due to disc surface, then

$$\mathbf{F}_{reaction} = \frac{mR\omega^2}{2}[e^{\omega t} - e^{-\omega t}]\hat{\mathbf{j}} + mg\,\hat{\mathbf{k}}$$

6. (6) From mass conservation,

$$\rho \cdot \frac{4}{3}\pi R^3 = \rho \cdot K \cdot \frac{4}{3}\pi r^3$$

$$\Rightarrow \quad R = K^{1/3} r$$

$$\therefore \quad \Delta U = T\Delta A = T(K \cdot 4\pi r^2 - 4\pi R^2)$$

$$= T(K \cdot 4\pi R^2 K^{-2/3} - 4\pi R^2)$$

$$\Delta U = 4\pi R^2 T[K^{1/3} - 1]$$

Putting the values, we get

$$10^{-3} = \frac{10^{-1}}{4\pi} \times 4\pi \times 10^{-4}[K^{1/3} - 1]$$

$$100 = K^{1/3} - 1$$

$$\Rightarrow \quad K^{1/3} \cong 100 = 10^2$$

Given that $K = 10^\alpha$

$$\therefore \quad 10^{\alpha/3} = 10^2$$

$$\Rightarrow \quad \frac{\alpha}{3} = 2 \Rightarrow \alpha = 6$$

7. (d)

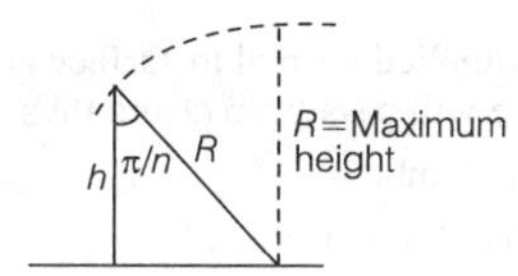

$$\cos\left(\frac{\pi}{n}\right) = \frac{h}{R}$$

$$\Delta = R - h = \frac{h}{\cos\left(\frac{\pi}{n}\right)} - h$$

$$= h\left[\frac{1}{\cos\left(\frac{\pi}{n}\right) - 1}\right]$$

8. (a) $m = \frac{4\pi R^3}{3} \times \rho$

On taking log both sides, we have

$$\ell n(m) = \ell n\left(\frac{4\pi}{3}\right) + \ell n(\rho) + 3\ell n(R)$$

On differentiating with respect to time,

$$0 = 0 + \frac{1}{\rho}\frac{d\rho}{dt} + \frac{3}{R}\frac{dR}{dt}$$

$$\Rightarrow \quad \left(\frac{dR}{dt}\right) = v \propto -R \times \frac{1}{\rho}\left(\frac{d\rho}{dt}\right)$$

$$v \propto R$$

9. (c) Given, $v_e = 11.2 \text{ km/s} = \sqrt{\frac{2GM_e}{R_e}}$

From energy conservation,

$$K_i + U_i = K_f + U_f$$

$$\frac{1}{2}mv_s^2 - \frac{GM_s m}{r} - \frac{GM_e m}{R_e} = 0 + 0$$

Here, r = distane of rocket from sun

$$\Rightarrow \quad v_s = \sqrt{\frac{2GM_e}{R_e} + \frac{2GM_s}{r}}$$

Given, $M_s = 3 \times 10^5 \, M_e$

and $\quad r = 2.5 \times 10^4 \, R_e$

$$\Rightarrow \quad v_s = \sqrt{\frac{2GM_e}{R_e} + (2G)\left(\frac{3\times 10^5 M_e}{2.5\times 10^4 R_e}\right)}$$

$$= \sqrt{\frac{2GM_e}{R_e}\left(1 + \frac{3\times 10^5}{2.5\times 10^4}\right)}$$

$$= \sqrt{\frac{2GM_e}{R_e} \times 13}$$

$$\Rightarrow \quad v_s \simeq 42 \text{ km/s}$$

10. (a,c)

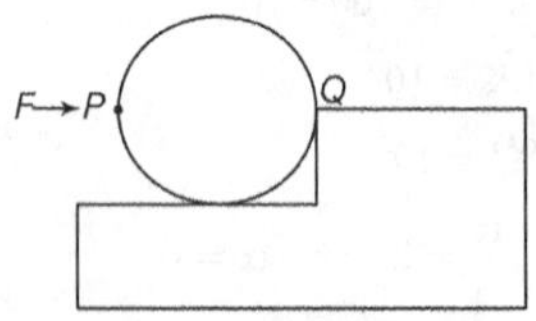

(a) If force is applied normal to surface at P, then line of action of force will pass from Q and thus $\tau = 0$.

(b) Wheel can climb.

(c) $\tau = F(2R\cos\theta) - mgR\cos\theta$

$\tau \propto \cos\theta$

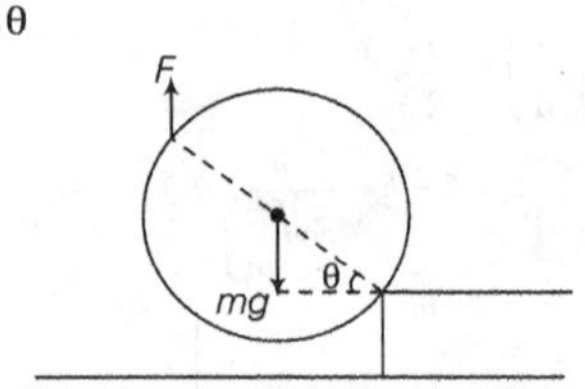

Hence, as θ increases, τ decreases.

So its correct.

(d)

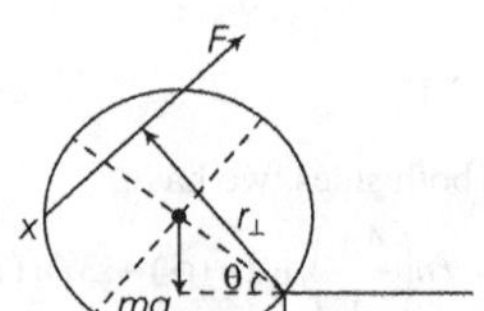

$\tau = Fr_\perp - mg\cos\theta$; τ increases with θ.

11. (a,c,d) When the bar makes an angle θ, the height of its COM (mid-point) is $\frac{L}{2}\cos\theta$.

$$\therefore \quad \text{Displacement} = L - \frac{L}{2}\cos\theta$$

$$= \frac{L}{2}(1 - \cos\theta)$$

Since, force on COM is only along the vertical direction, hence COM is falling vertically downward. Instantaneous torque about point of contact is

$$\tau = mg \times \frac{L}{2}\sin\theta \quad \text{or} \quad \tau \propto \sin\theta$$

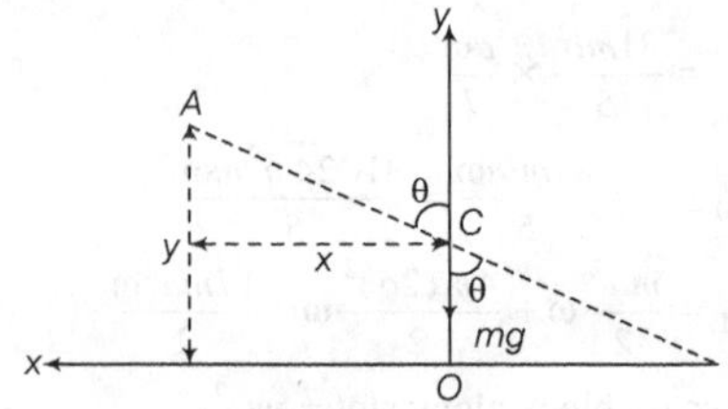

Now, $\quad x = \frac{L}{2}\sin\theta$

$$y = L\cos\theta$$

$$\frac{x^2}{(L/2)^2} + \frac{y^2}{L^2} = 1$$

Path of A is an ellipse.

12. (*) Question is not very clear.

13. (c) If height of the cone $h >> r$

Then, $\quad \mu N = mg$

$$\mu m(R - r)\omega_0^2 = mg$$

$$\omega_0 = \sqrt{\frac{g}{\mu(R - r)}}$$

14. (a,b) Δx_{cm} of the block and point mass system = 0

$$\therefore \quad m(x + R) + Mx = 0$$

where, x is displacement of the block.

Solving this equation, we get

$$x = -\frac{mR}{M + m}$$

From conservation of momentum and mechanical energy of the combined system

$$0 = mv - MV$$

$$mgR = \frac{1}{2}mv^2 + \frac{1}{2}MV^2$$

Solving these two equations, we get

$$v = \sqrt{\frac{2gR}{1 + \frac{m}{M}}}$$

15. (b, c) $V = \frac{Kr^2}{2}$

$$F = -\frac{dV}{dr} = -Kr \qquad \text{(towards centre)}$$

$$\left[F = -\frac{dV}{dr}\right]$$

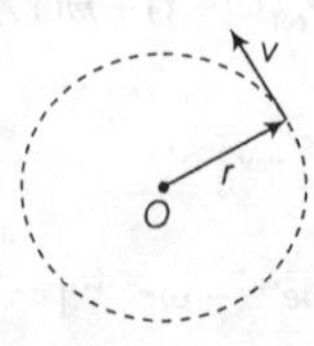

$$kR = \frac{mv^2}{R} \qquad \text{(Centripetal force)}$$

$$v = \sqrt{\frac{kR^2}{m}} = \sqrt{\frac{k}{m}}R$$

$$\Rightarrow \quad L = mvR = \sqrt{\frac{k}{m}}R^2$$

16. (a, c) $h=\dfrac{2\sigma\cos\theta}{r\rho g}$

(a) $\rightarrow h \propto \dfrac{1}{r}$

(b) h depends upon σ.

(c) If lift is going up with constant acceleration.

$$g_{\text{eff}} = (g+a) \quad \Rightarrow h = \frac{2\sigma\cos\theta}{r\rho(g+a)}$$

It means h decreases.

(d) h is proportional to $\cos\theta$.

17. (0.75 m) $a = \dfrac{g\sin\theta}{1+\dfrac{I}{MR^2}}$

$$a_{\text{ring}} = \frac{g\sin\theta}{2} \qquad (I = MR^2)$$

$$a_{\text{disc}} = \frac{2g\sin\theta}{3} \qquad \left(I = \frac{MR^2}{2}\right)$$

$$s = \frac{h}{\sin\theta} = \frac{1}{2}at^2 = \frac{1}{2}\left(\frac{g\sin\theta}{2}\right)t_1^2$$

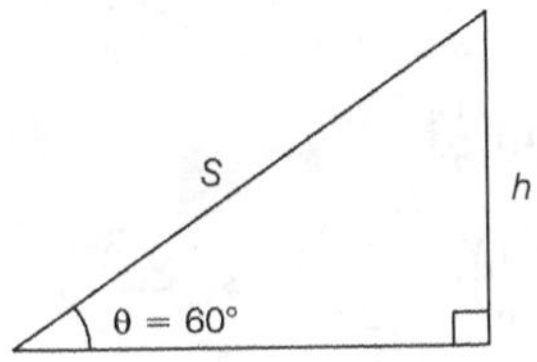

$$\Rightarrow \qquad t_1 = \sqrt{\frac{4h}{g\sin^2\theta}} = \sqrt{\frac{16h}{3g}}$$

$$s = \frac{h}{\sin\theta} = \frac{1}{2}at^2 = \frac{1}{2}\left(\frac{2g\sin\theta}{3}\right)t_2^2$$

$$\Rightarrow \qquad t_2 = \sqrt{\frac{3h}{g\sin^2\theta}} = \sqrt{\frac{4h}{g}}$$

$$t_2 - t_1 = \sqrt{\frac{16h}{3g}} - \sqrt{\frac{4h}{g}} = \frac{2-\sqrt{3}}{\sqrt{10}}$$

$$\sqrt{h}\left[\frac{4}{\sqrt{3}} - 2\right] = 2 - \sqrt{3}$$

Soving this equation we get, $h = 0.75$ m.

18. (a, c, d)

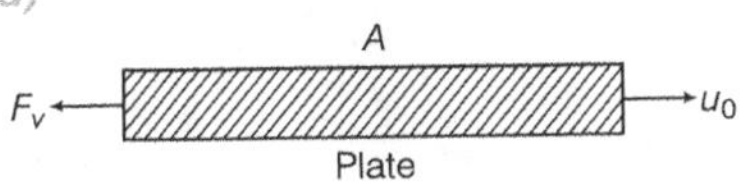

$$F_v = -\eta A\left(\frac{dv}{dy}\right)$$

Since, height h of the liquid in tank is very small.

$$\Rightarrow \qquad \frac{dv}{dy} = \frac{\Delta v}{\Delta y} = \left(\frac{u_0}{h}\right)$$

$$F_v = -(\eta)A\left(\frac{u_0}{h}\right)$$

$$F_v \propto \left(\frac{1}{h}\right), F_v \propto u_0, F \propto A, F_v \propto \eta$$

19. (b) $v = \sqrt{\dfrac{GM}{R}}$

Let $R_1 = R$, then $R_2 = 4R$

If $m_2 = m$, then $m_1 = 2m$

List-I

(P) $\dfrac{v_1}{v_2} = \sqrt{\dfrac{R_2}{R_1}} = \sqrt{\dfrac{4R}{R}} = 2:1$

(Q) $L = mvR$

$$\frac{L_1}{L_2} = \frac{R(2m)v_1}{4R(m)v_2} = \frac{1}{2}(2) = 1:1$$

(R) $\dfrac{K_1}{K_2} = \dfrac{\frac{1}{2}(2m)v_1^2}{\frac{1}{2}(m)v_2^2} = 2(4) = 8:1$

(S) $\dfrac{T_1}{T_2} = \left(\dfrac{R_1}{R_2}\right)^{3/2} = \left(\dfrac{1}{4}\right)^{3/2} = 1:8$

20. (a) When force $F = 0 \Rightarrow$ potential energy U = constant

$F \neq 0 \Rightarrow$ force is conservative $\Rightarrow$ Total energy E = constant

List-I

(P) $\mathbf{r}(t) = \alpha t\hat{\mathbf{i}} + \beta t\hat{\mathbf{j}}$

$$\frac{d\mathbf{r}}{dt} = \mathbf{v} = \alpha\hat{\mathbf{i}} + \beta\hat{\mathbf{j}} = \text{constant}$$

$\Rightarrow$ $\mathbf{p}$ = constant

$|\mathbf{v}| = \sqrt{\alpha^2 + \beta^2}$ = constant

$\Rightarrow$ K = constant

$\dfrac{d\mathbf{v}}{dt} = \mathbf{a} = 0$

$\Rightarrow$ $F = 0 \Rightarrow U$ = constant

$E = U + K$ = constant

$\mathbf{L} = m(\mathbf{r}\times\mathbf{v}) = 0$

$\mathbf{L}$ = constant

$P \rightarrow 1, 2, 3, 4, 5$

(Q) $\mathbf{r}(t) = \alpha\cos\omega t\hat{\mathbf{i}} + \beta\sin\omega t\hat{\mathbf{j}}$

$$\frac{d\mathbf{r}}{dt} = \mathbf{v} = \alpha\omega\sin\omega t(-\hat{\mathbf{i}}) + \beta\omega\cos\omega t\hat{\mathbf{j}} \neq \text{constant}$$

$\Rightarrow$ $\mathbf{p} \neq$ constant

$|\mathbf{v}| = \omega\sqrt{(\alpha\sin\omega t)^2 + (\beta\cos\omega t)^2} \neq$ constant

$\Rightarrow K \neq$ constant

$$\mathbf{a} = \frac{d\mathbf{v}}{dt} = -\omega^2\mathbf{r} \neq 0$$

$\Rightarrow E$ = constant = $K + U$

But $K \neq$ constant $\Rightarrow U \neq$ constant

$\mathbf{L} = m(\mathbf{r}\times\mathbf{v}) = m\omega\alpha\beta\,(\hat{\mathbf{k}})$ = constant

$Q \rightarrow 2, 5$

(R) $\mathbf{r}(t) = \alpha\,(\cos\omega t\hat{\mathbf{i}} + \sin\omega t\hat{\mathbf{j}})$

$$\frac{d\mathbf{r}}{dt} = \mathbf{v} = \alpha\omega[\sin\omega t(-\hat{\mathbf{i}}) + \cos\omega t\hat{\mathbf{j}}]$$

$\neq$ constant

$\Rightarrow$ $\mathbf{p} \neq$ constant

$|\mathbf{v}| = \alpha\omega =$ constant $\Rightarrow K =$ constant

$\mathbf{a} = \dfrac{d\mathbf{v}}{dt} = -\omega^2\mathbf{r} \neq 0 \Rightarrow E =$ constant, $U =$ constant

$\mathbf{L} = m(\mathbf{r} \times \mathbf{v}) = m\omega\alpha^2\hat{\mathbf{k}} =$ constant

$R \rightarrow 2, 3, 4, 5$

(S) $\mathbf{r}(t) = \alpha t\hat{\mathbf{i}} + \dfrac{\beta}{2}t^2\hat{\mathbf{j}}$

$\dfrac{d\mathbf{r}}{dt} = \mathbf{v} = \alpha\hat{\mathbf{i}} + \beta t\hat{\mathbf{j}} \neq$ constant

$\Rightarrow$ $\mathbf{p} \neq$ constant

$|\mathbf{v}| = \sqrt{\alpha^2 + (\beta t)^2} \neq$ constant

$\Rightarrow$ $K \neq$ constant

$\mathbf{a} = \dfrac{d\mathbf{v}}{dt} = \beta\hat{\mathbf{j}} \neq 0$

$\Rightarrow$ $E =$ constant $= K + U$

But $K \neq$ constant

$\therefore$ $U \neq$ constant

$\mathbf{L} = m(\mathbf{r} \times \mathbf{v}) = \dfrac{1}{2}\alpha\beta t^2\hat{\mathbf{k}} \neq$ constant

$S \rightarrow 5$

21. (a, c) $\mathbf{F} = (\alpha t)\hat{\mathbf{i}} + \beta\hat{\mathbf{j}}$ [at $t=0, v=0, \mathbf{r}=0$]

$\alpha = 1, \beta = 1$

$\mathbf{F} = t\hat{\mathbf{i}} + \hat{\mathbf{j}}$

$$m\frac{d\mathbf{v}}{dt} = t\hat{\mathbf{i}} + \hat{\mathbf{j}}$$

On integrating, $m\mathbf{v} = \dfrac{t^2}{2}\hat{\mathbf{i}} + t\hat{\mathbf{j}}$ [$m = 1$ kg]

$\dfrac{d\mathbf{r}}{dt} = \dfrac{t^2}{2}\hat{\mathbf{i}} + t\hat{\mathbf{j}}$ [$\mathbf{r}=0$ at $t=0$]

Again, on integrating,

$$\mathbf{r} = \frac{t^3}{6}\hat{\mathbf{i}} + \frac{t^2}{2}\hat{\mathbf{j}}$$

At $t = 1$ s, $\tau = (\mathbf{r} \times \mathbf{F}) = \left(\dfrac{1}{6}\hat{\mathbf{i}} + \dfrac{1}{2}\hat{\mathbf{j}}\right) \times (\hat{\mathbf{i}} + \hat{\mathbf{j}})$

$$\tau = \frac{1}{3}\hat{\mathbf{k}}$$

$$v = \frac{t^3}{2}\hat{\mathbf{i}} + t\hat{\mathbf{j}}$$

At $t = 1$ s, $\mathbf{v} = \left(\dfrac{1}{2}\hat{\mathbf{i}} + \hat{\mathbf{j}}\right) = \dfrac{1}{2}(\hat{\mathbf{i}} + 2\hat{\mathbf{j}})$ m/s

At $t = 1$ s, $\mathbf{r}_1 - \mathbf{r}_0 = \left[\dfrac{1}{6}\hat{\mathbf{i}} + \dfrac{1}{2}\hat{\mathbf{j}}\right] - [0]$

$$\mathbf{s} = \frac{1}{6}\hat{\mathbf{i}} + \frac{1}{2}\hat{\mathbf{j}}$$

$$|\mathbf{s}| = \sqrt{\left(\frac{1}{6}\right)^2 + \left(\frac{1}{2}\right)^2} \Rightarrow \frac{\sqrt{10}}{6}\text{m}$$